Mathematical Surveys and Monographs
Volume 83

Groups and Geometric Analysis

Integral Geometry, Invariant Differential Operators, and Spherical Functions

Sigurdur Helgason

AMERICAN MATHEMATICAL SOCIETY
Providence, Rhode Island

2000 *Mathematics Subject Classification.* Primary 22E30, 22–02, 43A85, 53–02, 53C65, 22E46, 53C35, 58C35, 43A77, 43A90, 35C15, 44A12, 51M10, 58J70.

Library of Congress Cataloging-in-Publication Data

Helgason, Sigurdur, 1927–
Groups and geometric analysis : integral geometry, invariant differential operators, and spherical functions / Sigurdur Helgason.
p. cm. — (Mathematical surveys and monographs ; v. 83)
Originally published: Orlando : Academic Press, c1984.
Includes bibliographical references and index.
ISBN 0-8218-2673-5 (alk. paper) | Softcover ISBN: 978-0-8218-3211-0
1. Lie groups. 2. Geometry, Differential. I. Title. II. Mathematical surveys and monographs ; no. 83.
QA387.H45 2000
512′.55—dc21 00-034997

Printed in the United States of America.
Reprinted by the American Mathematical Society, 2022
♾ The paper used in this book is acid-free and falls within the guidelines established to ensure permanence and durability.
Visit the AMS home page at URL: http://www.ams.org/

11 10 9 8 7 6 5 4 3 27 26 25 24 23 22

Selected Titles in This Series

83 **Sigurdur Helgason,** Groups and geometric analysis: Integral geometry, invariant differential operators, and spherical functions, 2000

82 **Goro Shimura,** Arithmeticity in the theory of automorphic forms, 2000

81 **Michael E. Taylor,** Tools for PDE: Pseudodifferential operators, paradifferential operators, and layer potentials, 2000

80 **Lindsay N. Childs,** Taming wild extensions: Hopf algebras and local Galois module theory, 2000

79 **Joseph A. Cima and William T. Ross,** The backward shift on the Hardy space, 2000

78 **Boris A. Kupershmidt,** KP or mKP: Noncommutative mathematics of Lagrangian, Hamiltonian, and integrable systems, 2000

77 **Fumio Hiai and Dénes Petz,** The semicircle law, free random variables and entropy, 2000

76 **Frederick P. Gardiner and Nikola Lakic,** Quasiconformal Teichmüller theory, 2000

75 **Greg Hjorth,** Classification and orbit equivalence relations, 2000

74 **Daniel W. Stroock,** An introduction to the analysis of paths on a Riemannian manifold, 2000

73 **John Locker,** Spectral theory of non-self-adjoint two-point differential operators, 2000

72 **Gerald Teschl,** Jacobi operators and completely integrable nonlinear lattices, 1999

71 **Lajos Pukánszky,** Characters of connected Lie groups, 1999

70 **Carmen Chicone and Yuri Latushkin,** Evolution semigroups in dynamical systems and differential equations, 1999

69 **C. T. C. Wall (A. A. Ranicki, Editor),** Surgery on compact manifolds, second edition, 1999

68 **David A. Cox and Sheldon Katz,** Mirror symmetry and algebraic geometry, 1999

67 **A. Borel and N. Wallach,** Continuous cohomology, discrete subgroups, and representations of reductive groups, second edition, 2000

66 **Yu. Ilyashenko and Weigu Li,** Nonlocal bifurcations, 1999

65 **Carl Faith,** Rings and things and a fine array of twentieth century associative algebra, 1999

64 **Rene A. Carmona and Boris Rozovskii, Editors,** Stochastic partial differential equations: Six perspectives, 1999

63 **Mark Hovey,** Model categories, 1999

62 **Vladimir I. Bogachev,** Gaussian measures, 1998

61 **W. Norrie Everitt and Lawrence Markus,** Boundary value problems and symplectic algebra for ordinary differential and quasi-differential operators, 1999

60 **Iain Raeburn and Dana P. Williams,** Morita equivalence and continuous-trace C^*-algebras, 1998

59 **Paul Howard and Jean E. Rubin,** Consequences of the axiom of choice, 1998

58 **Pavel I. Etingof, Igor B. Frenkel, and Alexander A. Kirillov, Jr.,** Lectures on representation theory and Knizhnik-Zamolodchikov equations, 1998

57 **Marc Levine,** Mixed motives, 1998

56 **Leonid I. Korogodski and Yan S. Soibelman,** Algebras of functions on quantum groups: Part I, 1998

55 **J. Scott Carter and Masahico Saito,** Knotted surfaces and their diagrams, 1998

54 **Casper Goffman, Togo Nishiura, and Daniel Waterman,** Homeomorphisms in analysis, 1997

53 **Andreas Kriegl and Peter W. Michor,** The convenient setting of global analysis, 1997

52 **V. A. Kozlov, V. G. Maz'ya, and J. Rossmann,** Elliptic boundary value problems in domains with point singularities, 1997

(Continued in the back of this publication)

To Thor and Annie

CONTENTS

INTRODUCTION

Geometric Fourier Analysis on Spaces of Constant Curvature

CHAPTER I

Integral Geometry and Radon Transforms

CHAPTER II

Invariant Differential Operators

CHAPTER III

Invariants and Harmonic Polynomials

CHAPTER IV

Spherical Functions and Spherical Transforms

CHAPTER V

Analysis on Compact Symmetric Spaces

PREFACE

This volume is intended as an introduction to group-theoretic methods in analysis on spaces that possess certain amounts of mobility and symmetry.

The role of group theory in elementary classical analysis is a rather subdued one; the motion group of $\boldsymbol{R}^3$ enters rather implicitly in standard vector analysis, and the conformal groups of the sphere and of the unit disk become important tools primarily after the Riemann mapping theorem has been established.

In contrast, our point of view here is to place a natural transformation group of a given space in the foreground. We use this group as a guide for the principal concepts (like that of an invariant differential operator) and as motivation for the leading problems in analysis on the space. For examples of problems that arise naturally in such a framework, we call attention here to the Eigenfunction Problems (A, B, and C in Introduction §1, No. 1, p. 2), the Radon-inversion and Orbital Integral Problems (A–D in Chapter I, §3, p. 147), and the Invariant Differential Operator Problems (A–E in Chapter II, §4, p. 277).

The present volume is intended as a textbook and a reference work on the three topics in the subtitle. Its length is caused by a rather leisurely style with many applications and digressions, sometimes in the form of exercises (with solutions).

The introductory chapter deals with the two-dimensional case, requiring only elementary methods and no Lie group theory. While this example has considerable interest in itself, it serves here to connect our main topics to classical analysis. Its primary purpose, however, is to introduce techniques and theories that generalize to semisimple Lie groups and symmetric spaces, including Harish-Chandra's c-functions (treated in generality in Chapter IV), the author's work on harmonic analysis on symmetric spaces (to be treated in generality in another volume), and group-theoretic analysis of the eigenfunctions of the Laplacian, including classical boundary value properties of harmonic functions.

In Chapter I we give an exposition of invariant integration on homogeneous spaces and relate this to the structure theory of semisimple Lie groups. We formulate general analytic integral–geometric problems (Radon transforms and orbital integrals) for a double fibration of a homogeneous space and develop their solutions in an elementary fashion for some simple classes of homogeneous spaces.

In Chapter II we discuss the effect on differential operators of a group action on a manifold. This gives rise to a separation of variables for differential operators, projections, transversal parts, orbital parts, and radial parts of differential operators; all these are useful for problems with built-in invariance conditions.

Chapter III deals with linear group action on a vector space and with the corresponding invariant polynomials and harmonic polynomials. The results, together with a description of the orbits, find analytic applications in the case of the linear isotropy representation of a symmetric space.

In Chapter IV we study the (zonal) spherical functions, that is, the K-invariant eigenfunctions of the G-invariant differential operators on the Riemannian homogeneous space G/K. Their theory, as well as that of the corresponding spherical transform, is worked out in considerable detail for the case of symmetric space (and its tangent space). Here the connection with representation theory is particularly simple and important.

In the investigation of the spherical function's behavior at ∞, Harish-Chandra's beautiful $\mathbf{c}$-function emerges. Eventually this function is expressed by the formula

$$\mathbf{c}(\lambda) = c_0 \prod_{\alpha \in \Sigma_0^+} \frac{2^{-\langle i\lambda, \alpha_0 \rangle}\Gamma(\langle i\lambda, \alpha_0 \rangle)}{\Gamma(\frac{1}{2}(\frac{1}{2}m_\alpha + 1 + \langle i\lambda, \alpha_0 \rangle))\Gamma(\frac{1}{2}(\frac{1}{2}m_\alpha + m_{2\alpha} + \langle i\lambda, \alpha_0 \rangle))},$$

the notation being explained at the end of §6. As indicated there, each detail in this formula has its own significance. In particular, the location of the singularities of $\mathbf{c}(\lambda)$ is crucial for the proof of the Paley–Wiener-type theorem for the spherical transform, which in turn enters into the proof of the corresponding inversion and Plancherel formula.

Chapter V deals with harmonic analysis on compact homogeneous spaces U/K. Because of the intimate connection with finite-dimensional representation theory for U, the chapter begins with a detailed exposition of the weight theory and the character theory for U.

Since this book is in part intended as a textbook, we now give some description of its level. The first third of the book is introductory and has on occasion been used as a textbook for first-year graduate students without background in Lie group theory. The remainder of the book requires some standard functional analysis results. The Lie-theoretic tools needed can for example be found in my book "Differential Geometry, Lie Groups, and Symmetric Spaces" (abbreviated [DS]), of which the present book can be considered an "analytic continuation." Thus, our aim has been to provide complete proofs of all the results in the book. Although this process of unification and consolidation has at times led to some simplifications of proofs, we have in the exposition been more concerned with clarity than brevity.

Each chapter begins with a short summary and ends with historical notes giving references to source material: an effort has been made to give appropriate credit to authors of individual results. At the same time we have tried to make these notes reflect the fact that the logical order of the exposition often differs drastically from the order of the historical development.

Much of the material in this book has been the subject of lectures at the Massachusetts Institute of Technology in recent years; some of the content of my lecture notes [1980c, 1981] has been incorporated in the Introduction and in Chapter I. Parts of the manuscript have been read and commented on by M. Baum, M. Cowling, M. Flensted-Jensen, F. Gonzales, A. G. Helmink, G. F. Helmink, B. Hoogenboom, A. Korányi, M. Mazzarello, J. Orloff, F. Richter, H. Schlichtkrull, and G. Travaglini. I am particularly indebted to T. Koornwinder for his numerous useful suggestions.

PREFACE TO THE 2000 PRINTING

The manuscript for this book was completed in late 1982 and was published by Academic Press in 1984. The book has been unavailable for a long time and I am happy to have this new printing published by the American Mathematical Society.

The book is the forerunner to my book "Geometric Analysis on Symmetric Spaces" published in 1994 by the American Mathematical Society as Vol. 39 in the Series "Mathematical Surveys and Monographs".

In this new printing I have taken the opportunity to correct a few inaccuracies and also to add a number of footnotes which are collected towards the end of the book in a section entitled "Some Details". The place of the footnotes is indicated by a dagger †. Some of these footnotes furnish alternative, more elementary proofs and some are clarifications of occasional tough spots in the exposition. In addition, an errata list appears after the Index.

Many of these corrections and footnotes are based on suggestions by readers and I would like to express my appreciation to A. Korányi, A. Mattuck, F. Rouvière, H. Schlichtkrull, G. Shimura, H. Stetkaer, and particularly P. Kuchment who translated the book into Russian.

SUGGESTIONS TO THE READER

Since this book is intended for readers with varied backgrounds and since it is written with several objectives in mind, we give some suggestions for its use.

The primary purpose has been to provide readers having a modest Lie-theoretic background with a self-contained account of the three topics in the subtitle. However, the various chapters are largely independent of each other and could individually serve as textbooks for one- or two-semester courses.

(i) The introductory chapter is so elementary that it could be used for an advanced undergraduate course.

(ii) Chapter I is an independent account of group-invariant integration and analytic integral geometry. A reader wishing only an elementary treatment of the Radon transform with some natural generalizations could read this chapter and skip §1 and §5.

(iii) Chapter II (together with §1 and §5 of Chapter I) provides an independent treatment of invariant differential operators. Some of the most natural problems for these operators are discussed in the introduction to §4, Chapter II.

(iv) Chapter III deals with invariants, particularly Weyl group invariants, harmonic polynomials, and orbit decomposition for the complex isotropy representation associated with a symmetric space. This chapter is almost entirely independent of the previous ones.

(v) Chapter IV, when preceded by Chapter I (§1 and §5) and Chapter II, provides an independent account of the theory of spherical functions with a certain degree of completeness.

(vi) Chapter V (together with Chapter IV, §1, No. 2) gives an independent exposition of analysis on compact homogeneous spaces with emphasis on the symmetric ones. The chapter starts with a brief account of the basic representation theory of compact Lie groups.

Exercises and Further Results. Each chapter ends with a few exercises; some of these are routine applications and others develop the general theory further. Stating such results with no hint of proof or reference I consider counterproductive; hence solutions are provided at the end of the book. These solutions are often fairly concise and some of them (indicated with a star) rely on other references. Hopefully such exercises can furnish suitable topics for student seminars.

A SEQUEL TO THE PRESENT VOLUME

GEOMETRIC ANALYSIS ON SYMMETRIC SPACES

was published in 1994 as

Volume 39 in the series "Mathematical Surveys and Monographs".
American Mathematical Society.

Contents:

Chapter I. A Duality in Integral Geometry

1. Generalities. 2. The Radon Transform for Points and Hyperplanes. 3. Homogeneous Spaces in Duality. Exercises and Further Results.

Chapter II. A Duality for Symmetric Spaces

1. The Space of Horocycles. 2. Invariant Differential Operators. 3. The Radon Transform and its Dual. 4. Finite-dimensional Spherical and Conical Representations. 5. Conical Distributions. 6. Some Rank-One Results. Exercises and Further Results.

Chapter III. The Fourier Transform on a Symmetric Space

1. The Inversion and the Plancherel Formula. 2. Generalized Spherical Functions (Eisenstein Integrals). 3. The Q^{δ}-matrices. 4. The Simplicity Criterion. 5. The Paley-Wiener Theorem for the Fourier Transform on $X = G/K$. 6. Eigenfunctions and Eigenspace Representations. 7. Tangent Space Analysis. 8. Eigenfunctions and Eigenspace Representations on X_o. 9. The Compact Case. 10. Elements of $\mathbf{D}(G/K)$ as Fractions. 11. The Rank One Case. 12. The Spherical Transform Revisited. Exercises and Further Results.

Chapter IV. The Radon Transform on X and on X_o. Range Questions

1. The Support Theorem. 2. The Ranges of $\mathcal{D}(X)^{\wedge}, \mathcal{E}'(X)^{\wedge}$ and $\mathcal{E}(\Xi)^{\vee}$. 3. The Range and Kernel for K-types. 4. The Radon Transform and its Dual for K-invariants. 5. The Radon Transform on X_o. Exercises and Further Results.

INTRODUCTION

GEOMETRIC FOURIER ANALYSIS ON SPACES OF CONSTANT CURVATURE

In this introductory chapter we formulate what might reasonably be called harmonic analysis on homogeneous spaces of Lie groups; the examples of the plane $\boldsymbol{R}^2$, the sphere $\boldsymbol{S}^2$, and the hyperbolic plane $\boldsymbol{H}^2$ are worked out in detail. For the last case the proofs exemplify those which generalize to symmetric spaces of the noncompact type. Since this chapter is limited to these basic examples, it is elementary and self-contained and does not require familiarity with Lie group theory.

After the generalities of §1 we turn to the case $\boldsymbol{R}^2$ as a homogeneous space of its group of isometries. Here the Laplacian generates the algebra of invariant differential operators; we describe its eigenfunctions by means of "entire functionals" and determine the irreducibility of the eigenspace representations. We also state a variation of the classical Paley–Wiener theorem; its proof is postponed to §3, where the necessary tools from the theory of spherical harmonics are developed.

The principal section (§4) of this chapter deals with the hyperbolic plane. The non-Euclidean Laplacian is essentially the only invariant differential operator. We give an integral representation of its eigenfunctions and determine the (exceptional) eigenvalues for which the corresponding eigenspace representation is not irreducible. A basic tool is a certain Fourier transform on the hyperbolic plane; when viewed in non-Euclidean terms, the classical Poisson integral formula for bounded harmonic functions becomes a result in this non-Euclidean Fourier analysis. The integral representation mentioned generalizes this formula to all eigenfunctions of the non-Euclidean Laplacian.

The radial eigenfunctions are called spherical functions; their behavior at ∞ is determined by the so-called $\boldsymbol{c}$-function of Harish-Chandra which enters in the spectral decomposition of the Laplacian. The knowledge of the singularities of the $\boldsymbol{c}$-function is also important in the proof of the Paley–Wiener-type theorem for the Fourier transform mentioned; this proof occupies a good part of this section.

§1. Harmonic Analysis on Homogeneous Spaces

1. General Problems

Let X be a locally compact Hausdorff space acted on transitively by a locally compact group G. We assume G leaves invariant a positive measure μ on X. Let T_X denote the (unitary) representation of G on the Hilbert space $L^2(X)$ defined by $(T_X(g)f)(x) = f(g^{-1} \cdot x)$ for $g \in G$,

$f \in L^2(X)$, $x \in X$. By harmonic analysis on X is frequently meant the decomposition (in the sense of the so-called direct integral theory) of T_X into irreducible representations. The corresponding decomposition of any $f \in L^2(X)$ will then resemble the classical Fourier integral decomposition. The main tools in the analysis are the measure μ and the theory of operator algebras.

If $X = G/K$ is a homogeneous space of a Lie group G, K a closed subgroup, the situation changes drastically because the machinery of differential calculus becomes available. To exploit this we consider the algebra $\boldsymbol{D}(G/K)$ of all differential operators on G/K which are *invariant* under the translations $\tau(g): xK \to gxK$ from G (cf. Chapter II, §4). A function on G/K which is an eigenfunction of each $D \in \boldsymbol{D}(G/K)$ will be called a *joint eigenfunction* of $\boldsymbol{D}(G/K)$. Given a homomorphism

$$\chi: \boldsymbol{D}(G/K) \to \boldsymbol{C}$$

the space

$$E_\chi(X) = \{f \in C^\infty(X) : Df = \chi(D)f \text{ for all } D \in \boldsymbol{D}(G/K)\}$$

is called a *joint eigenspace*. Let T_χ denote the natural representation of G on $E_\chi(X)$, that is, $(T_\chi(g)f)(x) = f(g^{-1} \cdot x)$. These representations are called *eigenspace representations*. By *harmonic analysis* on G/K we shall mean "answers" to the following problems:

A. *Decompose "arbitrary" functions on $X = G/K$ into joint eigenfunctions of $\boldsymbol{D}(G/K)$.*

B. *Describe the joint eigenspaces $E_\chi(X)$ of $\boldsymbol{D}(G/K)$.*

C. *Determine for which χ the eigenspace representation T_χ is irreducible.*

The space $C^\infty(X)$ has a standard topology (Chapter II, §2), $E_\chi(X)$ is given the induced topology, and irreducibility above means that there are no closed nontrivial invariant subspaces.

It may happen that $\boldsymbol{D}(G/K)$ has no nontrivial operators, in which case the above considerations have no interest. But we shall see that for the symmetric spaces G/K, Problems A, B, and C are reasonable and interesting. We shall now discuss them in detail for $\boldsymbol{R}^2$, $\boldsymbol{S}^2$, and $\boldsymbol{H}^2$, the three simply connected two-dimensional Riemannian manifolds of constant curvature.

2. Notation and Preliminaries

As usual, $\boldsymbol{R}$ and $\boldsymbol{C}$ will denote the fields of real and complex numbers, respectively, and $\boldsymbol{Z}$ the ring of integers. We write Re c and Im c, respectively, for the real and imaginary parts of a complex number c. Let

$$\boldsymbol{R}^+ = \{t \in \boldsymbol{R} : t \geq 0\}, \qquad \boldsymbol{Z}^+ = \boldsymbol{Z} \cap \boldsymbol{R}^+.$$

If X is a topological space and $A \subset X$, then $\mathring{A}$ denotes the interior of A and $\bar{A}$ or $\mathrm{Cl}(A)$ denotes the closure of A. Also $C(X)$ [resp. $C_c(X)$] denotes the space of complex-valued continuous functions (resp. of compact support) on X.

Let $\boldsymbol{R}^n = \{x = (x_1, \ldots, x_n) : x_i \in \boldsymbol{R}\}$ and let ∂_i denote the partial derivative $\partial/\partial x_i$. If $\alpha = (\alpha_1, \ldots, \alpha_n)$ is an n-tuple of integers $\alpha_i \geq 0$, we put

$$D^\alpha = \partial_1^{\alpha_1} \cdots \partial_n^{\alpha_n}, \qquad x^\alpha = x_1^{\alpha_1} \cdots x_n^{\alpha_n}, \qquad |\alpha| = \alpha_1 + \cdots + \alpha_n.$$

Let $C^m(\boldsymbol{R}^n)$ denote the space of complex-valued functions with continuous partial derivatives of order $\leq m$. Let $C^\infty(\boldsymbol{R}^n)$ or $\mathscr{E}(\boldsymbol{R}^n)$ denote the space of complex-valued C^∞-functions f on $\boldsymbol{R}^n$. Given $m \in \boldsymbol{Z}^+$ and a compact subset $K \subset \boldsymbol{R}^n$, let

$$\|f\|_m^K = \sum_{|\alpha| \leq m} \sup_{x \in K} |(D^\alpha f)(x)|. \tag{1}$$

The space $\mathscr{E}(\boldsymbol{R}^n)$ is topologized by means of the seminorms $\|f\|_m^K$. It is then a Fréchet space.

Let $\mathscr{D}(\boldsymbol{R}^n)$ or $C_c^\infty(\boldsymbol{R}^n)$ denote the space $C_c(\boldsymbol{R}^n) \cap C^\infty(\boldsymbol{R}^n)$. Sometimes this notation $C(\boldsymbol{R}^n)$, $\mathscr{E}(\boldsymbol{R}^n)$, and $\mathscr{D}(\boldsymbol{R}^n)$ will be used even when the functions are assumed to be real-valued.

If $K \subset \boldsymbol{R}^n$ is compact, let $\mathscr{D}_K(\boldsymbol{R}^n)$ denote the space of C^∞-functions with support contained in K; this space has a topology given by the norms $\| \ \|_m^K$, where $m = 0, 1, 2, \ldots$. A linear functional T on $\mathscr{D}(\boldsymbol{R}^n)$ is called a *distribution* if its restriction to each $\mathscr{D}_K(\boldsymbol{R}^n)$ is continuous. Let $\mathscr{D}'(\boldsymbol{R}^n)$ denote the space of all distributions on $\boldsymbol{R}^n$. A locally integrable function F on $\boldsymbol{R}^n$ gives rise to a distribution T_F: $f \to \int f(x)F(x)\,dx$ on $\boldsymbol{R}^n$.

A distribution $T \in \mathscr{D}'(\boldsymbol{R}^n)$ is said to be 0 on an open set $U \subset \boldsymbol{R}^n$ if $T(\phi) = 0$ for each $\phi \in \mathscr{D}(\boldsymbol{R}^n)$ whose support is contained in U. If V is the union of all open subsets $U_\alpha \subset \boldsymbol{R}^n$ on which $T = 0$, then a partition of unity argument shows that $T = 0$ on V. The complement of V is called the *support* of T.

A distribution T of *compact support* extends to a continuous linear form on $\mathscr{E}(\boldsymbol{R}^n)$ by putting

$$T(f) = T(f_0 f), \qquad f \in \mathscr{E}(\boldsymbol{R}^n),$$

if f_0 is any function in $\mathscr{D}(\boldsymbol{R}^n)$ which is identically 1 on a neighborhood of the support of T. The choice of f_0 is immaterial. In this way the dual space $\mathscr{E}'(\boldsymbol{R}^n)$ is identified with the space of distributions of compact support.

In Chapter II we shall define these notions for manifolds M. Here we need them only when M is an open subset of $\boldsymbol{R}^n$ and when M is the n-dimensional sphere $\boldsymbol{S}^n$.

In Chapter I, §2, No. 8 we review some further notions from distribution theory, in particular, differentiation, convolution, and Fourier transforms of distributions.

§2. The Euclidean Plane R^2

1. Eigenfunctions and Eigenspace Representations

If we view $\boldsymbol{R}^2$ as a homogeneous space of $\boldsymbol{R}^2$ acting on itself by translations, Problems A, B, and C boil down to ordinary Fourier analysis. In fact, the invariant differential operators are just the differential operators with constant coefficients, and the joint eigenfunctions are the constant multiples of the exponential functions. The eigenspace representations are one-dimensional, hence irreducible.

However, we can also consider $\boldsymbol{R}^2$ as the homogeneous space $\boldsymbol{R}^2 = \boldsymbol{M}(2)/\boldsymbol{O}(2)$ of the group $\boldsymbol{M}(2)$ of all isometries of $\boldsymbol{R}^2$ over the orthogonal group $\boldsymbol{O}(2)$ [leaving $(0, 0)$ fixed]. Then it is easy to see that $\boldsymbol{D}(G/K)$ consists of the polynomials in the Laplacian $L_{\boldsymbol{R}^2} = \partial^2/\partial x_1^2 + \partial^2/\partial x_2^2$. If $\lambda \in \boldsymbol{C}$ and ω a unit vector, the function $x \to e^{i\lambda(x,\omega)}$, where $(\ ,\ )$ is the usual inner product, is an eigenfunction of $L_{\boldsymbol{R}^2}$ with eigenvalue $-\lambda^2$.

We write the Fourier transform $\tilde{f}$ of a function f on $\boldsymbol{R}^2$ in the form

$$\tilde{f}(\lambda\omega) = \int_{\boldsymbol{R}^2} f(x) e^{-i\lambda(x,\omega)}\, dx. \tag{1}$$

The Plancherel formula now reads

$$\int_{\boldsymbol{R}^2} |f(x)|^2\, dx = \frac{1}{(2\pi)^2} \int_{S^1 \times \boldsymbol{R}^+} |\tilde{f}(\lambda\omega)|^2 \lambda\, d\lambda\, d\omega.$$

The Fourier inversion formula, valid, for example, if $f \in \mathscr{D}(\boldsymbol{R}^2)$, can be written

$$f(x) = \frac{1}{(2\pi)^2} \int_{S^1} \int_{\boldsymbol{R}^+} \tilde{f}(\lambda\omega) e^{i\lambda(x,\omega)} \lambda\, d\lambda\, d\omega, \tag{2}$$

and gives an explicit answer to Problem A. Here $d\omega$ denotes the circular measure on S^1.

This formula suggests that general eigenfunctions should be obtained from the functions $e^{i\lambda(x,\omega)}$ by some kind of superposition. We shall now prove a precise result of this kind.

Given $a, b \geq 0$ let $E_{a,b}$ denote the space of holomorphic functions on $\boldsymbol{C} - \{0\}$ satisfying

$$\|f\|_{a,b} = \sup_z (|f(z)| e^{-a|z| - b|z|^{-1}}) < \infty. \tag{3}$$

Then $E_{a,b}$ is a Banach space with the norm $\| \ \|_{a,b}$. Also $E_{a,b} \subset E_{a',b'}$ if and only if $a \le a'$, $b \le b'$, and in this case $\|f\|_{a,b} \ge \|f\|_{a',b'}$, so that the injection of $E_{a,b}$ into $E_{a',b'}$ is continuous. We can give the union

$$E = \bigcup_{a,b} E_{a,b},$$

the *inductive limit topology*. This means that a fundamental system of neighborhoods of 0 in E is given by the *convex* sets W such that for each (a, b), $W \cap E_{a,b}$ is a neighborhood of 0 in $E_{a,b}$. We identify the members of E with their restrictions to the unit circle $\boldsymbol{S}^1$ and call the members of the dual space E' *entire functionals* on $\boldsymbol{S}^1$. Since these generalize measures it is convenient to write

$$T(f) = \int_{\boldsymbol{S}^1} f(\omega)\, dT(\omega), \qquad f \in E, \quad T \in E'.$$

The following result gives an answer to Problem B.

Theorem 2.1. *The eigenfunctions of the Laplacian on $\boldsymbol{R}^2$ are precisely the harmonic functions and the functions*

$$f(x) = \int_{\boldsymbol{S}^1} e^{i\lambda(x,\omega)}\, dT(\omega), \tag{4}$$

where $\lambda \in \boldsymbol{C} - \{0\}$ and T is an entire functional on $\boldsymbol{S}^1$.

We note first that the right-hand side of (4) is well-defined because if $x = (x_1, x_2)$, the integrand is the restriction to $\boldsymbol{S}^1$ of the function

$$z \to \exp[\tfrac{1}{2}(i\lambda)x_1(z + z^{-1}) + \tfrac{1}{2}\lambda x_2(z - z^{-1})],$$

which indeed does belong to E.

Now we need to characterize the members of E in terms of their Laurent series expansions.

Lemma 2.2. *Let f be holomorphic in $\boldsymbol{C} - \{0\}$ and let*

$$f(z) = \sum_n \alpha_n z^n$$

be its Laurent expansion. Then

$$f \in E_{a+\varepsilon, b+\varepsilon} \text{ for all } \varepsilon > 0$$

if and only if

$$\alpha_n = O\left(\frac{(a+\delta)^n}{n!}\right), \qquad \alpha_{-n} = O\left(\frac{(b+\delta)^n}{n!}\right), \qquad n \ge 0,$$

for each $\delta > 0$.

Consider the decomposition

$$f(z) = \sum_{0}^{\infty} \alpha_n z^n + \sum_{-\infty}^{-1} \alpha_n z^n = f_+(z) + f_-(z); \tag{5}$$

since $f_+ = f - f_-$ and since f_- is bounded for $|z| \geq 1$, (5) shows that $E_{a,b} = E_{a,0} + E_{0,b}$. Thus it suffices to show that the function

$$f(z) = \sum_{0}^{\infty} \alpha_n z^n$$

belongs to $E_{a+\varepsilon,0}$ for each $\varepsilon > 0$ if and only if

$$\alpha_n = O\left(\frac{(a+\delta)^n}{n!}\right), \qquad n \geq 0,$$

for each $\delta > 0$. Assume first this last condition is satisfied. Then for each $\delta > 0$ we have a constant $C_\delta > 0$ such that

$$|\alpha_n| \leq C_\delta \frac{(a+\delta)^n}{n!}.$$

Then

$$|f(z)| \leq \sum_{0}^{\infty} |\alpha_n| |z|^n \leq C_\delta e^{(a+\delta)|z|}$$

so $f \in E_{a+\delta,0}$.

On the other hand, suppose $f \in E_{a+\varepsilon}$ for all $\varepsilon > 0$. Put

$$\nu = \limsup_n (n|\alpha_n|^{1/n}).$$

First take $\nu > 0$, and let $0 < \delta < \nu$. Then for infinitely many n,

$$|\alpha_n| \geq \left(\frac{\nu - \delta}{n}\right)^n$$

(where $\nu - \delta$ is to be interpreted as an arbitrary large number if $\nu = \infty$). Using

$$|\alpha_n| = |f^{(n)}(0)/n!| \leq (2\pi)^{-1} \int_{|z|=r} |z|^{-n-1} |f(z)| \, |dz|,$$

so

$$|\alpha_n| \leq r^{-n} \sup_{|z|=r} |f(z)|,$$

we have for the sequence $r = r_n = ne/(\nu - \delta)$ (n as in the inequality above)

$$\sup_{|z|=r} |f(z)| \geq |\alpha_n| r^n \geq e^n = \exp\left(\frac{(\nu - \delta)r}{e}\right).$$

It follows that for each $\varepsilon > 0$,

$$\frac{\nu}{e} \le a + \varepsilon,$$

so $\nu \le ae$. Given $\varepsilon > 0$ this means

$$|\alpha_n| \le \left(\frac{ae + \varepsilon}{n}\right)^n$$

for all n sufficiently large, which by Stirling's formula

$$n! = n^n e^{-n}(2\pi n)^{1/2} e^{\rho/12n} \qquad (0 < \rho < 1)$$

implies

$$\alpha_n = O\left(\frac{(a + \varepsilon)^n}{n!}\right),$$

for $\nu > 0$. The case $\nu = 0$ gives the result more directly.

We can also formulate the result by saying that $f(z) = \sum_0^\infty \alpha_n z^n$ belongs to $E_{a+\varepsilon,0}$ for all $\varepsilon > 0$ if and only if its Laplace transform

$$F(z) = \int_0^\infty f(t)e^{-zt}\, dt \qquad (\text{Re } z > a)$$

has its Laurent series

(6) $$F(z) = \sum_0^\infty n!\alpha_n z^{-n-1}$$

convergent for $|z| > a$.

Given an entire functional $T \in E'$ we associate with it a Fourier series

(7) $$T \sim \sum_n a_n e^{in\theta},$$

where, by definition,

$$a_n = \int_{S^1} e^{-in\theta}\, dT(\theta).$$

Proposition 2.3. *A series $\sum_n a_n e^{in\theta}$ is the Fourier series for an entire functional if and only if*

(8) $$\sum_n |a_n| \left(\frac{r^{|n|}}{|n|!}\right) < \infty \qquad \textit{for} \quad r \ge 0.$$

Proof. Let $T \in E'$ have the Fourier series (7). Then for each $a, b > 0$ and each $\varepsilon > 0$, T has a continuous restriction to $E_{a+\varepsilon,0}$ and to $E_{0,b+\varepsilon}$.

The series $\sum_0^\infty a^n z^n/n!$ converges to e^{az} in the topology of $E_{a+\varepsilon,0}$. In fact, if

$$f_N(z) = \sum_{N+1}^{\infty} \frac{a^n z^n}{n!}, \qquad g_N(z) = f_N(z)e^{-(a+\varepsilon)|z|},$$

then $|g_N(z)| \le e^{-\varepsilon|z|}$, so for $R > 0$

$$\begin{aligned}\sup_z |g_N(z)| &\le \sup_{|z|\le R} |g_N(z)| + \sup_{|z|>R} |g_N(z)| \\ &\le \sup_{|z|\le R} |f_N(z)| + \sup_{|z|>R} e^{-\varepsilon|z|}.\end{aligned}$$

By the uniform convergence of f_N to 0 on compact sets this implies $f_N \to 0$ in $E_{a+\varepsilon,0}$, as claimed. Consequently,

$$\int_{S^1} e^{ae^{i\theta}}\, dT(\theta) = \lim_{N\to\infty} \sum_0^N \frac{a^n a_{-n}}{n!}.$$

Using a similar argument for $E_{0,b+\varepsilon}$, the convergence (8) follows.

On the other hand, suppose (8) holds. Then we can define a linear form T on E by

$$T(f) = \sum_n a_n \alpha_{-n}$$

if $f = \sum_n \alpha_n z^n$. By the definition of the inductive limit, T is continuous provided that for each $a, b \ge 0$ the restriction $T|E_{a,b}$ is continuous. Now it is easily seen by means of the closed graph theorem that the map

$$f \in E_{a,b} \to f_+ \in E_{a,0},$$

with the notation of (5), is continuous. Thus it suffices to prove that for each $a > 0$ the restriction $T|E_{a,0}$ is continuous. For $f \in E_{a,0}$ let $f(z) = \sum_0^\infty \alpha_n z^n$ denote its Taylor series and F as in (6) its Laplace transform. Let $(f_k) \subset E_{a,0}$ be a sequence converging to 0. Then

(9) $$|f_k(z)| \le A_k e^{a|z|},$$

where $A_k \to 0$. The Laplace transforms $F_k(z)$ then satisfy for $x > a$

(10) $$|F_k(x)| \le A_k \int_0^\infty e^{(a-x)t}\, dt = A_k(x-a)^{-1}.$$

But since the right-hand side of (9) is rotation-invariant and since the function $z \to f(e^{i\theta}z)$ has Laplace transform $z \to e^{-i\theta}F(e^{-i\theta}z)$ [because of (6)], the estimate in (10) implies

(11) $$|F_k(z)| \le A_k(|z| - a)^{-1} \qquad \text{for} \quad |z| > a.$$

If $\alpha_n^{(k)}$ are the Taylor coefficients of $f_k(z)$, we have

$$n!\,\alpha_n^{(k)} = (2\pi i)^{-1} \int F_k(z) z^n\, dz,$$

with the integration taken, for example, over the circle $|z| = a + 1$. But then by (11)

$$n!|\alpha_n^{(k)}| \le A_k(a + 1)^{n+1},$$

so that

$$|T(f_k)| \le \left| \sum_0^{\infty} a_{-n}\alpha_n^{(k)} \right| \le A_k \sum_0^{\infty} |a_{-n}|(a + 1)^{n+1}(n!)^{-1}.$$

Using our assumption (8), we deduce $T(f_k) \to 0$, proving the continuity.

Lemma 2.4. *Let $\lambda \neq 0$, $n \in \boldsymbol{Z}$. The solutions f to $L_{\boldsymbol{R}^2} f = -\lambda^2 f$ satisfying $f(e^{i\theta}z) = e^{in\theta}f(z)$ are the constant multiples of the function*

$$\phi_{\lambda,n}(x) = \frac{1}{2\pi} \int_{S^1} e^{i\lambda(x,\,\omega)} \chi_n(\omega)\, d\omega,$$

where $\chi_n(e^{i\theta}) = e^{in\theta}$.

Proof. It is clear that two such functions f are proportional on the unit circle; hence by Theorem 2.7 in Chapter II they are proportional on $\boldsymbol{R}^2$. A more direct proof is obtained by noting that in polar coordinates,

$$L_{\boldsymbol{R}^2} = \frac{\partial^2}{\partial r^2} + \frac{1}{r}\frac{\partial}{\partial r} + \frac{1}{r^2}\frac{\partial^2}{\partial \theta^2},$$

so

$$\frac{d^2 f}{dr^2} + \frac{1}{r}\frac{df}{dr} - \frac{n^2}{r^2} f = -\lambda^2 f.$$

Because of the ellipticity of $L_{\boldsymbol{R}^2}$, the eigenfunction f is analytic.† Expanding it in a power series in r, the differential equation gives a recursion formula for the coefficients which readily shows that all the solutions are proportional.

To conclude the proof of the lemma it just remains to remark that $\phi_{\lambda,n} \not\equiv 0$. This is a special case of the elementary Lemma 2.7.

Proposition 2.5. *A function f satisfying $L_{\boldsymbol{R}^2} f = -\lambda^2 f$ satisfies the functional equation*

$$(12) \qquad \frac{1}{2\pi} \int_0^{2\pi} f(z + e^{i\theta}w)\, d\theta = f(z)\phi_{\lambda,0}(w), \qquad z, w \in \boldsymbol{C}.$$

Conversely a continuous function f satisfying (12) *is automatically of class C^∞ and satisfies $L_{\boldsymbol{R}^2} f = -\lambda^2 f$.*

†See "Some Details," p. 611.

Proof. If f is an eigenfunction, so is the function

$$F: w \to \int_0^{2\pi} f(z + e^{i\theta}w)\, d\theta.$$

In addition, F is radial, so by Lemma 2.4 it equals the function $\phi_{\lambda,0}(w)$ multiplied by some factor. This factor is obtained by putting $w = 0$; (12) follows.

On the other hand, suppose f is a continuous function satisfying (12). We multiply (12) by a function $h(w) \in \mathcal{D}(\mathbf{R}^2)$ and integrate with respect to w. By the invariance of dw under $\mathbf{M}(2)$ we get

$$\frac{1}{2\pi}\int_0^{2\pi} d\theta \int_{\mathbf{R}^2} f(u)h(e^{-i\theta}(u - z))\, du = f(z) \int_{\mathbf{R}^2} \phi_{\lambda,0}(w)h(w)\, dw$$

and this shows that f is of class C^∞.

Now apply $L = L_{\mathbf{R}^2}$ to (12) as a function of w. By its invariance under $\mathbf{M}(2)$ we get

$$\frac{1}{2\pi}\int_0^{2\pi} (Lf)(z + e^{i\theta}w)\, d\theta = f(z)(-\lambda^2)\phi_{\lambda,0}(w).$$

Putting $w = 0$ we get $Lf = -\lambda^2 f$, so the lemma is proved.

We can now prove Theorem 2.1. Let f satisfy $L_{\mathbf{R}^2} f = -\lambda^2 f$, where $\lambda \in \mathbf{C} - \{0\}$. We expand the function $\theta \to f(e^{i\theta}z)$ in an absolutely convergent Fourier series

$$f(e^{i\theta}z) = \sum_n c_n(z)e^{in\theta}, \tag{13}$$

where

$$c_n(z) = (2\pi)^{-1} \int_0^{2\pi} f(e^{i\theta}z)e^{-in\theta}\, d\theta.$$

But this is an eigenfunction of the Laplacian satisfying the homogeneity assumption of Lemma 2.4. Hence

$$c_n(z) = a_n \phi_{\lambda,n}(z), \qquad a_n \in \mathbf{C}.$$

Also, substituting $\theta = (\pi/2) - \phi$, we get

$$\phi_{\lambda,n}(r) = i^n(2\pi)^{-1} \int_0^{2\pi} e^{i\lambda r \sin\phi} e^{-in\phi}\, d\phi.$$

This shows incidentally that

$$\phi_{\lambda,-n}(r) = (-1)^n \phi_{\lambda,n}(-r).$$

Consider now the Bessel function

$$J_n(z) = \sum_{k=0}^{\infty} (-1)^k (\tfrac{1}{2}z)^{2k+n} [k!\,\Gamma(n+k+1)]^{-1},$$

defined for all $n, z \in \boldsymbol{C}$. (For $-n \in \boldsymbol{Z}^+$ the first coefficients vanish.) If in the integral for $\phi_{\lambda,n}$ we expand $e^{i\lambda r \sin\phi}$ in a power series and integrate term by term, we obtain

(13a) $$\phi_{\lambda,n}(r) = i^n J_n(\lambda r).$$

On the other hand, we have for $n \in \boldsymbol{Z}^+$

$$\Gamma(n+1)(\tfrac{1}{2}z)^{-n} J_n(z) = 1 + h(z),$$

where

$$|h(z)| = \left| \sum_{r=1}^{\infty} (-1)^r (\tfrac{1}{2}z)^{2r} \frac{1}{r!} \frac{\Gamma(n+1)}{\Gamma(n+r+1)} \right|$$
$$\leq \sum_{r=1}^{\infty} (\tfrac{1}{2}|z|)^{2r} \frac{1}{r!(n+1)^r} = \exp((\tfrac{1}{2}|z|)^2/(n+1)) - 1$$

so

$$\lim_{n\to+\infty} n!(\tfrac{1}{2}z)^{-n} J_n(z) = 1.$$

Hence the relation

$$\sum |a_n J_n(\lambda r)| < \infty,$$

which follows from (13) and the formulas

$$J_{-n}(\lambda r) = J_n(-\lambda r) = (-1)^n J_n(\lambda r),$$

which follow from (13a), imply

$$\sum_n |a_n| (r^{|n|}/|n|!) < \infty.$$

Thus by Prop. 2.3 there exists a $T \in E'$ such that

$$T \sim \sum a_n e^{in\theta}.$$

The formula for $\phi_{\lambda,n}$ shows that

$$e^{i\lambda(x,e^{i\theta})} = \sum_n \phi_{\lambda,n}(x) e^{-in\theta},$$

which, if $x = (x_1, x_2)$, gives the Laurent expansion

(14) $$\exp[\tfrac{1}{2}i\lambda x_1(z + z^{-1}) + \tfrac{1}{2}\lambda x_2(z - z^{-1})] = \sum_n \phi_{\lambda,n}(x) z^{-n}.$$

As we saw for the Taylor series for e^{az}, the series (14) converges in the topology of E. Thus we can apply T to it term by term. This gives

$$\int_{S^1} e^{i\lambda(x,\omega)}\,dT(\omega) = \sum_n \phi_{\lambda,n}(x)a_n = f(x),$$

which is the desired representation of f.

On the other hand, let $T \in E'$ and consider the "integral"

$$f(x) = \int_{S^1} e^{i\lambda(x,\omega)}\,dT(\omega)$$

as well as the Fourier expansion

$$T \sim \sum a_n e^{in\theta}.$$

Applying T to the expansion (14), we get

$$f(x) = \sum_n a_n \phi_{\lambda,n}(x); \tag{15}$$

by (8) and the above estimates for $J_n(\lambda r)$ this series converges uniformly on compact sets, so f at least is continuous. The functions $\phi_{\lambda,n}$ all satisfy (12), and so does f because of the uniform convergence. Hence it is an eigenfunction of $L_{\mathbf{R}^2}$ and Theorem 2.1 is proved.

Problem C in §1 will now be answered with the following result. For $\lambda \in \mathbf{C}$ let $\mathscr{E}_\lambda(\mathbf{R}^2)$ denote the eigenspace

$$\mathscr{E}_\lambda(\mathbf{R}^2) = \{f \in \mathscr{E}(\mathbf{R}^2) : L_{\mathbf{R}^2} f = -\lambda^2 f\}$$

and let T_λ denote the natural representation of $\mathbf{M}(2)$ on $\mathscr{E}_\lambda(\mathbf{R}^2)$; that is, $(T_\lambda(g)f)(x) = f(g^{-1} \cdot x)$. The topology of $\mathscr{E}(\mathbf{R}^2)$ was explained in §1, No. 2; $\mathscr{E}_\lambda(\mathbf{R}^2)$ is a closed subspace.

Theorem 2.6. *The eigenspace representation T_λ is irreducible if and only if $\lambda \neq 0$.*

We first have to prove a couple of lemmas. Given $F \in L^2(S^1)$, consider the function

$$f(x) = \frac{1}{2\pi}\int_{S^1} e^{i\lambda(x,\omega)}F(\omega)\,d\omega. \tag{16}$$

Lemma 2.7. *Let $\lambda \neq 0$. Then the mapping $F \to f$ defined by* (16) *is one-to-one.*

Proof. Let $p(\zeta) = p(\zeta_1, \zeta_2)$ be a polynomial and D the corresponding constant coefficient differential operator on $\mathbf{R}^2$ such that

$$D_x(e^{i(x,\zeta)}) = p(\zeta)e^{i(x,\zeta)}$$

for $\zeta \in C^2$. If $f \equiv 0$ in (16), we deduce from the above equation that

$$\int_{S^1} p(\lambda\omega_1, \lambda\omega_2)F(\omega)\,d\omega = 0.$$

Since p is arbitrary and $\lambda \neq 0$, this implies $F \equiv 0$.

For $\lambda \neq 0$ let $\mathscr{H}_\lambda$ denote the space of functions f as defined by (16); $\mathscr{H}_\lambda$ is a Hilbert space if the norm of f is defined as the L^2-norm of F on S^1. Because of Lemma 2.7 this is well-defined.

Lemma 2.8. *Let $\lambda \neq 0$. Then the space $\mathscr{H}_\lambda$ is dense in $\mathscr{E}_\lambda(R^2)$.*

Proof. Let $f \in \mathscr{E}_\lambda(R^2)$ and let $f^\theta(z) = f(e^{i\theta}z)$. Consider the mapping $\Phi: \theta \to f^\theta$ of S^1 into the Fréchet space $\mathscr{E}_\lambda(R^2)$. Using the seminorms (1) in §1 on $\mathscr{E}(R^2)$, it is clear what is meant by a differentiable mapping of a manifold into $\mathscr{E}(R^2)$. In that sense the map Φ is differentiable. The series (13) can be written in the form

$$\Phi(\theta) = \sum_n a_n \phi_{\lambda,n} e^{in\theta},$$

so $a_n \phi_{\lambda,n}$ is the Fourier coefficient of the vector-valued function Φ. Since the series (13) can be differentiated at will with respect to θ, we obtain for the kth derivative $\Phi^{(k)}$ of Φ,

$$a_n \phi_{\lambda,n} = (in)^{-k}(2\pi)^{-1} \int_0^{2\pi} \Phi^{(k)}(\theta)e^{-in\theta}\,d\theta.$$

This gives an estimate $\|a_n \phi_{\lambda,n}\| \leq (\text{const.})n^{-k}$ for each of the indicated seminorms $\| \ \|$ of $\mathscr{E}(R^2)$, so by the completeness the series

$$\sum_n a_n \phi_{\lambda,n} e^{in\theta}$$

converges absolutely in the topology of $\mathscr{E}(R^2)$ to some limit $\Phi_0(\theta)$. But then Φ and Φ_0 will have the same Fourier coefficients, so $\Phi_0 = \Phi$ and

$$\lim_{N\to\infty} \sum_{-N}^{N} a_n \phi_{\lambda,n} = \Phi(0) = f$$

in the topology of $\mathscr{E}(R^2)$. This proves the lemma.

We can now prove Theorem 2.6. If $\lambda = 0$, T_λ is obviously not irreducible since the constants form a nontrivial invariant subspace. Assuming now $\lambda \neq 0$, we first prove that $M(2)$ acts irreducibly on $\mathscr{H}_\lambda$. Let $V \neq 0$ be a closed invariant subspace of the Hilbert space $\mathscr{H}_\lambda$. Then there exists an $h \in V$ such that $h(0) = 1$. We write

$$h(x) = \frac{1}{2\pi} \int_{S^1} e^{i\lambda(x,\omega)} H(\omega)\,d\omega.$$

Because of Lemma 2.4 the average

$$h^\natural(z) = \frac{1}{2\pi}\int_0^{2\pi} h(e^{i\theta}z)\,d\theta$$

is then given by

$$h^\natural(x) = \phi_{\lambda,0}(x) = \frac{1}{2\pi}\int_{S^1} e^{i\lambda(x,\omega)}\,d\omega.$$

If f in (16) lies in the annihilator V^0 of V in $\mathscr{H}_\lambda$, the functions F and H are orthogonal on S^1. Since V^0 is $O(2)$-invariant, this remains true for H replaced by its average over S^1; in other words, $\phi_{\lambda,0}$ belongs to the double annihilator $(V^0)^0$, which, by Hilbert space theory, equals V. But then, since V is invariant under translations, it follows that for each $t \in \mathbf{R}^2$ the function

$$x \to \int_{S^1} e^{i\lambda(x,\omega)}e^{i\lambda(t,\omega)}\,d\omega$$

belongs to V. Lemma 2.7 then implies that $V^0 = \{0\}$, whence the irreducibility of $\mathbf{M}(2)$ on $\mathscr{H}_\lambda$.

Passing now to $\mathscr{E}_\lambda$, let $W \subset \mathscr{E}_\lambda$ be a closed invariant subspace. Then $W \cap \mathscr{H}_\lambda$ is an invariant subspace of $\mathscr{H}_\lambda$. Let (f_n) be a sequence in $W \cap \mathscr{H}_\lambda$ converging to $f \in \mathscr{H}_\lambda$ (in the topology of $\mathscr{H}_\lambda$). From (16) for each f_n we see, using Schwarz's inequality, that $f_n \to f$ in the topology of $\mathscr{E}_\lambda$. Thus $f \in W$, so $W \cap \mathscr{H}_\lambda$ is closed in $\mathscr{H}_\lambda$. Hence, by the above, $W \cap \mathscr{H}_\lambda$ is $\mathscr{H}_\lambda$ or $\{0\}$. In the first case, $W = \mathscr{E}_\lambda$ because of Lemma 2.8. In the second case, consider for each $f \in W$ the expansion (15). Each term $a_n\phi_{\lambda,n}$ is given by

$$a_n\phi_{\lambda,n}(z) = \frac{1}{2\pi}\int_0^{2\pi} f(e^{i\theta}z)e^{-in\theta}\,d\theta.$$

This belongs to $W \cap \mathscr{H}_\lambda$ because the left-hand side is in $\mathscr{H}_\lambda$ and the right-hand side belongs to W as a limit [in $\mathscr{E}(\mathbf{R}^2)$] of translates of members of W. It follows that $f = 0$ and the irreducibility is proved.

Remarks. (i) Theorem 2.6 generalizes to arbitrary dimensions (cf. Exercise A2).

(ii) For the exceptional value $\lambda = 0$ the solution space $\mathscr{E}_0(\mathbf{R}^2)$ consists of the space of harmonic functions. In this case there is a bigger group acting on the solution space, namely, the conformal group (or rather its Lie algebra), and again a certain irreducibility holds (cf. Exercise A4).

2. A Theorem of Paley–Wiener Type

A holomorphic function F on C^n is said to have *exponential type* $A > 0$ if for each $N \in Z^+$ there exists a constant C_N such that

$$|F(\zeta)| \leq C_N(1 + |\zeta|)^{-N} e^{A|\operatorname{Im}\zeta|}, \qquad \zeta \in C^n.$$

Here $|\zeta|^2 = |\zeta_1|^2 + \cdots + |\zeta_n|^2$ and Im ζ denotes the imaginary part of the vector $\zeta = (\zeta_1, \ldots, \zeta_n)$. Let $\mathscr{H}^A(C^n)$ denote the space of holomorphic functions F satisfying the condition above and let

$$\mathscr{H}(C^n) = \bigcup_{A>0} \mathscr{H}^A(C^n).$$

For simplicity we write $\mathscr{D}_A(R^n)$ for the space $\mathscr{D}_{\bar{B}_A(0)}(R^n)$ of C^∞-functions on R^n with support in the ball $|x| \leq A$. Then we have the following classical result.

Theorem 2.9. (Paley–Wiener) *The Fourier–Laplace transform $f \to \tilde{f}$, where*

$$\tilde{f}(\zeta) = \int_{R^n} f(x) e^{-i(x,\zeta)}\, dx, \qquad f \in \mathscr{D}(R^n), \tag{17}$$

is a bijection of $\mathscr{D}_A(R^n)$ onto $\mathscr{H}^A(C^n)$.

Our viewpoint of harmonic analysis on the homogeneous space $R^2 = M(2)/O(2)$, and more generally on $R^n = M(n)/O(n)$, relates functions f on R^n to functions ϕ on $R \times S^{n-1}$ via the Fourier transform

$$\phi(\lambda, \omega) = \tilde{f}(\lambda\omega) = \int_{R^n} f(x) e^{-i\lambda(x,\omega)}\, dx. \tag{18}$$

(We pass from $n = 2$ to general n because the proof of the result below requires a kind of an induction on the dimension). While Theorem 2.9 characterizes the holomorphic functions $\tilde{f}(\zeta)$ it is more appropriate in the present context to ask for an intrinsic characterization of the functions $\phi(\lambda, \omega)$ in (18) as f runs through $\mathscr{D}(R^n)$. (The similar problem for $f \in L^2$ offers no difficulties.) We shall now obtain such a characterization. Let Im c denote the imaginary part of a complex number c. A vector $a = (a_1, \ldots, a_n) \in C^n$ is called *isotropic* if $a_1^2 + \cdots + a_n^2 = 0$.

Theorem 2.10. *The mapping $f \to \tilde{f}$ maps $\mathscr{D}(R^n)$ onto the set of functions $\tilde{f}(\lambda\omega) = \phi(\lambda, \omega) \in C^\infty(R \times S^{n-1})$ satisfying:*

(i) *There exists a constant* $A > 0$ *such that for each* ω *the function* $\lambda \to \phi(\lambda, \omega)$ *extends to a holomorphic function on* $\mathbf{C}$ *with the property*

$$\sup_{\lambda, \omega} |\phi(\lambda, \omega)(1 + |\lambda|)^N e^{-A|\mathrm{Im}\,\lambda|}| < \infty \tag{19}$$

for each $N \in \mathbf{Z}$.

(ii) *For each* $k \in \mathbf{Z}^+$ *and each isotropic vector* $a \in \mathbf{C}^n$ *the function*

$$\lambda \to \lambda^{-k} \int_{S^{n-1}} \phi(\lambda, \omega)(a, \omega)^k \, d\omega$$

is even and holomorphic on $\mathbf{C}$.

Since the proof of this result uses several facts concerning spherical harmonics, we postpone it until the next section. In the process we shall also give a proof of Theorem 2.9.

§3. The Sphere S^2

1. Spherical Harmonics

In this section we consider Problems A, B, and C in §1 for the sphere S^2 viewed as a homogeneous space $\mathbf{O}(3)/\mathbf{O}(2)$, $\mathbf{O}(n)$ denoting the orthogonal group in $\mathbf{R}^n$. Using the spherical polar coordinates

$$x_1 = r \cos \psi \sin \theta, \; x_2 = r \sin \psi \sin \theta, \; x_3 = r \cos \theta$$

on $\mathbf{R}^3$, the Laplacian takes the form

$$L_{\mathbf{R}^3} = \frac{\partial^2}{\partial r^2} + \frac{2}{r}\frac{\partial}{\partial r} + \frac{1}{r^2} L, \tag{1}$$

where

$$L = \frac{\partial^2}{\partial \theta^2} + \cot \theta \frac{\partial}{\partial \theta} + \sin^{-2} \theta \frac{\partial^2}{\partial \psi^2}. \tag{2}$$

More generally, the Laplacian $L_{\mathbf{R}^n}$ $(n > 1)$ has the form

$$L_{\mathbf{R}^n} = \frac{\partial^2}{\partial r^2} + \frac{n-1}{r}\frac{\partial}{\partial r} + \frac{1}{r^2} L, \tag{3}$$

where L is the Laplacian on S^{n-1} (cf., Chapter II, §5). The operator L generates the algebra of $\mathbf{O}(n)$-invariant differential operators on $S^{n-1} = \mathbf{O}(n)/\mathbf{O}(n-1)$ (cf. Chapter II, Cor. 4.11). Let $(s_1, \ldots, s_n)$ denote the Cartesian coordinates of points on S^{n-1}.

Theorem 3.1.

(i) *The eigenspaces of L on* S^{n-1} *are of the form*

$$E_k = \text{span of } \{f_{a,k}(s) = (a_1 s_1 + \cdots + a_n s_n)^k, s \in S^{n-1}\}.$$

Here $a = (a_1, \ldots, a_n) \in C^n$ *is an isotropic vector and* $k \in Z^+$. *The eigenvalue is* $-k(k+n-2)$.

(ii) *Each eigenspace representation is irreducible.*

(iii) $L^2(S^{n-1}) = \sum_0^\infty E_k$ *(orthogonal Hilbert space decomposition).*

Proof. Let P_k denote the space of homogeneous polynomial functions $p(x_1, \ldots, x_n)$ of degree k on R^n and $H_k \subset P_k$ the subspace of harmonic polynomials. Consider the bilinear form $\langle\ ,\ \rangle$ on $P = \sum_0^\infty P_k$ given by

$$\langle p, q \rangle = \left\{p\left(\frac{\partial}{\partial x_1}, \ldots, \frac{\partial}{\partial x_n}\right)q\right\}(0).$$

A simple computation shows that $\langle\ ,\ \rangle$ is strictly positive definite and that $\langle p, q\rangle = \langle q, p\rangle$. Also, if $p, q, r \in P$ and if we put

$$\partial(p) = p\left(\frac{\partial}{\partial x_1}, \ldots, \frac{\partial}{\partial x_n}\right),$$

then

$$\langle p, qr\rangle = (\partial(qr)p)(0) = (\partial(r)\partial(q)p)(0) = \langle \partial(q)p, r\rangle,$$

so that the operators $r \to qr$ and $r \to \partial(q)r$ are adjoint operators. In particular, if $p \in P_{k-2}$, $q = x_1^2 + \cdots + x_n^2 = |x|^2$,

$$\langle qp, h\rangle = \langle p, \partial(q)h\rangle,$$

so that H_k is the orthogonal complement of qP_{k-2} in P_k; hence

$$P_k = qP_{k-2} + H_k \tag{4}$$

and by iteration

$$P_k = H_k + |x|^2 H_{k-2} + \cdots + |x|^{2m} H_{k-2m}, \qquad m = [\tfrac{1}{2}k]. \tag{5}$$

On the other hand, if $c = (c_1, \ldots, c_n) \in C^n$ satisfies

$$(c, c) = c_1^2 + \cdots + c_n^2 = 0,$$

then the polynomial $h_c(x) = (c_1 x_1 + \cdots + c_n x_n)^k$ belongs to H_k^c. Let H_k^0 be the span of the h_c for $(c, c) = 0$ and suppose $h \in H_k$ is orthogonal to H_k^0. A simple computation shows for $p \in P_k$,

$$\partial(p)(c_1 x_1 + \cdots + c_n x_n)^m$$
$$= m(m-1)\cdots(m-k+1)p(c)(c_1 x_1 + \cdots + c_n x_n)^{m-k},$$

so that, in particular,

$$h(c) = 0 \qquad \text{if} \quad (c, c) = 0.$$

In other words, h vanishes identically on the variety $\{c \in \boldsymbol{C}^n : (c, c) = 0\}$. By Hilbert's *Nullstellensatz* some power h^m belongs to the ideal

$$(x_1^2 + \cdots + x_n^2)P.$$

If $n \geq 3$, a simple computation shows that $x_1^2 + \cdots + x_n^2$ is irreducible, the ideal is a prime ideal, so h itself is divisible by $x_1^2 + \cdots + x_n^2$, whence by (4), $h = 0$. Consequently, if $n \geq 3$

(6) $$H_k^c = \text{span of } \{(c_1 x_1 + \cdots + c_n x_n)^k : (c, c) = 0\}.$$

This holds also for $n = 2$; in fact, we can represent a real harmonic polynomial $u(x_1, x_2)$ as the real part of a holomorphic function

$$f(x_1 + ix_2) = u(x_1, x_2) + iv(x_1, x_2).$$

By the Cauchy–Riemann equations, v is a polynomial in x_1, x_2. The function f, being holomorphic, is therefore a polynomial in z. Hence $u = \frac{1}{2}(f + \bar{f})$ is a linear combinations of powers $(x_1 \pm ix_2)^k$ as claimed.

By definition, and by (6), E_k consists of the restrictions $H_k | \boldsymbol{S}^{n-1}$. It follows from (3) that each $h \in E_k$ satisfies $Lh = -k(k + n - 2)h$. The eigenvalues $-k(k + n - 2)$ being different for different k, the symmetry of L (Chapter II, Prop. 2.3) shows that the spaces E_k are mutually orthogonal. By the Stone–Weierstrass theorem the algebraic direct sum

$$\sum_0^\infty P_k | \boldsymbol{S}^{n-1},$$

which by (5) equals $\sum_0^\infty E_k$, is dense in $C(\boldsymbol{S}^{n-1})$ in the uniform norm; this implies (iii).

Now consider the normalized measure $d'\omega = \Omega_n^{-1}\, d\omega$ on $\boldsymbol{S}^{n-1}$ (Ω_n being the area of $\boldsymbol{S}^{n-1}$) and let $\langle\ ,\ \rangle$ denote the inner product on $L^2(\boldsymbol{S}^{n-1})$ corresponding to the measure $d'\omega$. Let S_{km} $[1 \leq m \leq d(k)]$ be an orthonormal basis of E_k, and let F be any eigenfunction of L, $LF = cF$ $(c \in \boldsymbol{C})$. We have the expansion

(7) $$F \sim \sum_{k,m} a_{km} S_{km}, \qquad a_{km} = \langle F, S_{km} \rangle.$$

But then, using the symmetry of L,

$$c \sum_{k,m} a_{km} S_{km} \sim cF = LF \sim \sum_{k,m} (-k)(k + n - 2) a_{km} S_{km},$$

so c is one of the eigenvalues $-k_0(k_0 + n - 2)$ and E_{k_0} is the corresponding eigenspace. This proves (i).

For (ii) suppose $E_k = E' \oplus E''$ is an orthogonal decomposition of E_k into two nonzero invariant subspaces. By the $\mathbf{O}(n)$ invariance each of the spaces E', E'' contains a function which equals 1 at the point $o = (0, \ldots, 0, 1)$. By averaging over $\mathbf{O}(n-1)$, the isotropy group at o, we would obtain functions $\phi' \in E'$, $\phi'' \in E''$, with $\phi'(o) = \phi''(o) = 1$, depending only on the distance θ from o. Using Chapter II, Theorem 2.7, we deduce that $\phi' = \phi''$. A more direct proof of the identity $\phi' = \phi''$ can be obtained by observing that on functions on S^{n-1} depending only on θ, L has the form $d^2/d\theta^2 + (n-2)\cot\theta\, d/d\theta$ (Chapter II, Proposition 5.26). Thus ϕ' and ϕ'' both satisfy the singular differential equation

$$\frac{d^2\phi}{d\theta^2} + (n-2)\cot\theta\,\frac{d\phi}{d\theta} = -k(k+n-2)\phi. \tag{8}$$

Using a power series expansion† $\phi(\theta) = \sum_0^\infty a_m \sin^m \theta$ for small θ, one obtains the formulas

$$a_1 = 0, \qquad (m+2)(m+n-1)a_{m+2} = [m(m+n-2) - k(k+n-2)]a_m.$$

Thus ϕ is determined up to a constant factor. Hence $\phi' = \phi''$ and Theorem 3.1 is proved. Note that ϕ' must be real-valued.

Remark. The eigenfunctions of L on S^{n-1} are called *spherical harmonics*. The name is derived from the fact that they coincide with the set of restrictions $\bigcup_{k \geq 0}(H_k | S^{n-1})$. It is useful to remark that the restriction mapping $p \to p|S^{n-1}$ of H_k onto E_k is one-to-one.

According to Theorem 3.1 each $f \in L^2(S^{n-1})$ can be expanded in a series of spherical harmonics. We shall now need some refinements of this expansion. Classically this is done by a direct investigation of the functions in E_k. A simpler method (which is also closer to the spirit of this book) is to represent the members of E_k as representation coefficients (Chapter IV, §1) of the orthogonal group $\mathbf{O}(n)$. For example, with this approach, the formula from Prop. 3.2 below,

$$p(u \cdot o) = d(\delta)\langle p, \delta(u)\phi_k\rangle,$$

replaces the classical formula

$$Y_k(s) = c \int_{S^{n-1}} P_k^{(1/2)(n-2)}((s, s'))\,Y_k(s')\,d\omega(s')$$

(Heine [1878]), where $Y_k \in E_k$,

$$c = \tfrac{1}{2}\Gamma(\tfrac{1}{2}n - 1)(\tfrac{1}{2}n + k - 1)\pi^{-(1/2)n},$$

†See "Some Details," p. 611.

and $P_k^{(1/2)(n-2)}$ is the ultraspherical polynomial defined by

$$(1 - 2zt + z^2)^{1-(1/2)n} = \sum_0^\infty z^k P_k^{(1/2)(n-2)}(t). \tag{8a}$$

The group-theoretic method is still quite elementary because the only required tools are the Haar measure on $\mathbf{O}(n)$ and the Schur orthogonality relations (Chapter IV, §1, No. 2).

If $s \in S^{n-1}, u \in \mathbf{O}(n)$, let $u \cdot s$ denote the image of s under u and let the Haar measure du on $\mathbf{O}(n)$ be normalized by

$$\int_{\mathbf{O}(n)} F(u \cdot o)\, du = \int_{S^{n-1}} F(s)\, d'\omega(s), \tag{9}$$

(cf. Chapter I, §1).

Proposition 3.2. *For a fixed $k \in \mathbf{Z}^+$ let δ be the natural representation of $\mathbf{O}(n)$ on E_k and $d(\delta) = d(k)$ its degree. Let $\phi_k \in E_k$ be the unique element which is invariant under the isotropy group $\mathbf{O}(n-1)$ at o, normalized by $\phi_k(o) = 1$. Then*

(i) $p(u \cdot o)\langle \phi_k, \phi_k \rangle = \langle p, \delta(u)\phi_k \rangle$ *for* $p \in E_k$;
(ii) $\langle \phi_k, \phi_k \rangle = 1/d(\delta)$, $\phi_k(u^{-1} \cdot o) = \phi_k(u \cdot o)$, $|\phi_k(s)| \leq 1$.

Proof. (i) Let $M = \mathbf{O}(n-1)$ and dm its normalized Haar measure. Since $\phi = \phi_k$ is M-invariant, we have for $m \in M$,

$$\begin{aligned} \int_{S^{n-1}} p(s)\phi(s)\, d'\omega(s) &= \int_{S^{n-1}} p(m \cdot s)\phi(s)\, d'\omega(s) \\ &= \int_{S^{n-1}} \phi(s)\left(\int_M p(m \cdot s)\, dm\right) d'\omega(s) \\ &= \int_{S^{n-1}} \phi(s)p(o)\phi(s)\, d'\omega(s), \end{aligned}$$

so that, since ϕ is real,

$$\langle p, \phi \rangle = p(o)\langle \phi, \phi \rangle.$$

Replacing $p(s)$ by $(\delta(u)p)(s) = p(u^{-1}s)$, relation (i) follows.

Using Schur's orthogonality relations (Chapter IV, §1, No. 2) on relation (i), we obtain

$$\langle p, p \rangle \langle \phi, \phi \rangle^2 = \frac{1}{d(\delta)} \langle p, p \rangle \langle \phi, \phi \rangle,$$

which proves $\langle \phi, \phi \rangle = d(\delta)^{-1}$. Since ϕ_k is real-valued (by the uniqueness) we have by (i)

$$\phi_k(u \cdot o) = d(\delta)\langle \phi_k, \delta(u)\phi_k \rangle = d(\delta)\langle \delta(u)\phi_k, \phi_k \rangle = \phi_k(u^{-1} \cdot o)$$

and the relation $|\phi_k| \leq 1$ follows from Schwarz's inequality.

Corollary 3.3. *For each $s_0 \in \mathbf{S}^{n-1}$ and each $p \in E_k$, $F \in C(\mathbf{S}^{n-1})$, the function*

$$g: s \to \int_{\mathbf{O}(n)} F(u \cdot s)p(u \cdot s_0)\, du, \qquad s \in \mathbf{S}^{n-1},$$

belongs to E_k.

For this it is sufficient to consider the case $s_0 = o$. Writing $s = v \cdot o$, the integral is

$$g(v \cdot o) = d(\delta) \int_{\mathbf{O}(n)} F(w \cdot o)\langle p, \delta(wv^{-1})\phi_k \rangle\, dw$$

$$= d(\delta) \int_{\mathbf{O}(n)} F(w \cdot o)\langle \delta(w^{-1})p, \delta(v^{-1})\phi_k \rangle\, dw = q(v^{-1} \cdot o)$$

if

$$q = \int_{\mathbf{O}(n)} F(w \cdot o)\delta(w^{-1})p\, dw.$$

Clearly $q \in E_k$. Taking v as a rotation in the two-plane through 0, s, and o, we see that $g(s) = q(\sigma \cdot s)$, where σ is the symmetry of $\mathbf{S}^{n-1}$ with respect to o. Hence $g \in E_k$, as claimed.

Theorem 3.4. *For $k \in \mathbf{Z}^+$ let S_{km} $[1 \leq m \leq d(k)]$ be an orthonormal basis of E_k. Let $F \in \mathscr{E}(\mathbf{S}^{n-1})$ and $a_{km} = \langle F, S_{km} \rangle$. Then*

(i) *$F = \sum_{k,m} a_{km} S_{km}$, the series being absolutely and uniformly convergent.*

(ii) *The mapping $F \to \{a_{km}\}$ maps $\mathscr{E}(\mathbf{S}^{n-1})$ onto the set of all sequences a_{km} $[k \in \mathbf{Z}^+, 1 \leq m \leq d(k)]$ satisfying*

$$\sup_{k,m} |a_{km}| k^q < \infty. \tag{10}$$

for each $q \in \mathbf{Z}^+$.

Proof. By Hilbert space theory,

$$\sum_{k,m} |a_{km}|^2 = \int_{\mathbf{S}^{n-1}} |F(s)|^2\, d'\omega(s).$$

Replacing here F by $L^q F$, inequality (10) follows. Putting $p = S_{km}$ in Prop. 3.2, we obtain by Schwarz's inequality

$$|S_{km}(s)| \leq d(\delta)^{1/2}, \tag{11}$$

which by (4) is bounded by a fixed power of k. This proves (i).

It remains to prove that a sequence $\{a_{km}\}$ satisfying (10) gives a sum

$$F = \sum_{k,m} a_{km} S_{km}, \tag{11a}$$

which is necessarily smooth on S^{n-1}. For this we define for

$$F_1, F_2 \in C(S^{n-1})$$

the convolution $F_1 * F_2$ by

$$(F_1 * F_2)(s) = \int_{O(n)} F_1(u \cdot o) F_2(u^{-1} \cdot s)\, du. \tag{12}$$

Now, as well as later, we need the following simple lemma.

Lemma 3.5. *If $p \in E_k$, we have*

$$p * \phi_k = d(k)^{-1} p, \qquad \phi_k * p = p(o) d(k)^{-1} \phi_k. \tag{13}$$

In the notation of Theorem 3.4,

$$d(k) F * \phi_k = \sum_{1 \leq m \leq d(k)} a_{km} S_{km}, \tag{14}$$

so that (i) *reads*

$$F = \sum_0^\infty d(k) F * \phi_k. \tag{15}$$

Proof. We have by Prop. 3.2,

$$(p * \phi_k)(v \cdot o) = \int_{O(n)} d(k)^2 \langle p, \delta(u)\phi_k \rangle \langle \phi_k, \delta(u^{-1} v)\phi_k \rangle \, du,$$

which by the orthogonality relations equals

$$\langle p, \delta(v)\phi_k \rangle = d(k)^{-1} p(v \cdot o).$$

Second, $\phi_k * p$ is $O(n-1)$-invariant, hence proportional to ϕ_k, so evaluating at o and using Proposition 3.2, we obtain (13).

If F_1 is orthogonal to E_k, the definition (12) shows $F_1 * \phi_k = 0$. Hence (13) implies (14) and (15).

Finally, to prove that F in (11a) is smooth, we consider for $q \in \boldsymbol{Z}^+$ the functions

$$F_q(s) = \sum_{k,m} a_{km}(k+1)^q S_{km}(s),$$

$$\psi_q(s) = \sum_k (k+1)^{-q} d(k)\phi_k(s),$$

which by (13) satisfy

$$F = F_q * \psi_q.$$

Let $n = 3$. Using part (i) of Theorem 3.1 and the invariance of ϕ_k under $\boldsymbol{O}(n-1)$, we conclude that ϕ_k is the restriction to $\boldsymbol{S}^2$ of the polynomial

$$\frac{1}{2\pi}\int_{-\pi}^{\pi} (ix_1 \cos\phi + ix_2 \sin\phi + x_3)^k \, d\phi.$$

Hence

$$\phi_k(s) = \frac{1}{\pi}\int_0^{\pi} (\cos\theta + i\sin\theta\cos\phi)^k \, d\phi, \qquad s \in \boldsymbol{S}^2,$$

with $s_3 = \cos\theta$. For n arbitrary we have similarly [for example, by verifying (8)]

$$\phi_k(s) = \frac{\Omega_{n-2}}{\Omega_{n-1}}\int_0^{\pi} (\cos\theta + i\sin\theta\cos\phi)^k \sin^{n-3}\phi \, d\phi$$

if $s_n = \cos\theta$. Given $N \in \boldsymbol{Z}^+$ we can therefore choose q so large that the series for ψ_q can be differentiated term by term N times, so ψ_q has continuous derivatives up to order N. The function F_q is continuous [by (10) and (11)], so, using $F = F_q * \psi_q$, we conclude that F is N times differentiable. Thus $F \in \mathscr{E}(\boldsymbol{S}^{n-1})$, so Theorem 3.4 is proved.

2. Proof of Theorem 2.10

We first prove that conditions (i) and (ii) are satisfied for each $\tilde{f}$ if $f \in \mathscr{D}(\boldsymbol{R}^n)$. First note that (18) in §2 can be written

$$\phi(\lambda, \omega) = \tilde{f}(\lambda\omega) = \int_{\boldsymbol{R}} \hat{f}(\omega, p) e^{-i\lambda p} \, dp, \tag{16}$$

where $\hat{f}(\omega, p)$ is the integral of f over the hyperplane $(x, \omega) = p$. Since the function $p \to \hat{f}(\omega, p)$ belongs to $\mathscr{D}(\boldsymbol{R})$, property (i) follows immediately. Also (18) in §2 implies that $d^k\tilde{f}/d\lambda^k$ at $\lambda = 0$ is a homogeneous kth-degree polynomial in $(\omega_1, \ldots, \omega_n)$, which according to (4) can be expressed as a harmonic polynomial of degree $\leq k$. Expanding the function $\lambda \to \phi(\lambda, \omega) = \tilde{f}(\lambda\omega)$ in a convergent Taylor series around $\lambda = 0$ and integrating against $(a, \omega)^k$, property (ii) follows since E_k and E_l are orthogonal for $k > l$.

Conversely, suppose ϕ satisfies (i) and (ii) and define $f \in \mathscr{E}(\boldsymbol{R}^n)$ by

$$f(x) = (2\pi)^{-n} \int_{\boldsymbol{R}^+ \times \boldsymbol{S}^{n-1}} \phi(\lambda, \omega) e^{i\lambda(x, \omega)} \lambda^{n-1} \, d\lambda \, d\omega. \tag{17}$$

We shall show that $f(x) = 0$ for $|x| > A$ and then the inversion formula for the Fourier transform implies $\tilde{f} = \phi$. Since the assumptions (i) and (ii) are invariant under rotations around 0, it suffices to prove $f(x) = 0$ for $x = re_n$, where $e_n = (0, \ldots, 0, 1)$ and $r > A$. Let $U = \boldsymbol{O}(n)$ and M $[=\boldsymbol{O}(n-1)]$ be the subgroup leaving e_n (i.e., the north pole o) fixed. We can apply (15) to the function $F(u \cdot o) = f(u \cdot x)$. Evaluating at $u = e$, we obtain

$$f(x) = \sum_0^\infty d(k) \int_U f(u \cdot x) \phi_k(u^{-1} \cdot o) \, du. \tag{18}$$

Using (17) we get

$$d(k) \int_U f(u \cdot x) \phi_k(u^{-1} \cdot o) \, du$$

$$= d(k)(2\pi)^{-n} \int_{\boldsymbol{R}^+ \times \boldsymbol{S}^{n-1}} \left[\int_U \phi(\lambda, u \cdot \omega) \phi_k(u^{-1} \cdot o) \, du \right] e^{i\lambda(x, \omega)} \lambda^{n-1} d\lambda \, d\omega.$$

By Corollary 3.3 the function in the bracket belongs to E_k, so

$$d(k) \int_U \phi(\lambda, u \cdot \omega) \phi_k(u^{-1} \cdot o) \, du = \sum_{1 \leq m \leq d(k)} \phi_{km}(\lambda) S_{km}(\omega),$$

where (using $\phi_k(u^{-1} \cdot o) = \phi_k(u \cdot o)$)

$$\phi_{km}(\lambda) = d(k) \int_U \int_U \phi(\lambda, uv \cdot o) \phi_k(u^{-1} \cdot o) \overline{S_{km}(v \cdot o)} \, du \, dv$$

$$= d(k) \int_{\boldsymbol{S}^{n-1}} \phi(\lambda, \omega)(\phi_k * \overline{S_{km}})(\omega) \, d'\omega,$$

so by (13)

$$\phi_{km}(\lambda) = \overline{S_{km}(o)} \int_{S^{n-1}} \phi(\lambda, \omega) \phi_k(\omega) \, d'\omega.$$

We now have

$$f(x) = \sum_{km} f_{km}(x), \qquad x = re_n, \quad r \in \boldsymbol{R}, \tag{19}$$

where

(20) $$f_{km}(x) = (2\pi)^{-n} \int_{\mathbf{R}^+ \times \mathbf{S}^{n-1}} \phi_{km}(\lambda) S_{km}(\omega) e^{i\lambda(x,\omega)} \lambda^{n-1} \, d\lambda \, d\omega.$$

Because of our assumptions about ϕ, the function $\phi_{km}(\lambda)\lambda^{-k}$ is even and holomorphic and

(21) $$|\phi_{km}(\lambda)| \leq C_N(1 + |\lambda|)^{-N} e^{A|\mathrm{Im}\,\lambda|}.$$

In order to deal with (20) we shall use the following classical lemma (cf., Bochner [1955], p. 37, Erdèlyi *et al.* [1953], Vol. II, p. 247, or Vilenkin [1968], p. 554). We shall indicate a group-theoretic proof later.

Lemma 3.6. *For each* $p \in E_k$, $\eta \in \mathbf{S}^{n-1}$,

$$\int_{\mathbf{S}^{n-1}} e^{i\lambda(\eta,\omega)} p(\omega) \, d\omega = c_{n,k} \, p(\eta) \frac{J_{k+(1/2)n-1}(\lambda)}{\lambda^{(1/2)n-1}},$$

where $c_{n,k} = (2\pi)^{(1/2)n} i^k$.

Because of this lemma and (19), (20), our problem is reduced to proving

(22) $$\int_0^\infty \phi_{km}(\lambda)(\lambda r)^{1-(1/2)n} J_{k+(1/2)n-1}(\lambda r) \lambda^{n-1} \, d\lambda = 0$$

for $r > A$.

First let us assume $\phi(\lambda, \omega) = \phi(\lambda)$ is independent of ω. We "extend" ϕ to an invariant holomorphic function Φ of n variables by putting

$$\Phi(\zeta) = \Phi(\zeta_1, \ldots, \zeta_n) = \phi(\lambda)$$

if $\lambda^2 = \zeta_1^2 + \cdots + \zeta_n^2$. Since ϕ is even, this is possible. Writing

$$\lambda = \mu + i\nu, \qquad \zeta = \xi + i\eta, \qquad \xi, \eta \in \mathbf{R}^n,$$

we have

(23) $$\mu^2 - \nu^2 = |\xi|^2 - |\eta|^2, \; \mu^2\nu^2 = (\xi \cdot \eta)^2,$$

whence

(24) $$|\lambda|^4 = (|\xi|^2 - |\eta|^2)^2 + 4(\xi \cdot \eta)^2$$

and

$$2|\mathrm{Im}\,\lambda|^2 = |\eta|^2 - |\xi|^2 + [(|\xi|^2 - |\eta|^2)^2 + 4(\xi \cdot \eta)^2]^{1/2}.$$

In particular, since $|(\xi \cdot \eta)| \leq |\xi||\eta|$, we have

(25) $$|\mathrm{Im}\,\lambda| \leq |\eta|.$$

Now ϕ satisfies for each $N \in \boldsymbol{Z}^+$

$$|\phi(\lambda)| \leq C_N(1 + |\lambda|)^{-N} e^{A|\text{Im}\,\lambda|} \tag{26}$$

and (17) can be written

$$f(x) = (2\pi)^{-n} \int_{\boldsymbol{R}^n} \Phi(\xi_1, \ldots, \xi_n) e^{i(x,\xi)}\, d\xi.$$

By a suitable variation of the classical Paley–Wiener theorem argument we shall prove $f \in \mathcal{D}(\boldsymbol{R}^n)$.

If $\eta \in \boldsymbol{R}^n$ is any fixed vector, we can by Cauchy's theorem shift the integration into the complex domain, giving

$$f(x) = (2\pi)^{-n} \int_{\boldsymbol{R}^n} \Phi(\xi + i\eta) e^{i(x,\xi + i\eta)}\, d\xi. \tag{27}$$

In fact, (24) and (26) show that, η being fixed, $\Phi(\xi + it\eta)$ is rapidly decreasing in ξ, uniformly for $0 \leq t \leq 1$. Also, by (25), and (24), we have

$$\begin{aligned} |\Phi(\xi + i\eta)| &\leq C_N(1 + |\lambda|)^{-N} e^{A|\eta|} \\ &\leq C_N e^{A|\eta|}(1 + ||\xi|^2 - |\eta|^2|^{1/2})^{-N}. \end{aligned}$$

Taking $N = n + 1$ we have

$$\int_{\boldsymbol{R}^n} (1 + ||\xi|^2 - |\eta|^2|^{1/2})^{-N}\, d\xi \leq a|\eta|^n + b,$$

where a and b are constants, as we see by breaking the integral up into the regions $|\xi| \leq 2|\eta|$ and $|\xi| > 2|\eta|$. Thus (27) implies

$$|f(x)| \leq (2\pi)^{-n} C_{n+1}(a|\eta|^n + b) e^{A|\eta| - (x,\eta)}, \tag{28}$$

valid for all $x, \eta \in \boldsymbol{R}^n$. Now fix x with $|x| > A$ and put $\eta = tx$ in (28). Since

$$(a|x|^n t^n + b) e^{t|x|(A - |x|)} \to 0$$

as $t \to +\infty$ we obtain

$$f(x) = 0 \qquad \text{for} \quad |x| > A. \tag{29}$$

Consider now the general $\phi(\lambda, \omega)$. We recall that $\phi_{km}(\lambda)\lambda^{-k}$ is even and holomorphic and ϕ_{km} satisfies (21). We now write the expression in (22) as r^k times

$$r^{-((1/2)n + k) + 1} \int_0^\infty (\phi_{km}(\lambda)\lambda^{-k}) J_{k + (1/2)n - 1}(\lambda r) \lambda^{(1/2)n + k}\, d\lambda. \tag{30}$$

But as a special case[†] of Lemma 3.6

$$\int_{S^{n-1+2k}} e^{i\lambda(x,\omega)}\, d\omega = c\,\frac{J_{k+(1/2)n-1}(\lambda r)}{(\lambda r)^{(1/2)n+k-1}},$$

where $r = |x|$ and c is a constant. Consequently, the expression (30) is up to a constant factor the Fourier transform in $\boldsymbol{R}^{n+2k}$ of the radial function

$$\Phi_{km}(\xi_1, \ldots, \xi_{n+2k}) = \phi_{km}(|\xi|)|\xi|^{-k}, \qquad \xi \in \boldsymbol{R}^{n+2k}.$$

Letting $\phi_{km}(\lambda)\lambda^{-k}$ play the role of $\phi(\lambda)$ above, the function

$$\Phi(\zeta_1, \ldots, \zeta_{n+2k})$$

restricts to Φ_{km} on $\boldsymbol{R}^{n+2k}$. By the conclusion (29) we therefore find that the integral (30) vanishes for $r > A$, as desired.

Finally, we give a proof of Lemma 3.6. Consider the function

$$(31) \qquad \Phi_{\lambda,\delta}(x) = \int_U e^{i\lambda(x, ue_n)}\langle v, \delta(u)v\rangle\, du, \qquad x \in \boldsymbol{R}^n,$$

v denoting the unit vector $d(\delta)^{1/2}\phi_k \in E_k$.

Lemma 3.7. *Let* $w \in E_k$ *and*

$$x = l \cdot re_n, \qquad l \in U, \quad r \in \boldsymbol{R}.$$

Then

$$(32) \qquad \int_U e^{i\lambda(x, ue_n)}\langle w, \delta(u)v\rangle\, du = \langle w, \delta(l)v\rangle\Phi_{\lambda,\delta}(re_n).$$

Proof. Define the map $F\colon \boldsymbol{R}^n \to E_k$ by

$$F(x) = \int_U e^{-i\bar{\lambda}(x, ue_n)}\delta(u)v\, du.$$

Then the left-hand side of (32) equals $\langle w, F(x)\rangle$. Also,

$$F(u \cdot x) = \delta(u)F(x),$$

so in particular, the vector $F(re_n)$ is M-invariant and hence proportional to v. Thus

$$F(re_n) = (\Phi_{\lambda,\delta}(re_n))^{-} v,$$

so that

$$\langle w, F(l \cdot re_n)\rangle = \langle w, \delta(l)F(re_n)\rangle = \langle w, \delta(l)v\rangle\Phi_{\lambda,\delta}(re_n),$$

proving the lemma.

[†]See "Some Details," p. 611.

It remains to compute $\Phi_{\lambda,\delta}(x)$. We have of course

$$L_{\boldsymbol{R}^n}\Phi_{\lambda,\delta} = -\lambda^2\Phi_{\lambda,\delta}, \tag{33}$$

and we shall now use (3) to separate the variables. Since $\Phi_{\lambda,\delta}$ is M-invariant, it is not hard to see that $\Phi_{\lambda,\delta}(l \cdot e_n) = \Phi_{\lambda,\delta}(l^{-1}e_n)$. Hence

$$\Phi_{\lambda,\delta}(rle_n) = \Phi_{\lambda,\delta}(rl^{-1}e_n) = \int_U e^{i\lambda r(e_n,\, lue_n)}\langle v, \delta(u)v\rangle\, du,$$

so that

$$\Phi_{\lambda,\delta}(rle_n) = \int_U e^{i\lambda r(e_n,\, ue_n)}\langle \delta(l)v, \delta(u)v\rangle\, du.$$

Thus for a fixed r, the function $lM \to \Phi_{\lambda,\delta}(rle_n)$ on S^{n-1} is an eigenfunction of L with eigenvalue $-k(k+n-2)$. Combining this with (33) and (3) we deduce that the function $\psi(r) = \Phi_{\lambda,\delta}(re_n)$ satisfies the differential equation

$$\psi'' + \frac{n-1}{r}\psi' - \frac{k(k+n-2)}{r^2}\psi = -\lambda^2\psi. \tag{34}$$

But the constant multiples of the function $J_{k+(1/2)n-1}(\lambda r)/(\lambda r)^{(1/2)n-1}$ are the smooth solutions to this differential equation. This, combined with Lemma 3.7, gives the formula in Lemma 3.6 up to a constant factor, which we shall not determine since it is not necessary for the application at hand.[†]

At any rate we have proved $f_{km}(x) = 0$ for $|x| > A$, so by (19) $f(x) = 0$ for $x = re_n$, $r > A$. This proves Theorem 2.10.

Remark. While the inclusion

$$(\mathscr{D}_A)^{\sim} \subset \mathscr{H}^A$$

in Theorem 2.9 is obvious, it should be noted that the surjectivity $(\mathscr{D}_A)^{\sim} = \mathscr{H}^A$ follows from Theorem 2.10. In fact, if $F \in \mathscr{H}^A$, the function $\phi(\lambda, \omega) = F(\lambda\omega)$ certainly satisfies (19) in §2; also, expanding F in power series around 0, we obtain a power series

$$\phi(\lambda, \omega) = F(\lambda\omega) = \sum_{l \geq 0} \lambda^l P_l(\omega),$$

where P_l is a polynomial in $(\omega_1, \ldots, \omega_n)$ of degree l. If $l < k$, it is clear from (5) that P_l is orthogonal to the function $\omega \to (a, \omega)^k$ on S^{n-1} if the vector $a \in \boldsymbol{C}^n$ is isotropic. Thus condition (ii) of Theorem 2.10 is fulfilled, so there exists an $f \in \mathscr{D}_A(\boldsymbol{R}^n)$ such that $\phi(\lambda, \omega) = \tilde{f}(\lambda\omega)$. But then $F = \tilde{f}$, so Theorem 2.9 is also proved. Naturally, it can be proved much more directly.

[†]See "Some Details," p. 611.

§4. The Hyperbolic Plane H^2

1. Non-Euclidean Fourier Analysis. Problems and Results

Let D be the open disk $|z| < 1$ in $\boldsymbol{R}^2$ with the usual manifold structure but given the Riemannian structure (cf. [DS], Chapter I, §9)

$$\langle u, v\rangle_z = \frac{(u, v)}{(1 - |z|^2)^2} \tag{1}$$

if u and v are any tangent vectors at $z \in D$. Here (,) denotes the usual inner product on $\boldsymbol{R}^2$. This Riemannian manifold is usually called the *Poincaré model of the hyperbolic plane* $\boldsymbol{H}^2$. We shall now recall briefly some of its geometric properties (cf., [DS], pp. 93, 548). Since

$$\frac{\langle u, v\rangle^2}{\langle u, u\rangle\langle v, v\rangle} = \frac{(u, v)^2}{(u, u)(v, v)},$$

the angle between vectors in this Riemannian structure coincides with the Euclidean angle.

The length of a curve $\gamma(t)$ $(\alpha \le t \le \beta)$ in D is, according to the general definition for a Riemannian manifold, defined by

$$L(\gamma) = \int_\alpha^\beta \langle \gamma'(t), \gamma'(t)\rangle^{1/2}\, dt$$

and the distance between any two points $z, w \in D$ defined by

$$d(z, w) = \inf_\gamma L(\gamma),$$

the infimum being taken over all curves joining z and w. If $\gamma(t) = (x(t), y(t))$ and $s(\tau)$ is the arc length of the segment $\gamma(t)$ $(\alpha \le t \le \tau)$, we get from (1)

$$\left(\frac{ds}{d\tau}\right)^2 = (1 - x(\tau)^2 - y(\tau)^2)^{-2}\left[\left(\frac{dx}{d\tau}\right)^2 + \left(\frac{dy}{d\tau}\right)^2\right],$$

which is the meaning of the customary terminology

$$ds^2 = \frac{dx^2 + dy^2}{(1 - x^2 - y^2)^2}. \tag{2}$$

In particular, if $\gamma(\alpha) = o$ (the origin), $\gamma(\beta) = x$ (point on the x-axis) and we denote by γ_0 the line segment from o to x, we get from

$$\frac{x'(\tau)^2}{(1 - x(\tau)^2)^2} \le \frac{x'(\tau)^2 + y'(\tau)^2}{(1 - x(\tau)^2 - y(\tau)^2)^2}$$

the inequality[†]

$$L(\gamma_0) \leq L(\gamma).$$

This shows that straight lines through the origin are geodesics. Also,

(3) $$d(o, x) = L(\gamma_0) = \int_0^1 \frac{|x|}{1 - t^2 x^2}\, dt = \frac{1}{2} \log \frac{1 + |x|}{1 - |x|}.$$

Consider now the group

$$\boldsymbol{SU}(1, 1) = \left\{ \begin{pmatrix} a & b \\ \bar{b}, & \bar{a} \end{pmatrix} : |a|^2 - |b|^2 = 1 \right\},$$

which acts on D by means of the maps

(4) $$g : z \to \frac{az + b}{\bar{b}z + \bar{a}}, \qquad (z \in D).$$

The action is transitive; the subgroup fixing o is $\boldsymbol{SO}(2)$, so we have the identification

(5) $$D = \boldsymbol{SU}(1, 1)/\boldsymbol{SO}(2).$$

The Riemannian structure (1) is preserved by the maps (4). In fact, let $z(t)$ be a curve with $z(0) = z$, $z'(0) = u$. Then

$$g \cdot u = \left\{ \frac{d}{dt} g(z(t)) \right\}_{t=0} = \text{the vector } \frac{z'(0)}{(\bar{b}z + \bar{a})^2} \quad \text{at} \quad g(z),$$

and the desired relation

$$\langle g \cdot u, g \cdot u \rangle_{g(z)} = \langle u, u \rangle_z$$

follows by a simple computation.

The mappings (4) are conformal and map circles (and lines) into circles and lines. It follows that the *geodesics in D are the circular arcs perpendicular to the boundary* $|z| = 1$.

If z_1, z_2 are any points in D, the isometry

$$z \to \frac{z - z_1}{1 - \bar{z}_1 z}$$

maps z_1 to o and z_2 to the point $(z_2 - z_1)/(1 - \bar{z}_1 z_2)$. Following this with a rotation around o, this point is mapped into $|z_2 - z_1|/|1 - \bar{z}_1 z_2|$, so we deduce from (3)

(6) $$d(z_1, z_2) = \frac{1}{2} \log \frac{|1 - \bar{z}_1 z_2| + |z_2 - z_1|}{|1 - \bar{z}_1 z_2| - |z_2 - z_1|}, \qquad z_1, z_2 \in D.$$

[†]See "Some Details," p. 611.

Writing the Riemannian structure (2) in the general form (cf., Chapter II, §2, No. 4)

$$ds^2 = \sum_{i,j} g_{ij}\, dx_i\, dx_j, \qquad g_{ij} = (1 - |z|^2)^{-2}\delta_{ij}$$

and putting as usual $\bar{g} = |\det(g_{ij})|$, $g^{ij} = (g_{ij})^{-1}$ (inverse matrix), the Riemannian measure

$$f \to \int f \sqrt{\bar{g}}\, dx_1 \cdots dx_n,$$

and the Laplace–Beltrami operator

$$L : f \to \frac{1}{\sqrt{\bar{g}}} \sum_k \partial_k \left(\sum_i g^{ik} \sqrt{\bar{g}}\, \partial_i f \right)$$

become, respectively,

(7) $$dz = (1 - x^2 - y^2)^{-2}\, dx\, dy,$$

(8) $$L = (1 - x^2 - y^2)^2 \left(\frac{\partial^2}{\partial x^2} + \frac{\partial^2}{\partial y^2} \right).$$

By general theory (Chapter II, §2) they are invariant under all isometries; for the isometries (4) above (which together with $z \to \bar{z}$ generate all isometries of D) this can of course be verified directly. It is also easy to prove directly the special case of Chapter II, Corollary 4.11, that each $\boldsymbol{SU}(1, 1)$-invariant differential operator on D is a polynomial in L.

With these preparations we now consider Problems A, B, and C. We shall define some eigenfunctions of L, motivated by the Euclidean case. If $\mu \in \boldsymbol{C}$ and $\omega \in \boldsymbol{S}^{n-1}$, the function

(9) $$x \to e^{\mu(x, \omega)} \qquad (x \in \boldsymbol{R}^n)$$

has the following properties:

(i) It is an eigenfunction of the Laplacian $L_{\boldsymbol{R}^n}$ on $\boldsymbol{R}^n$.

(ii) It is a *plane wave with normal* ω; that is, it is constant on each hyperplane perpendicular to ω.

A hyperplane in $\boldsymbol{R}^n$ is orthogonal to a family of parallel lines. The geometric analog for D is a *horocycle*, i.e., a circle in D tangential to the boundary $B = \partial D$; in fact such a circle ξ is orthogonal to all geodesics in D tending to the point of contact b. If z is a point on ξ we put

(10) $\langle z, b \rangle =$ distance from o to ξ (with sign; to be taken negative if o lies inside ξ).

This "inner product" $\langle z, b\rangle$ is in fact the non-Euclidean analog of (x, ω), which geometrically means the (signed) distance from $0 \in \boldsymbol{R}^n$ to the hyperplane through x with normal ω. We shall write $\xi(z, b)$ for the above horocycle through z and b.

By analogy with (9) we consider the function

$$e_{\mu, b}: z \to e^{\mu\langle z, b\rangle}, \qquad z \in D. \tag{11}$$

Lemma 4.1. *For the Laplacian L we have*

$$Le_{\mu, b} = \mu(\mu - 2)e_{\mu, b}.$$

Proof. For $t \in \boldsymbol{R}$ let

$$a_t = \begin{pmatrix} \cosh t & \sinh t \\ \sinh t & \cosh t \end{pmatrix} \in \boldsymbol{SU}(1, 1).$$

Then by (3) we find

$$d(o, a_t \cdot o) = d(o, \tanh t) = t. \tag{12}$$

If b_0 is the point of B on the positive x-axis, we therefore have

$$e_{\mu, b_0}(\tanh t) = e^{\mu t}. \tag{13}$$

The horocycles tangential to B at b_0 are the orbits of the group

$$N = \left\{ n_s = \begin{pmatrix} 1 + is & -is \\ is & 1 - is \end{pmatrix} : s \in \boldsymbol{R} \right\}.$$

In fact, the orbit $N \cdot o$ is such a horocycle and so is the orbit $Na_t \cdot o = a_t N \cdot o$ (a_t normalizes N). Consider now functions on D which, like e_{μ, b_0}, are constant on each of these horocycles. Because of its N-invariance, L maps the class of such functions into itself. The restriction of L to such functions is a differential operator L_0 in the variable t. Since the isometry a_t satisfies $a_t \cdot \tanh \tau = \tanh(t + \tau)$, the invariance of L under a_t means that L_0 is invariant under the translation $\tau \to \tau + t$. Thus L_0 has constant coefficients, so $e^{\mu t}$ is an eigenfunction for it; hence by (13), e_{μ, b_0} is an eigenfunction of L. The eigenvalue can be calculated by expressing L in the coordinates $a_t n_s \cdot o \to (t, s)$; in fact, one finds

$$L_0 = \frac{d^2}{dt^2} - 2\frac{d}{dt}.$$

A simpler computation of the eigenvalue is suggested later (see also Exercise B2).

Reformulating the lemma, we have

$$L_z(e^{(i\lambda + 1)\langle z, b\rangle}) = -(\lambda^2 + 1)e^{(i\lambda + 1)\langle z, b\rangle}, \qquad \lambda \in \boldsymbol{C}.$$

We now define a Fourier transform on D motivated by the Euclidean case (1) in §2. Again dz will denote the surface element (7).

Definition. If f is a complex-valued function on D, its *Fourier transform* is defined by

$$\tilde{f}(\lambda, b) = \int_D f(z) e^{(-i\lambda+1)\langle z, b\rangle}\, dz \tag{14}$$

for all $\lambda \in \boldsymbol{C}$, $b \in B$, for which this integral exists.

In order to state the next theorem we call a C^∞-function $\psi(\lambda, b)$ on $\boldsymbol{C} \times B$, which is holomorphic in λ, a *holomorphic function of uniform exponential type R* if for each $N \in \boldsymbol{Z}^+$

$$\sup_{\lambda \in \boldsymbol{C},\, b \in B} e^{-R|\mathrm{Im}\,\lambda|}(1 + |\lambda|)^N |\psi(\lambda, b)| < \infty.$$

Here Im λ denotes the imaginary part of λ.

We can now state rather explicit answers to Problems A, B, and C.

Theorem 4.2.

(i) *If* $f \in \mathscr{D}(D)$, *then*

$$f(z) = \frac{1}{4\pi} \int_{\boldsymbol{R}} \int_B \tilde{f}(\lambda, b) e^{(i\lambda+1)\langle z, b\rangle} \lambda \tanh\left(\frac{\pi\lambda}{2}\right) d\lambda\, db, \tag{15}$$

where db is the circular measure on B normalized by $\int db = 1$.

(ii) *The mapping* $f \to \tilde{f}$ *is a bijection of* $\mathscr{D}(D)$ *onto the space of holomorphic functions* $\psi(\lambda, b)$ *of uniform exponential type satisfying the functional equation*

$$\int_B e^{(i\lambda+1)\langle z, b\rangle} \psi(\lambda, b)\, db = \int_B e^{(-i\lambda+1)\langle z, b\rangle} \psi(-\lambda, b)\, db.$$

(iii) *The mapping* $f \to \tilde{f}$ *extends to an isometry of* $L^2(D)$ *onto*

$$L^2(\boldsymbol{R}^+ \times B, (2\pi)^{-1}\lambda \tanh(\tfrac{1}{2}\pi\lambda)\, d\lambda\, db).$$

Remarks. The three parts of this theorem are non-Euclidean analogs of the inversion formula, the Paley–Wiener theorem, and the Plancherel formula, respectively, for the Fourier transform on $\boldsymbol{R}^n$. There are however some interesting differences:

(a) While the functional equation in (ii) gives genuine restrictions on the Fourier tansform $\psi(\lambda, b) = \tilde{f}(\lambda, b)$, the analogous condition for the Euclidean Fourier transform $\tilde{f}$ would be

$$\int_{S^1} e^{i\lambda(x, \omega)} \tilde{f}(\lambda\omega)\, d\omega = \int_{S^1} e^{-i\lambda(x, \omega)} \tilde{f}(-\lambda\omega)\, d\omega.$$

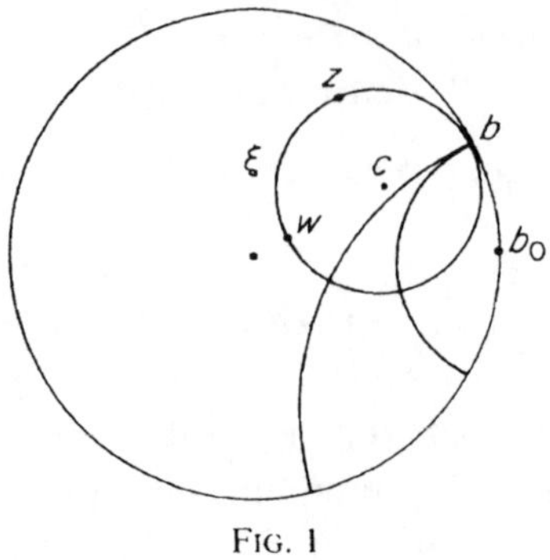

FIG. 1

This condition is however quite trivial so imposes no restrictions on $\tilde{f}$. The appropriate Euclidean analog is condition (ii) in Theorem 2.10.

(b) Comparing formulas (14) and (15) with their Euclidean analogs (1), (2) in §2 we see the kernel $e^{2\langle z, b\rangle}$, which does not have any Euclidean analog. If w is the point on the horocycle $\xi(z, b)$ closest to o, we have from (3),

$$e^{2\langle z, b\rangle} = \frac{1 + |w|}{1 - |w|}.$$

But $|w|$ can be found by using the cosine relation on the triangles (o, z, b) and (o, z, c), where c is the center of the horocycle. We obtain (see Fig. 1)

$$\cos(zob) = \frac{1 + |z|^2 - |z - b|^2}{2|z|} = \frac{|z|^2 + (\frac{1}{2}(1 + |w|))^2 - (\frac{1}{2}(1 - |w|))^2}{|z|(1 + |w|)},$$

so

$$e^{2\langle z, b\rangle} = \frac{1 - |z|^2}{|z - b|^2}, \tag{16}$$

which is just the classical Poisson kernel

$$P(z, b) = \frac{1 - |z|^2}{|z - b|^2} = \frac{1 - r^2}{1 - 2r\cos(\phi - \theta) + r^2}$$

if $z = re^{i\theta}$, $b = e^{i\phi}$.

The Poisson representation

$$u(z) = \int_B \frac{1 - |z|^2}{|z - b|^2} F(b)\, db$$

of a harmonic function $u(z)$ with continuous boundary values F can thus be written

$$u(z) = \int_B e^{2\langle z,b\rangle} F(b)\, db$$

and can therefore, in accordance with (15), *be viewed as a formula in non-Euclidean Fourier analysis.*

We shall see later that this leads to a simple proof of Schwarz's limit theorem

$$\lim_{z \to b} u(z) = F(b)$$

at each point b of continuity for F (Theorem 4.25). It also leads to group-theoretic proof of the classical Fatou theorem that a bounded harmonic function on D has boundary values along almost all radii (cf. Theorem 4.26).

It should be observed that because of (8) the Euclidean and the non-Euclidean harmonic functions coincide. We also note that (16) can be used to give a slightly simpler proof of Lemma 4.1.

In order to describe the solution to Problem **B** we need the concept of an *analytic functional* (*hyperfunction*) on a compact analytic manifold.

Let $\mathscr{A}(B)$ denote the space of analytic functions on the boundary B, considered as an analytic manifold. For each such manifold B, $\mathscr{A}(B)$ carries a certain natural topology; in the case of the circle it can be described rather simply as follows. Let U be an open annulus containing B, $\mathscr{H}(U)$ the space of holomorphic functions on U topologized by uniform convergence on compact subsets. Since each analytic function on B extends to a function in $\mathscr{H}(U)$ for a suitably chosen U, we can identify $\mathscr{A}(B)$ with the union $\bigcup_u \mathscr{H}(U)$ and give it the inductive limit topology. The elements of the dual space $\mathscr{A}'(B)$ are called *analytic functionals* (or *hyperfunctions*). Since the elements of $\mathscr{A}'(B)$ generalize measures, it is convenient to write

$$T(f) = \int_B f(b)\, dT(b), \qquad f \in \mathscr{A}(B), \quad T \in \mathscr{A}'(B),$$

For $\lambda \in \boldsymbol{C}$, let $\mathscr{E}_\lambda(D)$ denote the eigenspace

$$\mathscr{E}_\lambda(D) = \{f \in \mathscr{E}(D) : Lf = -(\lambda^2 + 1)f\}$$

with the topology induced by that of $\mathscr{E}(D)$.

Theorem 4.3. *The eigenfunctions of the Laplace–Beltrami operator L on D are precisely the functions*

$$f(z) = \int_B e^{(i\lambda+1)\langle z, b\rangle}\, dT(b),$$

where $\lambda \in \boldsymbol{C}$ and $T \in \mathscr{A}'(B)$. Moreover, if $i\lambda \neq -1, -3, -5, \ldots$, then the mapping $T \to f$ is a bijection of $\mathscr{A}'(B)$ onto $\mathscr{E}_\lambda(D)$.

We shall see later (85) that the hyperfunction T can be recovered from f by a suitable limit process at least for $\operatorname{Re}(i\lambda) > 0$. We shall also prove (Theorem 4.24) that T is a distribution on B if and only if f grows at most exponentially with the distance $d(o, z)$.

For $\lambda \in \boldsymbol{C}$ let T_λ denote the representation of $\boldsymbol{SU}(1, 1)$ on the eigenspace $\mathscr{E}_\lambda(D)$. For Problem C we then have the following result.

Theorem 4.4. *The eigenspace representation T_λ is irreducible if and only if*

$$i\lambda + 1 \notin 2\boldsymbol{Z}.$$

In the remainder of this section we give proofs of these theorems.

In addition to the subgroup $K = \boldsymbol{SO}(2) \subset \boldsymbol{SU}(1, 1) = G$ we have considered the subgroups

$$A = \left\{a_t = \begin{pmatrix} \operatorname{ch} t & \operatorname{sh} t \\ \operatorname{sh} t & \operatorname{ch} t \end{pmatrix} : t \in \boldsymbol{R}\right\},$$

$$N = \left\{n_x = \begin{pmatrix} 1 + ix & -ix \\ ix & 1 - ix \end{pmatrix} : x \in \boldsymbol{R}\right\}.$$

It is clear from (12) that the mapping $t \to a_t \cdot o$ is an isometry of $\boldsymbol{R}$ onto the geodesic $A \cdot o$. Let us now verify that the mapping $\phi : x \to n_x \cdot o$ is an isometry of $\boldsymbol{R}$ onto the horocycle $N \cdot o$. For this note that

$$\phi(x) = \frac{x}{x + i}, \qquad \frac{d\phi}{dx} = \frac{i}{(x + i)^2},$$

such that

$$\left\langle \frac{d\phi}{dx}, \frac{d\phi}{dx} \right\rangle_{x(x+i)^{-1}} = (x^2 + 1)^{-4} \Big/ \left(1 - \left|\frac{x}{x + i}\right|^2\right)^2 = 1,$$

proving the statement.

Let b_0 be the point of intersection of B with the positive x-axis. Consider the horocycles $\xi_t = Na_t \cdot o$, $\xi_{t+\Delta t} = Na_{t+\Delta t} \cdot o$ and the geodesics $n_x A \cdot o$ and $n_{x+\Delta x} A \cdot o$, indicated in Figure 2. Note (by matrix computation) that $a_t^{-1} n_x a_t = n_{xe^{-2t}}$ and that a_t is an isometry of D mapping

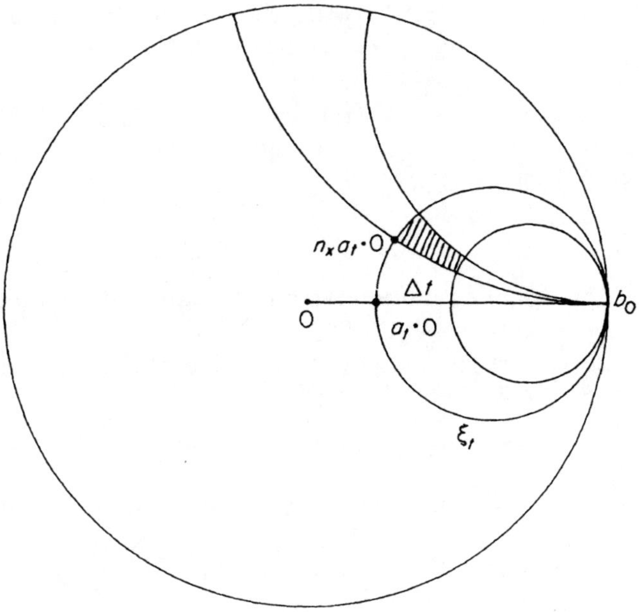

FIG. 2

$N \cdot o$ onto $Na_t \cdot o$. It follows that the shaded "rectangle" in Figure 2 has sides

$$\Delta t, \qquad e^{-2t}\Delta x, \qquad \Delta t, \qquad e^{-2(t+\Delta t)}\Delta x$$

such that the surface element dz is given by

$$\int_D f(z)\,dz = \int_{\mathbf{R}\times\mathbf{R}} f(n_x a_t \cdot o)e^{-2t}\,dx\,dt$$

and the arc element $d\omega_t$ on the horocycle ξ_t is given by

$$d\omega_t = e^{-2t}\,dx.$$

The formula for dz can also be proved by writing

$$\zeta = n_x a_t \cdot o = \frac{\operatorname{sh} t - ixe^{-t}}{\operatorname{ch} t - ixe^{-t}},$$

which gives

$$dz = \frac{i}{2}\frac{d\zeta \wedge d\bar{\zeta}}{(1-\zeta\bar{\zeta})^2} = e^{-2t}\,dx\,dt.$$

The integral formula above can also be written

$$\int_D f(z)\,dz = \int_{\mathbf{R}}\left(\int_{\xi_t} f(z)\,d\omega_t(z)\right)dt.$$

Rotating by θ we can therefore write the Fourier transform (14) in the form

$$\tilde{f}(\lambda, e^{i\theta}) = \int_{\mathbf{R}} e^{(-i\lambda+1)t}\left(\int_{\xi(\tanh t\, e^{i\theta},\, e^{i\theta})} f(z)\, d\omega_t(z)\right) dt,$$

and this will be useful later.

2. The Spherical Functions and Spherical Transforms

A *spherical function* on D is by definition a radial eigenfunction of L. If $z \in D$, $r = d(o, z)$, we know from (12) that

$$z = |z|e^{i\theta} = \tanh r\, e^{i\theta}.$$

In the coordinates (r, θ) (geodesic polar coordinates) L has the form

$$L = \frac{\partial^2}{\partial r^2} + 2\coth 2r \frac{\partial}{\partial r} + 4\sinh^{-2}(2r)\frac{\partial^2}{\partial \theta^2}. \tag{17}$$

Thus a spherical function ϕ satisfies

$$\frac{\partial^2 \phi}{\partial r^2} + 2\coth 2r \frac{\partial \phi}{\partial r} = -(\lambda^2 + 1)\phi. \tag{18}$$

By expanding in a power series in $\sinh 2r$ or by Theorem 2.7, Chapter II, we see that all smooth solutions of (18) are proportional—hence the constant multiples of the function

$$\phi_\lambda(z) = \int_B e^{(i\lambda+1)\langle z, b\rangle}\, db. \tag{19}$$

In particular, spherical functions are $\neq 0$ at $z = o$; normalizing spherical functions ϕ so that $\phi(o) = 1$, we see that (19) gives all the spherical functions. Also, $\phi_\lambda = \phi_{-\lambda}$, since both functions satisfy (18).

If we substitute $z = \tanh r\, e^{i\phi}$, $b = e^{i\theta}$ into (16), we obtain from (19),

$$\phi_\lambda(a_r \cdot o) = \frac{1}{2\pi}\int_{-\pi}^{\pi} (\cosh 2r - \sinh 2r \cos \theta)^{-(1/2)(i\lambda+1)}\, d\theta. \tag{20}$$

Now let $G = \boldsymbol{SU}(1, 1)$, $K = \boldsymbol{SO}(2)$ have Haar measures dg and dk normalized by

$$\int_G f(g \cdot o)\, dg = \int_D f(z)\, dz, \qquad \int_K dk = 1, \qquad f \in C_c(D)$$

(cf., Chapter I, §1). The spherical function ϕ_λ satisfies the functional equation

$$\int_K \phi_\lambda(gk \cdot z)\, dk = \phi_\lambda(g \cdot o)\phi_\lambda(z), \qquad z \in D,\, g \in G. \tag{21}$$

In fact, as a function of z the left-hand side is a radial eigenfunction of L. Hence it is a constant multiple of $\phi_\lambda(z)$. The constant is found by putting $z = o$.

Definition. If f is a radial function on D, its *spherical transform* is defined by

$$\tilde{f}(\lambda) = \int_D f(z)\phi_{-\lambda}(z)\, dz, \tag{22}$$

whenever this integral exists. Let $\mathscr{D}^\natural(D)$ denote the space of radial functions in $\mathscr{D}(D)$.

We shall now prove an inversion formula and a Plancherel formula for the spherical transform. The spectral theory of the ordinary differential operator (18) (cf., Weyl [1910], Coddington and Levinson [1955], pp. 251–256) does indeed suggest an inversion formula

$$f(z) = \int \tilde{f}(\lambda)\phi_\lambda(z)\, d\mu(\lambda),$$

where the "spectral measure" is determined by the asymptotic behavior of ϕ_λ at ∞. Specifically this is given by Harish-Chandra's $\mathbf{c}$-function $\lambda \to \mathbf{c}(\lambda)$ displayed in the next theorem.

Theorem 4.5. *If* $\operatorname{Re}(i\lambda) > 0$, *then the limit*

$$\mathbf{c}(\lambda) = \lim_{r \to +\infty} e^{(-i\lambda + 1)r}\phi_\lambda(a_r \cdot o)$$

exists and

$$\mathbf{c}(\lambda) = \pi^{-1/2} \frac{\Gamma(\frac{1}{2}i\lambda)}{\Gamma(\frac{1}{2}(i\lambda + 1))}.$$

Proof. Since $\phi_\lambda(a_{-r} \cdot o) = \phi_\lambda(a_r \cdot o) = \phi_{-\lambda}(a_r \cdot o)$, we can replace λ and r in the integral (20) by $-\lambda$ and $-r$. Using the substitution

$$u = \tan \tfrac{1}{2}\theta, \qquad \tfrac{1}{2}\, d\theta = (1 + u^2)^{-1}\, du,$$

we obtain

$$\begin{aligned} \phi_\lambda(a_r \cdot o) &= \frac{1}{\pi}\int_{-\infty}^{\infty} \left(\cosh 2r + \sinh 2r \frac{1 - u^2}{1 + u^2}\right)^{(1/2)(i\lambda - 1)} \frac{du}{1 + u^2} \\ &= \frac{1}{\pi} e^{(i\lambda - 1)r} \int_{-\infty}^{\infty} (1 + e^{-4r}u^2)^{(1/2)(i\lambda - 1)}(1 + u^2)^{-(1/2)(i\lambda + 1)}\, du. \end{aligned} \tag{23}$$

Assuming $\operatorname{Re}(i\lambda) > 0$ we shall show that the integral tends to a limit as $r \to +\infty$. Let $\lambda = \xi + i\eta$, so that $\operatorname{Re}(i\lambda) = -\eta > 0$. Select ε $(0 < \varepsilon < \frac{1}{2})$ so small that $1 + 2\varepsilon\eta > 0$. Then the integrand above has the majorization

$$\begin{aligned}(1 + e^{-4r}u^2)^{-(1/2)\eta - 1/2}&(1 + u^2)^{(1/2)\eta - 1/2}\\ &\le (1 + e^{-4r}u^2)^{-(1/2)\eta + \varepsilon\eta}(1 + u^2)^{(1/2)\eta - 1/2}\\ &\le (1 + u^2)^{-(1/2)\eta + \varepsilon\eta}(1 + u^2)^{(1/2)\eta - 1/2} = (1 + u^2)^{-(1/2) + \varepsilon\eta},\end{aligned}$$

and the last expression is integrable. This gives the formula

$$\lim_{r \to +\infty} e^{(-i\lambda + 1)r}\phi_\lambda(a_r \cdot o) = \frac{1}{\pi}\int_{-\infty}^{\infty} (1 + u^2)^{-(1/2)(i\lambda + 1)}\, du.$$

Putting $t = (1 + u^2)^{-1}$, this becomes

$$\frac{1}{\pi}\int_0^1 t^{(1/2)(i\lambda + 1)}t^{-(1/2)3}(1 - t)^{-1/2}\, dt = \frac{1}{\pi}\frac{\Gamma(\frac{1}{2}i\lambda)\Gamma(\frac{1}{2})}{\Gamma(\frac{1}{2}(i\lambda + 1))}$$

and the result follows.

Remark. The formula defines c as a meromorphic function on $\boldsymbol{C}$.

Using now the integral formula at the end of No. 1 we obtain from (19) the following expression for the spherical transform (22):

$$\tilde{f}(\lambda) = \int_{\boldsymbol{R}} e^{-i\lambda t}\left(e^{-t}\int_{\boldsymbol{R}} f(n_x a_t \cdot o)\, dx\right) dt. \tag{24}$$

Theorem 4.6. *The spherical transform $f \to \tilde{f}$ is inverted by the formula*

$$f(z) = \frac{1}{2\pi^2}\int_{\boldsymbol{R}} \tilde{f}(\lambda)\phi_\lambda(z)|c(\lambda)|^{-2}\, d\lambda, \qquad f \in \mathscr{D}^\natural(D). \tag{25}$$

Moreover,

$$\int_D |f(z)|^2\, dz = \frac{1}{2\pi^2}\int_{\boldsymbol{R}} |\tilde{f}(\lambda)|^2|c(\lambda)|^{-2}\, d\lambda. \tag{26}$$

Proof. In Chapter IV we shall give a proof of this theorem valid for a general symmetric space. This is based on detailed study of the spherical function, including Theorem 4.5. For the present case, $G = \boldsymbol{SU}(1, 1)$, we give a different proof, which, in addition to brevity, has some interesting features which we shall comment on later.

The starting point is formula (24). The function f is radial, so that with ζ as above

$$f(\zeta) = f(\tanh s)$$

if

$$(\tanh s)^2 = \zeta\bar{\zeta} = \frac{(\operatorname{sh} t)^2 + x^2 e^{-2t}}{(\operatorname{ch} t)^2 + x^2 e^{-2t}}.$$

This relation holds if

$$(\operatorname{ch} s)^2 = (\operatorname{ch} t)^2 + x^2 e^{-2t}.$$

Defining F on $[1, \infty)$ by

$$F((\operatorname{ch} s)^2) = f(\tanh s),$$

we have

$$F'((\operatorname{ch} s)^2) = f'(\tanh s)(2 \operatorname{sh} s \operatorname{ch}^3 s)^{-1},$$

so that, since $f'(o) = 0$, $\lim_{u\to 1} F'(u)$ exists. Furthermore,

$$\tilde{f}(\lambda) = \int_{\mathbf{R}} e^{-i\lambda t} e^{-t}\left(\int_{\mathbf{R}} F((\operatorname{ch} t)^2 + x^2 e^{-2t})\, dx\right) dt$$

$$= \int_{\mathbf{R}} e^{-i\lambda t}\left(\int_{\mathbf{R}} F((\operatorname{ch} t)^2 + y^2)\, dy\right) dt.$$

The integral equation

$$\phi(u) = \int_{\mathbf{R}} F(u + y^2)\, dy \qquad (u \geq 1)$$

(equivalent to Abel's integral equation) can be solved as follows:

$$\int_{\mathbf{R}} \phi'(u + z^2)\, dz = \int_{\mathbf{R}} \int_{\mathbf{R}} F'(u + y^2 + z^2)\, dy\, dz$$

$$= 2\pi \int_0^\infty F'(u + r^2) r\, dr = \pi \int_0^\infty F'(u + \rho)\, d\rho,$$

so that

$$-\pi F(u) = \int_{\mathbf{R}} \phi'(u + z^2)\, dz.$$

Thus

$$f(o) = F(1) = -\frac{1}{\pi} \int_{\mathbf{R}} \phi'(1 + z^2)\, dz$$

$$= -\frac{1}{\pi} \int_{\mathbf{R}} \phi'((\operatorname{ch} \tau)^2) \operatorname{ch} \tau\, d\tau.$$

On the other hand, we have by the Euclidean Fourier inversion formula

$$\phi((\operatorname{ch} t)^2) = \frac{1}{2\pi} \int_{\boldsymbol{R}} \tilde{f}(\lambda) e^{i\lambda t}\, d\lambda$$

$$= \frac{1}{2\pi} \int_{\boldsymbol{R}} \tilde{f}(\lambda) \cos \lambda t\, d\lambda,$$

since $\tilde{f}(\lambda)$ is even. Differentiating with respect to t, we obtain

$$-\phi'((\operatorname{ch} t)^2) 2 \operatorname{ch} t \operatorname{sh} t = \frac{1}{2\pi} \int_{\boldsymbol{R}} \tilde{f}(\lambda) \lambda \sin \lambda t\, d\lambda.$$

Using the formula

$$\int_{-\infty}^{\infty} \frac{\sin \lambda t}{\operatorname{sh} t}\, dt = \pi \tanh\left(\frac{\pi\lambda}{2}\right), \tag{27}$$

which is obtained by integrating the function $z \to e^{i\lambda z}/\operatorname{sh} z$ over the contour $\boldsymbol{R} \cup (\boldsymbol{R} + \pi i)$, we get from the above

$$f(o) = \frac{1}{2\pi^2} \int_{\boldsymbol{R}} \tilde{f}(\lambda) \frac{\lambda\pi}{2} \tanh\left(\frac{\pi\lambda}{2}\right) d\lambda.$$

But if $\lambda \in \boldsymbol{R}$, we have by simple identities for the gamma function,

$$|\boldsymbol{c}(\lambda)|^{-2} = \boldsymbol{c}(\lambda)^{-1} \boldsymbol{c}(-\lambda)^{-1} = \frac{\lambda\pi}{2} \tanh\left(\frac{\pi\lambda}{2}\right), \qquad \lambda \in \boldsymbol{R}, \tag{28}$$

so we have proved

$$f(o) = c \int_{\boldsymbol{R}} \tilde{f}(\lambda) |\boldsymbol{c}(\lambda)|^{-2}\, d\lambda, \qquad c = (2\pi^2)^{-1}. \tag{29}$$

Given $g \in \boldsymbol{SU}(1, 1)$ let us apply (29) to the average

$$F(z) = \int_K f(gk \cdot z)\, dk, \qquad z \in D.$$

Using the functional equation (21) we find

$$\tilde{F}(\lambda) = \int_K \left(\int_D f(gk \cdot z) \phi_{-\lambda}(z)\, dz \right) dk$$

$$= \int_D f(g \cdot z) \phi_{-\lambda}(z)\, dz = \int_D f(z) \phi_{-\lambda}(g^{-1} \cdot z)\, dz$$

$$= \int_D f(z) \left(\int_K \phi_{-\lambda}(g^{-1} k \cdot z)\, dk \right) dz = \phi_{-\lambda}(g^{-1} \cdot o) \tilde{f}(\lambda),$$

so that

$$\tilde{F}(\lambda) = \phi_\lambda(g \cdot o)\tilde{f}(\lambda).$$

Since $F(o) = f(g \cdot o)$, we get the desired formula,

(30) $$f(g \cdot o) = c \int_{\boldsymbol{R}} \tilde{f}(\lambda)\phi_\lambda(g \cdot o)|\boldsymbol{c}(\lambda)|^{-2}\, d\lambda.$$

For the second part of Theorem 4.6 consider for $f_1, f_2 \in \mathscr{D}^\natural(D)$ the convolution

$$(f_1 * f_2)(g \cdot o) = \int_G f_1(h \cdot o) f_2(h^{-1}g \cdot o)\, dh.$$

It has spherical transform

$$\begin{aligned}(f_1 * f_2)^\sim(\lambda) &= \int_G f_1(h \cdot o)\left(\int_G f_2(g \cdot o)\phi_{-\lambda}(hg \cdot o)\, dg\right) dh \\ &= \int_G f_1(h \cdot o)\left(\int_G f_2(g \cdot o)\left(\int_K \phi_{-\lambda}(hkg \cdot o)\, dk\right) dg\right) dh \\ &= \tilde{f}_1(\lambda)\tilde{f}_2(\lambda)\end{aligned}$$

because of (21). Applying (29) to the function

$$gK \to \int_D f(g \cdot z)\overline{f(z)}\, dz,$$

we obtain (26), so Theorem 4.6 is proved.

Ranges. In analogy to the subspace $\mathscr{D}^\natural(D) \subset \mathscr{D}(D)$ let $L^2_\natural(D)$ denote the space of radial functions in $L^2(D)$. Recall that a holomorphic function F on $\boldsymbol{C}$ is said to be an *entire function of exponential type* R if for each integer $N \geq 0$

(31) $$\sup_{\lambda \in \boldsymbol{C}} e^{-R|\operatorname{Im}\lambda|}(1 + |\lambda|)^N |F(\lambda)| < \infty.$$

Let $\mathscr{H}(\boldsymbol{C})$ denote the space of all such functions, $\mathscr{H}_e(\boldsymbol{C})$ the subspace of even functions in $\mathscr{H}(\boldsymbol{C})$. As special cases of Theorem 4.2 to be proved in the next subsection we have the following statements about the ranges of the spherical transform:

(32) $$\mathscr{D}^\natural(D)^\sim = \mathscr{H}_e(\boldsymbol{C});$$

(33) $$L^2_\natural(D)^\sim = L^2\left(\boldsymbol{R}^+, \frac{1}{2\pi}\lambda \tanh(\tfrac{1}{2}\lambda\pi)\, d\lambda\right).$$

In (32) f has support in $\mathrm{Cl}(B_R(o))$ if and only if $\tilde{f}$ has exponential type R; in (33) the mapping $f \to \tilde{f}$ is an isometry between the spaces indicated.

3. The Non-Euclidean Fourier Transform. Proof of the Main Result

We shall now prove Theorem 4.2, which includes the inversion formula, the Paley–Wiener theorem, and the Plancherel formula for the Fourier transform (14) on the hyperbolic plane D.

As before, let dg denote the Haar measure on the group $G = \boldsymbol{SU}(1, 1)$, normalized by

$$\int_G f(g \cdot o)\, dg = \int_D f(z)\, dz, \qquad f \in C_c(D).$$

Extending the convolution $*$ defined earlier, for two functions f_1, f_2 on D we put

$$(f_1 * f_2)(z) = \int_G f_1(g \cdot o) f_2(g^{-1} \cdot z)\, dg$$

whenever this integral exists. Equation (42) below shows how the Fourier transform behaves with respect to convolution.

Lemma 4.7. *Let $f \in C_c(D)$. Then*

$$f * \phi_\lambda(z) = \int_B e^{(i\lambda + 1)\langle z, b\rangle} \tilde{f}(\lambda, b)\, db.$$

Proof. We first observe the geometric identity

$$\langle g \cdot z, g \cdot b\rangle = \langle z, b\rangle + \langle g \cdot o, g \cdot b\rangle \tag{34}$$

for each $g \in \boldsymbol{SU}(1, 1)$. For this we consider the horocycles $\xi(g \cdot o, g \cdot b)$ and $\xi(g \cdot z, g \cdot b)$. As remarked earlier in connection with Figure 2, these horocycles cut segments of equal length of the parallel geodesics $(o, g \cdot b)$ and $(g \cdot o, g \cdot b)$. Now the identity (34) is clear from Figure 3. Taking $z = g^{-1} \cdot o$ the identity implies $\langle g^{-1} \cdot o, b\rangle = -\langle g \cdot o, g \cdot b\rangle$. Since

$$\begin{aligned}\langle g^{-1} \cdot z, b\rangle &= \langle z, g \cdot b\rangle + \langle g^{-1} \cdot o, b\rangle \\ &= \langle z, g \cdot b\rangle - \langle g \cdot o, g \cdot b\rangle,\end{aligned}$$

we obtain for the spherical function

$$\phi_\lambda(g^{-1} \cdot z) = \int_B e^{(i\lambda + 1)(\langle z, g \cdot b\rangle - \langle g \cdot o, g \cdot b\rangle)}\, db.$$

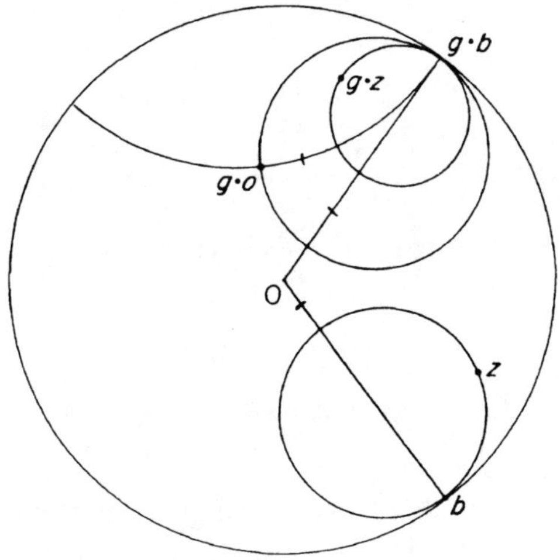

FIG. 3

On the other hand, when the conformal mapping g acts on the boundary B, the Jacobian is given by

$$\frac{d(g \cdot b)}{db} = P(g^{-1} \cdot o, b), \tag{34a}$$

where P is the Poisson kernel. In fact, if

$$e^{i\phi} = \frac{e^{i\theta} - a}{1 - \bar{a}e^{i\theta}},$$

a simple computation shows

$$\frac{d\phi}{d\theta} = \frac{1 - |a|^2}{|e^{i\theta} - a|^2}.$$

Using the non-Euclidean form of it given by (16), we get by the above

$$\frac{d(g \cdot b)}{db} = e^{2\langle g^{-1} \cdot o, b\rangle} = e^{-2\langle g \cdot o, g \cdot b\rangle}.$$

Hence, changing variables in the formula for ϕ_λ and putting $z = h \cdot o$, we obtain the symmetric formula

$$\phi_\lambda(g^{-1}h \cdot o) = \int_B e^{(-i\lambda+1)\langle g \cdot o, b\rangle} e^{(i\lambda+1)\langle h \cdot o, b\rangle}\, db. \tag{35}$$

The lemma follows by integrating against $f(g \cdot o)$.

Consider now for a fixed $g \in G$ the radial function

$$f_1(z) = \int_K f(gk \cdot z)\, dk.$$

From (29) we have

$$f_1(o) = c \int_{\boldsymbol{R}} \tilde{f}_1(\lambda) |\boldsymbol{c}(\lambda)|^{-2}\, d\lambda. \tag{36}$$

But

$$\begin{aligned} \tilde{f}_1(\lambda) &= \int_D \left(\int_K f(gk \cdot z)\, dk \right) \phi_{-\lambda}(z)\, dz \\ &= \int_D f(w) \phi_{-\lambda}(g^{-1} \cdot w)\, dw \\ &= \int_G f(h \cdot o) \phi_{-\lambda}(g^{-1} h \cdot o)\, dh = \int_G f(h \cdot o) \phi_\lambda(h^{-1} g \cdot o)\, dh, \end{aligned}$$

so

$$\tilde{f}_1(\lambda) = (f * \phi_\lambda)(g \cdot o).$$

Since $f_1(o) = f(g \cdot o)$, we get from Lemma 4.7 and (36)

$$f(z) = c \int_{\boldsymbol{R}} \left(\int_B \tilde{f}(\lambda, b) e^{(i\lambda + 1)\langle z, b \rangle}\, db \right) |\boldsymbol{c}(\lambda)|^{-2}\, d\lambda.$$

Using now formula (28) for $|\boldsymbol{c}(\lambda)|^{-2}$, we obtain the inversion formula in (15).

For the determination of the range of the Fourier transform $f(z) \to \tilde{f}(\lambda, b)$ [Theorem 4.2, (ii) and (iii)] we first introduce a definition.

Definition. The point $\lambda \in \boldsymbol{C}$ is called *simple* if the mapping $F \to f$ of $L^2(B)$ into $C^\infty(D)$ given by

$$f(z) = \int_B e^{(i\lambda + 1)\langle z, b \rangle} F(b)\, db \tag{37}$$

is one-to-one.

Proposition 4.8. *The point λ is nonsimple if and only if*

$$\lambda = i(1 + 2k), \qquad k \in \boldsymbol{Z}^+.$$

Remark. The "conceptual" version of this criterion is that the nonsimple points λ are the poles of the denominator of the $\boldsymbol{c}$-function.

We first prove a lemma (cf. Takahashi [1963]).

Lemma 4.9. *Let $F(\theta)$ be integrable on $0 \leq \theta \leq \pi$ and let*

$$H(t) = \int_0^{\pi} (\cosh t - \sinh t \cos \theta)^{-s} F(\theta)\, d\theta.$$

If $-s \notin \boldsymbol{Z}^+$, then $H \equiv 0$ if and only if $F \equiv 0$.

Proof. We have, by induction on p,

$$\frac{d^p}{dt^p}(\cosh t - \sinh t \cos \theta)^{-s}$$

$$= \sum_{q=0}^{p} c_q^p(s)(\cosh t - \sinh t \cos \theta)^{-s-q}(\sinh t - \cosh t \cos \theta)^q,$$

where the $c_q^p(s)$ are constants and

$$c_p^p(s) = (-1)^p s(s+1) \cdots (s+p-1).$$

If $-s \notin \boldsymbol{Z}^+$, then $c_p^p(s) \neq 0$ for all p. Thus if $H \equiv 0$ the equation $H^{(p)}(0) = 0$ implies successively

$$\int_0^{\pi} F(\theta) \cos^p \theta\, d\theta = 0, \qquad p = 0, 1, 2, \ldots,$$

so $F = 0$ (by the Stone–Weierstrass theorem), proving the lemma.

For Prop. 4.8 let $z = (\tanh r)e^{i\phi}$, $b = e^{i\theta}$. Then formula (37) reads

$$(38) \quad f(z) = \frac{1}{2\pi} \int_0^{2\pi} (\cosh 2r - \sinh 2r \cos \theta)^{-(1/2)(i\lambda+1)} F(e^{i(\phi+\theta)})\, d\theta.$$

Now if $\lambda = i(1+2k)$, $k \in \boldsymbol{Z}^+$, so that $-\frac{1}{2}(i\lambda + 1) = k$, the function $F(e^{i\theta}) = e^{(k+1)i\theta}$ gives $f \equiv 0$, so λ is not simple. On the other hand, suppose

$$-\tfrac{1}{2}(1 + i\lambda) \notin \boldsymbol{Z}^+.$$

If $f \equiv 0$ in (38), the proof of Lemma 4.9 implies

$$(39) \qquad \int_0^{2\pi} \cos^p \theta\, F(e^{i(\phi+\theta)})\, d\theta = 0, \qquad p \in \boldsymbol{Z}^+.$$

Expanding F in a Fourier series

$$F(e^{i(\phi+\theta)}) \sim \sum a_n e^{in\phi} e^{in\theta},$$

(39) implies successively

$$a_p e^{ip\phi} + a_{-p} e^{-ip\phi} = 0, \qquad p = 0, 1, 2, \ldots,$$

so $F = 0$, as desired.

Proposition 4.10. *If $-\lambda$ is simple, then the function space on B,*

$$\{\tilde{f}(\lambda, \cdot) : f \in \mathscr{D}(D)\}$$

is dense in $L^2(B)$.

Proof. If this were not so let $F \not\equiv 0$ in $L^2(B)$ satisfy

$$\int_B \tilde{f}(\lambda, b) F(b)\, db = 0 \qquad \text{for all} \quad f \in \mathscr{D}(D).$$

This implies

$$\int_B e^{(-i\lambda+1)\langle z, b\rangle} F(b) = 0 \qquad (z \in D),$$

contradicting the simplicity of $-\lambda$.

Passing now to the proof of Theorem 4.2, part (iii), we derive from Lemma 4.7 for λ real

$$\int_D f * \phi_\lambda(z) \overline{f(z)}\, dz = \int_B |\tilde{f}(\lambda, b)|^2\, db. \tag{40}$$

Here we multiply by $|c(\lambda)|^{-2}$, use Lemma 4.7 and the inversion formula, and obtain

$$\int_D |f(z)|^2\, dz = c \int_{R \times B} |\tilde{f}(\lambda, b)|^2 |c(\lambda)|^{-2}\, d\lambda\, db$$

$$= 2c \int_{R^+ \times B} |\tilde{f}(\lambda, b)|^2 |c(\lambda)|^{-2}\, d\lambda\, db,$$

the latter identity coming from the evenness in λ, which is clear from (40) and (28).

Now suppose $F \in L^2(R^+ \times B, |c(\lambda)|^{-2}\, d\lambda\, db)$ were orthogonal to the range, i.e.,

$$\int_{R^+ \times B} \tilde{f}(\lambda, b) F(\lambda, b) \lambda \tanh(\tfrac{1}{2}\pi\lambda)\, d\lambda\, db = 0, \qquad f \in \mathscr{D}(D). \tag{41}$$

If $\psi \in \mathscr{D}^\natural(D)$, we have

$$(f * \psi)^\sim(\lambda, b) = \tilde{f}(\lambda, b) \tilde{\psi}(\lambda). \tag{42}$$

In fact,

$$(f * \psi)^\sim(\lambda, b) = \int_D \int_G f(g \cdot o) \psi(g^{-1} \cdot z)\, dg\, e^{(-i\lambda+1)\langle z, b\rangle}\, dz$$

$$= \int_G f(g \cdot o) \left(\int_D \psi(z) e^{(-i\lambda+1)\langle g \cdot z, b\rangle}\, dz \right) dg.$$

Writing now

$$\langle g\cdot z, b\rangle = \langle z, g^{-1}\cdot b\rangle + \langle g\cdot o, b\rangle,$$

relation (42) follows.

The functions $\tilde{\psi}$ vanish at ∞ on $\boldsymbol{R}$ [by (24)]. They form an algebra closed under complex conjugation because

$$(\psi_1 * \psi_2)^\sim(\lambda) = \tilde{\psi}_1(\lambda)\tilde{\psi}_2(\lambda),$$
$$(\bar{\psi})^\sim(\lambda) = (\tilde{\psi}(\lambda))^-, \qquad \lambda\in\boldsymbol{R},$$

by $\phi_{-\lambda} = \phi_\lambda$. Finally, the algebra separates points on $\boldsymbol{R}/\boldsymbol{Z}_2$ because if $\tilde{\psi}(\lambda_1) = \tilde{\psi}(\lambda_2)$ for all $\psi\in\mathscr{D}^\natural(D)$, we have $\phi_{\lambda_1}\equiv\phi_{\lambda_2}$, so that applying L we get $\lambda_1^2+1=\lambda_2^2+1$. Thus the algebra $\mathscr{D}^\natural(D)^\sim$ is dense in the space of even continuous functions on $\boldsymbol{R}$, vanishing at ∞. Replacing $\tilde{f}(\lambda, b)$ in (41) by $\tilde{f}(\lambda, b)\tilde{\psi}(\lambda)$, we can therefore conclude that there exists a null set $N_f\subset\boldsymbol{R}^+$ (depending on f) such that if $\lambda\notin N_f$, then the function $b\to F(\lambda, b)$ lies in $L^2(B)$ and

$$\int_B F(\lambda, b)\tilde{f}(\lambda, b)\,db = 0. \tag{43}$$

For each $n\in\boldsymbol{Z}^+$ let $\phi_n\in\mathscr{D}(D)$ be the smoothed-out characteristic function of the ball $B_n(0)\subset D$, i.e., $\phi_n(z)=1$ for $d(o,z)<n-(1/n)$, $\phi_n(z)=0$ for $d(o,z)>n+(1/n)$, and $0\le\phi(z)\le 1$ for all z. Let $\mathfrak{M}$ denote the countable set of all functions on D of the form $\phi_n(z)p(x,y)$, where $n\in\boldsymbol{Z}^+$ and p is a polynomial with rational coefficients. Let $N=\bigcup_{f\in\mathfrak{M}}N_f$, so (43) holds for all $\lambda\notin N$ and all $f\in\mathfrak{M}$. Now if $f\in\mathscr{D}(D)$, we can choose $n\in\boldsymbol{Z}^+$ such that $f=\phi_n f$, and then a suitable subsequence $f_k=\phi_n f_k$ $(k=1,2,\ldots)$ from $\mathfrak{M}$ converges uniformly to f. It follows that for each $\lambda\in\boldsymbol{R}$, $\tilde{f}_k(\lambda, b)$ converges to $\tilde{f}(\lambda, b)$ uniformly on B. Thus (43) holds for all $f\in\mathscr{D}(D)$ and all $\lambda\in\boldsymbol{R}^+ - N$. But then Prop. 4.10 implies

$$\int_B F(\lambda, b)P(b)\,db = 0 \qquad \text{for all} \quad P\in L^2(B)$$

for $\lambda\in\boldsymbol{R}^+ - N$. Hence $\int_C F(\lambda, b)\,d\lambda\,db = 0$ for each "rectangle" C in $(\boldsymbol{R}^+ - N)\times B$, whence $F\equiv 0$ almost everywhere on $\boldsymbol{R}^+\times B$. This proves Theorem 4.2, part (iii).

Finally we prove Theorem 4.2, part (ii), characterizing the image $(\mathscr{D}(D))^\sim$. For this we must consider generalizations of the spherical function.

Proposition 4.11. *Let $m\in\boldsymbol{Z}$. The eigenfunctions f of L satisfying the homogeneity condition*

$$f(e^{i\theta}z) = e^{im\theta}f(z) \tag{44}$$

are the constant multiples of the functions

$$\Phi_{\lambda, m}(z) = \int_B e^{(i\lambda + 1)\langle z, b\rangle} \chi_m(b)\, db,$$

where $\lambda \in \mathbf{C}$ *and* $\chi_m(e^{i\phi}) = e^{im\phi}$.

Proof. In view of (17) and (44) the function $F(r) = f(\tanh r)$ satisfies

$$\frac{d^2F}{dr^2} + 2\coth(2r)\frac{dF}{dr} - 4m^2\sinh^{-2}(2r)F + (\lambda^2 + 1)F = 0 \tag{45}$$

for a suitable λ. Expanding F in a power series of $\sinh(2r)$, we can prove that the analytic solutions to (45) are proportional.[†] In fact, writing

$$F(r) = \sum_0^\infty a_k \sinh^k(2r),$$

we find the recursion formula

$$((k + 2)^2 - m^2)a_{k+2} = -[\tfrac{1}{4}(\lambda^2 + 1) + k(k + 1)]a_k,$$

from which the statement is readily derived. On the other hand, the functions $\Phi_{\lambda, m}(\tanh r)$ and $\Phi_{-\lambda, m}(\tanh r)$ are solutions, and by Prop. 4.8 they are not both 0. The analyticity of F being automatic, this proves the proposition. Another proof comes from Theorem 2.7 in Chapter II.

The function $\Phi_{\lambda, m}$ can be expressed in terms of the hypergeometric function. Using (16) and Erdélyi *et al.* [1953] (Vol. I, p. 81), we obtain, writing $v = \frac{1}{2}(i\lambda + 1)$,

$$\Phi_{\lambda, m}(|z|) = (1 - |z|^2)^v |z|^{|m|} \frac{\Gamma(|m| + v)}{\Gamma(v)|m|!} F(v, |m| + v; |m| + 1; |z|^2). \tag{46}$$

Using now the transformation formula

$$F(a, b; c; z) = (1 - z)^{c-a-b} F(c - a, c - b; c; z), \tag{47}$$

we obtain the following invariance property of $\Phi_{\lambda, m}$.

Proposition 4.12. *Let* p_m *denote the polynomial*

$$p_m(x) = (\tfrac{1}{2}(x + 1))(\tfrac{1}{2}(x + 1) + 1) \cdots (\tfrac{1}{2}(x + 1) + |m| - 1).$$

Then

$$\Phi_{\lambda, m}(z) = \Phi_{-\lambda, m}(z)\frac{p_m(i\lambda)}{p_m(-i\lambda)}.$$

† See "Some Details," p. 611.

Since the coefficients $2\coth(2r)$ and $-4m^2\sinh^{-2}(2r)$ in (45) have expansions

$$2\coth(2r) = -2 + 4\sum_0^\infty e^{-4nr}$$

and

$$-4m^2\sinh^{-2}(2r) = +16m^2 - 16m^2\sum_0^\infty e^{-4nr},$$

it is reasonable to try to find both solutions to (45) by means of such expansions. As $r \to +\infty$ the differential equation (45) approaches the equation

$$F'' + 2F' + (\lambda^2 + 1)F = 0,$$

which has solutions $e^{(\pm i\lambda - 1)r}$. This suggests finding a solution to (45) of the form

$$\Phi_\lambda(r) = e^{(i\lambda - 1)r}\sum_0^\infty \Gamma_n(\lambda)e^{-2nr}. \tag{48}$$

Substituting this into (45) and equating coefficients to $e^{(i\lambda - 1 - 2n)r}$, we are led to the recursion formula

$$n(n - i\lambda)\Gamma_n(\lambda) = \sum_{1 \le k \le [(1/2)n]} \Gamma_{n-2k}(\lambda)(2n - 4k - i\lambda + 1 + 4m^2k). \tag{49}$$

Putting $\Gamma_0 \equiv 1$, this relation defines $\Gamma_n(\lambda)$ recursively as a rational function on $\boldsymbol{C}$. Note that $\Gamma_n(\lambda) \equiv 0$ for n odd. In order to show that (48) converges we use (49) to estimate the growth of $\Gamma_n(\lambda)$ in n.

Lemma 4.13. *Let $i\lambda \notin 2\boldsymbol{Z}$ and let $h > 0$. Then there exists a constant $K_{\lambda,h}$ such that*

$$|\Gamma_n(\lambda)| \le K_{\lambda,h}e^{nh} \quad \textit{for} \quad n \in \boldsymbol{Z}^+. \tag{50}$$

Proof. (by induction on n). First we select constants $c_1 > 0$, $c_2 > 0$ such that for all even n

$$|n(n - i\lambda)| \ge c_1n^2, \qquad |2n + 1 + 2m^2n - i\lambda| \le c_2(n + 1).$$

Then by (49),

$$|\Gamma_n(\lambda)| \le \frac{c}{n}\sum_{k \ge 1} |\Gamma_{n-2k}(\lambda)|$$

with $c = \text{const}$. Let N_0 be an integer such that

$$c(1 - e^{-2h})^{-1} < N_0$$

and select $K = K_{\lambda, h}$ such that

$$|\Gamma_n(\lambda)| \le K e^{nh} \qquad \text{for} \quad n \le N_0.$$

Let $N \in \mathbf{Z}^+$, $N > N_0$. Assuming (50) holds for $n < N$, we have

$$\begin{aligned} |\Gamma_N(\lambda)| &\le \frac{c}{N} \sum_{k \ge 1} K e^{(N-2k)h} \\ &\le K e^{Nh} N^{-1} c(1 - e^{-2h})^{-1} \le K e^{Nh}, \end{aligned}$$

and this proves the lemma by induction.

Since $h > 0$ is arbitrary in Lemma 4.13, the estimate shows that the series for $\Phi_\lambda(r)$ converges, can be differentiated term by term, and provides a solution to (45). But then $\Phi_{-\lambda}$ is another solution, and if $i\lambda \notin 2\mathbf{Z}$, these solutions are linearly independent. It follows that for these λ, $\Phi_{\lambda,m}$ is a linear combination of Φ_λ and $\Phi_{-\lambda}$; say

$$\Phi_{\lambda, m} = \mathbf{C}_1(\lambda)\Phi_\lambda + \mathbf{C}_{-1}(\lambda)\Phi_{-\lambda}. \tag{51}$$

Here $\mathbf{C}_1$ and $\mathbf{C}_{-1}$ are holomorphic for $i\lambda \notin 2\mathbf{Z}$. In fact, we can evaluate the terms in (51) at points r_1, r_2 at which $\Phi_\lambda/\Phi_{-\lambda}$ takes different values and then solve with respect to $\mathbf{C}_1(\lambda)$ and $\mathbf{C}_{-1}(\lambda)$. To determine them we generalize Theorem 4.5 as follows.

Theorem 4.14. *If* $\mathrm{Re}(i\lambda) > 0$ *and* $F \in C(B)$, *then (with* b_0 *as in Fig. 2)*

$$\lim_{r \to +\infty} e^{(-i\lambda+1)r} \int_B e^{(i\lambda+1)\langle a_r \cdot o, b\rangle} F(b)\, db = \mathbf{c}(\lambda)\, F(b_0).$$

Proof. We have from (16), if $b = e^{i\theta}$, $z = a_r \cdot o = \tanh r$,

$$\begin{aligned} &\int_B e^{(i\lambda+1)\langle z, b\rangle} F(b)\, db \\ &\qquad = \frac{1}{2\pi} \int_{-\pi}^{\pi} (\cosh 2r - \sinh 2r \cos\theta)^{-(1/2)(i\lambda+1)} F(e^{i\theta})\, d\theta. \end{aligned}$$

Using the substitution

$$u = \tan(\tfrac{1}{2}\theta), \qquad \tfrac{1}{2} d\theta = (1 + u^2)^{-1}\, du,$$

the integral becomes for $g(u) = F(e^{i\theta})$,

$$\begin{aligned} &\frac{1}{\pi} \int_{-\infty}^{\infty} \left[\cosh 2r - \sinh 2r \frac{1 - u^2}{1 + u^2}\right]^{-(1/2)(i\lambda+1)} g(u) \frac{du}{(1 + u^2)} \\ &\qquad = \frac{1}{\pi} e^{(i\lambda+1)r} \int_{-\infty}^{\infty} (1 + e^{4r}u^2)^{-(1/2)(i\lambda+1)} (1 + u^2)^{(1/2)(i\lambda-1)} g(u)\, du, \end{aligned}$$

which by the substitution $v = e^{2r}u$ becomes

$$\frac{1}{\pi} e^{(i\lambda - 1)r} \int_{-\infty}^{\infty} (1 + v^2)^{-(1/2)(i\lambda + 1)}(1 + e^{-4r}v^2)^{(1/2)(i\lambda - 1)} g(e^{-2r}v)\, dv.$$

Because of the estimates in the proof of Theorem 4.5 we take $\lim_{r \to \infty}$ of this last integral by taking the limit under the integral sign. The result is

$$\int_{-\infty}^{\infty} (1 + v^2)^{-(1/2)(i\lambda + 1)} g(0)\, dv = \pi c(\lambda) g(0),$$

so that the theorem is proved.

Theorem 4.15. *The generalized spherical function* $\Phi_{\lambda, m}$ *is for* $i\lambda \notin 2\mathbf{Z}$ *given by the following expansion*:

$$\Phi_{\lambda, m}(\tanh r) = C_1(\lambda) \sum_0^{\infty} \Gamma_n(\lambda) e^{(i\lambda - 1 - 2n)r}$$
$$+ C_{-1}(\lambda) \sum_0^{\infty} \Gamma_n(-\lambda) e^{(-i\lambda - 1 - 2n)r},$$

where

$$C_1(\lambda) = c(\lambda), \qquad C_{-1}(\lambda) = c(-\lambda) \frac{p_m(i\lambda)}{p_m(-i\lambda)},$$

$\Gamma_0(\lambda) \equiv 1$, *and* $\Gamma_n(\lambda)$ *is given by the recursion formula* (49).

Proof. From (51) we see that

$$e^{(-i\lambda + 1)r} \Phi_{\lambda, m}(\tanh r) = C_1(\lambda) \sum_0^{\infty} \Gamma_n(\lambda) e^{-2nr}$$
$$+ C_{-1}(\lambda) e^{-2i\lambda r} \sum_0^{\infty} \Gamma_n(-\lambda) e^{-2nr}.$$

Letting $r \to +\infty$ and using Theorem 4.14, we obtain $C_1(\lambda) = c(\lambda)$ for $\operatorname{Re}(i\lambda) > 0$; by analytic continuation this holds for $i\lambda \notin 2\mathbf{Z}$. Handling $e^{(i\lambda + 1)(r)} \Phi_{-\lambda, m}(\tanh r)$ similarly and using Prop. 4.12, we obtain

$$C_{-1}(\lambda) \frac{p_m(-i\lambda)}{p_m(i\lambda)} = c(-\lambda) \qquad \text{for} \quad \operatorname{Re}(i\lambda) < 0,$$

so that the formula for $C_{-1}(\lambda)$ follows by analytic continuation.

In order to prove Theorem 4.2, part (ii), let $f \in \mathscr{D}(D)$. The functional equation

$$\int_B e^{(i\lambda + 1)\langle z, b \rangle} \tilde{f}(\lambda, b)\, db = \int_B e^{(-i\lambda + 1)\langle z, b \rangle} \tilde{f}(-\lambda, b)\, db$$

is immediate from Lemma 4.7. Moreover, as proved in No. 1,

$$\tilde{f}(\lambda, e^{i\theta}) = \int_{\boldsymbol{R}} e^{(-i\lambda+1)t}\left(\int_{\xi(\tanh t\, e^{i\theta},\, e^{i\theta})} f(z)\, d\omega_t\right) dt,$$

where $d\omega_t$ is the arc element on the horocycle indicated. This shows by the Paley–Wiener theorem on $\boldsymbol{R}$ that $\tilde{f}$ has uniform exponential type.

It now remains to prove that each ψ with the properties in Theorem 4.2, part (ii), has the form $\tilde{f}$ for some $f \in \mathscr{D}(D)$. For this purpose we define the function $f \in C(D)$ by

$$f(z) = \frac{1}{2\pi^2} \int_{\boldsymbol{R} \times B} \psi(\lambda, b) e^{(i\lambda+1)\langle z, b\rangle} |\boldsymbol{c}(\lambda)|^{-2}\, d\lambda\, db. \tag{52}$$

It is clear from (16) that the derivatives of $e^{(i\lambda+1)\langle z, b\rangle}$ with respect to z are bounded by a polynomial in λ [uniformly for (z, b) varying in a compact set]. Because of the rapid decrease of $\psi(\lambda, b)$ in λ, we have therefore $f \in \mathscr{E}(D)$. Assuming ψ of uniform exponential type R, we shall prove $f(z) = 0$ for $d(o, z) > R$. For z and λ fixed we expand the functions $\theta \to f(e^{i\theta}z)$, $\theta \to \psi(\lambda, e^{i\theta}b)$ into convergent Fourier series and evaluate at $\theta = 0$. This gives

$$f(z) = \sum_{m \in \boldsymbol{Z}} f_m(z), \qquad \psi(\lambda, e^{i\phi}) = \sum_{m \in \boldsymbol{Z}} \psi_m(\lambda) e^{im\phi},$$

where

$$f_m(z) = \frac{1}{2\pi} \int_0^{2\pi} f(e^{i\theta}z) e^{-im\theta}\, d\theta, \qquad \psi_m(\lambda) = \frac{1}{2\pi} \int_0^{2\pi} \psi(\lambda, e^{i\phi}) e^{-im\phi}\, d\phi.$$

In the first formula we substitute expression (52) for f. Then since $\psi(\lambda, b)$ is rapidly decreasing in λ, we can interchange the integrations $\int d\theta$ and $\int d\lambda\, db$. This leads to the integral

$$\int_0^{2\pi} \int_0^{2\pi} \psi(\lambda, e^{i\phi}) e^{(i\lambda+1)\langle e^{i\theta}z, e^{i\phi}\rangle} e^{-im\theta}\, d\theta\, d\phi.$$

But $\langle e^{i\theta}z, e^{i\phi}\rangle = \langle z, e^{i(\phi-\theta)}\rangle$, so putting $\theta = \phi - \zeta$ this integral can be written

$$\begin{aligned}\iint \psi(\lambda, e^{i\phi}) e^{(i\lambda+1)\langle z, e^{i\zeta}\rangle} e^{im\zeta}\, d\zeta\, e^{-im\phi}\, d\phi \\ = (2\pi)^2 \Phi_{\lambda, m}(z) \psi_m(\lambda).\end{aligned}$$

Thus we have obtained

$$f_m(z) = \frac{1}{2\pi^2} \int_{\boldsymbol{R}} \Phi_{\lambda, m}(z) \psi_m(\lambda) |\boldsymbol{c}(\lambda)|^{-2}\, d\lambda, \tag{53}$$

and since $f_m(e^{i\theta}z) = e^{im\theta}f_m(z)$, it suffices to prove $f_m(a_r \cdot o) = 0$ for $r > R$. For this we use the expansion of Theorem 4.15. The following lemma is crucial.

Lemma 4.16. *For $\varepsilon = \pm 1$ we have*

$$\int_{\boldsymbol{R}} C_\varepsilon(\lambda)\Gamma_n(\varepsilon\lambda)e^{\varepsilon i\lambda r}\psi_m(\lambda)|\boldsymbol{c}(\lambda)|^{-2}\, d\lambda = 0 \qquad \textit{for} \quad r > R.$$

Proof. We consider first the case $\varepsilon = +1$. Since $|\boldsymbol{c}(\lambda)|^2 = \boldsymbol{c}(\lambda)\boldsymbol{c}(-\lambda)$ for $\lambda \in \boldsymbol{R}$, the integral equals

$$\int_{\boldsymbol{R}} \psi_m(-\lambda)\Gamma_n(-\lambda)\boldsymbol{c}(\lambda)^{-1}e^{-i\lambda r}\, d\lambda. \tag{54}$$

If the function $\psi_m(-\lambda)\Gamma_n(-\lambda)\boldsymbol{c}(\lambda)^{-1}$ were an entire function of exponential type R [like $\psi_m(-\lambda)$], this integral would vanish for $r > R$ as a consequence of the classical Paley–Wiener theorem. However, both $\Gamma_n(-\lambda)$ and $\boldsymbol{c}(\lambda)^{-1}$ have poles. Fortunately there is a half-plane where both are holomorphic. In fact, by the recursion formula for the function $\Gamma_n(\lambda)$ its poles are among the points $i\lambda \in \boldsymbol{Z}^+$; in particular, the function $\lambda \to \Gamma_n(-\lambda)$ is holomorphic for λ in the lower half-plane. By Theorem 4.5 the poles of $\lambda \to \boldsymbol{c}(\lambda)^{-1}$ are the poles of $\Gamma(\frac{1}{2}(i\lambda + 1))$, so, it is also holomorphic in the lower half-plane. We therefore use Cauchy's theorem to shift the contour downward: For $\eta < 0$

$$\int_{\boldsymbol{R}} \psi_m(-\xi)\Gamma_n(-\xi)\boldsymbol{c}(\xi)^{-1}e^{-i\xi r}\, d\xi \tag{55}$$
$$= \int_{\boldsymbol{R}} \psi_m(-\xi - i\eta)\Gamma_n(-\xi - i\eta)\boldsymbol{c}(\xi + i\eta)^{-1}e^{-i\xi r}e^{\eta r}\, d\xi.$$

For this to be permissible we must check the behavior of the integrand at ∞. First, $\Gamma_n(\lambda)$ is rational. Second, $\boldsymbol{c}(\xi + i\eta)^{-1}$ can be estimated by means of the known asymptotics for the gamma function (Magnus and Oberhettinger [1948], p. 5); for example,

$$\left|\frac{\Gamma(z + \frac{1}{2})}{\Gamma(z)}\right| \le K_1 + K_2|z|^{1/2} \qquad \text{for} \quad \operatorname{Re} z \ge 0,$$

where K_1 and K_2 are constants. This implies by Theorem 4.5,

$$|\boldsymbol{c}(\xi + i\eta)|^{-1} \le C_1 + C_2|\xi + i\eta|^{1/2}, \qquad \xi \in \boldsymbol{R}, \quad \eta < 0, \tag{56}$$

where C_1 and C_2 are constants. But by (31) for $\psi_m(-\lambda)$ the rapid decrease of $\psi_m(-\lambda)$ cancels out the growth in Γ_n and $\boldsymbol{c}$ at ∞, so the contour shift above is permissible. Denoting the right-hand side of (55) by $Q(r)$, we have

$$|Q(r)| \leq Ce^{R|\eta|}e^{r\eta},$$

where C is a constant (independent of η). Letting $\eta \to -\infty$, we deduce $Q(r) = 0$ for $r > R$. This proves the lemma for the case $\varepsilon = +1$.

For $\varepsilon = -1$ the integral becomes

$$\int_{\mathbf{R}} \psi_m(\lambda)\Gamma_n(-\lambda)c(\lambda)^{-1} \frac{p_m(i\lambda)}{p_m(-i\lambda)} e^{-i\lambda r}\, d\lambda, \tag{57}$$

and again we would like to shift the integration downward. However, the denominator $p_m(-i\lambda)$ vanishes for $\lambda = -i, -3i, \ldots$, which at first seems to cause complications. Fortunately, $\psi_m(\lambda)$ turns out to have zeros at these points. We shall, in fact, prove

$$\psi_m(\lambda) = \psi_m(-\lambda)\frac{p_m(-i\lambda)}{p_m(i\lambda)}. \tag{58}$$

To see this we integrate the relation

$$e^{(i\lambda+1)\langle z, b\rangle} = \sum_m \Phi_{\lambda, m}(z)\chi_{-m}(b) \tag{59}$$

against $\psi(\lambda, b)$. This gives

$$\begin{aligned}\int_B \psi(\lambda, b)e^{(i\lambda+1)\langle z, b\rangle}\, db &= \sum_m \Phi_{\lambda, m}(z)\psi_m(\lambda)\\ &= \sum_m \Phi_{\lambda, m}(\tanh r)\psi_m(\lambda)e^{im\theta}\end{aligned}$$

if $z = \tanh r\, e^{i\theta}$. By our assumption about ψ, the left-hand side is even in λ, so that

$$\Phi_{\lambda, m}(\tanh r)\psi_m(\lambda) = \Phi_{-\lambda, m}(\tanh r)\psi_m(-\lambda).$$

Now (58) follows from Prop. 4.12. But then the integral (57) reduces to (54), which was already handled, so Lemma 4.16 is proved.

Now we substitute the expansion of $\Phi_{\lambda, m}(\tanh r)$ into the integral

$$f_m(\tanh r) = \frac{1}{2\pi^2}\int_{\mathbf{R}} \Phi_{\lambda, m}(\tanh r)\psi_m(\lambda)|c(\lambda)|^{-2}\, d\lambda.$$

Integrating term by term and using Lemma 4.16, we obtain the desired result, $f_m(\tanh r) = 0$ for $r > R$. The term-by-term integration is justified by the following lemma.

Lemma 4.17. *The Γ-coefficients satisfy*

$$|\Gamma_n(\lambda)| \leq c(1 + n^d), \qquad \lambda \in \mathbf{R}, \quad n \in \mathbf{Z}^+,$$

for suitable constants c and d.

Proof. The recursion formula (49) shows that $\Gamma_n(\lambda) \equiv 0$ if n is odd. Put $a_n(\lambda) = \Gamma_{2n}(\lambda)$, so that

$$2n(2n - i\lambda)a_n(\lambda) = \sum_{r=0}^{n-1} a_r(\lambda)(4r - i\lambda + 1 + 4m^2(n - r)).$$

We write this in the form

$$na_n(\lambda) = \sum_{r=0}^{n-1} a_r(\lambda)A_{r,n}(\lambda), \tag{60}$$

and then there exists an integer N such that

$$|A_{r,n}(\lambda)| \leq N, \qquad 0 \leq r < n, \quad n \in \mathbf{Z}^+, \quad \lambda \in \mathbf{R}. \tag{61}$$

Defining b_n by

$$nb_n = N \sum_{r=0}^{n-1} b_r, \qquad b_0 = 1, \tag{62}$$

we have from (60) and (61) by induction

$$|a_n(\lambda)| \leq b_n \qquad \text{for} \quad n \in \mathbf{Z}^+, \quad \lambda \in \mathbf{R}. \tag{63}$$

Now (62) implies

$$nb_n = N \sum_{0}^{n-2} b_r + Nb_{n-1} = (n - 1 + N)b_{n-1}, \tag{64}$$

so

$$b_n = \frac{(n + N - 1)!}{n!(N - 1)!}. \tag{65}$$

The lemma is now obvious from (63).

We have now proved that the function f defined by (52) satisfies

$$f(z) = 0 \qquad \text{for} \quad d(o, z) > R. \tag{66}$$

It remains to prove $\tilde{f} = \psi$. By the inversion formula [Theorem 4.2, part (i)] we have for $\phi = \tilde{f} - \psi$

$$\int_{\mathbf{R}} \left(\int_B \phi(\lambda, b) e^{(i\lambda + 1)\langle z, b\rangle} |c(\lambda)|^{-2} \, db \right) d\lambda \equiv 0.$$

Since the inner integral is even in λ, we can here replace $\boldsymbol{R}$ by $-\boldsymbol{R}^+$. Integrating the resulting relation against an arbitrary $F \in \mathscr{D}(D)$, we obtain

$$\int_{\boldsymbol{R}^+ \times B} \phi(-\lambda, b)\tilde{F}(\lambda, b)|c(\lambda)|^{-2}\, d\lambda\, db = 0.$$

But we have seen that the functions $\tilde{F}$ form a dense subset of $L^2(\boldsymbol{R}^+ \times B)$; hence $\phi \equiv 0$ on $(-\boldsymbol{R}^+) \times B$. But

$$\int_B \phi(\lambda, b)e^{(i\lambda+1)\langle z, b\rangle}\, db = \int_B \phi(-\lambda, b)e^{(-i\lambda+1)\langle z, b\rangle}\, db,$$

so, since each $\lambda \in \boldsymbol{R}$ is simple, $\phi \equiv 0$.

Theorem 4.2 is now completely proved.

4. Eigenfunctions and Eigenspace Representations. Proofs of Theorems 4.3 and 4.4

Given $\lambda \in \boldsymbol{C}$ we consider now the representation T_λ of $G = \boldsymbol{SU}(1, 1)$ on the eigenspace

$$\mathscr{E}_\lambda(D) = \{f \in C^\infty(D) : Lf = -(\lambda^2 + 1)f\}.$$

If $g \in G$, $T_\lambda(g)$ is the linear transformation of $\mathscr{E}_\lambda(D)$ given by

$$(T_\lambda(g)f)(z) = f(g^{-1} \cdot z).$$

We shall now prove Theorem 4.4:

(67) $\qquad T_\lambda$ is irreducible if and only if $i\lambda + 1 \notin 2\boldsymbol{Z}$.

Remark. T_{λ_0} is irreducible if and only if λ_0 is not a pole of the denominator of the function $c(\lambda)c(-\lambda)$.

Lemma 4.18. *The point $\lambda \in \boldsymbol{C}$ is simple if and only if the functions*

$$b \to \sum_k a_k e^{(i\lambda+1)\langle z_k, b\rangle}, \qquad a_k \in \boldsymbol{C}, \quad z_k \in D,$$

form a dense subspace of $L^2(B)$.

This is an immediate reformulation of the definition of simplicity.

Lemma 4.19. *The mapping $F \to f$ of $L^2(B)$ into $\mathscr{E}_\lambda(D)$ given by*

(68) $$f(z) = \int_B e^{(i\lambda+1)\langle z, b\rangle} F(b)\, db$$

is continuous.

The topology of $\mathscr{E}(D)$ is described in §1. The lemma follows immediately by using Schwarz's inequality.

For $\lambda \in \boldsymbol{C}$ simple let $\mathscr{H}_\lambda$ denote the space of functions f defined by (68) as F runs through $L^2(B)$. Then $\mathscr{H}_\lambda$ is a Hilbert space if f is given the L^2 norm of F.

Lemma 4.20. *Suppose $\lambda \in \boldsymbol{C}$ is simple. Then $\mathscr{H}_\lambda$ is dense in $\mathscr{E}_\lambda(D)$.*

The proof is analogous to that of Lemma 2.8, so we omit it.

Suppose now $i\lambda + 1 \notin 2\boldsymbol{Z}$. This means that both λ and $-\lambda$ are simple. Let $0 \neq V \subset \mathscr{E}_\lambda(D)$ be a closed invariant subspace. Then V contains an element f such that $f(o) \neq 0$; averaging over the rotations around o, we conclude $\phi_\lambda \in V$. Now by (35)

$$\sum_k a_k \phi_\lambda(g_k^{-1} \cdot z) = \int_B e^{(i\lambda+1)\langle z, b\rangle} \sum_k a_k e^{(-i\lambda+1)\langle g_k \cdot o, b\rangle}\, db. \tag{69}$$

Since $-\lambda$ is simple, we conclude from Lemmas 4.18 and 4.19 that the closed subspace of $\mathscr{E}_\lambda(D)$ generated by the various linear combinations (69) contains $\mathscr{H}_\lambda$. But then Lemma 4.20 shows $V = \mathscr{E}_\lambda(D)$. Hence T_λ is irreducible.

On the other hand, suppose T_λ is irreducible. Since $T_\lambda = T_{-\lambda}$, since the condition $i\lambda + 1 \notin 2\boldsymbol{Z}$ is invariant under $\lambda \to -\lambda$, and since either λ or $-\lambda$ is simple, we may assume that λ is simple.

Let $0 \neq E \subset \mathscr{H}_\lambda$ be a closed invariant subspace of the Hilbert space $\mathscr{H}_\lambda$. By the irreducibility of T_λ, E is dense in $\mathscr{E}_\lambda(D)$. Let

$$f * \chi_m(z) = \frac{1}{2\pi}\int_0^{2\pi} f(e^{-i\theta} \cdot z)\chi_m(e^{i\theta})\, d\theta.$$

Then E being closed in $\mathscr{H}_\lambda$ and invariant,

$$E * \chi_m \subset E. \tag{70}$$

Since E is dense in $\mathscr{E}_\lambda(D)$ and the map $f \to f * \chi_m$ is continuous

$$E * \chi_m \text{ is dense in } \mathscr{E}_\lambda(D) * \chi_m. \tag{71}$$

But by Prop. 4.11 this last space equals $\boldsymbol{C}\Phi_{\lambda,m}$. Thus (70) and (71) imply $\Phi_{\lambda,m} \in E$ for each m, whence $E = \mathscr{H}_\lambda$. Thus G acts irreducibly on $\mathscr{H}_\lambda$: hence the functions (69) are dense in $\mathscr{H}_\lambda$. Using Lemma 4.18, we conclude that $-\lambda$ is simple. This finishes the proof of Theorem 4.4.

Finally, we prove Theorem 4.3 giving the integral representation of the eigenfunctions of L by means of analytic functionals. Let f be any eigenfunction of L and select $\lambda \in \boldsymbol{C}$ simple such that

$$Lf = -(\lambda^2 + 1)f.$$

For fixed $z \in D$ we develop the function $\theta \to f(e^{i\theta}z)$ in an absolutely convergent Fourier series

$$f(e^{i\theta}z) = \sum_n c_n(z)e^{in\theta}, \tag{72}$$

where

$$c_n(z) = \frac{1}{2\pi}\int_0^{2\pi} f(e^{i\theta}z)e^{-in\theta}\,d\theta.$$

Then by Prop. 4.11

$$c_n(z) = a_n\Phi_{\lambda,n}(z), \quad a_n \in \boldsymbol{C}.$$

Using the transformation formula

$$F(a, b; c; z) = (1 - z)^{-a}F\left(a, c - b; c; \frac{z}{z - 1}\right)$$

for the hypergeometric function (Erdélyi *et al.* [1953], Vol. I, p. 64), we derive from (46) with $v = \frac{1}{2}(i\lambda + 1)$

$$\Phi_{\lambda,n}(r) = r^{|n|}\frac{\Gamma(|n| + v)}{\Gamma(v)|n|!}F\left(v, 1 - v; |n| + 1; \frac{r^2}{r^2 - 1}\right), \tag{73}$$

where $r = |z|$. We know from (72) that

$$\sum_n |a_n\Phi_{\lambda,n}(r)| < \infty \tag{74}$$

and we want to deduce from this that

$$\sum_n |a_n|r^{|n|} < \infty, \qquad 0 < r < 1. \tag{75}$$

Let us choose $k \in \boldsymbol{Z}^+$ such that $k > |v|$ and put $x = r^2(r^2 - 1)^{-1}$. Then for $n > 0$

$$F(v, 1 - v; n + 1; x) = p_k(n, x) + \rho_{k+1}(n, x),$$

where p_k is the kth Taylor polynomial

$$p_k(n, x) = 1 + \frac{v(1 - v)}{(n + 1)}x + \cdots \frac{v(v + 1)\cdots(v + k - 1)(1 - v)\cdots(1 - v + k - 1)}{(n + 1)\cdots(n + k)\,k!}x^k$$

and ρ_{k+1} is the remainder. Using the general formula

$$R_k(x) = \frac{1}{k!}\int_0^1 x^{k+1}(1 - s)^k f^{(k+1)}(xs)\,ds$$

for the remainder term in a Taylor series and the Euler integral formula for the hypergeometric function

$$F(a, b; c; z) = \frac{\Gamma(c)}{\Gamma(b)\Gamma(c-b)} \int_0^1 t^{b-1}(1-t)^{c-b-1}(1-tz)^{-a}\,dt$$

$$(\operatorname{Re} c > \operatorname{Re} b > 0),$$

we obtain for $n + 1 > \operatorname{Re}(1 - \nu) > 0$

$$\rho_{k+1}(n, x) = \frac{n!\,\Gamma(k+\nu+1)x^{k+1}}{\Gamma(1-\nu)\Gamma(n+\nu)\Gamma(\nu)\,k!} \int_0^1 \int_0^1 t^{1-\nu+k}(1-t)^{n+\nu-1}$$

$$\times (1-s)^k(1-stx)^{-\nu-k-1}\,ds\,dt.$$

The definition of ρ_{k+1} and formula (73) show that $\rho_{k+1}(n, x)$ is holomorphic in ν except for $\nu \in -\boldsymbol{Z}^+$ and these points are excluded, since λ is simple. Thus the last formula holds by analytic continuation for $n + \operatorname{Re} \nu > 0$, since $k > |\nu|$. Since $x < 0$, $0 \le s, t \le 1$, and $k > |\nu|$, we have

$$|(1-stx)^{-\nu-k-1}| \le 1,$$

so

$$|\rho_{k+1}(n, x)| \le |x|^{k+1} \frac{n!\,|\Gamma(k+\nu+1)|}{|\Gamma(1-\nu)\Gamma(n+\nu)\Gamma(\nu)|\,k!}$$

$$\times \frac{\Gamma(2 - \operatorname{Re}\nu + k)\Gamma(n + \operatorname{Re}\nu)}{\Gamma(2+n+k)}.$$

Using the asymptotic property

$$\lim_{|z|\to\infty} e^{-\alpha \log z} \frac{\Gamma(z+\alpha)}{\Gamma(z)} = 1$$

of the gamma function, we have

$$\lim_{n\to\infty} \frac{\Gamma(n + \operatorname{Re}\nu)}{|\Gamma(n+\nu)|} = 1,$$

and consequently,

$$|\rho_{k+1}(n, x)| \le C_k |x|^{k+1} n^{-(k+1)}, \tag{76}$$

where C_k is a constant. Now we put

$$b_n = |a_n| \frac{|\Gamma(n+\nu)|}{|\Gamma(\nu)|\,n!}, \qquad \varepsilon(n, x) = \frac{\rho_{k+1}(n, x)}{C_k |x|^{k+1}} n^{k+1}.$$

Then by (74)

$$\sum_0^\infty b_n r^n |p_k(n, x) + p_{k+1}(n, x)| < \infty,$$

whence

$$\sum_0^\infty b_n r^n n^{-(k+1)} \left| \frac{n^{k+1}}{|x|^{k+1} C_k} p_k(n, x) + \varepsilon(n, x) \right| < \infty. \tag{77}$$

We have ν and k fixed. Let $0 < r_0 < 1$ be arbitrary and put $x_0 = r_0^2/(r_0^2 - 1)$. Then select N_{r_0} such that

$$\left| \frac{n^{k+1}}{|x_0|^{k+1} C_k} p_k(n, x_0) \right| \geq 2 \qquad \text{for} \quad n \geq N_{r_0}.$$

This can be done since

$$\lim_{n \to \infty} p_k(n, x_0) = 1.$$

But $|\varepsilon(n, x)| \leq 1$, so by (77)

$$\sum_0^\infty |b_n r_0^n n^{-(k+1)}| < \infty.$$

Since r_0 was arbitrary, this proves

$$\sum_0^\infty |a_n| r^n < \infty, \qquad 0 \leq r < 1.$$

Now (75) follows since n appears in (73) only as $|n|$.

Now given a hyperfunction $T \in \mathscr{A}'(B)$ we define its Fourier series by

$$T \sim \sum_n \alpha_n e^{in\theta} \qquad \text{if} \quad \alpha_n = \int_0^{2\pi} e^{-in\theta}\, dT(\theta).$$

Then we have the following result.

Lemma 4.21. *A hyperfunction is uniquely determined by its Fourier series. The series*

$$\sum_n \alpha_n e^{in\theta}$$

is the Fourier series of a hyperfunction $T \in \mathscr{A}'(B)$ if and only if

$$\sum_n |\alpha_n| r^{|n|} < \infty \qquad \textit{for all} \quad 0 \leq r < 1. \tag{78}$$

Proof. Suppose we have a Laurent series

$$F(z) = \sum_n b_n z^n \tag{79}$$

converging in an annulus containing $B: |z| = 1$. This is equivalent to both series

$$\sum_0^\infty b_n z^n, \qquad \sum_0^\infty b_{-n} z^n$$

having radii of convergence >1 or equivalently,

$$\text{(80)} \qquad \limsup_n |b_n|^{1/n} = \beta < 1, \qquad \limsup_n |b_{-n}|^{1/n} = \gamma < 1;$$

all such numbers β, γ, $0 \leq \beta, \gamma < 1$, can occur.

By the definition of the inductive limit, the series (79) converges in the topology of $\mathscr{A}(B)$. Thus if $T \in \mathscr{A}'(B)$ has Fourier coefficients (α_n), we have

$$T(F) = \sum_n b_n \alpha_{-n}.$$

In particular, T is determined by its Fourier series. Because of (80) and the arbitrariness of β and γ, we deduce

$$\sum_n |\alpha_n| r^{|n|} < \infty, \qquad 0 \leq r < 1.$$

On the other hand, if this condition is satisfied, we can define a linear form T on $\mathscr{A}(B)$ by

$$T(F) = \sum_n \alpha_n b_{-n}.$$

To see that T is continuous we represent $\mathscr{A}(B)$ as the union

$$\mathscr{A}(B) = \bigcup_{n=1}^\infty \mathscr{H}_n,$$

where $\mathscr{H}_n$ is the set of holomorphic functions in the annulus $1 - (1/n) < |z| < 1 + (1/n)$. By the definition of the topology of $\mathscr{A}(B)$ we must prove that for each n, the restriction $T|\mathscr{H}_n$ is continuous (for the topology of uniform convergence on compact sets). Because of (78) the functions

$$f_1(z) = \sum_0^\infty \alpha_k z^k, \qquad f_2(z) = \sum_1^\infty \alpha_{-k} z^k$$

are holomorphic in the unit disk $|z| < 1$. For the curves

$$C_1: |z| = 1 - \frac{1}{2n}, \qquad C_2: |z| = 1 + \frac{1}{2n}$$

we have

$$\sum_0^\infty \alpha_k b_{-k} = \frac{1}{2\pi i}\int_{C_1} F(\zeta) f_1(\zeta)\frac{d\zeta}{\zeta}, \qquad F \in \mathscr{H}_n,$$

$$\sum_1^\infty \alpha_{-k} b_k = \frac{1}{2\pi i}\int_{C_2} F(\zeta) f_2\left(\frac{1}{\zeta}\right)\frac{d\zeta}{\zeta}, \qquad F \in \mathscr{H}_n,$$

from which the continuity of $T|\mathscr{H}_n$ is obvious. This proves the lemma.

Going back to the proof of Theorem 4.3, we know now that there exists a hyperfunction T with Fourier series

$$T \sim \sum_n a_n e^{in\theta}.$$

But given $z \in D$ the function†

$$\zeta \to \left(\frac{1 - z\bar{z}}{(\zeta - z)(\zeta^{-1} - \bar{z})}\right)^\nu, \qquad \nu = \tfrac{1}{2}(i\lambda + 1),$$

is holomorphic in an annulus U containing B and can be expanded there in a Laurent series. The restriction of this series to B is the Fourier series (59). Since the Laurent series converges uniformly on compact subsets and since the injection $\mathscr{H}(U) \to \mathscr{A}(B)$ is continuous it follows that (59) converges in the topology of $\mathscr{A}(B)$. Hence we can apply T to it term by term and deduce from (72)

$$\int_B e^{(i\lambda + 1)\langle z, b\rangle}\, dT(b) = \sum_m \Phi_{\lambda, m}(z) a_m = f(z),$$

which is the desired integral representation of f.

On the other hand, we must verify that if $T \in \mathscr{A}'(B)$, the function

(81) $$f(z) = \int_B e^{(i\lambda + 1)\langle z, b\rangle}\, dT(b)$$

is an eigenfunction of L. The continuity of f is clear, since $z_n \to z_0$ implies $e^{(i\lambda+1)\langle z_n, b\rangle} \to e^{(i\lambda+1)\langle z_0, b\rangle}$ in the topology† of $\mathscr{A}(B)$. Next we use the following result whose proof is entirely analogous to that of Prop. 2.5. Let dk be the normalized Haar measure in $K = \mathbf{SO}(2)$.

Proposition 4.22. *A function f satisfying $Lf = -(\lambda^2 + 1)f$ satisfies the functional equation*

(82) $$\int_K f(gk \cdot z)\, dk = f(g \cdot o)\phi_\lambda(z), \qquad g \in G, \quad z \in D.$$

†See "Some Details," p. 611.

Conversely, a continuous function f satisfying this functional equation is automatically of class C^∞ and satisfies $Lf = -(\lambda^2+1)f$.

The integrand in (81) satisfies the functional equation (for each b); hence to show that f does too, it suffices to show the commutation

$$\int_K \left(\int_B e^{(i\lambda+1)\langle gk\cdot z, b\rangle}\, dT(b)\right) dk = \int_B \left(\int_K e^{(i\lambda+1)\langle gk\cdot z, b\rangle}\, dk\right) dT(b).$$

Thinking of the integral over K as a limit, this follows just as the continuity of f above.[†]

To conclude the proof of Theorem 4.3 it remains to remark that if λ is simple, then the mapping

$$T \in \mathscr{A}'(B) \to f \in \mathscr{E}_\lambda(B)$$

given by (81) is one-to-one. For this note that f and T are related by

$$(83) \qquad f(z) = \sum_n a_n \Phi_{\lambda,n}(z), \qquad T \sim \sum a_n e^{in\theta},$$

and

$$(84) \qquad a_n \Phi_{\lambda,n}(z) = \frac{1}{2\pi}\int_0^{2\pi} f(e^{i\theta}z)e^{-in\theta}\, d\theta.$$

By the simplicity of λ each $\Phi_{\lambda,n}$ is $\not\equiv 0$, so that f determines the sequence (a_n) which by Lemma 4.21 determines T. This finishes the proof.

For $\mathrm{Re}(i\lambda) > 0$ the hyperfunction T can be recovered from f somewhat more directly; in fact we have as a special case of Theorem 4.14,

$$(85) \qquad \lim_{t\to+\infty} e^{(1-i\lambda)t} a_n \Phi_{\lambda,n}(e^{i\theta}\tanh t) = c(\lambda) a_n e^{in\theta}.$$

Under a mild condition on λ we shall now prove that the eigenfunction $f(z)$ grows at most exponentially with the distance $d(o, z)$ if and only if the hyperfunction T is a distribution. We shall need the following analog of Lemma 4.21.

Lemma 4.23. *The series*

$$(86) \qquad \sum_n \alpha_n e^{in\theta}$$

is the Fourier series of a distribution T on B if and only if

$$(87) \qquad |\alpha_n| \le C(1+|n|)^l, \qquad n \in \mathbf{Z},$$

for some constants C and l.

[†]See "Some Details," p. 611.

Proof. If T is a distribution, it is continuous on $\mathscr{D}(B)$ in the topology of the norms $\|F\|_k = \sup_{\theta, 0 \leq l \leq k} |d^l F/d\theta^l|$, $k = 0, 1, 2, \ldots$. Hence there exists an $l \geq 0$ and a constant C such that $|T(F)| \leq C\|F\|_l$. This proves (87). On the other hand, if $\sum \alpha_n e^{in\theta}$ is a series satisfying (87), then for $k \in \boldsymbol{Z}^+$ large enough the series

$$\sum_n \alpha_n (1 + n^2)^{-k} e^{in\theta}$$

is absolutely convergent to a continuous function G. The mapping

$$T\colon F \to \frac{1}{2\pi} \int_0^{2\pi} G(\theta)(1 - L)^k F(\theta)\, d\theta$$

is then a distribution with Fourier series $\sum \alpha_n e^{in\theta}$.

Given $\lambda \in \boldsymbol{C}$, let $\mathscr{E}_\lambda^*(D)$ denote the subspace of functions $f \in \mathscr{E}_\lambda(D)$ satisfying an inequality

(88) $$|f(z)| \leq C e^{a\, d(o, z)} \qquad (z \in D)$$

for some constants C and a.

Theorem 4.24. *Assume $i\lambda \notin \boldsymbol{Z}$. Then the mapping $T \to f$, where*

$$f(z) = \int_B e^{(i\lambda + 1)\langle z, b\rangle}\, dT(b),$$

is a bijection of the space $\mathscr{D}'(B)$ (of distributions on B) onto $\mathscr{E}_\lambda^*(D)$.

Proof. Consider the Fourier expansion (59). If L_B is the Laplacian $d^2/d\theta^2$ on B, we have

$$(-L_B)_b^p (e^{(i\lambda + 1)\langle z, b\rangle}) = \sum_n \Phi_{\lambda, n}(z) n^{2p} \chi_{-n}(b),$$

so that

(89) $$\Phi_{\lambda, n}(z) n^{2p} = \int_B (-L_B)_b^p (e^{(i\lambda + 1)\langle z, b\rangle}) \chi_n(b)\, db.$$

If $z = \tanh t\, e^{i\phi}$, $b = e^{i\theta}$, we have by (16)

(90) $$e^{2\nu\langle z, b\rangle} = [\mathrm{ch}(2t) - \mathrm{sh}(2t)\cos(\phi - \theta)]^{-\nu}.$$

From the definition of $\langle z, b\rangle$ it is easily seen that

$$|\langle z, b\rangle| \leq d(o, z) \qquad \text{for all} \quad z, b.$$

Thus if we apply L_B^p to (90), it follows quickly that

$$|L_B^p(e^{2\nu\langle z, b\rangle})| \leq C e^{a\, d(o, z)},$$

where C and a are constants depending on p and ν. Thus by (89)

$$|\Phi_{\lambda,n}(z)n^{2p}| \leq Ce^{a\,d(o,z)}. \tag{91}$$

Suppose now $T \in \mathscr{D}'(B)$ has Fourier coefficients a_n. Since (59) converges in the topology of $\mathscr{E}(B)$, we can apply T to it term by term and deduce that

$$f(z) = \sum_n \Phi_{\lambda,n}(z)a_n. \tag{92}$$

But by Lemma 4.23 a_n is bounded by a power of n and since (91) holds for each $p \in \mathbf{Z}^+$, it follows that $f \in \mathscr{E}_\lambda^*(D)$.

Conversely, suppose λ satisfies the conditions of the theorem and suppose $f \in \mathscr{E}_\lambda^*(D)$. Let $T \in \mathscr{A}'(B)$ represent f, in the sense that

$$f(z) = \int_B e^{(i\lambda+1)\langle z,b\rangle}\,dT(b).$$

Since λ is simple, we know by a previous remark that T is unique.

In order to prove that the a_n satisfy the condition of Lemma 4.23 we shall establish a modified version of (85). For this we need the following relation:

$$F(a,b;c;x) = \frac{\Gamma(c)\Gamma(c-a-b)}{\Gamma(c-a)\Gamma(c-b)} F(a,b;a+b-c+1;1-x)$$

$$+ (1-x)^{c-a-b}\frac{\Gamma(a+b-c)\Gamma(c)}{\Gamma(a)\Gamma(b)} F(c-a,c-b;c-a-b+1;1-x)$$

(cf., Erdélyi *et al.* [1953], Vol. I, p. 108) valid for $-c \notin \mathbf{Z}^+$, $a+b-c \notin \mathbf{Z}$. Using this on (46) we obtain after simple manipulations [with $\nu = \frac{1}{2}(i\lambda+1)$]

$$\Phi_{\lambda,m}(\tanh t)$$

$$= \tanh^{|m|} t\, ch^{-2\nu} t \frac{\Gamma(1-2\nu)\Gamma(|m|+\nu)}{\Gamma(\nu)\Gamma(1-\nu)\Gamma(|m|+1-\nu)} F(|m|+\nu,\nu;2\nu;\mathrm{ch}^{-2} t)$$

$$+ \tanh^{|m|} t\,\mathrm{ch}^{2\nu-2} t \frac{\Gamma(2\nu-1)}{\Gamma(\nu)^2} F(|m|-\nu+1,1-\nu;2-2\nu;\mathrm{ch}^{-2} t)$$

valid if $2\nu \notin \mathbf{Z}$ (which is equivalent to our assumption about λ).

Moreover, if ${}_1F_1$ denotes, as usual, the *confluent hypergeometric function*, we have for each $|x| < 1$

$$\lim_{m\to+\infty} F(a+m,b;c;x/m) = {}_1F_1(b,c,x) \tag{93}$$

(Erdélyi *et al.* [1953], Vol. I, p. 248). We construct now the sequence $t_m \to \infty$ by

(94) $$\operatorname{ch}^2 t_m = |m|/a,$$

where $a > 0$ is a number to be determined later. Using (93) and the relations

$$\lim_{m \to +\infty} \frac{\Gamma(x+m)}{\Gamma(m)} m^{-x} = 1,$$

$$\lim_{m \to +\infty} (\tanh t_m)^m = e^{-(1/2)a},$$

we deduce from the formula for $\Phi_{\lambda, m}$,

$$\lim_{|m| \to \infty} (\operatorname{ch} t_m)^{2-2\nu} \Phi_{\lambda, m}(\tanh t_m)$$

$$= e^{-(1/2)a} \frac{1}{\Gamma(\nu)} \left[\frac{\Gamma(1-2\nu)}{\Gamma(1-\nu)} a^{2\nu-1} {}_1F_1(\nu, 2\nu, a) + \frac{\Gamma(2\nu-1)}{\Gamma(\nu)} {}_1F_1(1-\nu, 2-2\nu, a) \right].$$

Since (cf. loc. cit. p. 278)

$$\lim_{x \to +\infty} {}_1F_1(b, c, x) e^{-x} x^{b-c} = \Gamma(c)/\Gamma(b)$$

and since $2\nu \neq 1$, the two terms inside the square brackets behave differently for a large a. Hence we can select a such that the last limit is $\neq 0$. Then for a suitable constant $C_0 \neq 0$ we have

(95) $$|\Phi_{\lambda, m}(\tanh t_m| \geq C_0 |m|^{(1/2)(\mathrm{Re}(i\lambda) - 1)}.$$

On the other hand, putting $z = \tanh t_m e^{i\theta}$ in (92), we obtain

(96) $$\sum_n |\Phi_{\lambda, n}(\tanh t_m)|^2 |a_n|^2 = \frac{1}{2\pi} \int_0^{2\pi} |f(\tanh t_m e^{i\theta})|^2 \, d\theta,$$

which by the assumption $f \in \mathscr{E}_\lambda^*(D)$ has a bound $C|m|^b$ for some constants C and b. But (95) and (96) then imply

(97) $$|a_m|^2 \leq (C/C_0)|m|^{b + (1/2)(1 - \mathrm{Re}(i\lambda))},$$

so by Lemma 4.23, T is a distribution and the theorem is proved.

Remark. While the case $i\lambda = 1$ was excluded, the theorem is valid for this case. The proof is even easier. In fact, we have here

$$\Phi_{\lambda, m}(\tanh t) = \tanh^{|m|} t,$$

so that

$$\lim_{|m| \to \infty} \Phi_{\lambda, m}(\tanh t_m) = e^{-(1/2)a}.$$

Hence (95) is obvious and (97) follows as before. Thus the ordinary harmonic functions

$$u(z) = \int_B \frac{1 - |z|^2}{|z - b|^2}\, dT(b),$$

which are Poisson integrals of distributions, are characterized by the growth condition

$$|u(z)| \leq Ce^{a\, d(o, z)}.$$

It is of interest to compare this with the classical characterization (Herglotz, Evans, Charathéodory, cf., Nevanlinna [1936]) of Poisson integrals of measures as the harmonic functions u satisfying

$$\sup_{r<1} \int_0^{2\pi} |u(re^{i\phi})|\, d\phi < \infty.$$

5. Limit Theorems

To conclude this section we prove two classical theorems from potential theory.

Theorem 4.25. *Let F be continuous on B and define the harmonic function u on D by the Poisson integral*

$$u(z) = \int_B \frac{1 - |z|^2}{|z - b|^2}\, F(b)\, db.$$

Then

$$\lim_{z \to b_0} u(z) = F(b_0) \qquad \textit{for} \quad b_0 \in B.$$

This classical theorem of H. A. Schwarz is usually proved by reducing it to the case $F(b_0) = 0$; then B is divided up into an ε-arc B_0 around b_0 and its complement. Then in similar fashion u decomposes into $u = u_0 + (u - u_0)$. The first term is small near b_0 because of the continuity of F at b_0; the second term is small since the kernel is small for z near b_0, $|b - b_0| \geq \varepsilon > 0$.

Because of formula (16) for the Poisson kernel, the theorem is a special case of Theorem 4.14 (with $i\lambda = 1$). For this case we can give an even more elementary proof. Because of (34a) u can be written

$$u(g \cdot o) = \int_B F(g \cdot b)\, db.$$

It suffices to consider the case $z \to b_0$, $b_0 = 1$.

Let a_t be the one-parameter group considered before and k_t rotation by the angle θ_t, where $\theta_t \to 0$ for $t \to +\infty$. Then

$$u(k_t a_t \cdot o) = u(e^{i\theta_t} \tanh t) = \frac{1}{2\pi} \int_0^{2\pi} F(k_t a_t \cdot e^{i\theta})\, d\theta$$

$$= \frac{1}{2\pi} \int_0^{2\pi} F\left(e^{i\theta_t} \cdot \frac{e^{i\theta} + \tanh t}{\tanh t\, e^{i\theta} + 1} \right) d\theta.$$

Letting $t \to +\infty$, we get by the dominated convergence theorem

$$\lim_{t \to +\infty} u(k_t a_t \cdot o) = \frac{1}{2\pi} \int_0^{2\pi} F(1)\, d\theta = F(b_0),$$

as desired.

The proof works because the one-parameter group a_t pulls the entire boundary B (except for the point -1) to the point b_0.

We shall now prove the classical theorem of Fatou that a bounded harmonic function on D has radial boundary values almost everywhere on B. While the proof is hardly simpler than the classical one, its group-theoretic features serve as a basis for the generalization to symmetric spaces. The theorem generalizes Theorem 4.25 in that continuity of F is dropped.

Theorem 4.26. (Fatou) *Let u be a bounded harmonic function on D. Then for almost all θ*

$$\lim_{r \to 1} u(re^{i\theta}) \qquad \textit{exists.}$$

Proof. It is well known that u is the Poisson integral of a function $F \in L^\infty(B)$. Consider now the subgroup Ξ of matrices

$$\Xi = \left\{ \xi_x = \begin{pmatrix} 1 + ix & ix \\ -ix & 1 - ix \end{pmatrix} : x \in \boldsymbol{R} \right\}$$

and the mapping

$$\xi_x \to \xi_x \cdot 1 = \frac{1 + 2ix}{1 - 2ix}$$

of Ξ into B. It is a bijection of Ξ onto $B - \{-1\}$. Writing

$$e^{i\theta} = (1 + 2ix)(1 - 2ix)^{-1},$$

we obtain

$$d\theta = \frac{4\, dx}{1 + 4x^2},$$

so that

$$\int_B f(b)\,db = \frac{2}{\pi}\int_{\mathbf{R}} f(\xi_x \cdot 1) \cdot \frac{dx}{1+4x^2}.$$

Now fix $\xi_y \in \Xi$ and use this relation on the function $f(b) = F(\xi_y a_t \cdot b)$. Then, since $a_t \xi_x \cdot 1 = a_t \xi_x a_t^{-1} \cdot 1$, we have

$$\begin{aligned} u(\xi_y a_t \cdot o) &= \int_B F(\xi_y a_t \cdot b)\,db \\ &= \frac{2}{\pi}\int_{\mathbf{R}} F(\xi_y a_t \xi_x \cdot 1)\frac{dx}{1+4x^2} \\ &= \frac{2}{\pi}\int_{\mathbf{R}} F(\xi_{y+xe^{-2t}} \cdot 1)\frac{dx}{1+4x^2}. \end{aligned}$$

Writing $\phi(y) = F(\xi_y \cdot 1)$, we deduce that

(98) $$|u(\xi_y a_t \cdot o) - F(\xi_y \cdot 1)| \leq \frac{2}{\pi}\int_{\mathbf{R}} |\phi(y + xe^{-2t}) - \phi(y)|\frac{dx}{1+4x^2}.$$

We now use the Lebesgue differentiation theorem, which states that

(99) $$\lim_{h \to 0} \frac{1}{2h}\int_{-h}^{h} |\phi(y+x) - \phi(y)|\,dx = 0$$

for almost all y. Given $\varepsilon > 0$ we select $c > 0$ such that, writing the right-hand-side of (98) as the sum

(100) $$\frac{2}{\pi}\int_{-c}^{c} + \frac{2}{\pi}\int_{|x|>c},$$

the latter term is $<\varepsilon$ for all y and all t. In the first term we put $x_1 = xe^{-2t}$ and then write $h = ce^{-2t}$. Then the first term in (100) becomes

$$\frac{4c}{\pi}\frac{1}{2h}\int_{-h}^{h} |\phi(y + x_1) - \phi(y)|\,dx_1.$$

Using (99), we can thus conclude that for almost all y

$$\lim_{t \to \infty} u(\xi_y a_t \cdot o) \qquad \text{exists.}$$

This means that u has a limit along the geodesic from $\xi_y \cdot o$ to the boundary point $\xi_y \cdot 1$. From this it is easy to deduce that u has the same limit along the radius $(o, \xi_y \cdot 1)$. For this we use the elementary inequality

$$|(\nabla u)(z)| \leq C/R \qquad \text{for} \quad |z| < 1 - R,$$

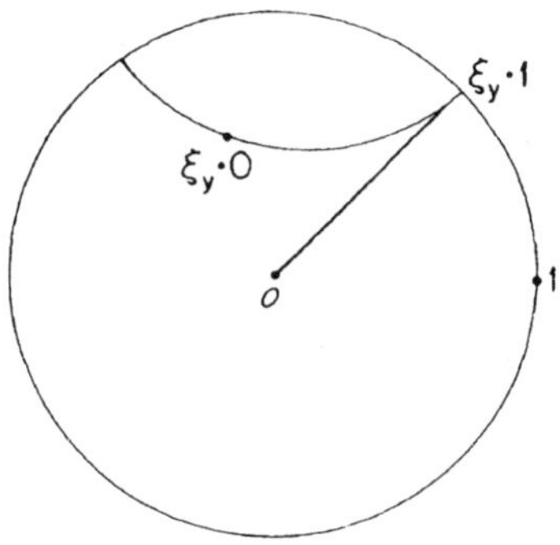

FIG. 4

∇u being the gradient of u and C being a constant. To prove this inequality we consider the Poisson formula

$$u(z) = \frac{1}{2\pi} \int_0^{2\pi} \frac{R^2 - |z|^2}{|Re^{i\phi} - z|^2} u(Re^{i\phi})\, d\phi$$

for the disk $|z| \leq R$, which implies by differentiation

$$\left(\frac{\partial u}{\partial x}\right)(o) = \frac{1}{R\pi} \int_0^{2\pi} u(Re^{i\phi}) \cos \phi \, d\phi,$$

so, using Schwarz's inequality,

$$|(\nabla u)(o)| \leq \frac{2}{R} \sup_{|\zeta| \leq R} |u(\zeta)|. \tag{101}$$

The desired inequality now follows by using (101) on the function $w \to u(w + z)$ in the disk $|w| \leq R$.

As R decreases to 0, the two geodesics above come together at a faster rate, so combining the inequality above with the mean value inequality $|u(z'') - u(z')| \leq |\nabla u||z'' - z'|$ we see that u has the same limit along both geodesics. This concludes the proof.

EXERCISES AND FURTHER RESULTS

A. The Spaces R^n and S^n

1. Let H be a group acting on a space X. If f is a function on X and $h \in H$, let $f^h(x) = f(h^{-1} \cdot x)$, $x \in X$. A continuous function f is said to be *H-finite* if the translated functions f^h all lie in a finite-dimensional space.

(i) For the sphere $S^{n-1} \subset \boldsymbol{R}^n$ show that a function f is $\boldsymbol{O}(n)$-finite if and only if f is the restriction of a polynomial on $\boldsymbol{R}^n$. Equivalently, f is a finite linear combination of spherical harmonics.

(ii)* Consider the orthogonal group $H = \boldsymbol{O}(p, q+1)$ acting on the quadric

$$C_{pq}: x_1^2 + \cdots + x_p^2 - x_{p+1}^2 - \cdots - x_{p+q+1}^2 = -1, \qquad p \geq 0, \quad q \geq 0,$$

in $\boldsymbol{R}^{p+q+1}$. Assume $(p, q) \neq (1, 0)$. Show that a function f on C_{pq} is H-finite if and only if it is the restriction of a polynomial on $\boldsymbol{R}^{p+q+1}$ (Helgason [1963], §3).

(iii) Show that the statement (ii) fails for $\boldsymbol{O}(1, 1)$ acting on $x_1^2 - x_2^2 = 1$.

(iv)* Let K be a compact group of linear transformations of a real vector space V, M the subgroup of K fixing a certain vector $v \neq 0$ in V and i the imbedding $kM \to k \cdot v$ of K/M into V. Then the K-finite functions on K/M are precisely $p \circ i$, where p is a polynomial function on V (ibid., §3).

(v) For a noncompact analog see Chapter IV, Exercise A5.

2. For $\lambda \in \boldsymbol{C}$ let $\mathscr{E}_\lambda(\boldsymbol{R}^n)$ denote the eigenspace

$$\mathscr{E}_\lambda(\boldsymbol{R}^n) = \{f \in \mathscr{E}(\boldsymbol{R}^n) : L_{\boldsymbol{R}^n} f = -\lambda^2 f\}$$

and let T_λ denote the representation of the group $\boldsymbol{M}(n)$ of all isometries of $\boldsymbol{R}^n$ on $\mathscr{E}_\lambda(\boldsymbol{R}^n)$ given by

$$(T_\lambda(g)f)(x) = f(g^{-1} \cdot x), \qquad g \in \boldsymbol{M}(n), \quad x \in \boldsymbol{R}^n.$$

Using Theorem 2.7, Chapter II show that if $\lambda \neq 0$ the $\boldsymbol{O}(n)$-finite functions in $\mathscr{E}_\lambda(\boldsymbol{R}^n)$ are precisely

$$f(x) = \int_{S^{n-1}} e^{i\lambda(x, \omega)} F(\omega) \, d\omega,$$

where F is an $\boldsymbol{O}(n)$-finite function on S^{n-1}. Deduce, generalizing the proof of Theorem 2.6, that

T_λ is irreducible if and only if $\lambda \neq 0$

(cf., Helgason [1974], §8).

3. For the representation T_0 (from Exercise 2) the space $\mathscr{H}_k = \sum_{i=0}^k H_i$ of harmonic polynomials of degree $\leq k$ is an invariant subspace; in particular, T_0 is not irreducible. Show, however, that the corresponding representation of $\boldsymbol{M}(n)$ on the factor space $\mathscr{H}_k/\mathscr{H}_{k-1}$ is irreducible.

*4. While the representation T_0 of $M(2)$ on the space $\mathscr{E}_0(\boldsymbol{R}^2)$ of harmonic functions on $\boldsymbol{R}^2$ is not irreducible, there is an irreducible action on $\mathscr{E}_0(\boldsymbol{R}^2)$ of a larger group (or rather its Lie algebra):

Let $G = \boldsymbol{SL}(2, \boldsymbol{C})$ denote the group of unimodular matrices acting on the one-point compactification of $\boldsymbol{C}$ by means of the maps

$$g: z \to \frac{az + b}{cz + d}, \qquad z \in \boldsymbol{C}$$

$(\begin{smallmatrix} a & b \\ c & d \end{smallmatrix})$ being an element of $\boldsymbol{SL}(2, \boldsymbol{C})$. If $L = L_{\boldsymbol{R}^2}$, consider the operator $L^g: \phi \to (L(\phi \circ g)) \circ \phi^{-1}$; show that L has the *quasi-invariance*

$$(L^g\phi)(z) = |cz - a|^4(L\phi)(z),$$

so ϕ harmonic $\Rightarrow \phi \circ g$ harmonic. Show that the corresponding action of the Lie algebra $\mathfrak{g}$ of G on $\mathscr{E}_0(\boldsymbol{R}^2)$ is *scalar irreducible*; that is, the only continuous operators commuting with the action are the scalar multiples of identity (cf., Helgason [1977a], also for a modified extension to n dimensions).

5. Let H_k be as in §3. Show that

(i) $\dim H_k = \binom{n+k-2}{k} + \binom{n+k-3}{k-1}$.

(ii) $L_{\boldsymbol{R}^n}(|x|^{2j}h) = c|x|^{2j-2}h$, $h \in H_k$, where c is the constant $2j(2j + 2k + n - 2)$. Deduce that if $h \in H_k$,

$$x_j h = h_{k+1} + (n + 2k - 2)^{-1}|x|^2 \frac{\partial h}{\partial x_j},$$

where $h_{k+1} \in H_{k+1}$.

(iii) Let $h \in H_k$. Using the mean-value theorem for harmonic functions, show that

$$\int_{\boldsymbol{R}^n} e^{-(1/2)|x|^2} h(x) e^{-i(x,y)}\, dx = (-i)^k (2\pi)^{(1/2)n} h(y) e^{-|y|^2/2}.$$

(iv) Using the fact that $|x|^{2-n}$ is a harmonic function in $\boldsymbol{R}^n - \{0\}$ $(n > 2)$, prove that

$$\partial(h)\left(\frac{1}{|x|^{n-2}}\right) = c_{k,n} \frac{1}{|x|^{n+2k-2}} h, \qquad (h \in H_k),$$

where

$$c_{k,n} = (-1)^k(n - 2)n(n + 2) \cdots (n + 2k - 4).$$

(Maxwell [1892], Hobson [1931], Coifman and Weiss [1971]).

B. The Hyperbolic Plane

1. Derive the Iwasawa decomposition $G = KAN$ for the group $G = \boldsymbol{SU}(1, 1)$ by means of Figure 2.

2. (i) Show that the transformation

$$c: z \to w = -i\frac{z+i}{z-i} \qquad (w = u + iv)$$

is an isometry of D onto the upper half-plane $H: v > 0$ with the metric

$$\frac{du^2 + dv^2}{4v^2}.$$

(ii) Show that if

$$g(z) = \frac{az + b}{\bar{b}z + \bar{a}}, \qquad |a|^2 - |b|^2 = 1,$$

then

$$cgc^{-1}(w) = \frac{[(a + \bar{a}) - i(b - \bar{b})]w + (b + \bar{b}) - i(a - \bar{a})}{[(b + \bar{b}) + i(a - \bar{a})]w + (a + \bar{a}) + i(b - \bar{b})}$$

and this can be written in the form

$$cgc^{-1}(w) = \frac{\alpha w + \beta}{\gamma w + \delta}, \qquad \text{where} \quad \begin{pmatrix} \alpha & \beta \\ \gamma & \delta \end{pmatrix} \in \boldsymbol{SL}(2, \boldsymbol{R}).$$

In this upper half-plane model the isometry group is generated by $\boldsymbol{SL}(2, \boldsymbol{R})$ and the map $w \to 1/\bar{w}$.

(iii) Show that the circle $S_r(i)$ in the Riemannian manifold H equals the circle $S_{\sinh 2r}(i \cosh 2r)$ in the flat Riemannian manifold $\boldsymbol{R}^2$.

(iv) Show that the Laplace–Beltrami operator on H is given by

$$L = 4v^2\left(\frac{\partial^2}{\partial u^2} + \frac{\partial^2}{\partial v^2}\right)$$

and that, by analogy with Lemma 4.1,

$$L(v^{(1/2)\mu}) = \mu(\mu - 2)v^{(1/2)\mu}.$$

3. Prove that the mapping $f \to F_f$, where

$$F_f(t) = e^{-t}\int_{\boldsymbol{R}} f(n_x a_t \cdot o)\, dx$$

is a bijection of $\mathscr{D}^\natural(D)$ onto the space $\mathscr{D}^\natural(\mathbf{R})$ of even C^∞-functions on $\mathbf{R}$ of compact support. Also,

$$F_{f*g} = F_f * F_g.$$

(Takahashi [1963]).

4. The spherical function ϕ_λ has the following properties: With $\lambda = \xi + i\eta$ ($\xi, \eta \in \mathbf{R}$),

(i) $\phi_\lambda(z)$ is real for all $z \Leftrightarrow \xi\eta = 0$,

(ii) ϕ_λ is bounded $\Leftrightarrow -1 \leq \eta \leq 1$.

5. Prove Prop. 4.12 by comparing the behavior of $\Phi_{\lambda,m}$ and $\Phi_{-\lambda,m}$ near $z = 0$.

6. Let f be a radial function in $L^1(D)$. Show that

$$\lim_{|\xi|\to\infty} \tilde{f}(\xi + i\eta) = 0 \qquad \text{uniformly} \quad \text{in } |\eta| \leq 1.$$

7. If f is a function on D and ξ a horocycle, put

$$\hat{f}(\xi) = \int_\xi f(x)\, d\omega(x),$$

$d\omega$ being the arc element on ξ. Writing

$$\hat{f}(e^{i\theta}, t) = \hat{f}(\xi(\tanh t\, e^{i\theta}, e^{i\theta})),$$

we have seen that

$$\tilde{f}(\lambda, e^{i\theta}) = \int_{\mathbf{R}} e^{(-i\lambda+1)t}\hat{f}(e^{i\theta}, t)\, dt.$$

Deduce from Theorem 4.2 and Prop. 4.12 the following:

The mapping $f \to \hat{f}$ is a bijection of $\mathscr{D}(D)$ onto the space of functions $\psi \in \mathscr{D}(\mathbf{S}^1 \times \mathbf{R})$, which when expanded,

$$\psi(e^{i\theta}, t) = \sum_{n\in\mathbf{Z}} \psi_n(t)e^{in\theta},$$

satisfy

$$\psi_n(t) = e^{-t}\left(\frac{d}{dt} - 1\right)\left(\frac{d}{dt} - 3\right)\cdots\left(\frac{d}{dt} - 2|n| + 1\right)\phi_n(t),$$

where $\phi_n \in \mathscr{D}(\mathbf{R})$ is even (cf., Helgason [1983a]).

8. Describe the orbits of A on D.

9. In the notation of Figure 5, where r and t denote non-Euclidean distances in D, show

$$x(r, t) = \frac{\operatorname{ch} r \operatorname{th} t\, i + \operatorname{sh} r}{\operatorname{sh} r \operatorname{th} t\, i + \operatorname{ch} r}$$

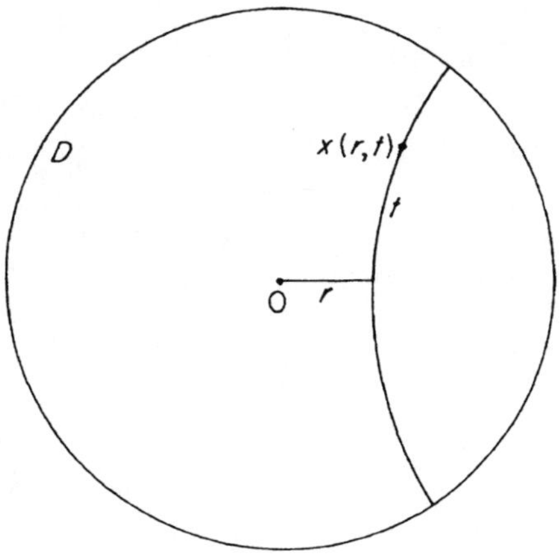

FIG. 5

and prove accordingly that

$$\int_D f(x)\,dx = \int_{\boldsymbol{R}} \int_{\boldsymbol{R}} f(x(r,t)) \cosh 2t\,dt\,dr.$$

Remark. The diffeomorphism $(r, t) \to x(r, t)$ is a special case of [DS], Theorem 14.6, Chapter I.

10. With the transform $f \to F_f$ as in Exercise 3 show that if

$$f(z) = \operatorname{ch} d(o, z)^{-i\lambda - 1}$$

then

$$F_f(t) = \pi \mathbf{c}(\lambda)(\operatorname{ch} t)^{-i\lambda}$$

(Koornwinder).

C. Fourier Analysis on the Sphere

1. (Cf., Sherman [1975]) Let $\boldsymbol{S}^n$ denote the unit sphere in $\boldsymbol{R}^{n+1}$. Fix $a = (0, \ldots, 0, 1)$ and $B = \{s \in \boldsymbol{S}^n : (a, s) = 0\}$ (a is the North Pole and B the Equator). By analogy with (11) in §4 consider for $k \in \boldsymbol{Z}^+$, $b \in B$ the functions

$$e_{b,k}(s) = (a + ib, s)^k, \qquad s \in \boldsymbol{S}^n,$$

$$f_{b,k}(s) = \operatorname{sgn}(s, a)^{n-1}(a + ib, s)^{-k-n+1}, \qquad s \in \boldsymbol{S}^n - B.$$

(i) Show that $e_{k,b}$ and $f_{k,b}$ are eigenfunctions of $L_{\boldsymbol{S}^n}$ with eigenvalue $-k(k + n - 1)$.

(ii) Let C_k^n be the constant multiple of the ultraspherical polynomial $P_k^{(1/2)(n-1)}$ in (8a) in §3 such that $C_k^n(1) = 1$. Then

$$C_k^n((s, a)) = \int_B e_{b,k}(s)\, db = \int_B f_{b,k}(s)\, db \qquad (s \notin B),$$

where db is the normalized invariant measure on B.

(iii)* In analogy with the formula

$$\phi_\lambda(g^{-1}h \cdot o) = \int_B e^{(-i\lambda+1)\langle g\cdot o, b\rangle} e^{(i\lambda+1)\langle h\cdot o, b\rangle}\, db$$

[cf. (35) in §4], prove that

$$C_k^n((s, s')) = \int_B e_{b,k}(s) f_{b,k}(s')\, db.$$

(iv) Given $f \in \mathscr{E}(S^n)$ vanishing in a neighborhood of B, define its Fourier transform

$$\tilde{f}(b, k) = \int_{S^n} f(s) f_{b,k}(s)\, ds.$$

Then [in analogy with Theorem 4.2, part (i)]

$$f(s) = \int_{B\times Z^+} e_{b,k}(s)\tilde{f}(b, k)\, d\mu(b, k),$$

where $d\mu(b, k) = \dim H_k(\boldsymbol{R}^{n+1})\, db$ in the notation of (6) in §3.

NOTES

The representation (Theorem 4.3) of the eigenfunctions of the Laplace–Beltrami operator on $\boldsymbol{H}^2$ was proved by the author ([1970a], Chapter IV, §1, and [1974]). This latter paper also contains the Euclidean analog, Theorem 2.1, which was extended to higher dimensions by Morimoto [1981]. Less specific version for $\boldsymbol{R}^n$ had been given by Hashizume *et al.* [1972]. Different integral representations for constant coefficient operators on $\boldsymbol{R}^n$ are given by Ehrenpreis [1969], Chapter VII, and Palamodov [1970] (Chapter VI). Lemma 2.2 is an immediate generalization of a classical Taylor series characterization of entire functions of exponential type (Boas [1954]). The proof incorporates some simplifications kindly communicated by M. Cowling. Propositions 2.5 and 4.22 are special cases of a mean value theorem for homogeneous spaces (Helgason [1962a], Chapter X, Corollary 7.4; the case of zero eigenvalue had been proved by Godement [1952a]). The eigenspace irreducibility (Theorem 2.6 and Exercise A2) is from the author's paper [1974]. Theorem 2.9 goes back to Paley and Wiener [1934]; see also Hörmander [1963]. The modified version in Theorem 2.10 is from Helgason [1976], §11. (The theorem is misstated in this paper: "polynomial" in Theorem 11.1 should read "harmonic polynomial.")

The expansion of a function into its "Laplace series" of spherical harmonics is classical (see Hobson [1931], and references there). For convergence results in the spirit of Theorem 3.4 see, for example, Calderón and Zygmund [1957], Clerc [1972]. Inspired by Peter-Weyl [1927], Cartan in [1929] interpreted the spherical harmonics group-theoretically and generalized the Laplace series to compact symmetric spaces. Further generalization was given by Weyl [1934] (see Chapter V, §§3-4, of this book). Another approach, more in line with our Fourier transform on $\boldsymbol{H}^2$, was initiated by Sherman [1975, 1977]; see Exercise C1. The classical decompositions (5) and (6) for harmonic functions will be generalized much further in Chapter III.

Theorem 4.2 is a special case of results from the author's papers [1965b, 1970a], (Chapter III, §5), and [1973a], §8, where a Fourier transform for symmetric spaces is defined and developed. The definition is sketched here at the end of Chapter IV, §7 (see Remark 3); the Paley-Wiener theorem for this transform is used in the answer to Problem B in Chapter II, §4. The integral representation in Theorem 4.3 and the irreducibility criterion in Theorem 4.4 (both from Helgason [1970a], Chapter IV) have natural generalizations, some of which are described in Chapter II, §4, No. 1.

The proof of the inversion formula, (25) in §4, for the spherical transform is from Godement [1957a]; the more general Plancherel theorem for the group $G = \boldsymbol{SL}(2, \boldsymbol{R})$ was already given by Harish-Chandra [1952], whereas the basic representation theory for $\boldsymbol{SL}(2, \boldsymbol{R})$ was developed by Bargmann [1947]. The general theory of spherical functions is the subject of Chapter IV, and we refer to the notes to that chapter for further references to the literature.

The expansion in Theorem 4.15 is a special case of an expansion due to Harish-Chandra valid for all semisimple Lie groups (cf., Warner [1972], Vol. II, Theorem 9.1.5.1). The functional equation in Prop. 4.12 is a special case of Lemma 6.2 in Helgason [1973a]; the proof of the Paley-Wiener-type theorem [Theorem 4.2, part (ii) and its generalization] also comes from this paper. An alternative proof was given by Torasso [1977].

Theorem 4.24 on the exponentially bounded eigenfunctions is due to Lewis [1978], as well as its extension to symmetric spaces of rank one. Lewis proved also for symmetric spaces of any rank a generalization of the inclusion $P_\lambda(\mathscr{D}'(B)) \subset \mathscr{E}_\lambda^*(D)$ for the Poisson transform $P_\lambda\colon T \to f$. The converse inclusion was proved by Oshima-Sekiguchi [1980] (for λ outside some hyperplanes).

The Schwarz limit theorem was extended by Karpelevič [1965]; see also Lowdenslager [1958] and Moore [1964] for the case of bounded symmetric domains. Furstenberg's basic paper [1963] (as well as Karpelevič [1965]) gives a Poisson integral representation of bounded harmonic functions on a noncompact symmetric space X. In a *Mathematical Reviews* article on Furstenberg's paper the author posed the problem of generalizing Fatou's theorem (Theorem 4.26) to the symmetric space X in the following form: *Let u be a bounded solution of Laplace's equation $Lu = 0$ on X and let $p \in X$ be any point. Then for almost all geodesics γ starting at p,*

$$\lim_{t \to \infty} u(\gamma(t)) \qquad \textit{exists.}$$

This was proved by Helgason-Korányi [1968]. Many variations of this result have been proved, relaxing the boundedness condition and allowing more general convergence to infinity, see Korányi [1970, 1979], for a survey.

Some of the material of this chapter was published earlier in Helgason [1981] and some in lecture notes in DeMichele and Ricci [1982].

CHAPTER I

INTEGRAL GEOMETRY AND RADON TRANSFORMS

This chapter deals with various features of group-invariant integration. In §1 it is described how an n-form on an n-dimensional manifold M gives rise to a measure on M. If M is a Riemannian manifold, orientable or not, this method gives a canonical measure on M. If M is a Lie group, or more generally a homogeneous space (satisfying a mild assumption), the process leads to an essentially unique invariant measure. In §5 we prove various integral theorems related to the Cartan, Iwasawa, and Bruhat decompositions of a semisimple Lie group.

It was proved by Funk [1916] and Radon [1917], respectively, that a symmetric function on S^2 can be determined by its great circle integrals and that a function on $\mathbf{R}^2$ can be determined by its line integrals. In §2 we give a self-contained elementary exposition of the Radon transform on $\mathbf{R}^n$. The emphasis is on inversion formulas and correspondence theorems for naturally defined function spaces. We also indicate applications to partial differential equations and to tomography.

The theory of the Radon transform on $\mathbf{R}^n$ suggest the general problem of determining a function on a manifold by means of its integrals over certain submanifolds. The diversity of results known for such problems suggests that a general theory would not go very far. However, the quoted examples $\mathbf{S}^2$ and $\mathbf{R}^2$ suggest the following viewpoint for generalizations.

The set of points on $\mathbf{S}^2$ and the set of great circles on $\mathbf{S}^2$ are both homogeneous spaces of the orthogonal group $\mathbf{O}(3)$. Similarly the set of points in $\mathbf{R}^2$ and the set of lines in $\mathbf{R}^2$ are both homogeneous spaces of the group $\mathbf{M}(2)$ of rigid motions of $\mathbf{R}^2$. This motivates our general Radon transform definition, which forms the framework of §3. Given two homogeneous spaces G/K and G/H of the same group G, the Radon transform $f \to \hat{f}$ associates to a function f on G/K a function $\hat{f}$ on G/H in a canonical fashion. The theory of this transform has in certain cases significant applications to differential equations, group representations, tomography, and differential geometry. Another general problem indicated in §3 is the orbital integral problem for a homogeneous space G/H (H noncompact), special cases of which have played a long-term role in representation theory. In §6 we give a solution to this problem in the simplest cases $\mathbf{O}(1, n+1)/\mathbf{O}(1, n)$, $\mathbf{O}(2, n)/\mathbf{O}(1, n)$, and $(L \times L)/\Delta L$, where $L = \mathbf{SL}(2, \mathbf{R})$ and ΔL denotes the diagonal in $L \times L$.

§1. Integration on Manifolds

1. Integration of Forms. Riemannian Measure

Let S be a locally compact Hausdorff space. The set of real-valued continuous functions on S will be denoted by $C(S)$, and $C_c(S)$ shall denote the set of functions in $C(S)$ of compact support. A *measure* on S is by definition a linear mapping $\mu: C_c(S) \to \boldsymbol{R}$ with the property that for each compact subset $K \subset S$ there exists a constant M_K such that

$$|\mu(f)| \leq M_K \sup_{x \in S} |f(x)|$$

for all $f \in C_c(S)$ whose support is contained in K. We recall that a linear mapping $\mu: C_c(S) \to \boldsymbol{R}$ which satisfies $\mu(f) \geq 0$ for $f \geq 0$, $f \in C_c(S)$, is a measure on S. Such a measure is called a *positive*† *measure*. For a manifold M we put $\mathscr{E}(M) = C^\infty(M)$, $\mathscr{D}(M) = \mathscr{E}(M) \cap C_c(M)$. Suppose M is an orientable m-dimensional manifold and let $(U_\alpha, \phi_\alpha)_{\alpha \in A}$ be a collection of local charts on M by which M is oriented (cf., [DS], Chapter VIII, §2). Let ω be an m-form on M. We shall define the integral $\int_M f\omega$ for each $f \in C_c(M)$. First we assume that f has compact support contained in a coordinate neighborhood U_α and let

$$\phi_\alpha(q) = (x_1(q), \ldots, x_m(q)), \qquad q \in U_\alpha.$$

On U_α, ω has an "expression" (cf., [DS], Chapter I, §2, No. 4)

$$\omega_{U_\alpha} = F_\alpha(x_1, \ldots, x_m)\, dx_1 \wedge \cdots \wedge dx_m, \tag{1}$$

and we set

$$\int_M f\omega = \int_{\phi_\alpha(U_\alpha)} (f \circ \phi_\alpha^{-1})(x_1, \ldots, x_m) F_\alpha(x_1, \ldots, x_m)\, dx_1 \cdots dx_m.$$

On using the transformation formula for multiple integrals we see that if f has compact support inside the intersection $U_\alpha \cap U_\beta$ of two coordinate neighborhoods, then the right-hand side in the formula above is

$$\int_{\phi_\beta(U_\beta)} (f \circ \phi_\beta^{-1})(y_1, \ldots, y_m) F_\beta(y_1, \ldots, y_m)\, dy_1 \cdots dy_m,$$

if $F_\beta\, dy_1 \wedge \cdots \wedge dy_m$ is the expression for ω on U_β. Thus $\int_M f\omega$ is well-defined. Next, let f be an arbitrary function in $C_c(M)$. Then f vanishes outside a paracompact open submanifold of M, and by Theorem 1.3, Chapter I, in [DS] f can be expressed as a finite sum $f = \sum_i f_i$, where

† See "Some Details," p. 611.

each f_i has compact support inside some neighborhood U_α from our covering. We put

$$\int_M f\omega = \sum_i \int_M f_i\omega.$$

Here it has to be verified (Chevalley [1946], p. 163) that the right-hand side is independent of the chosen decomposition $f = \sum_i f_i$ of f. Let $f = \sum_j g_j$ be another such decomposition and select $\phi \in C_c(M)$ such that $\phi = 1$ on the union of the supports of all f_i and g_j. Then $\phi = \sum \phi_\alpha$ (finite sum), where each ϕ_α has support inside a coordinate neighborhood from our covering. We have

$$\sum_i f_i\phi_\alpha = \sum_j g_j\phi_\alpha,$$

and since each summand has support inside a fixed coordinate neighborhood,

$$\sum_i \int (f_i\phi_\alpha)\omega = \sum_j \int (g_j\phi_\alpha)\omega.$$

For the same reason the formulas

$$f_i = \sum_\alpha f_i\phi_\alpha, \qquad g_j = \sum_\alpha g_j\phi_\alpha$$

imply that

$$\int f_i\omega = \sum_\alpha \int (f_i\phi_\alpha)\omega, \qquad \int g_j\omega = \sum_\alpha \int (g_j\phi_\alpha)\omega,$$

from which we derive the desired relation

$$\sum_i \int f_i\omega = \sum_j \int g_j\omega.$$

The integral $\int f\omega$ is now well-defined and the mapping

$$f \to \int_M f\omega, \qquad f \in C_c(M),$$

is a measure on M. We have obviously

Lemma 1.1. *If $\int_M f\omega = 0$ for all $f \in C_c(M)$, then $\omega = 0$.*

Definition. The m-form ω is said to be *positive* if for each $\alpha \in A$, the function F_α in (1) is >0 on $\phi_\alpha(U_\alpha)$.

If ω is a positive m-form on M, then it follows readily from Theorem 1.3, Chapter I, in [DS] that $\int_M f\omega \geq 0$ for each nonnegative function $f \in C_c(M)$. Thus, a positive m-form gives rise to a positive measure.

Suppose M and N are two oriented manifolds and let Φ be a diffeomorphism of M onto N. We assume that Φ is *orientation preserving*, that is, if the collection of local charts $(U_\alpha, \phi_\alpha)_{\alpha \in A}$ defines the orientation on M, then the collection $(\Phi(U_\alpha), \phi_\alpha \circ \Phi^{-1})_{\alpha \in A}$ of local charts on N defines the orientation on N. Let m denote the dimension of M and N.

Let ω be an m-form on N and $\Phi^*\omega$ its *transform* or *pullback* by Φ ([DS], p. 25). Then the formula

$$\int_M f\Phi^*\omega = \int_N (f \circ \Phi^{-1})\omega \tag{2}$$

holds for all $f \in C_c(M)$. In fact, it suffices to verify (2) in the case when f has compact support inside a coordinate neighborhood U_α. If we evaluate the left-hand side of (2) by means of the coordinate system ϕ_α and the right-hand side of (2) by means of the coordinate system $\phi_\alpha \circ \Phi^{-1}$, both sides of (2) reduce to the same integral.

If M is a pseudo-Riemannian manifold, orientable or not, a measure can be defined on M as follows. Consider a local chart (U_α, ϕ_α) on M and, as before, let $\phi_\alpha(q) = (x_1, \ldots, x_m)$, $q \in U_\alpha$. Let g denote the pseudo-Riemannian structure and generalizing the definition in [DS], Chapter VIII, §2, put

$$g_{ij} = g\left(\frac{\partial}{\partial x_i}, \frac{\partial}{\partial x_j}\right), \qquad \bar{g} = |\det(g_{ij})|.$$

For each function f in $C_c(U_\alpha)$ put

$$\mu(f) = \int_{\phi_\alpha(U_\alpha)} (f \circ \phi_\alpha^{-1})(x_1, \ldots, x_m)\sqrt{\bar{g}}\, dx_1\, dx_2 \cdots dx_m.$$

The expression on the right is invariant under coordinate changes. Using partition of unity as before, $\mu(f)$ can be defined for all $f \in C_c(M)$. The result is a positive measure on M, which we shall refer to as the *Riemannian measure* on M. We sometimes denote it by

$$\sqrt{\bar{g}}\, dx_1 \cdots dx_m.$$

Example. Let $\boldsymbol{H}^2$ denote the hyperbolic plane, that is, the unit disk $x^2 + y^2 < 1$ with the Riemannian structure

$$g = \frac{dx^2 + dy^2}{(1 - x^2 - y^2)^2}$$

(cf., [DS], Exercise G, Chapter I). The Riemannian measure is given by

$$(1 - x^2 - y^2)^{-2}\, dx\, dy.$$

In the geodesic polar coordinates (r, θ) (where $x = \tanh r \cos\theta$, $y = \tanh r \sin\theta$), it is given by

$$\tfrac{1}{2}\sinh(2r)\, dr\, d\theta.$$

For an oriented Riemannian manifold we shall now describe the Riemannian measure a little differently.

Lemma 1.2. *Let M be a Riemannian manifold, oriented by means of a collection $(U_\alpha, \phi_\alpha)_{\alpha\in A}$ of local charts. Let $\phi_\alpha(q) = (x_1, \dots, x_m)$ for $q \in U_\alpha$, and let $\bar{g}$ be defined as above. There exists a unique m-form ω on M such that*

$$f \to \int_M f\omega$$

is the Riemannian measure. On U_α, ω is given by

$$\omega = \sqrt{\bar{g}}\, dx_1 \wedge \cdots \wedge dx_m. \tag{3}$$

Proof. We can define an m-form ω on M by

$$\omega_q(X_1, \dots, X_m) = 1/m! \tag{4}$$

if $q \in U_\alpha$ and (X_i) is any orthonormal basis of M_q for which the matrix (a_{ij}) given by $(\partial/\partial x_j)_q = \sum_i a_{ij} X_i$ has positive determinant. Clearly ω_q does not depend on the choice of (X_i). We have $g_{ij} = \sum_k a_{ki} a_{kj}$, so that by the positivity, $\det(a_{ij}) = \sqrt{\bar{g}}$. Moreover,

$$\omega_q\left(\frac{\partial}{\partial x_1}, \dots, \frac{\partial}{\partial x_m}\right) = \sum a_{1i_1} \cdots a_{mi_m} \omega_q(X_{i_1}, \dots, X_{i_m})$$

$$= \frac{1}{m!}\det(a_{ij}) = \sqrt{\bar{g}}(dx_1 \wedge \cdots \wedge dx_m)_q\left(\frac{\partial}{\partial x_1}, \dots, \frac{\partial}{\partial x_m}\right).$$

This ω defined by (4) satisfies (3). Also (2) implies for $f \in C_c(U_\alpha)$

$$\int_{U_\alpha} f\omega = \int_{\phi_\alpha(U_\alpha)} (f \circ \phi_\alpha^{-1})(x_1, \dots, x_m)\sqrt{\bar{g}}\, dx_1 \cdots dx_m,$$

so that ω gives the Riemannian measure. The uniqueness being obvious, the lemma is proved.

Proposition 1.3. *Let M and N be Riemannian manifolds and Φ a diffeomorphism of M onto N. For $p \in M$ let $|\det d\Phi_p|$ denote the absolute* *value of the determinant of the linear isomorphism* $d\Phi_p: M_p \to N_{\Phi(p)}$ *when expressed in terms of any orthonormal bases. Then for* $F \in \mathscr{D}(N)$

$$(5) \qquad \int_N F(q)\, dq = \int_M F(\Phi(p))|\det d\Phi_p|\, dp,$$

if dp and dq denote the Riemannian measures on M and N, respectively.

Proof. By partition of unity it suffices to prove (5) when F has support in a coordinate neighborhood. We may therefore assume M and N to be oriented and that Φ is orientation preserving. In accordance with Lemma 1.2 let ω and θ be the m-forms on M and N, respectively, giving the Riemannian measures dp and dq. Then by (2), if $f = F \circ \Phi$,

$$\int_N F(q)\, dq = \int_N (f \circ \Phi^{-1})\theta = \int_M (F \circ \Phi)\Phi^*\theta.$$

But $\Phi^*\theta = \det(d\Phi)\, \omega$ because of (4), and now (5) follows.

Remark. Proposition 1.3 shows, in particular, that the Riemannian measure is invariant under isometries.

2. Invariant Measures on Coset Spaces

Let M be a manifold and Φ a diffeomorphism of M onto itself. We recall that a differential form ω on M is called invariant under Φ if

$$\Phi^*\omega = \omega.$$

Let G be a Lie group with Lie algebra $\mathfrak{g}$. A differential form ω on G is called *left-invariant* if $L_x^*\omega = \omega$ for all $x \in G$, L_x [or $L(x)$] denoting the left translation $g \to xg$ on G. Also, R_x [or $R(x)$] denotes the right translation $g \to gx$ on G and *right-invariant* differential forms on G can be defined. If $X \in \mathfrak{g}$, let $\tilde{X}$ denote the corresponding left-invariant vector field on G. Let $X_1, \ldots, X_n$ be a basis of $\mathfrak{g}$. The equations $\omega^i(\tilde{X}_j) = \delta^i_j$ determine uniquely n 1-forms ω^i on G. These are clearly left-invariant and the exterior product $\omega = \omega^1 \wedge \cdots \wedge \omega^n$ is a left-invariant n-form on G. Each 1- form on G can be written $\sum_{i=1}^n f_i\omega^i$, where $f_i \in \mathscr{E}(G)$; it follows that each n-form can be written $f\omega$, where $f \in \mathscr{E}(G)$. Thus, except for a constant factor, ω is the only left-invariant n-form on G. Let

$$\phi: x \to (x_1(x), \ldots, x_n(x))$$

be a system of canonical coordinates with respect to the basis $X_1, \ldots, X_n$ of $\mathfrak{g}$, valid on a connected open neighborhood U of e (cf. [DS], Chapter II, §1). On U, ω has an expression

$$\omega_U = F(x_1, \ldots, x_n)\, dx_1 \wedge \cdots \wedge dx_n$$

and $F > 0$. Now, if $g \in G$, the pair $(L_g U, \phi \circ L_{g^{-1}})$ is a local chart on a connected neighborhood of g. We put $(\phi \circ L_{g^{-1}})(x) = (y_1(x), \ldots, y_n(x))$ $(x \in L_g U)$. Since $y_i(gx) = x_i(x)$ $(x \in U \cap L_g U)$, the mapping

$$L_g : U \to L_g U$$

has coordinate expression ([DS] Chapter I, §3, No. 1) given by

$$(y_1, \ldots, y_n) = (x_1, \ldots, x_n).$$

On $L_g U$, ω has an expression

$$\omega_{L_g U} = G(y_1, \ldots, y_n)\, dy_1 \wedge \cdots \wedge dy_n,$$

so that the invariance condition $\omega_x = L_g^* \omega_{gx}$ $(x \in U \cap L_g U)$ can be written

$$\begin{aligned} &G(y_1(x), \ldots, y_n(x))(dy_1 \wedge \cdots \wedge dy_n)_x \\ &\quad = G(x_1(x), \ldots, x_n(x))(dx_1 \wedge \cdots \wedge dx_n)_x. \end{aligned}$$

Hence $F(x_1(x), \ldots, x_n(x)) = G(x_1(x), \ldots, x_n(x))$ and

$$F(x_1(x), \ldots, x_n(x)) = F(y_1(x), \ldots, y_n(x)) \frac{\partial(y_1(x), \ldots, y_n(x))}{\partial(x_1(x), \ldots, x_n(x))}$$

for $x \in U \cap L_g U$, which shows that the Jacobian of $(\phi \circ L_{g^{-1}}) \circ \phi^{-1}$ is >0. Consequently, the collection $(L_g U, \phi \circ L_{g^{-1}})_{g \in G}$ of local charts turns G into an oriented manifold and each left translation is orientation preserving. The orientation of G depends on the choice of basis of $\mathfrak{g}$. If $X'_1, \ldots, X'_n$ is another basis, then the resulting orientation of G is the same as that before if and only if the linear transformation

$$X_i \to X'_i \qquad (1 \leq i \leq n)$$

has positive determinant.

The form ω is a positive left-invariant n-form on G and except for a constant positive factor, ω is uniquely determined by these properties. We shall denote it by $d_l g$. The linear mapping of $C_c(G)$ into $\boldsymbol{R}$ given by $f \to \int f\, d_l g$ is a *measure* on G, which we denote by μ_l. This measure is positive; moreover, it is *left-invariant* in the sense that $\mu_l(f \circ L_x) = \mu_l(f)$ for $x \in G$, $f \in C_c(G)$.

Similarly, G can be turned into an oriented manifold such that each R_g $(g \in G)$ is orientation preserving. There exists a right-invariant positive n-form $d_r g$ on G and this is unique except for a constant positive factor. We define the *right-invariant* positive measure μ_r on G by

$$\mu_r(f) = \int f \, d_r g, \qquad f \in C_c(G).$$

The group G has been oriented in two ways. The left-invariant orientation is invariant under all right translations R_x ($x \in G$) if and only if it is invariant under all $I(x) = L_x \circ R_{x^{-1}}$ ($x \in G$). Since the differential $dI(x)_g$ satisfies

$$dI(x)_g = dL_{xgx^{-1}} \circ \mathrm{Ad}(x) \circ dL_{g^{-1}},$$

the necessary and sufficient condition is $\det \mathrm{Ad}(x) > 0$ for all $x \in G$. This condition is always fulfilled if G is connected.

Lemma 1.4. *With the notation above we have*

$$d_r g = c \det \mathrm{Ad}(g) \, d_l g,$$

where c is a constant.

Proof. Let $\theta = \det \mathrm{Ad}(g) \, d_l g$ and let $x \in G$. Then

$$(R_{x^{-1}})^* \theta = \det \mathrm{Ad}(gx^{-1})(R_{x^{-1}})^* \, d_l g = \det \mathrm{Ad}(gx^{-1}) I(x)^* \, d_l g.$$

At the point $g = e$ we have

$$(I(x)^*(d_l g))_e = \det \mathrm{Ad}(x)(d_l g)_e.$$

Consequently,

$$(R_{x^{-1}}{}^* \theta)_e = \det \mathrm{Ad}(e)(d_l g)_e = \theta_e.$$

Thus, θ is right-invariant and therefore proportional to $d_r g$.

Remark. If G is connected it can be oriented in such a way that all left and right translations are orientation preserving. If $d_r g$ and $d_l g$ are defined by means of this orientation, Lemma 1.4 holds with $c > 0$.

Corollary 1.5. *Let $x, y \in G$ and put $d_l(ygx) = (L_y R_x)^* \, d_l g$, $d_r(xgy) = (L_x R_y)^* \, d_r g$. Moreover, if J denotes the mapping $g \to g^{-1}$, put $d_l(g^{-1}) = J^*(d_l g)$. Then*

$$d_l(gx) = \det \mathrm{Ad}(x^{-1}) \, d_l(g), \qquad d_r(xg) = \det \mathrm{Ad}(x) \, d_r g,$$

$$d_l(g^{-1}) = (-1)^{\dim G} \det \mathrm{Ad}(g) \, d_l g.$$

In fact, the lemma implies that

$$c \det \mathrm{Ad}(g) \, d_l g = d_r g = d_r(gx) = c \det \mathrm{Ad}(gx) \, d_l(gx),$$

$$d_r(xg) = c \det \mathrm{Ad}(xg) \, d_l(xg) = c \det \mathrm{Ad}(xg) \, d_l g.$$

Finally, since $JR_x = L_{x^{-1}}J$, we have

$$(R_x)^* d_l(g^{-1}) = (R_x)^* J^* d_l g = (JR_x)^* d_l g = (L_{x^{-1}} J)^* d_l g = J^* d_l g,$$

so that $d_l(g^{-1})$ is right-invariant, hence proportional to $d_r g$. But obviously

$$(d_l(g^{-1}))_e = (-1)^{\dim G}(d_l g)_e,$$

so that the corollary is verified.

Definition. A Lie group G is called *unimodular* if the left invariant measure μ_l is also right-invariant.

In view of Corollary 1.5 we have by (2)

$$\mu_l(f \circ R_x) = |\det \mathrm{Ad}(x)| \mu_l(f). \tag{6}$$

It follows that G is unimodular if and only if $|\det \mathrm{Ad}(x)| = 1$ for all $x \in G$. If this condition is satisfied, the measures μ_l and μ_r coincide except for a constant factor.

Proposition 1.6. *The following Lie groups are unimodular:*

(i) *Lie groups G for which* $\mathrm{Ad}(G)$ *is compact;*

(ii) *semisimple Lie groups;*

(iii) *connected nilpotent Lie groups.*

Proof. In the case (i), the group $\{|\det \mathrm{Ad}(x)| : x \in G\}$ is a compact subgroup of the multiplicative group of positive real numbers. This subgroup necessarily consists of one element, so that G is unimodular. In the case (ii), each $\mathrm{Ad}(x)$ leaves invariant a nondegenerate bilinear form (namely, the Killing form). It follows that $(\det \mathrm{Ad}(x))^2 = 1$. Finally, let N be a connected nilpotent Lie group with Lie algebra $\mathfrak{n}$. If $X \in \mathfrak{n}$, then $\mathrm{ad}\, X$ is nilpotent, so that $\mathrm{Tr}(\mathrm{ad}\, X) = 0$. Since

$$\det e^A = r^{\mathrm{Tr}\, A}$$

for an arbitrary linear transformation A, we obtain

$$\det \mathrm{Ad}(\exp X) = e^{\mathrm{Tr}(\mathrm{Ad}\, X)} = 1.$$

This proves (iii).

Notation. In the sequel we shall mostly use the left invariant measure μ_l. The measure dg is usually called *Haar measure* on G. For simplicity we shall write μ instead of μ_l and dg instead of $d_l g$.

Let G be a Lie group with Lie algebra $\mathfrak{g}$; let H be a closed subgroup with Lie algebra $\mathfrak{h} \subset \mathfrak{g}$. Each $x \in G$ gives rise to an analytic diffeomorphism $\tau(x): gH \to xgH$ of G/H onto itself. Let π denote the natural mapping of G onto G/H and put $o = \pi(e)$. If $h \in H$, $(d\tau(h))_o$ is an endomorphism of the tangent space $(G/H)_o$. For simplicity, we shall write $d\tau(h)$ instead of $(d\tau(h))_o$ and $d\pi$ instead of $(d\pi)_e$.

Lemma 1.7.

$$\det(d\tau(h)) = \frac{\det \mathrm{Ad}_G(h)}{\det \mathrm{Ad}_H(h)} \qquad (h \in H).$$

Proof. It was shown in [DS], Chapter II, §4, that the differential $d\pi$ is a linear mapping of $\mathfrak{g}$ onto $(G/H)_o$ and has kernel $\mathfrak{h}$. Let $\mathfrak{m}$ be any subspace of $\mathfrak{g}$ such that $\mathfrak{g} = \mathfrak{h} + \mathfrak{m}$ (direct sum). Then $d\pi$ induces an isomorphism of $\mathfrak{m}$ onto $(G/H)_o$. Let $X \in \mathfrak{m}$. Then

$$\mathrm{Ad}_G(h)\, X = dR_{h^{-1}} \circ dL_h(X).$$

Since $\pi \circ R_h = \pi$, $(h \in H)$ and $\pi \circ L_g = \tau(g) \circ \pi$, $(g \in G)$, we obtain

$$d\pi \circ \mathrm{Ad}_G(h)\, X = d\tau(h) \circ d\pi(X), \qquad h \in H, \quad X \in \mathfrak{m}. \tag{7}$$

The vector $\mathrm{Ad}_G(h)\, X$ decomposes according to $\mathfrak{g} = \mathfrak{h} + \mathfrak{m}$,

$$\mathrm{Ad}_G(h)\, X = X(h)_{\mathfrak{h}} + X(h)_{\mathfrak{m}}.$$

The endomorphism $A_h\colon X \to X(h)_{\mathfrak{m}}$ of $\mathfrak{m}$ satisfies

$$d\pi \circ A_h(X) = d\tau(h) \circ d\pi(X), \qquad X \in \mathfrak{m},$$

so that $\det A_h = \det(d\tau(h))$. On the other hand,

$$\exp \mathrm{Ad}_G(h) tT = h \exp tT\, h^{-1} = \exp \mathrm{Ad}_H(h)\, tT$$

for $t \in R$, $T \in \mathfrak{h}$. Hence $\mathrm{Ad}_G(h)\, T = \mathrm{Ad}_H(h)\, T$ so that

$$\det \mathrm{Ad}_G(h) = \det A_h \det \mathrm{Ad}_H(h),$$

and the lemma is proved.

Proposition 1.8. *Let* $m = \dim G/H$. *The following conditions are equivalent*:

(i) *G/H has a nonzero G-invariant m-form ω;*
(ii) $\det \mathrm{Ad}_G(h) = \det \mathrm{Ad}_H(h)$ *for* $h \in H$.

If these conditions are satisfied, then G/H has a G-invariant orientation and the G-invariant m-form ω is unique up to a constant factor.

Proof. Let ω be a G-invariant m-form on G/H, $\omega \neq 0$. Then the relation $\tau(h)^*\omega = \omega$ at the point o implies $\det(d\tau(h)) = 1$, so (ii) holds. On the other hand, let $X_1, \ldots, X_m$ be a basis of $(G/H)_o$ and let $\omega^1, \ldots, \omega^m$ be the linear functions on $(G/H)_o$ determined by $\omega^i(X_j) = \delta_{ij}$. Consider the element $\omega^1 \wedge \cdots \wedge \omega^m$ in the Grassmann algebra of the tangent space $(G/H)_o$. Condition (ii) implies that $\det(d\tau(h)) = 1$ and the element $\omega^1 \wedge \cdots \wedge \omega^m$ is invariant under the linear transformation $d\tau(h)$. It follows that there exists a unique G-invariant m-form ω on G/H such that $\omega_o = \omega^1 \wedge \cdots \wedge \omega^m$. If ω^* is another G-invariant m-form on

G/H, then $\omega^* = f\omega$, where $f \in \mathscr{E}(G/H)$. Owing to the G-invariance, $f =$ constant.

Assuming (i), let $\phi: p \to (x_1(p), \ldots, x_m(p))$ be a system of coordinates on an open connected neighborhood U of $o \in G/H$ on which ω has an expression

$$\omega_U = F(x_1, \ldots, x_m)\, dx_1 \wedge \cdots \wedge dx_m,$$

with $F > 0$. The pair $(\tau(g)U,\ \phi \circ \tau(g^{-1}))$ is a local chart on a connected neighborhood of $g \cdot o \in G/H$. We put $(\phi \circ \tau(g^{-1}))(p) = (y_1(p), \ldots, y_m(p))$ for $p \in \tau(g)U$. Then the mapping $\tau(g): U \to \tau(g)U$ has expression ([DS], Chapter I, §3.1) $(y_1, \ldots, y_m) = (x_1, \ldots, x_m)$. On $\tau(g)U$, ω has an expression

$$\omega_{\tau(g)U} = G(y_1, \ldots, y_m)\, dy_1 \wedge \cdots \wedge dy_m$$

and since $\omega_q = \tau(g)^*\omega_{\tau(g)q}$ we have for $q \in U \cap \tau(g)U$

$$\begin{aligned} \omega_q &= G(y_1(q), \ldots, y_m(q))(dy_1 \wedge \cdots \wedge dy_m)_q \\ &= G(x_1(q), \ldots, x_m(q))(dx_1 \wedge \cdots \wedge dx_m)_q. \end{aligned}$$

Hence $F(x_1(q), \ldots, x_m(q)) = G(x_1(q), \ldots, x_m(q))$ and

$$F(x_1(q), \ldots, x_m(q)) = F(y_1(q), \ldots, y_m(q)) \frac{\partial(y_1(q), \ldots, y_m(q))}{\partial(x_1(q), \ldots, x_m(q))},$$

which shows that the Jacobian of the mapping $(\phi \circ \tau(g^{-1})) \circ \phi^{-1}$ is >0. Consequently, the collection $(\tau(g)U,\ \phi \circ \tau(g^{-1}))_{g \in G}$ of local charts turns G/H into an oriented manifold and each $\tau(g)$ is orientation preserving.

The G-invariant form ω now gives rise to an integral $\int f\omega$ which is invariant in the sense that

$$\int_{G/H} f\omega = \int_{G/H} (f \circ \tau(g))\omega, \qquad g \in G.$$

However, just as the Riemannian measure did not require orientability, an invariant measure can be constructed on G/H under a condition which is slightly more general than (ii). The projective space $\boldsymbol{P}^2(\boldsymbol{R})$ will, for example, satisfy this condition but it does not satisfy (ii). We recall that a measure μ on G/H is said to be invariant (or more precisely G-invariant) if $\mu(f \circ \tau(g)) = \mu(f)$ for all $g \in G$.

Theorem 1.9. *Let G be a Lie group and H a closed subgroup. The relation*

$$|\det \mathrm{Ad}_G(h)| = |\det \mathrm{Ad}_H(h)|, \qquad h \in H, \tag{8}$$

is a necessary and sufficient condition for the existence of a G-invariant measure >0 on G/H. This measure dg_H is unique (up to a constant factor) and

$$\int_G f(g)\,dg = \int_{G/H}\left(\int_H f(gh)\,dh\right)dg_H, \qquad f \in C_c(G), \tag{9}$$

if the left-invariant measures dg and dh are suitably normalized.

Formula (9) is illustrated in Fig. 6, where $\pi: G \to G/H$ is the natural mapping.

We begin by proving a simple lemma.

Lemma 1.10. *Let G be a Lie group and H a closed subgroup. Let dh be a left-invariant measure >0 on H and put*

$$\tilde{f}(gH) = \int_H f(gh)\,dh, \qquad f \in C_c(G).$$

Then the mapping $f \to \tilde{f}$ is a linear mapping of $C_c(G)$ onto $C_c(G/H)$.

Proof. Let $F \in C_c(G/H)$; we have to prove that there exists a function $f \in C_c(G)$ such that $F = \tilde{f}$. Let C be a compact subset of G/H outside which F vanishes and let C' be a compact subset of G whose image is C under the natural mapping $\pi: G \to G/H$. Let C_H be a compact subset of H of positive measure and put $\tilde{C} = C' \cdot C_H$. Then $\pi(\tilde{C}) = C$. Select $f_1 \in C_c(G)$ such that $f_1 \geq 0$ on G and $f_1 > 0$ on $\tilde{C}$. Then $\tilde{f}_1 > 0$ on C (since C_H has positive measure) and the function

$$f(g) = \begin{cases} f_1(g)\dfrac{F(\pi(g))}{\tilde{f}_1(\pi(g))} & \text{if } \pi(g) \in C \\ 0 & \text{if } \pi(g) \notin C \end{cases}$$

belongs to $C_c(G)$ and $\tilde{f} = F$.

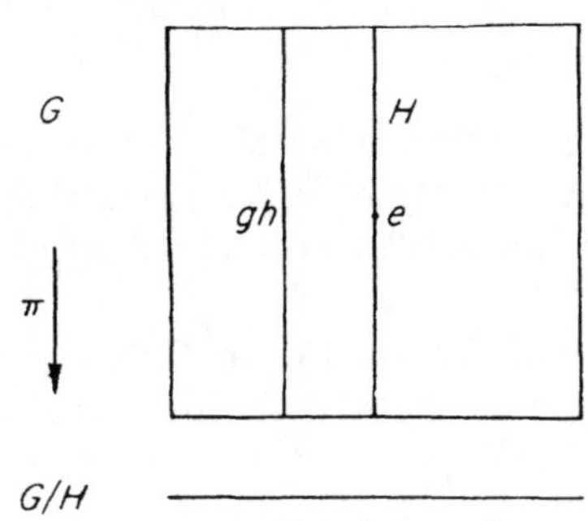

FIG. 6

Now in order to prove Theorem 1.9 suppose first that the relation

$$|\det \mathrm{Ad}_G(h)| = |\det \mathrm{Ad}_H(h)|, \qquad h \in H,$$

holds. Let $\phi \in C_c(G)$. Since we are dealing with measures rather than differential forms, we have by Cor. 1.5

$$\begin{aligned}\int_G \phi(g)\left(\int_H f(gh)\,dh\right) dg &= \int_H dh \int_G \phi(g) f(gh)\,dg \\ &= \int_H dh \int_G \phi(gh^{-1}) f(g)\,|\det \mathrm{Ad}_G(h)|\,dg \\ &= \int_G f(g)\,dg \int_H \phi(gh^{-1})\,|\det \mathrm{Ad}_G(h)|\,dh.\end{aligned}$$

But the relation (8) and the last part of Corollary 1.5 shows that

$$\int_H \phi(gh^{-1})\,|\det \mathrm{Ad}_G(h)|\,dh = \int_H \phi(gh)\,dh,$$

so that

$$\int_G \phi(g)\,dg \int_H f(gh)\,dh = \int_G f(g)\,dg \int_H \phi(gh)\,dh.$$

Taking ϕ such that $\int \phi(gh)\,dh = 1$ on the support of f, we conclude that

$$\int_G f(g)\,dg = 0 \qquad \text{if } \tilde{f} \equiv 0.$$

In view of the lemma we can therefore define a linear mapping $\mu: C_c(G/H) \to \boldsymbol{R}$ by

$$\mu(F) = \int_G f(g)\,dg \qquad \text{if} \quad F = \tilde{f}.$$

Since $\mu(F) \geq 0$ if $F \geq 0$, μ is a positive measure on G/H; moreover,

$$\mu((\tilde{f})^{\tau(x)}) = \int_G f^{L(x)}(g)\,dg = \int_G f(g)\,dg = \mu(\tilde{f}),$$

so that μ is invariant.

In order to prove the converse we shall make use of the theorem that the group G has (up to a constant factor) a unique† positive left invariant measure. For this "uniqueness of Haar measure" see e.g., Weil's book [1940].

If μ is a positive invariant measure on G/H, the mapping $f \to \mu(\tilde{f})$ is a positive left invariant measure on G. Owing to the uniqueness mentioned,

$$\int_G f(g)\,dg = \mu(\tilde{f}).$$

†See "Some Details," p. 611.

In view of the lemma this proves the uniqueness of μ as well as (9). In order to derive (8), replace $f(g)$ by $f(gh_1)$ in (9). Owing to Corollary 1.5 the left-hand side is multiplied by $|\det \mathrm{Ad}_G(h_1)|$ and the right-hand side is multiplied by $|\det \mathrm{Ad}_H(h_1)|$. This finishes the proof of Theorem 1.9.

Remark. If H is compact, condition (8) is satisfied; hence in this case G/H has a G-invariant measure.

In the remainder of this section we shall often have to calculate how invariant measures transform under mappings. To a certain extent these calculations can be reduced to the following general lemma.

Lemma 1.11. *Let G and S be Lie groups and $H \subset G$ and $T \subset S$ closed subgroups. Suppose that the coset spaces G/H and S/T have the same dimension m and that they possess positive invariant differential m-forms, denoted dg_H and ds_T. Let $o = \{H\}$, $o' = \{T\}$.*

Let ϕ be a differentiable mapping of G/H into S/T such that $\phi \cdot \{H\} = \{T\}$. Then

$$\phi^*(ds_T) = D\, dg_H,$$

where D is a function on G/H computed as follows: Let $X_1, \ldots, X_m$ and $Y_1, \ldots, Y_m$ be fixed (but arbitrary) bases of the tangent spaces $(G/H)_o$ and $(S/T)_{o'}$, respectively, such that

$$dg_H(X_1, \ldots, X_m) = ds_T(Y_1, \ldots, Y_m).$$

Let $g \in G$ and select $s \in \phi(gH)$. Consider the linear mapping

$$A(g) = d\tau(s^{-1}) \circ d\phi_{gH} \circ d\tau(g)_o$$

of $(G/H)_o$ into $(S/T)_{o'}$ and put $A(g)X_j = \sum_i a_{ij}(g) Y_i$. Then

$$D(gH) = \det(a_{ij}(g)).$$

Proof. Let $\omega^1, \ldots, \omega^m$ be the linear functions on $(G/H)_{o'}$ determined by $\omega^i(X_j) = \delta_{ij}$; let $\theta^1, \ldots, \theta^m$ be the linear functions on $(S/T)_{o'}$ determined by $\theta^i(Y_j) = \delta_{ij}$. Then the dual mapping

$${}^t(A(g)) = \tau(g)^* \circ \phi^* \circ \tau(s^{-1})^*,$$

which maps the dual of $(S/T)_{o'}$ into the dual of $(G/H)_o$ satisfies

$${}^t(A(g))\theta^i = \sum_j a_{ij}(g)\omega^j.$$

Now, due to the assumption about dg_H and ds_T,

$$(dg_H)_{gH} = c\tau(g^{-1})^*(\omega^1 \wedge \cdots \wedge \omega^m),$$

$$(ds_T)_{sT} = c\tau(s^{-1})^*(\theta^1 \wedge \cdots \wedge \theta^m),$$

where c is a constant. Consequently,

$$\begin{aligned}(\phi^*(ds_T))_{gH} &= c\phi^* \circ \tau(s^{-1})^*(\theta^1 \wedge \cdots \wedge \theta^m)\\ &= c\tau(g^{-1})^{*t}A(g)(\theta^1 \wedge \cdots \wedge \theta^m)\\ &= c\det(a_{ij}(g))\tau(g^{-1})^*(\omega^1 \wedge \cdots \wedge \omega^m)\\ &= \det(a_{ij}(g))(dg_H)_{gH}.\end{aligned}$$

Next we calculate the behavior of the Haar measure under the decomposition of a group into a product.

Proposition 1.12. *Let U be a Lie group with Lie algebra $\mathfrak{u}$. Suppose $\mathfrak{u}$ is a direct sum $\mathfrak{u} = \mathfrak{m} + \mathfrak{h}$, where $\mathfrak{m}$ and $\mathfrak{h}$ are subalgebras of $\mathfrak{u}$ (not necessarily ideals). Let M and H denote the analytic subgroups of U with Lie algebras $\mathfrak{m}$ and $\mathfrak{h}$, respectively. Suppose the mapping $\alpha: (m, h) \to mh$ is a one-to-one mapping of $M \times H$ onto U.*

Then the positive left-invariant measures dh, dm, du, can be normalized in such a way that

$$\int_U f(u)\, du = \int_{M\times H} f(mh) \frac{\det \mathrm{Ad}_H(h)}{\det \mathrm{Ad}_U(h)}\, dm\, dh$$

for all $f \in C_c(U)$.

Proof. From Lemma 5.2, Chapter VI [DS] we know that α is a diffeomorphism and

$$d\alpha_{(m,h)}(dL_m Y, dL_h Z) = dL_{mh}(\mathrm{Ad}_U(h^{-1})Y + Z)$$

for $m \in M$, $h \in H$, $Y \in \mathfrak{m}$, and $Z \in \mathfrak{h}$. Using Lemma 1.11 we see that

$$\alpha^*(du) = D(m, h)\, dm\, dh,$$

where $dm\, dh$ is the invariant measure on the product group $M \times H$ and $D(m, h)$ is the determinant of the linear mapping

$$A(m, h): (Y, Z) \to \mathrm{Ad}_U(h^{-1})Y + Z$$

of $\mathfrak{m} \times \mathfrak{h}$ into $\mathfrak{g}$. [Since we identify $\mathfrak{m} \times \mathfrak{h}$ with $\mathfrak{g}$ we can indeed regard $A(m, h)$ as an endomorphism of $\mathfrak{g}$ and thus can speak of its determinant.] Since

$$\mathrm{Ad}_U(h^{-1})Y + Z = \mathrm{Ad}_U(h^{-1})(Y + \mathrm{Ad}_H(h)Z),$$

it is clear that

$$\det A(m, h) = \frac{\det \mathrm{Ad}_H(h)}{\det \mathrm{Ad}_U(h)}$$

and the lemma follows from (2).

We conclude this section with a useful "chain rule" version of Theorem 1.9.

Proposition 1.13. *Let G be a Lie group, H and N closed subgroups such that $H \subset N \subset G$. Assume that G/H and G/N have positive G-invariant measures dg_H and dg_N. Then N/H has an N-invariant positive measure dn_H which (suitably normalized) satisfies*

$$\int_{G/H} f(gH)\, dg_H = \int_{G/N}\left(\int_{N/H} f(gnH)\, dn_H\right) dg_N \tag{10}$$

for all $f \in C_c(G/H)$. Moreover, if for some $f \in C(G/H)$ one side is absolutely convergent, then so is the other and both sides are equal.

Proof. By Theorem 1.9, the assumptions mean that

$$|\det \mathrm{Ad}_G(h)| = |\det \mathrm{Ad}_H(h)|, \qquad h \in H,$$

$$|\det \mathrm{Ad}_G(n)| = |\det \mathrm{Ad}_N(n)|, \qquad n \in N.$$

Hence $|\det \mathrm{Ad}_N(h)| = |\det \mathrm{Ad}_H(h)|$, so that dn_H exists. Let

$$\pi_H : G \to G/H, \qquad \pi_N : G \to G/N$$

be the natural maps. Let $f \in C_c(G/H)$, $g \in G$, $n_0 \in N$. Since

$$N/H = G/H - \pi_H(G - N),$$

the subset $N/H \subset G/H$ is closed, so that the function $nH \to f(gnH)$ on N/H has compact support. The integral

$$\int_{N/H} f(gnH)\, dn_H$$

is thus well-defined, and since it is invariant under $g \to gn_0$, it can be regarded as a function F on G/N,

$$F(gN) = \int_{N/H} f(gnH)\, dn_H, \tag{11}$$

which is continuous. To see that F has compact support let $C \subset G/H$ be a compact set outside of which f vanishes. Select a compact set $\tilde{C} \subset G$ such that $\pi_H(\tilde{C}) = C$. The integral (11) vanishes if $g^{-1} \cdot C \cap N/H = \emptyset$; thus it vanishes if $\tilde{C}H \cap gN = \emptyset$. Hence F vanishes outside the compact set $\pi_N(\tilde{C})$. Thus we can form

$$I(f) = \int_{G/N}\left(\int_{N/H} f(gnH)\, dn_H\right) dg_N.$$

The functional $f \to I(f)$ is a positive measure on G/H which is clearly G-invariant. Also, $I(f) > 0$ if $f \geq 0$ and >0 near the origin in G/H. Thus $I(f)$ is a positive multiple of dg_H, so (10) follows, with a suitable normalization of dn_H. By standard measure theory, (10) extends to the case when f is the characteristic function of a Borel set; approximating by an increasing sequence of linear combinations of such functions, we see that (10) holds for all positive $f \in C(G/H)$ (possibly with both sides $+\infty$). Decomposing $f \in C(G/H)$ into two such functions, the proposition follows.

3. Haar Measure in Canonical Coordinates

Let G be a Lie group with Lie algebra $\mathfrak{g}$. Select neighborhoods N_0 of 0 in $\mathfrak{g}$ and N_e of e in G such that the exponential mapping $\exp: \mathfrak{g} \to G$ gives a diffeomorphism of N_0 onto N_e. Fix a Euclidean measure dX on $\mathfrak{g}$ and let dg denote the left-invariant form on G such that $(dg)_e = dX$.

Theorem 1.14. *With dg and dX as above we have for the pullback by* exp

$$(\exp)^*(dg) = \det\left(\frac{1 - e^{-\operatorname{ad} X}}{\operatorname{ad} X}\right) dX. \tag{12}$$

If $f \in C(G)$ has compact support contained in the canonical coordinate neighborhood N_e, then

$$\int_G f(g)\, dg = \int_{\mathfrak{g}} f(\exp X) \det\left(\frac{1 - e^{-\operatorname{ad} X}}{\operatorname{ad} X}\right) dX. \tag{13}$$

Proof. Since dg is left-invariant, formula (12) is an immediate consequence of Theorem 1.7, Chapter II in [DS]. Then (13) follows from (2) in §1 used on the function $f \circ \exp$.

§2. The Radon Transform on R^n

1. Introduction

It was proved by J. Radon in 1917 that a differentiable function on $\boldsymbol{R}^3$ can be determined explicitly by means of its integrals over the planes in $\boldsymbol{R}^3$. Let $J(\omega, p)$ denote the integral of f over the hyperplane $(x, \omega) = p$, ω denoting a unit vector and $(\ ,\ )$ the inner product. Then

$$f(x) = -\frac{1}{8\pi^2} L_x\left(\int_{S^2} J(\omega, (\omega, x))\, d\omega\right),$$

where L is the Laplacian on $\boldsymbol{R}^3$ and $d\omega$ the area element on the sphere $\boldsymbol{S}^2$ (cf. Theorem 2.13).

We observe that the formula above contains two integrations dual to each other: first one integrates over the set of points in a hyperplane, then one integrates over the set of hyperplanes passing through a given point. This suggests considering the transforms $f \to \hat{f}$ $\phi \to \check{\phi}$ defined below.

The formula has another interesting feature. For a fixed ω the integrand $x \to J(\omega, (\omega, x))$ is a *plane wave*, that is, a function constant on each plane perpendicular to ω. Ignoring the Laplacian, the formula gives a continuous decomposition of f into plane waves. Since a plane wave amounts to a function of just one variable (along the normal to the planes), this decomposition can sometimes reduce a problem for $\boldsymbol{R}^3$ to a similar problem for $\boldsymbol{R}$. This principle has been particularly useful in the theory of partial differential equations.

The analog of the formula above for line integrals is of importance in radiography, where the objective is the description of a density function by means of certain line integrals.

In this section we discuss relationships between a function on $\boldsymbol{R}^n$ and its integrals over k-dimensional planes in $\boldsymbol{R}^n$. The case $k = n - 1$ will be the one of primary interest.

At some stages we use some elementary facts concerning distributions and Riesz potentials. Since this material will be familiar to many readers it is placed in an appendix to this section.

2. The Radon Transform of the Spaces $\mathscr{D}(\boldsymbol{R}^n)$ and $\mathscr{S}(\boldsymbol{R}^n)$. The Support Theorem

Let f be a function on $\boldsymbol{R}^n$, integrable on each hyperplane in $\boldsymbol{R}^n$. Let $\boldsymbol{P}^n$ denote the space of all hyperplanes in $\boldsymbol{R}^n$, $\boldsymbol{P}^n$ being furnished with the obvious topology. The *Radon transform* of f is defined as the function $\hat{f}$ on $\boldsymbol{P}^n$ given by

$$\hat{f}(\xi) = \int_\xi f(x)\, dm(x),$$

where dm is the Euclidean measure on the hyperplane ξ. Along with the transformation $f \to \hat{f}$ we consider also the *dual transform* $\phi \to \check{\phi}$, which to a continuous function ϕ on $\boldsymbol{P}^n$ associates the function $\check{\phi}$ on $\boldsymbol{R}^n$ given by

$$\check{\phi}(x) = \int_{x \in \xi} \phi(\xi)\, d\mu(\xi),$$

where $d\mu$ is the measure on the compact set $\{\xi \in \boldsymbol{P}^n : x \in \xi\}$ which is invariant under the group of rotations around x and for which the measure of the whole set is 1. We shall relate certain function spaces on $\boldsymbol{R}^n$ and on $\boldsymbol{P}^n$ by means of the transforms $f \to \hat{f}$, $\phi \to \check{\phi}$; later we obtain explicit inversion formulas.

Each hyperplane $\xi \in \boldsymbol{P}^n$ can be written $\xi = \{x \in \boldsymbol{R}^n : (x, \omega) = p\}$, where $(\, , \,)$ is the usual inner product, $\omega = (\omega_1, \ldots, \omega_n)$ a unit vector, and $p \in \boldsymbol{R}$. Note that the pairs (ω, p) and $(-\omega, -p)$ give the same ξ; the mapping $(\omega, p) \to \xi$ is a double covering of $\boldsymbol{S}^{n-1} \times \boldsymbol{R}$ onto $\boldsymbol{P}^n$. Thus $\boldsymbol{P}^n$ has a canonical manifold structure with respect to which this covering map is differentiable and regular. We thus identify continuous (differentiable) functions ϕ on $\boldsymbol{P}^n$ with continuous (differentiable) functions ϕ on $\boldsymbol{S}^{n-1} \times \boldsymbol{R}$ satisfying $\phi(\omega, p) = \phi(-\omega, -p)$. Writing $\hat{f}(\omega, p)$ instead of $\hat{f}(\xi)$ and f_t for the translated function $x \to f(t + x)$, we have

$$\hat{f}_t(\omega, p) = \int_{(x, \omega) = p} f(x + t)\, dm(x) = \int_{(y, \omega) = p + (t, \omega)} f(y)\, dm(y),$$

so

$$\hat{f}_t(\omega, p) = \hat{f}(\omega, p + (t, \omega)). \tag{1}$$

Taking limits, we see that if $\partial_i = \partial/\partial x_i$,

$$(\partial_i f)\hat{}\,(\omega, p) = \omega_i \frac{\partial \hat{f}}{\partial p}(\omega, p). \tag{2}$$

Let L denote the Laplacian $\sum_i \partial_i^2$ on $\boldsymbol{R}^n$ and let $\Box$ denote the operator $\phi(\omega, p) \to (\partial^2/\partial p^2)\phi(\omega, p)$, which is a well-defined operator on $\mathscr{E}(\boldsymbol{P}^n)$. By analogy with the Laplacian L (cf. Chapter II, Corollary 4.11) it can be shown (cf. Exercise C2 in Chapter II) that if $\boldsymbol{M}(n)$ is the group of isometries of $\boldsymbol{R}^n$, then $\Box$ generates the algebra of $\boldsymbol{M}(n)$-invariant differential operators on $\boldsymbol{P}^n$.

Lemma 2.1. *The transforms $f \to \hat{f}$, $\phi \to \check{\phi}$ intertwine L and $\Box$, i.e.,*

$$(Lf)\hat{} = \Box(\hat{f}), \qquad (\Box\phi)\check{} = L\check{\phi}.$$

Proof. The first relation follows from (2) by iteration. For the second we just note that for a constant c

$$\check{\phi}(x) = c \int_{\boldsymbol{S}^{n-1}} \phi(\omega, (x, \omega))\, d\omega, \tag{3}$$

where $d\omega$ is the usual measure on $\boldsymbol{S}^{n-1}$.

The Radon transform is closely connected with the Fourier transform

$$\tilde{f}(u) = \int_{\boldsymbol{R}^n} f(x) e^{-i(x, u)}\, dx, \qquad u \in \boldsymbol{R}^n.$$

In fact, if $s \in \boldsymbol{R}$, ω a unit vector, then

$$\tilde{f}(s\omega) = \int_{-\infty}^{\infty} dr \int_{(x,\omega)=r} f(x) e^{-is(x,\omega)}\, dm(x),$$

so

$$\tilde{f}(s\omega) = \int_{-\infty}^{\infty} \hat{f}(\omega, r) e^{-isr}\, dr. \tag{4}$$

This means that the n-dimensional Fourier transform is the 1-dimensional Fourier transform of the Radon transform. From (4), or directly, it follows that the Radon transform of the convolution

$$f(x) = \int_{\boldsymbol{R}^n} f_1(x - y) f_2(y)\, dy$$

is the convolution

$$\hat{f}(\omega, p) = \int_{\boldsymbol{R}} \hat{f}_1(\omega, p - q) \hat{f}_2(\omega, q)\, dq. \tag{5}$$

Although slightly greater generality is possible, we shall work with the space $\mathscr{S}(\boldsymbol{R}^n)$ of complex-valued rapidly decreasing functions on $\boldsymbol{R}^n$. We recall that $f \in \mathscr{S}(\boldsymbol{R}^n)$ if and only if for each polynomial P and each integer $m \geq 0$,

$$\sup_x \big| |x|^m P(\partial_1, \ldots, \partial_n) f(x) \big| < \infty, \tag{6}$$

$|x|$ denoting the norm of x. We now formulate this in a more invariant fashion.

Lemma 2.2. *A function $f \in \mathscr{E}(\boldsymbol{R}^n)$ belongs to $\mathscr{S}(\boldsymbol{R}^n)$ if and only if for each pair $k, l \in \boldsymbol{Z}^+$*

$$\sup_{x \in \boldsymbol{R}^n} |(1 + |x|)^k (L^l f)(x)| < \infty.$$

This is easily proved just by using Fourier transforms.

By analogy with $\mathscr{S}(\boldsymbol{R}^n)$ we define $\mathscr{S}(\boldsymbol{S}^{n-1} \times \boldsymbol{R})$ as the space of C^∞-functions ϕ on $\boldsymbol{S}^{n-1} \times \boldsymbol{R}$ which for any integers $k, l \geq 0$ and any differential operator D on $\boldsymbol{S}^{n-1}$ satisfy

$$\sup_{\omega \in \boldsymbol{S}^{n-1}, r \in \boldsymbol{R}} \left| (1 + |r|^k) \frac{d^l}{dr^l} (D\phi)(\omega, r) \right| < \infty. \tag{7}$$

The space $\mathscr{S}(\boldsymbol{P}^n)$ is then defined as the set of $\phi \in \mathscr{S}(\boldsymbol{S}^{n-1} \times \boldsymbol{R})$ satisfying $\phi(\omega, p) = \phi(-\omega, -p)$.

Lemma 2.3. *For each $f \in \mathscr{S}(\boldsymbol{R}^n)$ the Radon transform $\hat{f}(\omega, p)$ satisfies the following condition: For $k \in \boldsymbol{Z}^+$ the integral*

$$\int_{\boldsymbol{R}} \hat{f}(\omega, p) p^k \, dp$$

can be written as a kth-degree homogeneous polynomial in $\omega_1, \ldots, \omega_n$.

Proof. This is immediate from the relation

$$(8) \quad \int_{\boldsymbol{R}} \hat{f}(\omega, p) p^k \, dp = \int_{\boldsymbol{R}} p^k \, dp \int_{(x, \omega) = p} f(x) \, dm(x) = \int_{\boldsymbol{R}^n} f(x)(x, \omega)^k \, dx.$$

In accordance with this lemma we define the space

$$\mathscr{S}_H(\boldsymbol{P}^n) = \left\{ \phi \in \mathscr{S}(\boldsymbol{P}^n) : \begin{array}{l} \text{For each } k \in \boldsymbol{Z}^+, \int_{\boldsymbol{R}} \phi(\omega, p) p^k \, dp \text{ is a} \\ k\text{th-degree homogeneous polynomial in } \omega_1, \ldots, \omega_n \end{array} \right\}.$$

With the notation $\mathscr{D}(\boldsymbol{P}^n) = C_c^\infty(\boldsymbol{P}^n)$ we write

$$\mathscr{D}_H(\boldsymbol{P}^n) = \mathscr{S}_H(\boldsymbol{P}^n) \cap \mathscr{D}(\boldsymbol{P}^n).$$

According to Schwartz [1966] (p. 249), the Fourier transform $f \to \tilde{f}$ maps the space $\mathscr{S}(\boldsymbol{R}^n)$ onto itself. We shall now, by analogy with the Schwartz theorem, determine the image of $\mathscr{S}(\boldsymbol{R}^n)$ under the Radon transform.

Theorem 2.4. *The Radon transform $f \to \hat{f}$ is a linear one-to-one mapping of $\mathscr{S}(\boldsymbol{R}^n)$ onto $\mathscr{S}_H(\boldsymbol{P}^n)$.*

Proof. Since

$$\frac{d}{ds} \tilde{f}(s\omega) = \sum_{i=1}^{n} \omega_i (\partial_i \tilde{f}),$$

it is clear from (4) that for each fixed ω the function $r \to \hat{f}(\omega, r)$ lies in $\mathscr{S}(\boldsymbol{R})$. For each $\omega_0 \in \boldsymbol{S}^{n-1}$ a subset of $\{\omega_1, \ldots, \omega_n\}$ will serve as local coordinates on a neighborhood of ω_0 in $\boldsymbol{S}^{n-1}$. To see that $\hat{f} \in \mathscr{S}(\boldsymbol{P}^n)$, it therefore suffices to verify (7) for $\phi = \hat{f}$ on an open subset $N \subset \boldsymbol{S}^{n-1}$, where ω_n is bounded away from 0 and $\omega_1, \ldots, \omega_{n-1}$ serve as coordinates, in terms of which D is expressed. Since

$$(9) \quad u_1 = s\omega_1, \quad \ldots, \quad u_{n-1} = s\omega_{n-1}, \quad u_n = s(1 - \omega_1^2 - \cdots - \omega_{n-1}^2)^{1/2}$$

we have

$$\frac{\partial}{\partial \omega_i} (\tilde{f}(s\omega)) = s \frac{\partial \tilde{f}}{\partial u_i} - s\omega_i (1 - \omega_1^2 - \cdots - \omega_{n-1}^2)^{-1/2} \frac{\partial \tilde{f}}{\partial u_n}.$$

It follows that if D is any differential operator on $\boldsymbol{S}^{n-1}$ and if $k, l \in \boldsymbol{Z}^+$, then

$$\sup_{\omega \in N, s \in \boldsymbol{R}} \left| (1 + s^{2k}) \frac{d^l}{ds^l} (D\tilde{f})(\omega, s) \right| < \infty. \tag{10}$$

We can therefore apply D under the integral sign in the inversion formula to (4),

$$\hat{f}(\omega, r) = \frac{1}{2\pi} \int_{\boldsymbol{R}} \tilde{f}(s\omega) e^{isr} \, ds,$$

and obtain

$$(1 + r^{2k}) \frac{d^l}{dr^l} (D_\omega(\hat{f}(\omega, r))) = \frac{1}{2\pi} \int \left(1 + (-1)^k \frac{d^{2k}}{ds^{2k}} \right) ((is)^l D_\omega(\tilde{f}(s\omega))) e^{isr} \, ds.$$

Now (10) shows that $\hat{f} \in \mathscr{S}(\boldsymbol{P}^n)$, so that by Lemma 2.3, $\hat{f} \in \mathscr{S}_H(\boldsymbol{P}^n)$.

Because of (4) and the fact that the Fourier transform is one-to-one, it only remains to prove the surjectivity in Theorem 2.4. Let $\phi \in \mathscr{S}_H(\boldsymbol{P}^n)$. In order to prove $\phi = \hat{f}$ for some $f \in \mathscr{S}(\boldsymbol{R}^n)$, we put

$$\Phi(s, \omega) = \int_{-\infty}^{\infty} \phi(\omega, r) e^{-irs} \, dr.$$

Then $\Phi(s, \omega) = \Phi(-s, -\omega)$ and $\Phi(0, \omega)$ is a homogeneous polynomial of degree 0 in $\omega_1, \ldots, \omega_n$, and hence is constant. Thus there exists a function F on $\boldsymbol{R}^n$ such that

$$F(s\omega) = \int_{\boldsymbol{R}} \phi(\omega, r) e^{-irs} \, dr.$$

While F is clearly smooth away from the origin, we shall now prove it to be smooth at the origin too; this is where the homogeneity condition in the definition of $\mathscr{S}_H(\boldsymbol{P}^n)$ enters decisively. Consider the coordinate neighborhood $N \subset \boldsymbol{S}^{n-1}$ above, and if $h \in C^\infty(\boldsymbol{R}^n - \{0\})$, let $h^*(\omega_1, \ldots, \omega_{n-1}, s)$ be the function obtained from h by means of the substitution (9). Then

$$\frac{\partial h}{\partial u_i} = \sum_{j=1}^{n-1} \frac{\partial h^*}{\partial \omega_j} \frac{\partial \omega_j}{\partial u_i} + \frac{\partial h^*}{\partial s} \cdot \frac{\partial s}{\partial u_i} \qquad (1 \le i \le n)$$

and

$$\frac{\partial \omega_j}{\partial u_i} = \frac{1}{s} \left(\delta_{ij} - \frac{u_i u_j}{s^2} \right) \qquad (1 \le i \le n, \ 1 \le j \le n-1),$$

$$\frac{\partial s}{\partial u_i} = \omega_i \qquad (1 \le i \le n-1), \qquad \frac{\partial s}{\partial u_n} = (1 - \omega_1^2 - \cdots - \omega_{n-1}^2)^{1/2}.$$

Hence

$$\frac{\partial h}{\partial u_i} = \frac{1}{s}\frac{\partial h^*}{\partial \omega_i} + \omega_i\left(\frac{\partial h^*}{\partial s} - \frac{1}{s}\sum_{j=1}^{n-1}\omega_j\frac{\partial h^*}{\partial \omega_j}\right) \qquad (1 \le i \le n-1),$$

$$\frac{\partial h}{\partial u_n} = (1 - \omega_1^2 - \cdots - \omega_{n-1}^2)^{1/2}\left(\frac{\partial h^*}{\partial s} - \frac{1}{s}\sum_{j=1}^{n-1}\omega_j\frac{\partial h^*}{\partial \omega_j}\right).$$

In order to use this for $h = F$ we write

$$F(s\omega) = \int_{-\infty}^{\infty}\phi(\omega, r)\,dr + \int_{-\infty}^{\infty}\phi(\omega, r)(e^{-irs} - 1)\,dr.$$

By assumption the first integral is independent of ω. Thus using (7) we have for a constant $K > 0$

$$\left|\frac{1}{s}\frac{\partial}{\partial \omega_i}(F(s\omega))\right| \le K\int(1 + r^4)^{-1}s^{-1}|e^{-isr} - 1|\,dr \le K\int\frac{|r|}{1 + r^4}\,dr,$$

and a similar estimate is obvious for $\partial F(s\omega)/\partial s$. The formulas above therefore imply that all the derivatives $\partial F/\partial u_i$ are bounded in a punctured ball $0 < |u| < \varepsilon$, so that (since $n > 1$) we can conclude that F is uniformly continuous on $0 < |u| < \varepsilon$, hence continuous at $u = 0$.

More generally, we prove by induction that

$$(11)\qquad \frac{\partial^q h}{\partial u_{i_1}\cdots\partial u_{i_q}} = \sum_{1\le i+j\le q,\ 1\le k_1,\ldots,k_i\le n-1} A_{j,k_1,\ldots,k_i}(\omega, s)\frac{\partial^{i+j}h^*}{\partial\omega_{k_1}\cdots\partial\omega_{k_i}\partial s^j},$$

where the A have the form

$$(12)\qquad A_{j,k_1,\ldots,k_i}(\omega, s) = a_{j,k_1,\ldots,k_i}(\omega)s^{j-q}.$$

For $q = 1$ this is in fact proved above. Assuming (11) for q, we calculate

$$\frac{\partial^{q+1}h}{\partial u_{i_1}\cdots\partial u_{i_{q+1}}},$$

using the above formulas for $\partial/\partial u_i$. If $A_{j,k_1,\ldots,k_i}(\omega, s)$ is differentiated with respect to $u_{i_{q+1}}$ we get a formula like (12) with q replaced by $q + 1$. If, on the other hand, the $(i + j)$th derivative of h^* in (11) is differentiated with respect to $u_{i_{q+1}}$, we get a combination of terms

$$s^{-1}\frac{\partial^{i+j+1}h^*}{\partial\omega_{k_1}\cdots\partial\omega_{k_{i+1}}\partial s^j}, \qquad \frac{\partial^{i+j+1}h^*}{\partial\omega_{k_1}\cdots\partial\omega_{k_i}\partial s^{j+1}},$$

and in both cases we get coefficients satisfying (12) with q replaced by $q + 1$. This proves (11) in general. Now

$$(13)\qquad F(s\omega) = \int_{-\infty}^{\infty}\phi(\omega, r)\sum_0^{q-1}\frac{(-isr)^k}{k!}\,dr + \int_{-\infty}^{\infty}\phi(\omega, r)e_q(-irs)\,dr,$$

where

$$e_q(t) = \frac{t^q}{q!} + \frac{t^{q+1}}{(q+1)!} + \cdots .$$

Our assumption on ϕ implies that the first integral in (13) is a polynomial in $u_1, \ldots, u_n$ of degree $\le q - 1$ and is therefore annihilated by the differential operator (11). If $0 \le j \le q$, we have

$$\left| s^{j-q} \frac{\partial^j}{\partial s^j} (e_q(-irs)) \right| = |(-ir)^q (-irs)^{j-q} e_{q-j}(-irs)| \le K_j r^q, \tag{14}$$

where K_j is a constant because the function $t \to (it)^{-p} e_p(it)$ is obviously bounded on $\boldsymbol{R}$ $(p \ge 0)$. Since $\phi \in \mathscr{S}(\boldsymbol{P}^n)$, it follows from (11), (13), and (14) that each qth-order derivative of F with respect to $u_1, \ldots, u_n$ is bounded in a punctured ball $0 < |u| < \varepsilon$. Thus we have proved $F \in C^\infty(\boldsymbol{R}^n)$. That F is rapidly decreasing is now clear from (7) and (11). Finally, if f is the function in $\mathscr{S}(\boldsymbol{R}^n)$ whose Fourier transform is F, then

$$\tilde{f}(s\omega) = F(s\omega) = \int_{-\infty}^{\infty} \phi(\omega, r) e^{-irs} \, dr;$$

hence by (4), $\hat{f} = \phi$ and the theorem is proved.

To make further progress we introduce some useful notation. Let $S_r(x)$ denote the sphere $\{y : |y - x| = r\}$ in $\boldsymbol{R}^n$ and $A(r)$ its area. Let $B_r(x)$ denote the open ball $\{y : |y - x| < r\}$. For a continuous function f on $S_r(x)$ let $(M^r f)(x)$ denote the mean value

$$(M^r f)(x) = \frac{1}{A(r)} \int_{S_r(x)} f(\omega) \, d\omega,$$

where $d\omega$ is the Euclidean measure. Let K denote the orthogonal group $\boldsymbol{O}(n)$, dk its Haar measure, normalized by $\int_K dk = 1$. If $y \in \boldsymbol{R}^n$, $r = |y|$, then

$$(M^r f)(x) = \int_K f(x + k \cdot y) \, dk. \tag{15}$$

In fact, for x, y fixed, both sides represent rotation-invariant measures on $S_r(x)$, having the same value for the function $f \equiv 1$. The rotations being transitive on $S_r(x)$, (15) follows from the uniqueness in Theorem 1.9. Formula (3) can similarly be written

$$\check{\phi}(x) = \int_K \phi(x + k \cdot \xi_0) \, dk \tag{16}$$

if ξ_0 is some fixed hyperplane through the origin. We see then that if $f \in \mathscr{S}(\boldsymbol{R}^n)$, Ω_k the area of the unit sphere in $\boldsymbol{R}^k$,

$$(\hat{f})^{\vee}(x) = \int_K \hat{f}(x + k \cdot \xi_0)\, dk = \int_K \left(\int_{\xi_0} f(x + k \cdot y)\, dm(y) \right) dk$$

$$= \int_{\xi_0} (M^{|y|}f)(x)\, dm(y) = \Omega_{n-1} \int_0^{\infty} r^{n-2} \left(\frac{1}{\Omega_n} \int_{S^{n-1}} f(x + r\omega)\, d\omega \right) dr,$$

so that

$$(\hat{f})^{\vee}(x) = \frac{\Omega_{n-1}}{\Omega_n} \int_{\boldsymbol{R}^n} |x - y|^{-1} f(y)\, dy. \tag{17}$$

We consider now the analog of Theorem 2.4 for the transform $\phi \to \check{\phi}$. But $\phi \in \mathscr{S}_H(\boldsymbol{P}^n)$ does not imply $\check{\phi} \in \mathscr{S}(\boldsymbol{R}^n)$. [If this were so and by Theorem 2.4 we wrote $\phi = \hat{f}$, $f \in \mathscr{S}(\boldsymbol{R}^n)$, then the inversion formula in Theorem 2.13 for $n = 3$ would imply $\int f(x)\, dx = 0$.] On a smaller space we shall obtain a more satisfactory result.

Let $\mathscr{S}^*(\boldsymbol{R}^n)$ denote the space of all functions $f \in \mathscr{S}(\boldsymbol{R}^n)$ which are orthogonal to all polynomials, i.e.,

$$\int_{\boldsymbol{R}^n} f(x)P(x)\, dx = 0 \qquad \text{for all polynomials } P.$$

Similarly, let $\mathscr{S}^*(\boldsymbol{P}^n) \subset \mathscr{S}(\boldsymbol{P}^n)$ be characterized by

$$\int_{\boldsymbol{R}} \phi(\omega, r)p(r)\, dr = 0 \qquad \text{for all polynomials } p.$$

Note that under the Fourier transform the space $\mathscr{S}^*(\boldsymbol{R}^n)$ corresponds to the subspace $\mathscr{S}_0(\boldsymbol{R}^n) \subset \mathscr{S}(\boldsymbol{R}^n)$ of functions all of whose derivatives vanish at 0.

Corollary 2.5. *The transforms $f \to \hat{f}$, $\phi \to \check{\phi}$ are bijections of $\mathscr{S}^*(\boldsymbol{R}^n)$ onto $\mathscr{S}^*(\boldsymbol{P}^n)$ and of $\mathscr{S}^*(\boldsymbol{P}^n)$ onto $\mathscr{S}^*(\boldsymbol{R}^n)$.*

The first statement is clear from (8) if we take into account the elementary fact that the polynomials $x \to (x, \omega)^k$ span the space of homogeneous polynomials of degree k (cf., Exercise C1). To see that $\phi \to \check{\phi}$ is a bijection of $\mathscr{S}^*(\boldsymbol{P}^n)$ onto $\mathscr{S}^*(\boldsymbol{R}^n)$ we use (17), knowing that $\phi = \hat{f}$ for some $f \in \mathscr{S}^*(\boldsymbol{R}^n)$. The right-hand side of (17) is the convolution of f with the tempered distribution $|x|^{-1}$ whose Fourier transform is by Lemma 2.41 a constant multiple of $|u|^{1-n}$. (Here we leave out the trivial case $n = 1$.) By the general theory of tempered distributions [cf. (69) later], this convolution is a tempered distribution whose Fourier transform is a constant multiple of $|u|^{1-n}\tilde{f}(u)$. But this lies in the space $\mathscr{S}_0(\boldsymbol{R}^n)$ since $\tilde{f}$ does. Now (17) implies that $\check{\phi} = (\hat{f})^{\vee} \in \mathscr{S}^*(\boldsymbol{R}^n)$ and that

$\check{\phi} \not\equiv 0$ if $\phi \not\equiv 0$. Finally, the mapping $\phi \to \check{\phi}$ is surjective because the function $((\hat{f})^{\vee})^{\sim}(u) = c|u|^{1-n}\tilde{f}(u)$ (c a constant) runs through $\mathscr{S}_0(\boldsymbol{R}^n)$ as f runs through $\mathscr{S}^*(\boldsymbol{R}^n)$.

We now turn to the space $\mathscr{D}(\boldsymbol{R}^n) = C_c^\infty(\boldsymbol{R}^n)$ and its image under the Radon transform.

Theorem 2.6. (The Support Theorem) *Let $f \in C(\boldsymbol{R}^n)$ satisfy the following conditions:*

(i) *For each integer $k > 0$, $|x|^k f(x)$ is bounded.*

(ii) *There exists a constant $A > 0$ such that*

$$\hat{f}(\xi) = 0 \qquad \textit{for} \quad d(0, \xi) > A,$$

d denoting distance.

Then

$$f(x) = 0 \qquad \textit{for} \quad |x| > A.$$

Proof. Replacing f by the convolution $\phi * f$, where ϕ is a radial C^∞-function with support in a small ball $B_\varepsilon(0)$, we see that it suffices to prove the theorem for $f \in \mathscr{E}(\boldsymbol{R}^n)$. In fact, $\phi * f$ is smooth, it satisfies (i), and by (5) it satisfies (ii) with A replaced by $A + \varepsilon$. Assuming the theorem for the smooth case, we deduce that support $(\phi * f) \subset \mathrm{Cl}(B_{A+\varepsilon}(0))$ so letting $\varepsilon \to 0$, we obtain support $(f) \subset \mathrm{Cl}(B_A(0))$.

To begin with we assume that f is a radial function. Then $f(x) = F(|x|)$, where $F \in \mathscr{E}(\boldsymbol{R})$ and even. Then $\hat{f}$ has the form $\hat{f}(\xi) = \hat{F}(d(0, \xi))$, where $\hat{F}$ is given by

$$\hat{F}(p) = \int_{\boldsymbol{R}^{n-1}} F((p^2 + |y|^2)^{1/2})\, dm(y) \qquad (p \geq 0),$$

because of the definition of the Radon transform. In particular, F being even, $\hat{F}$ extends to an even function on $\boldsymbol{R}$. Using polar coordinates in $\boldsymbol{R}^{n-1}$, we obtain

$$\hat{F}(p) = \Omega_{n-1} \int_0^\infty F((p^2 + t^2)^{1/2}) t^{n-2}\, dt. \tag{18}$$

Here we substitute $s = (p^2 + t^2)^{-1/2}$ and then put $u = p^{-1}$. Then (18) becomes

$$u^{n-3}\hat{F}(u^{-1}) = \Omega_{n-1} \int_0^u (F(s^{-1})s^{-n})(u^2 - s^2)^{(1/2)(n-3)}\, ds.$$

We write this equation for simplicity

$$h(u) = \int_0^u g(s)(u^2 - s^2)^{(1/2)(n-3)}\, ds. \tag{19}$$

This integral equation is very close to Abel's integral equation (Whittaker and Watson [1927]) and can be inverted as follows. Multiplying both sides by $u(t^2 - u^2)^{(1/2)(n-3)}$ and integrating over $0 \leq u \leq t$, we obtain

$$\begin{aligned}
&\int_0^t h(u)(t^2 - u^2)^{(1/2)(n-3)} u \, du \\
&= \int_0^t \left\{ \int_0^u g(s)[(u^2 - s^2)(t^2 - u^2)]^{(1/2)(n-3)} \, ds \right\} u \, du \\
&= \int_0^t g(s) \left\{ \int_{u=s}^t u[(t^2 - u^2)(u^2 - s^2)]^{(1/2)(n-3)} \, du \right\} ds.
\end{aligned}$$

The substitution $(t^2 - s^2)v = (t^2 + s^2) - 2u^2$ gives an explicit evaluation of the inner integral, and we obtain

$$\int_0^t h(u)(t^2 - u^2)^{(1/2)(n-3)} u \, du = C \int_0^t g(s)(t^2 - s^2)^{n-2} \, ds,$$

where C is a constant. Here we apply the operator $d/d(t^2) = (1/2t)(d/dt)$ $(n-1)$ times, whereby the right-hand side gives a constant multiple of $t^{-1}g(t)$. Hence we obtain with another constant c

$$F(t^{-1})t^{-n} = ct \left(\frac{d}{d(t^2)} \right)^{n-1} \int_0^t (t^2 - u^2)^{(1/2)(n-3)} u^{n-2} \hat{F}(u^{-1}) \, du. \tag{20}$$

By assumption (ii) we have $\hat{F}(u^{-1}) = 0$ if $u^{-1} \geq A$, that is, if $u \leq A^{-1}$. But then (20) implies $F(t^{-1}) = 0$ if $t < A^{-1}$, that is, if $t^{-1} > A$. This proves the theorem for the case when f is radial.

We consider next the case of a general f. Fix $x \in \boldsymbol{R}^n$ and consider the function

$$g_x(y) = \int_K f(x + k \cdot y) \, dk$$

as in (15). Then g_x satisfies (i) and

$$\hat{g}_x(\xi) = \int_K \hat{f}(x + k \cdot \xi) \, dk, \tag{21}$$

$x + k \cdot \xi$ denoting the translate of the hyperplane $k \cdot \xi$ by x. The triangle inequality shows that

$$d(0, x + k \cdot \xi) \geq d(0, \xi) - |x|, \qquad x \in \boldsymbol{R}^n, \quad k \in K.$$

Hence we conclude from assumption (i) and (21) that

$$\hat{g}_x(\xi) = 0 \qquad \text{if} \quad d(0, \xi) > A + |x|. \tag{22}$$

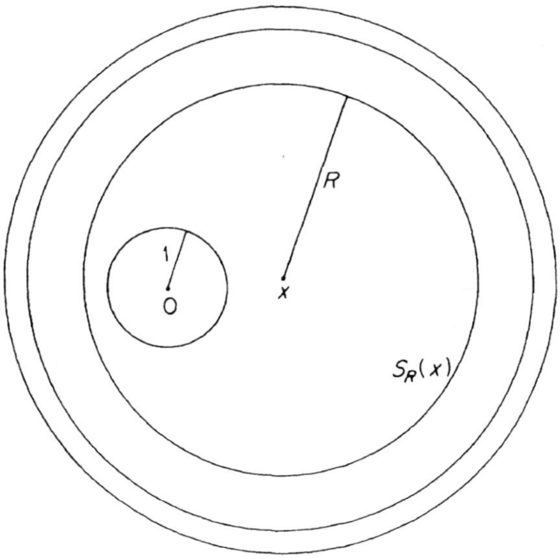

FIG. 7

But g_x is a radial function, so that (22) implies by the first part of the proof that

$$\int_K f(x + k \cdot y)\, dk = 0 \qquad \text{if} \quad |y| > A + |x|. \tag{23}$$

Geometrically, this formula reads: The surface integral of f over $S_{|y|}(x)$ is 0 if the ball $B_{|y|}(x)$ contains the ball $B_A(0)$. The theorem is therefore a consequence of the following lemma.

Lemma 2.7. *Let $f \in C(\boldsymbol{R}^n)$ be such that for each integer $k > 0$,*

$$\sup_{x \in \boldsymbol{R}^n} |x|^k |f(x)| < \infty.$$

Suppose f has surface integral 0 over every sphere S which encloses the unit ball. Then $f(x) \equiv 0$ for $|x| > 1$.

Proof. The idea is to perturb S in the relation

$$\int_S f(s)\, d\omega(s) = 0 \tag{24}$$

slightly and differentiate with respect to the parameter of perturbation, thereby obtaining additional relations. Replacing, as above, f with a suitable convolution $\phi * f$, we see that it suffices to prove the lemma for f in $\mathscr{E}(\boldsymbol{R}^n)$. Writing $S = S_R(x)$ and viewing the exterior of the ball $B_R(x)$ as a union of spheres with center x (see Fig. 7) we have by the assumptions

$$\int_{B_R(x)} f(y)\, dy = \int_{\boldsymbol{R}^n} f(y)\, dy,$$

which is a constant. Differentiating with respect to x_i, we obtain

$$\int_{B_R(0)} (\partial_i f)(x + y)\, dy = 0. \tag{25}$$

We use now the divergence theorem

$$\int_{B_R(0)} (\operatorname{div} F)(y)\, dy = \int_{S_R(0)} (F, \boldsymbol{n})(s)\, d\omega(s) \tag{26}$$

for a vector field F on $\boldsymbol{R}^n$, $\boldsymbol{n}$ denoting the outgoing unit normal and $d\omega$ the surface element on $S_R(0)$. For the vector field $F(y) = f(x + y)(\partial/\partial y_i)$ we obtain from (25) and (26), since $\boldsymbol{n} = R^{-1}(s_1, \ldots, s_n)$,

$$\int_{S_R(0)} f(x + s)s_i\, d\omega(s) = 0. \tag{27}$$

But by (24)

$$\int_{S_R(0)} f(x + s)x_i\, d\omega(s) = 0.$$

Adding the two last relations we get

$$\int_S f(s)s_i\, d\omega(s) = 0.$$

This means that the hypotheses of the lemma hold for $f(x)$ replaced by the function $x_i f(x)$. By iteration

$$\int_S f(s)P(s)\, d\omega(s) = 0$$

for any polynomial P, so $f \equiv 0$ on S. This proves the lemma as well as Theorem 2.6.

Corollary 2.8. *Let $f \in C(\boldsymbol{R}^n)$ satisfy* (i) *and assume*

$$\hat{f}(\xi) = 0$$

for all hyperplanes ξ disjoint from a certain compact convex set C. Then

$$f(x) = 0 \qquad \textit{for} \quad x \notin C. \tag{28}$$

In fact, if B is a closed ball containing C we have by Theorem 2.6, $f(x) = 0$ for $x \notin B$. But C is the intersection of such balls, so (28) follows.

Remark 2.9. While condition (i) of rapid decrease entered in the proof of Lemma 2.7 [we used $|x|^k f(x) \in L^1(\boldsymbol{R}^n)$ for each $k > 0$], one may wonder whether it could not be weakened in Theorem 2.6 and perhaps even dropped in Lemma 2.7.

As an example, showing that the condition of rapid decrease cannot be dropped in either result, consider for $n = 2$ the function

$$f(x, y) = (x + iy)^{-5}$$

made smooth in $\boldsymbol{R}^2$ by changing it in a small disk around 0. Using Cauchy's theorem for a large semicircle, we have $\int_l f(x)\, dm(x) = 0$ for every line l outside the unit circle. Thus (ii) is satisfied in Theorem 2.6. Hence (i) cannot be dropped or weakened substantially.

This same example works for Lemma 2.7. In fact, let S be a circle $|z - z_0| = r$ enclosing the unit disk. Then

$$d\omega(s) = -ir\frac{dz}{z - z_0},$$

so by expanding S, or by residue calculus,

$$\int_S z^{-5}(z - z_0)^{-1}\, dz = 0,$$

(the residues at $z = 0$ and $z = z_0$ cancel), so that we have in fact

$$\int_S f(s)\, d\omega(s) = 0.$$

We recall now that by its definition

$$\mathscr{D}_H(\boldsymbol{P}^n) = \left\{\phi \in \mathscr{D}(\boldsymbol{P}^n) : \begin{array}{l}\text{For each } k \in \boldsymbol{Z}^+, \int_{\boldsymbol{R}} \phi(\omega, p)p^k\, dp \text{ is a homogeneous} \\ k\text{th-degree polynomial in } \omega_1, \ldots, \omega_n.\end{array}\right\}.$$

Combining Theorems 2.4 and 2.6, we obtain the following characterization of the Radon transform of the space $\mathscr{D}(\boldsymbol{R}^n) = C_c^\infty(\boldsymbol{R}^n)$. This can be regarded as the analog for the Radon transform of the Paley–Wiener theorem for the Fourier transform.

Theorem 2.10. *The Radon transform is a bijection of $\mathscr{D}(\boldsymbol{R}^n)$ onto $\mathscr{D}_H(\boldsymbol{P}^n)$.*

We conclude this subsection with a variation and a consequence of Theorem 2.6.

Lemma 2.11. *Let $f \in C_c(\boldsymbol{R}^n)$, $A > 0$, ω_0 a fixed unit vector, and $N \subset \boldsymbol{S}^{n-1}$ a neighborhood of ω_0 in the unit sphere $\boldsymbol{S}^{n-1} \subset \boldsymbol{R}^n$. Assume*

$$\hat{f}(\omega, p) = 0 \qquad \textit{for} \quad \omega \in N, \quad p > A.$$

Then

(29) $$f(x) = 0 \qquad \textit{in the half-space } (x, \omega_0) > A.$$

Proof. Let B be a closed ball around the origin containing the support of f. Let $\varepsilon > 0$ and let H_ε be the union of the half-spaces $(x, \omega) > A + \varepsilon$ as ω runs through N. Then by our assumption

(30) $$\hat{f}(\xi) = 0 \qquad \text{if} \quad \xi \subset H_\varepsilon.$$

Now choose a ball B_ε with center on the ray from 0 through $-\omega_0$, with the point $(A + 2\varepsilon)\omega_0$ on the boundary, and with radius so large that any hyperplane ξ intersecting B but not B_e must lie in H_ε. Then by (30)

$$\hat{f}(\xi) = 0 \quad \text{whenever} \quad \xi \in \boldsymbol{P}^n, \quad \xi \cap B_\varepsilon = \varnothing.$$

Hence by Theorem 2.6, $f(x) = 0$ for $x \notin B_\varepsilon$. In particular, $f(x) = 0$ for $(x, \omega_0) > A + 2\varepsilon$; since $\varepsilon > 0$ is arbitrary, the lemma follows.

Corollary 2.12. *Let N be any open subset of the unit sphere $\boldsymbol{S}^{n-1}$. If $f \in C_c(\boldsymbol{R}^n)$ and*

$$\hat{f}(\omega, p) = 0 \qquad \textit{for} \quad p \in \boldsymbol{R}, \quad \omega \in N,$$

then $f \equiv 0$.

Since $\hat{f}(-\omega, -p) = \hat{f}(\omega, p)$, this is obvious from Lemma 2.11.

3. The Inversion Formulas

We shall now establish explicit inversion formulas for the Radon transform $f \to \hat{f}$ and its dual $\varphi \to \check{\varphi}$.

Theorem 2.13. *The function f can be recovered from its Radon transform by means of the following inversion formula:*

$$cf = (-L)^{(n-1)/2}((\hat{f})^{\vee}), \quad f \in \mathscr{S}(\boldsymbol{R}^n),$$

where c is the constant

$$c = (4\pi)^{(n-1)/2} \frac{\Gamma(n/2)}{\Gamma(1/2)}.$$

Here L is the Laplacian on $\boldsymbol{R}^n$. For n even the fractional power $L^{(n-1)/2}$ requires a definition which will be given in course of the proof.

To begin with, we recall the familiar fact that if $f \in C^2(\boldsymbol{R}^n)$ is a radial function, i.e., $f(x) = F(r)$, $r = |x|$, then

(31) $$(Lf)(x) = \frac{d^2F}{dr^2} + \frac{n-1}{r}\frac{dF}{dr}.$$

This is immediate from the relations

$$\frac{\partial^2 f}{\partial x_i^2} = \frac{\partial^2 f}{\partial r^2}\left(\frac{\partial r}{\partial x_i}\right)^2 + \frac{\partial f}{\partial r}\frac{\partial^2 r}{\partial x_i^2}.$$

Lemma 2.14.

(i) *$LM^r = M^r L$ for each $r > 0$.*

(ii) *For $f \in C^2(\boldsymbol{R}^n)$ the mean value $(M^r f)(x)$ satisfies the "Darboux equation"*

$$L_x((M^r f)(x)) = \left(\frac{\partial^2}{\partial r^2} + \frac{n-1}{r}\frac{\partial}{\partial r}\right)(M^r f(x));$$

that is, the function $F(x, y) = (M^{|y|} f)(x)$ satisfies

$$L_x(F(x, y)) = L_y(F(x, y)).$$

Proof. We prove this group-theoretically, using expression (15) for the mean value. For $z \in \boldsymbol{R}^n$, $k \in K$, let T_z denote the translation $x \to x + z$ and R_k the rotation $x \to k \cdot x$. Since L is invariant under these transformations (cf. Chapter II, §2, No. 1), we have, if $r = |y|$,

$$(LM^r f)(x) = \int_K L_x(f(x + k \cdot y))\,dk = \int_K (Lf)(x + k \cdot y)\,dk$$

$$= (M^r L f)(x) = \int_K [(Lf) \circ T_x \circ R_k](y)\,dk$$

$$= \int_K [L(f \circ T_x \circ R_k)](y)\,dk = L_y\left(\int_K f(x + k \cdot y)\,dk\right),$$

which proves the lemma.

In order to prove Theorem 2.13 let $f \in \mathscr{S}(\boldsymbol{R}^n)$. Fix a hyperplane ξ_0 through 0 and an isometry $g \in \boldsymbol{M}(n)$. As k runs through $\boldsymbol{O}(n)$, $gk \cdot \xi_0$ runs through the set of hyperplanes through $g \cdot 0$, and we have

$$\check{\phi}(g \cdot 0) = \int_K \phi(gk \cdot \xi_0)\,dk,$$

so that

$$(\hat{f})^{\vee}(g \cdot 0) = \int_K \left(\int_{\xi_0} f(gk \cdot y)\,dm(y)\right) dk$$

$$= \int_{\xi_0} dm(y) \int_K f(gk \cdot y)\,dk = \int_{\xi_0} (M^{|y|} f)(g \cdot 0)\,dm(y).$$

Hence

$$(\hat{f})^{\vee}(x) = \Omega_{n-1} \int_0^{\infty} (M^r f)(x) r^{n-2}\, dr, \tag{32}$$

where Ω_{n-1} is the area of the unit sphere in $\boldsymbol{R}^{n-1}$, i.e.,

$$\Omega_{n-1} = \frac{2\pi^{(n-1)/2}}{\Gamma(\frac{1}{2}(n-1))}.$$

Applying L to (32), using (31) and Lemma 2.14, we obtain

$$L((\hat{f})^{\vee}) = \Omega_{n-1} \int_0^{\infty} \left(\frac{d^2F}{dr^2} + \frac{n-1}{r}\frac{dF}{dr}\right) r^{n-2}\, dr, \tag{33}$$

where $F(r) = (M^r f)(x)$. Integrating by parts and using $F(0) = f(x)$, $\lim_{r\to\infty} r^k F^{(m)}(r) = 0$, we get

$$L((\hat{f})^{\vee}) = \begin{cases} -\Omega_{n-1} f(x) & \text{if } n = 3 \\ -\Omega_{n-1}(n-3) \displaystyle\int_0^{\infty} F(r) r^{n-4}\, dr & (n > 3). \end{cases}$$

More generally,

$$L_x\left(\int_0^{\infty} (M^r f)(x) r^k\, dr\right) = \begin{cases} -(n-2) f(x) & \text{if } k = 1 \\ -(n-1-k)(k-1) \displaystyle\int_0^{\infty} F(r) r^{k-2}\, dr & (k > 1). \end{cases}$$

If n is odd, the formula in Theorem 2.13 follows by iteration.

We now pass to the case of even n and use the definition of the fractional power $L^{(n-1)/2}$ in terms of Riesz potentials I^{γ} in (81) and (82). Using (17) in the present section, we have

$$(\hat{f})^{\vee} = 2^{n-1}\pi^{(n/2)-1}\Gamma\left(\frac{n}{2}\right) I^{n-1} f. \tag{34}$$

Using Prop. 2.38 and the definition of $L^{(n-1)/2}$ ((85) No. 8), we obtain the desired formula,

$$(-L)^{(n-1)/2}((\hat{f})^{\vee}) = cf, \tag{35}$$

in Theorem 2.13.

We shall now prove a similar inversion formula for the dual transform $\phi \to \check{\phi}$ on the subspace $\mathscr{S}^*(\boldsymbol{P}^n) \subset \mathscr{S}(\boldsymbol{P}^n)$.

Theorem 2.15. *We have*

$$c\phi = (-\square)^{(n-1)/2}((\check{\phi})^{\wedge}), \qquad \phi \in \mathscr{S}^*(\boldsymbol{P}^n),$$

where c is the constant

$$c = (4\pi)^{(n-1)/2} \frac{\Gamma(n/2)}{\Gamma(1/2)}.$$

Here $\square$ denotes as before the operator d^2/dp^2, and its fractional powers are again defined in terms of Riesz's potentials on the one-dimensional p-space.

If n is odd our inversion formula follows from the odd-dimensional case in Theorem 2.13 if we put $f = \check{\phi}$ and take Lemma 2.1 and Corollary 2.5 into account.

Suppose now n is even. We claim that

$$((-L)^{(n-1)/2}f)\hat{} = (-\square)^{(n-1)/2}\hat{f}, \qquad f \in \mathscr{S}^*(\boldsymbol{R}^n). \tag{36}$$

In fact, by Lemma 2.37 and Cor. 2.5 both sides belong to $\mathscr{S}^*(\boldsymbol{P}^n)$. Taking the 1-dimensional Fourier transform of $((-L)^{(n-1)/2}f)\hat{}$, we get by (4),

$$((-L)^{(n-1)/2}f)\tilde{}(s\omega) = |s|^{n-1}\tilde{f}(s\omega).$$

This coincides with the Fourier transform of $(-\square)^{(n-1)/2}\hat{f}$, so that (36) is proved. Now Theorem 2.15 follows from (35) if we put in (36)

$$\phi = \hat{g}, \qquad f = (\hat{g})\check{}, \qquad g \in \mathscr{S}^*(\boldsymbol{R}^n).$$

Because of its theoretical importance we now prove the inversion theorem (2.13) in a different form. The proof is less geometric and involves just the one-variable Fourier transform.

Let $\mathscr{H}$ denote the Hilbert transform

$$(\mathscr{H}F)(t) = \frac{i}{\pi}\int_{-\infty}^{\infty} \frac{F(p)}{t-p}\,dp, \qquad F \in \mathscr{S}(\boldsymbol{R}),$$

the integral being considered as the Cauchy principal value. For $\phi \in \mathscr{S}(\boldsymbol{P}^n)$ let $\Lambda\phi$ be defined by

$$(\Lambda\phi)(\omega, p) = \begin{cases} \dfrac{d^{n-1}}{dp^{n-1}}\phi(\omega, p), & n \text{ odd} \\[2ex] \mathscr{H}_p \dfrac{d^{n-1}}{dp^{n-1}}\phi(\omega, p), & n \text{ even.} \end{cases} \tag{37}$$

Note that in both cases $(\Lambda\phi)(-\omega, -p) = (\Lambda\phi)(\omega, p)$, so that $\Lambda\phi$ is a function on $\boldsymbol{P}^n$.

Theorem 2.16. *Let Λ be as defined by* (37). *Then*

$$cf = (\Lambda\hat{f})\check{}, \qquad f \in \mathscr{S}(\boldsymbol{R}^n),$$

where

$$c = i^{n-1}(4\pi)^{(n-1)/2}\frac{\Gamma(n/2)}{\Gamma(1/2)}.$$

Proof. By the inversion formula for the Fourier transform and by (4)

$$f(x) = (2\pi)^{-n}\int_{S^{n-1}} d\omega \int_0^\infty \left(\int_{-\infty}^\infty e^{-isp}\hat{f}(\omega, p)\, dp\right) e^{is(x,\omega)} s^{n-1}\, ds,$$

which we write as

$$f(x) = (2\pi)^{-n}\int_{S^{n-1}} F(\omega, x)\, d\omega$$

$$= (2\pi)^{-n}\int_{S^{n-1}} \tfrac{1}{2}(F(\omega, x) + F(-\omega, x))\, d\omega.$$

Using $\hat{f}(-\omega, -p) = \hat{f}(\omega, p)$, this gives the formula

$$(38)\quad f(x) = \tfrac{1}{2}(2\pi)^{-n}\int_{S^{n-1}} d\omega \int_{-\infty}^\infty |s|^{n-1} e^{is(x,\omega)}\, ds \int_{-\infty}^\infty e^{-isp}\hat{f}(\omega, p)\, dp.$$

If n is odd the absolute value on s can be dropped. The factor s^{n-1} can be removed by replacing $\hat{f}(\omega, p)$ by $(-i)^{n-1}(d^{n-1}/dp^{n-1})\hat{f}(\omega, p)$. The inversion formula for the Fourier transform on $\boldsymbol{R}$ then gives

$$f(x) = \tfrac{1}{2}(2\pi)^{-n}(2\pi)^{+1}(-i)^{n-1}\int_{S^{n-1}} \left\{\frac{d^{n-1}}{dp^{n-1}}\hat{f}(\omega, p)\right\}_{p=(x,\omega)} d\omega,$$

as desired.

Supposing now n is even, we let

$$\operatorname{sgn} s = \begin{cases} 1 & \text{if } s \geq 0 \\ -1 & \text{if } s < 0. \end{cases}$$

Then for a constant c_0,

$$(39)\quad f(x) = c_0\int_{S^{n-1}} d\omega \int_{\boldsymbol{R}} (\operatorname{sgn} s) e^{is(x,\omega)}\, ds \int_{\boldsymbol{R}} \frac{d^{n-1}}{dp^{n-1}}\hat{f}(\omega, p) e^{-isp}\, dp.$$

The Cauchy principal value

$$\psi \to \lim_{\varepsilon \to 0}\int_{|x| \geq \varepsilon} \frac{\psi(x)}{x}\, dx$$

is a tempered distribution whose Fourier transform is $-\pi i \operatorname{sgn} s$ (cf., Proposition 2.39). Hence

$$(40)\qquad (\mathscr{H}F)^{\sim}(s) = \operatorname{sgn} s\, \tilde{F}(s).$$

Thus sgn s can be removed[†] in the integral above if we replace

$$(d^{n-1}/dp^{n-1})\hat{f}(\omega, p) \quad \text{by} \quad \Lambda\hat{f}(\omega, p).$$

The inversion formula for the one-dimensional Fourier transform now gives the result.

4. The Plancherel Formula

We recall that the functions on $\boldsymbol{P}^n$ have been identified with the functions ϕ on $\boldsymbol{S}^{n-1} \times \boldsymbol{R}$ which are even: $\phi(-\omega, -p) = \phi(\omega, p)$. The functional

$$\phi \to \int_{S^{n-1}} \int_{\boldsymbol{R}} \phi(\omega, p)\, d\omega\, dp, \qquad \phi \in C_c(\boldsymbol{P}^n), \tag{41}$$

is therefore a well-defined measure on $\boldsymbol{P}^n$, denoted $d\omega\, dp$. The group $\boldsymbol{M}(n)$ of rigid motions of $\boldsymbol{R}^n$ acts transitively on $\boldsymbol{P}^n$; it also leaves the measure $d\omega\, dp$ invariant. It suffices to verify this latter statement for the translations T in $\boldsymbol{M}(n)$ because $\boldsymbol{M}(n)$ is generated by them together with the rotations around 0, and these rotations clearly leave $d\omega\, dp$ invariant. But

$$(\phi \circ T)(\omega, p) = \phi(\omega, p + q(\omega, T)),$$

where $q(\omega, T) \in \boldsymbol{R}$ is independent of p, so

$$\begin{aligned} \iint (\phi \circ T)(\omega, p)\, d\omega\, dp &= \iint \phi(\omega, p + q(\omega, T))\, dp\, d\omega \\ &= \iint \phi(\omega, p)\, dp\, d\omega, \end{aligned}$$

proving the invariance.

In accordance with (81) and (82) the fractional power $\square^k$ is defined on $\mathscr{S}(\boldsymbol{P}^n)$ by

$$(-\square)^k \phi(\omega, p) = \frac{1}{H_1(-2k)} \int_{\boldsymbol{R}} \phi(\omega, q)|p - q|^{-2k-1}\, dq \tag{42}$$

and then the one-dimensional Fourier transform satisfies

$$((-\square)^k \phi)^{\sim}(\omega, s) = |s|^{2k} \tilde{\phi}(\omega, s). \tag{43}$$

Now, if $f \in \mathscr{S}(\boldsymbol{R}^n)$, we have by (4)

$$\hat{f}(\omega, p) = (2\pi)^{-1} \int \tilde{f}(s\omega) e^{isp}\, ds$$

[†]See "Some Details," p. 611.

and

$$(-\square)^{(n-1)/4}\hat{f}(\omega, p) = (2\pi)^{-1}\int_{\boldsymbol{R}} |s|^{(n-1)/2}\tilde{f}(s\omega)e^{isp}\, ds. \tag{44}$$

Theorem 2.17. *The mapping $f \to \square^{(n-1)/4}\hat{f}$ extends to an isometry of $L^2(\boldsymbol{R}^n)$ onto the space $L^2_e(\boldsymbol{S}^{n-1} \times \boldsymbol{R})$ of even functions in $L^2(\boldsymbol{S}^{n-1} \times \boldsymbol{R})$, the measure on $\boldsymbol{S}^{n-1} \times \boldsymbol{R}$ being*

$$\tfrac{1}{2}(2\pi)^{1-n}\, d\omega\, dp.$$

Proof. From (44) we have from the Plancherel formula on $\boldsymbol{R}$

$$(2\pi)\int_{\boldsymbol{R}} |(-\square)^{(n-1)/4}\hat{f}(\omega, p)|^2\, dp = \int_{\boldsymbol{R}} |s|^{n-1}|\tilde{f}(s\omega)|^2\, ds,$$

so that by integration over $\boldsymbol{S}^{n-1}$ and use of the Plancherel formula for $f(x) \to \tilde{f}(s\omega)$ we obtain

$$\int_{\boldsymbol{R}^n} |f(x)|^2\, dx = \tfrac{1}{2}(2\pi)^{1-n}\int_{\boldsymbol{S}^{n-1}\times\boldsymbol{R}} |\square^{(n-1)/4}\hat{f}(\omega, p)|^2\, d\omega\, dp.$$

It remains to prove that the mapping is surjective. For this it would suffice to prove that if $\phi \in L^2(\boldsymbol{S}^{n-1} \times \boldsymbol{R})$ is even and satisfies

$$\int_{\boldsymbol{S}^{n-1}}\int_{\boldsymbol{R}} \phi(\omega, p)(-\square)^{(n-1)/4}\hat{f}(\omega, p)\, d\omega\, dp = 0$$

for all $f \in \mathscr{S}(\boldsymbol{R}^n)$, then $\phi = 0$. Taking Fourier transforms, we must prove that if $\psi \in L^2(\boldsymbol{S}^{n-1} \times \boldsymbol{R})$ is even and satisfies

$$\int_{\boldsymbol{S}^{n-1}}\int_{\boldsymbol{R}} \psi(\omega, s)|s|^{(n-1)/2}\tilde{f}(s\omega)\, ds\, d\omega = 0 \tag{45}$$

for all $f \in \mathscr{S}(\boldsymbol{R}^n)$, then $\psi = 0$. Using the condition $\psi(-\omega, -s) = \psi(\omega, s)$, we see that

$$\int_{\boldsymbol{S}^{n-1}}\int_{-\infty}^{0} \psi(\omega, s)|s|^{(1/2)(n-1)}\tilde{f}(s\omega)\, ds\, d\omega$$
$$= \int_{\boldsymbol{S}^{n-1}}\int_{0}^{\infty} \psi(\omega, t)|t|^{(1/2)(n-1)}\tilde{f}(t\omega)\, dt\, d\omega,$$

so (45) holds with $\boldsymbol{R}$ replaced by the positive axis $\boldsymbol{R}^+$. But then the function

$$\Psi(u) = \psi\left(\frac{u}{|u|}, |u|\right)|u|^{-(n-1)/2}, \qquad u \in \boldsymbol{R}^n - \{0\},$$

satisfies

$$\int_{\boldsymbol{R}^n} \Psi(u)\tilde{f}(u)\, du = 0, \qquad f \in \mathscr{S}(\boldsymbol{R}^n),$$

so $\Psi = 0$ almost everywhere, whence $\psi = 0$.

Remark. Combining Theorem 2.16 with (46), we have an alternative Plancherel formula:

$$(45') \qquad c\int_{\boldsymbol{R}^n} f(x)g(x)\,dx = \int_{\boldsymbol{P}^n} (\Lambda\hat{f})(\xi)\hat{g}(\xi)\,d\xi, \qquad f, g \in \mathscr{D}(\boldsymbol{R}^n).$$

5. The Radon Transform of Distributions

It will be proved in a general context in §3 (Proposition 3.6) that

$$(46) \qquad \int_{\boldsymbol{P}^n} \hat{f}(\xi)\phi(\xi)\,d\xi = \int_{\boldsymbol{R}^n} f(x)\check{\phi}(x)\,dx$$

for $f \in C_c(\boldsymbol{R}^n)$, $\phi \in C(\boldsymbol{P}^n)$ if $d\xi$ is a suitable fixed $\boldsymbol{M}(n)$-invariant measure on $\boldsymbol{P}^n$. Thus $d\xi = \gamma\,d\omega\,dp$, where γ is a constant, independent of f and ϕ. With applications to distributions in mind we shall prove (46) in a somewhat stronger form. Since composite functions will be considered, it is more convenient to work with Borel measurable functions.

Lemma 2.18. *Let f and ϕ be Borel-measurable functions on $\boldsymbol{R}^n$ and $\boldsymbol{P}^n$, respectively. Formula (46) holds (with $\hat{f}$ and $\check{\phi}$ existing almost everywhere) in the following two situations:*

(a) *$f \in L^1(\boldsymbol{R}^n)$ vanishing outside a compact set; $\phi \in C(\boldsymbol{P}^n)$.*

(b) *$f \in C_c(\boldsymbol{R}^n)$, ϕ locally integrable.*

Also, $d\xi = \Omega_n^{-1}\,d\omega\,dp$.

Proof. We shall use the Fubini theorem repeatedly both on the product $\boldsymbol{R}^n \times \boldsymbol{S}^{n-1}$ and on the product $\boldsymbol{R}^n = \boldsymbol{R} \times \boldsymbol{R}^{n-1}$. Since $f \in L^1(\boldsymbol{R}^n)$, we have for each $\omega \in \boldsymbol{S}^{n-1}$ that $\hat{f}(\omega, p)$ exists for almost all p and that

$$\int_{\boldsymbol{R}^n} f(x)\,dx = \int_{\boldsymbol{R}} \hat{f}(\omega, p)\,dp.$$

Also, since for each ω, $\hat{f}(\omega, p)$ exists for almost all p, the Fubini theorem implies that $\hat{f}(\omega, p)$ exists for almost all $(\omega, p) \in \boldsymbol{S}^{n-1} \times \boldsymbol{R}$. Next we consider the Borel-measurable function $(x, \omega) \to f(x)\phi(\omega, (\omega, x))$ on $\boldsymbol{R}^n \times \boldsymbol{S}^{n-1}$. We have

$$\begin{aligned}
&\int_{\boldsymbol{S}^{n-1}\times\boldsymbol{R}^n} |f(x)\phi(\omega, (\omega, x))|\,d\omega\,dx \\
&= \int_{\boldsymbol{S}^{n-1}} \left(\int_{\boldsymbol{R}^n} |f(x)\phi(\omega, (\omega, x))|\,dx\right) d\omega \\
&= \int_{\boldsymbol{S}^{n-1}} \left(\int_{\boldsymbol{R}} |f|\hat{\ }(\omega, p)|\phi(\omega, p)|\,dp\right) d\omega,
\end{aligned}$$

which in both cases is finite. Thus $f(x)\phi(\omega, (\omega, x))$ is integrable on $\boldsymbol{R}^n \times \boldsymbol{S}^{n-1}$, and its integral can be calculated by removing the absolute values above. This gives the left-hand side of (46). Reversing the integrations, we conclude that $\check{\phi}(x)$ exists for almost all x and that the double integral reduces to the right-hand side of (46).

The formula (46) dictates how to define the Radon transform and its dual for distributions. In order to make the definitions formally consistent with those for functions we would require $\hat{S}(\phi) = S(\check{\phi})$, $\check{\Sigma}(f) = \Sigma(\hat{f})$ if S and Σ are distributions on $\boldsymbol{R}^n$ and $\boldsymbol{P}^n$, respectively. But while $f \in \mathscr{D}(\boldsymbol{R}^n)$ implies $\hat{f} \in \mathscr{D}(\boldsymbol{P}^n)$, a similar implication does not hold for ϕ; we do not even have $\check{\phi} \in \mathscr{S}(\boldsymbol{R}^n)$ for $\phi \in \mathscr{D}(\boldsymbol{P}^n)$; hence $\hat{S}$ cannot be defined by the formula above, even if S is assumed to be tempered. Using the notation $\mathscr{E}$ (resp., $\mathscr{D}$) for the space of C^∞-functions (resp., of compact support) and $\mathscr{D}'$ (resp., $\mathscr{E}'$) for the space of distributions (resp., of compact support), we make the following definition.

Definition. For $S \in \mathscr{E}'(\boldsymbol{R}^n)$ we define the functional $\hat{S}$ by

$$\hat{S}(\phi) = S(\check{\phi}) \qquad \text{for} \quad \phi \in \mathscr{E}(\boldsymbol{P}^n).$$

For $\Sigma \in \mathscr{D}'(\boldsymbol{P}^n)$ we define the functional $\check{\Sigma}$ by

$$\check{\Sigma}(f) = \Sigma(\hat{f}) \qquad \text{for} \quad f \in \mathscr{D}(\boldsymbol{R}^n).$$

Lemma 2.19.

(i) *For each $\Sigma \in \mathscr{D}'(\boldsymbol{P}^n)$ we have $\check{\Sigma} \in \mathscr{D}'(\boldsymbol{R}^n)$.*

(ii) *For each $S \in \mathscr{E}'(\boldsymbol{R}^n)$ we have $\hat{S} \in \mathscr{E}'(\boldsymbol{P}^n)$.*

Moreover,

(iii) $(LS)^{\wedge} = \square \hat{S}$, $(\square \Sigma)^{\vee} = L\check{\Sigma}$.

Proof. For $A > 0$ let $\mathscr{D}_A(\boldsymbol{R}^n)$ denote the set of functions $f \in \mathscr{D}(\boldsymbol{R}^n)$ with support in the closure of $B_A(0)$. Similarly, let $\mathscr{D}_A(\boldsymbol{P}^n)$ denote the set of functions $\phi \in \mathscr{D}(\boldsymbol{P}^n)$ with support in the closure of the "ball"

$$\beta_A(0) = \{\xi \in \boldsymbol{P}^n : d(0, \xi) < A\}.$$

The mapping $f \to \hat{f}$ being continuous from $\mathscr{D}_A(\boldsymbol{R}^n)$ to $\mathscr{D}_A(\boldsymbol{P}^n)$ (with the topologies from §1 of the Introduction), the restriction of $\check{\Sigma}$ to each $\mathscr{D}_A(\boldsymbol{R}^n)$ is continuous, so (i) follows. That $\hat{S}$ is a distribution is clear from (3). Concerning its support select $R > 0$ such that S has support inside $B_R(0)$. Then if $\phi(\omega, p) = 0$ for $p \leq R$, we have $\check{\phi}(x) = 0$ for $|x| \leq R$, whence $\hat{S}(\phi) = S(\check{\phi}) = 0$.

For (iii) we note by Lemma 2.1,

$$(LS)^{\widehat{}}(\phi) = (LS)(\check{\phi}) = S(L\check{\phi}) = S((\Box\phi)^{\vee})$$
$$= \hat{S}(\Box\phi) = (\Box\hat{S})(\phi).$$

The other relation is proved in the same manner.

We shall now prove an analog of the support theorem (Theorem 2.6) for distributions. For $A > 0$ let $\beta_A(0)$ be defined as above.

Theorem 2.20. *Let $T \in \mathscr{E}'(\boldsymbol{R}^n)$ satisfy the condition*

$$\operatorname{supp} \hat{T} \subset \operatorname{Cl}(\beta_A(0)).$$

Then

$$\operatorname{supp}(T) \subset \operatorname{Cl}(B_A(0)).$$

Proof. For $f \in \mathscr{D}(\boldsymbol{R}^n)$, $\phi \in \mathscr{D}(\boldsymbol{P}^n)$ we can consider the *mixed convolution*

$$(f \times \phi)(\xi) = \int_{\boldsymbol{R}^n} f(y)\phi(\xi - y)\, dy,$$

where for $\xi \in \boldsymbol{P}^n$, $\xi - y$ denotes the translate of the hyperplane ξ by $-y$. Then

$$(f \times \phi)^{\vee} = f * \check{\phi}.$$

In fact, if ξ_0 is any hyperplane through 0,

$$(f \times \phi)^{\vee}(x) = \int_K dk \int_{\boldsymbol{R}^n} f(y)\phi(x + k \cdot \xi_0 - y)\, dy$$
$$= \int_K dk \int_{\boldsymbol{R}^n} f(x - y)\phi(y + k \cdot \xi_0)\, dy$$
$$= (f * \check{\phi})(x).$$

By the definition of $\hat{T}$, the support assumption on $\hat{T}$ is equivalent to

$$T(\check{\phi}) = 0$$

for all $\phi \in \mathscr{D}(\boldsymbol{P}^n)$ with support in $\boldsymbol{P}^n - \operatorname{Cl}(\beta_A(0))$. Let $\varepsilon > 0$, let $f \in \mathscr{D}(\boldsymbol{R}^n)$ be a symmetric function with support in $\operatorname{Cl}(B_\varepsilon(0))$ and let $\phi \in \mathscr{D}(\boldsymbol{P}^n)$ have support contained in $\boldsymbol{P}^n - \operatorname{Cl}(\beta_{A+\varepsilon}(0))$. Since $d(0, \xi - y) \leq d(0, \xi) + |y|$, it follows that $f \times \phi$ has support in $\boldsymbol{P}^n - \operatorname{Cl}(\beta_A(0))$; thus by the formulas above and the symmetry of f,

$$(f * T)(\check{\phi}) = T(f * \check{\phi}) = T((f \times \phi)^{\vee}) = 0.$$

But then

$$(f * T)^{\widehat{}}(\phi) = (f * T)(\check{\phi}) = 0,$$

which means that $(f * T)^{\widehat{}}$ has support in $\mathrm{Cl}(\beta_{A+\varepsilon}(0))$. But now Theorem 2.6 implies that $f * T$ has support in $\mathrm{Cl}(B_{A+\varepsilon}(0))$. Letting $\varepsilon \to 0$, we obtain the desired conclusion, $\mathrm{supp}(T) \subset \mathrm{Cl}(B_A(0))$.

Remark. For a strengthening of this result see Exercise B5.

We can now extend the inversion formulas for the Radon transform to distributions. First we observe that the Hilbert transform $\mathscr{H}$ can be extended to distributions T on $\boldsymbol{R}$ of compact support. It suffices to put

$$\mathscr{H}(T)(F) = -T(\mathscr{H}F), \qquad F \in \mathscr{D}(\boldsymbol{R}).$$

In fact, $\mathscr{H}$ being the convolution with a (tempered) distribution, the mapping $F \to \mathscr{H}F$ is a continuous mapping of $\mathscr{D}(\boldsymbol{R})$ into $\mathscr{E}(\boldsymbol{R})$ (cf. No. 8 below). In particular, $\mathscr{H}(T) \in \mathscr{D}'(\boldsymbol{R})$.

Theorem 2.21. *The Radon transform $S \to \hat{S}$ $(S \in \mathscr{E}'(\boldsymbol{R}^n))$ is inverted by the following formula*:

$$cS = (\Lambda\hat{S})^{\vee}, \qquad S \in \mathscr{E}'(\boldsymbol{R}^n),$$

where the constant c equals

$$c = (-4\pi)^{(n-1)/2}\frac{\Gamma(n/2)}{\Gamma(1/2)}.$$

In the case when n is odd we have also

$$cS = L^{(1/2)(n-1)}((\hat{S})^{\vee}).$$

Remark. Since $\hat{S}$ has compact support and since Λ is defined by means of the Hilbert transform, the remarks above show that $\Lambda\hat{S} \in \mathscr{D}'(\boldsymbol{P}^n)$. Hence the right-hand side is well-defined.

Proof. Using Theorem 2.16, we have

$$\begin{aligned}(\Lambda\hat{S})^{\vee}(f) &= (\Lambda\hat{S})(\hat{f}) = \hat{S}(\Lambda\hat{f}) \\ &= S((\Lambda\hat{f})^{\vee}) = cS(f).\end{aligned}$$

The other inversion then follows, using Lemma 2.19.

Let M be a manifold and $d\mu$ a measure such that on each local coordinate patch with coordinates $(t_1, \ldots, t_n)$ the Lebesgue measure $dt_1 \cdots dt_n$ and $d\mu$ are absolutely continuous with respect to each other. If h is a function on M locally integrable with respect to $d\mu$, the distribution $\phi \to \int h\phi \, d\mu$ will be denoted T_h.

Proposition 2.22.

(a) *Let $f \in L^1(\boldsymbol{R}^n)$ vanish outside a compact set. Then the distribution T_f has Radon transform given by*

$$\hat{T}_f = T_{\hat{f}}. \tag{47}$$

(b) *Let ϕ be a locally integrable function on $\boldsymbol{P}^n$. Then*

$$(T_\phi)^{\vee} = T_{\check{\phi}}. \tag{48}$$

Proof. The existence and local integrability of $\hat{f}$ and $\check{\phi}$ was established during the proof of Lemma 2.18. The two formulas now follow directly from Lemma 2.18.

As a result of this proposition the smoothness assumption can be dropped in the inversion formulas. In particular, we can state the following result.

Corollary 2.23. (n odd) *The inversion formula*

$$cf = L^{(n-1)/2}((\hat{f})^{\vee}), \qquad c = (-4\pi)^{(n-1)/2}\frac{\Gamma(n/2)}{\Gamma(1/2)},$$

holds for all $f \in L^1(\boldsymbol{R}^n)$ of compact support, the derivative interpreted in the sense of distributions.

Examples. If μ is a measure (or a distribution) on a submanifold S of a manifold M, the distribution on M given by $\phi \to \mu(\phi|S)$ will also be denoted by μ.

(a) Let δ_0 be the delta distribution $f \to f(0)$ on $\boldsymbol{R}^n$. Then

$$\hat{\delta}_0(\phi) = \delta_0(\check{\phi}) = \Omega_n^{-1}\int_{\boldsymbol{S}^{n-1}} \phi(\omega, 0)\, d\omega,$$

so that

$$\hat{\delta}_0 = \Omega_n^{-1} m_{\boldsymbol{S}^{n-1}}, \tag{49}$$

the normalized measure on $\boldsymbol{S}^{n-1}$ considered as a distribution on $\boldsymbol{S}^{n-1} \times \boldsymbol{R}$.

(b) Let ξ_0 denote the hyperplane $x_n = 0$ in $\boldsymbol{R}^n$, and δ_{ξ_0} the delta distribution $\phi \to \phi(\xi_0)$ on $\boldsymbol{P}^n$. Then

$$\check{\delta}_{\xi_0}(f) = \int_{\xi_0} f(x)\, dm(x),$$

so that

$$\check{\delta}_{\xi_0} = m_{\xi_0}, \tag{50}$$

the Euclidean measure of ξ_0.

(c) Let χ_B be the characteristic function of the unit ball $B \subset \boldsymbol{R}^n$. Then by (47)

$$(\chi_B)\widehat{}\,(\omega, p) = \begin{cases} \Omega_{n-1}(1 - p^2)^{(1/2)(n-1)}, & |p| \leq 1 \\ 0, & |p| > 1. \end{cases}$$

(d) Let Ω be a convex region in $\boldsymbol{R}^n$ whose boundary is a smooth surface. We shall obtain a formula for the volume of Ω in terms of the areas of its hyperplane sections. For simplicity we assume n odd. The characteristic function χ_Ω is a distribution of compact support and $(\chi_\Omega)\widehat{}$ is thus well-defined. Approximating χ_Ω in the L^2-norm by a sequence $(\psi_n) \subset \mathscr{D}(\Omega)$, we see from Theorem 2.17 that $\partial_p^{(1/2)(n-1)}\hat{\psi}_n(\omega, p)$ converges in the L^2-norm on $\boldsymbol{P}^n$. Since

$$\int \hat{\psi}_n(\xi)\phi(\xi)\, d\xi = \int \psi_n(x)\check{\phi}(x)\, dx,$$

it follows from Schwarz's inequality that $\hat{\psi}_n \to (\chi_\Omega)\widehat{}$ in the sense of distributions, and accordingly $\partial^{(1/2)(n-1)}\hat{\psi}_n$ converges as a distribution to $\partial^{(1/2)(n-1)}((\chi_\Omega)\widehat{})$. Since the L^2-limit is also a limit in the sense of distributions, this last function equals the L^2-limit of the sequence $\partial^{(1/2)(n-1)}\hat{\psi}_n$. From Theorem 2.17 and Corollary 2.23 we therefore have the following result.

Theorem 2.24. *Let $\Omega \subset \boldsymbol{R}^n$ (n odd) be a convex region as above and $V(\Omega)$ its volume. Let $A(\omega, p)$ denote the $(n-1)$-dimensional area of the intersection of Ω with the hyperplane $(x, \omega) = p$. Then*

$$(51) \qquad c_0\, \chi_\Omega(x) = L_x^{(1/2)(n-1)}\left(\int_{\boldsymbol{S}^{n-1}} A(\omega, (x, \omega))\, d\omega\right), \qquad c_0 = 2(2\pi i)^{n-1},$$

in the sense of distributions.

Furthermore,

$$V(\Omega) = \int A(\omega, p)dp.$$

6. Integration over d-Planes. X-Ray Transforms

Let d be a fixed integer in the range $0 < d < n$. Since a hyperplane in $\boldsymbol{R}^n$ can be viewed as a disjoint union of parallel d-planes, parametrized by $\boldsymbol{R}^{n-1-d}$, it is obvious from (4) that if $f \in \mathscr{S}(\boldsymbol{R}^n)$ has 0 integral over each d-plane in $\boldsymbol{R}^n$, then it is identically 0. Similarly, we can deduce the following consequence of Theorem 2.6.

Corollary 2.25. *Let $f \in C(\boldsymbol{R}^n)$ satisfy the following conditions:*

(i) *For each integer $m > 0$, $|x|^m f(x)$ is bounded on $\boldsymbol{R}^n$.*

(ii) *For each d-plane ξ_d outside a ball $|x| < A$, we have*

$$\int_{\xi_d} f(x)\, dm(x) = 0,$$

dm being the Euclidean measure. Then

$$f(x) = 0 \qquad \text{for} \quad |x| > A.$$

We now define the *d-dimensional Radon transform* $f \to \hat{f}$ by

$$\hat{f}(\xi) = \int_{\xi} f(x)\, dm(x), \qquad \xi \quad \text{a } d\text{-plane.} \tag{52}$$

Because of the applications to radiology indicated in No. 7b), the one-dimensional Radon transform is often called the *X-ray transform.* We can then reformulate Corollary 2.25 as follows.

Corollary 2.26. *Let $f, g \in C(\boldsymbol{R}^n)$ satisfy the rapid decrease condition:*
For each $m > 0$, $|x|^m f(x)$ and $|x|^m g(x)$ are bounded on $\boldsymbol{R}^n$.
Assume for the d-dimensional Radon transforms

$$\hat{f}(\xi) = \hat{g}(\xi)$$

whenever the d-plane ξ lies outside the unit ball. Then

$$f(x) = g(x) \qquad \textit{for} \quad |x| > 1.$$

We shall now generalize the inversion formula in Theorem 2.13. If ϕ is a continuous function on the space of d-planes in $\boldsymbol{R}^n$, we denote by $\check{\phi}$ the point function

$$\check{\phi}(x) = \int_{x \in \xi} \phi(\xi)\, d\mu(\xi),$$

where μ is the unique measure on the (compact) space of d-planes passing through x, invariant under all rotations around x, and with total measure 1. If σ is a fixed d-plane through the origin, we have by analogy with (16),

$$\check{\phi}(x) = \int_K \phi(x + k \cdot \sigma)\, dk. \tag{53}$$

Theorem 2.27. *The d-dimensional Radon transform in* $\mathbf{R}^n$ *is inverted by the formula*

$$cf = (-L)^{d/2}((\hat{f})^{\vee}), \qquad f \in \mathscr{S}(\mathbf{R}^n), \tag{54}$$

where

$$c = \frac{\Gamma(\frac{1}{2}n)}{\Gamma(\frac{1}{2}(n-d))}(4\pi)^{(1/2)d}.$$

Proof. We have by analogy with (32)

$$\begin{aligned}(\hat{f})^{\vee}(x) &= \int_K \left(\int_\sigma f(x + k \cdot y)\, dm(y)\right) dk \\ &= \int_\sigma dm(y) \int_K f(x + k \cdot y)\, dk = \int_\sigma (M^{|y|}f)(x)\, dm\,(y).\end{aligned}$$

Hence

$$(\hat{f})^{\vee}(x) = \Omega_d \int_0^\infty (M^r f)(x)\, r^{d-1}\, dr,$$

so, using polar coordinates around x,

$$(\hat{f})^{\vee}(x) = \frac{\Omega_d}{\Omega_n} \int_{\mathbf{R}^n} |x - y|^{d-n} f(y)\, dy. \tag{55}$$

The theorem now follows from Prop. 2.38.

As a corollary of Theorem 2.10 we now obtain a generalization, characterizing the image of the space $\mathscr{D}(\mathbf{R}^n)$ under the d-dimensional Radon transform.

The set $\mathbf{G}(d, n)$ of d-planes in $\mathbf{R}^n$ is a manifold, in fact, a homogeneous space of the group $\mathbf{M}(n)$ of all isometries of $\mathbf{R}^n$. Let $\mathbf{G}_{d,n}$ denote the manifold of all d-dimensional subspaces (d-planes through 0) of $\mathbf{R}^n$. The parallel translation of a d-plane to one through 0 gives a mapping π of $\mathbf{G}(d, n)$ onto $\mathbf{G}_{d,n}$. The inverse image $\pi^{-1}(\sigma)$ of a member $\sigma \in \mathbf{G}_{d,n}$ is naturally identified with the orthogonal complement $\sigma^\perp$. Let us write $\xi = (\sigma, x'')$ if $\sigma = \pi(\xi)$ and $x'' = \sigma^\perp \cap \xi$. Then (52) can be written

$$\hat{f}(\sigma, x'') = \int_\sigma f(x' + x'')\, dx'. \tag{56}$$

For $k \in \mathbf{Z}^+$ we consider the polynomial

$$P_k(u) = \int_{\mathbf{R}^n} f(x)(x, u)^k\, dx. \tag{57}$$

If $u = u'' \in \sigma^\perp$, this can be written

$$\int_{R^n} f(x)(x, u'')^k\, dx = \int_{\sigma^\perp}\int_\sigma f(x' + x'')(x'', u'')^k\, dx'\, dx'',$$

so the polynomial

$$P_{\sigma,k}(u'') = \int_{\sigma^\perp} \hat{f}(\sigma, x'')(x'', u'')^k\, dx''$$

is the restriction to $\sigma^\perp$ of the polynomial P_k.

By analogy with the space $\mathscr{D}_H(\boldsymbol{P}^n)$ in No. 2 we define the space $\mathscr{D}_H(\boldsymbol{G}(d, n))$ as the set of C^∞-functions $\phi(\xi) = \phi_\sigma(x'')$ on $\boldsymbol{G}(d, n)$ of compact support satisfying the following condition.

(H): *For each $k \in \boldsymbol{Z}^+$ there exists a homogeneous kth-degree polynomial P_k on $\boldsymbol{R}^n$ such that for each $\sigma \in \boldsymbol{G}_{d,n}$ the polynomial*

$$P_{\sigma,k}(u'') = \int_{\sigma^\perp} \phi_\sigma(x'')(x'', u'')^k\, dx'', \qquad u'' \in \sigma^\perp,$$

coincides with the restriction $P_k|\sigma^\perp$.

Corollary 2.28. *The d-dimensional Radon transform is a bijection of $\mathscr{D}(\boldsymbol{R}^n)$ onto $\mathscr{D}_H(\boldsymbol{G}(d, n))$.*

Proof. For $d = n - 1$ this is Theorem 2.10. We shall now reduce the case of general d to the case $d = n - 1$. It remains just to prove the surjectivity in Corollary 2.28.

Let $\phi \in \mathscr{D}_H(\boldsymbol{G}(d, n))$. Let $\omega \in \boldsymbol{R}^n$ be a unit vector. Choose a d-dimensional subspace σ perpendicular to ω and consider the $(n - d - 1)$-dimensional integral

$$\Psi_\sigma(\omega, p) = \int_{(\omega, x'') = p,\, x'' \in \sigma^\perp} \phi_\sigma(x'')\, d_{n-d-1}(x''), \qquad p \in \boldsymbol{R}. \tag{58}$$

We claim that this is independent of the choice of σ. In fact,

$$\int_{\boldsymbol{R}} \Psi_\sigma(\omega, p)p^k\, dp = \int_{\boldsymbol{R}} p^k\left(\int \phi_\sigma(x'')\, d_{n-d-1}(x'')\right) dp$$

$$= \int_{\sigma^\perp} \phi_\sigma(x'')(x'', \omega)^k\, dx'' = P_k(\omega).$$

If we had chosen another σ, say σ_1, perpendicular to ω, then by the above, $\Psi_\sigma(\omega, p) - \Psi_{\sigma_1}(\omega, p)$ would have been orthogonal to all polynomials in p; having compact support it would have been identically O.

Thus we have a well-defined function $\Psi(\omega, p) = \Psi_\sigma(\omega, p)$ to which Theorem 2.10 applies. (It should be noted that Ψ is smooth; in fact, for ω near a fixed ω_0, we can let σ depend smoothly on ω, so that $\Psi_\sigma(\omega, p)$ is smooth.) From Theorem 2.10 we get a function $f \in \mathscr{D}(\boldsymbol{R}^n)$ such that

$$\Psi(\omega, p) = \int_{(x,\omega)=p} f(x)\, dm(x). \tag{59}$$

It remains to prove that

$$\phi_\sigma(x'') = \int_\sigma f(x' + x'')\, dx'. \tag{60}$$

But as x'' runs through an arbitrary hyperplane in $\sigma^\perp$ it follows from (58) and (59) that both sides of (60) have the same integral. By the injectivity of the $(n-d-1)$-dimensional Radon transform on $\sigma^\perp$, Eq. (60) follows. This proves Corollary 2.28.

7. Applications

A. *Partial Differential Equations*

The inversion formula in Theorem 2.13 is very well suited for applications to partial differential equations. To explain the underlying principle we write the inversion formula in the form

$$f(x) = \gamma L_x^{(n-1)/2}\left(\int_{S^{n-1}} \hat{f}(\omega, (x, \omega))d\omega\right), \tag{61}$$

where the constant γ equals $\frac{1}{2}(2\pi i)^{1-n}$. Note that the function $f_\omega(x) = \hat{f}(\omega, (x, \omega))$ is a *plane wave with normal* ω; that is, it is constant on each hyperplane perpendicular to ω.

Consider now a differential operator

$$D = \sum_{(k)} a_{k_1 \cdots k_n} \partial_1^{k_1} \cdots \partial_n^{k_n}$$

with constant coefficients $a_{k_1 \ldots k_n}$, and suppose we want to solve the differential equation

$$Du = f, \tag{62}$$

where f is a given function in $\mathscr{S}(\boldsymbol{R}^n)$. To simplify the use of (61) we assume n to be odd. We begin by considering the differential equation

$$Dv = f_\omega, \tag{63}$$

where f_ω is the plane wave defined above and we look for a solution v which is also a plane wave with normal ω. But a plane wave with normal ω is just a function of one variable; also, if v is a plane wave with normal ω, so is the function Dv. The differential equation (63) (with v a plane wave) is therefore an *ordinary* differential equation with constant coefficients. Suppose $v = u_\omega$ is a solution and assume that this choice can be made smoothly in ω. Then the function

$$u = \gamma L^{(n-1)/2} \int_{S^{n-1}} u_\omega \, d\omega \tag{64}$$

is a solution to the differential equation (62). In fact, since D and $L^{(n-1)/2}$ commute, we have

$$Du = \gamma\, L^{(n-1)/2} \int_{S^{n-1}} Du_\omega \, d\omega = \gamma\, L^{(n-1)/2} \int_{S^{n-1}} f_\omega \, d\omega = f.$$

This method only assumes that the plane wave solution u_ω to the ordinary differential equation $Dv = f_\omega$ exists and can be chosen so as to depend smoothly on ω. This cannot always be done because D might annihilate all plane waves with normal ω. [For example, take $D = \partial^2/(\partial x_1\, \partial x_2)$ and $\omega = (1, 0)$.] However, if this restriction to plane waves is never 0 it follows from a theorem of Trèves [1963] that the solution u_ω can be chosen to depend smoothly on ω. Thus we can state

Theorem 2.29. *Assuming the restriction D_ω of D to the space of plane waves with normal ω is $\neq 0$ for each ω formula* (64) *gives a solution to the differential equation $Du = f(f \in \mathscr{S}(\boldsymbol{R}^n))$.*

The method of plane waves can also be used to solve the Cauchy problem for hyperbolic differential equations with constant coefficients. We illustrate the method by means of the wave equation in $\boldsymbol{R}^n$,

$$Lu = \frac{\partial u^2}{\partial t^2}, \qquad u(x, 0) = 0, \qquad \frac{\partial u}{\partial t}(x, 0) = f(x), \tag{65}$$

$f \in \mathscr{S}(\boldsymbol{R}^n)$ being a given function.

Lemma 2.30. *For each $x \in \boldsymbol{R}^n$*

$$\int_{S^{n-1}} |(\omega, x)|^k \, d\omega = \frac{2\pi^{(1/2)(n-1)}\Gamma(\frac{1}{2}(k+1))}{\Gamma(\frac{1}{2}(k+n))} |x|^k. \tag{66}$$

Proof. The hyperplane perpendicular to x with distance p from 0 intersects $\boldsymbol{S}^{n-1}$ in an $(n-2)$-sphere of radius $(1 - p^2)^{1/2}$. The integrand is constant $|x|p$ on this sphere and from Fig. 8 we have

$$\frac{ds}{dp} = (1 - p^2)^{-1/2}.$$

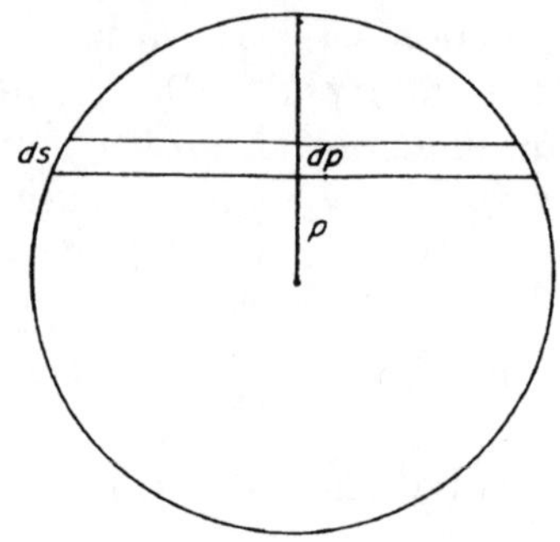

FIG. 8

It follows that

$$\int_{S^{n-1}} |(\omega, x)|^k \, d\omega = 2\int_0^1 \Omega_{n-1}(1-p^2)^{(n-2)/2}(p|x|)^k(1-p^2)^{-1/2}\, dp,$$

which implies the lemma.

Using (66) and the identities (88) and (89) in No. 8 we can derive the identity

$$(67) \qquad L_x^{(n+1)/2} \int_{R^n} \left(\int_{S^{n-1}} f(y)|(x-y, \omega)| \, d\omega \right) dy = 4(2\pi i)^{n-1} f(x)$$

for n odd.

Suppose u_ω is a plane wave with normal ω satisfying the wave equation

$$Lu_\omega = \frac{\partial^2}{\partial t^2} u_\omega,$$

ignoring the initial condition. Then the function

$$Z(x, t) = \int_{S^{n-1}} u_\omega(x, t)\, d\omega$$

is also a solution and so is the convolution

$$v(x, t) = \int_{R^n} f(y) Z(x-y, t)\, dy.$$

The time derivative satisfies

$$v_t(x, 0) = \int_{R^n} f(y) Z_t(x-y, 0)\, dy,$$

so in view of (67) we try to determine u_ω such that

$$Z_t(x, 0) = \int_{S^{n-1}} |(x, \omega)| \, d\omega.$$

For this we put

$$u_\omega(x, t) = \tfrac{1}{2}((x, \omega) + t)^2 \operatorname{sgn}((x, \omega) + t),$$

and have then obtained the following result.

Theorem 2.31. *Let n be odd. A solution to the Cauchy problem* (65) *is given by*

$u(x, t) =$

$$\frac{1}{8(2\pi i)^{n-1}} L_x^{(n+1)/2} \left[\int_{R^n} f(x - y) \int_{S^{n-1}} ((y, \omega) + t)^2 \operatorname{sgn}((y, \omega) + t) \, d\omega \, dy \right].$$

We shall now put this into another form for $n = 3$. Since u_ω satisfies the wave equation, the formula can be written

$$u(x, t) = -\frac{1}{16\pi^2} \frac{\partial}{\partial t} \int_{R^3} f(y) \left[\frac{\partial}{\partial t} \int_{S^2} \operatorname{sgn}\{(x - y, \omega) + t\} \, d\omega \right] dy.$$

As for Lemma 2.30 we compute the S^2 integral by slices perpendicular to $x - y$. This gives

$$\int_{S^2} \operatorname{sgn}\{(x - y, \omega) + t\} \, d\omega = 2\pi \int_{-1}^{1} \operatorname{sgn}\{|x - y| p + t\} \, dp,$$

which, since $t \geq 0$, equals

$$2\pi \int_{-1}^{1} \operatorname{sgn}\{p + |x - y|^{-1} t\} \, dp = \begin{cases} 4\pi |x - y|^{-1} t & \text{if } t < |x - y| \\ 4\pi & \text{if } t > |x - y|. \end{cases}$$

Hence

$$\begin{aligned} u(x, t) &= -\frac{1}{4\pi} \frac{\partial}{\partial t} \int_{|x - y| > t} f(y) |x - y|^{-1} \, dy \\ &= -\frac{1}{4\pi} \frac{\partial}{\partial t} \int_t^\infty d\rho \int_{S_\rho(x)} \rho^{-1} f(y) \, d\omega(y) \\ &= \frac{1}{4\pi} (4\pi t^2) t^{-1} (M^t f)(x). \end{aligned}$$

This proves the following result of Poisson.

Corollary 2.32. *For $n = 3$ the solution to the Cauchy problem* (65) *is given by*

$$u(x, t) = t(M^t f)(x).$$

Remarks. (i) A similar formula holds for arbitrary n; the solution to (65) is given by

$$u(x, t) = \frac{1}{(n-2)!} \frac{\partial^{n-2}}{\partial t^{n-2}} \int_0^t (M^\rho f)(x)\rho(t^2 - \rho^2)^{(1/2)(n-3)} \, d\rho;$$

(cf. Tedone [1898], Ásgeirsson [1937], John [1955], Chapter II). A formula of this type is derived in Exercise F1, Chapter II.

(ii) *Huygens's Principle.* The formula in Corollary 2.32 shows that $u(x, t)$ is determined by the values of f on the sphere $S_t(x)$ in $\boldsymbol{R}^3$. This phenomenon, called *Huygens's principle*, remains true for $\boldsymbol{R}^n$ (n *odd*) in a slightly weaker form. In fact, if the differentiations $\partial^{n-2}/\partial t^{n-2}$ are carried out, we end up with a linear combination with polynomial coefficients

$$\sum_k p_k(t) \frac{\partial^k (M^t f(x))}{\partial t^k},$$

which shows the following:

For n odd, the solution $u(x, t)$ to the Cauchy problem (65) *is determined by the values of f in an arbitrarily thin shell around $S_t(x)$.*

For even n the integral will remain, so $u(x, t)$ will require knowing f in the ball $B_t(x)$.

B. *Radiography*

The classic interpretation of an X-ray picture is an attempt at reconstructing properties of a three-dimensional body by means of the X-ray projection on a plane.

In modern X-ray technology the picture is given a more refined mathematical interpretation. Let $B \subset \boldsymbol{R}^3$ be a body (for example, a part of a human body) and let $f(x)$ denote its density at a point x. Let ξ be a line in $\boldsymbol{R}^3$ and suppose a thin beam of X rays is directed at B along ξ. Let I_0 and I denote respectively, the intensity of the beam before entering B and after leaving B. For physical reasons we then have the formula (Cormack [1963]),

$$\log(I_0/I) = \int_\xi f(x) \, dm(x),$$

the integral $\hat{f}(\xi)$ of f along ξ. Since the left-hand side is determined by the X-ray picture, the X-ray reconstruction problem amounts to the determination of the function f by means of its line integrals $\hat{f}(\xi)$. *The inversion formula in Theorem* 2.27 *gives an explicit solution of this problem.*

If $B_0 \subset B$ is a convex subset (for example, the heart), it may be of interest to determine the density f outside B_0 using only X rays which do not intersect B_0. *The support theorem (Theorem 2.6 and Corollaries 2.8 and 2.26) implies that f is determined outside B_0 on the basis of the integrals $\hat{f}(\xi)$ for which ξ does not intersect B_0.*

8. Appendix. Distributions and Riesz Potentials

In this subsection we collect a few notions and facts from distribution theory as well as some results about Riesz potentials which have been used in this section.

In the Introduction, §1, we reviewed the definition of a distribution on $\boldsymbol{R}^n$ and of the support of a distribution. If F is a locally integrable function on $\boldsymbol{R}^n$, the distribution $\phi \to \int F(x)\phi(x)\,dx$ ($\phi \in \mathcal{D}(\boldsymbol{R}^n)$) is denoted T_F.

If $T \in \mathcal{D}'(\boldsymbol{R})$, the *derivative* T' is defined as the distribution

$$\phi \to -T(\phi'), \qquad \phi \in \mathcal{D}(\boldsymbol{R}).$$

If $F \in C^1(\boldsymbol{R})$, then by partial integration, the distributions $T_{F'}$ and $(T_F)'$ are seen to coincide.

A linear form on the space $\mathcal{S}(\boldsymbol{R})$ is called a *tempered distribution* if it is continuous in the topology given by the norms (6). The restriction of a tempered distribution to $\mathcal{D}(\boldsymbol{R})$ is a distribution, and since the subspace $\mathcal{D}(\boldsymbol{R}) \subset \mathcal{S}(\boldsymbol{R})$ is dense, two tempered distributions coincide if they coincide on $\mathcal{D}(\boldsymbol{R})$. A distribution of compact support is tempered. Let $\mathcal{S}'(\boldsymbol{R})$ denote the space of tempered distributions.

If $T \in \mathcal{S}'(\boldsymbol{R})$, its *Fourier transform* $\tilde{T}$ is the linear form on $\mathcal{S}(\boldsymbol{R})$ defined by

$$\tilde{T}(\phi) = T(\tilde{\phi}), \qquad \phi \in \mathcal{S}(\boldsymbol{R}),$$

where

$$\tilde{\phi}(u) = \int_{\boldsymbol{R}} \phi(x) e^{-ixu}\,dx.$$

Since $\phi \to \tilde{\phi}$ is a homeomorphism of $\mathcal{S}(\boldsymbol{R})$ onto itself, $\tilde{T} \in \mathcal{S}'(\boldsymbol{R})$. Moreover, since $\int F\tilde{\phi} = \int \tilde{F}\phi$ [$\phi, F \in \mathcal{S}(\boldsymbol{R})$], the distributions $(T_F)^{\sim}$ and $T_{\tilde{F}}$ coincide. Since $(\tilde{\phi})^{\sim}(x) = 2\pi\phi(-x)$, we have $\tilde{1} = 2\pi\delta$. We also have the rules

$$(\tilde{T})^{\sim} = 2\pi\check{T},\ (T')^{\sim} = ix\tilde{T},\ (uT)^{\sim} = i(\tilde{T})', \tag{68}$$

where $\check{T}$ is the image of T under the image $x \to -x$.

Since distributions generalize measures, it is sometimes convenient to write

$$T(\phi) = \int_{\boldsymbol{R}} \phi(x)\,dT(x)$$

for the value of a distribution T on the function ϕ. If $T \in \mathscr{E}'(\boldsymbol{R})$, then $\tilde{T}$, as defined above, is the distribution given by the function $u \to \int e^{-ixu}\, dT(x)$.

If S and T are two distributions at least one of which is of compact support, their *convolution* is the distribution defined by

$$\phi \to \int_{\boldsymbol{R}}\int_{\boldsymbol{R}} \phi(x + y)\, dS(x)\, dT(y), \qquad \phi \in \mathscr{D}(\boldsymbol{R}).$$

This distribution is denoted $S * T$. If $f \in \mathscr{D}(\boldsymbol{R})$, $T \in \mathscr{D}'(\boldsymbol{R})$, the distribution $T_f * T$ has the form T_g, where

$$g(x) = \int_{\boldsymbol{R}} f(x - y)\, dT(y),$$

so we write for simplicity $g = f * T$. The mapping $f \to f * T$ is continuous from $\mathscr{D}(\boldsymbol{R})$ into $\mathscr{E}(\boldsymbol{R})$.

If S and T both have compact supports, so does $S * T$. Also, $\tilde{S}$ and $\tilde{T}$ have the form $\tilde{S} = T_s$, $\tilde{T} = T_t$, where $s, t \in \mathscr{E}(\boldsymbol{R})$, and, in addition, $(S * T)^{\sim} = T_{st}$. We express this in the form

$$(S * T)^{\sim} = \tilde{S}\tilde{T}. \tag{69}$$

This formula holds also in the cases

$$S \in \mathscr{S}(\boldsymbol{R}), \qquad T \in \mathscr{S}'(\boldsymbol{R}),$$
$$S \in \mathscr{E}'(\boldsymbol{R}), \qquad T \in \mathscr{S}'(\boldsymbol{R}),$$

and in both cases $S * T$ is tempered.

These notions (distributions, tempered distributions, derivative, Fourier transform, and convolution) generalize in an obvious manner to several variables. We thus have function spaces and distribution spaces satisfying the following canonical inclusions:

$$\begin{array}{ccccc} \mathscr{D}(\boldsymbol{R}^n) & \subset & \mathscr{S}(\boldsymbol{R}^n) & \subset & \mathscr{E}(\boldsymbol{R}^n), \\ \cap & & \cap & & \cap \\ \mathscr{E}'(\boldsymbol{R}^n) & \subset & \mathscr{S}'(\boldsymbol{R}^n) & \subset & \mathscr{D}'(\boldsymbol{R}^n). \end{array}$$

We shall now study some useful examples in detail.

If $\alpha \in \boldsymbol{C}$ satisfies $\operatorname{Re}\alpha > -1$, the functional

$$x_+^\alpha : \phi \to \int_0^\infty x^\alpha \phi(x)\, dx, \qquad \phi \in \mathscr{S}(\boldsymbol{R}),$$

is a well-defined distribution. The mapping $\alpha \to x_+^\alpha$ can be extended to a holomorphic distribution-valued function in the region

$$\operatorname{Re}\alpha > -n - 1, \qquad \alpha \neq -1, -2, \ldots, -n,$$

by means of the formula

$$(70)\quad x_+^\alpha(\phi) = \int_0^1 x^\alpha\left[\phi(x) - \phi(0) - x\phi'(0) - \cdots - \frac{x^{n-1}}{(n-1)!}\phi^{(n-1)}(0)\right]dx$$
$$+ \int_1^\infty x^\alpha\,\phi(x)\,dx + \sum_{k=1}^{n}\frac{\phi^{(k-1)}(0)}{(k-1)!(\alpha+k)}.$$

In this manner $\alpha \to x_+^\alpha$ is a meromorphic distribution-valued function on C, with simple poles at $\alpha = -1, -2, \ldots$. We note that the residue at $\alpha = -k$ is given by

$$(71)\qquad \operatorname*{Res}_{\alpha=-k} x_+^\alpha = \lim_{\alpha\to -k}(\alpha+k)x_+^\alpha = \frac{(-1)^{k-1}}{(k-1)!}\delta^{(k-1)}.$$

Here $\delta^{(h)}$ is the hth derivative of the delta distribution δ. We note that x_+^α is always a tempered distribution.

Next we consider for $\operatorname{Re}\alpha > -n$ the distribution r^α on R^n given by

$$r^\alpha\colon \phi \to \int_{R^n}\phi(x)|x|^\alpha\,dx, \qquad \phi\in\mathscr{D}(R^n).$$

Lemma 2.33. *The mapping $\alpha \to r^\alpha$ extends uniquely to a meromorphic mapping from C to the space $\mathscr{S}'(R^n)$ of tempered distributions. The poles are the points*

$$\alpha = -n-2h \qquad (h\in Z^+),$$

and they are all simple.

Proof. We have for $\operatorname{Re}\alpha > -n$

$$(72)\qquad r^\alpha(\phi) = \Omega_n\int_0^\infty (M^t\phi)(0)t^{\alpha+n-1}\,dt.$$

Next we note that the mean value function $t \to (M^t\phi)(0)$ extends to an even C^∞-function on R, and its odd-order derivatives at the origin vanish. Each even-order derivative is nonzero if ϕ is suitably chosen. Since by (72)

$$(73)\qquad r^\alpha(\phi) = \Omega_n t_+^{\alpha+n-1}(M^t\phi),$$

the first statement of the lemma follows. The possible (simple) poles of r^α are by the remarks about x_+^α given by $\alpha + n - 1 = -1, -2, \ldots$. However, if $\alpha + n - 1 = -2, -4, \ldots$, formula (71) shows, since $(M^t\phi(0))^{(h)} = 0$ (h odd) that $r^\alpha(\phi)$ is holomorphic at the points

$$\alpha = -n-1, -n-3, \ldots.$$

The remark about the even derivatives of $M^t\phi$ shows, on the other hand, that the points $\alpha = -n - 2h$ ($h \in \mathbf{Z}^+$) are genuine poles. We note also from (71) and (73) that

(74) $$\operatorname*{Res}_{\alpha=-n} r^\alpha = \lim_{\alpha\to -n} (\alpha + n)\, r^\alpha = \Omega_n \delta.$$

We recall now that the Fourier transform $T \to \tilde{T}$ of a tempered distribution T on $\boldsymbol{R}^n$ is defined by

$$\tilde{T}(\phi) = T(\tilde{\phi}), \qquad \phi \in \mathscr{S}(\boldsymbol{R}^n).$$

We shall now calculate the Fourier transforms of these tempered distributions r^α.

Lemma 2.34. *We have the identities*

(75) $$(r^\alpha)^\sim = 2^{n+\alpha}\pi^{n/2}\frac{\Gamma(\frac{1}{2}(n+\alpha))}{\Gamma(-\frac{1}{2}\alpha)} r^{-\alpha-n}, \qquad \alpha,\ -\alpha - n \notin 2\mathbf{Z}^+,$$

(76) $$(r^{2h})^\sim = (2\pi)^n(-L)^h\delta, \qquad h \in \mathbf{Z}^+.$$

Proof. We use the fact that if $\psi(x) = e^{-(1/2)|x|^2}$, then

$$\tilde{\psi}(u) = (2\pi)^{(n/2)} e^{-(1/2)|u|^2},$$

so that by the formula $\int f\tilde{g} = \int \tilde{f}g$ we obtain for $\phi \in \mathscr{S}(\boldsymbol{R}^n)$, $t > 0$,

(77) $$\int \tilde{\phi}(x) e^{-(1/2)t|x|^2}\, dx = (2\pi)^{n/2}\, t^{-n/2} \int \phi(u)\, e^{-(1/2)t^{-1}|u|^2}\, du.$$

We multiply this equation by $t^{-(1/2)\alpha-1}$ and integrate with respect to t. On the left we obtain, using the formula

$$\int_0^\infty e^{-(1/2)t|x|^2}\, t^{-(1/2)\alpha-1}\, dt = \Gamma\left(-\frac{\alpha}{2}\right) 2^{-\alpha/2}\, |x|^\alpha,$$

the expression

$$\Gamma\left(-\frac{\alpha}{2}\right) 2^{-\alpha/2} \int \tilde{\phi}(x) |x|^\alpha\, dx.$$

On the right we similarly obtain

$$(2\pi)^{n/2}\, \Gamma\left(\frac{n+\alpha}{2}\right) 2^{(n+\alpha)/2} \int \phi(u) |u|^{-\alpha-n}\, du.$$

The interchange of the integrations is valid for α in the strip $-n < \operatorname{Re}\alpha < 0$, so that (75) is proved for these α. For the remaining ones it follows by analytic continuation. Finally, (76) is immediate from the definitions.

Lemma 2.35. *The action of the Laplacian is given by*

(78) $$Lr^\alpha = \alpha(\alpha + n - 2)r^{\alpha-2}, \qquad -\alpha - n + 2 \notin 2\mathbf{Z}^+,$$

(79) $$Lr^{2-n} = (2 - n)\Omega_n\delta, \qquad n \neq 2.$$

For $n = 2$ this "Poisson equation" is replaced by

(80) $$L(\log r) = 2\pi\delta.$$

Proof. For Re α sufficiently large, (78) is obvious by computation. For the remaining ones it follows by analytic continuation. For (79) we use the Fourier transform and the fact that for a tempered distribution S,

$$(-LS)\tilde{} = r^2\tilde{S}.$$

Hence, by (75),

$$(-Lr^{2-n})\tilde{} = 4\frac{\pi^{n/2}}{\Gamma((n/2) - 1)} = \frac{2\pi^{n/2}}{\Gamma(n/2)}(n - 2)\,\tilde{\delta}.$$

Finally, we prove (80). If $\phi \in \mathscr{D}(\mathbf{R}^2)$, we have, putting $F(r) = (M^r\phi)(0)$,

$$(L(\log r))(\phi) = \int_{\mathbf{R}^2} \log r\,(L\phi)(x)\,dx = \int_0^\infty \log r\,(2\pi r)(M^rL\phi)(0)\,dr.$$

Using Lemma 2.14, this becomes

$$\int_0^\infty \log r\; 2\pi r(F''(r) + r^{-1}F'(r))\,dr,$$

which by integration by parts reduces to

$$[\log r\,(2\pi r)F'(r)]_0^\infty - 2\pi\int_0^\infty F'(r)\,dr = 2\pi\,F(0).$$

[See Exercise D4 for a simpler proof.]

We shall now define fractional powers of L, motivated by the relation

$$(-Lf)\tilde{}(u) = |u|^2\tilde{f}(u),$$

so that formally we should like to have a relation

$$((-L)^pf)\tilde{}(u) = |u|^{2p}\tilde{f}(u).$$

Since the Fourier transform of a convolution is the product of the Fourier transforms, formula (75) (for $2p = -\alpha - n$) suggests defining

(81) $$(-L)^pf = I^{-2p}(f),$$

where I^γ is the *Riesz potential*

(82) $$(I^\gamma f)(x) = \frac{1}{H_n(\gamma)}\int_{\mathbf{R}^n} f(y)|x - y|^{\gamma-n}\,dy$$

with

$$H_n(\gamma) = 2^\gamma \pi^{n/2} \frac{\Gamma(\frac{1}{2}\gamma)}{\Gamma(\frac{1}{2}(n-\gamma))}. \tag{83}$$

Writing (82) as $H_n(\gamma)^{-1}(f * r^{\gamma-n})(x)$ and assuming $f \in \mathscr{S}(\boldsymbol{R}^n)$, we see that the poles of $r^{\gamma-n}$ are cancelled by the poles of $\Gamma(\frac{1}{2}\gamma)$, so that $(I^\gamma f)(x)$ extends to a holomorphic function in the set $C_n = \{\gamma \in \boldsymbol{C} : \gamma - n \notin 2\boldsymbol{Z}^+\}$. We also have by (74) and the formula for Ω_n

$$I^0 f = \lim_{\gamma \to 0} I^\gamma f = f. \tag{84}$$

Furthermore, by (78) and by analytic continuation,

$$I^\gamma L f = L I^\gamma f = I^{\gamma-2} f, \qquad f \in \mathscr{S}(\boldsymbol{R}^n), \quad \gamma \in C_n. \tag{85}$$

We now prove an important property of the Riesz potentials.

Proposition 2.36. *The following identity holds:*

$$I^\alpha(I^\beta f) = I^{\alpha+\beta} f \qquad \textit{for} \quad f \in \mathscr{S}(\boldsymbol{R}^n), \quad \operatorname{Re}\alpha, \operatorname{Re}\beta > 0, \quad \operatorname{Re}(\alpha+\beta) < n.$$

Proof. We have

$$\begin{aligned} I^\alpha(I^\beta f)(x) &= \frac{1}{H_n(\alpha)} \int |x-z|^{\alpha-n} \left(\frac{1}{H_n(\beta)} \int f(y)|z-y|^{\beta-n}\, dy \right) dz \\ &= \frac{1}{H_n(\alpha) H_n(\beta)} \int f(y) \left(\int |x-z|^{\alpha-n} |z-y|^{\beta-n}\, dz \right) dy. \end{aligned}$$

The substitution $v = (x-z)/|x-y|$ reduces the inner integral to the form

$$|x-y|^{\alpha+\beta-n} \int_{\boldsymbol{R}^n} |v|^{\alpha-n} |w-v|^{\beta-n}\, dv, \tag{86}$$

where w is the unit vector $(x-y)/|x-y|$. Using a rotation around the origin, we see that the integral in (86) equals the number

$$c_n(\alpha, \beta) = \int_{\boldsymbol{R}^n} |v|^{\alpha-n} |e_1 - v|^{\beta-n}\, dv, \tag{87}$$

where $e_1 = (1, 0, \ldots, 0)$. The assumptions made on α and β ensure that this integral converges. By the Fubini theorem the exchange of order of integrations above is permissible and

$$I^\alpha(I^\beta f) = \frac{H_n(\alpha+\beta)}{H_n(\alpha) H_n(\beta)} c_n(\alpha, \beta) I^{\alpha+\beta} f. \tag{88}$$

It remains to calculate $c_n(\alpha, \beta)$. For this we use the following lemma. As in Corollary 2.5 let $\mathscr{S}^*(\boldsymbol{R}^n)$ denote the set of functions in $\mathscr{S}(\boldsymbol{R}^n)$ which are orthogonal to all polynomials.

Lemma 2.37. *Each I^α leaves the space $\mathscr{S}^*(\boldsymbol{R}^n)$ invariant.*

Proof. By continuity it suffices to prove this for those α for which $\alpha - n$ satisfies the assumptions of Lemma 2.34, that is, $\alpha - n, -\alpha \notin 2\boldsymbol{Z}^+$. But then, if $f \in \mathscr{S}^*$,

$$(I^\alpha f)^\sim(u) = \frac{1}{H_n(\alpha)}(f * r^{\alpha-n})^\sim = \tilde{f}(u)|u|^{-\alpha}, \tag{89}$$

since $(\phi * S)^\sim = \tilde{\phi}\tilde{S}$ for $\phi \in \mathscr{S}$, $S \in \mathscr{S}'$. But $\tilde{f}$ has all derivatives 0 at 0 and so does $\tilde{f}(u)|u|^{-\alpha}$, so the lemma is proved.

We can now finish the proof of Proposition 2.36. Taking $f_0 \in \mathscr{S}^*$ we can put $f = I^\beta f_0$ in (89) and then

$$\begin{aligned}(I^\alpha(I^\beta f_0))^\sim &= (I^\beta f_0)^\sim(u)|u|^{-\alpha} = \tilde{f}_0(u)|u|^{-\alpha-\beta}\\ &= (I^{\alpha+\beta} f_0)^\sim(u).\end{aligned}$$

This shows that the scalar factor in (88) equals 1, so Proposition 2.36 is proved.

In the process we have obtained the evaluation

$$\int_{\boldsymbol{R}^n} |v|^{\alpha-n}|e_1 - v|^{\beta-n}\,dv = \frac{H_n(\alpha)H_n(\beta)}{H_n(\alpha+\beta)}.$$

We now prove a variation of Proposition 2.36 needed in the theory of the Radon transform.

Proposition 2.38. *Let $0 < k < n$. Then*

$$I^{-k}(I^k f) = f \qquad \text{for all } f \in \mathscr{S}(\boldsymbol{R}^n).$$

Proof. By Proposition 2.36 we have

$$I^\alpha(I^k f) = I^{\alpha+k} f \qquad \text{for} \quad 0 < \operatorname{Re}\alpha < n - k. \tag{90}$$

We shall now prove, following a suggestion of R. Seeley that the function $\phi = I^k f$ satisfies

$$\sup_x |\phi(x)|\,|x|^{n-k} < \infty. \tag{91}$$

For each N we have an estimate $|f(y)| \leq C_N(1 + |y|)^{-N}$, where C_N is a constant. Thus we have

$$\left| \int_{\mathbf{R}^n} f(y)|x - y|^{k-n}\, dy \right| \leq C_N \int_{|x-y| \geq |x|/2} (1 + |y|)^{-N}|x - y|^{k-n}\, dy$$

$$+ C_N \int_{|x-y| \leq |x|/2} (1 + |y|)^{-N}|x - y|^{k-n}\, dy.$$

In the first integral we have $|x - y|^{k-n} \leq |\frac{1}{2}x|^{k-n}$ and in the second we have the inequality

$$|y| \geq |x|/2.$$

Taking N large enough, both integrals on the right-hand side will satisfy (91), so (91) is proved.

We claim now that $I^\alpha(\phi)(x)$, which by (90) is holomorphic for $0 < \operatorname{Re} \alpha < n - k$, extends to a holomorphic function in the half-plane $\operatorname{Re} \alpha < n - k$. It suffices to prove this for $x = 0$. We decompose $\phi = \phi_1 + \phi_2$, where ϕ_1 is a smooth function identically 0 in a neighborhood $|x| < \varepsilon$ of 0 and $\phi_2 \in \mathscr{S}(\mathbf{R}^n)$. Since ϕ_1 satisfies (91), we have for $\operatorname{Re} \alpha < n - k$, putting $\beta = \operatorname{Re} \alpha$,

$$\left| \int \phi_1(x)|x|^{\beta-n}\, dx \right| \leq C \int_\varepsilon^\infty |x|^{k-n}|x|^{\beta-n}|x|^{n-1}\, d|x|$$

$$= C \int_\varepsilon^\infty |x|^{\beta+k-n-1}\, d|x| < \infty,$$

so $I^\alpha \phi_1$ is holomorphic in this half-plane. On the other hand, $I^\alpha \phi_2$ is holomorphic for all $\alpha \in C_n$. Now we can put $\alpha = -k$ in (90) and the proposition is proved.

Proposition 2.39. *Let S denote the Cauchy principal value*

$$S: \psi \to \lim_{\varepsilon \to 0} \int_{|x| \geq \varepsilon} \frac{\psi(x)}{x}\, dx.$$

Then S is a tempered distribution and $\tilde{S}$ is the function

$$\tilde{S}(s) = -\pi i \operatorname{sgn} s = \begin{cases} -\pi i, & s \geq 0 \\ \pi i, & s < 0. \end{cases}$$

Proof. We have $xS = 1$, so by (68),

$$2\pi\delta = \tilde{1} = (xS)^\sim = i(\tilde{S})',$$

But $(\operatorname{sgn} s)' = 2\delta$, so

$$\tilde{S} = -\pi i \operatorname{sgn} s + C \qquad (C = \text{constant}).$$

But $\tilde{S}$ and sgn s are odd, so that the result follows.

For later use we formulate now a result for the analogs of the Riesz potential corresponding to the quadratic form $Q(X) = x_1^2 - \cdots - x_n^2$ $(X = (x_1, \ldots, x_n))$ on $\boldsymbol{R}^n$. Let D denote the *retrograde cone*

$$D = \{x \in \boldsymbol{R}^n : Q(X) > 0, x_1 < 0\}$$

and for $f \in \mathscr{D}(\boldsymbol{R}^n)$ consider the integral

$$J^\gamma f = \frac{1}{K_n(\gamma)} \int_D f(x) Q(x)^{(1/2)(\gamma - n)}\, dX, \tag{92}$$

where

$$K_n(\gamma) = \pi^{(1/2)(n-2)} 2^{\gamma - 1} \Gamma\left(\frac{\gamma}{2}\right) \Gamma\left(\frac{\gamma + 2 - n}{2}\right). \tag{93}$$

The following result is proved in Riesz [1949], Chapter III.

Theorem 2.40. *The integral $J^\gamma f$ which converges for* $\operatorname{Re} \gamma \geq n$ *extends to a holomorphic function $\gamma \to J^\gamma f$ on all of $\boldsymbol{C}$. Moreover,*

$$J^0 f = f(0).$$

Remark. The result remains valid if f, together with each of its partial derivatives, depends holomorphically on γ (cf., loc. cit., Lemma IV, p. 62).

§3. A Duality in Integral Geometry. Generalized Radon Transforms and Orbital Integrals

1. A Duality for Homogeneous Spaces

The inversion formulas in Theorems 2.13, 2.15, 2.16, and 2.19 suggest the general problem of determining a function on a manifold by means of its integrals over certain submanifolds. In order to provide a natural framework for such problems, we consider the Radon transform $f \to \hat{f}$ on $\boldsymbol{R}^n$ and its dual $\phi \to \check{\phi}$ from a group-theoretic point of view, motivated by the fact that the isometry group $\boldsymbol{M}(n)$ acts transitively on both $\boldsymbol{R}^n$ and on the hyperplane space $\boldsymbol{P}^n$. Thus

$$\boldsymbol{R}^n = \boldsymbol{M}(n)/\boldsymbol{O}(n), \quad \boldsymbol{P}^n = \boldsymbol{M}(n)/\boldsymbol{Z}_2 \times \boldsymbol{M}(n-1), \tag{1}$$

where $\boldsymbol{O}(n)$ is the orthogonal group fixing the origin $0 \in \boldsymbol{R}^n$ and $\boldsymbol{Z}_2 \times \boldsymbol{M}(n-1)$ is the subgroup of $\boldsymbol{M}(n)$ leaving a certain hyperplane ξ_0 through 0 stable ($\boldsymbol{Z}_2$ consists of the identity and the reflection in this hyperplane).

We observe now that a point $g_1 \boldsymbol{O}(n)$ in the first coset space above lies on a plane $g_2(\boldsymbol{Z}_2 \times \boldsymbol{M}(n-1))$ in the second if and only if these cosets, considered as subsets of $\boldsymbol{M}(n)$, have a point in common. This leads to the following general setup.

Let G be a locally compact group, X and Ξ two left coset spaces of G,

$$X = G/H_X, \qquad \Xi = G/H_\Xi, \tag{2}$$

where H_X and H_Ξ are closed subgroups of G. It will be convenient to make the following assumptions:

(i) The groups G, H_X, H_Ξ, $H_X \cap H_\Xi$ are all unimodular (the left invariant Haar measures are right-invariant).

(ii) If $h_X H_\Xi \subset H_\Xi H_X$, then $h_X \in H_\Xi$. If $h_\Xi H_X \subset H_X H_\Xi$, then $h_\Xi \in H_X$.

(iii) The set $H_X H_\Xi \subset G$ is closed.

We note that (iii) is satisfied if one of the subgroups H_X, H_Ξ is compact.

All the assumptions are satisfied for the pair of coset spaces in (1). Let us, for example, check the first part of (ii). If $h_X H_\Xi \subset H_\Xi H_X$, we obtain by applying both sides to the origin, $h_X \cdot \xi_0 \subset \xi_0$, so that $h_X \in H_\Xi$.

Two elements $x \in X$, $\xi \in \Xi$ are said to be *incident* if as cosets in G they intersect. We put

$$\check{x} = \{\xi \in \Xi : x \text{ and } \xi \text{ incident}\},$$

$$\hat{\xi} = \{x \in X : x \text{ and } \xi \text{ incident}\}.$$

Using the notation $A^g = gAg^{-1}$ for $g \in G$, $A \subset G$, we have the following lemma.

Lemma 3.1. *Let $g \in G$, $x = gH_X$, $\xi = gH_\Xi$. Then*

(a) *$\check{x}$ is an orbit of $(H_X)^g$ and we have the coset space identification*

$$\check{x} = (H_X)^g/(H_X \cap H_\Xi)^g;$$

(b) *$\hat{\xi}$ is an orbit of $(H_\Xi)^g$ and*

$$\hat{\xi} = (H_\Xi)^g/(H_X \cap H_\Xi)^g.$$

Proof. By definition

$$\check{x} = \{\gamma H_\Xi : \gamma H_\Xi \cap gH_X \neq \emptyset\},$$

which can be written

$$\check{x} = \{gh_X H_\Xi : h_X \in H_X\}. \tag{3}$$

This is the orbit of the point gH_Ξ in Ξ under the group gH_Xg^{-1}. The subgroup leaving the point gH_Ξ fixed is $(gH_\Xi g^{-1}) \cap (gH_Xg^{-1})$. This proves (a); (b) follows in the same way.

Let $x_0 = \{H_X\}$ and $\xi_0 = \{H_\Xi\}$ denote the origins in X and Ξ, respectively. Then by (3)

$$\check{x} = g \cdot \check{x}_0, \qquad \hat{\xi} = g \cdot \hat{\xi}_0,$$

where $\cdot$ denotes the action of G on X and on Ξ.

Lemma 3.2. *The maps $x \to \check{x}$ and $\xi \to \hat{\xi}$ are one-to-one.*

Proof. Suppose $x_1, x_2 \in X$ and $\check{x}_1 = \check{x}_2$. Let $g_1, g_2 \in G$ be such that $x_1 = g_1H_X$, $x_2 = g_2H_X$. Then by (3) $g_1 \cdot \check{x}_0 = g_2 \cdot \check{x}_0$, so that writing $g = g_1^{-1}g_2$, we have $g \cdot \check{x}_0 = \check{x}_0$. In particular, $g \cdot \xi_0 \in \check{x}_0$, so that, since $\check{x}_0$ is the orbit of $\xi_0 \in \Xi$ under H_X, we have $g \cdot \xi_0 = h_X \cdot \xi_0$ for some $h_X \in H_X$, whence $h_X^{-1}g = h_\Xi \in H_\Xi$. It follows that $h_\Xi \cdot \check{x}_0 = \check{x}_0$, so that

$$h_\Xi H_X \cdot \xi_0 = H_X \cdot \xi_0,$$

that is, $h_\Xi H_X \subset H_X H_\Xi$. By assumption (ii) $h_\Xi \in H_X$, which gives $x_1 = x_2$. This proves the lemma.

In view of this lemma, X and Ξ are homogeneous spaces of the same group G in which each point in Ξ can be viewed as a subset of X and each point of X can be viewed as a subset of Ξ. We say X and Ξ are *homogeneous spaces in duality.* The terminology is suggested by the familiar duality between $\boldsymbol{R}^n$ and $\boldsymbol{P}^n$ in projective geometry.

The maps $x \to \check{x}$ and $\xi \to \hat{\xi}$ are also conveniently described by means of the following double fibration,

$$\begin{array}{ccc} & G/(H_X \cap H_\Xi) & \\ & {\scriptstyle p}\swarrow \quad \searrow{\scriptstyle \pi} & \\ X = G/H_X & & \Xi = G/H_\Xi, \end{array} \tag{4}$$

where the maps p and π are given by $p(gH_X \cap H_\Xi) = gH_X$, $\pi(gH_X \cap H_\Xi) = gH_\Xi$. Then by (3) we have

$$\check{x} = \pi(p^{-1}(x)), \qquad \hat{\xi} = p(\pi^{-1}(\xi)). \tag{5}$$

Lemma 3.3. *Each $\check{x} \subset \Xi$ is closed and each $\hat{\xi} \subset X$ is closed.*

Proof. If $p_\Xi : G \to G/H_\Xi$ is the natural mapping, we have

$$\begin{aligned}(p_\Xi)^{-1}(\Xi - (\check{x}_0)) &= \{g : gH_\Xi \notin H_X \cdot H_\Xi\} \\ &= G - H_XH_\Xi.\end{aligned}$$

In particular, $p_\Xi(G - H_X H_\Xi) = \Xi - \check{x}_0$, so that by using (ii) and the fact that p_Ξ is an open mapping, we deduce that $\check{x}_0$ is closed. By translation each $\check{x}$ is closed and similarly each $\hat{\xi}$ is closed.

Examples. (i) *Points Outside Hyperplanes.* We saw before that if in the coset space representation (1) $\boldsymbol{O}(n)$ is viewed as the isotropy group of 0 and $\boldsymbol{Z}_2\boldsymbol{M}(n-1)$ is viewed as the isotropy group of a hyperplane *through* 0, then the abstract incidence notion is equivalent to the naive one: $x \in \boldsymbol{R}^n$ is incident to $\xi \in \boldsymbol{P}^n$ if and only if $x \in \xi$.

On the other hand, we can also view $\boldsymbol{Z}_2\boldsymbol{M}(n-1)$ as the isotropy group of a hyperplane ξ_δ at a distance $\delta > 0$ from 0. [This amounts to a different imbedding of the group $\boldsymbol{Z}_2\boldsymbol{M}(n-1)$ into $\boldsymbol{M}(n)$.] Then we have the following generalization.

Proposition 3.4. *The point $x \in \boldsymbol{R}^n$ and the hyperplane $\xi \in \boldsymbol{P}^n$ are incident if and only if distance $(x, \xi) = \delta$.*

Proof. Let $x = gH_X$, $\xi = \gamma H_\Xi$, where $H_X = \boldsymbol{O}(n)$, $H_\Xi = \boldsymbol{Z}_2\boldsymbol{M}(n-1)$. Then if $gH_X \cap \gamma H_\Xi \neq \emptyset$, we have $gh_X = \gamma h_\Xi$ for some $h_X \in H_X$, $h_\Xi \in H_\Xi$. Now the orbit $H_\Xi \cdot 0$ consists of the two planes ξ'_δ and ξ''_δ parallel to ξ_δ at a distance δ from ξ_δ. The relation $g \cdot 0 = \gamma h_\Xi \cdot 0 \in \gamma \cdot (\xi'_\delta \cup \xi''_\delta)$ together with the fact that g and γ are isometries shows that x has distance δ from $\gamma \cdot \xi_\delta = \xi$.

On the other hand, if distance $(x, \xi) = \delta$, we have $g \cdot 0 \in \gamma \cdot (\xi'_\delta \cup \xi''_\delta) = \gamma H_\Xi \cdot 0$, which means $gH_X \cap \gamma H_\Xi \neq \emptyset$.

(ii) *Unit Spheres.* Let σ_0 be a sphere in $\boldsymbol{R}^n$ of radius one passing through the origin. Denoting by Σ the set of all *unit* spheres in $\boldsymbol{R}^n$, we have the dual homogeneous spaces

$$\boldsymbol{R}^n = \boldsymbol{M}(n)/\boldsymbol{O}(n), \qquad \Sigma = \boldsymbol{M}(n)/\boldsymbol{O}^*(n), \tag{6}$$

where $\boldsymbol{O}^*(n)$ is the set of rotations around the center of σ_0. Here a point $x = g\boldsymbol{O}(n)$ is incident to the sphere $\sigma = \gamma\boldsymbol{O}^*(n)$ if and only if $x \in \sigma$.

(iii) *Complex Flag Manifolds.* Consider the complex n-space $\boldsymbol{C}^n$ and a fixed set of integers $0 < d_1 < \cdots < d_r < n$. A *flag* in $\boldsymbol{C}^n$ of type $(d_1, \ldots, d_r)$ is a sequence of subspaces $L_1 \subset L_2 \subset \cdots \subset L_r \subset \boldsymbol{C}^n$, where $\dim L_i = d_i$ $(1 \leq i \leq r)$. The set $F_{d_1, \ldots, d_r}$ of all flags in $\boldsymbol{C}^n$ of type $(d_1, \ldots, d_r)$ is a complex manifold in a natural way. In fact, the general linear group $\boldsymbol{GL}(n, \boldsymbol{C})$ acts transitively on it with a complex isotropy group.

Consider, in particular, the manifolds F_{12} of flags of type (1, 2) in $\boldsymbol{C}^4$, the complex projective space $F_1 = \boldsymbol{P}_3(\boldsymbol{C})$ (of complex lines in $\boldsymbol{C}^4$), and the Grassmann manifold $F_2 = \boldsymbol{G}_{2,4}(\boldsymbol{C})$ of complex two-dimensional subspaces of $\boldsymbol{C}^4$. The group $\boldsymbol{U}(4)$ acts transitively on these manifolds F_{12}, F_1, and F_2, and our double fibration (4) becomes

$$\begin{array}{ccc} & F_{12} = \boldsymbol{U}(4)/(\boldsymbol{U}(1) \times \boldsymbol{U}(1) \times \boldsymbol{U}(2)) & \\ \swarrow & & \searrow \\ F_1 = \boldsymbol{U}(4)/(\boldsymbol{U}(1) \times \boldsymbol{U}(3)) & & F_2 = \boldsymbol{U}(4)/(\boldsymbol{U}(2) \times \boldsymbol{U}(2)) \end{array}$$

and our maps $x \to \check{x}$, $\xi \to \hat{\xi}$ mean the following:

(A) The map $x \to \check{x}$ associates to a complex line the set of two-dimensional complex subspaces $\subset \boldsymbol{C}^4$ containing it.

(B) The map $\xi \to \hat{\xi}$ associates to a two-dimensional complex subspace of $\boldsymbol{C}^4$ the set of complex lines contained in it.

These maps A, B have been utilized by Penrose; they are called *Penrose correspondences* in the paper Wells [1979], to which the reader should refer for an account of their applications.

2. The Radon Transform for the Double Fibration

In accordance with the unimodularity assumption (i), we fix invariant measures dg, dh_X, dh_Ξ, and dh on the groups

$$G,\ H_X,\ H_\Xi, \text{ and } H = H_X \cap H_\Xi,$$

respectively. By the existence theorem for invariant measures on homogeneous spaces (Theorem 1.9) there exists a unique H_X-invariant measure $d\mu = d(h_X)_H$ on $\check{x}_0 = H_X/H$ satisfying

$$\int_{H_X} f(h_X)\, dh_X = \int_{\check{x}_0} \left(\int_H f(h_X h)\, dh \right) d\mu(h_X H) \tag{7}$$

for all $f \in C_c(H_X)$. Similarly, there is defined an H_Ξ-invariant measure dm on $\hat{\xi}_0 = H_\Xi/H$. We shall now see that by translations we obtain consistently defined invariant measures on each $\check{x}$.

Lemma 3.5. *There exists a nonzero measure on each $\check{x}$, coinciding with $d\mu$ on $\check{x}_0$, such that whenever $g \cdot \check{x}_1 = \check{x}_2$, the measures correspond under g. A similar statement holds for $\hat{\xi}$.*

Proof. If $\check{x} = g \cdot \check{x}_0$, we transfer the measure $d\mu$ on $\check{x}_0$ over to $\check{x}$ by means of the homeomorphism $\xi \to g \cdot \xi$. This gives an $(H_X)^g$-invariant measure on $\check{x}$, but since such measures are only determined up to a constant factor, we must still prove independence of the choice of g. But if

$g' \cdot \check{x}_0 = g \cdot \check{x}_0$, we have $(g \cdot x_0)\check{} = (g' \cdot x_0)\check{}$, so that by Lemma 3.2, $g \cdot x_0 = g' \cdot x_0$, whence $g \in g'H_X$. Since $d\mu$ on $\check{x}_0$ is H_X-invariant, the lemma follows.

The measures on $\check{x}$ and $\hat{\xi}$ defined by Lemma 3.5 will be denoted $d\mu$ and dm, respectively. We also denote by dg_{H_X} and dg_{H_Ξ}, respectively, the G-invariant measures on X and Ξ which are normalized by the relations

(8) $$\int_G F(g)\,dg = \int_X \left(\int_{H_X} F(gh_X)\,dh_X\right) dg_{H_X},$$

(9) $$\int_G \Phi(g)\,dg = \int_\Xi \left(\int_{H_\Xi} \Phi(gh_\Xi)\,dh_\Xi\right) dg_{H_\Xi}.$$

We shall also put $dx = dg_{H_X}$, $d\xi = dg_{H_\Xi}$ for simplicity. We now define the *Radon transform* $f \to \hat{f}$ *and its dual* $\phi \to \check{\phi}$ by the formulas

(10) $$\hat{f}(\xi) = \int_{\hat{\xi}} f(x)\,dm(x), \qquad \check{\phi}(x) = \int_{\check{x}} \phi(\xi)\,d\mu(\xi)$$

for $f \in C_c(X)$, $\phi \in C_c(\Xi)$. Since $\hat{\xi}$ is closed, the restriction $f|\hat{\xi}$ belongs to $C_c(\hat{\xi})$. Thus the integrals (10) are well-defined. Formulas (10) can also be written in group-theoretic terms:

(11)
$$\hat{f}(gH_\Xi) = \int_{H_\Xi/H} f(gh_\Xi H_X)\,d(h_\Xi)_H, \qquad \check{\phi}(gH_X) = \int_{H_X/H} \phi(gh_X H_\Xi)\,d(h_X)_H.$$

Proposition 3.6. *Let* $f \in C_c(X)$, $\phi \in C_c(\Xi)$. *Then* $\hat{f}$ *and* $\check{\phi}$ *are continuous and*

$$\int_X f(x)\check{\phi}(x)\,dx = \int_\Xi \hat{f}(\xi)\phi(\xi)\,d\xi.$$

Proof. The continuity of $\hat{f}$ (and of $\check{\phi}$) is immediate from (11). Next we consider the double fibration (4), where $p(gH) = gH_X$, $\pi(gH) = gH_\Xi$. We fix a G-invariant measure dg_H on G/H such that for all $F \in C_c(G)$

(12) $$\int_G F(g)\,dg = \int_{G/H} \left(\int_H F(gh)\,dh\right) dg_H.$$

By the "chain rule" for invariant integration (cf. Proposition 1.13 for Lie groups), we have

(13) $$\int_{G/H} Q(gH)\,dg_H = c\int_{G/H_X} dg_{H_X} \int_{H_X/H} Q(gh_X H)\,d(h_X)_H,$$

where $Q \in C_c(G/H)$ is arbitrary and c is a constant independent of Q. Combining (12) and (13), we obtain

$$(14) \qquad \int_G F(g)\,dg = c \int_{G/H_X} dg_{H_X} \int_{H_X/H} d(h_X)_H \int_H F(gh_X h)\,dh,$$

which by (7) equals

$$c \int_{G/H_X} dg_{H_X} \int_{H_X} F(gh_X)\,dh_X.$$

By comparing this with (8) we obtain $c = 1$.

We consider now the function

$$P = (f \circ p)(\phi \circ \pi)$$

on G/H. We integrate it over G/H in two ways corresponding to the two fibrations in (4). This amounts to using (13) and its analog for H_Ξ. The result is

$$(15) \qquad \int_{G/H} P(gH)\,dg_H = \int_{G/H_X} dg_{H_X} \int_{H_X/H} P(gh_X H)\,d(h_X)_H,$$

$$(16) \qquad \int_{G/H} P(gH)\,dg_H = \int_{G/H} dg_{H_\Xi} \int_{H_\Xi/H} P(gh_\Xi H)\,d(h_\Xi)_H,$$

and as in Proposition 1.13 absolute convergence of one side implies that of the other. But

$$P(gh_X H) = f(gH_X)\phi(gh_X H_\Xi)$$

and

$$\int_{H_X/H} \phi(gh_X H_\Xi)\,d(h_X)_H = \check{\phi}(gH_X),$$

so that the right-hand side of (15) reduces to $\int f(x)\check{\phi}(x)\,dx$. Treating (16) similarly, we obtain the proposition. Figure 9 illustrates the equality of (15) and (16).

The result shows how to define the Radon transform and its dual for measures and, in case G is a Lie group, for distributions.

Definition. Let s be a measure on X of compact support. Its Radon transform is the functional $\hat{s}$ on $C_c(\Xi)$ defined by

$$(17) \qquad \hat{s}(\phi) = s(\check{\phi}).$$

Similarly, $\check{\sigma}$ is defined by

$$(18) \qquad \check{\sigma}(f) = \sigma(\hat{f}), \qquad f \in C_c(X),$$

if σ is a compactly supported measure on Ξ.

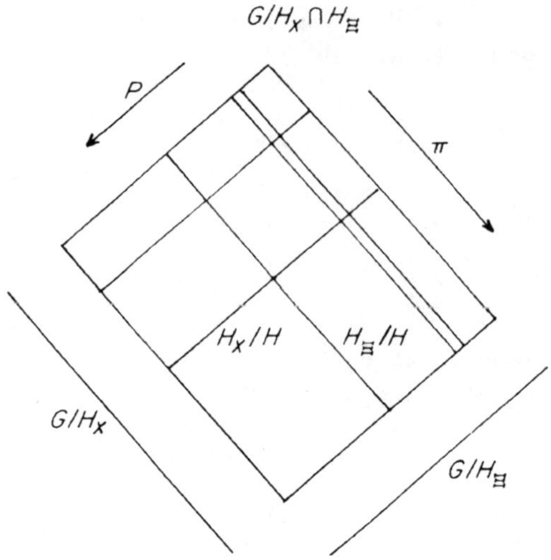

FIG. 9

Lemma 3.7. (i) *If s is a compactly supported measure on X, $\hat{s}$ is a measure on Ξ.*

(ii) *If s is a bounded measure on X and $\check{x}_0$ has finite measure, then $\hat{s}$ as defined by* (17) *is a bounded measure.*

Proof. (i) The measure s can be written as a difference $s = s^+ - s^-$ of two positive measures, each of compact support. Then $\hat{s} = \widehat{s^+} - \widehat{s^-}$ is a difference of two *positive* functionals on $C_c(\Xi)$. Since a positive functional is necessarily a measure, $\hat{s}$ is a measure.

(ii) We have

$$\sup|\check{\phi}(x)| \leq \sup|\phi(\xi)|\,\mu(\check{x}_0),$$

so for a constant K,

$$|\hat{s}(\phi)| = |s(\check{\phi})| \leq K \sup|\check{\phi}| \leq K\mu(\check{x}_0) \sup|\phi|,$$

and the boundedness of $\hat{s}$ follows.

If G is a Lie group, then (17), (18) with $f \in \mathscr{D}(X)$, $\phi \in \mathscr{D}(\Xi)$ serve to define the Radon transform $s \to \hat{s}$ and the dual $\sigma \to \check{\sigma}$ for distributions s and σ of compact support. We consider the spaces $\mathscr{D}(X)$ and $\mathscr{E}(X)$ [$= C^\infty(X)$] with their customary topologies (Chapter II, §2, No. 2). The duals $\mathscr{D}'(X)$ and $\mathscr{E}'(X)$ then consists of the distributions on X and the distributions on X of compact support, respectively.

Proposition 3.8. *The mappings*

$$f \in \mathscr{D}(X) \to \hat{f} \in \mathscr{E}(\Xi),$$

$$\phi \in \mathscr{D}(\Xi) \to \check{\phi} \in \mathscr{E}(X)$$

are continuous. In particular,

$$s \in \mathscr{E}'(X) \Rightarrow \hat{s} \in \mathscr{D}'(\Xi),$$

$$\sigma \in \mathscr{E}'(\Xi) \Rightarrow \check{\sigma} \in \mathscr{D}'(X).$$

Proof. We have

$$\hat{f}(g \cdot \xi_0) = \int_{\xi_0} f(g \cdot x)\, dm(x). \tag{19}$$

Let g run through a local cross section through e in G over a neighborhood of ξ_0 in Ξ. If $(t_1, \ldots, t_n)$ are coordinates of g and $(x_1, \ldots, x_m)$ the coordinates of $x \in \xi_0$, then (19) can be written in the form

$$\hat{F}(t_1, \ldots, t_n) = \int F(t_1, \ldots, t_n; x_1, \ldots, x_m)\, dx_1 \cdots dx_m.$$

Now it is clear that $\hat{f} \in \mathscr{E}(\Xi)$ and that $f \to \hat{f}$ is continuous, proving the proposition.

The results for the Radon transform on $\boldsymbol{R}^n$ in §2 now suggest the following general problems:

(A) *Relate function spaces on X and on Ξ by means of the integral transforms $f \to \hat{f}$, $\phi \to \check{\phi}$. In particular, relate the supports of f and of $\hat{f}$.*

(B) *Relate directly the functions f and $(\hat{f})^{\vee}$ on X, the functions ϕ and $(\check{\phi})^{\wedge}$ on Ξ.*

(C) In case G is a Lie group, let $\boldsymbol{D}(X)$ and $\boldsymbol{D}(\Xi)$ denote the algebras of G-invariant differential operators on X and Ξ, respectively.

Does there exist a map $D \to \hat{D}$ of $\boldsymbol{D}(X)$ into $\boldsymbol{D}(\Xi)$ and a map $E \to \check{E}$ of $\boldsymbol{D}(\Xi)$ into $\boldsymbol{D}(X)$ such that

$$(Df)^{\wedge} = \hat{D}\hat{f}, \qquad (E\phi)^{\vee} = \check{E}\check{\phi}?$$

Theorems 2.4, 2.6, 2.10, 2.13, 2.15, Lemma 2.1, and Corollary 2.5 give answers to these questions for $X = \boldsymbol{R}^n$, $\Xi = \boldsymbol{P}^n$. But while the problems can be posed for the general double fibration (4), one cannot expect complete solutions in this generality. For example, if $f \to \hat{f}$ is the Radon transform on $\boldsymbol{S}^2$ defined by integration over geodesics, the function $(\hat{f})^{\vee}$ will, in contrast to f, necessarily be a symmetric function. Here however, assumption (ii) is not satisfied.

In the next section we consider Problems A and B in detail for the case when X is a two-point homogeneous space, or equivalently, an isotropic Riemannian manifold.

Examples. (i) Let G denote the group $\boldsymbol{SL}(2, \boldsymbol{R})$ of 2×2 matrices of determinant one and Γ the *modular group* $\boldsymbol{SL}(2, \boldsymbol{Z})$. Let N denote the unipotent group $\left(\begin{smallmatrix} 1 & n \\ 0 & 1 \end{smallmatrix}\right)$, where $n \in \boldsymbol{R}$, and consider the homogeneous spaces

$$X = G/N, \qquad \Xi = G/\Gamma. \tag{20}$$

Under the usual action of G on $\boldsymbol{R}^2$ N is the isotropy subgroup of $(1, 0)$, so that X can be identified with $\boldsymbol{R}^2 - (0)$, whereas Ξ is of course three-dimensional.

In number theory one is interested in decomposing the space $L^2(G/\Gamma)$ into G-invariant irreducible subspaces. We now give a rough description of this in terms of the transforms $f \to \hat{f}$ and $\phi \to \check{\phi}$.

As customary we put $\Gamma_\infty = \Gamma \cap N$; our transforms (10) then take the form

$$\hat{f}(g\Gamma) = \sum_{\Gamma/\Gamma_\infty} f(g\gamma), \qquad \check{\phi}(gN) = \int_{N/\Gamma_\infty} \phi(gn\Gamma)\, dn_{\Gamma_\infty}.$$

Since N/Γ_∞ is the circle group, $\check{\phi}(gN)$ is just the constant term in the Fourier expansion of the function $n\Gamma_\infty \to \phi(gn\Gamma)$. The null space $L^2_{\mathrm{d}}(G/\Gamma)$ in $L^2(G/\Gamma)$ of the operator $\phi \to \check{\phi}$ is called the space of *cusp forms*. According to Proposition 3.6 they constitute the orthogonal complement of the image $C_c(X)^\wedge$.

We have now the G-invariant decomposition

$$L^2(G/\Gamma) = L^2_{\mathrm{c}}(G/\Gamma) \oplus L^2_{\mathrm{d}}(G/\Gamma), \tag{21}$$

where ($-$ denoting closure)

$$L^2_{\mathrm{c}}(G/\Gamma) = (C_c(X)^\wedge)^- \tag{22}$$

and as mentioned above,

$$L^2_{\mathrm{d}}(G/\Gamma) = (C_c(X)^\wedge)^\perp. \tag{23}$$

It is known (cf., Selberg [1962], Godement [1966]) that the representation of G on $L^2_{\mathrm{c}}(G/\Gamma)/\boldsymbol{C}$ is the *continuous* direct sum of the irreducible representations of G from the principal series, whereas the representation of G on $L^2_{\mathrm{d}}(G/\Gamma)$ is the *discrete* direct sum of irreducible representations each occurring with finite multiplicity.

(ii) The determination of a function in $\boldsymbol{R}^n$ in terms of its integrals over unit spheres (John [1955]) can be regarded as a solution to the first half of Problem B for the double fibration (6).

3. Orbital Integrals

As before, let $X = G/H_X$ be a homogeneous space with origin $o = \{H_X\}$. Given $x_0 \in X$ let G_{x_0} denote the subgroup of G leaving x_0 fixed, i.e., the isotropy subgroup of G at x_0.

Definition. A *generalized sphere* is an orbit $G_{x_0} \cdot x$ in X of some point $x \in X$ under the isotropy subgroup at some point $x_0 \in X$.

Examples. (i) If $X = \mathbf{R}^n$, $G = \mathbf{M}(n)$, then the generalized spheres are just the spheres.

(ii) Let X be a locally compact group L and G the product group $L \times L$ acting on L on the right and left, the element $(l_1, l_2) \in L \times L$ inducing the action $l \to l_1 l l_2^{-1}$ on L. Let ΔL denote the diagonal in $L \times L$, so that $L = (L \times L)/\Delta L$. If $l_0 \in L$, then the isotropy subgroup of l_0 is given by

$$(L \times L)_{l_0} = (l_0, e)\Delta L(l_0^{-1}, e),$$

and the orbit of l under it by

$$(L \times L)_{l_0} \cdot l = l_0(l_0^{-1} l)^L,$$

that is, the left translate by l_0 of the conjugacy class of the element $l_0^{-1} l$. Thus the *generalized spheres in the group L are the left (or right) translates of its conjugacy classes.*

Coming back to the general case $X = G/H_X = G/G_0$ we assume that G_0, and therefore each G_{x_0}, is unimodular. Writing the orbit $G_{x_0} \cdot x$ as $G_{x_0}/(G_{x_0})_x$, we see that if $(G_{x_0})_x$ is unimodular, the orbit $G_{x_0} \cdot x$ has an invariant measure determined up to a constant factor. We can then consider the following general problem (following A, B, C above).

(D) *The Orbital Integral Problem. Determine a function f on X in terms of its integrals over generalized spheres.*

Remarks. In this problem it is of course significant how the invariant measures on the various orbits are normalized.

(a) If G_0 is compact, the problem above is rather trivial because each orbit $G_{x_0} \cdot x$ has finite invariant measure, so that $f(x_0)$ is given as the limit as $x \to x_0$ of the average of f over $G_{x_0} \cdot x$.

(b) Suppose that for each $x_0 \in X$ there is a G_{x_0}-invariant open set $C_{x_0} \subset X$ containing x_0 in its closure such that for each $x \in C_{x_0}$ the isotropy group $(G_{x_0})_x$ is compact. The invariant measure on the orbits $G_{x_0} \cdot x$ $(x_0 \in X, x \in C_{x_0})$ can then be consistently normalized as follows: Fix a Haar measure dg_0 on G_0. If $x_0 = g \cdot o$, we have $G_{x_0} = gG_0g^{-1}$ and can carry dg_0 over to a measure dg_{x_0} on G_{x_0} by means of the conjugation $z \to gzg^{-1}$ ($z \in G_0$). Since dg_0 is biinvariant, dg_{x_0} is independent of the choice of g satisfying $x_0 = g \cdot o$ and is biinvariant. Since $(G_{x_0})_x$ is compact, it has a unique Haar measure $dg_{x_0,x}$ with total measure 1, and now dg_{x_0} and $dg_{x_0,x}$ determine canonically an invariant measure μ on the orbit $G_{x_0} \cdot x = G_{x_0}/(G_{x_0})_x$ (Theorem 1.9). We can therefore state Problem D in a more specific form.

(D′) *Express $f(x_0)$ in terms of the integrals*

$$\int_{G_{x_0}} f(p)\, d\mu(p), \qquad x \in C_{x_0}.$$

For the case when X is a Lorentzian manifold of constant curvature, the assumptions above are satisfied (with C_{x_0} consisting of the "timelike" rays from x_0) and an explicit solution to Problem D′ can be given (cf. Theorem 6.17 in this chapter).

(c) If in Example (ii) above L is a semisimple Lie group, Problem D is a basic step in proving the Plancherel formula for the Fourier transform on L (see notes to §6 at the end of this chapter).

§4. The Radon Transform on Two-Point Homogeneous Spaces. The X-Ray Transform

Let X be a complete Riemannian manifold, x a point in X and X_x the tangent space to X at x. Let Exp_x denote the mapping of X_x into X given by $\mathrm{Exp}_x(u) = \gamma_u(1)$, where $t \to \gamma_u(t)$ is the geodesic in X through x with tangent vector u at x.

A connected submanifold S of a Riemannian manifold X is said to be *totally geodesic* if each geodesic in X which is tangential to S at a point lies entirely in S.

The totally geodesic submanifolds of $\boldsymbol{R}^n$ are the planes in $\boldsymbol{R}^n$. Therefore, in generalizing the Radon transform to Riemannian manifolds, it is natural to consider integration over totally geodesic submanifolds. In order to have enough totally geodesic submanifolds at our disposal, we consider in this section Riemannian manifolds X which are *two-point homogeneous* in the sense that for any two-point pairs $p, q \in X$, $p', q' \in X$ satisfying $d(p, q) = d(p', q')$ (where d = distance), there exists an isometry g of X such that $g \cdot p = p'$, $g \cdot q = q'$. A Riemannian manifold X is said to be *isotropic* if for each $p \in X$ and each pair of unit tangent vectors u, v to X at p there exists an isometry of X leaving p fixed and mapping

u to v. It is well known that *a Riemannian manifold is two-point homogeneous if and only if it is isotropic* (see, e.g., [DS], pp. 535, 585). We now start with the class of Riemannian manifolds with the richest supply of totally geodesic submanifolds, namely, the spaces of constant curvature.

1. Spaces of Constant Curvature

Let X be a simply connected complete Riemannian manifold of dimension $n \geq 2$ and constant sectional curvature.

Lemma 4.1. *Let $x \in X$, V a subspace of the tangent space X_x. Then* $\operatorname{Exp}_x(V)$ *is a totally geodesic submanifold of X.*

Proof. For this we choose a specific imbedding of X into $\mathbf{R}^{n+1}$, and assume for simplicity that the curvature is ε $(= \pm 1)$. Consider the quadratic form

$$B_\varepsilon(x) = x_1^2 + \cdots + x_n^2 + \varepsilon x_{n+1}^2$$

and the quadric Q_ε given by $B_\varepsilon(x) = \varepsilon$. The orthogonal group $\mathbf{O}(B_\varepsilon)$ acts transitively on Q_ε. The form B_ε is positive definite on the tangent space $\mathbf{R}^n \times (0)$ to Q_ε at $x^0 = (0, \ldots, 0, 1)$; by the transitivity B_ε induces a positive definite quadratic form at each point of Q_ε, turning Q_ε into a Riemannian manifold, on which $\mathbf{O}(B_\varepsilon)$ acts as a transitive group of isometries.

The isotropy subgroup at the point x_0 is isomorphic to $\mathbf{O}(n)$ and it acts transitively on the set of two-dimensional subspaces of the tangent space $(Q_\varepsilon)_{x^0}$. It follows that all sectional curvatures at x^0 are the same, namely, ε, so by the homogeneity, Q_ε has constant curvature ε. In order to work with connected manifolds, we replace Q_{-1} by its intersection Q_{-1}^+ with the half-space $x_{n+1} > 0$. Then Q_{+1} and Q_{-1}^+ are simply connected complete Riemannian manifolds of constant curvature. Since such manifolds are uniquely determined by the dimension and the curvature (see, e.g., [DS] pp. 227, 250, 564), it follows that we can identify X with Q_{+1} or Q_{-1}^+.

The geodesic in X through x^0 with tangent vector $(1, 0, \ldots, 0)$ will be left pointwise fixed by the isometry

$$(x_1, x_2, \ldots, x_n, x_{n+1}) \to (x_1, -x_2, \ldots, -x_n, x_{n+1}).$$

This geodesic is therefore the intersection of X with the two-plane $x_2 = \cdots = x_n = 0$ in $\mathbf{R}^{n+1}$. By the transitivity of $\mathbf{O}(n)$ all geodesics in X through x^0 are intersections of X with two-planes through 0. By the transivity of $\mathbf{O}(Q_\varepsilon)$ it then follows that *the geodesics in X are precisely the nonempty intersections of X with two-planes through the origin.*

Now if $V \subset X_{x^0}$ is a subspace, $\mathrm{Exp}_{x^0}(V)$ is by the above the intersection of X with the subspace of $\mathbf{R}^{n+1}$ spanned by V and x^0. Thus $\mathrm{Exp}_{x^0}(V)$ is a quadric in $V + \mathbf{R}x^0$, and its Riemannian structure induced by X is the same as induced by the restriction $B_\varepsilon|(V + \mathbf{R}x^0)$. Thus, by the above, the geodesics in $\mathrm{Exp}_{x^0}(V)$ are obtained by intersecting it with two-planes in $V + \mathbf{R}x^0$ through 0. Consequently, the geodesics in $\mathrm{Exp}_{x^0}(V)$ are geodesics in X, so $\mathrm{Exp}_{x^0}(V)$ is a totally geodesic submanifold of X. By the homogeneity of X this holds with x^0 replaced by an arbitrary point $x \in X$. The lemma is proved.

In accordance with the viewpoint of §3 we consider X as a homogeneous space of the identity component G of the group $\mathbf{O}(Q_\varepsilon)$. Let H_X denote the isotropy subgroup of G at the point $x^0 = (0, \ldots, 0, 1)$. Then H_X can be identified with the special orthogonal group $\mathbf{SO}(n)$. Let k be a fixed integer, $1 \leq k \leq n-1$; let $\xi_0 \subset X$ be a fixed totally geodesic submanifold of dimension k passing through x^0 and let H_Ξ be the subgroup of G leaving ξ_0 invariant. We have then

$$(1) \qquad X = G/H_X, \qquad \Xi = G/H_\Xi,$$

Ξ denoting the set of totally geodesic k-dimensional submanifolds of X. Since $x^0 \in \xi_0$, it is clear that the abstract incidence notion boils down to the naive one, in other words: The cosets $x = gH_X$ $\xi = \gamma H_\Xi$ have a point in common if and only if $x \in \xi$.

A. *The Hyperbolic Space*

We take first the case of negative curvature, that is, $\varepsilon = -1$. The transform $f \to \hat{f}$ is now given by

$$(2) \qquad \hat{f}(\xi) = \int_\xi f(x)\, dm(x),$$

ξ being any k-dimensional totally geodesic submanifold of X $(1 \leq k)$ with the induced Riemannian structure and dm the corresponding measure. From our description of the geodesics in X it is clear that any two points in X can be joined by a unique geodesic. Let d be the distance function on X, and for simplicity we write o for the origin x^0 in X. Consider now geodesic polar coordinates for X at o; this is a mapping

$$\mathrm{Exp}_o\, Y \to (r, \theta_1, \ldots, \theta_{n-1}),$$

where Y runs through the tangent space X_0, $r = |Y|$ (the norm given by the Riemannian structure), and $(\theta_1, \ldots, \theta_{n-1})$ are coordinates of the unit vector $Y/|Y|$. Then the Riemannian structure of X is given by

$$(3) \qquad ds^2 = dr^2 + (\sinh r)^2\, d\sigma^2,$$

where $d\sigma^2$ is the Riemannian structure

$$\sum_{i,j=1}^{n-1} g_{ij}(\theta_1, \ldots, \theta_{n-1})\, d\theta_i\, d\theta_j.$$

on the unit sphere in X_o. This is easy to see by using the model (15) below for X. Then a point with distance r from the origin o is given by

$$x_i = \tanh \frac{r}{2}\, \omega_i \qquad \text{where} \quad \sum_1^n \omega_i^2 = 1$$

[cf. Introduction, Eq. (12), §4, where the curvature is -4]. Substituting this and

$$dx_i = \frac{1}{2}\left(\cosh \frac{r}{2}\right)^{-2} \omega_i\, dr - \tanh \frac{r}{2}\, d\omega_i$$

into (15) and using $\sum \omega_i\, d\omega_i = 0$, we obtain the expression (3). The surface area $A(r)$ and volume $V(r) = \int_0^r A(t)\, dt$ of a sphere in X of radius r are thus given by

$$(4) \qquad A(r) = \Omega_n (\sinh r)^{n-1}, \qquad V(r) = \Omega_n \int_0^r \sinh^{n-1} t\, dt,$$

so $V(r)$ increases like $e^{(n-1)r}$. This explains the growth condition in the next result, where $d(o, \xi)$ denotes the distance of o to the manifold ξ.

Theorem 4.2. (The Support Theorem) *Suppose $f \in C(X)$ satisfies*

(i) *For each integer $m > 0$, $f(x)e^{md(o,x)}$ is bounded.*
(ii) *There exists a number $R > 0$ such that*

$$\hat{f}(\xi) = 0 \qquad \textit{for} \quad d(o, \xi) > R.$$

Then

$$f(x) = 0 \qquad \textit{for} \quad d(o, x) > R.$$

Taking $R \to 0$, we obtain the following consequence.

Corollary 4.3. *The Radon transform $f \to \hat{f}$ is one-to-one on the space of continuous functions on X satisfying condition* (i) *of "exponential decrease."*

Proof of Theorem 4.2. Using smoothing of the form

$$\int_G \phi(g) f(g^{-1} \cdot x)\, dg$$

[$\phi \in \mathscr{D}(G)$, dg being the Haar measure on G], we can (as in Theorem 2.6) assume that $f \in \mathscr{E}(X)$.

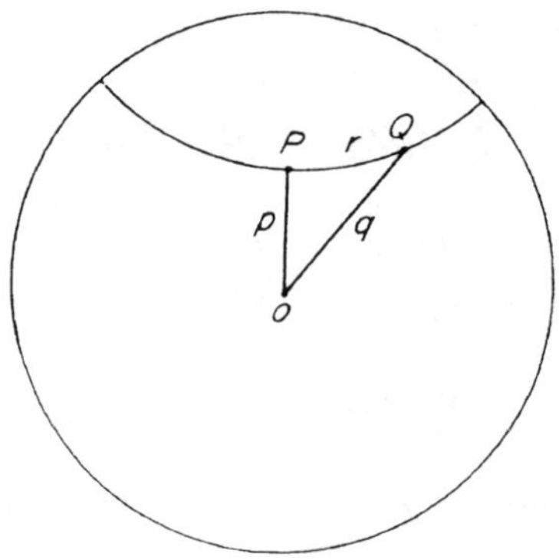

FIG. 10

We first consider the case when f in (2) is a radial function. Let P denote the point in ξ at the minimum distance $p = d(o, \xi)$ from o, let $Q \in \xi$ be arbitrary, and let

$$q = d(o, Q), \qquad r = d(P, Q).$$

Since ξ is totally geodesic, $d(P, Q)$ is also the distance between P and Q in ξ. Consider now the totally geodesic plane π through the geodesics oP and oQ as given by Lemma 4.1. Since a totally geodesic submanifold contains the geodesic joining any two of its points, π contains the geodesic PQ. The angle oPQ being 90° (see, e.g., [DS], Chapter I, Lemma 13.6), we conclude by hyperbolic trigonometry (cf. Exercise C2 and Fig. 10)

$$\cosh q = \cosh p \cosh r. \tag{5}$$

Since f is radial, it follows from (5) that the restriction $f \mid \xi$ is constant on spheres in ξ with center P. Since these have area $\Omega_k(\sinh r)^{k-1}$, formula (2) takes the form

$$\hat{f}(\xi) = \Omega_k \int_0^\infty f(Q)(\sinh r)^{k-1}\, dr. \tag{6}$$

Since f is a radial function, it is invariant under the subgroup $H_X \subset G$ which fixes o. But H_X is not only transitive on each sphere $S_r(o)$ with center o, it is for each fixed k transitive on the set of k-dimensional totally geodesic submanifolds which are tangent to $S_r(o)$. Consequently, $\hat{f}(\xi)$ depends only on the distance $d(o, \xi)$. Thus we can write

$$f(Q) = F(\cosh q), \qquad \hat{f}(\xi) = \hat{F}(\cosh p)$$

for certain 1-variable functions F and $\hat{F}$, so that by (5) we obtain

$$\hat{F}(\cosh p) = \Omega_k \int_0^\infty F(\cosh p \cosh r)(\sinh r)^{k-1}\, dr. \tag{7}$$

Writing here $t = \cosh p$, $s = \cosh r$, this reduces to

$$\hat{F}(t) = \Omega_k \int_1^{\infty} F(ts)(s^2 - 1)^{(1/2)k-1}\,ds. \tag{8}$$

Here we substitute $u = (ts)^{-1}$ and then put $v = t^{-1}$. Then (8) becomes

$$v^{-1}\hat{F}(v^{-1}) = \Omega_k \int_0^{v} \{F(u^{-1})u^{-k}\}(v^2 - u^2)^{(1/2)k-1}\,du.$$

This integral equation is of the form (19) in §2, so that we get the following analog of (20) in §2:

$$F(u^{-1})u^{-k} = cu\left(\frac{d}{d(u^2)}\right)^k \int_0^u (u^2 - v^2)^{(1/2)k-1}\hat{F}(v^{-1})\,dv. \tag{9}$$

Now by assumption (ii), $\hat{F}(\cosh p) = 0$ if $p > R$. Thus $\hat{F}(v^{-1}) = 0$ if $0 < v < (\cosh R)^{-1}$. From (9) we can then conclude that $F(u^{-1}) = 0$ if $u < (\cosh R)^{-1}$, which means $f(x) = 0$ for $d(o, x) > R$. This proves the theorem for f radial.

Next we consider an arbitrary $f \in \mathscr{E}(X)$ satisfying (i) and (ii). Fix $x \in X$, and if dk is the normalized Haar measure on H_X, consider the integral

$$F_x(y) = \int_{H_X} f(gk \cdot y)\,dk, \qquad y \in X,$$

where $g \in G$ is an element such that $g \cdot o = x$. By Theorem 1.9, F_x is the average of f on the sphere with center x passing through $g \cdot y$. The function F_x satisfies the decay condition (i) and it is radial. Moreover,

$$\hat{F}_x(\xi) = \int_{H_X} \hat{f}(gk \cdot \xi)\,dk. \tag{10}$$

We now need the following estimate:

$$d(o, gk \cdot \xi) \geq d(o, \xi) - d(o, g \cdot o). \tag{11}$$

For this let x_0 be a point on ξ closest to $k^{-1}g^{-1} \cdot o$. Then by the triangle inequality

$$\begin{aligned} d(o, gk \cdot \xi) &= d(k^{-1}g^{-1} \cdot o, \xi) \\ &\geq d(o, x_0) - d(o, k^{-1}g^{-1} \cdot o) \\ &\geq d(o, \xi) - d(o, g \cdot o). \end{aligned}$$

Thus it follows by (ii) that

$$\hat{F}_x(\xi) = 0 \qquad \text{if} \quad d(o, \xi) > d(o, x) + R.$$

Since F_x is radial, this implies by the first part of the proof that

$$\int_{H_X} f(gk \cdot y)\, dk = 0 \tag{12}$$

if

$$d(o, y) > d(o, g \cdot o) + R. \tag{13}$$

But the set $\{gk \cdot y : k \in H_X\}$ is the sphere $S_{d(o,y)}(g \cdot o)$ with center $g \cdot o$ and radius $d(o, y)$; furthermore, the inequality in (13) implies the inclusion relation

$$B_R(o) \subset B_{d(o,y)}(g \cdot o) \tag{14}$$

for the balls. But considering the part in $B_R(o)$ of the geodesic through o and $g \cdot o$, we see that conversely relation (14) implies (13). Theorem 4.2 will therefore be proved if we establish the following lemma.

Lemma 4.4. *Let $f \in C(X)$ satisfy the conditions:*

(i) *For each integer $m > 0$, $f(x)e^{md(o,x)}$ is bounded.*
(ii) *There exists a number $R > 0$ such that the surface integral*

$$\int_S f(s)\, d\omega(s) = 0$$

whenever the sphere S encloses the ball $B_R(o)$.

Then

$$f(x) = 0 \qquad \textit{for} \quad d(o, x) > R.$$

Proof. This lemma is the exact analog of Lemma 2.7, whose proof, however, used the vector space structure of $\boldsymbol{R}^n$. By using a special model of the hyperbolic space we shall nevertheless adapt the proof to the present situation. As before we may assume f is smooth, i.e., $f \in \mathscr{E}(X)$.

Consider the unit ball $\{x \in \boldsymbol{R}^n : \sum_1^n x_i^2 < 1\}$ with the Riemannian structure

$$ds^2 = \rho(x_1, \ldots, x_n)^2(dx_1^2 + \cdots + dx_n^2), \tag{15}$$

where

$$\rho(x_1, \ldots, x_n) = 2(1 - x_1^2 - \cdots - x_n^2)^{-1}.$$

We know from Exercise C1 that this Riemannian manifold has constant curvature -1, so we can use it for a model of X. This model is useful here because the spheres in X are the ordinary Euclidean spheres inside the ball. This fact is obvious for the spheres Σ with center o. For the general statement it suffices to prove that if T is the geodesic symmetry with respect to a point (which we can take on the x_1-axis), then $T(\Sigma)$ is a Euclidean sphere. The unit disk D in the x_1x_2-plane is totally geodesic in X, hence invariant under T. Now the isometries of the non-Euclidean disk D are generated by the complex conjugation

$$x_1 + ix_2 \to x_1 - ix_2$$

and fractional linear transformations, so they map Euclidean circles into Euclidean circles. In particular, $T(\Sigma \cap D) = T(\Sigma) \cap D$ is a Euclidean circle. But T commutes with the rotations around the x_1-axis. Thus $T(\Sigma)$ is invariant under such rotations and intersects D in a circle; hence it is a Euclidean sphere.

The totally geodesic hypersurfaces in X are the spherical caps perpendicular to the boundary.

After these preliminaries we pass to the proof of Lemma 4.4. Let $S = S_r(y)$ be a sphere in X enclosing $B_R(o)$, and let $B_r(y)$ denote the corresponding ball. Expressing the exterior $X - B_r(y)$ as a union of spheres in X with center y, we deduce from assumption (ii)

$$\int_{B_r(y)} f(x)\,dx = \int_X f(x)\,dx, \tag{16}$$

which is a constant for small variations in r and y. The Riemannian measure dx is given by

$$dx = \rho^n\,dx_0, \tag{17}$$

where $dx_0 = dx_1 \cdots dx_n$ is the Euclidean volume element. Let r_0 and y_0, respectively, denote the Euclidean radius and Euclidean center of $S_r(y)$. Then $S_{r_0}(y_0) = S_r(y)$, $B_{r_0}(y_0) = B_r(y)$ set-theoretically, and by (16) and (17)

$$\int_{B_{r_0}(y_0)} f(x_0)\rho(x_0)^n\,dx_0 = \text{const.} \tag{18}$$

for small variations in r_0 and y_0; thus by differentiation with respect to r_0

$$\int_{S_{r_0}(y_0)} f(s_0)\rho(s_0)^n\,d\omega_0(s_0) = 0, \tag{19}$$

where $d\omega_0$ is the Euclidean surface element. Putting

$$f^*(x) = f(x)\rho(x)^n,$$

we have by (18)

$$\int_{B_{r_0}(y_0)} f^*(x_0)\,dx_0 = \text{const.}$$

so, differentiating with respect to y_0, we get

$$\int_{B_{r_0}(o)} (\partial_i f^*)(y_0 + x_0)\,dx_0 = 0.$$

By using the divergence theorem (26) in §2 on the vector field $F(x_0) = f^*(y_0 + x_0)\,\partial_i$ defined in a neighborhood of $B_{r_0}(o)$, the last equation implies

$$\int_{S_{r_0}(o)} f^*(y_0 + s)s_i\,d\omega_0(s) = 0,$$

which in combination with (19) gives

$$\int_{S_{r_0}(y_0)} f^*(s)s_i\,d\omega_0(s) = 0. \tag{20}$$

The Euclidean and the non-Euclidean Riemannian structures on $S_{r_0}(y_0)$ differ by the factor ρ^2. It follows that $d\omega = \rho(s)^{n-1}\,d\omega_0$, so that (20) takes the form

$$\int_{S_r(y)} f(s)\rho(s)s_i\,d\omega(s) = 0. \tag{21}$$

We have thus proved that the function $x \to f(x)\rho(x)x_i$ satisfies the assumptions of the theorem. By iteration we obtain

$$\int_{S_r(y)} f(s)\rho(s)^k s_{i_1} \cdots s_{i_k}\,d\omega(s) = 0. \tag{22}$$

In particular, this holds for $y = 0$ and $r > R$. Then $\rho(s) = \text{constant}$ and (22) gives $f \equiv 0$ outside $B_R(o)$ by the Weierstrass approximation theorem. Thus Theorem 4.2 is proved.

Now let L denote the Laplace–Beltrami operator on X. (See Chapter II §2, for the definition.) Because of formula (3) for the Riemannian structure of X, L is given by

$$L = \frac{\partial^2}{\partial r^2} + (n-1)\coth r\,\frac{\partial}{\partial r} + (\sinh r)^{-2}L_S, \tag{23}$$

where L_S is the Laplace–Beltrami operator on the unit sphere in X_o.

(See also Chapter II, Proposition 5.26.) We consider also for each $r \geq 0$ the mean value operator M^r defined by

$$(M^r f)(x) = \frac{1}{A(r)} \int_{S_r(x)} f(s)\, d\omega(s).$$

As we saw before this can also be written

$$(M^r f)(g \cdot o) = \int_{H_x} f(gk \cdot y)\, dk \tag{24}$$

if $g \in G$ is arbitrary and $y \in X$ is such that $r = d(o, y)$. From Prop. 4.12 in Chapter II (which could easily be proved at the present stage) we have the commutativity

$$M^r L = L M^r \tag{25}$$

and the analog of the Darboux equation

$$L_x(F(x, y)) = L_y(F(x, y)) \tag{26}$$

for the function $F(x, y) = (M^{d(o, y)} f)(x)$.

For a fixed integer k $(1 \leq k \leq n-1)$ let Ξ denote the manifold of all k-dimensional totally geodesic submanifolds of X. If ϕ is a continuous function on Ξ, we denote by $\check{\phi}$ the point function

$$\check{\phi}(x) = \int_{x \in \xi} \phi(\xi)\, d\mu(\xi),$$

where μ is the unique measure on the (compact) space of ξ passing through x, invariant under all rotations around x and having total measure one.

Theorem 4.5. (The Inversion Formula). *For k even let Q_k denote the polynomial*

$$Q_k(z) = [z + (k-1)(n-k)][z + (k-3)(n-k+2)] \cdots [z + 1 \cdot (n-2)]$$

of degree $k/2$. The k-dimensional Radon transform on X is then inverted by the formula

$$cf = Q_k(L)(\hat{f})^{\vee}, \qquad f \in \mathscr{D}(X).$$

Here c is the constant

$$c = \frac{\Gamma(\frac{1}{2}n)}{\Gamma(\frac{1}{2}(n-k))} (-4\pi)^{(1/2)k}. \tag{27}$$

The formula holds also for f rapidly decreasing in the sense of condition (i) *in Theorem* 4.2.

Proof. Fix $\xi \in \Xi$ passing through the origin $o \in X$. If $x \in X$, fix $g \in G$ such that $g \cdot o = x$. As k runs through H_X, $gk \cdot \xi$ runs through the set of all totally geodesic submanifolds of X passing through x and

$$\check{\phi}(g \cdot o) = \int_K \phi(gk \cdot \xi)\, dk.$$

Hence

$$\begin{aligned}(\hat{f})^{\vee}(g \cdot o) &= \int_K \left(\int_\xi f(gk \cdot y)\, dm(y) \right) dk \\ &= \int_\xi (M^r f)(g \cdot o)\, dm(y),\end{aligned}$$

where $r = d(o, y)$. But since ξ is totally geodesic in X, it has also constant curvature -1 and two points in ξ have the same distance in ξ as in X. Thus we have

$$(\hat{f})^{\vee}(x) = \Omega_k \int_0^\infty (M^r f)(x)(\sinh r)^{k-1}\, dr. \tag{28}$$

We apply L to both sides and use (23). Then

$$(L(\hat{f})^{\vee})(x) = \Omega_k \int_0^\infty (\sinh r)^{k-1} L_r(M^r f(x))\, dr, \tag{29}$$

where L_r is the "radial part"

$$\frac{\partial^2}{\partial r^2} + (n-1)\coth r \frac{\partial}{\partial r}$$

of L. Putting now $F(r) = (M^r f)(x)$, we have the following result.

Lemma 4.6. *Let m be an integer, $0 < m < n = \dim X$. Then*

$$\int_0^\infty \sinh^m r\, L_r F\, dr$$

$$= (m + 1 - n)\left[m \int_0^\infty \sinh^m r\, F(r)\, dr + (m-1) \int_0^\infty \sinh^{m-2} r\, F(r)\, dr \right].$$

If $m = 1$, the term $(m-1)\int_0^\infty \sinh^{m-2} r\, F(r)\, dr$ should be replaced by $F(0)$.

This follows by repeated integration by parts.

From this lemma combined with the Darboux equation (26) in the form

$$L_x(M^r f(x)) = L_r(M^r f(x)) \tag{30}$$

we deduce

$$[L_x + m(n - m - 1)] \int_0^\infty \sinh^m r\, (M^r f)(x)\, dr$$

$$= -(n - m - 1)(m - 1) \int_0^\infty \sinh^{m-2} r\, (M^r f)(p)\, dr.$$

Applying this repeatedly to (29) we obtain Theorem 4.5.

B. *The Spheres and the Elliptic Spaces*

Now let X be the unit sphere $S^n(0) \subset \mathbf{R}^{n+1}$ and Ξ the set of k-dimensional totally geodesic submanifolds of X. Each $\xi \in \Xi$ is a k-sphere. We shall now invert the Radon transform

$$\hat{f}(\xi) = \int_\xi f(x)\, dm(x), \qquad f \in \mathscr{E}(X),$$

where dm is the measure on ξ given by the Riemannian structure induced by that of X. In contrast to the hyperbolic space, each geodesic in X through a point x also passes through the antipodal point Ax. As a result, $\hat{f} = (f \circ A)^\wedge$ and our inversion formula will reflect this fact. Although we state our result for the sphere, it is really a result for the *elliptic space*, that is, the sphere with antipodal points identified. The functions on this space are naturally identified with symmetric functions on the sphere.

Again let

$$\check{\phi}(x) = \int_{x \in \xi} \phi(\xi)\, d\mu(\xi)$$

denote the average of a continuous function on Ξ over the set of ξ passing through x.

Theorem 4.7. *Let k be an integer, $1 \leq k < n = \dim X$.*

(i) *The mapping $f \to \hat{f}$ ($f \in \mathscr{E}(X)$) has a kernel consisting of the skew functions (the functions f satisfying $f + f \circ A = 0$).*

(ii) *Assume k even and let P_k denote the polynomial*

$$P_k(z) = [z - (k-1)(n-k)][z - (k-3)(n-k+2)] \cdots [z - 1(n-2)]$$

of degree $k/2$. The k-dimensional Radon transform on X is then inverted by the formula

$$c(f + f \circ A) = P_k(L)((\hat{f})^\vee), \qquad f \in \mathscr{E}(X),$$

where c is the constant in (27).

Proof. We first prove (ii) in a similar way to that in the noncompact case. The Riemannian structure in (3) is now replaced by

$$ds^2 = dr^2 + \sin^2 r\, d\sigma^2,$$

the Laplace–Beltrami operator is now given by

$$(31)\qquad L = \frac{\partial^2}{\partial r^2} + (n-1)\cot r \frac{\partial}{\partial r} + (\sin r)^{-2} L_S$$

instead of (23), and

$$(\hat{f})^{\vee}(x) = \Omega_k \int_0^{\pi} (M^r f)(x) \sin^{k-1} r\, dr.$$

For a fixed x we put $F(r) = (M^r f)(x)$. The analog of Lemma 4.6 now reads as follows.

Lemma 4.8. *Let m be an integer, $0 < m < n = \dim X$. Then*

$$\int_0^{\pi} \sin^m r\, L_r F\, dr$$
$$= (n-m-1)\left[m \int_0^{\pi} \sin^m r\, F(r)\, dr - (m-1) \int_0^{\pi} \sin^{m-2} r\, F(r)\, dr\right].$$

If $m = 1$, the term $(m-1)\int_0^{\pi} \sin^{m-2} r\, F(r)\, dr$ should be replaced by $F(0) + F(\pi)$.

Since (30) is still valid, the lemma implies

$$[L_x - m(n-m-1)] \int_0^{\pi} \sin^m r\, (M^r f)(x)\, dr$$
$$= -(n-m-1)(m-1) \int_0^{\pi} \sin^{m-2} r\, (M^r f)(x)\, dr$$

and the desired inversion formula follows by iteration, since $F(0) + F(\pi) = f(x) + f(Ax)$.

In the case when k is even, part (i) follows from (ii). Next suppose $k = n-1$, n even. For each ξ there are exactly two points x and Ax at maximum distance, namely, $\pi/2$, from ξ, and we write

$$\hat{f}(x) = \hat{f}(Ax) = \hat{f}(\xi).$$

We have then

$$(32)\qquad \hat{f}(x) = \Omega_n (M^{\pi/2} f)(x).$$

Next we recall the decomposition

$$(33)\qquad L^2(X) = \sum_0^{\infty} E_s$$

from §3 of the Introduction. Here the space E_s consists of the restrictions to X of the homogeneous harmonic polynomials on $\mathbf{R}^{n+1}$ of degree s. The space is the eigenspace of L for the eigenvalue $-s(s+n-1)$ $(s \geq 0)$. As proved there, E_s contains a unique function ϕ_s which is invariant under the group K of rotations around the vertical axis (the x_{n+1}-axis in $\boldsymbol{R}^{n+1}$), and satisifies $\phi_s(o) = 1$, o being the north pole.

Since the mean value operator $M^{\pi/2}$ is invariant under $\boldsymbol{O}(n+1)$, which in turn acts irreducibly on E_s, we see from Schur's lemma that $M^{\pi/2}$ acts as a scalar c_s on E_s. The equations

$$M^{\pi/2}\phi_s = c_s\phi_s, \qquad \phi_s(o) = 1$$

then imply

$$c_s = \phi_s(\pi/2). \tag{34}$$

Lemma 4.9. *The scalar $\phi_s(\pi/2)$ is zero if and only if s is odd.*

Proof. Let H_s be the K-invariant homogeneous harmonic polynomial whose restriction to X equals ϕ_s. Then H_s is a polynomial in

$$(x_1^2 + \cdots + x_n^2) \quad \text{and} \quad x_{n+1}.$$

Thus we see that if the degree s is odd, x_{n+1} occurs in each term; hence $\phi_s(\pi/2) = H_s(1, 0, \ldots, 0, 0) = 0$. If s is even, say $s = 2d$, we write

$$H_s = a_0(x_1^2 + \cdots + x_n^2)^d + a_1x_{n+1}^2(x_1^2 + \cdots + x_n^2)^{d-1} + \cdots + a_dx_{n+1}^{2d}.$$

By using $L_{n+1} = L_n + \partial^2/\partial x_{n+1}^2$ and formula (31) in §2, the equation $L_{n+1}H_s \equiv 0$ gives the recursion formula

$$a_i(2d - 2i)(2d - 2i + n - 2) + a_{i+1}(2i + 2)(2i + 1) = 0$$

$(0 \leq i < d)$. Hence $H_s(1, 0, \ldots, 0)$, which equals a_0, is $\neq 0$. This proves the lemma.

Now by Theorem 3.3 in the Introduction, each $f \in \mathscr{E}(X)$ has a uniformly convergent expansion,

$$f = \sum_0^\infty h_s \qquad (h_s \in E_s),$$

and by (32)

$$\hat{f} = \Omega_n \sum_0^\infty c_s h_s.$$

If $\hat{f} \equiv 0$, then by Lemma 4.9, $h_s = 0$ for s even, so f is skew. Conversely, $\hat{f} = 0$ if f is skew; thus Theorem 4.7 is proved for the case $k = n - 1$, n even.

If k is odd, $0 < k < n - 1$, the proof just carried out shows that $\hat{f}(\xi) = 0$ for all $\xi \in \Xi$ implies that f has integral 0 over every $(k+1)$-dimensional sphere with radius 1 and center 0. Since $k + 1$ is even and $<n$, we conclude by (ii) that $f + f \circ A \equiv 0$, so that the theorem is proved.

2. Compact Two-Point Homogeneous Spaces

We shall now extend the inversion formula in Theorem 4.7 to compact two-point homogeneous spaces X of dimension $n > 1$. By virtue of Wang's classification [1952] these are also the compact symmetric spaces of rank one, so their geometry can be described very explicitly. Here we shall use some geometric and group-theoretic properties of these spaces [(i)–(vii) below] and refer to [DS], Chapter VII, §10, for their proofs.

Let U denote the group $I(X)$ of isometries of X. Fix an origin $o \in X$ and let K denote the isotropy subgroup U_o. Let $\mathfrak{k}$ and $\mathfrak{u}$ be the Lie algebras of K and U, respectively. Then $\mathfrak{u}$ is semisimple. Let $\mathfrak{p}$ be the orthogonal complement of $\mathfrak{k}$ in $\mathfrak{u}$ with respect to the Killing form B of $\mathfrak{u}$. Changing the distance function on X by a constant factor, we may assume that the differential of the mapping $u \to u \cdot 0$ of U onto X gives an isometry of $\mathfrak{p}$ (with the metric of $-B$) onto the tangent space X_o. This is the canonical metric on X which we shall use.

Let L denote the diameter of X, that is, the maximal distance between any two points. If $x \in X$, let A_x denote the set of points in X of distance L from x. By the two-point homogeneity the isotropy subgroup U_x acts transitively on A_x; thus $A_x \subset X$ is a submanifold, *the antipodal manifold* associated to x.

(i) *Each A_x is a totally geodesic submanifold of X; with the Riemannian structure induced by that of X it is another two-point homogeneous space.*

(ii) *Let Ξ denote the set of all antipodal manifolds in X; since U acts transitively on Ξ, the set Ξ has a natural manifold structure. Then the mapping $j: x \to A_x$ is a one-to-one diffeomorphism; also, $x \in A_y$ if and only if $y \in A_x$.*

(iii) *Each geodesic in X has period $2L$. If $x \in X$, the mapping Exp_x: $X_x \to X$ gives a diffeomorphism of the ball $B_L(0)$ onto the open set $X - A_x$.*

Fix a vector $H \in \mathfrak{p}$ of length L [i.e., $L^2 = -B(H, H)$]. For $Z \in \mathfrak{p}$ let T_Z denote the linear transformation $Y \to [Z, [Z, Y]]$ of $\mathfrak{p}$, $[\ ,\ ]$ denoting the Lie bracket in $\mathfrak{u}$. For simplicity we now write Exp instead of Exp_o. A point $Y \in \mathfrak{p}$ is said to be *conjugate* to o if the differential $d\,\mathrm{Exp}$ is singular at Y.

The line $\mathfrak{a} = \mathbf{R}H$ is a maximal abelian subspace of $\mathfrak{p}$. The eigenvalues of T_H are 0, $\alpha(H)^2$, and possibly $(\frac{1}{2}\alpha(H))^2$, where $\pm\alpha$ (and possibly $\pm\frac{1}{2}\alpha$) are the roots of $\mathfrak{u}$ with respect to $\mathfrak{a}$. Let

$$\mathfrak{p} = \mathfrak{a} + \mathfrak{p}_\alpha + \mathfrak{p}_{(1/2)\alpha} \tag{35}$$

be the corresponding decomposition of $\mathfrak{p}$ into eigenspaces; the dimensions $q = \dim(\mathfrak{p}_\alpha)$, $p = \dim(\mathfrak{p}_{(1/2)\alpha})$ are called the *multiplicities* of α and $\frac{1}{2}\alpha$, respectively.

(iv) *Suppose H is conjugate to o. Then* $\operatorname{Exp}(\mathfrak{a} + \mathfrak{p}_\alpha)$, *with the Riemannian structure induced by that of X, is a sphere, totally geodesic in X, having o and* $\operatorname{Exp} H$ *as antipodal points, and having curvature π^2/L^2. Moreover,*

$$A_{\operatorname{Exp} H} = \operatorname{Exp}(\mathfrak{p}_{(1/2)\alpha}).$$

(v) *If H is not conjugate to o, then* $\mathfrak{p}_{(1/2)\alpha} = 0$ and

$$A_{\operatorname{Exp} H} = \operatorname{Exp} \mathfrak{p}_\alpha.$$

(vi) *The differential at Y of* Exp *is given by*

$$d \operatorname{Exp}_Y = d\tau(\exp Y) \circ \sum_0^\infty \frac{(T_Y)^k}{(2k+1)!},$$

where for $u \in U$, $\tau(u)$ is the isometry $x \to u \cdot x$.

(vii) *By analogy with* (23) *the Laplace–Beltrami operator L on X has the expression*

$$L = \frac{\partial^2}{\partial r^2} + \frac{1}{A(r)} A'(r) \frac{\partial}{\partial r} + L_{S_r},$$

where L_{S_r} is the Laplace–Beltrami operator on $S_r(o)$ and $A(r)$ its area (Chapter II, Proposition 5.26).

Lemma 4.10. *The surface area $A(r)$ $(0 < r < L)$ is given by*

$$A(r) = \Omega_n \lambda^{-p} (2\lambda)^{-q} \sin^p(\lambda r) \sin^q(2\lambda r),$$

where p and q are the multiplicities above and $\lambda = |\alpha(H)|/2L$.

Proof. Because of (iii) and (vi), the surface area of $S_r(o)$ is given by

$$A(r) = \int_{|Y|=r} \det\left(\sum_0^\infty \frac{T_Y^k}{(2k+1)!}\right) d\omega_r(Y),$$

where $d\omega_r$ is the surface element on the sphere $|Y| = r$ in $\mathfrak{p}$. Because of the two-point homogeneity, the integrand depends on r only, so that we can evaluate it for $Y = H_r = (r/L)H$. Since the nonzero eigenvalues of T_{H_r} are $\alpha(H_r)^2$ with multiplicity q and $(\frac{1}{2}\alpha(H_r))^2$ with multiplicity p; a trivial computation gives the lemma.

We consider now Problems A, B, and C from §3, No. 2, for the homogeneous spaces X and Ξ, which are acted on transitively by the same group U. Fix an element $\xi_o \in \Xi$ passing through the origin $o \in X$. If $\xi_o =$

$A_{o'}$ then an element $u \in U$ leaves ξ_o invariant if and only if it lies in the isotropy subgroup $K' = U_{o'}$; we have the identifications

$$X = U/K, \qquad \Xi = U/K',$$

and $x \in X$ and $\xi \in \Xi$ are incident if and only if $x \in \xi$.

On Ξ we now choose a Riemannian structure such that the diffeomorphism $j: x \to A_x$ from (ii) is an isometry. Let L and Λ denote the Laplacians on X and Ξ, respectively. With $\check{x}$ and $\hat{\xi}$ defined as in §3, No. 1 we have

$$\hat{\xi} = \xi, \qquad \check{x} = \{j(y) : y \in j(x)\};$$

the first relation amounts to the incidence description above, and the second is a consequence of the property $x \in A_y \Leftrightarrow y \in A_x$ listed under (ii).

The sets $\check{x}$ and $\hat{\xi}$ will be given the measures $d\mu$ and dm, respectively, induced by the Riemannian structures of Ξ and X. The Radon transform and its dual are then given by

$$\hat{f}(\xi) = \int_\xi f(x)\, dm(x), \qquad \check{\phi}(x) = \int_{\check{x}} \phi(\xi)\, d\mu(\xi).$$

But

$$\check{\phi}(x) = \int_{\check{x}} \phi(\xi)\, d\mu(\xi) = \int_{y \in j(x)} \phi(j(y))\, d\mu(j(y)) = \int_{j(x)} (\phi \circ j)(y)\, dm(y),$$

so that

$$\check{\phi} = (\phi \circ j)\hat{} \circ j. \tag{36}$$

Because of this correspondence between the transforms $f \to \hat{f}$, $\phi \to \check{\phi}$, it suffices to consider the first one. Let $\boldsymbol{D}(X)$ denote the algebra of differential operators on X invariant under U. It can be shown (cf. Chapter II, §4) that $\boldsymbol{D}(X)$ is generated by L. Similarly, $\boldsymbol{D}(\Xi)$ is generated by Λ.

Theorem 4.11. (i) *The mapping $f \to \hat{f}$ is a linear one-to-one mapping of $\mathscr{E}(X)$ onto $\mathscr{E}(\Xi)$ and*

$$(Lf)\hat{} = \Lambda \hat{f}.$$

(ii) *Except for the case when X is an even-dimensional elliptic space,*

$$f = P(L)((\hat{f})\check{}\,), \qquad f \in \mathscr{E}(X),$$

where P is a polynomial, independent of f, explicitly given below, (44)–(50). In all cases

$$\text{degree}(P) = \tfrac{1}{2}\ \textit{dimension of the antipodal manifold.}$$

Proof. We first prove (ii). Let dk be the Haar measure on K such that $\int dk = 1$ and let Ω_X denote the total measure of an antipodal manifold in X. Then $u(\check{o}) = m(A_o) = \Omega_X$ and if $u \in U$,

$$\check{\phi}(u \cdot o) = \Omega_X \int_K \phi(uk \cdot \xi_o)\, dk.$$

Hence

$$(\hat{f})\check{}(u \cdot o) = \Omega_X \int_K \left(\int_{\xi_o} f(uk \cdot y)\, dm(y) \right) dk$$
$$= \Omega_X \int_{\xi_o} (M^r f)(u \cdot o)\, dm(y),$$

where r is the distance $d(o, y)$ in the space X between o and y. If $d(o, y) < L$, there is a unique geodesic in X of length $d(o, y)$ joining o to y, and since ξ_0 is totally geodesic, $d(o, y)$ is also the distance in ξ_0 between o and y. Thus using geodesic polar coordinates in ξ_o in the last integral, we obtain

$$(\hat{f})\check{}(x) = \Omega_X \int_0^L (M^r f)(x) A_1(r)\, dr, \tag{37}$$

where $A_1(r)$ is the area of a sphere of radius r in ξ_o. By Lemma 4.10 we have

$$A_1(r) = C_1 \sin^{p_1}(\lambda_1 r) \sin^{q_1}(2\lambda_1 r), \tag{38}$$

where C_1 and λ_1 are constants and p_1, q_1 are the multiplicities for the antipodal manifold. In order to prove (ii) on the basis of (37), we need the following complete list of the compact symmetric spaces of rank one and their corresponding antipodal manifolds:

X		A_o
Spheres	S^n $(n = 1, 2, \ldots)$	Point
Real projective spaces	$P^n(R)$ $(n = 2, 3, \ldots)$	$P^{n-1}(R)$
Complex projective spaces	$P^n(C)$ $(n = 4, 6, \ldots)$	$P^{n-2}(C)$
Quaternion projective spaces	$P^n(H)$ $(n = 8, 12, \ldots)$	$P^{n-4}(H)$
Cayley plane	$P^{16}(Cay)$	S^8

Here the superscripts denote real dimension. For the lowest dimension, note that

$$P^1(R) = S^1, \qquad P^2(C) = S^2, \qquad P^4(H) = S^4.$$

For the case S^n, (ii) is trivial, and the case $X = \boldsymbol{P}^n(\boldsymbol{R})$ was already done in Theorem 4.7. For the remaining cases, α and $\frac{1}{2}\alpha$ are both roots, so by (v) H is conjugate to o; hence we have the properties in (iv). Since a closed geodesic in A_o is a closed geodesic in X, we have

$$\begin{aligned} L &= \text{diameter } X = \text{diameter } A_x \\ &= \text{distance from } o \text{ to the nearest conjugate point in } X_o \\ &= \text{smallest number } M \text{ such that } \lim_{r \to M} A(r) = 0. \end{aligned}$$

The multiplicities p and q for the rank-one symmetric spaces were determined by Cartan [1927] (see also Araki [1962], or [DS], p. 532) and we can now derive the following list:

$X = \boldsymbol{P}^n(\boldsymbol{C})$:

$$p = n - 2, q = 1, \lambda = \pi/2L,$$

$$A(r) = \tfrac{1}{2}\Omega_n \lambda^{-n+1} \sin^{(n-2)}(\lambda r) \sin(2\lambda r),$$

$$A_1(r) = \tfrac{1}{2}\Omega_{n-2} \lambda^{-n+3} \sin^{(n-4)}(\lambda r) \sin(2\lambda r).$$

$X = \boldsymbol{P}^n(\boldsymbol{H})$:

$$p = n - 4, q = 3, \lambda = \pi/2L,$$

$$A(r) = \tfrac{1}{8}\Omega_n \lambda^{-n+1} \sin^{(n-4)}(\lambda r) \sin^3(2\lambda r),$$

$$A_1(r) = \tfrac{1}{8}\Omega_{n-4} \lambda^{-n+5} \sin^{n-8}(\lambda r) \sin^3(2\lambda r).$$

$X = \boldsymbol{P}^{16}(\boldsymbol{Cay})$:

$$p = 8, q = 7, \lambda = \pi/2L,$$

$$A(r) = (1/2^7)\Omega_{16} \lambda^{-15} \sin^8(\lambda r) \sin^7(2\lambda r),$$

$$A_1(r) = \Omega_8 \sin^7(2\lambda r).$$

In all cases we have

$$\Omega_X = m(A_0) = \int_0^L A_1(r)\, dr. \tag{39}$$

Thus $A(r)$ and $A_1(r)$ are in all cases expressed in terms of n and L. But L can also be expressed in terms of n. In fact, the Killing form metric has the property

$$B(H, H) = \sum_\beta \beta(H)^2$$

as β runs over the roots. Thus

$$-L^2 = -|H|^2 = 2p(\tfrac{1}{2}\alpha(H))^2 + 2q(\alpha(H))^2.$$

But by (iii) H is the first conjugate point in the ray $\mathbf{R}^+H$, so that $|\alpha(H)| = \pi$ (see [DS], p. 294). Thus we get the formula

$$L^2 = p(\pi^2/2) + 2q\pi^2. \tag{40}$$

Lemma 4.12. *In the Killing form metric the diameters L of the projective spaces*

$$\mathbf{P}^n(\mathbf{C}), \qquad \mathbf{P}^n(\mathbf{H}), \quad \mathbf{P}^{16}(\mathbf{Cay})$$

are given by, respectively,

$$\left(\frac{n}{2}+1\right)^{1/2}\pi, \qquad \left(\frac{n}{2}+4\right)^{1/2}\pi, \qquad 3\sqrt{2}\pi.$$

As already used for spaces of constant curvature we have here the commutativity [cf. (25)]

$$M^rL = LM^r \tag{41}$$

which implies [cf. (26)]

$$L_x((M^rf)(x)) = L_r((M^rf)(x)), \tag{42}$$

where by (vii) and Lemma 4.10

$$L_r = \frac{\partial^2}{\partial r^2} + \lambda\{p\cot(\lambda r) + 2q\cot(2\lambda r)\}\frac{\partial}{\partial r} \qquad (0 < r < L).$$

We now apply the Laplacian to (37) and use (42) in order to simplify the right-hand side. The result is reduced by repeated use of the following three lemmas.

Lemma 4.13. *Let* $X = \mathbf{P}^n(\mathbf{C})$, $f \in C^\infty(X)$. *If m is an even integer,* $0 \le m \le n-4$, *then*

$$(L_x - \lambda^2(n-m-2)(m+2))\int_0^L \sin^m(\lambda r)\sin(2\lambda r)\,[M^rf](x)\,dr$$

$$= -\lambda^2(n-m-2)m\int_0^L \sin^{m-2}(\lambda r)\sin(2\lambda r)\,[M^rf](x)\,dr.$$

For $m = 0$ *the right-hand side should be replaced by*

$$-2\lambda(n-2)f(x).$$

Lemma 4.14. *Let* $X = \boldsymbol{P}^n(\boldsymbol{H})$, $f \in C^\infty(X)$. *Let* m *be an even integer,* $0 < m \leq n - 8$. *Then*

$$(L_x - \lambda^2(n - m - 4)(m + 6)) \int_0^L \sin^m(\lambda r) \sin^3(2\lambda r) \, [M^r f](x) \, dr$$

$$= -\lambda^2(n - m - 4)(m + 2) \int_0^L \sin^{m-2}(\lambda r) \sin^3(2\lambda r)[M^r f](x) \, dr.$$

Also,

$$(L_x - 6\lambda^2(n - 4))(L_x - 4\lambda^2(n - 2)) \int_0^L \sin^3(2\lambda r) \, [M^r f](x) \, dr$$

$$= 16\lambda^3(n - 2)(n - 4) f(x).$$

Lemma 4.15. *Let* $X = \boldsymbol{P}^{16}(\boldsymbol{Cay})$, $f \in C^\infty(X)$. *Let* $m > 1$ *be an integer. Then*

$$(L_x - 4\lambda^2 m(11 - m)) \int_0^L \sin^m(2\lambda r) \, [M^r f](x) \, dr$$

$$= -32\lambda^2(m - 1) \int_0^L \sin^{m-2}(2\lambda r) \cos^2(\lambda r) \, [M^r f](x)$$

$$+ 4\lambda^2(m - 1)(m - 7) \int_0^L \sin^{m-2}(2\lambda r) \, [M^r f](x) \, dr;$$

$$(L_x - 4\lambda^2(m + 1)(10 - m)) \int_0^L \sin^m(2\lambda r) \cos^2(\lambda r) \, [M^r f](x) \, dr$$

$$= 4\lambda^2(3m - 5) \int_0^L \sin^m(2\lambda r) \, [M^r f](x) \, dr$$

$$+ 4\lambda^2(m - 1)(m - 15) \int_0^L \sin^{m-2}(2\lambda r) \cos^2(\lambda r) \, [M^r f](x) \, dr.$$

Also,

$$(L_x - 72\lambda^2) \int_0^L \sin(2\lambda r) \cos^2(\lambda r) \, (M^r f)(x) \, dr \tag{43}$$

$$= -8\lambda^2 \int_0^L \sin(2\lambda r) \, (M^r f)(r) - 28\lambda f(x).$$

These lemmas are proved by means of long computations. Since the methods are similar for all cases, let us just verify the last formula (43). Here we have

$$L_r = \frac{\partial^2}{\partial r^2} + \lambda\{8 \cot(\lambda r) + 14 \cot(2\lambda r)\} \frac{\partial}{\partial r}.$$

Hence putting $F(r) = (M^r f)(x)$, we have by (42)

$$L_x \int_0^L \sin(2\lambda r) \cos^2(\lambda r) (M^r f)(x) \, dr$$

$$= \int_0^L [\sin(2\lambda r) \cos^2(\lambda r) F''(r) + (44\lambda \cos^4(\lambda r) - 14\lambda \cos^2(\lambda r))F'(r)] \, dr$$

$$= \int_0^L [36\lambda \cos^4(\lambda r) - 8\lambda \cos^2(\lambda r)]F'(r) \, dr$$

$$= -28\lambda F(0) - \lambda^2 \int_0^L F(r)[\sin(2\lambda r) (8 - 72 \cos^2(\lambda r))] \, dr,$$

which gives formula (43).

We can now prove (ii) in Theorem 4.11. Consider first the case $X = \boldsymbol{P}^{16}(\boldsymbol{Cay})$. We have

$$(\hat{f})^{\vee}(x) = \Omega_8 \Omega_X \int_0^L (M^r f)(x) \sin^7(2\lambda r) \, dr.$$

Here

$$L^2 = 18\pi^2, \qquad \Omega_X = \Omega_8 \int_0^L \sin^7(2\lambda r) \, dr, \qquad \lambda = \pi/2L.$$

Taking $m = 7$ in Lemma 4.15, we get

$$(L_x - 112\lambda^2) \int_0^L (M^r f)(x) \sin^7(2\lambda r) \, dr$$

$$= -192\lambda^2 \int_0^L (M^r f)(x) \sin^5(2\lambda r) \cos^2(\lambda r) \, dr,$$

and then taking $m = 5$ we get

$$(L_x - 120\lambda^2)(L_x - 112\lambda^2) \int_0^L (M^r f)(x) \sin^7(2\lambda r) \, dr$$

$$= (-192\lambda^2)(40\lambda^2) \left[\int_0^L (M^r f)(x) \sin^5(2\lambda r) \, dr \right.$$

$$\left. -4 \int_0^L (M^r f)(x) \sin^3(2\lambda r) \cos^2(\lambda r) \, dr \right].$$

Taking $m = 5$ again, we get

$$(L_x - 120\lambda^2)^2(L_x - 112\lambda^2)\int_0^L (M^r f)(x)\sin^7(2\lambda r)\,dr$$

$$= (-192\lambda^2)(40\lambda^2)\Big[(-128\lambda^2)\int_0^L \sin^3(2\lambda r)\cos^2(\lambda r)\,(M^r f)(x)\,dr$$

$$-32\lambda^2\int_0^L \sin^3(2\lambda r)\,(M^r f)(x)\,dr$$

$$-4(L_x - 120\lambda^2)\int_0^L (M^r f)(x)\sin^3(2\lambda r)\cos^2(\lambda r)\,dr\Big].$$

Taking $m = 3$, the last term is found to be

$$-4(L_x - 112\lambda^2 - 8\lambda^2)\int_0^L (M^r f)(x)\sin^3(2\lambda r)\cos^2(\lambda r)\,dr$$

$$= -64\lambda^2\int_0^L (M^r f)(x)\sin^3(2\lambda r)\,dr$$

$$+ 384\lambda^2\int_0^L (M^r f)(x)\sin(2\lambda r)\cos^2(\lambda r)\,dr$$

$$+ 32\lambda^2\int_0^L (M^r f)(x)\sin^3(2\lambda r)\cos^2(\lambda r)\,dr.$$

Hence

$$(L_x - 120\lambda^2)^2(L_x - 112\lambda^2)\int_0^L (M^r f)(x)\sin^7(2\lambda r)\,dr$$

$$= 192\cdot 40\cdot 96\cdot\lambda^6\Big[\int_0^L (M^r f)(x)\sin^3(2\lambda r)\cos^2(\lambda r)\,dr$$

$$+\int_0^L (M^r f)(x)\sin^3(2\lambda r)\,dr$$

$$-4\int_0^L (M^r f)(x)\sin(2\lambda r)\cos^2(\lambda r)\,dr\Big].$$

Finally, we apply the operator $(L_x - 112\lambda^2)$ to both sides. Taking $m=3$ in Lemma 4.15, we get

$$(L_x - 112\lambda^2)\int_0^L (M^r f)(x)\sin^3(2\lambda r)\cos^2(\lambda r)\,dr$$

$$= 16\lambda^2\int_0^L (M^r f)(x)\sin^3(2\lambda r)\,dr$$

$$-96\lambda^2\int_0^L (M^r f)(x)\sin(2\lambda r)\cos^2(\lambda r)\,dr,$$

$$(L_x - 112\lambda^2)\int_0^L (M^r f)(x)\sin^3(2\lambda r)\,dr$$

$$= -16\lambda^2\int_0^L (M^r f)(x)\sin^3(2\lambda r)\,dr$$

$$-64\lambda^2\int_0^L (M^r f)(x)\sin(2\lambda r)\cos^2(\lambda r)\,dr$$

$$-32\lambda^2\int_0^L (M^r f)(x)\sin(2\lambda r)\,dr$$

$$-4(L_x - 112\lambda^2)\int_0^L (M^r f)(x)\sin(2\lambda r)\cos^2(\lambda r)\,dr$$

$$= 160\lambda^2\int_0^L (M^r f)(x)\sin(2\lambda r)\cos^2(\lambda r)\,dr$$

$$+ 32\lambda^2\int_0^L (M^r f)(x)\sin(2\lambda r)\,dr$$

$$+ 112\lambda f(x).$$

Fortunately, all terms except the last one cancel out and we obtain

$$(L_x - 112\lambda^2)^2(L_x - 120\lambda^2)^2\int_0^L (M^r f)(x)\sin^7(2\lambda r)\,dr$$

$$= 192\cdot 40\cdot 96\cdot 112\cdot \lambda^7 f(x).$$

If we now substitute the values $\lambda^{-2} = 72$, we get

$$
\begin{aligned}
(L_x - \tfrac{14}{9})^2(L_x - \tfrac{15}{9})^2(\hat{f})^{\vee}(x)
&= \Omega_8^2 \frac{1}{2\lambda}\left(\int_0^\pi \sin^7 s\, ds\right)192 \cdot 40 \cdot 96 \cdot 112\lambda^7 f(x) \\
&= \frac{\pi^8}{9}\left(\int_0^\pi \sin^7 s\, ds\right)2^8 \cdot 3^{-4} \cdot 5 \cdot 7 f(x) = \frac{\pi^8 2^{13}}{3^6} f(x).
\end{aligned}
$$

Thus we have proved for $X = \boldsymbol{P}^{16}(\boldsymbol{Cay})$,

$$f = P(L)(\hat{f})^{\vee}, \qquad f \in C^\infty(X),$$

where

$$P(L) = \frac{3^6}{\pi^8 2^{13}}(L - \tfrac{14}{9})^2(L - \tfrac{15}{9})^2. \tag{44}$$

For $X = \boldsymbol{P}^n(\boldsymbol{C})$ we find similarly from Lemma 4.13 the formula

$$f = P(L)((\hat{f})^{\vee}), \qquad f \in \mathscr{E}(X)$$

where, since $\lambda^{-2} = 2(n + 2)$,

$$P(L) = c\left(L - \frac{(n-2)2}{2(n+2)}\right)\left(L - \frac{(n-4)4}{2(n+2)}\right)\cdots\left(L - \frac{2(n-2)}{2(n+2)}\right), \tag{45}$$

with

$$c = [-8\pi^2(n + 2)]^{1-(n/2)}. \tag{46}$$

For $X = \boldsymbol{P}^n(\boldsymbol{H})$ we derive from Lemma 4.14 the formula

$$f = P(L)((\hat{f})^{\vee}), \qquad f \in \mathscr{E}(X),$$

where, since $\lambda^{-2} = 2(n + 8)$,

$$P(L) = c\left(L - \frac{(n-2)4}{2(n+8)}\right)\left(L - \frac{(n-4)6}{2(n+8)}\right)\cdots\left(L - \frac{4(n-2)}{2(n+8)}\right), \tag{47}$$

with

$$c = \tfrac{1}{2}[-4\pi^2(n + 8)]^{2-(n/2)}. \tag{48}$$

Finally, we determine $P(L)$ for the case $X = \boldsymbol{P}^n(\boldsymbol{R})$ because now the metric on X is normalized by means of the Killing form of $U = I(X)$ rather than by the curvature $+1$ condition in Theorem 4.7. Instead of the functions on $\boldsymbol{P}^n(\boldsymbol{R})$ we shall deal with even functions f on the sphere

$\boldsymbol{S}^n$ and define $\hat{f}(\xi)$ as the integral of f over the totally geodesic $(n-1)$-sphere ξ. In order to preserve (36) we define $\check{\phi}(x)$ as an integral over an $(n-1)$-sphere (of unit normal vectors to hyperplanes through x).

The Killing form metric on $\boldsymbol{S}^n$ is obtained by multiplying the usual Riemannian structure (curvature $+1$) by $2n$; (see [DS], solution to Exercise 7, Chapter V). The new Laplacian is therefore $1/2n$ times the Laplacian in Theorem 4.7. Thus we have by Theorem 4.7, with $k = n-1$, n odd,

$$P(L)(\hat{f})^{\vee} = f,$$

where

$$P(L) = c\left(L - \frac{(n-2)1}{2n}\right)\left(L - \frac{(n-4)3}{2n}\right)\cdots\left(L - \frac{1(n-2)}{2n}\right), \tag{49}$$

c a constant. To find c we apply the formula to the function $f \equiv 1$. Since

$$\hat{1} \equiv \Omega_n (2n)^{(1/2)(n-1)}, \qquad (\hat{1})^{\vee} = \Omega_n^2 (2n)^{n-1},$$

we have

$$c = \tfrac{1}{4}(-4\pi^2 n)^{(1/2)(1-n)}. \tag{50}$$

This concludes the proof of part (ii) of Theorem 4.11.

That $f \to \hat{f}$ is injective as follows from (ii) except for the case

$$X = \boldsymbol{P}^n(\boldsymbol{R}), \qquad n \text{ even}.$$

But in this exceptional case the injectivity follows from Theorem 4.7.

For the surjectivity we use once more the fact that the mean value operator M^r commutes with the Laplacian [cf., (41)]. We have

$$\hat{f}(j(x)) = c(M^L f)(x), \tag{51}$$

where c is a constant. Thus by (36)

$$(\hat{f})^{\vee}(x) = (\hat{f} \circ j)^{\wedge}(j(x)) = cM^L(\hat{f} \circ j)(x),$$

so

$$(\hat{f})^{\vee} = c^2 M^L M^L f. \tag{52}$$

Thus if X is not an even-dimensional projective space f is a constant multiple of $M^L P(L) M^L f$, which by (51) shows $f \to \hat{f}$ surjective. For the case $X = \boldsymbol{P}^n(\boldsymbol{R})$, n even, we use Theorem 4.7. If the set $\mathscr{E}(X)^{\wedge}$ were not dense there would by (51) be a nonzero distribution T on X such that

$$T(M^L f) = 0, \qquad f \in \mathscr{E}(X). \tag{53}$$

Now take f to be an eigenfunction of L; then, as used earlier, f is an eigenfunction of M^L and, by the injectivity, with a nonzero eigenvalue. Thus (53) implies† the contradiction $T = 0$.

It remains to prove $(Lf)\hat{} = \Lambda\hat{f}$. For this we use (36), (41), and (51). By the definition of Λ we have

$$(\Lambda\phi)(j(x)) = L(\phi \circ j)(x), \qquad x \in X, \quad \phi \in \mathscr{E}(\Xi).$$

Thus

$$\begin{aligned}(\Lambda\hat{f})(j(x)) &= (L(\hat{f} \circ j))(x) = cL(M^L f)(x) = cM^L(Lf)(x) \\ &= (Lf)\hat{}(j(x)).\end{aligned}$$

This finishes the proof of Theorem 4.11.

Corollary 4.16. *Let B be an open set in $\boldsymbol{R}^{n+1}$, symmetric and star shaped with respect to 0, bounded by a hypersurface. Assume for a fixed k $(1 \leq k < n)$*

$$\text{(54)} \qquad \operatorname{Area}(B \cap P) = \text{const.}$$

for all $(k + 1)$-planes P through 0. Then B is an open ball.

In fact, we know from Theorem 4.7 that if f is a symmetric function on $X = \boldsymbol{S}^n$ with $\hat{f}(\boldsymbol{S}^n \cap P)$ constant (for all P), then f is a constant. We apply this to the function

$$f(\theta) = \rho(\theta)^{k+1}, \qquad \theta \in \boldsymbol{S}^n,$$

if $\rho(\theta)$ is the distance from the origin to each of the two points of intersection of the boundary of B with the line through 0 and θ; f is well-defined, since B is symmetric. If $\theta = (\theta_1, \ldots, \theta_k)$ runs through the k-sphere $\boldsymbol{S}^n \cap P$, then the point

$$x = r\theta \qquad (0 \leq r < \rho(\theta))$$

runs through the set $B \cap P$ and

$$\operatorname{Area}(B \cap P) = \int_{\boldsymbol{S}^n \cap P} d\omega(\theta) \int_0^{\rho(\theta)} r^k \, dr.$$

It follows that $\operatorname{Area}(B \cap P)$ is a constant multiple of $\hat{f}(\boldsymbol{S}^n \cap P)$, so (54) implies that f is constant. This proves the corollary.

Remark. R. Michel has in [1972] and [1973] used Theorem 1.11 to prove certain infinitesimal rigidity properties of the canonical metric on the real and complex projective spaces.

† See "Some Details," p. 611.

3. Noncompact Two-Point Homogeneous Spaces

Theorem 4.11 has an analog for noncompact two-point homogeneous spaces, which we shall now describe. By Tits's classification [1955] of isotropic homogeneous spaces (Riemannian or not) these spaces are the Euclidean spaces and the noncompact symmetric spaces $X = G/K$ of rank one. (This is proved without classification in Nagano [1959] and Helgason [1959].) Let d denote the distance in X and o the origin, that is, the coset $\{K\}$.

Let $\mathfrak{g} = \mathfrak{k} + \mathfrak{p}$ be the direct decomposition of the Lie algebra $\mathfrak{g}$ of G into the Lie algebra $\mathfrak{k}$ of K and its orthogonal complement $\mathfrak{p}$ (with respect to the Killing form of $\mathfrak{g}$). Fix a one-dimensional subspace $\mathfrak{a} \subset \mathfrak{p}$ and let

$$\mathfrak{p} = \mathfrak{a} + \mathfrak{p}_\alpha + \mathfrak{p}_{(1/2)\alpha} \tag{55}$$

be the decomposition of $\mathfrak{p}$ into eigenspaces of T_H [by analogy with (35)]. Let ξ_o denote the totally geodesic submanifold $\mathrm{Exp}(\mathfrak{p}_{(1/2)\alpha})$; in the case $\mathfrak{p}_{(1/2)\alpha} = 0$ we put $\xi_o = \mathrm{Exp}(\mathfrak{p}_\alpha)$. By the classification and duality for symmetric spaces we have the following complete list of the spaces G/K.

X		ξ_o
Real hyperbolic spaces	$\boldsymbol{H}^n(\boldsymbol{R})$ $(n = 2, 3, \ldots)$	$\boldsymbol{H}^{n-1}(\boldsymbol{R})$
Complex hyperbolic spaces	$\boldsymbol{H}^n(\boldsymbol{C})$ $(n = 4, 6, \ldots)$	$\boldsymbol{H}^{n-2}(\boldsymbol{C})$
Quaternion hyperbolic spaces	$\boldsymbol{H}^n(\boldsymbol{H})$ $(n = 8, 12, \ldots)$	$\boldsymbol{H}^{n-4}(\boldsymbol{H})$
Cayley hyperbolic space	$\boldsymbol{H}^{16}(\boldsymbol{Cay})$	$\boldsymbol{H}^8(\boldsymbol{R})$

Here the superscript denotes the real dimension; for the lowest dimension note that

$$\boldsymbol{H}^1(\boldsymbol{R}) = \boldsymbol{R}, \qquad \boldsymbol{H}^2(\boldsymbol{C}) = \boldsymbol{H}^2(\boldsymbol{R}), \qquad \boldsymbol{H}^4(\boldsymbol{H}) = \boldsymbol{H}^4(\boldsymbol{R}).$$

Let Ξ denote the set of submanifolds $g \cdot \xi_o$ of X as g runs through G; Ξ is given the canonical differentiable structure of a homogeneous space. Each $\xi \in \Xi$ has a measure m induced by the Riemannian structure of X, and the Radon transform on X is defined by

$$\hat{f}(\xi) = \int_\xi f(x)\, dm(x), \qquad f \in C_c(X).$$

The dual transform $\varphi \to \check{\varphi}$ is defined by

$$\check{\varphi}(x) = \int_{\xi \ni x} \varphi(\xi)\, d\mu(\xi), \qquad \varphi \in C(\Xi),$$

where μ is the invariant average on the set of ξ passing through x. Let L denote the Laplace–Beltrami operator on X, the Riemannian structure being that given by the Killing form of $\mathfrak{g}$.

Theorem 4.17. *The Radon transform $f \to \hat{f}$ is a one-to-one mappint of $\mathcal{D}(X)$ into $\mathcal{D}(\Xi)$ and, except for the case $X = \boldsymbol{H}^n(\boldsymbol{R})$, n even, is inverted by the formula*

$$f = Q(L)((\hat{f})^{\vee}),$$

the polynomial Q being given as follow:

$X = \boldsymbol{H}^n(\boldsymbol{R})$, n odd:

$$Q(L) = \gamma\left(L + \frac{(n-2)1}{2n}\right)\left(L + \frac{(n-4)3}{2n}\right)\cdots\left(L + \frac{1(n-2)}{2n}\right).$$

$X = \boldsymbol{H}^n(\boldsymbol{C})$:

$$Q(L) = \gamma\left(L + \frac{(n-2)2}{2(n+2)}\right)\left(L + \frac{(n-4)4}{2(n+2)}\right)\cdots\left(L + \frac{2(n-2)}{2(n+2)}\right).$$

$X = \boldsymbol{H}^n(\boldsymbol{H})$:

$$Q(L) = \gamma\left(L + \frac{(n-2)4}{2(n+8)}\right)\left(L + \frac{(n-4)6}{2(n+8)}\right)\cdots\left(L + \frac{4(n-2)}{2(n+8)}\right).$$

$X = \boldsymbol{H}^{16}(\boldsymbol{Cay})$:

$$Q(L) = \gamma(L + \tfrac{14}{9})^2(L + \tfrac{15}{9})^2.$$

The constants γ are obtained from the constants c in (44), (46), (48), *and* (50) *by multiplication by the factor Ω_X in* (39).

We omit the proof since it is analogous to that of Theorem 4.11. The adjustment from the constants c to the constants γ is caused by the difference in the normalizations of the dual Radon transform in the compact case and the noncompact case.

4. The X-Ray Transform on a Symmetric Space

Let X be a complete Riemannian manifold of dimension >1 in which any two points can be joined by a unique geodesic. By analogy with $\boldsymbol{R}^n$ we can define the *X-ray transform* $f \to \hat{f}$ on X by the formula

$$\hat{f}(\gamma) = \int_{\gamma} f(x)\, ds(x), \tag{56}$$

γ being any complete geodesic in X, ds the element of arc length, and f

any continuous function on X for which the integral converges. By analogy with the X-ray reconstruction problem on $\boldsymbol{R}^n$ (§2, No. 7b) one can consider the problem of inverting the X-ray transform $f \to \hat{f}$ on X. With d denoting the distance in X and $o \in X$ some fixed point, we now define two subspaces of $C(X)$. Let

$$F(X) = \left\{ f \in C(X) : \sup_x d(o, x)^k |f(x)| < \infty \text{ for each } k \geq 0 \right\},$$

$$\mathscr{F}(X) = \left\{ f \in C(X) : \sup_x e^{kd(o, x)} |f(x)| < \infty \text{ for each } k \geq 0 \right\}.$$

Because of the triangle inequality these spaces do not depend on the choice of o. We can informally refer to $F(X)$ as the space of continuous *rapidly decreasing functions* and to $\mathscr{F}(X)$ as the space of continuous *exponentially decreasing functions.* We shall now prove the analog of the support theorems (Theorems 2.6 and 4.2) for the X-ray transform on a symmetric space of the noncompact type. This general analog turns out to be a direct corollary of the Euclidean case and the hyperbolic case, already done.

Corollary 4.18. *Let X be a symmetric space of the noncompact type, B any ball in M.*

(i) *If a function $f \in \mathscr{F}(X)$ satisfies*

$$\hat{f}(\gamma) = 0 \qquad \textit{whenever} \quad \gamma \cap B = \emptyset, \tag{57}$$

then

$$f(x) = 0 \qquad \textit{for} \quad x \notin B. \tag{58}$$

In particular, the X-ray transform is one-to-one on $\mathscr{F}(X)$.

(ii) *If X has rank greater than one, statement* (i) *holds with $\mathscr{F}(X)$ replaced by $F(X)$.*

Proof. Let o be the center of B, r its radius, and let γ be an arbitrary geodesic in X through o.

Assume first X has rank greater than one. By a standard conjugacy theorem for symmetric spaces γ lies in a two-dimensional, flat, totally geodesic submanifold of X. Using Theorem 2.6, on this Euclidean plane we deduce $f(x) = 0$ if $x \in \gamma$, $d(o, x) > r$. Since γ is arbitrary, (58) follows.

Next suppose X has rank one. Identifying $\mathfrak{p}$ with the tangent space X_0, let $\mathfrak{a}$ be the tangent line to γ. We can then consider the eigenspace decomposition (55). If $\mathfrak{b} \subset \mathfrak{p}_\alpha$ is a line through the origin, then

$$S = \operatorname{Exp}(\mathfrak{a} + \mathfrak{b})$$

is a totally geodesic submanifold of X. To see this we verify that the subspace $\mathfrak{a} + \mathfrak{b} \subset \mathfrak{p}$ is a Lie triple system. In fact, if $H \in \mathfrak{a}$, $Y \in \mathfrak{b}$ are any nonzero vectors, we have for the Cartan involution θ of $\mathfrak{g}$ with respect to k, $Y = Z - \theta Z$, where $[H, Z] = \alpha(H)Z$. If follows that

$$[Y, [Y, H]] \subset \mathfrak{a},$$

which together with $[H, [H, Y]] = \alpha(H)^2 Y$ implies that $\mathfrak{a} + \mathfrak{b}$ is a Lie triple system. Hence by [DS], Theorem 7.2, Chapter IV, the set

$$S = \operatorname{Exp}(\mathfrak{a} + \mathfrak{b})$$

is a totally geodesic submanifold of X. Being two-dimensional, nonflat, and simply connected, S is necessarily a hyperbolic space. From Theorem 4.2 we can therefore conclude that $f(x) = 0$ for $x \in \gamma$, $d(o, x) > r$. Now (58) follows since γ is arbitrary.

For a Riemannian manifold all of whose geodesics are closed the analog of (56) can be considered (with the integration taken over the finite length of the geodesic), and again we can consider the injectivity question for the transform $f \to \hat{f}$.

Corollary 4.19. *Let X be a compact symmetric space of rank one. Let f be a continuous function on X satisfying*

$$\int_\gamma f(x)\, ds(x) = 0$$

for each (closed) geodesic γ in X, ds being the element of arc length.

(i) *If X is a sphere, f is skew.*
(ii) *If X is not a sphere, $f \equiv 0$.*

Taking a convolution with f we may assume f smooth. Part (i) is already contained in Theorem 4.7. For part (ii) we use the classification. For $X = \boldsymbol{P}^{16}(\boldsymbol{Cay})$ the antipodal manifolds are totally geodesic spheres; thus, using part (i), we conclude that $\hat{f} \equiv 0$, so that by Theorem 4.11, $f \equiv 0$. For the remaining cases $\boldsymbol{P}^n(\boldsymbol{C})$ $(n = 4, 6, \ldots)$ and $\boldsymbol{P}^n(\boldsymbol{H})$ $(n = 8, 12, \ldots)$, (ii) follows similarly by induction because the initial antipodal manifolds, namely, $\boldsymbol{P}^2(\boldsymbol{C})$ and $\boldsymbol{P}^4(\boldsymbol{H})$, are totally geodesic spheres.

§5. Integral Formulas

In this section we derive some integral formulas for semisimple Lie groups; the formulas, which to a large extent are due to Harish-Chandra relate the Haar measures of the groups entering in the Iwasawa, Cartan, and Bruhat decompositions. Sometimes we use the general tools developed in §1; however, it is often better to use ad hoc invariance considerations.

1. Integral Formulas Related to the Iwasawa Decomposition

Let G be a connected semisimple Lie group, $\mathfrak{g}$ its Lie algebra. Let θ be a Cartan involution of $\mathfrak{g}$, $\mathfrak{g} = \mathfrak{k} + \mathfrak{p}$ the corresponding Cartan decomposition, $\mathfrak{a} \subset \mathfrak{p}$ a maximal abelian subspace, and Σ the corresponding set of restricted roots (cf. [DS], Chapter IX). Fix a Weyl chamber $\mathfrak{a}^+ \subset \mathfrak{a}$; let Σ^+ denote the corresponding set of positive roots and $\mathfrak{n} = \sum_{\alpha \in \Sigma^+} \mathfrak{g}_\alpha$ the nilpotent algebra spanned by the root subspaces for the positive restricted roots. Let $m_\alpha = \dim(\mathfrak{g}_\alpha)$ and put $\rho = \frac{1}{2} \sum_{\alpha \in \Sigma^+} (m_\alpha)\alpha$.

If A and N denote the analytic subgroups of G with Lie algebras $\mathfrak{a}$ and $\mathfrak{n}$, respectively, we have the Iwasawa decompositions

$$\mathfrak{g} = \mathfrak{k} + \mathfrak{a} + \mathfrak{n} \qquad \text{(direct vector space sum)},$$

$$G = KAN,$$

that is, the mapping $(k, a, n) \to kan$ is a diffeomorphism of $K \times A \times N$ onto G. If $a \in A$, let $\log a$ denote the unique element $H \in \mathfrak{a}$ such that $\exp H = a$. As usual, M will denote the centralizer of A in K, and $\mathfrak{m}$ its Lie algebra.

Proposition 5.1. *Let $G = KAN$ be an Iwasawa decomposition of a connected semisimple Lie group G. Let dk, da, and dn be left-invariant measures on K, A, and N, respectively. Then the left-invariant measure dg on G can be normalized so that*

$$\int_G f(g)\, dg = \int_{K \times A \times N} f(kan) e^{2\rho(\log a)}\, dk\, da\, dn \qquad (f \in C_c(G)).$$

Proof. While this can be derived from Proposition 1.12, we proceed more directly. Since the mapping $(k, a, n) \to kan$ is a diffeomorphism, there exists a function $D(k, a, n)$ on $K \times A \times N$ such that

$$(1) \qquad \int_G f(g)\, dg = \int_{K \times A \times N} f(kan) D(k, a, n)\, dk\, da\, dn, \qquad f \in C_c(G).$$

The groups G, K, A, N are all unimodular, (Proposition 1.6). In particular, the left-hand side of (1) does not change if we replace $f(g)$ by

$f(k_1 g n_1)$ $(k_1 \in K, n_1 \in N)$. It follows that $D(k_1^{-1}k, a, nn_1^{-1}) \equiv D(k, a, n)$, so $D(k, a, n)$ is a function $\delta(a)$ of a alone. Next, let $a_1 \in A$. Then

$$\int_G f(g)\,dg = \int_G f(ga_1)\,dg = \int_{KAN} f(kana_1)\delta(a)\,dk\,da\,dn$$

$$= \int_{KAN} f(kaa_1(a_1^{-1}na_1)\delta(a)\,dk\,da\,dn$$

$$= \int_{KAN} f(ka(a_1^{-1}na_1))\delta(aa_1^{-1})\,dk\,da\,dn,$$

which by (2) in §1 equals

$$\int_{KAN} f(kan)\delta(aa_1^{-1})\,dk\,da\,(I(a_1)^*(dn)).$$

Here $I(a_1)$ denotes the automorphism $n \to a_1 n a_1^{-1}$ of N. It maps the Haar measure dn into another Haar measure. This new Haar measure is a multiple of dn by the factor

$$\det(dI(a_1)_e) = \det(\mathrm{Ad}(a_1)|\mathfrak{n}) = e^{2\rho(\log a_1)}. \tag{2}$$

Comparing with (1), we obtain

$$\delta(a) = \delta(aa_1^{-1})e^{2\rho(\log a_1)},$$

so that putting $a = e$, the proposition follows.

Corollary 5.2. *If $F \in C_c(AN)$, we have for $a \in A$,*

$$\int_N F(na)\,dn = e^{2\rho(\log a)} \int_N F(an)\,dn.$$

This is a consequence of (2); equivalently we can proceed as follows: The left-hand side is

$$\int_N (F \circ R(a))(n)\,dn = \int_N (F \circ R(a))(ana^{-1})\,d(ana^{-1})$$

$$= \int_N (F \circ R(a))(ana^{-1}) \frac{d(ana^{-1})}{dn}\,dn = \int_N F(an)e^{2\rho(\log a)}\,dn.$$

Corollary 5.3. *For $f \in C_c(G)$ we have*

$$\int_G f(g)\,dg = \int_{KNA} f(kna)\,dk\,dn\,da = \int_{ANK} f(ank)\,da\,dn\,dk.$$

The first identity follows from Proposition 5.1 and Corollary 5.2. By replacing f by the function $\check{f}: g \to f(g^{-1})$, the second identity follows from the first one.

Let $\mathfrak{a}' \subset \mathfrak{a}$ denote the subset of *regular* elements, that is, elements H such that $\alpha(H) \neq 0$ for all $\alpha \in \Sigma$. Let $A' = \exp \mathfrak{a}'$.

Lemma 5.4. *Let $H \in \mathfrak{a}'$ and $h = \exp H$. Then the mapping*

$$\xi : n \to h^{-1}nhn^{-1}$$

is a diffeomorphism of N onto N.

Proof. Clearly ξ maps N into N. Next we calculate the differential $d\xi_n$ at $n \in N$. Let $X \in \mathfrak{n}$. Then

$$\begin{aligned}\xi(n \exp tX) &= h^{-1}n \exp tX\, hn^{-1}n \exp(-tX)\, n^{-1} \\ &= h^{-1}nhn^{-1} \exp(\mathrm{Ad}(nh^{-1})tX) \exp \mathrm{Ad}(n)\,(-tX).\end{aligned}$$

Thus

$$d\xi_n(dL(n)X) = dL(h^{-1}nhn^{-1})\, \mathrm{Ad}(n)\, (\mathrm{Ad}(h^{-1}) - 1)\, X. \tag{3}$$

Since H is regular, $\mathrm{Ad}(h^{-1})$ has no nonzero fixed points on $\mathfrak{n}$, so ξ is everywhere regular. Thus there exist open neighborhoods U and V of e in N such that ξ is a diffeomorphism of U onto V. We have

$$U = \exp U_0,\ V = \exp V_0,$$

where U_0 and V_0 are neighborhoods of 0 in $\mathfrak{n}$. Fix $H_0 \in \mathfrak{a}^+$ and put $a_t = \exp tH_0$ $(t \in \boldsymbol{R})$. Then

$$\xi(a_t n a_t^{-1}) = a_t \xi(n) a_t^{-1}. \tag{4}$$

Since the decomposition $\mathfrak{n} = \sum_{\alpha > 0} \mathfrak{g}_\alpha$ diagonalizes $\mathrm{Ad}(a_t)$ (with diagonal elements $e^{t\alpha(H_0)}$), it is obvious that

$$\bigcup_{t>0} \mathrm{Ad}(a_t)V_0 = \mathfrak{n}.$$

Since the map $\exp : \mathfrak{n} \to N$ is surjective, it follows that $\bigcup_{t>0} a_t V a_t^{-1} = N$. But then (4) implies $\xi(N) = N$. Finally, ξ is one-to-one: if in fact $\xi(n_1) = \xi(n_2)$, (4) implies $\xi(a_t n_1 a_t^{-1}) = \xi(a_t n_2 a_t^{-1})$. Choosing for t a negative number so large that both $a_t n_1 a_t^{-1}$ and $a_t n_2 a_t^{-1}$ belong to U, if follows that $a_t n_1 a_t^{-1} = a_t n_2 a_t^{-1}$, so $n_1 = n_2$.

Corollary 5.5. *Let $H \in \mathfrak{a}'$ and $h = \exp H$. Then*

$$\int_N f(n)\, dn = \prod_{\alpha \in \Sigma^+} |1 - e^{-\alpha(H)}|^{m_\alpha} \int_N f(h^{-1}nhn^{-1})\, dn.$$

In fact, Lemma 1.11 and (3) above show that

$$\xi^*(dn) = \det(\mathrm{Ad}(n)\, (\mathrm{Ad}(h^{-1}) - 1))\, dn,$$

whence the result.

Proposition 5.6. *Assume G above has finite center (so K is compact). For $a = \exp H$, $H \in \mathfrak{a}$, let*

$$D(a) = \prod_{\alpha \in \Sigma^+} (\sinh(\tfrac{1}{2}\alpha(H)))^{m_\alpha}.$$

Let $a \in A$ be such that $D(a) \neq 0$. Then for a suitable normalization of the invariant measure dg_A on G/A we have

$$|D(a)| \int_{G/A} f(gag^{-1})\, dg_A = e^{\rho(\log a)} \int_{K \times N} f(kank^{-1})\, dk\, dn$$

for all $f \in C_c(G)$.

Proof. We must first verify that the integral on the left is defined. Let C denote the support of f and put

$$C_A = \{gA \in G/A : gag^{-1} \in C\}.$$

By writing $g \in G$ as $g = kna_1$, it is clear that

$$gag^{-1} \in C \Leftrightarrow knan^{-1}k^{-1} \in C \Rightarrow nan^{-1} \in kCk.$$

Thus if $gag^{-1} \in C$, then nan^{-1} lies in a compact subset of G, so by Lemma 5.4, n lies in a compact subset of N. Hence kn lies in a compact subset of G, so C_A is compact.

By Theorem 1.9 and the first relation in Corollary 5.3 we have for a suitable constant c

$$(5) \qquad \int_{G/A} F(gA)\, dg_A = c \int_{K \times N} F(knA)\, dk\, dn, \qquad F \in C_c(G/A).$$

Hence by Corollary 5.5,

$$\int_{G/A} f(gag^{-1})\, dg_A = c \int_{K \times N} f(knan^{-1}k^{-1})\, dk\, dn$$

$$= c \prod_{\alpha \in \Sigma^+} |1 - e^{-\alpha(H)}|^{-m_\alpha} \int_{K \times N} f(kank^{-1})\, dk\, dn,$$

so the result follows by writing

$$1 - e^{-\alpha(H)} = 2 \sinh(\tfrac{1}{2}\alpha(H))\, e^{-(1/2)\alpha(H)}.$$

This proposition has an interesting functional equation as a consequence. Let M' denote the normalizer of A in K and $W = M'/M$ the Weyl group.

Theorem 5.7. (Harish-Chandra) *Retaining the notation in Proposition 5.6, let $W = W(\mathfrak{g}, \mathfrak{a})$ denote the Weyl group of G/K and for $s \in W$, $H \in \mathfrak{a}$, $a = \exp H$, put $a^s = \exp s(H)$.*

Let $f \in C_c(G)$ satisfy $f(kgk^{-1}) \equiv f(g)$ $(k \in K)$. Then the function

$$F_f(a) = e^{\rho(\log a)} \int_N f(an)\, dn, \qquad a \in A, \tag{6}$$

satisfies the functional equation

$$F_f(a^s) = F_f(a), \qquad a \in A,$$

for each $s \in W$.

Proof. For $g \in G$, let $a^{g_A} = gag^{-1}$. First assume $D(a) \neq 0$. Then by Proposition 5.6

$$F_f(a) = |D(a)| \int_{G/A} f(a^{g_A})\, dg_A. \tag{7}$$

We shall now prove that the right-hand side of this relation is invariant under $a \to a^s$. For this purpose f can be arbitrary in $C_c(G)$. First, since $s \in W$ permutes the roots, $|D(a)| = (D(a)^2)^{1/2}$ is invariant under s. For the invariance of the integral select $u \in K$ such that $\mathrm{Ad}_G(u)$ coincides with s on $\mathfrak{a}$. Then $uAu^{-1} = A$, so the mapping $\phi: gA \to ugu^{-1}A$ is a well-defined mapping of G/A onto itself. Put $f^u(g) = f(ugu^{-1})$, and as in Lemma 1.10,

$$\bar{f}(gA) = \int_A f(ga)\, da.$$

Then the mapping $f \to \bar{f}$ maps $C_c(G)$ onto $C_c(G/A)$, and with dg suitably normalized,

$$\int_{G/A} \bar{f}(gA)\, dg_A = \int_G f(g)\, dg = \int_G f^u(g)\, dg = \int_{G/A} \overline{f^u}(gA)\, dg_A. \tag{8}$$

On the other hand,

$$\overline{f^u}(gA) = \int_A f(ugau^{-1})\, da = \int_A f(ugu^{-1}a^s)\, da = \int_A f(ugu^{-1}a)\, da,$$

so $\overline{f^u} = (\bar{f})\phi^{-1}$. Substituting this in (8), we deduce $\phi^*(dg_A) = dg_A$. Thus, invoking (5), we have

$$\begin{aligned}
\int_{G/A} f((a^s)^{g_A})\, dg_A &= \int_{G/A} f((a^s)^{(ugu^{-1})_A})\, dg_A \\
&= \int_{G/A} f(ugag^{-1}u^{-1})\, dg_A \\
&= \int_{K \times N} f(uknan^{-1}k^{-1}u^{-1})\, dk\, dn.
\end{aligned}$$

Here u can be cancelled, so this equals

$$\int_{G/A} f(a^{g_A})\, dg_A.$$

This proves $F_f(a^s) = F_f(a)$ for all a in a dense subset of A; by continuity it holds for all $a \in A$.

2. Integral Formulas for the Cartan Decomposition

A. *The Noncompact Case*

For the symmetric space $X = G/K$ as above, consider as in [DS] (Chapter IX), the set $A' = \exp \mathfrak{a}'$ ($\mathfrak{a}'$ = the set of regular elements in $\mathfrak{a}$), $G' = KA'K$, and the open dense subset $X' = G' \cdot o$ of X which is diffeomorphic to $(K/M) \times A^+$ under the "polar coordinate map"

$$\Phi\colon (kM, a) \to kaK$$

of $(K/M) \times A^+$ onto X'. Here o is the origin $\{K\}$ in X. Let dx denote the volume element on X corresponding to the canonical Riemannian structure $\langle\ ,\ \rangle$ induced by the Killing form B of $\mathfrak{g}$ (also denoted $\langle\ ,\ \rangle$) restricted to $\mathfrak{p}$. Let θ denote the Cartan involution of $\mathfrak{g}$ corresponding to the Cartan decomposition $\mathfrak{g} = \mathfrak{k} + \mathfrak{p}$.

Theorem 5.8. *For a suitable constant* c

$$\int_X f(x)\, dx = c \int_{K/M} \left(\int_{A^+} f(ka \cdot o)\delta(a)\, da \right) dk_M, \qquad f \in C_c(X),$$

where dk_M *is the* K*-invariant measure on* K/M, *normalized by* $\int dk_M = 1$, *and*

$$\delta(\exp H) = \prod_{\alpha \in \Sigma^+} (\sinh \alpha(H))^{m_\alpha}, \qquad H \in \mathfrak{a}^+.$$

Proof. Let $\mathfrak{l}$ denote the orthogonal complement of $\mathfrak{m}$ in $\mathfrak{k}$ (with respect to B). Identifying $\mathfrak{l}$ with tangent space of K/M, the restriction $(-B)|\mathfrak{l}$ induces a K-invariant Riemannian structure on K/M, and dk_M is a constant multiple of the corresponding volume element. According to Prop. 1.3 we just have to calculate $\det(d\Phi_{(kM,a)})$ for the polar coordinate map Φ above.

We know from [DS] (Chapter VII, §11), that $\mathfrak{l}$ is the direct sum

$$\mathfrak{l} = \sum_{\alpha \in \Sigma^+} \mathfrak{k}_\alpha,$$

where

$$\mathfrak{k}_\alpha = \{T \in \mathfrak{k} : (\operatorname{ad} H)^2 T = \alpha(H)^2 T \text{ for all } H \in \mathfrak{a}\}.$$

Moreover, the mapping $X \to X + \theta X$ is a bijection of $\mathfrak{g}_\alpha$ onto $\mathfrak{k}_\alpha$; in particular, $\dim \mathfrak{k}_\alpha = m_\alpha$. Let $T_1^\alpha, \ldots, T_{m_\alpha}^\alpha$ be an orthonormal basis of $\mathfrak{k}_\alpha$ ($\alpha \in \Sigma^+$) and let $H_1, \ldots, H_l$ be an orthonormal basis of $\mathfrak{a}$. As usual, we denote by $\tau(b_0)$ the translation $bC \to b_0 bC$ on any homogeneous space B/C. Let $\pi: G \to G/K$ be the natural mapping, so $d\pi$ is the projection of $\mathfrak{g}$ onto the tangent space $(G/K)_o$ with kernel $\mathfrak{k}$.

The curve $t \to k \exp(tT_i^\alpha) M$ in K/M has tangent vector $d\tau(k)T_i^\alpha$ at $t = 0$. Then for $a = \exp H$ ($H \in \mathfrak{a}^+$),

$$\begin{aligned} &\Phi(k \exp(tT_i^\alpha) M, a) \\ &\qquad = k \exp(tT_i^\alpha)\, a \cdot o = ka \exp(t \operatorname{Ad}(a^{-1})\, T_i^\alpha) \cdot o, \\ &d\Phi_{(kM, a)}(d\tau(k)T_i^\alpha, 0) \\ &\qquad = d\tau(ka)\, d\pi(\operatorname{Ad}(a^{-1})\, T_i^\alpha) \\ &\qquad = d\tau(ka)\, d\pi\{\tfrac{1}{2}(\operatorname{Ad}(a^{-1})\, T_i^\alpha - \theta \operatorname{Ad}(a^{-1})\, T_i^\alpha)\} \\ &\qquad = d\tau(ka)\, d\pi\{\tfrac{1}{2}(e^{-\operatorname{ad} H} T_i^\alpha - e^{\operatorname{ad} H} T_i^\alpha)\} \\ &\qquad = d\tau(ka)\, d\pi(-\alpha(H)^{-1}[H, T_i^\alpha] \sinh \alpha(H)), \end{aligned}$$

where we used the relation $(\operatorname{ad} H)^2 T_i^\alpha = \alpha(H)^2 T_i^\alpha$. Also,

$$\Phi(kM, a \exp(tH_j)) = ka \exp(tH_j) \cdot o,$$

so

$$d\Phi_{(kM, a)}(0, d\tau(a)H_j) = d\tau(ka)\, d\pi(H_j).$$

The maps

$$d\pi: \mathfrak{p} \to (G/K)_o, \qquad d\tau(ka): (G/K)_o \to (G/K)_{ka \cdot o}$$

are isometries, and the vectors

$$H_j, \qquad \alpha(H)^{-1}[H, T_i^\alpha] \qquad (1 \le j \le l, \alpha \in \Sigma^+, 1 \le i \le m_\alpha)$$

form an orthonormal basis of $\mathfrak{p}$. The formulas above thus imply that

$$|\det(d\Phi_{(kM, a)})| = \prod_{\alpha \in \Sigma^+} (\sinh \alpha(H))^{m_\alpha}, \tag{9}$$

and this proves the theorem.

B. *The Compact Case*

The case of a compact symmetric space can be handled by similar methods although the global properties of the "polar coordinate" decomposition now have a new but minor complication.

With the Cartan decomposition as above consider the compact real form

$$\mathfrak{u} = \mathfrak{k} + \mathfrak{p}_*, \qquad \text{where} \quad \mathfrak{p}_* = i\mathfrak{p}$$

of the complexification $\mathfrak{g}^C$. Then the subspace $\mathfrak{a}_* = i\mathfrak{a}$ is a maximal abelian subspace of $\mathfrak{p}_*$. Let (U, K_1) be a symmetric pair such that U and K_1 have Lie algebras $\mathfrak{u}$ and $\mathfrak{k}$, respectively. Let M_1 denote the centralizer of $\mathfrak{a}_*$ in K_1. The group $A_* = \exp \mathfrak{a}_*$ is a closed subgroup of U ([DS], §6, Chapter V), hence compact. Let du, dk, dm, and da denote the invariant measures on the compact groups U, K_1, M_1, and A_* such that in each case the total measure is 1. Let du_{K_1} and dk_{M_1} be the corresponding invariant measures on U/K_1, K_1/M_1, respectively.

By analogy with the polar coordinate map Φ above we consider the mapping

$$\Psi: (K_1/M_1) \times A_* \to U/K_1$$

given by

$$\Psi(kM_1, a) = kaK_1,$$

which by [DS], Theorem 6.7, Chapter V, is surjective. By analogy with (9) we find

$$|\det(d\Psi_{(kM_1, a)})| = \prod_{\alpha \in \Sigma^+} |\sin \alpha(iH)|^{m_\alpha} \tag{10}$$

if $a = \exp H$ $(H \in \mathfrak{a}_*)$. Let $\delta_*(a)$ denote the right-hand side of (10). Using Prop. 1.3 we have on each open subset on which Ψ is one-to-one and regular,

$$\Psi^*(du_{K_1}) = c\delta_*(a)\, dk_{M_1}\, da, \tag{11}$$

where c is a constant. Let A'_* denote the set of points $a \in A_*$ for which $\delta_*(a) \neq 0$ and let $(U/K_1)_r$ denote the complement in U/K_1 of the singular set as defined in [DS], §3, Chapter VII. The space $J = K_1 \cap A_*$ is compact and discrete, hence finite. Let j denote its cardinality and let w denote the cardinality of the Weyl group $W = W(\mathfrak{g}, \theta)$.

Lemma 5.9. *The mapping Ψ is a regular wj-to-one mapping of*

$$(K_1/M_1) \times A'_* \quad \textit{onto} \quad (U/K_1)_r.$$

Proof. The regularity and surjectivity of Ψ being known from Lemma 8.1, Chapter VII, in [DS], it only remains to prove that each point in $(U/K_1)_r$ has exactly wj preimages. Suppose $a_1, a_2 \in A'_*$, k_1, $k_2 \in K_1$, such that $\Psi(k_1M_1, a_1) = \Psi(k_2M_1, a_2)$. On putting $k = k_2^{-1}k_1$, we have $ka_1 = a_2k'$ for a suitable $k' \in K_1$. We apply the involutive automorphism θ of U, eliminate k' and obtain $ka_1^2k^{-1} = a_2^2$. This implies $\mathrm{Ad}(k)\,Z_1 = Z_2$ if Z_i $(i = 1, 2)$ denotes the centralizer

$$Z_i = \{X \in \mathfrak{p}_* : \mathrm{Ad}(a_i^2)\,X = X\}.$$

Let $H \in \mathfrak{a}_*$, $X \in \mathfrak{p}_*$. Writing

$$X = X_0 + \sum_{\alpha \in \Sigma^+} a_\alpha(X_\alpha - \theta X_\alpha),$$

where $X_0 \in \mathfrak{a}_*$, $X_\alpha \in \mathfrak{g}_\alpha$, $a_\alpha \in i\boldsymbol{R}$, we have

$$\begin{aligned} \mathrm{Ad}(\exp H)\, X &= e^{\mathrm{ad}\, H}\, X \\ &= X_0 + \sum_{\alpha \in \Sigma^+} a_\alpha(e^{\alpha(H)} X_\alpha - e^{-\alpha(H)} \theta X_\alpha). \end{aligned}$$

If $a_1 = \exp H_1$, $X \in Z_1$, we deduce that

$$a_\alpha = a_\alpha e^{2\alpha(H_1)} \qquad (\alpha \in \Sigma^+),$$

so that $\alpha(H_1) \in \pi i \boldsymbol{Z}$ if $a_\alpha \neq 0$. But the assumption $\delta_*(a_1) \neq 0$ then shows $a_\alpha = 0$ for all $\alpha \in \Sigma^+$, that is, $Z_1 = \mathfrak{a}_*$. Similarly $Z_2 = \mathfrak{a}_*$. This proves that k lies in the normalizer M_1' of $\mathfrak{a}_*$ in K_1. We have by the above

$$a_2 = (k a_1 k^{-1}) k (k')^{-1}. \tag{12}$$

Now let $s \in W$, $a \in A_*$. We recall ([DS], Chapter VII, §2) that $W = M_1'/M_1$, even if K_1 is not necessarily connected. Thus if $m \in M_1'$ is any element such that s and $\mathrm{Ad}(m)$ coincide on $\mathfrak{a}_*$, we can write

$$m a m^{-1} = a^s.$$

It is also clear that $\delta_*(a^s) = \delta_*(a)$. Let $b \in J$ and select $H \in \mathfrak{a}_*$ such that $b = \exp H$. Since $b^2 = e$, we have $\alpha(H) \in \pi i \boldsymbol{Z}$ for all $\alpha \in \Sigma^+$, so that

$$\delta_*(a_0 b) = \delta_*(a_0)$$

for all $a_0 \in A_*$. Formula (12) shows that $a_2 = a_1^t b$, where $t \in W$, $b \in J$. Also, $k_2 M_1 = k_1 k^{-1} M_1$.

On the other hand, if b runs through the set J and k runs through a complete set of representatives of M_1' mod M_1, then the elements

$$(k_1 k^{-1} M_1, (k a_1 k^{-1}) b) \in (K_1/M_1) \times A_*'$$

(for $(k_1 M_1, a_1)$ fixed) are all different. Hence they make up the complete inverse image $\Psi^{-1}(k_1 a_1 K_1)$. This proves the lemma.

New A_*' fills up A_* except for a set of measure 0 and $(U/K_1)_r$ fills up U/K_1 except for a set of measure 0. Since the density δ_*, as noted, has the same value at all the points $k a_1 k^{-1} b$ as $k \in M_1'$, $b \in J$, we can derive the following result from (11) and Lemma 5.9.

Theorem 5.10. *Let (U, K_1) be a Riemannian symmetric pair of the compact type. Let du_{K_1}, dk_{M_1}, da be the invariant measures on U/K_1, K_1/M_1, and A_* normalized by*

$$\int_{U/K_1} du_{K_1} = \int_{K_1/M_1} dk_{M_1} = \int_A da = 1.$$

Then for a suitable constant c_*,

$$\int_{U/K_1} f(uk_1)\, du_{K_1} = c_* \int_{K_1/M_1} \left(\int_{A_*} f(kaK_1)\delta_*(a)\, da \right) dk_{M_1} \tag{13}$$

for all $f \in C(U/K_1)$, $\delta_*(a)$ denoting the right-hand side of (10).

In the case when U is simply connected, K_1 connected, we can put the result in a more precise form. As in [DS] (Chapter VII, §3), let $D(\mathfrak{u}, \theta)$ denote the *diagram*

$$\{H \in \mathfrak{a}_* : \alpha(H) \in \pi i \mathbf{Z} \text{ for same } \alpha \in \Sigma^+\};$$

$\mathfrak{a}_r$ is the complement $\mathfrak{a}_* - D(\mathfrak{u}, \theta)$, and Q_0 is any component of $\mathfrak{a}_r$ whose closure $\bar{Q}_0$ contains the origin. We have then the following *global polar coordinate representation of* U/K_1.

Theorem 5.11. (U *is simply connected,* K_1 *connected.*) *The mapping*

$$\psi : (kM_1, H) \to k(\exp H)K_1$$

maps $(K_1/M_1) \times \bar{Q}_0$ onto U/K_1 and gives a bijection of $(K_1/M_1) \times Q_0$ onto $(U/K_1)_r$.

Proof. That

$$\psi(K_1/M_1 \times \bar{Q}_0) = U/K_1,$$
$$\psi(K_1/M_1 \times Q_0) = (U/K_1)_r$$

is clear from Chapter VII, §8, in [DS]. For the remaining (injectivity) statement suppose $H_1, H_2 \in Q_0$, $k_1, k_2 \in K_1$, such that

$$k_1(\exp H_1)K_1 = k_2(\exp H_2)K_1.$$

By (12) we have

$$\exp H_2 = \exp \mathrm{Ad}(k)\, H_1 b, \tag{14}$$

where $k = k_2^{-1}k_1 \in M'_1$ and $b \in J$. Since $b = \exp H_0$ $(H_0 \in \mathfrak{a}_*)$, it follows that H_2 and H_1 are congruent under the group Γ_{K_1} generated by $W(\mathfrak{u}, \theta)$ and the lattice $\mathfrak{a}_{K_1} = \{H \in \mathfrak{a}_* : \exp H \in K_1\}$. But then Chapter VII, §8 in [DS] implies that $H_1 = H_2$. Equation (14) then implies that

$$H_1 = sH_1 + H_*, \tag{15}$$

where $H_* \in \mathfrak{a}_{K_1}$ and s is the Weyl group element induced by Ad(k). But since Γ_{K_1} permutes the components of $\mathfrak{a}_r$ simply transitively, (15) implies $s = e$. Thus $k_1 M_1 = k_2 M_1$, proving the injectivity.

Let us now assume U/K_1 in Theorem 5.11 to be irreducible. Let $\mu_1, \ldots, \mu_l$ denote the simple restricted roots, δ the highest root, $\delta = \sum_1^l d_i \mu_i$. Then the component Q_0 can be taken as the polyhedron

$$Q_0 = \{H \in \mathfrak{a}_* : (1/i)\mu_j(H) > 0 \ (1 \leq j \leq l), (1/i)\delta(H) < \pi\}.$$

We can then state Theorem 5.10 in a more precise form.

Corollary 5.12. *With U, K_1 as in Theorem 5.11, U/K_1 irreducible, let dH be the Euclidean measure on $\mathfrak{a}_*$ induced by the Killing form. Then*

$$\int_{U/K_1} f(uK_1)\, du_{K_1} = c_0 \int_{K/M_1} dk_{M_1}\left(\int_{Q_0} f(k(\exp H)K_1)\delta_*(\exp H)\, dH\right),$$

where

$$c_0^{-1} = \int_{Q_0} \prod_{\alpha \in \Sigma^+} (\sin \alpha(i^{-1}H))^{m_\alpha}\, dH.$$

We now restate Theorem 5.11 for the case when the symmetric space is a group.

Let U be a compact semisimple Lie group, $T \subset U$ a maximal torus. As in [DS] (Chapter VII), §§4–7, let $\mathfrak{t}_0$ and $\mathfrak{u}$ denote the respective Lie algebras, $\mathfrak{t}$ and $\mathfrak{g}$ the complexifications. Let Δ denote the set of roots of $\mathfrak{g}$ with respect to $\mathfrak{t}$. Let $D(\mathfrak{u})$ denote the diagram

$$\{H \in \mathfrak{t}_0 : \alpha(H) \in 2\pi i \mathbf{Z} \text{ for some } \alpha \in \Delta\},$$

let $\mathfrak{t}_r = \mathfrak{t}_0 - D(\mathfrak{u})$ and fix a component P_0 of $\mathfrak{t}_r$ whose closure $\bar{P}_0$ contains the origin. As in [DS], Chapter VII, §4, we view U as the symmetric space $U \times U/U$. Then δ^* becomes a function on the set of pairs $\{(t, t^{-1}) : t \in T\}$. Writing $\delta(\exp 2H) = \delta^*(\exp H, \exp(-H))$ $(H \in \mathfrak{t}_0)$, we have (up to a constant factor)

$$\delta(t) = \left| \prod_{\alpha \in \Delta} 2 \sin \frac{\alpha(iH)}{2} \right| \quad \text{if} \quad t = \exp H, \quad H \in \mathfrak{t}_0.$$

The product is taken over all of Δ, not just over the positive roots, because here each restricted root has multiplicity 2. Let du be the Haar measure on U normalized by $\int du = 1$ and let dH be the Euclidean measure on $\mathfrak{t}_0$ induced by the Killing form of $\mathfrak{u}$. Then we have from Theorem 5.11 and its corrollary the following result.

Theorem 5.13. *Let U be a compact semisimple simply connected Lie group. The mapping*

$$\psi: (uT, H) \to u \exp H\, u^{-1}$$

maps $(U/T) \times \bar{P}_0$ onto U and gives a bijection of $(U/T) \times P_0$ onto U_r. Also,

$$\int_U f(u)\, du = c \int_{U/T} du_T \int_{P_0} f(u \exp H\, u^{-1}) \delta(\exp H)\, dH,$$

where c is a constant.

It is convenient to insert here a simple result about reduced root systems. Let E be a vector space over $\boldsymbol{R}$, V the dual of E, and R a reduced root system in V ([DS], Chapter X, §3). Let $\alpha_1, \ldots, \alpha_l$ be a basis of R and R^+ the corresponding set of positive roots. Let $\langle\ ,\ \rangle$ denote the Killing form on V ([DS], p. 523) and C the Weyl chamber corresponding to R^+, that is,

$$C = \{\gamma \in V : \langle \alpha, \gamma \rangle > 0 \text{ for } \alpha \in R^+\} = \{\gamma \in V : \langle \alpha_i, \gamma \rangle > 0\ (1 \leq i \leq l)\}.$$

As in [DS], Chapter X, §3 let $T(R)$ denote the subgroup of V generated by the vectors in R and, putting $\check{\alpha} = 2\alpha/\langle \alpha, \alpha \rangle$, let

$$\tilde{T}(R) = \{\omega \in V : \langle \omega, \check{\alpha} \rangle \in \boldsymbol{Z} \text{ for } \alpha \in R\}.$$

Let $\omega_1, \ldots, \omega_l$ be the "dual basis" of V given by $\langle \omega_i, \check{\alpha}_j \rangle = \delta_{ij}$; then (loc. cit.)

$$\tilde{T}(R) = \sum_1^l \boldsymbol{Z}\omega_i, \qquad T(R) = \sum_1^l \boldsymbol{Z}\alpha_i.$$

It follows that

$$\tilde{T}(R) \cap C = \{m_1\omega_1 + \cdots + m_l\omega_l : m_1 > 0, \ldots, m_l > 0\}. \tag{16}$$

The element $\rho = \frac{1}{2}\sum_{\alpha \in R^+} \alpha$ belongs to $\tilde{T}(R) \cap C$; in fact ([DS], Lemma 3.11, Chapter X)

$$\rho = \omega_1 + \cdots + \omega_l.$$

Consider now the algebra F of functions on E generated over $\boldsymbol{R}$ by the exponentials e^{ω} ($\omega \in \tilde{T}(R)$). The Weyl group $W(R)$ operates on F by

$$s(e^{\omega}) = e^{s\omega}.$$

An element $a \in F$ is called *invariant* if $sa = a$ ($s \in W$) and *skew* if $sa = (\det s)a$ ($s \in W$), det denoting determinant. The *alternation* $A: F \to F$ given by

$$A(e^{\omega}) = \sum_{s \in W(R)} (\det s) e^{s\omega}$$

maps F onto the set of skew elements in F (cf. [DS], Chapter I, §2, No. 3).

Lemma 5.14. *The elements $A(e^{\omega})$, $\omega \in \tilde{T}(R) \cap C$, form a basis of the space of skew elements in F.*

Proof. Since $W(R)$ permutes the Weyl chambers simply transitively (cf. the proof of Corollary 7.4 in [DS], Chapter VII), the elements $s\omega$ ($s \in W(R)$, $\omega \in \tilde{T}(R) \cap C$) are all different, so the linear independence of the elements $A(e^{\omega})$ follows.

Next let $f \in F$ be skew, $f = \sum f_{\omega} e^{\omega}$. Then if $f_{\omega} \neq 0$, ω is regular. In fact, suppose $s\omega_0 = \omega_0$ for some ω_0 and some reflection $s \in W(R)$. Since f is skew,

$$sf = (\det s)f = -f, \qquad \sum f_{\omega} e^{s\omega} = -\Sigma\, f_{\omega} e^{\omega},$$

whence $f_{\omega_0} = 0$. But each regular element is $W(R)$-conjugate to an element in C. Thus f has a unique representation

$$f = \sum_{\lambda \in \tilde{T}(R) \cap C,\, s \in W(R)} f_{\lambda, s} e^{s\lambda}. \tag{17}$$

But if $t \in W(R)$,

$$(\det t)f = tf = \sum_{\omega} f_{\omega} e^{t\omega},$$

so, by the linear independence of the $e^{s\lambda}$ in (17),

$$f_{\lambda, s} = (\det s)^{-1} f_{\lambda} = (\det s) f_{\lambda},$$

whence

$$f = \sum_{\lambda \in \tilde{T}(R) \cap C} f_{\lambda} A(e^{\lambda}).$$

Proposition 5.15.

(i) *The following identity holds in F:*

$$e^{\rho} \prod_{\alpha \in R^+} (1 - e^{-\alpha}) = \sum_{s \in W(R)} (\det s) e^{s\rho}. \tag{18}$$

(ii) *The element given by (18) is skew and divides every skew element of F.*

Proof. The left-hand side of (18), say D, can be written in the form

$$\prod_{\alpha \in R^+} (e^{(1/2)\alpha} - e^{-(1/2)\alpha}).$$

If α_i is a simple root, the corresponding reflection s_{α_i} permutes $R^+ - \{\alpha_i\}$ (cf. [DS], Chapter X, Lemma 3.11). It follows that $s_{\alpha_i} D = -D =$

$\det(s_{\alpha_i})\, D$, so that, since the s_{α_i} generate $W(R)$, D is skew. Thus, by Lemma 5.14, D is a linear combination

$$D = a_\rho A(e^\rho) + \sum_{\omega \in \tilde{T}(R) \cap C,\, \omega \neq \rho} c_\omega A(e^\omega).$$

On the other hand, by definition,

$$D = e^\rho + \sum_\lambda d_\lambda e^{\rho - \lambda} \qquad (d_\lambda \in \mathbf{Z}), \tag{19}$$

where λ is a positive integral linear combination of the α_i. Now if $c_\omega \neq 0$, then by (16) $\omega - \rho \in \bar{C}$ (closure) and, by (19), $\omega = \rho - \lambda$ for some λ. But this would give $-\lambda \in \bar{C}$, contradicting Lemma 2.20(iii) in [DS], (Chapter VII). (This lemma holds for abstract root systems because of Lemma 3.4, Chapter X.) Thus all c_ω vanish and $a_\rho = 1$, proving part (i).

For part (ii) we consider ω *regular* in $\tilde{T}(R)$ and let $\alpha \in R^+$. Then

$$\langle \omega, \check{\alpha} \rangle \in \mathbf{Z} - \{0\} \text{ and } e^\omega - s_\alpha e^\omega = e^\omega - e^{\omega - \langle \omega, \check{\alpha} \rangle \alpha},$$

so

$$e^\omega - s_\alpha e^\omega = \begin{cases} e^\omega(1 - e^{(-\alpha)\langle \omega, \check{\alpha} \rangle}) & \text{if } \langle \omega, \check{\alpha} \rangle > 0 \\ e^\omega(1 - e^{\alpha(-\langle \omega, \check{\alpha} \rangle)}) & \text{if } \langle \omega, \check{\alpha} \rangle < 0. \end{cases}$$

In the first case $e^\omega - s_\alpha e^\omega$ is divisible by $(1 - e^{-\alpha})$, in the second case by $(1 - e^\alpha) = e^\alpha(e^{-\alpha} - 1)$. Thus in both cases $(1 - e^{-\alpha})$ divides $e^\omega - s_\alpha e^\omega$ and hence divides each $A(e^\omega)$ $(= \frac{1}{2}(A(e^\omega) - s_\alpha A(e^\omega)))$. The factors $(1 - e^{-\alpha})$ $(\alpha \in R^+)$ being relatively prime† and F a unique factorization domain, D divides $A(e^\omega)$ and, by Lemma 5.14, each skew element in F.

Using Proposition 5.15, we can state the group case of Theorem 5.10 in an improved form.

Corollary 5.16. *Let U be a compact semisimple Lie group, $T \subset U$ a maximal torus, $\delta(t)$ given as above, and the Haar measures dt and du normalized by $\int dt = \int du = 1$. Then*

$$\int_U f(u)\, du = \frac{1}{|W|} \int_T \delta(t)\, dt \int_U f(utu^{-1})\, du,$$

where $|W|$ is the order of the Weyl group W of U.

It only remains to verify $\int_T \delta(t)\, dt = |W|$. For this we note that if $t = \exp H$, then

$$\delta(t) = \prod_{\alpha > 0} (e^{(1/2)\alpha} - e^{-(1/2)\alpha})(H) \operatorname{conj}\left(\prod_{\alpha > 0} (e^{(1/2)\alpha} - e^{-(1/2)\alpha})(H) \right),$$

which by Proposition 5.15 equals

$$\delta(t) = \sum_{s, \sigma \in W(R)} (\det s)(\det \sigma) e^{s\rho - \sigma\rho}(H).$$

†See "Some Details," p. 611.

If γ is a sum of roots, we can define a character e^γ of T by $e^\gamma(t) = e^{\gamma(H)}$ if $t = \exp H$ ($H \in \mathfrak{t}_0$). In fact, if $\exp H = e$, then $\gamma(H) \in 2\pi i \boldsymbol{Z}$. But $s\rho - \sigma\rho$ is a sum of roots and is $\neq 0$ unless $s = \sigma$. Hence $\int_T \delta(t)\, dt = |W|$, as stated.

C. *The Lie Algebra Case*

We shall now state and prove the infinitesimal version of Theorem 5.8 corresponding to the decomposition of $\mathfrak{p}$ into K-orbits. We retain the notation from the proof of Theorem 5.8.

Consider the mapping

$$\phi\colon (kM, H) \to \mathrm{Ad}(k)\, H, \qquad k \in K, \quad H \in \mathfrak{a}, \tag{20}$$

of $K/M \times \mathfrak{a}$ onto $\mathfrak{p}$. It is shown in [DS], Chapter VII, §3, that its differential at a point $(k_0 M, H_0) \in K/M \times \mathfrak{a}$ is given by

$$\begin{aligned} &d\phi_{(k_0 M, H_0)}(d\tau(k_0)\, L, d\tau(H_0)\, H) \\ &\quad = d\tau(\mathrm{Ad}(k_0)\, H_0)\, \mathrm{Ad}(k_0)\, ([L, H_0] + H) \end{aligned} \tag{21}$$

for $L \in \mathfrak{l}$, $H \in \mathfrak{a}$. As in the proof of (9) consider the orthonormal basis

$$H_1, \dots, H_l, T_i^\alpha \qquad (1 \leq i \leq m_\alpha, \alpha \in \Sigma^+)$$

of $\mathfrak{a} + \mathfrak{l}$; then if $H_0 \in \mathfrak{a}$ is regular,

$$H_1, \dots, H_l, \alpha(H_0)^{-1}[H_0, T_i^\alpha] \qquad (1 \leq i \leq m_\alpha, \alpha \in \Sigma^+)$$

is an orthonormal basis of $\mathfrak{p}$. Expressing $d\phi$ in terms of these bases, we see that

$$|\det(d\phi_{(k_0 M, H_0)})| = \prod_{\alpha \in \Sigma^+} |\alpha(H_0)|^{m_\alpha}. \tag{22}$$

Let $\mathfrak{p}'$ denote the set of regular elements in $\mathfrak{p}$ (elements whose centralizer in $\mathfrak{p}$ is abelian) and let dX and dH denote the Euclidean measures on $\mathfrak{p}$ and $\mathfrak{a}$ corresponding to the Killing form metric.

Theorem 5.17. *The mapping ϕ of $K/M \times \mathfrak{a}$ onto $\mathfrak{p}$ is a diffeomorphism of $K/M \times \mathfrak{a}^+$ onto $\mathfrak{p}'$; moreover,*

$$\int f(X)\, dX = c \int_{K/M} dk_M \left(\int_{\mathfrak{a}^+} f(\mathrm{Ad}(k)\, H) \prod_{\alpha \in \Sigma^+} \alpha(H)^{m_\alpha}\, dH \right), \qquad f \in C_c(\mathfrak{p}),$$

where c is a constant.

Only the injectivity on $K/M \times \mathfrak{a}^+$ remains to be verified. But suppose $\mathrm{Ad}(k_1)\, H_1 = \mathrm{Ad}(k_2)\, H_2$, where $k_1, k_2 \in K$, $H_1, H_2 \in \mathfrak{a}^+$. Let $k = k_1^{-1} k_2$, so that $\mathrm{Ad}(k)\, H_1 = H_2$. Then $\mathrm{Ad}(k)$ maps the centralizer of H_1 in $\mathfrak{p}$ onto the centralizer of H_2 in $\mathfrak{p}$. These being equal to $\mathfrak{a}$, we have $k \in M'$; but by the simple transitivity of the Weyl group on the set of Weyl chambers we conclude that $k \in M$, as desired.

We now prove a simple result showing once again how $\mathfrak{a}^+$ behaves like a multidimensional radius vector. For $X \in \mathfrak{p}$, let $\mathfrak{a}(X)$ denote the *unique point*

$$\mathfrak{a}(X) = (K \cdot X) \cap \overline{\mathfrak{a}^+}$$

(cf. [DS], Chapter IX, §1). Let d denote the distance function on $\mathfrak{p}$ and define distance between subsets as usual.

Proposition 5.18. *Let $X, Y \in \mathfrak{p}$. Then*

$$d(\mathrm{Ad}(K)\,X, \mathrm{Ad}(K)\,Y) = d(\mathfrak{a}(X), \mathfrak{a}(Y)).$$

Proof. Let $\mathrm{Ad}(k_0)\,Y$ be the point on the orbit $\mathrm{Ad}(K)\,Y$ which minimizes the square distance

$$F(k) = \langle X - \mathrm{Ad}(k)\;Y, X - \mathrm{Ad}(k)\;Y\rangle.$$

The relation

$$\left\{\frac{d}{dt} F(\exp tT \cdot k_0)\right\}_{t=0} = 0 \qquad (T \in \mathfrak{k})$$

gives

$$\langle X, [T, \mathrm{Ad}(k_0)\;Y]\rangle = 0, \qquad T \in \mathfrak{k},$$

which implies $[X, \mathrm{Ad}(k_0)\;Y] = 0$. Thus we may assume that X and $\mathrm{Ad}(k_0)\;Y$ are in $\mathfrak{a}$. But the proof of Theorem 2.12 in [DS] (Chapter VII), then shows that X and $\mathrm{Ad}(k_0)\;Y$ are in the same closed Weyl chamber, so the result follows.

3. Integral Formulas for the Bruhat Decomposition

We retain the notation from the beginning of this section and put $\bar{N} = \theta N$ As proved in [DS] (Chapter IX, §1), the mapping

(23) $$(\bar{n}, m, a, n) \to \bar{n}man$$

is a bijection of $\bar{N} \times M \times A \times N$ onto an open submanifold

$$\bar{N}MAN \subset G$$

whose complement has Haar measure 0. Furthermore, if we write $g = k(g) \exp H(g)\, n(g)$ $(k(g) \in K,\ H(g) \in \mathfrak{a},\ n(g) \in N)$, according to the Iwasawa decomposition $G = KAN$ the mapping

(24) $$\bar{n} \to k(\bar{n})M$$

is a diffeomorphism of $\bar{N}$ onto an open subset of K/M whose complement is a null set for the invariant measure dk_M. We shall now compute the Jacobian for these maps (23) and (24). As before, let ρ denote half the sum of the positive restricted roots (with multiplicity).

Lemma 5.19. *Let $g \in G$. The mapping $T_g: k \to k(gk)$ is a diffeomorphism of K onto itself and*

$$\int_K F(k(g^{-1}k))\,dk = \int_K F(k)e^{-2\rho(H(gk))}\,dk, \qquad F \in C(K).$$

Proof. Let $x \in G$. By Proposition 5.1 we have

$$\int f(kan)e^{2\rho(\log a)}\,dk\,da\,dn = \int f(g)\,dg = \int f(xg)\,dg.$$

But if $g = kan$,

$$\begin{aligned} xg &= xkan = k(xk)\exp H(xk)\,n(xk)an \\ &= k(xk)\exp H(xk)\,a(a^{-1}n(xk)an) = k_1 a_1 n_1, \end{aligned}$$

so that our integral is

$$\int f(k_1 a_1 n_1)e^{2\rho(\log a)}\,dk\,da\,dn.$$

But the translation

$$a \to \exp H(xk)\,a = a_1$$

preserves da, and the translation

$$n \to (a^{-1}n(xk)a)n = n_1$$

preserves dn. Thus the integral equals

$$\int f(k(xk)a_1 n_1)e^{2\rho(\log a_1)}e^{-2\rho(H(xk))}\,dk\,da_1\,dn_1.$$

Taking f of the form $f(kan) = F(k)F_1(a)F_2(n)$, we get

$$\int_K F(k)\,dk = \int_K F(k(xk))e^{-2\rho(H(xk))}\,dk, \tag{25}$$

and the lemma follows if we replace F with the function $F \circ T_{x^{-1}}$.

We can also write the lemma in the form

$$(T_g)^*(dk) = e^{-2\rho(H(gk))}\,dk. \tag{26}$$

Theorem 5.20. *The function* $\bar{n} \to e^{-2\rho(H(\bar{n}))}$ *is integrable on* $\bar{N}$ *and if* $d\bar{n}$ *and* dk_M *are normalized by*

$$\int_{K/M} dk_M = \int_{\bar{N}} e^{-2\rho(H(\bar{n}))}\, d\bar{n} = 1,$$

we have

$$\int_{K/M} F(kM)\, dk_M = \int_{\bar{N}} F(k(\bar{n})M) e^{-2\rho(H(\bar{n}))}\, d\bar{n}, \qquad F \in C(K/M).$$

Proof. We know from (24) that there exists a function $\psi \in \mathscr{E}(\bar{N})$ such that

$$\int_{K/M} F(kM)\, dk_M = \int_{\bar{N}} F(k(\bar{n})M)\psi(\bar{n})\, d\bar{n}. \tag{27}$$

Fix $x \in \bar{N}$. By Lemma 5.19 we have

$$\int_K F(k(x^{-1}k)M)\, dk = \int_K F(kM) e^{-2\rho(H(xk))}\, dk.$$

But $k(x^{-1}k(\bar{n})) = k(x^{-1}\bar{n})$, so by (27) the left-hand side is

$$\int_{\bar{N}} F(k(\bar{n})M)\psi(x\bar{n})\, d\bar{n},$$

whereas the right-hand side is

$$\int_{\bar{N}} F(k(\bar{n})M) e^{-2\rho(H(xk(\bar{n})))}\psi(\bar{n})\, d\bar{n}.$$

Thus

$$\psi(x\bar{n}) = e^{-2\rho(H(xk(\bar{n})))}\psi(\bar{n}),$$

and the theorem follows by putting $\bar{n} = e$.

Proposition 5.21. *Let* $d\bar{n}$, dm, da, *and* dn *be Haar measures on* $\bar{N}$, M, A, *and* N, *respectively. Then the Haar measure* dg *in* G *can be normalized so that*

$$\int_G f(g)\, dg = \int_{\bar{N} \times M \times A \times N} f(\bar{n}man) e^{2\rho(\log a)}\, d\bar{n}\, dm\, da\, dn \tag{28}$$

for $f \in C_c(G)$.

Proof. Because of the properties stated for the map (23), formula (28) holds with $e^{2\rho(\log a)}$ replaced by some unknown function $D(\bar{n}, m, a, n)$. Then D can be determined by the method of Proposition 5.1. Replacing

$f(g)$ by $f(\bar{n}_1 gn)$, we see that D is independent of $\bar{n}$ and n. Replacing $f(g)$ by $f(m_1 g)$ and noting that, by the compactness of M, $d\bar{n}$ is invariant under $\bar{n} \to m_1 \bar{n} m_1^{-1}$, we see that D is independent of m. Finally, D's dependence on a is found just as in Proposition 5.1.

§6. Orbital Integrals

In this section we give examples of solutions to Problem D in §3, No. 3, which amounts to the determination of a function in terms of its integrals over generalized spheres.

1. Pseudo-Riemannian Manifolds of Constant Curvature

Let X be a manifold. We recall that a *pseudo-Riemannian structure* of signature (p, q) is a smooth assignment $y \to g_y$, where g_y is a symmetric nondegenerate bilinear form on $X_y \times X_y$ of signature (p, q). This means that for a suitable basis $Y_1, \ldots, Y_{p+q}$ of X_y we have

$$g_y(Y, Y) = y_1^2 + \cdots + y_p^2 - y_{p+1}^2 - \cdots - y_{p+q}^2 \qquad \left(Y = \sum_1^{p+q} y_i Y_i\right).$$

If $q = 0$, we speak of a Riemannian structure, and if $p = 1$, we speak of a *Lorentzian* structure. Connected manifolds with such structures g are called pseudo-Riemannian (resp. Riemannian, Lorentzian) manifolds.

An isometry of a pseudo-Riemannian manifold X is a diffeomorphism preserving g. Let $I(X)$ denote the group of all isometries of X. For $y \in X$ let $I(X)_y$ denote the subgroup of $I(X)$ fixing y (the isotropy subgroup at y) and let H_y denote the group of linear transformations of the tangent space X_y induced by the action of $I(X)_y$. For each $a \in \boldsymbol{R}$ let $\Sigma_a(y)$ denote the "sphere"

$$\Sigma_a(y) = \{Z \in X_y : g_y(X, X) = a,\ Z \neq 0\}.$$

Definition. A pseudo-Riemannian manifold X is said to be *isotropic* if for each $a \in \boldsymbol{R}$ and each $y \in X$ the group H_y acts transitively on $\Sigma_a(y)$. (Note that the case $a = 0$ is included.)

An isotropic pseudo-Riemannian manifold X is necessarily homogeneous. For this let y, z be any two points in X and join them by a curve consisting of finitely many geodesic segments γ_i. For each γ_i there exists an isometry of X fixing the midpoint of γ_i but reversing the direction of γ_i. Their product will map y to z.

Let X be a manifold with a pseudo-Riemannian structure g and curvature tensor R. Let $y \in X$, $S \subset X_y$ a two-dimensional subspace on which g_y is nondegenerate. The curvature of X along the section S is defined by

$$K(S) = -\frac{g_y(R_y(X, Y)X, Y)}{g_y(X, X)g_y(Y, Y) - g_y(X, Y)^2}.$$

The denominator is in fact $\neq 0$, and the expression is independent of the choice of X and Y (cf., [DS], p. 250).

We shall now construct pseudo-Riemannian manifolds of signature (p, q) and constant curvature. Consider the space $\boldsymbol{R}^{p+q+1}$ with the flat pseudo-Riemannian structure

$$B_e(Y) = y_1^2 + \cdots + y_p^2 - y_{p+1}^2 - \cdots - y_{p+q}^2 + ey_{p+q+1}^2 \qquad (e = \pm 1).$$

Let Q_e denote the quadric in $\boldsymbol{R}^{p+q+1}$ given by

$$B_e(Y) = e. \tag{1}$$

The orthogonal group $\boldsymbol{O}(B_e)[=\boldsymbol{O}(p, q+1)$ or $\boldsymbol{O}(p+1, q)]$ acts transitively on Q_e; the isotropy subgroup at $o = (0, \ldots, 0, 1)$ is identified with $\boldsymbol{O}(p, q)$.

Theorem 6.1. (i) *The restriction of B_e to the tangent spaces to Q_e gives a pseudo-Riemannian structure g_e on Q_e of signature (p, q).*

(ii) *We have*

$$Q_{-1} \cong \boldsymbol{O}(p, q+1)/\boldsymbol{O}(p, q) \qquad (\textit{diffeomorphism}) \tag{2}$$

and the pseudo-Riemannian structure g_{-1} on Q_{-1} has constant curvature -1.

(iii) *We have*

$$Q_{+1} \cong \boldsymbol{O}(p+1, q)/\boldsymbol{O}(p, q) \qquad (\textit{diffeomorphism}) \tag{3}$$

and the pseudo-Riemannian structure g_{+1} on Q_{+1} has constant curvature $+1$.

(iv) *The flat space $\boldsymbol{R}^{p+q}$ with the quadratic form $g_0(Y) = \sum_1^p y_i^2 - \sum_{1+p}^{p+q} y_j^2$ and the spaces*

$$\boldsymbol{O}(p, q+1)/\boldsymbol{O}(p, q), \qquad \boldsymbol{O}(p+1, q)/\boldsymbol{O}(p, q)$$

are isotropic, and they exhaust the class of pseudo-Riemannian manifolds of constant curvature and signature (p, q) except for local isometry and multiplication of the pseudo-Riemannian structure by a constant factor.

Proof. If s_0 denotes the linear transformation

$$(y_1, \ldots, y_{p+q}, y_{p+q+1}) \to (-y_1, \ldots, -y_{p+q}, y_{p+q+1}),$$

then the mapping $\sigma: g \to s_0 g s_0$ is an involutive automorphism of $\mathbf{O}(p, q+1)$ whose differential $d\sigma$ has fixed point set $\mathfrak{o}(p, q)$ (the Lie algebra of $\mathbf{O}(p, q)$). The (-1)-eigenspace of $d\sigma$, say $\mathfrak{m}$, is spanned by the vectors

$$(4) \qquad Y_i = E_{i,p+q+1} + E_{p+q+1,i} \qquad (1 \leq i \leq p),$$

$$(5) \qquad Y_j = E_{j,p+q+1} - E_{p+q+1,j} \qquad (p+1 \leq j \leq p+q).$$

Here E_{ij} denotes a square matrix with entry 1 where the ith row and the jth column meet, all other entries being 0.

The mapping $\psi: g\mathbf{O}(p, q) \to g \cdot o$ has a differential $d\psi$ which maps $\mathfrak{m}$ bijectively onto the tangent plane $y_{p+q+1} = 1$ to Q_{-1} at o and $d\psi(X) = X \cdot o$ $(X \in \mathfrak{m})$. Thus

$$d\psi(Y_k) = (\delta_{1k}, \dots, \delta_{p+q+1,k}) \qquad (1 \leq k \leq p+q).$$

Thus $B_{-1}(d\psi(Y_k)) = 1$ if $1 \leq k \leq p$ and -1 if $p+1 \leq k \leq p+q$, proving (i). Next, since the space (2) is symmetric, its curvature tensor satisfies ([DS], Chapter IV, Exercise A1),

$$R_0(X, Y)(Z) = -[[X, Y], Z],$$

where [,] is the Lie bracket. A simple computation then shows

$$K(\mathbf{R}Y_k + \mathbf{R}Y_l) = -1 \qquad (1 \leq k, l \leq p+q),$$

and this implies (ii). Part (iii) is proved in the same way. For (iv) we first verify that the spaces listed are isotropic. Since the isotropy action of $\mathbf{O}(p, q+1)_0 = \mathbf{O}(p, q)$ on $\mathfrak{m}$ is the ordinary action of $\mathbf{O}(p, q)$ on $\mathbf{R}^{p+q}$, it suffices to verify that $\mathbf{R}^{p+q}$ with the quadratic form g_0 is isotropic. But we know that $\mathbf{O}(p, q)$ is transitive on $g_e = +1$ and on $g_e = -1$, so it remains to show that $\mathbf{O}(p, q)$ transitive on the cone

$$\{Y \neq 0 : g_e(Y) = 0\}.$$

By rotation in $\mathbf{R}^p$ and in $\mathbf{R}^q$, it suffices to verify the statement for $p = q = 1$. But for this case it is obvious. The uniqueness in (iv) follows from the general fact that a symmetric space is determined locally by its pseudo-Riemannian structure and curvature tensor at a point (cf., [DS], pp. 200–201). This finishes the proof.

The spaces (2) and (3) are the pseudo-Riemannian analogs of the spaces $\mathbf{O}(p, 1)/\mathbf{O}(p)$, $\mathbf{O}(p+1)/\mathbf{O}(p)$ from §4, No. 1. For the pseudo-Riemannian analogs of the other two-point homogeneous spaces, see Wolf [1967].

We shall later need a lemma about the connectivity of the groups $\mathbf{O}(p, q)$. Let $I_{p,q}$ denote the diagonal matrix (d_{ij}) with $d_{ii} = 1$ $(1 \leq i \leq p)$,

$d_{jj} = -1$ $(p+1 \le j \le p+q)$, so a matrix g with transpose ${}^t g$ belongs to $\boldsymbol{O}(p, q)$ if and only if

$$ {}^t g I_{p,q} g = I_{p,q}. \tag{6}$$

If $y \in \boldsymbol{R}^{p+q}$, let $y^T = (y_1, \ldots, y_p, 0 \cdots 0)$, $y^S = (0, \ldots, 0, y_{p+1}, \ldots, y_{p+q})$, and for $g \in \boldsymbol{O}(p, q)$, let g_T and g_S denote the matrices

$$(g_T)_{ij} = g_{ij} \qquad (1 \le i, j \le p),$$

$$(g_S)_{kl} = g_{kl} \qquad (p+1 \le k, l \le p+q).$$

If $g_1, \ldots, g_{p+q}$ denote the column vectors of the matrix g, then (6) means for the scalar products that

$$g_i^T \cdot g_i^T - g_i^S \cdot g_i^S = 1, \qquad 1 \le i \le p,$$

$$g_j^T \cdot g_j^T - g_j^S \cdot g_j^S = -1, \qquad p+1 \le j \le p+q,$$

$$g_j^T \cdot g_k^T = g_j^S \cdot g_k^S, \quad j \ne k.$$

Lemma 6.2. *We have for each* $g \in \boldsymbol{O}(p, q)$

$$|\det(g_T)| \ge 1, \qquad |\det(g_S)| \ge 1.$$

The components of $\boldsymbol{O}(p, q)$ *are obtained by*

(7) $\det g_T \ge 1, \quad \det g_S \ge 1$ *(identity component)*;

(8) $\det g_T \le -1, \quad \det g_S \ge 1$;

(9) $\det g_T \ge 1, \quad \det g_S \le -1$;

(10) $\det g_T \le -1, \quad \det g_S \le -1$.

Thus $\boldsymbol{O}(p, q)$ *has* 4 *components if* $p \ge 1$, $q \ge 1$, 2 *components if* p *or* $q = 0$.

Proof. Consider the Gram determinant

$$\det \begin{pmatrix} g_1^T \cdot g_1^T & g_1^T \cdot g_2^T & \cdots & g_1^T \cdot g_p^T \\ g_2^T \cdot g_1^T & & & \\ \vdots & & & \\ g_p^T \cdot g_1^T & & \cdots & g_p^T \cdot g_p^T \end{pmatrix},$$

which equals $(\det g_T)^2$. By using the relations above it can also be written

$$\det \begin{pmatrix} 1 + g_1^S \cdot g_1^S & g_1^S \cdot g_2^S & \cdots & g_1^S \cdot g_p^S \\ g_2^S \cdot g_1^S & & & \\ \vdots & & & \\ g_p^S \cdot g_1^S & & \cdots & 1 + g_p^S \cdot g_p^S \end{pmatrix},$$

which equals 1 plus a sum of lower-order Gram determinants each of which is still positive. Thus $(\det g_T)^2 \geq 1$, and similarly, $(\det g_S)^2 \geq 1$. Assuming now $p \geq 1$, $q \geq 1$, consider the decomposition of $\mathbf{O}(p, q)$ into the four pieces (7)–(10). Each of these is $\neq \emptyset$ because (8) is obtained from (7) by multiplication by $I_{1,p+q-1}$, etc. On the other hand, since the functions $g \to \det(g_T)$, $g \to \det(g_S)$ are continuous on $\mathbf{O}(p, q)$, the four pieces above belong to different components of $\mathbf{O}(p, q)$. However, by [DS], Chapter X, Lemma 2.3, $\mathbf{O}(p, q)$ is homeomorphic to the product of $\mathbf{O}(p, q) \cap \mathbf{U}(p + q)$ with a Euclidean space. Since

$$\mathbf{O}(p, q) \cap \mathbf{U}(p + q) = \mathbf{O}(p, q) \cap \mathbf{O}(p + q)$$

is homeomorphic to $\mathbf{O}(p) \times \mathbf{O}(q)$, it just remains to remark that $\mathbf{O}(n)$ has two components.

Now let X be a Lorentzian manifold of constant curvature. According to Theorem 6.1, X is (up to multiplication of the Lorentzian structure by a positive constant) locally isometric to one of the spaces

$$\mathbf{R}^{1+q} \qquad \text{(flat, signature } (1, q))$$

$$Q_{-1} = \mathbf{O}(1, q + 1)/\mathbf{O}(1, q): \qquad y_1^2 - y_2^2 - \cdots - y_{q+2}^2 = -1$$

$$Q_1 = \mathbf{O}(2, q)/\mathbf{O}(1, q): \qquad y_1^2 - y_2^2 - \cdots - y_{q+1}^2 + y_{q+2}^2 = 1$$

the Lorentzian structure being induced (in the two last cases) by the form

$$y_1^2 - y_2^2 - \cdots - y_{q+1}^2 \mp y_{q+2}^2.$$

2. Orbital Integrals for the Lorentzian Case

We shall now define the Lorentzian analog of the spherical averaging operator M^r from §2, No. 2, and §4. We start with some geometric preparation.

For manifolds X with a Lorentzian structure g we adopt the following customary terminology: If $y \in X$, the cone

$$C_y = \{Y \in X_y : g_y(Y, Y) = 0\}$$

is called the *null cone* (or the *light cone*) in X_y with vertex y. A nonzero vector $Y \in X_y$ is said to be *timelike*, *isotropic*, or *spacelike* if $g_y(Y, Y)$ is positive, 0, or negative, respectively. Similar designations apply to geodesics according to the type of their tangent vectors.

While the geodesics in $\mathbf{R}^{1+q}$ are just the straight lines, the geodesics in Q_{-1} and Q_{+1} can be found by the method of §4, No. 1.

Proposition 6.3. *The geodesics in the Lorentzian quadrics Q_{-1} and Q_{+1} have the following properties:*

(i) *The geodesics are the nonempty intersections of the quadrics with two-planes in $\boldsymbol{R}^{2+q}$ through the origin.*

(ii) *For Q_{-1} the spacelike geodesics are closed; for Q_{+1} the timelike geodesics are closed.*

(iii) *The isotropic geodesics are certain straight lines in $\boldsymbol{R}^{2+q}$.*

Proof. Part (i) follows by the symmetry considerations in §4, No. 1. For part (ii) consider the intersection of Q_{-1} with the two-plane $y_1 = y_4 = \cdots = y_{q+2} = 0$. The intersection is the circle $y_2 = \cos t$, $y_3 = \sin t$ whose tangent vector $(0, -\sin t, \cos t, 0, \ldots, 0)$ is clearly spacelike. Since $\boldsymbol{O}(1, q+1)$ permutes the spacelike geodesics transitively, the first statement in (ii) follows. For Q_{+1} we intersect similarly with the two-plane $y_2 = \cdots = y_{q+1} = 0$. For (iii) we note that the two-plane

$$\boldsymbol{R}(1, 0, \ldots, 0, 1) + \boldsymbol{R}(0, 1, \ldots, 0)$$

intersects Q_{-1} in a pair of straight lines

$$y_1 = t, \, y_2 = \pm 1, \, y_3 = \cdots = y_{q+1} = 0, \, y_{q+2} = t,$$

which clearly are isotropic. The transitivity of $\boldsymbol{O}(1, q+1)$ on the set of isotropic geodesics (Theorem 6.1) then implies that each of these is a straight line. The argument for Q_{+1} is similar.

Lemma 6.4. *The quadrics Q_{-1} and Q_{+1} $(q \geq 1)$ are connected.*

Proof. The q-sphere being connected, the point $(y_1, \ldots, y_{q+2})$ on $Q_{\mp 1}$ can be moved continuously on $Q_{\mp 1}$ to the point

$$(y_1, (y_2^2 + \cdots + y_{q+1}^2)^{1/2}, 0, \ldots, 0, y_{q+2}),$$

so the statement follows from the fact that the hyperboloids $y_1^2 - y_2^2 \mp y_3^2 = \mp 1$ are connected.

Lemma 6.5. *The identity components of $\boldsymbol{O}(1, q+1)$ and $\boldsymbol{O}(2, q)$ act transitively on Q_{-1} and Q_{+1}, respectively, and the isotropy subgroups are connected.*

Proof. The first statement comes from the general fact (cf., [DS], Chapter II, §4) that when a separable Lie group acts transitively on a connected manifold, then so does its identity component. For the isotropy groups we use the description (7) of the identity component. This shows quickly that

$$\boldsymbol{O}_0(1, q+1) \cap \boldsymbol{O}(1, q) = \boldsymbol{O}_0(1, q),$$

$$\boldsymbol{O}_0(2, q) \cap \boldsymbol{O}(1, q) = \boldsymbol{O}_0(1, q),$$

the subscript 0 denoting identity component. Thus we have

$$Q_{-1} = \boldsymbol{O}_0(1, q+1)/\boldsymbol{O}_0(1, q),$$
$$Q_{+1} = \boldsymbol{Q}_0(2, q)/\boldsymbol{O}_0(1, q),$$

proving the lemma.

We now write our three spaces $\boldsymbol{R}^{1+q}$, Q_{-1}, Q_{+1} in the form $X = G/H$, where $H = \boldsymbol{O}_0(1, q)$ and G is either $G^0 = \boldsymbol{R}^{1+q} \cdot \boldsymbol{O}_0(1, q)$ (semi-direct product, $\boldsymbol{R}^{1+q}$ normal), $G^- = \boldsymbol{O}_0(1, q+1)$ or $G^+ = \boldsymbol{O}_0(2, q)$. Let o denote the origin $\{H\}$ in X; that is,

$$o = \begin{cases} (0, \ldots, 0) & \text{if } X = \boldsymbol{R}^{1+q} \\ (0, \ldots, 0, 1) & \text{if } X = Q_{-1} \text{ or } Q_{+1}. \end{cases}$$

In the cases $X = Q_{-1}$, $X = Q_{+1}$, the tangent space X_o is the hyperplane $\{y_1, \ldots, y_{q+1}, 1\} \subset \boldsymbol{R}^{2+q}$.

The timelike vectors at o fill up the "interior" $\mathring{C}_o$ of the cone C_o. The set $\mathring{C}_o$ consists of two components. The components which contain the timelike vector

$$v_o = (-1, 0, \ldots, 0),$$

in the cases G^0/H, G^-/H, G^+/H, respectively, will be called the *retrograde cone* in X_o. It will be denoted D_o. The component of the hyperboloid $g_o(Y, Y) = r^2$ which lies in D_o will be denoted $S_r(o)$. If y is any other point of X, we define C_y, D_y, $S_r(y) \subset X_y$ by

$$C_y = g \cdot C_o, \qquad D_y = g \cdot D_o, \qquad S_r(y) = g \cdot S_r(o)$$

if $g \in G$ is chosen such that $g \cdot o = y$. This is a valid definition because the connectedness of H implies that $h \cdot D_o \subset D_o$. We also define

$$B_r(y) = \{Y \in D_y : 0 < g_y(Y, Y) < r^2\}.$$

If Exp denotes the exponential mapping of X_y into X, mapping rays through 0 onto geodesics through y, we put

$$\boldsymbol{D}_y = \operatorname{Exp} D_y, \qquad \boldsymbol{C}_y = \operatorname{Exp} C_y,$$
$$\boldsymbol{S}_r(y) = \operatorname{Exp} S_r(y), \qquad \boldsymbol{B}_r(y) = \operatorname{Exp} B_r(y).$$

Again $\boldsymbol{C}_y$ and $\boldsymbol{D}_y$ are called, respectively, the *light cone* and *retrograde cone* in X with vertex y. For the spaces $X = Q_{+1}$ we always assume $r < \pi$ in order that Exp will be one-to-one on $B_r(y)$ in view of Part (ii) of Proposition 6.3.

Lemma 6.6. *The negative of the Lorentzian structure on $X = G/H$ induces on each $\boldsymbol{S}_r(y)$ a Riemannian structure of constant negative curvature $(q > 1)$.*

Proof. The manifold X being isotropic, the group† $H = \mathbf{O}_0(1, q)$ acts transitively on $S_r(o)$. The subgroup leaving fixed the geodesic from o with tangent vector v_o is $\mathbf{O}_0(q)$. This implies the lemma.

Lemma 6.7. *The timelike geodesics from y intersect $S_r(y)$ under a right angle.*

Proof. By the group invariance it suffices to prove this for $y = o$ and the geodesic with tangent vector v_o. For this case the statement is obvious.

Let $\tau(g)$ denote the translation $xH \to gxH$ on G/H and for $Y \in \mathfrak{m}$ let T_Y denote the linear transformation $Z \to [Y, [Y, Z]]$ of $\mathfrak{m}$ into itself. As usual, we identify $\mathfrak{m}$ with $(G/H)_o$.

Lemma 6.8. *The exponential mapping* $\operatorname{Exp}: \mathfrak{m} \to G/H$ *has differential*

$$d\operatorname{Exp}_Y = d\tau(\exp Y) \circ \sum_0^\infty \frac{T_Y^n}{(2n+1)!} \qquad (Y \in \mathfrak{m}).$$

Proof. If $\pi: G \to G/H$ is the natural map, we have $\pi(\exp Y) = \operatorname{Exp} Y$ ([DS], Chapter IV, Exercise A1), so Theorem 4.1 (loc. cit.) is still valid, giving the lemma.

Lemma 6.9. *The linear transformation*

$$A_Y = \sum_0^\infty \frac{T_Y^n}{(2n+1)!}$$

has determinant given by

$$\det A_Y = \left\{\frac{\sinh(g(Y, Y))^{1/2}}{(g(Y, Y))^{1/2}}\right\}^q \qquad \text{for } Q_{-1},$$

$$\det A_Y = \left\{\frac{\sin(g(Y, Y))^{1/2}}{(g(Y, Y))^{1/2}}\right\}^q \qquad \text{for } Q_{+1}$$

for Y timelike.

Proof. Consider the case of Q_{-1}. Since $\det(A_Y)$ is invariant under H, it suffices to verify this for $Y = cY_1$ in (4), where $c \in \mathbf{R}$. We have $c^2 = g(Y, Y)$ and $T_{Y_1}(Y_j) = Y_j$ $(2 \leq j \leq q + 1)$. Thus T_Y has the eigenvalues 0 and $g(Y, Y)$; the latter is a q-tuple eigenvalue. This implies the formula for the determinant. The case Q_{+1} is treated in the same way.

From this lemma and the description of the geodesics in Proposition 6.3 we can now conclude the following result.

†See "Some Details," p. 611.

Proposition 6.10. (i) *The mapping* $\operatorname{Exp}: \mathfrak{m} \to Q_{-1}$ *is a diffeomorphism of* D_o *onto* $\mathbf{D}_0$.

(ii) *The mapping* $\operatorname{Exp}: \mathfrak{m} \to Q_{+1}$ *is a diffeomorphism of* $B_\pi(o)$ *onto* $\mathbf{B}_\pi(o)$.

Let dh denote a biinvariant measure on the unimodular group H. Let $u \in \mathscr{D}(X)$, $y \in X$ and $r > 0$. Select $g \in G$ such that $g \cdot o = y$ and select $x \in \mathbf{S}_r(o)$. Consider the integral

$$\int_H u(gh \cdot x)\, dh.$$

Since the subgroup $K \subset H$ leaving x fixed is compact, it is easy to see that the set

$$C_{g,x} = \{h \in H : gh \cdot x \in \operatorname{support}(u)\}$$

is compact; thus the integral above converges. By the biinvariance of dh it is independent of the choice of g (satisfying $g \cdot o = y$) and of the choice of $x \in \mathbf{S}_r(o)$. By analogy with the Riemannian case [§4, (24)] we can thus define the operator M^r (*the orbital integral*) by

$$(11) \qquad (M^r u)(y) = \int_H u(gh \cdot x)\, dh.$$

If g and x run through suitable compact neighborhoods, the sets $C_{g,x}$ are enclosed in a fixed compact subset of H, so $(M^r u)(y)$ depends smoothly on both r and y. It is also clear from (11) that the operator M^r is invariant under the action of G: if $l \in G$ and $\tau(l)$ denotes the transformation $nH \to lnH$ of G/H onto itself, then

$$M^r(u \circ \tau(l)) = (M^r u) \circ \tau(l).$$

If dk denotes the normalized Haar measure on K, we have by Theorem 1.9

$$\int_H u(h \cdot x)\, dh = \int_{H/K} d\dot{h} \int_K u(hk \cdot x)\, dk = \int_{H/K} u(h \cdot x)\, d\dot{h},$$

where $d\dot{h}$ is an H-invariant measure on H/K. But if $d\omega_r$ is the Riemannian measure on $\mathbf{S}_r(o)$ (cf. Lemma 6.6), we have by the uniqueness of H-invariant measures on the space $H/K \approx \mathbf{S}_r(o)$ that

$$(12) \qquad \int_H u(h \cdot x)\, dh = \frac{1}{A(r)} \int_{\mathbf{S}_r(o)} u(z)\, d\omega_r(z),$$

where $A(r)$ is a positive scalar. But since g is an isometry, we deduce from (12) that

$$(M^r u)(y) = \frac{1}{A(r)} \int_{\mathbf{S}_r(y)} u(z)\, d\boldsymbol{\omega}_r(z).$$

Now, we have to determine $A(r)$.

Lemma 6.11. *For a suitable fixed normalization of the Haar measure dh on H we have*

$$A(r) = r^q, \qquad (\sinh r)^q, \qquad (\sin r)^q$$

for the cases

$$\boldsymbol{R}^{1+q}, \qquad \boldsymbol{O}(1, q+1)/\boldsymbol{O}(1, q), \qquad \boldsymbol{O}(2, q)/\boldsymbol{O}(1, q),$$

respectively.

Proof. The relations above show that $dh = A(r)^{-1}\, d\boldsymbol{\omega}_r\, dk$. The mapping $\operatorname{Exp}: D_o \to \boldsymbol{D}_o$ preserves length on the geodesics through o and maps $S_r(o)$ onto $\boldsymbol{S}_r(o)$. Thus if $z \in S^r(o)$ and Z denotes the vector from 0 to z in X_o, the ratio of the volume elements of $\boldsymbol{S}_r(o)$ and $S_r(o)$ at z is given by $\det(d\operatorname{Exp}_Z)$. Because of Lemmas 6.8 and 6.9 this equals

$$1, \qquad \left(\frac{\sinh r}{r}\right)^q, \qquad \left(\frac{\sin r}{r}\right)^q$$

for the three respective cases. But the volume element $d\omega_r$ on $S_r(o)$ equals $r^q\, d\omega_1$. Thus we can write in the three respective cases

$$dh = \frac{r^q}{A(r)}\, d\omega_1\, dk, \qquad \frac{\sinh^q r}{A(r)}\, d\omega_1\, dk, \qquad \frac{\sin^q r}{A(r)}\, d\omega_1\, dk.$$

But we can once and for all normalize dh by $dh = d\omega_1\, dk$, and for this choice our formulas for $A(r)$ hold.

Let $\square$ denote the *wave operator* on $X = G/H$, that is, the Laplace–Beltrami operator for the Lorentzian structure g on X as defined in Chapter II, §2, No. 4.

Lemma 6.12. *Let $y \in X$. On the retrograde cone $\boldsymbol{D}_y$ the wave operator $\square$ can be written*

$$\square = \frac{\partial^2}{\partial r^2} + \frac{1}{A(r)} \frac{dA}{dr} \frac{\partial}{\partial r} - L_{\boldsymbol{S}_r(y)},$$

where $L_{\boldsymbol{S}_r(y)}$ is the Laplace–Beltrami operator on $\boldsymbol{S}_r(y)$.

Proof. We can take $y = o$. If $(\theta_1, \ldots, \theta_q)$ are coordinates on the "sphere" $S_1(o)$ in the flat space X_0, then $(r\theta_1, \ldots, r\theta_q)$ are coordinates on $S_r(o)$. The Lorentzian structure on D_o is therefore given by

$$dr^2 - r^2\, d\theta^2,$$

where $d\theta^2$ is the Riemannian structure of $S_1(o)$. Since A_Y in Lemma 6.9 is a diagonal matrix with eigenvalues 1 and $r^{-1}A(r)^{1/q}$ (q times), it follows from Lemma 6.8 that the image $\mathbf{S}_r(o) = \operatorname{Exp}(S_r(o))$ has Riemannian structure

$$r^2\, d\theta^2, \qquad \sinh^2 r\, d\theta^2, \qquad \text{and} \qquad \sin^2 r\, d\theta^2$$

in the cases $\boldsymbol{R}^{1+q}$, Q_{-1}, and Q_{+1}, respectively. By the perpendicularity in Lemma 6.7 it follows that the Lorentzian structure on $\boldsymbol{D}_o$ is given by

$$dr^2 - r^2\, d\theta^2, \qquad dr^2 - \sinh^2 r\, d\theta^2, \qquad dr^2 - \sin^2 r\, d\theta^2$$

in the three respective cases. Now the lemma follows immediately.

The operator M^r is of course the Lorentzian analog to the spherical mean value operator for isotropic Riemannian manifolds. We shall now prove that by analogy with the Riemannian case, the operator M^r commutes with the wave operator $\square$.

Theorem 6.13. *For each of the isotropic Lorentz spaces $X = G^-/H$, G^+/H, and G^0/H, the wave operator $\square$ and the orbital integral M^r commute*:

$$\square M^r u = M^r \square u \qquad \text{for} \quad u \in \mathscr{D}(X).$$

(For G^+/H we assume $r < \pi$.)

Given a function u on G/H we define the function $\tilde{u}$ on G by $\tilde{u}(g) = u(g \cdot o)$.

Lemma 6.14. *There exists a differential operator $\tilde{\square}$ on G invariant under all left and all right translations such that*

$$\tilde{\square}\tilde{u} = (\square u)^{\sim} \qquad \text{for} \quad u \in \mathscr{D}(X).$$

Proof. We consider first the case $X = G^-/H$. The bilinear form

$$K(Y, Z) = \tfrac{1}{2}\operatorname{Tr}(YZ)$$

on the Lie algebra $\mathfrak{o}(1, q + 1)$ of G^- is nondegenerate; in fact, K is nondegenerate on the complexification $\mathfrak{o}(q + 2, \boldsymbol{C})$ consisting of all complex skew symmetric matrices of order $q + 2$. A simple computation shows that in the notation of (4) and (5)

$$K(Y_1, Y_1) = 1, \qquad K(Y_j, Y_j) = -1 \qquad (2 \le j \le q + 1).$$

Since K is symmetric and nondegenerate, there exists a unique left-invariant pseudo-Riemannian structure $\tilde{K}$ on G^- such that $\tilde{K}_e = K$. Moreover, since K is invariant under the conjugation $Y \to gYg^{-1}$ of $\mathfrak{o}(1, q+1)$, $\tilde{K}$ is also right invariant. Let $\tilde{\square}$ denote the corresponding Laplace–Beltrami operator on G^-. Then $\tilde{\square}$ is invariant under all left and all right translations on G^-. Let $u \in \mathscr{D}(X)$. Since $\tilde{\square}\tilde{u}$ is invariant under all right translations from H, there is a unique function $v \in \mathscr{E}(X)$ such that $\tilde{\square}\tilde{u} = \tilde{v}$. The mapping $u \to v$ is a differential operator which at the origin must coincide with $\square$; that is, $\tilde{\square}\tilde{u}(e) = \square u(o)$. Since, in addition, both $\square$ and the operator $u \to v$ are invariant under the action of G^- on X, it follows that they coincide. This proves $\tilde{\square}\tilde{u} = (\square u)^{\sim}$.

The case $X = G^+/H$ is handled in the same manner. For the flat case $X = G^0/H$ let

$$Y_j = (0, \ldots, 1, \ldots, 0),$$

the jth coordinate vector on $\boldsymbol{R}^{1+q}$. Then $\square = Y_1^2 - Y_2^2 - \cdots - Y_{q+1}^2$. Since $\boldsymbol{R}^{1+q}$ is naturally imbedded in the Lie algebra of G^0, we can extend Y_j to a left-invariant vector field $\tilde{Y}_j$ on G^0. The operator $\tilde{\square} = \tilde{Y}_1^2 - \tilde{Y}_2^2 - \cdots - \tilde{Y}_{q+1}^2$ is then a left- and right-invariant differential operator on G^0, and again we have $\tilde{\square}\tilde{u} = (\square u)^{\sim}$. This proves the lemma.

We can now prove Theorem 6.13. If $g \in G$, let $L(g)$ and $R(g)$, respectively, denote the left and right translations $l \to gl$ and $l \to lg$ on G. If $l \cdot o = x$, $x \in \boldsymbol{S}_r(o)$ $(r > 0)$, and $g \cdot o = y$, then

$$(M^r u)(y) = \int_H \tilde{u}(ghl)\, dh$$

because of (11). As g and l run through sufficiently small compact neighborhoods, the integration takes place within a fixed compact subset of H as remarked earlier. Denoting by subscript the argument on which a differential operator is to act, we shall prove the following result.

Lemma 6.15.

$$\tilde{\square}_l\left(\int_H \tilde{u}(ghl)\, dh\right) = \int_H (\tilde{\square}\tilde{u})(ghl)\, dh = \tilde{\square}_g\left(\int_H \tilde{u}(ghl)\, dh\right).$$

Proof. The first equality sign follows from the left invariance of $\tilde{\square}$. In fact, the integral on the left is

$$\int_H (\tilde{u} \circ L(gh))(l)\, dh,$$

so

$$\tilde{\Box}_l\left(\int_H \tilde{u}(ghl)\,dh\right) = \int_H [\tilde{\Box}(\tilde{u} \circ L(gh))](l)\,dh$$
$$= \int_H [(\tilde{\Box}\tilde{u}) \circ L(gh)](l)\,dh = \int_H (\tilde{\Box}\tilde{u})(ghl)\,dh.$$

The second equality in the lemma follows similarly from the right invariance of $\tilde{\Box}$. But this second equality is just the commutativity statement in Theorem 6.13.

Lemma 6.15 also implies the following analog of the Darboux equation in Lemma 2.14.

Corollary 6.16. *Let $u \in \mathscr{D}(X)$ and put*

$$U(y, z) = (M^r u)(y) \qquad \textit{if} \quad z \in \boldsymbol{S}_r(o).$$

Then

$$\Box_y(U(y, z)) = \Box_z(U(y, z)).$$

Remark. In $\boldsymbol{R}^n$ the solutions to the Laplace equation $Lu = 0$ are characterized by the spherical mean value theorem $M^r u = u$ (all r). This can be stated in this equivalent form: $M^r u$ is constant in r. In this latter form the mean value theorem holds for the solutions of the wave equation $\Box u = 0$ in an isotropic Lorentzian manifold: *If u satisfies $\Box u = 0$ and if u is suitably small at ∞, then $(M^r u)(0)$ is constant in r.* For a precise statement and proof see Helgason [1959] (Chapter IV, 5). For $\boldsymbol{R}^2$ such a result was also noted by L. Ásgeirsson.

3. Generalized Riesz Potentials

In this section we generalize part of the theory of Riesz potentials (§2, No. 8) to Lorentzian manifolds of constant curvature.

Consider first the case

$$X = Q_{-1} = G^-/H = \boldsymbol{O}_0(1, n)/\boldsymbol{O}_0(1, n-1)$$

of dimension n and let $f \in \mathscr{D}(X)$ and $y \in X$. If $z = \operatorname{Exp}_y Y$ $(Y \in D_y)$, we put $r_{yz} = g(Y, Y)^{1/2}$ and consider the integral

$$(13) \qquad (I_-^\lambda f)(y) = \frac{1}{K_n(\lambda)} \int_{\boldsymbol{D}_y} f(z) \sinh^{\lambda-n}(r_{yz})\,dz,$$

where dz is the Riemannian measure on X, and

$$K_n(\lambda) = \pi^{(1/2)(n-2)}2^{\lambda-1}\Gamma\left(\frac{\lambda}{2}\right)\Gamma\left(\frac{\lambda+2-n}{2}\right). \tag{14}$$

The integral converges for $\operatorname{Re}\lambda \geq n$. We transfer the integral in (13) over to D_y by the diffeomorphism Exp ($=\operatorname{Exp}_y$). Since

$$dz = dr\, d\boldsymbol{\omega}_r = dr\left(\frac{\sinh r}{r}\right)^{n-1} d\omega_r$$

and since $dr\, d\omega_r$ equals the volume element dZ on D_y, we obtain

$$(I_-^\lambda f)(y) = \frac{1}{K_n(\lambda)}\int_{D_y}(f\circ \operatorname{Exp})(Z)\left(\frac{\sinh r}{r}\right)^{\lambda-1} r^{\lambda-n}\, dZ,$$

where $r = g(Z, Z)^{1/2}$. This has the form

$$\frac{1}{K_n(\lambda)}\int_{D_y} h(Z, \lambda) r^{\lambda-n}\, dZ, \tag{15}$$

where $h(Z, \lambda)$, as well as each of its partial derivatives with respect to the first argument, is holomorphic in λ and h has compact support in the first variable. Because of Theorem 2.40 and the subsequent remark, the function $\lambda \to (I_-^\lambda f)(y)$, which by its definition is holomorphic for $\operatorname{Re}\lambda > n$, admits a holomorphic continuation to the entire λ-plane. Its value at $\lambda = 0$ is $h(0, 0) = f(y)$. Denoting the holomorphic continuation of (13) by $(I_-^\lambda f)(y)$ we have thus obtained

$$I_-^0 f = f. \tag{16}$$

We would now like to differentiate (13) with respect to y. For this we write the integral in the form $\int_F f(z)K(y, z)\, dz$ over a bounded region F which properly contains the intersection of the support of f with the closure of $\boldsymbol{D}_y$. The kernel $K(y, z)$ is defined as $\sinh^{\lambda-n} r_{yz}$ if $z \in \boldsymbol{D}_y$, otherwise as 0. For $\operatorname{Re}\lambda$ sufficiently large, $K(y, z)$ is twice continuously differentiable in y, so we can deduce for such λ that $I_-^\lambda f$ is of class C^2 and that

$$(\square I_-^\lambda f)(y) = \frac{1}{K_n(\lambda)}\int_{D_y} f(z)\, \square_y(\sinh^{\lambda-n} r_{yz})\, dz. \tag{17}$$

Moreover, given $m \in \boldsymbol{Z}^+$, we can find k such that $I_-^\lambda f \in C^m$ for $\operatorname{Re}\lambda > k$ (and all f). Using Lemma 6.12 and the relation

$$\frac{1}{A(r)}\frac{dA}{dr} = (n-1)\coth r,$$

we find

$$\begin{aligned}\square_y(\sinh^{\lambda-n} r_{yz}) &= \square_z(\sinh^{\lambda-n} r_{yz})\\ &= (\lambda - n)(\lambda - 1)\sinh^{\lambda-n} r_{yz}\\ &\quad + (\lambda - n)(\lambda - 2)\sinh^{\lambda-n-2} r_{yz}.\end{aligned}$$

We also have

$$K_n(\lambda) = (\lambda - 2)(\lambda - n)K_n(\lambda - 2),$$

so by substituting into (17) we get

$$\square I_-^\lambda f = (\lambda - n)(\lambda - 1)I_-^\lambda f + I_-^{\lambda-2}f.$$

Still assuming $\operatorname{Re}\lambda$ large we can use Green's formula to express the integral

$$\int_{D_y} [f(z)\,\square_z(\sinh^{\lambda-n} r_{yz}) - \sinh^{\lambda-n} r_{yz}\,(\square f)(z)]\,dz \tag{18}$$

as a surface integral over a part of C_y (on which $\sinh^{\lambda-n} r_{yz}$ and its first-order derivatives vanish) together with an integral over a surface inside D_y (on which f and its derivatives vanish). Hence the expression (18) vanishes, so we have proved the relations

$$\square(I_-^\lambda f) = I_-^\lambda(\square f), \tag{19}$$

$$I_-^\lambda(\square f) = (\lambda - n)(\lambda - 1)I_-^\lambda f + I_-^{\lambda-2}f \tag{20}$$

for $\operatorname{Re}\lambda > k$, k being some number (independent of f).

Since both sides of (20) are holomorphic in λ, this relation holds for all $\lambda \in \boldsymbol{C}$. We shall now deduce that for each $\lambda \in \boldsymbol{C}$, $I_-^\lambda f \in \mathscr{E}(X)$ and (19) holds. For this we observe by iterating (20) that for each $p \in \boldsymbol{Z}^+$

$$I_-^\lambda f = I_-^{\lambda+2p}(Q_p(\square)f), \tag{21}$$

Q_p being a certain pth-degree polynomial. Choosing p arbitrarily large, we deduce from the remark following (17) that $I_-^\lambda f \in \mathscr{E}(X)$; second, (19) implies for $\operatorname{Re}\lambda + 2p > k$ that

$$\square I_-^{\lambda+2p}(Q_p(\square)f) = I_-^{\lambda+2p}(Q_p(\square)\,\square f).$$

By using (21) again, this means that (19) holds for all λ.

Putting $\lambda = 0$ in (20), we get

$$I^{-2}f = \square f - nf. \tag{22}$$

Remark. In Riesz's paper [1949] (p. 190), an analog I^α of the potentials (82) in §2, is defined for any analytic Lorentzian manifold. These potentials I^α are different, however, from our I_-^λ and satisfy the equation $I^{-2}f = \square f$ in contrast to (22).

We consider next the case

$$X = Q_{+1} = G^+/H = \boldsymbol{O}_0(2, n-1)/\boldsymbol{O}_0(1, n-1)$$

and define for $f \in \mathcal{D}(X)$

$$(I_+^\lambda f)(y) = \frac{1}{K_n(\lambda)} \int_{D_y} f(z) \sin^{\lambda-n}(r_{yz})\, dz. \tag{23}$$

Again, $K_n(\lambda)$ is given by (14) and dz is the volume element. In order to bypass the difficulties caused by the fact that the function $z \to \sin r_{yz}$ vanishes on $\boldsymbol{S}_\pi(y)$, we assume that f has support disjoint from $\boldsymbol{S}_\pi(o)$. Then the support of f is disjoint from $\boldsymbol{S}_\pi(y)$ for all y in some neighborhood of o in X. We can then prove just as before that

$$(I_+^0 f)(y) = f(y), \tag{24}$$

$$(\square I_+^\lambda f)(y) = (I_+^\lambda \square f)(y), \tag{25}$$

$$(I_+^\lambda \square f)(y) = -(\lambda - n)(\lambda - 1)(I_+^\lambda f)(y) + (I_+^{\lambda-2} f)(y) \tag{26}$$

for all $\lambda \in \boldsymbol{C}$. In particular,

$$I_+^{-2} f = \square f + nf. \tag{27}$$

Finally, we consider the flat case

$$X = \boldsymbol{R}^n = G^0/H = \boldsymbol{R}^n \cdot \boldsymbol{O}_0(1, n-1)/\boldsymbol{O}_0(1, n-1)$$

and define

$$(I_0^\lambda f)(y) = \frac{1}{K_n(\lambda)} \int_{D_y} f(z) r_{yz}^{\lambda-n}\, dz.$$

These are the potentials defined by Riesz in [1949] (p. 31), who proved

$$I_0^0 f = f, \qquad \square I_0^\lambda f = I_0^\lambda \square f = I_0^{\lambda-2} f. \tag{28}$$

4. Determination of a Function from Its Integrals over Lorentzian Spheres

In a Riemannian manifold a function is determined in terms of its spherical mean values by the simple relation $f = \lim_{r \to 0} M^r f$. We shall now solve the analogous problem for an even-dimensional Lorentzian manifold of constant curvature and express a function f in terms of its orbital integrals $M^r f$. Since the spheres $\boldsymbol{S}_r(y)$ do not shrink to a point as $r \to 0$ the formula (cf. Theorem 6.17) below is quite different.

For the solution of the problem we use the geometric description of the wave operator $\square$ developed in §6, No. 2, particularly its commutation with the orbital integral M^r, and combine this with the results about the generalized Riesz potentials established in §6, No. 3.

We consider first the negatively curved space $X = G^-/H$. Let $n = \dim X$ and assume n even. Let $f \in \mathcal{D}(X)$ and put $F(r) = (M^r f)(y)$. Since the volume element dz on $\boldsymbol{D}_y$ is given by $dz = dr\, d\boldsymbol{\omega}^r$, we obtain from (12) and Lemma 6.9

$$(I_-^\lambda f)(y) = \frac{1}{K_n(\lambda)} \int_0^\infty \sinh^{\lambda - 1} r\, F(r)\, dr. \tag{29}$$

Let $Y_1, \ldots, Y_n$ be a basis of X_y such that the Lorentzian structure is given by

$$g_y(Y, Y) = y_1^2 - y_2^2 - \cdots - y_n^2, \qquad Y = \sum_1^n y_i Y_i.$$

If $\theta_1, \ldots, \theta_{n-2}$ are geodesic polar coordinates on the unit sphere in $\boldsymbol{R}^{n-1}$, we put

$$\begin{aligned} y_1 &= -r \cosh \zeta \qquad (0 \le \zeta < \infty,\ 0 < r < \infty), \\ y_2 &= r \sinh \zeta \cos \theta_1, \\ &\ \vdots \\ y_n &= r \sinh \zeta \sin \theta_1 \cdots \sin \theta_{n-2}. \end{aligned}$$

Then $(r, \zeta, \theta_1, \ldots, \theta_{n-2})$ are coordinates on the retrograde cone D_y and the volume element on $S_r(y)$ is given by

$$d\omega_r = r^{n-1} \sinh^{n-2} \zeta\, d\zeta\, d\omega^{n-2},$$

where $d\omega^{n-2}$ is the volume element on the unit sphere in $\boldsymbol{R}^{n-1}$. It follows that (writing sh for sinh, ch for cosh)

$$d\boldsymbol{\omega}_r = \operatorname{sh}^{n-1} r \operatorname{sh}^{n-2} \zeta\, d\zeta\, d\omega^{n-2}$$

and therefore

$$\begin{aligned} F(r) = &\iint (f \circ \operatorname{Exp})(-r \operatorname{ch} \zeta, r \operatorname{sh} \zeta \cos \theta_1, \ldots, r \operatorname{sh} \zeta \sin \theta_1 \cdots \sin \theta_{n-2}) \\ &\times \operatorname{sh}^{n-2} \zeta\, d\zeta\, d\omega^{n-2}. \end{aligned} \tag{30}$$

Now select A such that $f \circ \operatorname{Exp}$ vanishes outside the sphere $y_1^2 + \cdots + y_n^2 = A^2$ in X_y. Then in the integral (30) the range of ζ is contained in the interval $(0, \zeta_0)$, where $r^2 \cosh^2 \zeta_0 + r^2 \sinh^2 \zeta_0 = A^2$. We see by the substitution $t = r \sinh \zeta$ that the integral expression (30) behaves for small r like

$$\int_0^K \phi(t)\left(\frac{t}{r}\right)^{n-2}(r^2 + t^2)^{-1/2}\,dt,$$

where ϕ is bounded. Therefore if $n > 2$, the limit

$$a = \lim_{r\to 0} \sinh^{n-2} r\, F(r) \qquad (n > 2) \tag{31}$$

exists. For $n = 2$ we find similarly by differentiating (30) that the limit

$$b = \lim_{r\to 0}(\sinh r)F'(r) \qquad (n = 2) \tag{32}$$

exists.

Consider now the case $n > 2$. We can rewrite (29) in the form

$$(I_-^\lambda f)(y) = \frac{1}{K_n(\lambda)}\int_0^A \sinh^{n-2} r\, F(r) \sinh^{\lambda-n+1} r\, dr,$$

where $F(A) = 0$. We now evaluate both sides for $\lambda = n - 2$. Since $K_n(\lambda)$ has a simple pole for $\lambda = n - 2$, the integral has at most a simple pole there and the residue of the integral is

$$\lim_{\lambda\to n-2} (\lambda - n + 2)\int_0^A \sinh^{n-2} r\, F(r) \sinh^{\lambda-n+1} r\, dr.$$

Here we can take λ real and greater than $n - 2$. This is convenient, since by (31) the integral is then absolutely convergent and we do not have to think of it as an implicitly given holomorphic extension. We split the integral into two terms:

$$(\lambda - n + 2)\int_0^A (\sinh^{n-2} r\, F(r) - a) \sinh^{\lambda-n+1} r\, dr$$

$$+\, a(\lambda - n + 2)\int_0^A \sinh^{\lambda-n+1} r\, dr.$$

For the last term we use the relation

$$\lim_{\mu\to 0+} \mu \int_0^A \sinh^{\mu-1} r\, dr$$

$$= \lim_{\mu\to 0+} \mu \int_0^{\sinh A} t^{\mu-1}(1 + t^2)^{-1/2}\, dt = 1$$

by (71) in §2. For the first term we can for each $\varepsilon > 0$ find a $\delta > 0$ such that

$$|\sinh^{n-2} r\, F(r) - a| < \varepsilon \qquad \text{for} \quad 0 < r < \delta.$$

If $N = \max|\sinh^{n-2} r\, F(r)|$, we have for $n - 2 < \lambda < n - 1$ the estimates

$$\left|(\lambda - n + 2)\int_\delta^A (\sinh^{n-2} r\, F(r) - a) \sinh^{\lambda-n+1} r\, dr\right|$$
$$\leq (\lambda - n + 2)(N + |a|)(A - \delta)(\sinh \delta)^{\lambda-n+1};$$
$$\left|(\lambda - n + 2)\int_0^\delta (\sinh^{n-2} r\, F(r) - a) \sinh^{\lambda-n+1} r\, dr\right|$$
$$\leq \varepsilon(\lambda - n + 2)\int_0^\delta r^{\lambda-n+1}\, dr = \varepsilon\delta^{\lambda-n+2}.$$

By taking $\lambda - (n - 2)$ small enough, the right-hand side of each of these inequalities is $<2\varepsilon$. We have therefore proved

$$\lim_{\lambda\to n-2} (\lambda - n + 2)\int_0^\infty \sinh^{\lambda-1} r\, F(r)\, dr$$
$$= \lim_{r\to 0} \sinh^{n-2} r\, F(r).$$

Taking into account the formula for $K_n(\lambda)$, we have proved for the integral (29)

$$I_-^{n-2}f = (4\pi)^{(1/2)(2-n)} \frac{1}{\Gamma(\frac{1}{2}(n-2))} \lim_{r\to 0} \sinh^{n-2} r\, M^r f. \tag{33}$$

On the other hand, using formula (20) recursively, we obtain for $u \in \mathscr{D}(X)$

$$I_-^{n-2}(Q(\square)u) = u,$$

where

$$Q(\square) = (\square + (n-3)2)(\square + (n-5)4)\cdots(\square + 1(n-2)).$$

We combine this with (33) and use the commutativity $\square M^r = M^r \square$. This gives

$$u = (4\pi)^{(1/2)(2-n)} \frac{1}{\Gamma(\frac{1}{2}(n-2))} \lim_{r\to 0} \sinh^{n-2} r\, Q(\square)M^r u. \tag{34}$$

Here we can replace $\sinh r$ by r and can replace $\square$ by the operator

$$\square_r = \frac{d^2}{dr^2} + (n-1)\coth r \frac{d}{dr}$$

because of Lemma 6.12 and Corollary 6.16.

For the case $n = 2$ we have by (29)

$$(I_-^2 f)(y) = \frac{1}{K_2(2)} \int_0^\infty \sinh r\, F(r)\, dr. \tag{35}$$

This integral, which in effect only goes from 0 to A, is absolutely convergent because our estimate of (30) shows (for $n = 2$) that $F(r) = O(\log r)$ near $r = 0$. But using (20), (35), Lemma 6.12, Theorem 6.13, and Corollary 6.16, we obtain for $u \in \mathscr{D}(X)$

$$\begin{aligned} u = I_-^2\, \square u &= \frac{1}{2} \int_0^\infty \sinh r\, M^r\, \square u\, dr \\ &= \frac{1}{2} \int_0^\infty \sinh r\, \square M^r u\, dr = \frac{1}{2} \int_0^\infty \sinh r \left(\frac{d^2}{dr^2} + \coth r \frac{d}{dr}\right) M^r u\, dr \\ &= \frac{1}{2} \int_0^\infty \frac{d}{dr}\left(\sinh r \frac{d}{dr} M^r u\right) dr = -\frac{1}{2} \lim_{r \to 0} \sinh r \frac{d(M^r u)}{dr}. \end{aligned}$$

This is the substitute for (34) in the case $n = 2$.

The spaces G^+/H and G^0/H can be treated in the same manner. We have thus proved the following result in response to Problem D (or D′) in §3.

Theorem 6.17. *Let X be one of the isotropic Lorentzian manifolds G^-/H, G^0/H, G^+/H. Let K denote the curvature of X ($K = -1, 0, +1$) and assume $n = \dim X$ to be even. Put*

$$Q(\square) = (\square - K(n-3)2)(\square - K(n-5)4) \cdots (\square - K \cdot 1(n-2)).$$

Then if $u \in \mathscr{D}(X)$,

$$u = (4\pi)^{(1/2)(2-n)} \frac{1}{\Gamma(\frac{1}{2}(n-2))} \lim_{r \to 0} r^{n-2}\, Q(\square_r)(M^r u), \qquad (n \neq 2) \tag{36}$$

$$u = -\frac{1}{2} \lim_{r \to 0} r \frac{d}{dr}(M^r u) \qquad (n = 2). \tag{37}$$

Here $\square$ is the Laplace–Beltrami operator and $\square_r$ its radial part:

$$\square_r = \frac{d^2}{dr^2} + \frac{1}{A(r)} \frac{dA}{dr} \frac{d}{dr}.$$

5. Orbital Integrals on $SL(2, R)$

In this subsection we solve Problem D (§3, No. 3) in another special case. We show for the group $G = \mathbf{SL}(2, \mathbf{R})$ that a function $f \in \mathscr{D}(G)$

is explicitly determined by its integrals over the conjugacy classes of elliptic elements (cf. [DS], Chapter IX, §7) and their translates.

Consider the Iwasawa decomposition $G = KAN$, where

$$A = \left\{a = a_t = \begin{pmatrix} e^t & 0 \\ 0 & e^{-t} \end{pmatrix} : t \in \boldsymbol{R}\right\}, \qquad A^+ = \{a_t : t > 0\},$$

$$N = \left\{n = n_x = \begin{pmatrix} 1 & x \\ 0 & 1 \end{pmatrix} : x \in \boldsymbol{R}\right\},$$

$$K = \left\{k = k_\theta = \begin{pmatrix} \cos\theta & \sin\theta \\ -\sin\theta & \cos\theta \end{pmatrix} : \theta \in \boldsymbol{R}\right\}.$$

Since $a_t n_x a_t^{-1} = n_{xe^{2t}}$, the positive root α is given by $\alpha(\log a_t) = 2t$, so that, because of $G = K\,\mathrm{Cl}(A^+)\,K$ and Theorem 5.8, the Haar measure dg can be normalized so that

$$\text{(38)} \qquad \int_G f(g)\,dg = \frac{1}{4\pi^2}\int_0^{2\pi}\int_0^{2\pi}\int_0^{\infty} f(k_\theta a_t k_\phi)\sinh(2t)\,d\theta\,d\phi\,dt.$$

Denoting by $d\dot{g}$ the invariant measure on G/K [normalized by dg and $dk = (1/(2\pi))\,d\theta$], we consider the *orbital integral*

$$\text{(39)} \qquad \Phi_f(\theta) = \int_{G/K} f(gk_\theta g^{-1})\,d\dot{g}, \qquad f \in \mathscr{D}(G),$$

of f over the conjugacy class k_θ^G of k_θ.

Theorem 6.18. *The integral* (39) *converges absolutely, and*

$$\lim_{\theta\to 0}\frac{d}{d\theta}(\theta\Phi_f(\theta)) = -\frac{1}{2}f(e).$$

Proof. We have

$$\text{(40)} \qquad a_t k_\theta a_t^{-1} = a_t \exp\begin{pmatrix} 0 & \theta \\ -\theta & 0 \end{pmatrix} a_t^{-1} = \exp\begin{pmatrix} 0 & \theta e^{2t} \\ -\theta_e^{-2t} & 0 \end{pmatrix}$$

$$= \begin{pmatrix} \cos\theta & e^{2t}\sin\theta \\ -e^{-2t}\sin\theta & \cos\theta \end{pmatrix}.$$

The last expression shows that k_θ^G is closed in G, so the integral (39) is absolutely convergent. Also, $\Phi_f(\theta)$ can be written as $\int_G f(gk_\theta g^{-1})\,dg$ and therefore equals $\Phi_{f^*}(\theta)$, where $f^*(g) = (1/(2\pi))\int f(k_\theta g k_\theta^{-1})\,d\theta$. For the proof of Theorem 6.18 we can therefore assume f to be invariant under conjugation by K. Then writing $\dot{g} = k_\phi a_t K$ so by (38)

$$d\dot{g} = (1/(2\pi))\sinh(2t)\,dt\,d\phi,$$

we have

$$\Phi_f(\theta) = \int_0^\infty f(a_t k_\theta a_t^{-1}) \sinh(2t)\, dt.$$

Here we use (40), replace $2t$ by t, and introduce the function

$$F(x, y) = (f \circ \exp)\begin{pmatrix} 0 & x \\ -y & 0 \end{pmatrix}, \qquad (x, y) \in \boldsymbol{R}^2.$$

Then $\theta\Phi_f(\theta) = \frac{1}{4}g(\theta)$, where

$$g(r) = r \int_0^\infty F(re^t, re^{-t})(e^t - e^{-t})\, dt, \qquad r \neq 0. \tag{41}$$

In (37) we represented $u(0, 0)$ by means of the integral

$$(M^r u)(0, 0) = \int_{-\infty}^\infty u(-r\cosh\zeta, r\sinh\zeta)\, d\zeta,$$

which is the invariant integral of u over the *semihyperbola* $x^2 - y^2 = r^2$, $x < 0$. The integral (41) represents the integral of the function

$$F(x, y)(x - y)$$

over the *quarter-hyperbola* $xy = r^2$, $x \geq r$. The following counterpart to (37) gives Theorem 6.18.

Lemma 6.19. *For $F \in \mathscr{D}(\boldsymbol{R}^2)$ define g by (41). Then*

$$\lim_{r \to 0} g'(r) = -2F(0, 0).$$

Proof. Let $F_1 = \partial F/\partial x$, $F_2 = \partial F/\partial y$. Then

$$g'(r) = \int_0^\infty (F + re^t F_1 + re^{-t} F_2)(re^t, re^{-t})(e^t - e^{-t})\, dt$$

$$= \int_0^\infty (e^t F + re^{2t} F_1)\, dt - \int_0^\infty e^{-t} F\, dt + \int_0^\infty (rF_2 - rF_1 - re^{-2t} F_2)\, dt.$$

By the dominating convergence theorem,

$$\lim_{r \to 0} \int_0^\infty e^{-t} F(re^t, re^{-t})\, dt = F(0, 0)$$

and

$$\lim_{r \to 0} r \int_0^\infty e^{-2t} F_2(re^t, re^{-t})\, dt = 0.$$

Since F_i has support contained in a ball $B_A(0)$, $F_i(re^t, re^{-t}) = 0$ for $t > \log(Ar^{-1})$. Hence

$$(42) \qquad \int F_i(re^t, re^{-t})\, dt = O(|\log r|),$$

so it remains to prove

$$(43) \qquad \lim_{r \to 0} \int_0^{\infty} [e^t F(re^t, re^{-t}) + re^{2t} F_1(re^t, re^{-t})]\, dt = -F(0,0).$$

The integrand, however, equals

$$\frac{d}{dt}(e^t F(re^t, re^{-t})) + rF_2(re^t, re^{-t}),$$

so (43) follows from (42).

EXERCISES AND FURTHER RESULTS

A. Invariant Measures

1. Let G be a Lie group and H a closed subgroup. Then

(i) If H is compact, G/H has an invariant measure.

(ii) If G is unimodular and H normal, then H is unimodular.

(iii) If G/H has a finite invariant measure and if H is unimodular, then G is unimodular.

2. For the group $\boldsymbol{O}(2)$ the element $g = \begin{pmatrix} 0 & 1 \\ 1 & 0 \end{pmatrix}$ satisfies $\operatorname{Ad}(g) = -I$.

3. Let G be a connected Lie group with Lie algebra $\mathfrak{g}$, and $H \subset G$ a closed analytic subgroup with Lie algebra $\mathfrak{h} \subset \mathfrak{g}$. Let $X_1, \ldots, X_n$ be a basis of $\mathfrak{g}$ such that $X_{r+1}, \ldots, X_n$ span $\mathfrak{h}$ and put

$$\mathfrak{m} = \boldsymbol{R}X_1 + \cdots + \boldsymbol{R}X_r.$$

Let c_{ij}^k be determined by $[X_i, X_j] = \sum_k c_{ij}^k X_k$.

(i) G is unimodular if and only if

$$\operatorname{Tr}_{\mathfrak{g}}(\operatorname{ad} X) = 0 \qquad \text{for} \quad X \in \mathfrak{g},$$

or, equivalently,

$$\sum_{k=1}^{n} c_{ik}^k = 0 \qquad \text{for} \quad \text{each } i, \quad 1 \le i \le n.$$

(ii) The space G/H has an invariant measure if and only if

$$\operatorname{Tr}(\operatorname{ad}_{\mathfrak{g}}(T)) = \operatorname{Tr}(\operatorname{ad}_{\mathfrak{h}}(T)) \qquad \text{for} \quad T \in \mathfrak{h},$$

or, equivalently,

$$\sum_{\alpha=1}^{r} c_{i\alpha}^{\alpha} = 0 \qquad \text{for} \quad r+1 \le i \le n$$

(Chern [1942]).

4. Show that the group $\boldsymbol{M}(n)$ of isometries of $\boldsymbol{R}^n$ is isomorphic to the group of matrices

$$g_{k,x} = \begin{pmatrix} & k & & x_1 \\ & & & \vdots \\ & & & x_n \\ 0 & \cdots & 0 & 1 \end{pmatrix},$$

where $k \in K = \boldsymbol{O}(n)$ and $x = (x_1, \ldots, x_n) \in \boldsymbol{R}^n$. A Haar measure dg on $\boldsymbol{M}(n)$ is then given by

$$\int_G f(g)\, dg = \int_{K \times \boldsymbol{R}^n} f(g_{k,x})\, dk\, dx, \qquad f \in C_c(\mathbf{M}(n)),$$

where dk is a Haar measure on K.

5. A biinvariant measure on the group $G = \boldsymbol{GL}(n, \boldsymbol{R})$ of nonsingular matrices $X = (x_{ij})$ is given by

$$f \to \int_G f(X) |\det X|^{-n} \prod_{i,j} dx_{ij}.$$

6. A biinvariant measure on the unimodular group $G = \boldsymbol{SL}(n, \boldsymbol{R})$ is given by

$$f \to \int_{G'} f(X) |\det X_{11}|^{-1} \prod_{(i,j) \neq (1,1)} dx_{ij}.$$

Here $X = (x_{ij})$, X_{ij} is the (i, j)-cofactor in X, and the x_{ij} (except for x_{11}) are taken as independent variables on the set G' given by $\det X_{11} \neq 0$.

7. Let $\boldsymbol{T}(n, \boldsymbol{R})$ denote the group of all $g \in \boldsymbol{GL}(n, \boldsymbol{R})$ which are upper triangular. A left-invariant measure on $\boldsymbol{T}(n, \boldsymbol{R})$ is given by

$$f \to \int_{\boldsymbol{T}(n, \boldsymbol{R})} f(t) t_{11}^{-n} t_{22}^{1-n} \cdots t_{nn}^{-1} \prod_{i \le j} dt_{ij}$$

and a right-invariant measure by

$$f \to \int_{\boldsymbol{T}(n, \boldsymbol{R})} f(t) t_{11}^{-1} t_{22}^{-2} \cdots t_{nn}^{-n} \prod_{i \le j} dt_{ij}.$$

8. Let $G = \boldsymbol{SL}(2, \boldsymbol{R})$ and let Γ be the *modular group* $\boldsymbol{SL}(2, \boldsymbol{Z})$. Prove that

$$\mu(G/\Gamma) < \infty$$

if μ is an invariant measure on G/Γ. [*Hint*: Let G act on the upper half-plane $H: y > 0$ by $z \to (az + b)/(cz + d)$. Since Γ contains the translation $T: z \to z + 1$ and the "inversion" $S: z \to -1/z$, show that each $z \in H$ can be mapped by Γ into a point in the set

$$D = \{z \in H : |z| \geq 1, |\mathrm{Re}\, z| \leq \tfrac{1}{2}\},$$

which in the G-invariant measure on H has finite area.]

9. Let $\boldsymbol{G}_{p,p+q} = \boldsymbol{O}(p + q)/\boldsymbol{O}(p) \times \boldsymbol{O}(q)$ denote the Grassmann manifold of p-spaces in $\boldsymbol{R}^{p+q}$. Let $e_1, \ldots, e_p$ and $f_1, \ldots, f_q$ be the canonical bases of the orthogonal subspaces

$$E_1 = \boldsymbol{R}^p \times 0, \qquad E_2 = 0 \times \boldsymbol{R}^q.$$

Let

$$\boldsymbol{G}'_{p,p+q} = \{P \in \boldsymbol{G}_{p,p+q} : P \cap E_2 = \{0\}\}.$$

(i) Let $P \in \boldsymbol{G}'_{p,p+q}$. For each i $(1 \leq i \leq p)$ P contains a unique vector of the form

$$e_i + \beta_{i1} f_1 + \cdots + \beta_{iq} f_q.$$

The mapping $P \to (\beta_{ij})_{1 \leq i \leq p,\, 1 \leq j \leq q}$ is a bijection of $\boldsymbol{G}'_{p,p+q}$ onto the space $\boldsymbol{M}_{p,q}(\boldsymbol{R})$ of $p \times q$ real matrices.

(ii) In the parametrization above an invariant measure on $\boldsymbol{G}'_{1,2}$ is given by

$$f \to \int (1 + \beta_{11}^2)^{-1} f \, d\beta_{11};$$

on $\boldsymbol{G}'_{1,3}$ by

$$f \to \int (1 + \beta_{11}^2 + \beta_{12}^2)^{-3/2} f \, d\beta_{11}\, d\beta_{12}$$

and on $\boldsymbol{G}'_{2,4}$ by

$$f \to \int \{1 + (ad - bc)^2 + a^2 + b^2 + c^2 + d^2\}^{-2} f \, da\, db\, dc\, dd,$$

where $a = \beta_{11}$, $b = \beta_{12}$, $c = \beta_{21}$, $d = \beta_{22}$. Generalization?

10. (i) The measure

$$d\mu = (dx_1 \cdots dx_{n-1})/|x_n|$$

is a rotation-invariant measure on the sphere $x_1^2 + \cdots + x_n^2 = 1$.

(ii) The measure

$$dv = (dx_1 \cdots dx_{p+q})/|x_{p+q+1}|$$

is an invariant measure on the quadric

$$x_1^2 + \cdots + x_p^2 - x_{p+1}^2 - \cdots - x_{p+q+1}^2 = -1$$

viewed as a homogeneous space $\boldsymbol{O}(p, q+1)/\boldsymbol{O}(p, q)$ (cf. [DS], Chapter V, Exercise 7).

11. Let $U = \boldsymbol{SU}(2)$ and σ the involution

$$u \to \begin{pmatrix} 1 & 0 \\ 0 & -1 \end{pmatrix} u \begin{pmatrix} 1 & 0 \\ 0 & -1 \end{pmatrix}$$

of U. Show that

(i) The eigenspaces of the automorphism $d\sigma$ of the Lie algebra of $\mathfrak{u}$ are given by

$$\mathfrak{k} = \boldsymbol{R}e_3, \qquad \mathfrak{p}_* = \boldsymbol{R}e_1 + \boldsymbol{R}e_2,$$

where

$$e_1 = \begin{pmatrix} 0 & i \\ i & 0 \end{pmatrix}, \qquad e_2 = \begin{pmatrix} 0 & 1 \\ -1 & 0 \end{pmatrix}, \qquad e_3 = \begin{pmatrix} i & 0 \\ 0 & -i \end{pmatrix}.$$

(ii) The adjoint representation Ad_U of U is a double covering of $\boldsymbol{SU}(2)$ onto the group $\boldsymbol{SO}(3)$ of proper rotations of the Euclidean space

$$\mathfrak{u} = \boldsymbol{R}e_1 + \boldsymbol{R}e_2 + \boldsymbol{R}e_3.$$

We have

$$\mathrm{Ad}_U(u)X = uXu^{-1}, \qquad u \in U, \quad X \in \mathfrak{u}.$$

(iii) Let

$$K = \left\{ k_\theta = \exp\left(\frac{\theta}{2} e_3\right) = \begin{pmatrix} e^{(i/2)\theta} & 0 \\ 0 & e^{-(i/2)\theta} \end{pmatrix} : \theta \in \boldsymbol{R} \right\},$$

$$A = \left\{ a_t = \exp\left(\frac{t}{2} e_1\right) = \begin{pmatrix} \cos\frac{t}{2} & i \sin\frac{t}{2} \\ i \sin\frac{t}{2} & \cos\frac{t}{2} \end{pmatrix} : t \in \boldsymbol{R} \right\}.$$

In terms of the basis (e_1, e_2, e_3) of $\mathfrak{u}$ we have

$$\mathrm{Ad}(k_\theta) = \begin{pmatrix} \cos\theta & -\sin\theta & 0 \\ \sin\theta & \cos\theta & 0 \\ 0 & 0 & 1 \end{pmatrix},$$

$$\mathrm{Ad}(a_t) = \begin{pmatrix} 1 & 0 & 0 \\ 0 & \cos t & \sin t \\ 0 & -\sin t & \cos t \end{pmatrix}.$$

Thus the Cartan decomposition $U = KAK$ ([DS], Chapter V, Theorem 6.7) gives for $u \in U$

$$u = k_\theta a_t k_\phi,$$

which under Ad_U gives the Euler angles decomposition of a rotation in $\boldsymbol{R}^3$.

(iv) Since

$$k_\theta a_t k_\phi = \begin{pmatrix} \alpha & \beta \\ -\bar{\beta} & \bar{\alpha} \end{pmatrix}$$

where

(1) $$\alpha = \cos\frac{t}{2}\, e^{(i/2)(\theta+\phi)}, \quad \beta = i \sin\frac{t}{2}\, e^{(i/2)(\theta-\phi)}$$

define

$$z = x_1 + ix_2 = r\cos\frac{t}{2}\, e^{(i/2)(\theta+\phi)}, \quad w = x_3 + ix_4 = ir\sin\frac{t}{2}\, e^{(i/2)(\theta-\phi)}.$$

Then (in the sense of exterior multiplication)

$$dz\, d\bar{z}\, dw\, d\bar{w} = \tfrac{1}{2} r^3 \sin t\, dr\, dt\, d\theta\, d\phi.$$

The group $\boldsymbol{SU}(2)$ is given by $r = 1$, i.e., the sphere $x_1^2 + x_2^2 + x_3^2 + x_4^2 = 1$. Thus

$$\sin t\, dt\, d\theta\, d\phi$$

is an invariant measure on $\boldsymbol{SU}(2)$. This confirms Theorem 5.10 in this case. More precisely, the normalized Haar measure du on $\boldsymbol{SU}(2)$ is given by

$$\int_{\boldsymbol{SU}(2)} f(u)\, du = \frac{1}{16\pi^2} \int_0^\pi \sin t\, dt \int_0^{2\pi} d\theta \int_{-2\pi}^{2\pi} f(k_\theta a_t k_\phi)\, d\phi.$$

B. Radon Transforms

1. Derive Theorem 2.10 in this chapter from Theorem 2.10 in the Introduction.

2. (i) The range $\mathscr{D}(\boldsymbol{R}^2)^\wedge$ of $\mathscr{D}$ under the Radon transform on $\boldsymbol{R}^2$ consists of the functions $\psi \in \mathscr{D}(\boldsymbol{P}^2)$ which when expanded in a Fourier series

$$\psi(e^{i\theta}, p) = \sum_{n \in \boldsymbol{Z}} \psi_n(p) e^{in\theta}$$

have the following property: For each $n \in \boldsymbol{Z}$

$$\psi_n(p) = \frac{d^{|n|}}{dp^{|n|}} \phi_n(p),$$

where $\phi_n \in \mathscr{D}(\boldsymbol{R})$ is even.

(ii) Generalize (i) to the Radon transform on $\boldsymbol{R}^n$.

3. Let $f \to \hat{f}$ denote the Radon transform on $\boldsymbol{R}^n$ and $\phi \to \check{\phi}$ its dual. Show that the mapping $T \to \check{T}$ is one-to-one on $\mathscr{E}'(\boldsymbol{P}^n)$.

4. Let $\mathscr{N}$ denote the kernel of the dual transform $\phi \to \check{\phi}$ on $\mathscr{E}(\boldsymbol{P}^n)$. Then $\mathscr{N}$ is the closed subspace of $\mathscr{E}(\boldsymbol{P}^n)$ generated by the spaces

$$E_k \otimes p^l, \qquad k, l \in \boldsymbol{Z}^+, \quad k - l > 0 \text{ and even.}$$

(Here $E_k \otimes p^l$ is the space of functions $\phi(\omega, p)$ of the form $\phi(\omega, p) = \psi(\omega)p^l$ where ψ belongs to the eigenspace E_k from Introduction §3.)

5*. The range $\mathscr{E}'(\boldsymbol{R}^n)^\wedge$ of $\mathscr{E}'(\boldsymbol{R}^n)$ under the Radon transform consists of the distributions $\Sigma \in \mathscr{E}'(\boldsymbol{P}^n)$ with the following property: For each $k \in \boldsymbol{Z}^+$ the distribution

$$\phi \in \mathscr{E}(\boldsymbol{S}^{n-1}) \to \int_{\boldsymbol{S}^{n-1} \times \boldsymbol{R}} \phi(\omega) p^k \, d\Sigma(\omega, p)$$

is a homogeneous polynomial in $\omega_1, \ldots, \omega_n$ of degree k (cf. Helgason [1983a]).

The result can also be stated (see **B4**)

$$\mathscr{E}'(\boldsymbol{R}^n)^\wedge = \{\Sigma \in \mathscr{E}'(\boldsymbol{P}^n) : \Sigma(\mathscr{N}) = 0\}.$$

In particular, $\mathscr{E}'(\boldsymbol{R}^n)^\wedge$ is closed in $\mathscr{E}'(\boldsymbol{P}^n)$. This implies by purely functional-analytic results (Hertle [1984]) that the mapping $\phi \to \check{\phi}$ of $\mathscr{E}(\boldsymbol{P}^n)$ into $\mathscr{E}(\boldsymbol{R}^n)$ is surjective.

C. Spaces of Constant Curvature

1. (i) (Quadric Model) The quadric Q_{-1} given by

$$B_{-1}(x) = x_1^2 + \cdots + x_n^2 - x_{n+1}^2 = -1, \qquad x_{n+1} > 0, \quad x = (x_1, \ldots, x_{n+1}),$$

with Riemannian structure induced by B_{-1} has constant curvature -1 (cf. [DS], Chapter V, Exercise 7).

(ii) (Unit Ball Model) The space

$$B_1(0) = \{y : |y| < 1\}, \qquad y = (y_1, \ldots, y_n), \quad |y|^2 = \sum_1^n y_i^2,$$

with the Riemannian structure

$$ds^2 = 4(1 - |y|^2)^{-2}(dy_1^2 + \cdots + dy_n^2),$$

has constant curvature -1.

(iii) (Half-Space Model) The space

$$(z_1, \ldots, z_n) \in \boldsymbol{R}^n, \qquad z_n > 0,$$

with the Riemannian structure

$$d\sigma^2 = (dz_1^2 + \cdots + dz_n^2)/z_n^2$$

has constant curvature -1.

(iv) The mapping $y = \Phi(x)$ $(x \in Q_{-1})$ given by

$$(y_1, \ldots, y_n) = \frac{1}{(x_{n+1} + 1)}(x_1, \ldots, x_n)$$

with inverse

$$(x_1, \ldots, x_n) = \frac{2}{1 - |y|^2}(y_1, \ldots, y_n), \qquad x_{n+1} = \frac{1 + |y|^2}{1 - |y|^2},$$

is an isometry of Q_{-1} onto $B_1(0)$.

(v) The mapping $z = \Psi(x)$ $(x \in Q_{-1})$ given by

$$(z_1, \ldots, z_n) = (x_{n+1} - x_n)^{-1}(x_1, \ldots, x_{n-1}, 1)$$

with inverse

$$(x_1, x_2, \ldots, x_n, x_{n+1}) = z_n^{-1}\left(z_1, z_2, \ldots, \frac{|z|^2 - 1}{2}, \frac{|z|^2 + 1}{2}\right),$$

where $|z| = (z_1^2 + \cdots + z_n^2)^{1/2}$, is an isometry of Q_{-1} onto H_n.

2. On the unit sphere S^2 we have the formulas

$$\frac{\sin a}{\sin A} = \frac{\sin b}{\sin B} = \frac{\sin c}{\sin C},$$

$$\cos a = \cos b \cos c + \sin b \sin c \cos A,$$

for a geodesic triangle with angles A, B, and C and sides of length a, b, c. The duality between compact and noncompact symmetric spaces ([DS], Chapter V, §§2–3) suggests the analogous formulas

$$\frac{\sinh a}{\sin A} = \frac{\sinh b}{\sin B} = \frac{\sinh c}{\sin C}, \tag{1}$$

$$\cosh a = \cosh b \cosh c - \sinh b \sinh c \cos A \tag{2}$$

on the hyperbolic plane with curvature -1.

(i) Verify (1) and (2).

(ii) Generalize (1) and (2) to the hyperbolic plane with curvature $-\varepsilon^2$.

(iii) What happens as $\varepsilon \to 0$?

D. Some Results in Analysis

1. Let $k \in \boldsymbol{Z}^+$. Show that the polynomials $x \to (x, \omega)^k$ $(\omega \in \boldsymbol{S}^{n-1})$ span the space of all homogeneous polynomials of degree k.

2. Show that the function

$$f(x) = \begin{cases} |x|^{-(1/2)n}(\log|x|)^{-1}, & |x| \geq 2 \\ 0, & |x| < 2 \end{cases}$$

belongs to $L^2(\boldsymbol{R}^n)$ but is not integrable over any plane of dimension $\geq \frac{1}{2}n$ (Solmon [1976]).

This example exhibits the difficulties in defining k-plane transforms directly on $L^p(\boldsymbol{R}^n)$; while the distributional definition essentially requires compact support, it is well suited to nonsmooth behavior.

3. Let $F \in \mathscr{D}(\boldsymbol{R}^2)$ and as in Lemma 6.19 define g by

$$g(r) = r \int_0^\infty F(re^t, re^{-t})(e^t - e^{-t})\, dt, \qquad r \neq 0.$$

Prove that

$$\lim_{r \to 0+} g(r) = \int_0^\infty F(x, 0)\, dx, \qquad \lim_{r \to 0-} g(r) = -\int_0^\infty F(x, 0)\, dx.$$

Thus g may not extend to a continuous function around $r = 0$ although g' does (Harish-Chandra (1964c]).

4. Give a proof of the Poisson equation in $\boldsymbol{R}^2$,

$$L(\log r) = 2\pi\delta,$$

by use of the formula $Lr^\alpha = \alpha^2 r^{\alpha - 2}$.

5. Let $m_1, \ldots, m_r$ be distinct real numbers and put

$$f(t) = \sum_{j=1}^{r} c_j e^{im_j t}, \qquad t \in \boldsymbol{R},$$

$c_1, \ldots, c_r$ being complex numbers. Using

$$\lim_{T \to +\infty} \frac{1}{T} \int_0^T |f(t)|^2\, dt = \sum_i |c_i|^2,$$

deduce that

$$(*) \qquad \limsup_{t \to +\infty} |f(t)| \geq \left\{ \sum_i |c_i|^2 \right\}^{1/2}.$$

Now let $k_1, \ldots, k_r$ be distinct complex numbers and $p_1, \ldots, p_r$ polynomials in t with complex coefficients such that all $p_i \neq 0$. Suppose

$$\limsup_{t \to +\infty} |p_0(t) + p_1(t)e^{k_1 t} + \cdots + p_r(t)e^{k_r t}| \leq a.$$

Using (*), deduce that p_0 is a constant and $|p_0| \leq a$. Moreover, if $\operatorname{Re} k_i \geq 0$ for each i, then each p_j is a constant and each k_i is purely imaginary (cf. Harish-Chandra [1958a], §15).

NOTES

With differential forms viewed as "infinitesimal volumes," the notion of their integrals (integral invariants) becomes a natural one (Crofton [1868], Poincaré [1887], Cartan [1896] [1922], Lie [1897]). In Chevalley [1946] differential forms appear as smooth families of multilinear functions on the tangent spaces. In §1 we follow his treatment of their integration. The Riemannian measure is usually presented under an unnecessary orientability condition.

The invariant integral on a compact Lie group was used by Hurwitz [1897], Schur [1924], and Weyl [1925]. The Haar measure was constructed by Haar [1933] on locally compact groups. Invariant measures on coset spaces occur in special cases in classical integral geometry (Crofton, Czuber, Deltheil, Herglotz, Blaschke), but Theorem 1.9 (for locally compact coset spaces) was first proved by Weil [1940] (see Chern [1942] for the Lie group case, described in Exercise A3). Proposition 1.12 is from Harish-Chandra [1953]; see also Mostow [1952]. This §1 differs little from Chapter X, §1, Nos. 1, 2, in Helgason [1962a].

§2, Nos. 1, 3, 4. The inversion formulas

(i) $$f(x) = \tfrac{1}{2}(2\pi i)^{1-n} L_x^{(1/2)(n-1)} \int_{S^{n-1}} J(\omega, (\omega, x))\, d\omega \qquad (n \text{ odd}),$$

(ii) $$f(x) = (2\pi i)^{-n} L_x^{(1/2)(n-2)} \int_{S^{n-1}} d\omega \int_{-\infty}^{\infty} \frac{dJ(\omega, p)}{p - (\omega, x)} \qquad (n \text{ even})$$

for a function $f \in \mathscr{D}(\boldsymbol{R}^n)$ in terms of its plane integrals $J(\omega, p)$ go back to Radon [1917] and John [1955]. According to Bockwinkel [1906], the case $n = 3$ had been proved already before 1906 by H. A. Lorentz. In John [1955] proofs are given based on the Poisson equation $Lu = f$. Alternative proofs, using distributions, were given by Gelfand and Schilov [1960]. The dual transforms $f \to \hat{f}$, $\phi \to \check{\phi}$, the unified inversion formula, and its dual

$$cf = L^{(1/2)(n-1)}((\hat{f})^{\vee}), \qquad c\phi = \square^{(1/2)(n-1)}((\check{\phi})^{\wedge})$$

were given by the author in [1964b]. The proofs from Helgason [1959] (Chapter IV), and [1965a] are based on the Darboux equation (Lemma 2.14) and therefore generalize to two-point homogeneous spaces. Formulas (17) and (55) were already given by Fuglede [1958]; according to Radon [1917], the first formula had even been observed by Herglotz. The modified inversion formula (Theorem 2.16) and Theorem 2.17 are proved in Ludwig [1966]; the latter result is attributed to Y. Reshetnyak in Gelfand *et al.* [1966].

§2. No. 2. The support theorem and the Paley–Wiener theorem (Theorems 2.4, 2.6, and 2.10) were first given in Helgason [1964b, 1965a]. The example in Remark 2.9 was also found by D. J. Newman, cf. B. Weiss's paper [1967], which gives another proof of the support theorem. The local result in Corollary 2.12 goes back to John [1935]; our derivation is suggested by the proof of a similar lemma in Flensted-Jensen [1977*b*], p. 81. Another proof is in Ludwig [1966].

Corollary 2.8 was derived by Ludwig [1966] in a different way. He proposes alternative proof of the Schwartz and Paley–Wiener theorems by expanding $\hat{f}(\omega, p)$ into spherical harmonics in ω. However, a crucial point in the proof is not established and seems difficult to settle in the context: the smoothness of the function $\tilde{f}(\xi)$ in (2.7), p. 57, at the point $\xi = 0$. (This is the function F in our proof of Theorem 2.4; here the smoothness of F at at 0 is the main point.)

Since the inversion formula (Theorem 2.13) is rather easy to prove for odd n, it is natural to try to prove Theorem 2.4 for n odd by showing directly that if $\phi \in \mathscr{S}_H(\boldsymbol{R}^n)$, then the function $f = L^{(1/2)(n-1)}(\check{\phi})$ belongs to $\mathscr{S}(\boldsymbol{R}^n)$ [in general, $\check{\phi} \notin \mathscr{S}(\boldsymbol{R}^n)$]. This approach is taken in Gelfand *et al.* [1966] (pp. 16–17); however, this method seems to me to offer some unresolved technical difficulties. Corollary 2.5 is stated in Semyanistyi [1960]. In [1984] Hertle has shown that $f \to \hat{f}$ is not a homeomorphism of $\mathscr{D}(\boldsymbol{R}^n)$ onto its image inside $\mathscr{D}(\boldsymbol{P}^n)$.

§2. Nos. 5, 6. The approach to the Radon transform of distributions adopted here is from the author's paper [1966a]. Other methods are proposed in Gelfand *et al.* [1966], Ludwig [1966] [where formula (46) is also proved], Ambrose (unpublished), and Guillemin and Sternberg [1977]. The d-plane transform, Theorem 2.27 and Corollary 2.28, characterizing the range are from Helgason [1959, 1980c]; an L^2 version of Corollary 2.28 was given by Solmon [1976], p. 77. A different characterization of the image for $d = 1$, $n = 3$ was given by John [1938]. For further work in this direction see Gelfand and Graev [1968] and Guillemin and Sternberg [1979].

§2, No. 7. The applications to differential equations given here are based on John [1955]. Other applications of the Radon transform to partial differential equations with constant coefficients can be found in Courant and Lax [1955], Gelfand and Shapiro [1955], John [1955], Borovikov [1959], Garding [1961], Ludwig [1966], and Lax and Phillips [1967].

Applications of the Radon transform in medicine were proposed by Cormack [1963, 1964] and in radioastronomy by Bracewell and Riddle [1967]. See Shepp and Kruskal [1978], Smith *et al.* [1977], and Shepp *et al.* [1983] for further account of applications.

§2, No. 8. For a thorough treatment of distribution theory on $\boldsymbol{R}^n$ and on manifolds see Schwartz [1966] or Trèves [1967]. A concise, but nevertheless self-contained, treatment of the basics is given in Hörmander [1963]. A more systematic study of the potentials I^γ and the distributions r^α, x_+^α is given in Riesz [1949], Schwartz [1966], and Gelfand and Schilov [1960]. A more general version of Proposition 2.38 is given in Ortner [1980].

§3. The notion of *incidence* in a pair of homogeneous spaces goes back to Chern [1942]. The *duality* between

$$\text{(i)} \qquad G/H_X \quad \text{and} \quad G/H_\Xi,$$

the *correspondences*

$$\text{(ii)} \qquad x \to \check{x}, \qquad \xi \to \hat{\xi},$$

and the *Radon transform for a double fibration* of a homogeneous space were introduced in the author's paper [1966a], from which most of this section comes. A further generalization, replacing the homogeneity assumption by postulates about the compatibility of the entering measures, was given by Gelfand *et al.* [1969].

For the case $G = \boldsymbol{U}(4)$, $H_X = \boldsymbol{U}(1) \times \boldsymbol{U}(3)$, $H_\Xi = \boldsymbol{U}(2) \times \boldsymbol{U}(2)$, the maps (ii) become Penrose correspondences (Penrose [1967]) described in more detail in Example (iii) in No. 1.

The example where X is the set of p-planes in $\boldsymbol{R}^n$, Ξ the set of q-planes ($n = p + q + 1$) is worked out in Helgason [1965a] with solutions to Problems A, B, and C. However, because of the convergence problems encountered there, it may be more satisfactory to adopt the group-theoretic incidence definition in the present section. If $x_0 \perp \xi_0, x_0 \cap \xi_0 \neq \emptyset$, then x and ξ are incident iff they intersect under a right angle. Here Problems **B** and C have been solved by Fulton Gonzalez.

§4. It was shown by Funk [1916] that a function f on the two-sphere, symmetric with respect to the center, can be determined by the integrals of f over the great circles. In [1917] Radon discussed this problem and the analogous one of determining a function on the non-Euclidean plane from its integrals over all geodesics.

The Radon transform on hyperbolic and on elliptic spaces corresponding to totally geodesic submanifolds was defined in Helgason [1959], where the inversion formulas in Theorems 4.5 and 4.7 were proved. A generalization was given by Semyanistyi [1961]. An alternative definition, with corresponding inversion formulas, was given in Gelfand *et. al.* [1966]. A support theorem and a Paley–Wiener theorem for this transform, valid for symmetric spaces, were proved by the author in [1973a] §8 (cf. specializations in [1983a, 1983b]). Some local support theorems and L^2 Paley–Wiener theorems for the case of hyperbolic spaces were proved by Lax and Phillips [1979, 1982].

The support theorem (Theorem 4.2) was proved by the author [1964b, 1980b] and its consequence, Corollary 4.18, in [1980d]. The theory of the Radon transform for antipodal manifolds in compact two-point homogeneous spaces (Theorem 4.11) is from Helgason [1965a]. Another proof of the inversion formula is given in Grinberg [1983]. R. Michel has in [1972] and [1973] used Theorem 4.11 in establishing certain infinitesimal rigidity properties of the canonical metrics on the real and complex projective spaces. Funk [1916], a paper noted earlier, showed that the standard metric on the projective plane cannot be deformed in such a way that the geodesics remain closed and of the same length. For $\boldsymbol{S}^2$, however, such a deformation is possible (cf. Guillemin [1976]).

§5. While the volume element on the orthogonal group goes back to Schur [1924], the general formula in Corollary 5.16 is due to Weyl [1925]. Its generalization, Theorem 5.10, is from Cartan [1929] (§23). The other integral formulas are for the most part due to Harish-Chandra [1953], p. 239, [1954d] p. 507, [1956a], §12, and [1958a], pp. 261, 287. For the case of the complex classical groups many such integral formulas are proved in Gelfand and Naimark [1957]. The proof of Lemma 5.4 (from Harish-Chandra [1954d] p. 507) is due to Wallach [1973] (p. 174), and so is Proposition 5.21. Proposition 5.18 is from Helgason [1980a], and the proofs of Harish-Chandra's results 5.8, 5.19, and 5.20 are from some old unpublished lectures by the author. The presentation of Lemma 5.14 and Proposition 5.15 is much influenced by that of Séminaire Sophus Lie [1955] (Exp. 19).

§6, No. 1. The construction of the constant curvature spaces (Theorem 6.1) is from Helgason [1959, 1961]. The proof of Lemma 6.2 on the connectivity is adapted from Boerner [1955]. For more information about these spaces as well as on isotropic manifolds see Wolf [1967].

§6. Nos. 2–4. The solution of the orbital integral problem on Lorentzian manifolds G/H of constant curvature given here is based on Helgason [1959]. For some related work for arbitrary signature see Faraut [1979].

§6, No. 5. The Plancherel formula for the Fourier transform on a complex semisimple Lie group G was given by Gelfand and Naimark [1948] for $G = \boldsymbol{SL}(n, \boldsymbol{C})$ (see also [1957]) and by Harish-Chandra [1951b, 1954d] for general G. In [1955], Gelfand and Graev observed that the Plancherel formula for complex G results from an expression of the value of a function $f \in \mathcal{D}(G)$ at e in terms of the integral of f over each conjugacy class. As we noted in §3, No. 3, this amounts to the orbital integral problem for $G \times G/\Delta G$, which, in the quoted paper, Gelfand and Graev solved when G is a complex classical group.

For G *real* and semisimple the solution of the orbital integral problem on $G \times G/\Delta G$ was given by Harish-Chandra [1957b]. Theorem 6.18 in the text is the special case $G = \boldsymbol{SL}(2, \boldsymbol{R})$. In contrast to the case of complex G, the orbital integral problem is only a partial step towards the full Plancherel formula (Harish-Chandra [1970]).

The orbital integrals considered in the above papers only involve orbits of semisimple $g \in G$. However, orbital integrals exist for arbitrary $g \in G$ (cf., Rao [1972]) and have been studied by Harish-Chandra, Barbash [1979], Barbash and Vogan [1980, 1982]; these offer new and interesting features.

As a synthesis of the results above it seems to be a promising research project to develop an orbital integral theory (cf., §3, Problem D) for affine symmetric spaces G/H (G semisimple) including a joint generalization of Theorem 6.17 and Theorem 6.18. For G/H of "rank one" this has been carried out by Jeremy Orloff.

CHAPTER II

INVARIANT DIFFERENTIAL OPERATORS

While applications to differential equations served as a motivation for Lie's theory of transformation groups, the theory of Lie groups and Lie algebras has grown into a force in its own right, exerting ever-increasing influence on many other fields of mathematics. In particular, with Lie group theory so highly developed, it is reasonable to reverse Lie's original viewpoint, that is, consider the group as the given object and investigate differential operators invariant under its action. Such invariance conditions are very natural, and the literature abounds with examples; see §4, No. 1 for a few elementary ones.

In §1 we review some structure theorems for $C^\infty(\mathbf{R}^n)$ and a coordinate-free characterization of differential operators. In §2 we discuss for a manifold M the spaces $\mathscr{D}(M) = C_c^\infty(M)$, $\mathscr{E}(M) = C^\infty(M)$, and their topologies, distributions, and differential operators on M with particular emphasis on the Laplace-Beltrami operator on pseudo-Riemannian manifolds.

In §3 we discuss projections of differential operators on submanifolds; also, when a group is acting on V we decompose an arbitrary differential operator on V into a "polynomial" in orbital operators with transversal operators as "coefficients." For the Laplacian this becomes a kind of a Pythagorean formula [see (21) in §3]. When a cross section W is selected for the group action on V, each differential operator on V has a radial part on W. When the orbits intersect W orthogonally, the radial part of the Laplacian has a simple geometric expression which in §3, No. 4 is worked out for some examples.

In §§4–5 we consider the invariant differential operators on a homogeneous space G/H and describe these in terms of the Lie algebras of G and of H. The case of a symmetric space is studied in more detail; we relate the operators to Weyl group invariants and for the rank-one case prove generalized forms of the Darboux equation and Ásgeirsson's mean value theorem and give an explicit solution of Poisson's equation. Symmetric spaces G/K with G complex have some special features worked out in No. 8–9.

§1. Differentiable Functions on R^n

Let $\mathbf{R}^n$ denote the n-dimensional Euclidean space and $x = (x_1, \ldots, x_n)$ an arbitrary point. Put $|x| = (x_1^2 + \cdots + x_n^2)^{1/2}$. If $V \subset \mathbf{R}^n$ is an open subset, let $\mathscr{E}(V)$ denote the space of complex-valued differentiable functions on V, and $\mathscr{D}(V)$ the space of functions in $\mathscr{E}(V)$ with compact support contained in V. While the space $\mathscr{E}(V)$ is the one of principal interest, $\mathscr{D}(V)$ is often more convenient to work with because $\mathscr{D}(U) \subset$

$\mathscr{D}(V)$ if $U \subset V$. Let ∂_i denote the partial differentiation $\partial/\partial x_i$, and if $\alpha = (\alpha_1, \ldots, \alpha_n)$ is an n-tuple of integers $\alpha_i \geq 0$, we put

$$D^\alpha = \partial_1^{\alpha_1} \cdots \partial_n^{\alpha_n}, \qquad x^\alpha = x_1^{\alpha_1} \cdots x_n^{\alpha_n},$$

$$|\alpha| = \alpha_1 + \cdots + \alpha_n, \qquad \alpha! = \alpha_1! \cdots \alpha_n!.$$

If $\beta = (\beta_1, \ldots, \beta_n)$ is another integral positive n-tuple and $\beta_j \leq \alpha_j$ for all j, we write $\beta \leq \alpha$ and put

$$\alpha - \beta = (\alpha_1 - \beta_1, \ldots, \alpha_n - \beta_n),$$

$$\binom{\alpha}{\beta} = \frac{\alpha!}{\beta!(\alpha - \beta)!}.$$

We have then the generalized Leibniz rule for the differentiation of the product of two functions f and g:

$$D^\alpha(fg) = \sum_{\mu + \nu = \alpha} \binom{\alpha}{\nu} (D^\nu f)(D^\mu g). \tag{1}$$

(Cf. Hörmander [1963], Chapter I; an identity of this type is in fact obvious and the coefficients can be determined by taking f and g to be exponentials.)

If $S \subset V$ is any subset and $m \in \boldsymbol{Z}^+$, we put

$$\|f\|_m^S = \sum_{|\alpha| \leq m} \sup_{x \in S} |(D^\alpha f)(x)|. \tag{2}$$

When $S = V$ we often drop the superscript S in the notation. Then we have for $f, g \in \mathscr{E}(V)$,

$$\|f + g\|_m^S \leq \|f\|_m^S + \|g\|_m^S. \tag{3}$$

Lemma 1.1. *Let $C \subset U \subset V \subset \boldsymbol{R}^n$, where U and V are open and C a closed subset of $\boldsymbol{R}^n$. Then there exists a function $\phi \in \mathscr{E}(V)$ such that $\phi \equiv 1$ on C, $\phi \equiv 0$ on $V - U$, and $0 \leq \phi \leq 1$ everywhere.*

For the proof see, e.g., [DS] (pp. 88 and 538), valid for V replaced by a paracompact manifold. Using the seminorms (2) we can prove a sharper version.

Proposition 1.2. *Let $C \subset \boldsymbol{R}^n$ be any closed subset. Then there exists a $\phi \in \mathscr{E}(\boldsymbol{R}^n)$ such that*

$$C = \{x \in \boldsymbol{R}^n : \phi(x) = 0\}.$$

Proof. If V_k is a $(1/k)$-neighborhood of C, then C is the intersection of the open sets V_k $(k \geq 1)$. Let $\phi_k \in \mathscr{E}(\boldsymbol{R}^n)$ such that $\phi_k \equiv 0$ on C, $\phi_k \equiv 1$ on $\boldsymbol{R}^n - V_k$, $0 \leq \phi_k \leq 1$. Let $B_r(0)$ denote the ball $|x| < r$ and put $\|f\|_m^r$ for the seminorm (2) when $S = B_r(0)$.

Choose $\varepsilon_k > 0$ such that

$$\sum_1^\infty \varepsilon_k \|\phi_k\|_k^k < \infty$$

and put

$$\psi_m = \sum_1^m \varepsilon_k \phi_k.$$

Then for a given k, if $m' \geq m > r$, k,

$$\|\psi_{m'} - \psi_m\|_k^r \leq \sum_{q>m} \varepsilon_q \|\phi_q\|_k^r \leq \sum_{q>m} \varepsilon_q \|\phi_q\|_q^q,$$

and this tends to 0 as $m \to \infty$. The limit $\phi = \lim_{m\to\infty} \psi_m$ clearly has the desired properties.

Lemma 1.3. *Let $m > 0$, and suppose $f \in \mathscr{E}(\boldsymbol{R}^n)$ has all derivatives of order $\leq m$ vanish at 0. Then given $\varepsilon > 0$ there exists a $g \in \mathscr{E}(\boldsymbol{R}^n)$ vanishing in a neighborhood of 0 and satisfying*

$$\|g - f\|_m < \varepsilon.$$

Proof. Let $\phi \in \mathscr{E}(\boldsymbol{R}^n)$ such that $\phi(x) = 0$ for $|x| \leq \frac{1}{2}$, $\phi(x) = 1$ for $|x| \geq 1$, and $0 \leq \phi \leq 1$ everywhere. For $\delta > 0$ put

$$g_\delta(x) = \phi(x/\delta) f(x).$$

Then $g_\delta \in \mathscr{E}(\boldsymbol{R}^n)$, $g_\delta = 0$ near $x = 0$, and $g_\delta(x) = f(x)$ for $|x| \geq \delta$. Thus it suffices to prove that if $|\alpha| \leq m$,

$$\sup_{|x|\leq\delta} |D^\alpha g_\delta(x) - D^\alpha f(x)| \to 0 \qquad \text{as} \quad \delta \to 0. \tag{4}$$

But by our assumption about f, $(D^\alpha f)(0) = 0$ for $|\alpha| \leq m$, so

$$\sup_{|x|\leq\delta} |(D^\alpha f)(x)| \to 0 \qquad \text{as} \quad \delta \to 0. \tag{5}$$

Also, by (1),

$$D^\alpha g_\delta(x) = \sum_{\mu+\nu=\alpha} \binom{\alpha}{\nu} \delta^{-|\nu|} (D^\nu \phi)\left(\frac{x}{\delta}\right)(D^\mu f)(x)$$

so for a constant C,

$$|D^\alpha g_\delta(x)| \leq C \sum_{\mu+\nu=\alpha} \delta^{-|\nu|} |(D^\mu f)(x)|, \qquad x \in \boldsymbol{R}^n. \tag{6}$$

But $D^\mu f$ has all derivatives up to order $m - |\mu|$ vanishing at $x = 0$. Thus

$$\sup_{|x|\leq\delta} |D^\mu f(x)| = o(\delta^{m-|\mu|}),$$

so by (6)

$$\sup_{|x| \leq \delta} |D^\alpha g_\delta(x)| = o\left(\sum_{\mu + \nu = \alpha} \delta^{m - |\mu| - |\nu|} \right) = o(\delta^{m - |\alpha|}),$$

and now (4) follows if we take (5) into account.

Definition. Let $V \subset \boldsymbol{R}^n$ be an open set. A *differential operator* on V is a linear mapping $D\colon \mathscr{D}(V) \to \mathscr{D}(V)$ with the following property:

For each relatively compact open set $U \subset V$ such that $\overline{U} \subset V$, there exists a finite family of functions $a_\alpha \in \mathscr{E}(U)$ $[\alpha = (\alpha_1, \ldots, \alpha_n),\ \alpha_i \in \boldsymbol{Z}^+]$ such that

$$D\phi = \sum_\alpha a_\alpha D^\alpha \phi, \qquad \phi \in \mathscr{D}(U). \tag{7}$$

It is clear that a differential operator D on V has the property

$$\operatorname{supp}(D\psi) \subset \operatorname{supp}(\psi), \qquad \psi \in \mathscr{D}(V),$$

supp denoting support. This property implies that D can be extended to a linear operator (also denoted D) from $\mathscr{E}(V)$ to $\mathscr{E}(V)$ by means of the formula

$$(Df)(x) = (D\phi)(x). \tag{8}$$

Here $x \in V$, $f \in \mathscr{E}(V)$ are arbitrary and ϕ is any function in $\mathscr{D}(V)$ which coincides with f in a neighborhood of x. The choice of ϕ is clearly immaterial.

Theorem 1.4. *Let $D\colon \mathscr{D}(V) \to \mathscr{D}(V)$ be a linear mapping satisfying the condition*

$$\operatorname{supp}(D\phi) \subset \operatorname{supp}(\phi), \qquad \phi \in \mathscr{D}(V). \tag{9}$$

Then D is a differential operator on V. Conversely, any differential operator on V satisfies (9).

Remark. If D is assumed continuous [in the topology on $\mathscr{D}(V)$ defined in §2, No. 2], then for $x \in V$ the mapping $\phi \to (D\phi)(x)$ is by (9) a distribution with point support. Hence (7) follows immediately from Schwartz's theorem that a distribution with point support is a linear combination of derivatives of the delta distribution. The main point of Theorem 1.4 is therefore that this continuity of D is automatic.

Proof. Suppose D: $\mathscr{D}(V) \to \mathscr{D}(V)$ satisfies (9) and extend D to $\mathscr{E}(V)$ by (8). Next we shall prove that for each point $a \in V$ there exists an open relatively compact neighborhood U of a, $(\overline{U} \subset V)$, an integer m, and a constant C such that

$$\|Du\|_0 \leq C\|u\|_m, \qquad u \in \mathscr{D}(U - \{a\}). \tag{10}$$

Suppose this were false for a point a. Let $U^0 \subset V$ be an open relatively compact neighborhood of a, $\overline{U}^0 \subset V$. Then there is a function

$$u_1 \in \mathscr{D}(U^0 - \{a\})$$

such that

$$\|Du_1\|_0 > 2^2\|u_1\|_1.$$

If $U_1 = \{x : u_1(x) \neq 0\}$, then $U^0 - \overline{U}_1$ is an open neighborhood of a, so by our assumption there is a function $u_2 \in \mathscr{D}(U^0 - \overline{U}_1 - \{a\})$ such that

$$\|Du_2\|_0 > 2^4\|u_2\|_2.$$

Let $U_2 = \{x : u_2(x) \neq 0\}$. By induction we obtain a sequence $U_1, U_2, \ldots$ of open sets such that

$$\overline{U}_k \subset U^0 - \{a\}, \qquad \overline{U}_k \cap \overline{U}_l = \varnothing \qquad \text{for} \quad k \neq l, \tag{11}$$

and functions

$$u_k \in \mathscr{D}(U^0 - \overline{U}_1 - \cdots - \overline{U}_{k-1} - \{a\}) \subset \mathscr{D}(V) \tag{12}$$

satisfying $U_k = \{x : u_k(x) \neq 0\}$ and

$$\|Du_k\|_0 > 2^{2k}\|u_k\|_k. \tag{13}$$

Because of (11) and (12) the sum

$$u = \sum_1^\infty \frac{2^{-k}u_k}{\|u_k\|_k}$$

is a well-defined function in $\mathscr{D}(V)$, and on U_k, u coincides with the function $2^{-k}\|u_k\|_k^{-1}u_k$. Hence if $|$ denotes restriction, (9) implies,

$$Du \,|\, U_k = 2^{-k}\|u_k\|_k^{-1}(Du_k) \,|\, U_k.$$

Thus, using (13), we conclude that there exists a point $x_k \in U_k$ such that

$$(Du)(x_k) > 2^k,$$

and this contradicts the boundedness of Du. This proves (10).

Lemma 1.5. *Let $D: \mathscr{D}(V) \to \mathscr{D}(V)$ satisfy (9). Let $U \subset V$ be any open set. Assume there exist constants $C > 0$, $m \in \boldsymbol{Z}^+$ such that*

$$\|Du\|_0 \leq C\|u\|_m \qquad \textit{for} \quad u \in \mathscr{D}(U). \tag{14}$$

Then there exist functions $a_\alpha \in \mathscr{E}(V)$ *such that*

$$(15) \qquad (Du)(x) = \sum_{|\alpha| \leq m} a_\alpha(x)(D^\alpha u)(x), \qquad x \in U, \quad u \in \mathscr{D}(U).$$

Proof. For each $a \in V$ let

$$Q_{\alpha, a}(x) = (x_1 - a_1)^{\alpha_1} \cdots (x_n - a_n)^{\alpha_n}.$$

Since this is a polynomial in $(a_1, \ldots, a_n)$ with coefficients which are polynomials in $(x_1, \ldots, x_n)$, the function

$$b_\alpha : a \to (DQ_{\alpha, a})(a)$$

belongs to $\mathscr{E}(V)$. Now let $u \in \mathscr{D}(U)$ and $a \in U$. Then the function

$$f = u - \sum_{|\alpha| \leq m} \frac{1}{\alpha!}(D^\alpha u)(a)Q_{\alpha, a}$$

satisfies $(D^\alpha f)(a) = 0$ for $|\alpha| \leq m$. Now we use Lemma 1.3 and its proof to approximate f in the seminorm $\| \ \|_m$ by functions g_ν which coincide with f outside some neighborhood N_ν of a, but vanish identically near a. From (9) [and (8)] we conclude that Dg_ν vanishes identically near a. Now using (14) on $u_\nu = f - g_\nu$, we conclude that $Du_\nu(a) \to 0$, so

$$(Df)(a) = 0.$$

This means

$$(Du)(a) = \sum_{|\alpha| \leq m} \frac{1}{\alpha!}(D^\alpha u)(a)(DQ_{\alpha, a})(a),$$

which gives the conclusion (15) with $\alpha!\, a_\alpha = b_\alpha$.

We can now prove Theorem 1.4. Let $U \subset V$ be open, $\overline{U}$ compact, $\overline{U} \subset V$. Applying (10) to each point of $\overline{U}$, we get a finite covering of $\overline{U}$ by open relatively compact sets $U_1, \ldots, U_r$, points $a_1 \in U_1, \ldots, a_r \in U_r$, and constants $C > 0$, $m \in \mathbf{Z}^+$ such that for each i $(1 \leq i \leq r)$

$$(16) \qquad \|Du\|_0 \leq C\|u\|_m$$

for $u \in \mathscr{D}(U_i - \{a_i\})$. Let $1 = \sum_1^{r+1} \phi_i$ be a partition of unity for the covering $U_1, \ldots, U_r, V - \overline{U}$ of V ([DS], p. 89) and let

$$u \in \mathscr{D}(U - \{a_1\} - \cdots - \{a_r\}).$$

Then $u = \sum_1^{r+1} \phi_i u = \sum_1^r \phi_i u$ and (16) holds for each $\phi_i u$. Using (1), we deduce that (16) holds for u itself (with another C). The lemma then implies that (15) holds for $x \in U$, $x \neq a_1, \ldots, a_r$. But then (15) holds for all $x \in U$, both sides being continuous. This proves Theorem 1.4.

§2. Differential Operators on Manifolds

1. Definition. The Spaces $\mathscr{D}(M)$ and $\mathscr{E}(M)$

Let M be a manifold. Motivated by Theorem 1.4, we define a *differential operator* D on M to be a linear mapping of $C_c^\infty(M)$ into itself which decreases supports:

$$\operatorname{supp}(Df) \subset \operatorname{supp}(f), \qquad f \in C_c^\infty(M).$$

If (U, ϕ) is a local coordinate system on M, the mapping

$$D^\phi: F \to (D(F \circ \phi)) \circ \phi^{-1}, \qquad F \in C_c^\infty(\phi(U)), \tag{1}$$

satisfies the assumption of Theorem 1.4. Using (7) in §1 we obtain for each open relatively compact set W, with $\overline{W} \subset U$, a finite family of functions $a_\alpha \in C^\infty(W)$ such that

$$Df = \sum_\alpha a_\alpha(D^\alpha(f \circ \phi^{-1})) \circ \phi, \qquad f \in C_c^\infty(W). \tag{2}$$

Thus the definition above of a differential operator coincides with the customary one. Just as for open sets in $\mathbf{R}^m$ [cf. (8) in §1] we can extend a differential operator to $C^\infty(M)$.

We also adopt Schwartz's notation

$$\mathscr{D}(M) = C_c^\infty(M), \qquad \mathscr{E}(M) = C^\infty(M),$$

and if $K \subset M$ is any compact subset, $\mathscr{D}_K(M)$ denotes the set of functions in $\mathscr{D}(M)$ with support in K.

For convenience we now make the assumption that M *has a countable base for the open sets.* We shall now give $\mathscr{E}(M)$ a topology.

2. Topology of the Spaces $\mathscr{D}(M)$ and $\mathscr{E}(M)$. Distributions

If $V \subset \mathbf{R}^m$ is an open set, $\mathscr{E}(V)$ is topologized by the seminorms $\|f\|_k^C$ [see (2) in §1] as C runs through the compact subsets of V and k runs through $\mathbf{Z}^+$. If (U, ϕ) is a local coordinate system on M, this gives a topology of $\mathscr{E}(U)$ with the property that a sequence (f_n) in $\mathscr{E}(U)$ converges to 0 if and only if for each differential operator D on U, $Df_n \to 0$ uniformly on each compact subset of U. In particular, the topology of $\mathscr{E}(U)$ is independent of the coordinate system.

The space $\mathscr{E}(M)$ is now provided with the weakest topology for which all the maps $f \to f \mid U$ ($\mid$ denotes restriction), as (U, ϕ) runs through all local coordinate systems on M, are continuous. Here we can, by the countability assumption, restrict the charts (U, ϕ) to a countable family (U_j, ϕ_j), $j = 1, 2, \ldots$. Since each $\mathscr{E}(U_j)$ is a Fréchet space, it follows that $\mathscr{E}(M)$ is also a Fréchet space, and again, a sequence (f_n) in $\mathscr{E}(M)$ converges to 0 if and only if for each differential operator D on M, $Df_n \to 0$ uniformly on each compact subset of M. Writing M as the union of an increasing sequence of compact sets, we see that *$\mathscr{D}(M)$ is dense in $\mathscr{E}(M)$*. We note also that a differential operator on M is automatically a continuous endomorphism of $\mathscr{E}(M)$.

For each compact set $K \subset M$, $\mathscr{D}_K(M)$ is given the topology induced by $\mathscr{E}(M)$. It is a closed subspace of $\mathscr{E}(M)$, hence a Fréchet space. A linear functional T on $\mathscr{D}(M)$ is called a *distribution* if for any compact set $K \subset M$ the restriction of T to $\mathscr{D}_K(M)$ is continuous. The set of all distributions on M is denoted $\mathscr{D}'(M)$. We often write $\int_M f(m)\, dT(m)$ instead of $T(f)$.

We now give $\mathscr{D}(M)$ the *inductive limit topology* of the spaces $\mathscr{D}_K(M)$ by taking as a fundamental system of neighborhoods of 0 the convex sets W such that for each compact subset $K \subset M$, $W \cap \mathscr{D}_K(M)$ is a neighborhood of 0 in $\mathscr{D}_K(M)$. With this topology of $\mathscr{D}(M)$ the continuous linear functionals T are precisely the distributions on M. In fact, if $r > 0$, $T^{-1}(B_r(0))$ is a convex set containing 0, so the continuity assumption amounts to the definition of a distribution. Thus $\mathscr{D}'(M)$ is just the dual space of $\mathscr{D}(M)$. A similar argument shows that a differential operator gives a continuous endomorphism of $\mathscr{D}(M)$.

A distribution T is said to vanish on an open set $V \subset M$ if $T(f) = 0$ for each $f \in \mathscr{D}(V)$ $(\subset \mathscr{D}(M))$. Let $\{U_\alpha\}_{\alpha \in A}$ denote the family of all open sets on which T vanishes and U their union. Then T vanishes on U. In fact, if $f \in \mathscr{D}(U)$, $\operatorname{supp}(f)$ can be covered by finitely many U_α, say $U_1, \ldots, U_r$. Then $\{U_1, \ldots, U_r, M - \operatorname{supp}(f)\}$ is a covering of M. If $1 = \sum_1^{r+1} \phi_i$ is a corresponding partition of unity, we have $f = \sum_1^r \phi_i f$, so $T(f) = 0$. The complement $M - U$ is called the *support of* T. Let $\mathscr{E}'(M)$ denote the set of distributions on M of compact support. We shall now prove that this set can be identified with the dual space of $\mathscr{E}(M)$.

Proposition 2.1. *The restriction of a functional from $\mathscr{E}(M)$ to $\mathscr{D}(M)$ gives a bijection of the dual of $\mathscr{E}(M)$ onto $\mathscr{E}'(M)$.*

Proof. Let τ be a continuous linear functional on $\mathscr{E}(M)$, T its restriction to $\mathscr{D}(M)$. Since T is continuous on each $\mathscr{D}_K(M)$, it is a distribution. Also, $\tau_1 \neq \tau_2$ implies $T_1 \neq T_2$ because $\mathscr{D}(M)$ is dense in $\mathscr{E}(M)$. Next we prove that T has compact support. For this let $K_1 \subset K_2 \subset \cdots$ be a sequence of compact subsets of M such that each compact set $K \subset M$ is contained in some K_i. If $\operatorname{supp}(T)$ were noncompact, we could find $\phi_i \in$

$\mathscr{E}(M)$ such that $\phi_i \equiv 0$ on K_i but $T(\phi_i) = 1$ $(i = 1, 2, \ldots)$. Then $\phi_i \to 0$ in $\mathscr{E}(M)$, which is a contradiction.

To see that the mapping $\tau \to T$ is surjective let S be a distribution of compact support. Fix a function $\phi \in \mathscr{D}(M)$ such that $\phi \equiv 1$ on a neighborhood of supp(S). The mapping $f \to \phi f$ is a continuous mapping of $\mathscr{E}(M)$ into itself; hence we can define a continuous linear functional σ on $\mathscr{E}(M)$ by

$$\sigma(f) = S(\phi f), \qquad f \in \mathscr{E}(M). \tag{3}$$

It is independent of the choice of ϕ because if $\phi_1 \in \mathscr{D}(M)$, $\phi_1 \equiv 1$ on a neighborhood of supp(S), then

$$S(\phi_1 f) - S(\phi f) = S((\phi_1 - \phi)f) = 0,$$

since $\text{supp}((\phi_1 - \phi)f) \subset M - \text{supp}(S)$. The same argument shows that S coincides with the restriction of σ to $\mathscr{D}(M)$.

3. Effect of Mappings. The Adjoint

Let $\mathbf{E}(M)$ denote the set of all differential operators on M. If $f \in \mathscr{E}(M)$, $D \in \mathbf{E}(M)$, the value $(Df)(p)$ of Df at p will sometimes be denoted $D_p(f(p))$. The composition of two differential operators D_1 and D_2 will often be denoted $D_1 \circ D_2$.

Let M and N be manifolds and $\phi: M \to N$ a differentiable mapping. The differential of ϕ at $p \in M$ which maps M_p into $N_{\phi(p)}$ is denoted $d\phi_p$. If ϕ is a diffeomorphism of M onto N and if $f \in \mathscr{D}(N)$, $g \in \mathscr{E}(N)$, $T \in \mathscr{D}'(M)$, $D \in \mathbf{E}(M)$, we put

$$g^{\phi^{-1}} = g \circ \phi, \qquad T^{\phi}(f) = T(f^{\phi^{-1}}), \qquad D^{\phi}(g) = (D(g^{\phi^{-1}}))^{\phi}. \tag{4}$$

Then $g^{\phi^{-1}} \in \mathscr{E}(M)$, $T^{\phi} \in \mathscr{D}'(N)$, and D^{ϕ} is a differential operator on N, the *image* of D under ϕ. Suppose ϕ is a diffeomorphism of M onto itself; D is said to be *invariant under* ϕ if $D^{\phi} = D$, that is, if

$$Dg = (D(g \circ \phi)) \circ \phi^{-1}$$

for all $g \in \mathscr{E}(M)$.

A measure μ on the manifold M is said to be *equivalent* to Lebesgue measure if on each coordinate neighborhood on M it is the multiple of the Lebesgue measure by a nowhere vanishing C^{∞}-function. In this case the *adjoint* D^* of $D \in \mathbf{E}(M)$ is defined as the operator on $\mathscr{D}'(M)$ which is the transpose of D:

$$(D^*T)(f) = T(Df), \qquad f \in \mathscr{D}(M), \quad T \in \mathscr{D}'(M). \tag{5}$$

The space $\mathscr{E}(M)$ is imbedded in $\mathscr{D}'(M)$ by means of μ if to $g \in \mathscr{E}(M)$ we associate the distribution

$$f \to \int_M fg \, d\mu, \qquad f \in \mathscr{D}(M),$$

on M. Thus the restriction $D^*|\mathscr{E}(M)$ is well-defined. Also, D^*, as an operator on $\mathscr{D}'(M)$, decreases supports, so if $\{x_1, \ldots, x_m\}$ is a coordinate system on an open set U and $g \in \mathscr{D}(U)$, the distribution D^*g is determined by its restriction to $\mathscr{D}(U)$. But if in terms of the local coordinates we calculate the usual adjoint D' of D *on* U, we have

$$(6) \qquad \int_U (D'g)(x)f(x) \, d\mu(x) = \int_U g(x)(Df)(x) \, d\mu(x), \qquad f \in \mathscr{D}(U),$$

so the distribution D^*g equals the function $D'g$ on U. This shows that D^* maps $\mathscr{E}(M)$ into itself and is a differential operator. We remark that while this definition of the adjoint D^* depends on the measure μ, it does not require orientation of M.

Now if $E \in \mathbf{E}(M)$, $T \in \mathscr{D}'(M)$, we define $ET \in \mathscr{D}'(M)$ by

$$(ET)(f) = T(E^*f), \qquad f \in \mathscr{D}(M).$$

If T is a function g, i.e.,

$$T(f) = \int fg \, d\mu,$$

then formula (6) shows that ET is the function Eg. If τ is a diffeomorphism of M leaving μ invariant, then

$$(E^*)^\tau = (E^\tau)^*, \qquad (ET)^\tau = E^\tau T^\tau$$

for $E \in \mathbf{E}(M)$ and $T \in \mathscr{D}'(M)$.

4. The Laplace–Beltrami Operator

A pseudo-Riemannian manifold M always possesses a differential operator of particular interest, the so-called *Laplace–Beltrami operator* which we shall now define. Let g denote the pseudo-Riemannian structure on M and let $\phi: q \to (x_1(q), \ldots, x_m(q))$ be a coordinate system valid on an open set $U \subset M$. As customary we define the functions g_{ij}, g^{ij}, $\bar{g}$ on U by

$$g_{ij} = g\left(\frac{\partial}{\partial x_i}, \frac{\partial}{\partial x_j}\right), \qquad \sum_j g_{ij} g^{jk} = \delta_{ik},$$

$$\bar{g} = |\det(g_{ij})|.$$

We often write $\langle\ ,\ \rangle$ in place of g and extend it bilinearly to complex vector fields.

Each C^∞-function f on M gives rise to a vector field grad f (gradient of f) defined by

$$\langle \operatorname{grad} f, X\rangle = Xf \tag{7}$$

for each vector field X. In terms of the coordinates on U, we have on U

$$\operatorname{grad} f = \sum_{i,j} g^{ij}\,\partial_i f\,\frac{\partial}{\partial x_j}, \tag{8}$$

where $\partial_i f = (\partial/\partial x_i)(f)$.

On the other hand, if X is a vector field on M, the *divergence* of X is the function on M which on U is given by

$$\operatorname{div} X = \frac{1}{\sqrt{\bar{g}}}\sum_i \partial_i(\sqrt{\bar{g}}\,X_i) \tag{9}$$

if $X = \sum_i X_i(\partial/\partial x_i)$ on U. To see that div X is well-defined (independent of the coordinate system), we interpret it in terms of the Riemannian measure. Let $\theta(X)$ denote the Lie derivative by X as defined in [DS, Exercise B, Chapter I]. We shall use some of its properties established there.

Lemma 2.2. *Suppose U is connected and let ω_U denote the differential form on U given by*

$$\omega_U = \sqrt{\bar{g}}\,dx_1 \wedge \cdots \wedge dx_m. \tag{10}$$

Then

$$\theta(X)\omega_U = (\operatorname{div} X)\omega_U. \tag{11}$$

Proof. Since $\theta(X)$ commutes with the exterior differentiation d, we have

$$\theta(X)\,dx_i = d\,\theta(X)x_i = d\,Xx_i = dX_i = \sum_j (\partial_j X_i)\,dx_j.$$

Since $\theta(X)$ is a derivation of the Grassmann algebra of U, we obtain

$$\begin{aligned}
&\theta(X)(\sqrt{\bar{g}}\,dx_1 \wedge \cdots \wedge dx_m)\\
&\quad = X(\sqrt{\bar{g}})\,dx_1 \wedge \cdots \wedge dx_m\\
&\quad\quad + \sqrt{\bar{g}}\sum_i dx_1 \wedge \cdots \wedge \left(\sum_r (\partial_r X_i)\,dx_r\right) \wedge \cdots \wedge dx_m,
\end{aligned}$$

where the parenthesis on the right occurs at the ith factor. It follows that the right-hand side equals

$$\Big\{\sum_i X_i\,\partial_i(\sqrt{\bar{g}}) + \sqrt{\bar{g}}\sum_i (\partial_i X_i)\Big\}\,dx_1 \wedge \cdots \wedge dx_m,$$

which, by comparison with (9), proves the lemma.

Now let $\{y_1, \ldots, y_m\}$ be another coordinate system on U and put

$$h_{ij} = g\left(\frac{\partial}{\partial y_i}, \frac{\partial}{\partial y_j}\right), \qquad (h^{ij}) = (h_{ij})^{-1},$$
$$\bar{h} = |\det(h_{ij})|.$$

Then

$$h_{ij} = \sum_{pq} \frac{\partial x_p}{\partial y_i}\frac{\partial x_q}{\partial y_j}\,g_{pq},$$

so if J denotes the Jacobi matrix

$$J = \frac{\partial(x_1, \ldots, x_m)}{\partial(y_1, \ldots, y_m)},$$

we have

$$\bar{h} = |\det J|^2 \bar{g}.$$

On the other hand, we have (cf. [DS], Chapter I, §2, No. 3)

$$dx_1 \wedge \cdots \wedge dx_m = \det J\, dy_1 \wedge \cdots \wedge dy_m.$$

Hence, if $\det J > 0$,

$$\bar{h}^{1/2}\, dy_1 \wedge \cdots \wedge dy_m = \bar{g}^{1/2}\, dx_1 \wedge \cdots \wedge dx_m.$$

Thus (11) shows that the right-hand side of (9) is the same for all coordinate systems on U which satisfy the positivity condition $\det J > 0$ relative to the original $\{x_1, \ldots, x_m\}$. But the right-hand side of (9) is obviously also invariant under the coordinate change $(x_1, x_2, \ldots, x_m) \to (x_2, x_1, \ldots, x_m)$; thus it is invariant under all coordinate changes. Hence $\operatorname{div} X$ is a well-defined function on M (without the customary assumption of orientability).

Remark. Another justification of the definition of $\operatorname{div} X$ comes from Lemma 2.5.

The Laplace–Beltrami operator L on M is now defined by

(12) $$Lf = \operatorname{div} \operatorname{grad} f, \qquad f \in \mathscr{E}(M).$$

In terms of local coordinates we have

$$(13) \qquad Lf = \frac{1}{\sqrt{\bar{g}}} \sum_k \partial_k \Big(\sum_i g^{ik} \sqrt{\bar{g}}\, \partial_i f \Big),$$

so L is a differential operator on M.

Proposition 2.3. *Let M be a pseudo-Riemannian manifold, L the Laplace–Beltrami operator on M. Then L is symmetric; that is,*

$$\int_M u(x)(Lv)(x)\, dx = \int_M (Lu)(x)v(x)\, dx, \qquad u \in \mathscr{D}(M), \quad v \in \mathscr{E}(M)$$

if dx is the Riemannian measure on M.

Proof. Let X be any vector field on M. Then we have from the definitions (8) and (9),

$$(14) \qquad \operatorname{div}(uX) = u \operatorname{div} X + Xu,$$

$$(15) \qquad \operatorname{grad} u\,(v) = \operatorname{grad} v\,(u) = \langle \operatorname{grad} u, \operatorname{grad} v \rangle.$$

Consequently,

$$uLv - vLu = \operatorname{div}(u \operatorname{grad} v) - \operatorname{grad} v\,(u) - \operatorname{div}(v \operatorname{grad} u) + \operatorname{grad} u\,(v),$$

so

$$uLv - vLu = \operatorname{div}(u \operatorname{grad} v - v \operatorname{grad} u).$$

It suffices therefore to prove

$$(16) \qquad \int_M (\operatorname{div} X)\, dx = 0$$

for any vector field X on M vanishing outside a compact subset. While this could be derived from (11) above [see also (2) in §1 of Chapter I and [DS], Exercise B3, Chapter I], we can proceed more directly as follows. Using partition of unity, we may assume that X vanishes outside a coordinate neighborhood U. Writing $X = \sum_i X_i\, \partial/\partial x_i$ on U, we have

$$\int_M (\operatorname{div} X)\, dx = \int \sum_i \partial_i(\sqrt{\bar{g}}\, X_i)\, dx_1 \cdots dx_m = 0,$$

proving (16) and the proposition.

In connection with (14) and (15) we mention the simple identity

$$(17) \qquad L(uv) = u(Lv) + 2\langle \operatorname{grad} u, \operatorname{grad} v \rangle + v(Lu),$$

whose proof is left to the reader. Next we prove a simple invariance property of the Laplacian. If (U, ϕ) is a local chart on M and $f \in \mathscr{E}(M)$, we often write f^* for the composite function $f \circ \phi^{-1}$.

Proposition 2.4. *Let Φ be a diffeomorphism of the pseudo-Riemannian manifold M. Then Φ leaves the Laplace–Beltrami operator L invariant if and only if it is an isometry.*

Proof. Let $p \in M$ and let (V, ψ) be a local chart around p. Then $(\Phi(V), \psi \circ \Phi^{-1})$ is a local chart around $\Phi(p)$. For $x \in V$, let $y = \Phi(x)$ and

$$\psi(x) = (x_1, \ldots, x_m), \qquad x \in V,$$
$$(\psi \circ \Phi^{-1})(y) = (y_1, \ldots, y_m), \qquad y \in \Phi(V).$$

Then

$$x_i(x) = y_i(\Phi(x)), \qquad d\Phi_x\left(\frac{\partial}{\partial x_i}\right)_x = \left(\frac{\partial}{\partial y_i}\right)_{\Phi(x)} \qquad (1 \leq i \leq m).$$

For each function $f \in \mathscr{E}(M)$,

$$(18) \quad ((Lf)^{\Phi^{-1}})(x) = (Lf)(\Phi(x)) = \frac{1}{\sqrt{\bar{g}(y)}} \sum_k \frac{\partial}{\partial y_k} \left(\sum_i g^{ik}(y) \sqrt{\bar{g}(y)} \frac{\partial f^*}{\partial y_i} \right),$$

$$(19) \qquad (Lf^{\Phi^{-1}})(x) = \frac{1}{\sqrt{\bar{g}(x)}} \sum_k \frac{\partial}{\partial x_k} \left(\sum_i g^{ik}(x) \sqrt{\bar{g}(x)} \frac{\partial (f \circ \Phi)^*}{\partial x_i} \right).$$

Because of the choice of coordinates we have

$$\frac{\partial f^*}{\partial y_i} = \frac{\partial (f \circ \Phi)^*}{\partial x_i}, \qquad \frac{\partial^2 f^*}{\partial y_i \, \partial y_j} = \frac{\partial^2 (f \circ \Phi)^*}{\partial x_i \, \partial x_j}, \qquad 1 \leq i, j \leq m.$$

Now if Φ is an isometry, then $g_{ij}(x) = g_{ij}(y)$ for all i, j. Thus the right-hand sides of (18) and (19) coincide and $L^\Phi = L$. On the other hand, if (18) and (19) agree, then we obtain, by equating coefficients, $g_{ij}(x) = g_{ij}(y)$, which shows that Φ is an isometry.

The Laplace–Beltrami operator was defined above in terms of the pseudo-Riemannian structure on M. We shall now describe it more directly in terms of the Riemannian connection ∇ (§9, Chapter I in [DS]).

Lemma 2.5. *If X is a vector field on M and $p \in M$, then*

$$(\operatorname{div} X)_p = \text{trace of the endomorphism } v \to \nabla_v(X) \text{ of } M_p.$$

Proof. Let $\{x_1, \ldots, x_m\}$ be a coordinate system as in Lemma 2.2 around p. Denoting the endomorphism $v \to \nabla_v(X)$ of M_p by D_X, we have by §5, Chapter I in [DS]

$$\text{Trace}(D_X) = \sum_i \left(\frac{\partial X^i}{\partial x_i} + \sum_j X^j \Gamma^i_{ij} \right) \tag{20}$$

if $X = \sum_i X^i(\partial/\partial x_i)$. On the other hand,

$$\operatorname{div} X = \sum_i \partial_i X^i + \sum_j \partial_j(\log \sqrt{\bar{g}}) X^j,$$

so it suffices to prove

$$\partial_j(\log \sqrt{\bar{g}}) = \sum_i \Gamma^i_{ij}. \tag{21}$$

The formula for ∇ in [DS] (Chapter I, §9), can be written

$$2 \sum_l g_{il} \Gamma^l_{jk} = \partial_j g_{ik} + \partial_k g_{ij} - \partial_i g_{jk}. \tag{22}$$

On the other hand, if G_{ij} is the cofactor of g_{ji} in the determinant $g_0 = \det(g_{ij})$, we have

$$G_{ij} = g_0 g^{ij}. \tag{23}$$

Applying to g_0 the rule for differentiating determinants, we have

$$\frac{\partial g_0}{\partial x_k} = \sum_{i,j} \partial_k(g_{ij})\, G_{ij} = g_0 \sum_{ij} g^{ij}\, \partial_k(g_{ij}). \tag{24}$$

Now (22) implies

$$\partial_k(g_{ij}) = \sum_l (\Gamma^l_{ik} g_{jl} + \Gamma^l_{jk} g_{il}), \tag{25}$$

whereby the right-hand side of (24) reduces to

$$2g_0\left(\sum_i \Gamma^i_{ik}\right).$$

Since $\bar{g} = |g_0|$, this proves (21) and the lemma.

Proposition 2.6. *In an arbitrary coordinate system, the Laplace–Beltrami operator is given by*

$$Lf = \sum_{i,j} g^{ij}\left(\partial_i \partial_j f - \sum_k \Gamma^k_{ij} \partial_k f\right). \tag{26}$$

Proof. Because of Lemma 2.5 and (20), we have

$$Lf = \sum_i \left[\partial_i \left(\sum_l g^{li} \, \partial_l f \right) + \sum_j \left(\sum_l g^{lj} \, \partial_l f \right) \Gamma^i_{ij} \right]. \tag{27}$$

But differentiating $\sum_j g^{ij} g_{jk} = \delta_{ik}$ and using (25), we derive

$$\partial_l (g^{ir}) = - \sum_j (\Gamma^r_{jl} g^{ij} + \Gamma^i_{jl} g^{jr}). \tag{28}$$

Carrying out the differentiation in (27) and using (28), we get the result

We consider now the case of a Riemannian manifold (i.e., g positive definite). Then

$$\langle \text{grad } f, \text{grad } \bar{f} \rangle = \sum g^{ij} \, \partial_i f \, \partial_j \bar{f} \geq 0.$$

Let us call a Riemannian manifold M *simply convex* if any two points in M can be joined by a unique geodesic. Every point in any Riemannian manifold has an open neighborhood, which with the induced Riemannian structure is simply convex (cf., [DS], Chapter I, Theorem 9.9).

Theorem 2.7. *Let M be an analytic simply convex Riemannian manifold and $p \in M$. Given $c \in \mathbf{C}$ there exists a sphere $S_R(p)$ around p such that if*

$$Lu = cu, \; u \in \mathscr{E}(M), \text{ and } u \equiv 0 \quad \text{on} \quad S_R(p),$$

then

$$u \equiv 0 \quad \text{on} \quad M.$$

Remark. The example $u(t) = \sin \lambda t$ in $\mathbf{R}$ which satisfies $Lu = -\lambda^2 u$ shows that the radius R will depend on c. The proof below will show, however, that for a given eigenvalue c all sufficiently small R will do.

Proof. Let N_0 be a normal neighborhood of the origin in M_p ([DS], Chapter I, §6), let $S_R(0) \subset N_0$ be a sphere inside N_0 with center 0, $B_R(0)$ the corresponding open ball, and put

$$S_R = \text{Exp}_p(S_R(0)), \; B_R = \text{Exp}_p(B_R(0)).$$

For $u \in \mathscr{E}(B_R)$ we put

$$\|u\| = \left[\int_{B_R} |u(x)|^2 \, dx + \int_{B_R} \langle \text{grad } u, \text{grad } \bar{u} \rangle (x) \, dx \right]^{1/2}$$

Let $u \in \mathscr{E}(M)$ such that $u \equiv 0$ on S_R. We shall prove that given $\varepsilon > 0$ there exists a $\phi \in \mathscr{D}(B_R)$ such that

$$\|u - \phi\| < \varepsilon. \tag{29}$$

For this let $(\theta_1, \ldots, \theta_m)$, where $\theta_m = r$, be geodesic polar coordinates around p (cf. [DS], p. 543). For $\delta > 0$ small, let α_δ be an even function in $\mathscr{E}(\mathbf{R})$ satisfying

1. $\alpha_\delta \equiv 1$ on $[0, R - 4\delta]$,
2. $\alpha_\delta \equiv 0$ on $[R - \delta, R]$,
3. $|\alpha'_\delta| \leq \delta^{-1}$,
4. $|\alpha_\delta| \leq 1$.

Let $\alpha \in \mathscr{E}(B_R)$ be defined by $\alpha(x) = \alpha_\delta(d(p, x))$, d denoting the distance. We shall prove that $\phi = \alpha u$ satisfies (29) provided δ is small enough. Since

$$\operatorname{grad} u = \sum_{i,j} g^{ij}(\partial_i u)\, \partial_j \qquad (r = \theta_m)$$

and since

$$\frac{\partial}{\partial\theta_i}(\alpha u - u) = (\alpha - 1)\frac{\partial u}{\partial\theta_i} + \frac{\partial\alpha}{\partial\theta_i} u,$$

it would be sufficient to prove

$$\int_{B_R} \left|\frac{\partial\alpha}{\partial r} u\right|^2 dx \to 0 \quad \text{as} \quad \delta \to 0. \tag{30}$$

But if $\bar{g}(r, \theta_1, \ldots, \theta_{m-1}) = |\det(g_{ij})|$, we have

$$\int_{B_R} \left|\frac{\partial\alpha}{\partial r} u\right|^2 dx \leq \frac{1}{\delta^2}\int_{R-4\delta}^{R-\delta} dr \int_{(\theta)} |u|^2 \bar{g}^{1/2}\, d\theta_1 \cdots d\theta_{m-1}.$$

Using now $u \equiv 0$ on S_R, we have

$$|u(r, \theta_1, \ldots, \theta_{m-1})|^2 = \left|\int_r^R \frac{\partial u}{\partial p} dp\right|^2 \leq \int_r^R dp \int_r^R \left|\frac{\partial u}{\partial p}\right|^2 dp.$$

Hence

$$|u(r, \theta_1, \ldots, \theta_{m-1})|^2 \leq 4\delta \int_{R-4\delta}^R \left|\frac{\partial u}{\partial p}\right|^2 dp \quad \text{if} \quad R - 4\delta \leq r \leq R.$$

Thus if $D = \max|\partial u/\partial p|^2$ on B_R, we obtain

$$\int_{B_R} \left|\frac{\partial\alpha}{\partial r} u\right|^2 dx \leq 16D \operatorname{Volume}(\Sigma_\delta),$$

where Σ_δ is the "shell" $R - 4\delta \leq d(p, x) \leq R - \delta$. This proves (30) and therefore also (29).

Now we choose an orthonormal basis of the tangent space M_p and let $\{x_1, \ldots, x_m\}$ be a system of normal coordinates with respect to this basis.

Let $\phi \in \mathscr{D}(B_R)$ be real. Using Schwarz's inequality on

$$\phi(x) = \int_{-R}^{x_j} (\partial_j \phi)(x_1, \ldots, t, \ldots, x_m)\, dt$$

and then integrating over B_R, we derive the Poincaré inequality,[†]

$$\int_{B_R} |\phi(x)|^2\, dx_1 \cdots dx_m \leq 4R^2 \int_{B_R} \sum_j (\partial_j \phi)^2\, dx_1 \cdots dx_m. \tag{31}$$

Now let

$$c_R = \inf_{i,\, x \in B_R} g^{ii}(x), \qquad d_R = \sup_{i \neq j,\, x \in B_R} |g^{ij}(x)|$$

$$e_R = \sup_{x \in B_R} \bar{g}^{1/2}(x), \qquad f_R = \inf_{x \in B_R} \bar{g}^{1/2}(x).$$

Then

$$\begin{aligned} \sum_{ij} g^{ij} \frac{\partial \phi}{\partial x_i} \frac{\partial \phi}{\partial x_j} &\geq c_R \sum_i \left(\frac{\partial \phi}{\partial x_i}\right)^2 - d_R \sum_{i \neq j} \left|\frac{\partial \phi}{\partial x_i} \frac{\partial \phi}{\partial x_j}\right| \\ &\geq (c_R - (n-1)d_R) \sum_i \left(\frac{\partial \phi}{\partial x_i}\right)^2. \end{aligned}$$

We assume R to be so small that $c_R - (n-1)d_R > 0$. Then, by (31),

$$\begin{aligned} \int_{B_R} |\phi(x)|^2\, dx &\leq e_R 4R^2 \int_{B_R} \sum_i (\partial_i \phi)^2\, dx_1 \cdots dx_m \\ &\leq 4e_R R^2 (c_R - (n-1)d_R)^{-1} \int_{B_R} \sum_{i,j} g^{ij} (\partial_i \phi)(\partial_j \phi)\, dx_1 \cdots dx_m \\ &\leq 4e_R f_R^{-1} R^2 (c_R - (n-1)d_R)^{-1} \int_{B_R} \sum_{i,j} g^{ij} (\partial_i \phi)(\partial_j \phi)\, dx. \end{aligned}$$

Thus we have for small $R > 0$, $\phi \in \mathscr{D}(B_R)$ real,

$$\int_{B_R} |\phi(x)|^2\, dx \leq g_R 4R^2 \int_{B_R} \langle \operatorname{grad} \phi, \operatorname{grad} \phi \rangle\, dx, \tag{32}$$

where g_R is a constant depending only on the Riemannian structure of M and $g_R \to 1$ as $R \to 0$.

To finish the proof of Theorem 2.7 suppose $Lu = cu$, where $c \in \boldsymbol{C}$ and $u \equiv 0$ on S_R. Let $\phi, \psi \in \mathscr{D}(B_R)$. By (17) we have

$$L(\phi\psi) = \phi L\psi + 2\langle \operatorname{grad} \phi, \operatorname{grad} \psi \rangle + \psi L\phi.$$

[†]See "Some Details," p. 611.

Integrating this over B_R and using (16) together with Proposition 2.3, we obtain

$$\int_{B_R} \phi L\psi \, dx = -\int_{B_R} \langle \text{grad } \phi, \text{grad } \psi \rangle \, dx.$$

Using (29), we deduce by approximation

$$\int_{B_R} uL\psi \, dx = -\int_{B_R} \langle \text{grad } u, \text{grad } \psi \rangle \, dx,$$

whence by Proposition 2.3 and $Lu = cu$

$$c \int_{B_R} u\psi \, dx = -\int_{B_R} \langle \text{grad } u, \text{grad } \psi \rangle \, dx.$$

Again, by (29) this implies, letting $\psi \to \bar{u}$,

$$c \int_{B_R} |u|^2 \, dx = -\int_{B_R} \langle \text{grad } u, \text{grad } \bar{u} \rangle \, dx. \tag{33}$$

Thus c is real (or $u \equiv 0$ on B_R), so we may assume u to be real. Again, by (29) we may replace ϕ by u in (32). But if $g_R(4R^2)(-c) < 1$, we can conclude from (33) that $u \equiv 0$ on B_R. But since the operator L has analytic coefficients and is elliptic, its eigenfunctions are analytic (see, for example, John [1955], p. 57). Hence $u \equiv 0$ and the theorem is proved.

Remark. The proof shows that if $c = 0$, we can take R arbitrary.

§3. Geometric Operations on Differential Operators

1. Projections of Differential Operators

Let V be a Riemannian manifold, D a differential operator on V. If $S \subset V$ is any submanifold, we shall now define the *projection of D on S*.

For each $s \in S$ consider the geodesics in V starting at s perpendicular to S. If we take sufficiently short pieces of these geodesics, their union is a submanifold $S_s^\perp$ of V. Putting this more precisely in terms of the exponential mapping Exp_s for V ([DS], Chapter I, §6), we let $V(0)$ be a spherical normal neighborhood of 0 in the tangent space V_s, let $V_s = S_s \oplus S_s'$ be the orthogonal decomposition, and put $S_s^\perp = \text{Exp}_s(V(0) \cap S_s')$. Fix $s_0 \in S$. Shrinking the $S_s^\perp$ further, we can assure that as s runs through a suitable neighborhood S_0 of s_0 in S the manifolds $S_s^\perp$ are disjoint and fill up a neighborhood V_0 (a "tubular neighborhood") of s_0 in V:

$$V_0 = \bigcup_{s \in S_0} S_s^{\perp}. \tag{1}$$

Given $F \in \mathscr{D}(S)$, we define a function $\tilde{F}$ on V_0 by making it constant on each $S_s^{\perp}$ and equal to F on S_0. Given the differential operator D on V, we define an operator D' on $\mathscr{D}(S)$ by

$$(D'F)(s_0) = (D\tilde{F})(s_0). \tag{2}$$

Since $\tilde{F}$ is C^{∞}, the right-hand side makes sense, and since D decreases supports, the right-hand side is independent of the choice of S_0 and V_0 as long as we have the smooth disjoint decomposition (1). This $D'F$ is a well-defined function on S. If we vary s_0 slightly, decomposition (1) can still be used and the right-hand side of (2) varies smoothly with s_0. Thus $D'F$ is a C^{∞}-function on S. It is also clear that the linear map $F \to D'F$ decreases support, so D' is a differential operator on S. Summarizing, we have proved the following result.

Proposition 3.1. *If D is a differential operator on V, formulas* (1) *and* (2) *define a differential operator D' on the submanifold S.*

Example. Consider the Laplacian L on $\mathbf{R}^3$ and let S denote the unit sphere $x_1^2 + x_2^2 + x_3^2 = 1$. If we express L in the spherical polar coordinates (r, θ, ϕ), we obtain the standard formula

$$L = \frac{\partial^2}{\partial r^2} + \frac{2}{r}\frac{\partial}{\partial r} + \frac{1}{r^2}\left(\frac{\partial^2}{\partial \theta^2} + \cot\theta\frac{\partial}{\partial\theta} + \sin^{-2}\theta\frac{\partial^2}{\partial\phi^2}\right). \tag{3}$$

The operator in the parentheses is clearly the projection L' of L on S. It is a familiar fact, which we generalize below, that L' is the Laplace–Beltrami operator on S.

Theorem 3.2. *Let V be a Riemannian manifold, $S \subset V$ a submanifold, and L_V and L_S the corresponding Laplace–Beltrami operators. Then L_S equals the projection of L_V on S:*

$$L_V' = L_S.$$

Proof. We use the coordinate representation of L_V given by Proposition 2.6. Now let $s_0 \in S$ be arbitrary and choose the local coordinates $(x_1, \ldots, x_n)$ on a neighborhood of s_0 in V in accordance with (1) as follows:

(i) The mapping

$$s \to (x_1(s), \ldots, x_r(s), 0, \ldots, 0)$$

is a system of local coordinates near s_0 on S.

(ii) For each $s \in S$ sufficiently close to s_0 and any constants $a_{r+1}, \ldots, a_n$, not all 0, the curve

(4) $$t \to (x_1(s), \ldots, x_r(s), a_{r+1}t, \ldots, a_n t)$$

is a geodesic in V, starting at s, perpendicular to S.

When writing out (2) in this coordinate system it is convenient to adopt the following range of indices:

$$1 \leq i, j, k \leq r, \qquad r+1 \leq \alpha, \beta, \gamma \leq n.$$

The geodesic (4) satisfies the differential equations

$$\frac{d^2 x_i}{dt^2} + \sum_{1 \leq p, q \leq n} \Gamma^i_{pq} \frac{dx_p}{dt} \frac{dx_q}{dt} = 0.$$

At the point s this reduces to

$$0 + \sum_{\alpha, \beta} \Gamma^i_{\alpha\beta} a_\alpha a_\beta + 0 = 0,$$

so that, since by (22) §2 $\Gamma^i_{\alpha\beta}$ is symmetric in α and β and a_α, a_β arbitrary,

(5) $$\Gamma^i_{\alpha\beta}(s) = 0.$$

Since the geodesic is perpendicular to S at s, we have

(6) $$g_{i\alpha}(s) = 0, \qquad g^{i\alpha}(s) = 0.$$

With F and $\tilde{F}$ related as above we have

(7) $$(L'_V F)(s_0) = (L_V \tilde{F})(s_0) = \sum_{p,q} g^{pq}\left(\partial_p \partial_q \tilde{F} - \sum_t \Gamma^t_{pq}\, \partial_t \tilde{F}\right)(s_0).$$

By definition, $\tilde{F}$ is constant in the $n - r$ last variables; because of this, (5) and (6), the right-hand side of (7) reduces to

(8) $$\sum_{i,j} g^{ij}\left(\partial_i \partial_j \tilde{F} - \sum_k \Gamma^k_{ij}\, \partial_k \tilde{F}\right)(s_0).$$

But $\tilde{F} = F$ on S and by (22) in §2 the Christoffel symbol Γ^k_{ij} on S is the restriction to S of the corresponding symbol for V. Thus expression (8) equals $(L_S F)(s_0)$, so the theorem is proved.

2. Transversal Parts and Separation of Variables for Differential Operators

Let V be a manifold satisfying the second axiom of countability and suppose H is a Lie transformation group of V. If $v \in V$, let H^v denote the subgroup of H leaving v fixed. As usual, the image of v under $h \in H$ will be denoted $h \cdot v$. Let $\mathfrak{h}$ denote the Lie algebra of H. For $X \in \mathfrak{h}$ let X^+ denote the vector field on V *induced by* X; that is,

$$(X^+f)(v) = \left\{\frac{d}{dt} f(\exp tX \cdot v)\right\}_{t=0} \qquad [v \in V, f \in \mathscr{E}(V)].$$

A C^∞-function f on an open subset of V is said to be *locally invariant* if $X^+f = 0$ for each $X \in \mathfrak{h}$. We recall that a submanifold $B \subset H$ is called a *local cross section* over an open set $U \subset H/H^v$ if the natural map $\pi\colon H \to H/H^v$ gives by restriction a diffeomorphism of B onto U.

Lemma 3.3. *Suppose W is a submanifold of V such that for each $w \in W$ the tangent spaces at w satisfy the following "transversality" condition:*

$$V_w = (H \cdot w)_w + W_w \qquad (\textit{direct sum}). \tag{9}$$

Fix $w_0 \in W$. Then there exists an open, relatively compact neighborhood W_0 of w_0 in W and a relatively compact submanifold B of H forming a local cross section through e over an open neighborhood U_0 of eH^{w_0} in H/H^{w_0} such that the mapping

$$\eta\colon (b, w) \to b \cdot w$$

is a diffeomorphism of $B \times W_0$ onto an open neighborhood of w_0 in V.

Proof. Let $\mathfrak{h}^0$ denote the Lie algebra of H^{w_0} and $\mathfrak{n} \subset \mathfrak{h}$ any subspace such that $\mathfrak{h} = \mathfrak{h}^0 \oplus \mathfrak{n}$ (direct sum). If $\mathfrak{n}_0$ is a sufficiently small neighborhood of 0 in $\mathfrak{n}$, $\exp \mathfrak{n}_0$ is a local cross section in H over a neighborhood of eH^{w_0} in H/H^{w_0} (cf. [DS], Lemma 4.1 in Chapter II). Therefore, it suffices to prove that the mapping $\phi\colon (X, w) \to \exp X \cdot w$ of $\mathfrak{n} \times W$ into V is regular at $(0, w_0)$. Since

$$(d\phi)_{(0, w_0)}(0, W_{w_0}) = W_{w_0},$$

it remains to verify that

$$d\phi_{(0, w_0)}(\mathfrak{n}, 0) = (H \cdot w_0)_{w_0}. \tag{10}$$

The mapping $h \to h \cdot w_0$ maps H onto the orbit $H \cdot w_0$; its differential maps $\mathfrak{h}$ onto the tangent space $(H \cdot w_0)_{w_0}$, has kernel $\mathfrak{h}^0$, and therefore gives a bijection of $\mathfrak{n}$ onto $(H \cdot w_0)_{w_0}$. This proves (10) and the lemma.

Now we assume V has a Riemannian structure g invariant under the action of H. We assume furthermore that all the orbits of H have the same dimension. [Even if this assumption is not satisfied for V, it may be satisfied for a suitable open submanifold; cf. $\boldsymbol{O}(n)$ acting on $\boldsymbol{R}^n$.] Let D be a differential operator on V. We shall associate to D a new differential operator D_T on V whose action is "transversal to the orbits."

Let $s_0 \in V$ and let S denote the orbit $H \cdot s_0$. We construct the manifolds $S_s^\perp$ as before with $s \in S$. Shrinking $S_{s_0}^\perp$ if necessary, we may assume that it satisfies the transversality condition (9) above for W. Put $w_0 = s_0$ and choose $W_0 \subset S_{s_0}^\perp$ as in the lemma. Its conclusion means that the disjoint union $\bigcup_{w \in W_0} B \cdot w$ is a neighborhood V_0 of s_0 in V. If $f \in \mathscr{E}(V)$, we consider its restriction to W_0, which we then extend to a function f_{s_0} on V_0 by

$$f_{s_0}(b \cdot w) = f(w), \qquad b \in B, \quad w \in W_0. \tag{11}$$

We then define D_T by

$$(D_T f)(s_0) = (Df_{s_0})(s_0). \tag{12}$$

Since f_{s_0} is smooth near s_0, the right-hand side makes sense; also, since $B \cdot w$ is a neighborhood of w in the orbit $H \cdot w$ (for reasons of dimensionality), the method for constructing B is immaterial. Thus $D_T f$ is a well-defined C^∞-function on V. If f vanishes on an open subset U of V, then so does $D_T f$. Hence the mapping $f \to D_T f$ decreases supports, so D_T is a differential operator. The operator D_T is called the *transversal part of* D. If $D = D_T$, then D is said to be transversal.

Observing now that $(f_{s_0})_{s_0} = f_{s_0}$, we have for $\phi \in \mathscr{E}(V)$

$$(\phi E_T)_T = \phi E_T. \tag{12'}$$

In fact,

$$\begin{aligned}((\phi E_T)_T f)(s_0) &= (\phi E_T)(f_{s_0})(s_0) = \phi(s_0)(E_T f_{s_0})(s_0) \\ &= \phi(s_0)(Ef_{s_0})(s_0) = \phi(s_0)(E_T f)(s_0) = (\phi E_T f)(s_0).\end{aligned}$$

We shall now prove a global "separation of variables" formula for an arbitrary differential operator E on V.

Theorem 3.4. *Let H be a Lie transformation group of isometries of a Riemannian manifold V; all orbits are assumed to have the same dimension. Let $X_1, \ldots, X_l$ be a basis of the Lie algebra $\mathfrak{h}$ of H and let $Y_i = X_i^+$ $(1 \le i \le l)$ denote the induced vector fields on V. Then each differential operator E on V can be written as a locally finite sum*

$$E = E_T + \sum_{(i)} E_{(i)} Y_{i_1} \cdots Y_{i_r}, \tag{13}$$

where E_T is the transversal part of E and each $E_{(i)}$ $((i) = (i_1, \ldots, i_r))$ is transversal.

Remark. Here "locally finite" means that each point of V has a neighborhood inside which all but a finite number of the terms in (13) vanish identically.

Example. Let V be the plane $\boldsymbol{R}^2$ and H the group of translations T_t: $(x, y) \to (x, y + t)$, $t \in \boldsymbol{R}$. In this case, the operators of the form

$$\sum_i b_i(x, y)(\partial/\partial x)^i$$

are the transversal operators, $\partial/\partial y$ the vector field induced by H, and (13) reduces to the representation

$$\sum_j \left(\sum_i a_{ij}(x, y)(\partial/\partial x)^i\right)(\partial/\partial y)^j$$

of an arbitrary differential operator on $\boldsymbol{R}^2$.

Proof. Let $s_0 \in V$ and let S, $S_s^\perp$, B, W_0, and V_0 be as above. Let $b \to (y_1(b), \ldots, y_r(b))$ be any coordinate system on B, $y_1(e) = \cdots = y_r(e) = 0$, and let $w \to (z_{r+1}(w), \ldots, z_n(w))$ be a coordinate system on W_0 such that the geodesics through s_0 correspond to the straight lines through 0. Then we define a coordinate system $\{x_1, \ldots, x_n\}$ on V_0 by

$$\begin{aligned}(x_1(b \cdot w), \ldots, x_r(b \cdot w),\ x_{r+1}(b \cdot w), \ldots, x_n(b \cdot w)) \\ = (y_1(b), \ldots, y_r(b), z_{r+1}(w), \ldots, z_n(w)).\end{aligned}$$

For each $v \in V_0$ the vector $(\partial/\partial x_i)_v$ $(1 \leq i \leq r)$ is tangential to the orbit $H \cdot v$, hence is a linear combination of the vectors $(Y_1)_v, \ldots, (Y_l)_v$. Pick r linearly independent among these vectors, say $(Y_{j_1})_v, \ldots, (Y_{j_r})_v$, and write $(\partial/\partial x_i)_v$ as a linear combination

$$\left(\frac{\partial}{\partial x_i}\right)_v = \sum_1^l a_{ij}(v)(Y_j)_v, \tag{14}$$

where $a_{ij}(v) = 0$ if $j \notin \{j_1, \ldots, j_r\}$. Since the vectors $(Y_{j_k})_n$ $(1 \leq k \leq r)$ remain linearly independent for u near v, (14) extends to a relation for vector fields,

$$\frac{\partial}{\partial x_i} = \sum_{j=1}^l a_{ij,v} Y_j,$$

valid on a relatively compact neighborhood N_v of v, the $a_{ij,v}$ being C^∞-functions and $a_{ij,v} \equiv 0$ if $j \notin \{j_1, \ldots, j_r\}$. The neighborhoods N_v form a covering of V_0. We can find a countable covering $\bigcup_{m=1}^\infty N_m$ of V_0 which is locally finite and such that for each m there exists an $N_{v(m)}$ such that $\bar{N}_m \subset N_{v(m)}$ (cf. [DS], Chapter I, §§1, 15). Let $\sum_{m=1}^\infty \phi_m$ be a corresponding partition of unity. Then we have

$$\phi_m \frac{\partial}{\partial x_i} = \sum_j a_{ij,v(m)} \phi_m Y_j$$

on $N_{v(m)}$, hence on all of V_0. Summing over m, we obtain on V_0

$$\frac{\partial}{\partial x_i} = \sum_{j=1}^{l} b_{ij} Y_j \qquad (1 \le i \le r), \tag{15}$$

where $b_{ij} = \sum_m a_{ij,\, v(m)} \phi_m$ lies in $\mathscr{E}(V_0)$, because of the local finiteness.
By definition, we have

$$(E - E_T)(\psi)(s_0) = 0 \tag{16}$$

whenever the function ψ satisfies

$$\psi(b \cdot w) = \psi(w), \qquad b \in B, \quad w \in W_0. \tag{17}$$

However, condition (17) just amounts to local invariance of ψ on V_0 ($\dim B$ = dimension of the H-orbits), so (16) holds not just for s_0 but

$$(E - E_T)(\psi)(s) = 0 \qquad \text{for} \quad s \text{ near } s_0 \text{ in } V_0. \tag{18}$$

In terms of our coordinate system (17) amounts to

$$\psi(x_1, \ldots, x_n) = \psi(0, \ldots, 0, x_{r+1}, \ldots, x_n),$$

so since $E - E_T$ annihilates all such functions, each term in its coordinate expression must include some ∂_i $(1 \le i \le r)$, whence by (15)

$$E - E_T = \sum_{1}^{l} E_i Y_i \tag{19}$$

in a neighborhood N_0 of s_0, each E_i being a differential operator on N_0. Again, we can form a locally finite covering $\{U_\alpha\}$ of V such that (19) holds on each member of the covering. Let $1 = \sum \phi_\alpha$ be a partition of unity subordinate to this covering. Then by (19)

$$\phi_\alpha E - (\phi_\alpha E)_T = \sum_{1}^{l} E_i^\alpha Y_i,$$

with $E_i^\alpha \in \mathbf{E}(U^\alpha)$. However, by (12′), $(\phi_\alpha E)_T = \phi_\alpha E_T$, so we obtain by summation over α,

$$E - E_T \in \mathbf{E}(V)\mathfrak{h}^+, \tag{20}$$

where, as before, $\mathbf{E}(V)$ denotes the set of all differential operators on V and $\mathfrak{h}^+$ denotes the set $\{X^+ : X \in \mathfrak{h}\}$ of vector fields induced by H. Now the theorem follows from (20) by iteration.

We shall now work out a more explicit form of (13) in the case when E is the Laplace–Beltrami operator on V.

Theorem 3.5. *Let V be a Riemannian manifold, H a Lie transformation group of isometries of V; all orbits are assumed to have the same* *dimension. Let S be any H-orbit and* $\bar{f}$ *the restriction to S of a function* $f \in \mathscr{E}(V)$*. Then the Laplace–Beltrami operators* L_S *on S,* L_V *on V satisfy*

$$(L_V f)^{-} = L_S(\bar{f}) + ((L_V)_T f)^{-}. \tag{21}$$

Remark. This formula generalizes decomposition (3) for the Laplacian on $\boldsymbol{R}^3$. Here $V = \boldsymbol{R}^3 - \{0\}$, H is the rotation group $\boldsymbol{O}(3)$, the operator $(\partial^2/\partial r^2) + (2/r)(\partial/\partial r)$ is the transversal part of $L_{\boldsymbol{R}^3}$, and the operator

$$r^{-2}\left(\frac{\partial^2}{\partial\theta^2} + \cot\theta\frac{\partial}{\partial\theta} + \sin^{-2}\theta\frac{\partial^2}{\partial\phi^2}\right)$$

is the Laplacian on the sphere $S_r(0)$.

For the proof of (21) let $s_0 \in S$ and let the notation be as in the last proof. Since H permutes the various $S_s^\perp$ $(s \in S)$, the coordinate system $\{x_1, \ldots, x_n\}$ in question has properties (i) and (ii) from the proof of Proposition 3.1. We saw before that if $\psi \in \mathscr{E}(V_0)$ satisfies

$$\psi(x_1, \ldots, x_n) = \psi(0, \ldots, 0, x_{r+1}, \ldots, x_n), \tag{22}$$

then

$$\psi = \psi_{s_0}, \qquad \bar{\psi} = \text{const.}, \tag{23}$$

so

$$((L_V)_T\psi)(s_0) = (L_V\psi)(s_0), \qquad (L_S\bar{\psi})(s_0) = 0. \tag{24}$$

On the other hand, if $\phi \in \mathscr{E}(V_0)$ satisfies

$$\phi(x_1, \ldots, x_n) = \phi(x_1, \ldots, x_r, 0, \ldots, 0), \tag{25}$$

then $\phi(b \cdot w) = \phi(b \cdot s_0)$ $(w \in W_0)$, and since H permutes the various $S_s^\perp$ $(s \in S)$, we have in a neighborhood of s_0

$$\phi = (\bar{\phi})^{\sim}, \qquad \phi_{s_0} = \text{const.}, \tag{26}$$

where $^{\sim}$ is defined in connection with Proposition 3.1. From Proposition 3.1 we therefore deduce

$$(L_V\phi)(s_0) = (L_V'\bar{\phi})(s_0) = (L_S\bar{\phi})(s_0), \qquad ((L_V)_T\phi)(s_0) = 0. \tag{27}$$

Now, by (17) in §2,

$$L_V(\phi\psi) = \phi L_V\psi + 2g(\text{grad}\,\phi, \text{grad}\,\psi) + \psi L_V\phi. \tag{28}$$

We have

$$\partial_i\psi = 0 \qquad (1 \le i \le r), \qquad \partial_\alpha\phi = 0 \qquad (r + 1 \le \alpha \le n),$$

so using the orthogonality (6) and the formula (8) in §2 for grad, we see that the middle term in (28) is 0 at the point s_0. But by (23) and (27) the last term at s_0 equals $L_s(\phi\bar{\psi})(s_0)$. Similarly, using (24), (26), and the definition of $(L_V)_T$, we see that the first term on the right in (28) equals $(L_V)_T(\phi\psi)(s_0)$ at the point s_0. Consequently, if $f = \phi\psi$,

$$(L_V f)(s_0) = L_S(\bar{f})(s_0) + ((L_V)_T f)(s_0).$$

It is an elementary fact (cf., Schwartz [1966], Chapter IV, §3) that the finite linear combinations of functions $\phi\psi$ [where ϕ satisfies (25) and ψ satisfies (22)] is dense in $\mathscr{E}(V_0)$. The theorem therefore follows by continuity.

Remark. Theorem 3.5 remains true with the same proof if V is a manifold with a pseudo-Riemannian structure g, provided g is nonsingular on S.

3. Radial Parts of a Differential Operator. General Theory

Again let V be a manifold satisfying the second axiom of countability, H a Lie transformation group of V, and $\mathfrak{h}$ its Lie algebra. As in Lemma 3.3 we assume $W \subset V$ is a submanifold satisfying this condition:

(29) $\qquad V_W = (H \cdot w)_w + W_w \qquad$ (direct sum) $\qquad$ for each $w \in W$.

Theorem 3.6. *Assume the transversality condition* (29). *Let D be a differential operator on V. Then there exists a unique differential operator $\Delta(D)$ on W such that*

(30) $$(Df)^{-} = \Delta(D)\bar{f}$$

for each locally invariant function f on an open subset of V, the bar denoting restriction to W.

Definition. The operator $\Delta(D)$ is called the *radial part* of D. It has an obvious connection to the transversal part defined earlier.

Proof. Let $w_0 \in W$ and select W_0, B, V_0, and η as in Lemma 3.3. Let $\phi \in \mathscr{E}(W_0)$ and define f on V_0 by

$$f(b \cdot w) = \phi(w) \qquad b \in B, \quad w \in W_0.$$

Since $(f \circ \eta)(b, w) = \phi(w)$, we have $f \in \mathscr{E}(V_0)$. Let $X \in \mathfrak{h}$. For reasons of dimensionality the tangent spaces $\eta(B \times w)_{b \cdot w}$ and $(H \cdot w)_{b \cdot w}$ coincide, whence $(X^+ f)(b \cdot w) = 0$, so f is locally invariant. Consider now the linear mapping

$$\Delta_{w_0, W_0, B}: \mathscr{E}(W_0) \to \mathscr{E}(W_0)$$

given by $\phi \to (Df)^{-}$. Since this mapping decreases supports, it is a differential operator on W_0. By the remarks above, the set $B \cdot w$ is an open submanifold of the orbit $H \cdot w$ $(w \in W_0)$; hence if B' is another local cross section with the same properties as B in Lemma 3.3, we have

$$\Delta_{w_0, W_0, B} = \Delta_{w_0, W_0, B'}.$$

Thus we can, for each $w_0 \in W$ and each open relatively compact neighborhood W_0 of w_0 in W for which a B with the properties of Lemma 3.3 exists, define $\Delta_{w_0, W_0} = \Delta_{w_0, W_0, B}$. It is obvious that if W' and W'' are two such neighborhoods, then

$$\Delta_{w_0, W'} = \Delta_{w_0, W''} \qquad \text{on} \quad \mathscr{E}(W' \cap W'').$$

Thus we can define the linear form

$$\Delta_{w_0}: \phi \to (\Delta_{w_0, W_0}(\phi))(w_0), \qquad \phi \in \mathscr{E}(W),$$

the choice of W_0 being immaterial. Finally, we define the mapping $\Delta(D)$ by

$$(\Delta(D)\phi)(w_0) = \Delta_{w_0}(\phi), \qquad w_0 \in W.$$

Then $(\Delta(D)\phi_1)(w_0) = (\Delta(D)\phi_2)(w_0)$ if ϕ_1 and ϕ_2 coincide near w_0; also, on the subspace $\mathscr{D}(W_0) \subset \mathscr{E}(W)$, $\Delta(D)$ coincides with the differential operator $\Delta_{w_0, W_0, B}$. It follows that $\Delta(D)$ is a differential operator on W, and its construction shows that it has the properties stated in the theorem. The operator $\Delta(D)$ is called *the radial part* of D.

We shall now give a formula for the radial part $\Delta(L_V)$ of the Laplacian for a suitable transversal submanifold W for an action by isometries.

Let V be a Riemannian manifold, $I(V)$ its group of isometries in the compact open topology. We shall now assume the result of Myers and Steenrod [1939] (cf., also Kobayashi and Nomizu [1963], Vol. I, Chapter VI) that $I(V)$ has an analytic structure compatible with this topology in which it is a Lie transformation group of V. (The case when V is symmetric is the one of principal interest for us; for this case a direct proof is given in [DS], Chapter IV, §3.)

Let $H \subset I(V)$ be a closed subgroup. With the induced topology H is a Lie transformation group of V ([DS], Chapter II, §2). If $v \in V$, the orbit $H \cdot v$ is a closed submanifold of V, the isotropy group H^v is compact, and the mapping $hH^v \to h \cdot v$ is a diffeomorphism of H/H^v onto $H \cdot v$ (with the topology induced by V). For proofs see [DS], Chapter IV, §2, Chapter II, Theorem 3.2, and Exercise C5.

We shall now introduce a function δ on V, *the density function*, which in a certain manner measures the relative "size" of the orbits of H. Let $v \in V$. The orbit $H \cdot v$ inherits a Riemannian structure from that of V.

The corresponding Riemannian measure $d\sigma_v$ is, of course, invariant under H. On the other hand, if we fix a left-invariant Haar measure dh on H and let dk denote the unique Haar measure on the compact group H^v normalized by $\int dk = 1$, we obtain an H-invariant measure $d\dot{h}$ on H/H^v by the formula

$$\int_H f(h)\,dh = \int_{H/H^v}\left(\int_{H^v} f(hk)\,dk\right)d\dot{h}, \qquad f \in C_c(H)$$

(cf. Chapter I, Theorem 1.9, and the remark following). Under the identification $H \cdot v = H/H^v$ the measures $d\sigma_v$ and $d\dot{h}$ must, by the uniqueness, be proportional. Hence there exists a function $\delta(v)$ on V, which we call the *density function*, such that

$$d\sigma_v = \delta(v)\,d\dot{h}, \qquad v \in V. \tag{31}$$

Example. For the usual action of $\boldsymbol{O}(n)$ on $\boldsymbol{R}^n$,

$$\delta(x) = c|x|^{n-1}, \qquad x \in \boldsymbol{R}^n,$$

where c is a constant.

We shall now determine the radial part of the Laplacian on V assuming the H orbits intersect the transversal submanifold orthogonally and in just one point.

Theorem 3.7. *Let V be a Riemannian manifold, H a closed unimodular subgroup of the isometry group $I(V)$. Assume that a submanifold $W \subset V$ satisfies the following transversality condition:*
For each $w \in W$,

$$(H \cdot w) \cap W = \{w\}, \qquad V_w = (H \cdot w)_w \oplus W_w \tag{32}$$

(orthogonal direct sum). Then the radial part of L_V is given by

$$\Delta(L_V) = \delta^{-1/2} L_W \circ \delta^{1/2} - \delta^{-1/2} L_W(\delta^{1/2}), \tag{33}$$

where $\circ$ denotes composition of differential operators and δ is the density function.

We shall see that (33) can also be written

$$\Delta(L_V) = L_W + \operatorname{grad}_W(\log \delta), \tag{33'}$$

where grad_W refers to the submanifold W and the vector field

$$\operatorname{grad}_W(\log \delta)$$

is viewed as a differential operator on W.

Proof. Put $V^* = H \cdot W$. The mapping $\pi: (h, w) \to h \cdot w$ of $H \times W$ onto V^* has a differential satisfying

$$\begin{aligned} d\pi_{(h,w)}(H_h \times W_w) &= d\pi_{(h,w)}(H_h \times \{0\}) + d\pi_{(h,w)}(\{0\} \times W_w) \\ &= (H \cdot w)_{h \cdot w} + (dh)_w(W_w) = (dh)_w((H \cdot w)_w + W_w). \end{aligned}$$

Because of (32) or (29), this image equals $V_{h \cdot w}$. This implies that V^* is open in V ([DS], Chapter I, Proposition 3.1). Next we prove the formula

$$\int_{V^*} F(v)\, dv = \int_W \delta(w)\left(\int_{H \cdot w} F(h \cdot w)\, d\dot{h}\right) dw, \qquad F \in \mathscr{D}(V^*), \tag{34}$$

where dv and dw are the Riemannian measures on V and W, respectively. For this let $w_0 \in W$ and select B, W_0, V_0, and η with the properties of Lemma 3.3. Assuming, further that $b \to (y_1, \ldots, y_r(b))$ is a coordinate system on B, with $y_1(e) = \cdots = y_r(e) = 0$, and the mapping

$$w \to (z_{r+1}(w), \ldots, z_n(w))$$

a coordinate system on W_0, we have a coordinate system $\tau = \{x_1, \ldots, x_n\}$ on V_0 by putting

$$\begin{aligned} &(x_1(b \cdot w), \ldots, x_r(b \cdot w), x_{r+1}(b \cdot w), \ldots, x_n(b \cdot w)) \\ &\quad = (y_1(b), \ldots, y_r(b), z_{r+1}(w), \ldots, z_n(w)). \end{aligned} \tag{35}$$

Once more we adopt the following range of indices: $1 \leq i, j, k \leq r$, $r + 1 \leq \alpha, \beta, \gamma \leq n$. We have the following coordinate expressions for the Riemannian measures:

$$dv = \bar{g}^{1/2}\, dx_1 \cdots dx_n, \qquad dw = \bar{\gamma}^{1/2}\, dx_{r+1} \cdots dx_n,$$

where

$$\bar{g} = |\det(g_{pq})_{1 \leq p,q \leq n}|, \qquad \bar{\gamma} = |\det g_{\alpha\beta}|. \tag{36}$$

Because of the orthogonality condition in (32) we have

$$g_{i\alpha}(w) = 0, \qquad w \in W_0. \tag{37}$$

Let $b \in B$. Since the mapping $w \to b \cdot w$ of W_0 onto $b \cdot W_0$ has coordinate expression

$$(x_{r+1}, \ldots, x_n) \to (x_{r+1}, \ldots, x_n),$$

it follows that

$$db\left(\frac{\partial}{\partial x_\alpha}\right)_w = \left(\frac{\partial}{\partial x_\alpha}\right)_{b \cdot w}. \tag{38}$$

On the other hand, the submanifold of V_0 given by fixing the x_α lies on an orbit of H. It follows that

$$db\left(\frac{\partial}{\partial x_i}\right)_w = \sum_{j=1}^{r} a_{ij}\left(\frac{\partial}{\partial x_j}\right)_{b \cdot w}, \tag{39}$$

where $a_{ij} \in \boldsymbol{R}$. By (37) and (38)

$$g_{\alpha\beta}(b \cdot w) = g_{\alpha\beta}(w), \qquad g_{i\alpha}(b \cdot w) = 0.$$

Hence, using (39), we obtain

$$\bar{g}(b \cdot w) = |\det(g_{ij}(b \cdot w))|\,\bar{\gamma}(w). \tag{40}$$

But the Riemannian measure $d\sigma_w$ on the orbit $H \cdot w$ has coordinate expression

$$d\sigma_w = |\det(g_{ij})(b \cdot w)|^{1/2}\, dx_1 \cdots dx_r(b \cdot w),$$

so by (40) we obtain for $F \in \mathscr{D}(V_0)$,

$$\int_V F(v)\, dv = \int_W \bar{\gamma}^{1/2}(w)\left(\int_{H \cdot w} F(p)\, d\sigma_w(p)\right) dx_{r+1} \cdots dx_n(w).$$

By using now the definition (31) of δ, formula (34) follows for $F \in \mathscr{D}(V_0)$. But then it holds also for $F \in \mathscr{D}(h \cdot V_0)$ for each $h \in H$. But as w_0 runs through W the sets $h \cdot V_0$ $(h \in H)$ form a covering of V^*. Passing to a locally finite refinement and a corresponding partition of unity, (34) follows in general.

If $f \in \mathscr{D}(V^*)$, let $\dot{f} \in \mathscr{D}(W)$ be determined by

$$\dot{f}(w) = \int_{H \cdot w} f(h \cdot w)\, d\dot{h}. \tag{41}$$

The mapping $f \to \dot{f}$ is surjective. In fact, let $F \in \mathscr{D}(W)$ and let $C \subset W$ be a compact subset outside which F vanishes. Let $\tilde{C}$ be a compact subset of V^* such that $(H \cdot w) \cap \tilde{C} = \varnothing$ for $w \notin C$ and such that for each $w \in C$, $(H \cdot w) \cap \tilde{C}$ has measure >0 (for $d\sigma_w$). For example, take $\tilde{C} = C_H \cdot C$, where C_H is a compact neighborhood of e in H. Let $f_1 \in \mathscr{D}(V^*)$ be such that $f_1 \geq 0$ on V^* and $f_1 > 0$ on $\tilde{C}$. Then $\dot{f}_1 > 0$ on C and the function

$$f(v) = \begin{cases} f_1(v)(F(w))/\dot{f}_1(w) & \text{if} \quad v \in H \cdot w, \quad w \in C \\ 0 & \text{otherwise} \end{cases}$$

belongs to $\mathscr{D}(V^*)$ and satisfies $\dot{f} = F$. This proves the surjectivity

$$\mathscr{D}(V^*)^{\boldsymbol{\cdot}} = \mathscr{D}(W). \tag{42}$$

Since the vector fields $\partial/\partial x_i$ $(1 \leq i \leq r)$ are tangential to the H-orbits, we have on V_0, just as in (15),

$$\frac{\partial}{\partial x_i} = \sum_1^l b_{ij} Y_i, \qquad b_{ij} \in \mathscr{E}(V_0). \tag{43}$$

Now we use the coordinate expression of L_V as given in Proposition 2.6. Combining it with (37) and (43), we deduce that

$$\Delta(L_V) = L_W + \text{lower-order terms}. \tag{44}$$

On the other hand, we have by the symmetry (Proposition 2.3)

$$\int_{V^*} (L_V f_1)(v) f_2(v)\, dv = \int_{V^*} f_1(v)(L_V f_2)(v)\, dv \tag{45}$$

for $f_1, f_2 \in \mathscr{D}(V^*)$. But then the relation holds for all $f_2 \in \mathscr{E}(V^*)$. In particular, let us take f_2 invariant under H. Then, by (34), the left-hand side of (45) is

$$\int_W \delta(w) f_2(w) \left(\int_{H \cdot w} (L_V f_1)(h \cdot w)\, d\dot{h} \right) dw. \tag{46}$$

However, H^v is compact for each $v \in V$, so the inner integral can be written

$$\int_H (L_V f_1)(h \cdot w)\, dh, \tag{47}$$

and for the same reason, since L_V is H-invariant,

$$(L_V)_v \left(\int_H f_1(h \cdot v)\, dh \right) = \int_H (L_V f_1)(h \cdot v)\, dh. \tag{48}$$

By the unimodularity of H the function $v \to \int_H f_1(h \cdot v)\, dh$ is H-invariant, so by (48) expression (47) equals $(\Delta(L_V)\bar{f}_1)(w)$, and the left-hand side of (45) reduces to

$$\int_W (\Delta(L_V)\bar{f}_1)(w) \bar{f}_2(w) \delta(w)\, dw.$$

Now, by using (34), the H-invariance of $L_V f_2$, and the definition of $\Delta(L_V)$, the right-hand side of (45) reduces to

$$\int_W \bar{f}_1(w)(\Delta(L_V)\bar{f}_2)(w)\delta(w)\, dw.$$

By (42) the functions $\bar{f}_1$ (and also $\bar{f}_2$) fill up $\mathscr{D}(W)$, so the equality of the last expressions means that $\Delta(L_V)$ is symmetric with respect to $\delta(w)\, dw$. But since L_W is symmetric with respect to dw (Proposition 2.3), the operator $\delta^{-1/2}L_W \circ \delta^{1/2}$ is symmetric with respect to $\delta(w)\,dw$, and, of course, it agrees with L_W up to lower-order terms. Thus by (44) the difference

$$\Delta(L_V) - \delta^{-1/2}L_W \circ \delta^{1/2},$$

which is symmetric (with respect to $\delta(w)\,dw$) has order ≤ 1. But no first-order differential operator can be symmetric (cf. Exercise A7), so this difference is a function; the desired formula now follows by applying the operators to the constant 1.

Remark 1. By using (7) and (17) in §2 and the formula

$$2\delta^{-1/2}\operatorname{grad}(\delta^{1/2}) = \delta^{-1}\operatorname{grad}(\delta) = \operatorname{grad}(\log\delta),$$

the formula for the radial part can be written

$$\Delta(L_V) = L_W + \operatorname{grad}_W(\log\delta), \tag{49}$$

the vector field $\operatorname{grad}_W(\log\delta)$ being viewed as a differential operator on W.

Remark 2. Theorem 3.7 remains true if V is a manifold with a pseudo-Riemannian structure g, provided the following two conditions are satisfied:

(i) For each $w \in W$ the orbit $H \cdot w$ is closed and g is nondegenerate on it and on W.

(ii) For each $w \in W$ the group H^w is compact.

In fact, by (i) the orbit $H \cdot w$ is locally compact and thus homeomorphic to H/H^w ([DS], Chapter I, Theorem 3.2). By (ii), H^v is compact for each $v \in V^*$, so no change is necessary in the proof.

The general formula (33) for the radial part is proved in Helgason [1972a,b] where a further variation in the assumptions (weakening of the compactness in (ii)) is also given.

4. Examples of Radial Parts

We shall now use Theorem 3.7 to determine the radial part of the Laplace–Beltrami operator for various examples.

(i) *$O(n)$ Acting on the Euclidean Space $\mathbf{R}^n$.*

Let $V = \mathbf{R}^n$, $H = O(n)$ the orthogonal group, acting on $\mathbf{R}^n$. Then the submanifold $W = \mathbf{R}^+ - \{0\}$ (the positive real axis) satisfies condition (32) in the theorem; hence we conclude [since $\delta(x) = c|x|^{n-1}$]

$$\Delta(L_{\mathbf{R}^n}) = r^{-(1/2)(n-1)}\frac{d^2}{dr^2} \circ r^{(1/2)(n-1)} - r^{-(1/2)(n-1)}\frac{d^2}{dr^2}(r^{(1/2)(n-1)}). \tag{50}$$

This can be reduced to the familiar expression [cf., (49)]

$$\Delta(L_{\boldsymbol{R}^n}) = \frac{d^2}{dr^2} + \frac{n-1}{r}\frac{d}{dr}.$$

Note that exactly in the case $n = 3$ does the last term in (50) disappear, and then the formula reduces to

$$\Delta(L_{\boldsymbol{R}^3}) = r^{-1}\frac{d^2}{dr^2} \circ r.$$

In this case, a radial function $f(x) = F(|x|)$ is an eigenfunction of $L_{\boldsymbol{R}^n}$ if and only if the function $rF(r)$ is an eigenfunction of d^2/dr^2. Thus $f(x)$ is an elementary function.

More generally the function

$$\int_{S^{n-1}} e^{i\lambda(x,\omega)}\, d\omega, \tag{51}$$

where (x, ω) is the inner product and $d\omega$ the Riemannian measure, is a radial eigenfunction of $L_{\boldsymbol{R}^n}$ on $\boldsymbol{R}^n$ for eigenvalue $-\lambda^2$. If $\lambda \in \boldsymbol{R}$, the function (51) is a constant multiple of

$$\frac{J_{((n/2)-1)}(\lambda|x|)}{(\lambda|x|)^{(n/2)-1}},$$

where J_ν is the Bessel function of order ν. Formula (50) therefore explains the classical fact that $J_{1/2}(r)$ is an elementary function.

(ii) *The Group N Acting on the Symmetric Space $X = G/K$.*

As usual, let $G = KAN$ be an Iwasawa decomposition of a connected semisimple Lie group G with finite center. We take $V = G/K$ and $H = N$ in Theorem 3.7, and adopt the notation from §5 in Chapter I. Then the submanifold $A \cdot o \subset V$ satisfies assumption (32) because of the uniqueness in the Iwasawa decomposition ([DS], Theorem 5.1, Chapter VI) and because of the orthogonality $N \cdot o \perp A \cdot o$ (loc. cit., Exercise B2), which implies the orthogonality $N \cdot (a \cdot o) \perp A \cdot o$.

Using Corollaries 5.2 and 5.3 in Chapter I, we obtain

$$\int_{G/K} f(gK)\, dg_K = \int_A e^{-2\rho(\log a)}\left(\int_N f(naK)\, dn\right) da,$$

so comparing with (34),

$$\delta(a \cdot 0) = e^{-2\rho(\log a)}, \qquad a \in A. \tag{52}$$

Now Theorem 3.7 implies the following result.

Proposition 3.8. *When N acts on the symmetric space $X = G/K$, the radial part for the transversal submanifold $A \cdot o$ is given by*

$$\Delta(L_X) = e^{\rho} L_A \circ e^{-\rho} - \langle \rho, \rho \rangle, \tag{53}$$

where L_A is the Laplacian on $A \cdot o$ and where e^{ρ} denotes the function $a \cdot o \to e^{\rho(\log a)}$ on $A \cdot o$.

Remark. One can compare this formula with the radial part of $L_{\boldsymbol{R}^n}$ when $\boldsymbol{R}^{n-k}$ acts on $\boldsymbol{R}^n$ by translation. Then the orthogonal complement $\boldsymbol{R}^k$ serves as a transversal submanifold and from (33) or directly we see that $\Delta(L_{\boldsymbol{R}^n}) = L_{\boldsymbol{R}^k}$. This corresponds to taking $\rho = 0$ in (53).

(iii) *The Group K Acting on $X = G/K$.*

Again, we follow the notation of §5 in Chapter I for the symmetric space $X = G/K$. We shall verify that the manifold $A^+ \cdot o$ satisfies the two transversality conditions in (32) for the group K playing the role of H. The first condition is guaranteed by the properties of the polar coordinate map $(kM, a) \to kaK$. To check the orthogonality condition

$$(G/K)_{a \cdot o} = (A^+ \cdot o)_{a \cdot o} \oplus (Ka \cdot o)_{a \cdot o}, \qquad a \in A^+, \tag{54}$$

let $T \in \mathfrak{k}$ and consider the curve $t \to \exp tTa \cdot o$ in the orbit $Ka \cdot o$. Its tangent vector for $t = 0$ equals, since

$$\exp tTa \cdot o = a \exp \operatorname{Ad}(a^{-1})\, tT \cdot o,$$

the vector $d\tau(a)\, d\pi(\operatorname{Ad}(a^{-1})\, T)$. [Here $\tau(a)$ is the translation $gK \to agK$ and $\pi: G \to G/K$ the natural mapping.] But if $H \in \mathfrak{a}$ (the tangent space to $A^+ \cdot o$), we have

$$\begin{aligned} &\langle d\tau(a)\, d\pi(\operatorname{Ad}(a^{-1})\, T), d\tau(a)\, H \rangle \\ &\quad = \langle \tfrac{1}{2}(\operatorname{Ad}(a^{-1})T - \theta \operatorname{Ad}(a^{-1})\, T), H \rangle = 0, \end{aligned}$$

verifying (54).

Proposition 3.9. *For K acting on $X = G/K$ with $A^+ \cdot o$ as a transversal manifold, the radial part of L_X is given by*

$$\Delta(L_X) = L_A + \sum_{\alpha \in \Sigma^+} m_\alpha (\coth \alpha) A_\alpha. \tag{55}$$

Here the vector $A_\alpha \in \mathfrak{a}$ is, as usual, determined by $\langle A_\alpha, H \rangle = \alpha(H)$ $(H \in \mathfrak{a})$, and in (55) it is considered as a differential operator on $A^+ \cdot o$. Using Theorem 5.8 in Chapter I, we find for each $H \in \mathfrak{a}$ that

$$H(\log \delta) = \sum_{\alpha \in \Sigma^+} m_\alpha (\coth \alpha) \alpha(H). \tag{56}$$

If $H_1, \ldots, H_l$ is an orthonormal basis of $\mathfrak{a}$, we obtain from (49) and (56)

$$\begin{aligned}\Delta(L_X) &= L_A + \sum_i H_i(\log \delta)H_i \\ &= L_A + \sum_{\alpha \in \Sigma^+} m_\alpha \coth \alpha \sum_i \alpha(H_i)H_i.\end{aligned}$$

But $\sum_i \alpha(H_i)H_i = \sum_i \langle A_\alpha, H_i \rangle H_i = A_\alpha$, so (55) is proved.

For the case when G is complex, the radial part can be put in a better form.

Proposition 3.10. *Suppose G is a complex semisimple Lie group. For K acting on $X = G/K$ with $A^+ \cdot o$ as a transversal manifold, the radial part of L_X is given by*

$$\Delta(L_X) = \delta^{-1/2}(L_A - \langle \rho, \rho \rangle) \circ \delta^{1/2}, \tag{57}$$

where

$$\delta^{1/2}(a) = \sum_{s \in W} (\det s) e^{s\rho(\log a)}.$$

This is an immediate consequence of Theorem 5.8 in Chapter I if we recall that each $\alpha \in \Sigma^+$ has multiplicity 2 and then use the formula

$$\prod_{\alpha \in \Sigma^+} (e^\alpha - e^{-\alpha}) = \sum_{s \in W} (\det s) e^{s\rho},$$

which follows from Proposition 5.15 in Chapter I for the root system $\{2\alpha : \alpha \in \Sigma\}$.

Remark. Consider, as in §4 in Chapter I, the hyperbolic space $X = \boldsymbol{H}^n$ with the Riemannian structure

$$ds^2 = dr^2 + (\sinh r)^2 \, d\sigma^2, \tag{58}$$

where $d\sigma^2$ is the Riemannian structure on the unit sphere in the tangent space X_o. Under the action of the isotropy group at o the radial part of the Laplacian is given by [cf. (23) in §4, Chapter I]

$$\Delta(L_X) = \frac{d^2}{dr^2} + (n-1)\coth r \frac{d}{dr}. \tag{59}$$

Viewing $\boldsymbol{H}^n$ as the symmetric space $\boldsymbol{O}_0(n, 1)/\boldsymbol{O}_0(n)$ (subscript indicating identity components), we can give it the canonical Killing form Riemannian structure, which by irreducibility is a constant multiple of (58). Thus (59) follows also from the general formula (55). We can also write (59) in the form

$$(\Delta(L_X) = \sinh^{-(1/2)(n-1)} r \left[\frac{d^2}{dr^2} \circ \sinh^{(1/2)(n-1)} r - \frac{d^2(\sinh^{(1/2)(n-1)} r)}{dr^2} \right]. \tag{60}$$

In analogy with what happened for $\boldsymbol{R}^n$ [cf., (i)], the case $n = 3$ is one of special simplicity; the zero-order term in (60) then reduces to a constant, and in fact

$$\Delta(L_X) = \sinh^{-1} r \left(\frac{d^2}{dr^2} - 1\right) \circ \sinh r.$$

This again implies that for dimension 3 the radial eigenfunctions of L_X are elementary functions. While for $\boldsymbol{R}^3$ we did not have a conceptual explanation of this fact (see, however, Proposition 3.13 for $G = \boldsymbol{SL}(2, \boldsymbol{C})$, $K = \boldsymbol{SU}(2)$), Proposition 3.10 gives such an explanation for $\boldsymbol{H}^3$: If $n = 3$, the Lorentz group $\boldsymbol{O}_0(n, 1)$ has a complex structure. In fact, it is locally isomorphic to $\boldsymbol{SL}(2, \boldsymbol{C})$ ([DS], Chapter V, §2).

(iv) *The Group K_1 Acting on the Compact Symmetric Space U/K_1.*

Consider now the simply connected compact symmetric space U/K_1 from Theorem 5.11 in Chapter I. Under the action of K_1 the submanifold $\operatorname{Exp} Q_0 = \exp Q_0 \cdot o$ serves as an orthogonal transversal submanifold for which Theorem 3.7 applies. [The orthogonality is proved in the same way as (54).] Now the Riemannian structure of U/K_1 is induced by $B_* = -B$, where B is the Killing form of $\mathfrak{g}^C$ restricted to $\mathfrak{u}$. Let $C_* = i\mathfrak{a}^+$, and for a root $\alpha \in \Sigma$ put $\alpha_* = (1/i)\alpha$ and $\Sigma_* = \{\alpha_* : \alpha \in \Sigma\}$, $\Sigma_*^+ = \{\alpha_* : \alpha \in \Sigma^+\}$. If $\alpha_* \in \Sigma_*$, $A_* \in \mathfrak{a}_*$ be determined by $B_*(A_*, H) \equiv \alpha_*(H)$. We call the α_* the roots of $\mathfrak{u}$ with respect to $\mathfrak{a}_*$. The roots in Σ_*^+ are the ones which are positive on the Weyl chamber C_*. We may assume the polyhedron Q_0 selected inside C_*. On $\operatorname{Exp} Q_0$ the density function is up to a factor given by

(61) $$\delta_*(\operatorname{Exp} H) = \prod_{\alpha_* \in \Sigma_*^+} (\sin \alpha_*(H))^{m_\alpha}.$$

Proposition 3.11. *For K_1 acting on the compact simply connected symmetric space U/K_1 with transversal manifold* $\operatorname{Exp} Q_0$, *the radial part of the Laplacian L_{U/K_1} is given by*

(62) $$\Delta(L_{U/K_1}) = L_{A_*} + \sum_{\alpha_* \in \Sigma_*^+} m_\alpha \cot \alpha_* \, A_{\alpha_*},$$

where L_{A_*} is the Laplacian on the flat manifold $A_* . o$.

The proof is entirely analogous to that of (55). In fact, if $H \in \mathfrak{a}_*$,

$$H(\log \delta_*) = \sum_{\alpha_* \in \Sigma_*^+} m_\alpha (\cot \alpha_*) \alpha_*(H)$$

and $A_{\alpha_*} = \sum_1^l \alpha_*(H_i) H_i$ if $H_1, \ldots, H_l$ is a basis of $\mathfrak{a}_*$ orthonormal with respect to B_*.

(v) *A Compact, Semisimple, Simply Connected Lie Group U Acting on Itself by Conjugacy.*

This situation is a special case of Proposition 3.11, so that the radial part can be found from (62); however, it is easier to use Theorem 5.13 in Chapter I. We use the notation prior to that theorem.

Let the Riemannian structure of U be given by B_*, the negative of the Killing form of $\mathfrak{u}$. Again, we put $\alpha_*(H) = (1/i)\alpha(H)$ for $H \in \mathfrak{t}_0$, $\alpha \in \Delta$. Let $C_0 \subset \mathfrak{t}_0$ denote the Weyl chamber containing P_0 and

$$\Delta^+ = \{\alpha_* : \alpha_*(C_0) \subset \boldsymbol{R}^+\}.$$

Let $\rho_* = \frac{1}{2}\sum_{\alpha_* \in \Delta^+} \alpha_*$ and let $|\rho_*|$ denote its norm in the metric B_*. Here we have

$$\delta(t)^{1/2} = \prod_{\alpha_* \in \Delta^+} 2 \sin \frac{\alpha_*(H)}{2}$$

if $t = \exp H$, $H \in P_0$. But using Proposition 5.15, Chapter I, we find that this reduces to

$$\delta(t)^{1/2} = \sum_{s \in W} (\det s) e^{is\rho_*(H)}. \tag{63}$$

Proposition 3.12. *For U acting on itself by conjugacy with transversal manifold* $\exp P_0$, *the radial part of the Laplacian L_U is given by*

$$\Delta(L_U) = \delta^{-1/2} L_T \circ \delta^{1/2} + |\rho_*|^2,$$

where L_T denotes the Laplacian on T.

This follows from Theorem 3.7 if we just take (63) into account.

(vi) *The Adjoint Action of K on $\mathfrak{p}$.*

This is the situation of Theorem 5.17 in Chapter I for the symmetric space $X = G/K$. The Weyl chamber $\mathfrak{a}^+$ can serve as an orthogonal transversal manifold. In fact, if $H_0 \in \mathfrak{a}^+$, the tangent space to the K-orbit $K \cdot H_0$ is, by (21) in §5, Chapter I, given by $[\mathfrak{l}, H_0] = [\mathfrak{k}, H_0]$ which is orthogonal to $\mathfrak{a}$ under the Killing form on $\mathfrak{p}$. Using the quoted theorem we thus derive the following result.

Proposition 3.13. *For the adjoint action of K on $\mathfrak{p}$ with transversal manifold $\mathfrak{a}^+$ the radial part of the Laplacian $L_{\mathfrak{p}}$ is given by*

$$\Delta(L_{\mathfrak{p}}) = L_{\mathfrak{a}} + \sum_{\alpha \in \Sigma^+} m_\alpha \frac{1}{\alpha} A_\alpha,$$

where $L_{\mathfrak{a}}$ is the Laplacian on $\mathfrak{a}$.

For the case when G is complex the formula can also be written

$$\Delta(L_{\mathfrak{p}}) = \pi^{-1} L_{\mathfrak{a}} \circ \pi,$$

where $\pi = \prod_{\alpha \in \Sigma^+} \alpha$.

In fact, the density is now proportional to $\prod_{\alpha \in \Sigma^+} \alpha(H)^{m_\alpha}$, so the first formula above follows from (49). The second formula is a consequence of (33) and the equation

$$L_{\mathfrak{a}}(\pi) = 0. \tag{64}$$

This equation is established as follows. If α is a simple root, the reflection s_α permutes the set $\Sigma^+ - \{\alpha\}$ and consequently $\pi^{s_\alpha} = -\pi$; it follows that π is a *skew* polynomial, i.e., $\pi^s = (\det s)\pi$ for $s \in W$. Since $L_{\mathfrak{a}}$ is invariant under the Weyl group, it follows that $L_{\mathfrak{a}}(\pi)$ is skew. For each reflection s_α $(\alpha \in \Sigma^+)$ the relation $(L_{\mathfrak{a}}\pi)^{s_\alpha} = -L_{\mathfrak{a}}\pi$ implies that $L_{\mathfrak{a}}\pi$ vanishes on the plane $\alpha = 0$. Thus the polynomial $L_{\mathfrak{a}}\pi$ is divisible by α, and since the positive roots are prime to each other (as polynomials), $L_{\mathfrak{a}}\pi$ is divisible by π. This proves (64).

(vii) *The Adjoint Action of a Complex Semisimple Lie Group G on Its Lie Algebra* $\mathfrak{g}$.

Let $\mathfrak{h}$ be an arbitrary Cartan subalgebra of $\mathfrak{g}$ and $\mathfrak{h}' \subset \mathfrak{h}$ the subset of regular elements. Let Δ denote the set of roots of $\mathfrak{g}$ with respect to $\mathfrak{h}$ and $\mathfrak{g} = \mathfrak{h} + \sum_{\alpha \in \Delta} \mathfrak{g}^\alpha$ the corresponding root space decomposition ([DS], Chapter III). If $X \in \mathfrak{g}$, the orbit $G \cdot X$ has tangent space $[\mathfrak{g}, X]$ at X; the root space decomposition therefore shows that the submanifold $\mathfrak{h}' \subset \mathfrak{g}$ satisfies the transversality condition (29).

The Killing form $\langle\ ,\ \rangle$ on $\mathfrak{g}$ is nondegenerate on both $\mathfrak{g}$ and on $\mathfrak{h}$, so that we can use it to identify $\mathfrak{g}$ and $\mathfrak{h}$ with their duals $\mathfrak{g}^*$ and $\mathfrak{h}^*$, respectively. If p is a polynomial function on $\mathfrak{g}$ (resp. $\mathfrak{h}$) let $\partial(p)$ denote the constant coefficient differential operator on $\mathfrak{g}$ (resp. $\mathfrak{h}$) which corresponds to p by means of this identification. If f is a function on $\mathfrak{g}$, let $\bar{f}$ denote its restriction to $\mathfrak{h}$ (or $\mathfrak{h}'$). Let $\omega(X) = \langle X, X\rangle$ $(X \in \mathfrak{g})$. We shall now determine the radial part of the "complex Laplacian" $\partial(\omega)$.

For $\alpha \in \Delta$ we denote by H_α the vector in $\mathfrak{h}$ satisfying $\langle H_\alpha, H\rangle = \alpha(H)$ $(H \in \mathfrak{h})$. Then the space $\mathfrak{h}_{\boldsymbol{R}} = \sum_{\alpha \in \Delta} \boldsymbol{R} H_\alpha$ is the subset of $\mathfrak{h}$ on which all the roots are real-valued. Fixing a Weyl chamber $\mathfrak{h}^+ \subset \mathfrak{h}_{\boldsymbol{R}}$, a root α is said to be *positive* if $\alpha(\mathfrak{h}^+) \subset \boldsymbol{R}^+$.

Proposition 3.14. *For the adjoint action of G on* $\mathfrak{g}$ *with transversal manifold* $\mathfrak{h}'$, *the radial part of the operator* $\partial(\omega)$ *is given by*

$$\Delta(\partial(\omega)) = \pi^{-1}\partial(\bar{\omega}) \circ \pi,$$

where π *is the product of the positive roots.*

Proof. Since we are dealing with a situation over $\mathbf{C}$ and since the isotropy groups are noncompact, Theorem 3.7 does not apply without some adjustment. We therefore adopt a more direct approach.

Choose elements $X_\alpha \in \mathfrak{g}^\alpha$ $(\alpha \in \Delta)$ such that $\langle X_\alpha, X_{-\alpha}\rangle = 1$. Then $[X_\alpha, X_{-\alpha}] = H_\alpha$ ([DS], Chapter III, §4(3)). Since $\langle X_\alpha, X_\beta\rangle = 0$ if $\alpha, \beta \in \Delta$, $\alpha + \beta \neq 0$, we have

$$\partial(\omega) = \partial(\bar{\omega}) + \sum_{\alpha\in\Delta} X_\alpha X_{-\alpha}. \tag{65}$$

Now let $H \in \mathfrak{h}'$ and suppose f is analytic and locally invariant in a neighborhood of H in $\mathfrak{g}$. Then if $\alpha \in \Delta$,

$$f(e^{\mathrm{ad}(sX_\alpha + tX_{-\alpha})}H) = f(H)$$

for s and t sufficiently small. Writing this as

$$f(H + X(s, t)) = f(H),$$

we have by Taylor's formula

$$\sum_{n=0}^{\infty} \frac{1}{n!}(X(s, t)^n f)(H) = f(H). \tag{66}$$

Here we collect the coefficients of st. Since

$$X(s, t) = [sX_\alpha + tX_{-\alpha}, H] + \tfrac{1}{2}[sX_\alpha + tX_{-\alpha}, [sX_\alpha + tX_{-\alpha}, H]] + \cdots$$

only terms with $n \leq 2$ in (66) give a contribution:

$X(s, t)$ gives coefficient $\frac{1}{2}[X_{-\alpha}, [X_\alpha, H]] + \frac{1}{2}[X_\alpha, [X_{-\alpha}, H]]$;

$\frac{1}{2}X(s, t)^2$ gives coefficient $[X_\alpha, H][X_{-\alpha}, H]$.

It follows that

$$[(\alpha(H)^2 X_\alpha X_{-\alpha} - \alpha(H)H_\alpha)f](H) = 0.$$

Thus by (65)

$$\Delta(\partial(\omega)) = \partial(\bar{\omega}) + 2\sum_{\alpha>0} \alpha^{-1} H_\alpha. \tag{67}$$

On the other hand, by (15) and (17) in §2 (which are also valid in the pseudo-Riemannian case), we have

$$\pi^{-1}\,\partial(\bar{\omega})(\pi f) = \partial(\bar{\omega})f + \pi^{-1}(\partial(\bar{\omega})\,\pi)f + 2(\mathrm{grad}(\log \pi))(f). \tag{68}$$

But

$$\mathrm{grad}(\log \pi) = \sum_{\alpha>0} \mathrm{grad}(\log \alpha) = \sum_{\alpha>0} \alpha^{-1} H_\alpha.$$

Finally, we have by analogy with (64) $\partial(\varpi)\pi = 0$, and now the proposition follows from (67) and (68).

(viii) *The Laplacian on X and on* $\mathfrak{p}$.

Continuing Example (iii) above, consider the diffeomorphism $\mathrm{Exp}: Y \to (\exp Y)K$ of the tangent space $\mathfrak{p}$ onto the symmetric space G/K and let J denote the corresponding volume element ratio; i.e.,

$$\int_{G/K} f(x)\,dx = \int_{\mathfrak{p}} f(\mathrm{Exp}\,X)J(X)\,dX, \qquad f \in C_c(G/K).$$

Then by [DS], Chapter IV, Theorem 4.1,

$$J(X) = \det\left(\left(\frac{\sinh \operatorname{ad} X}{\operatorname{ad} X}\right)_{\mathfrak{p}}\right), \tag{69}$$

when the subscript denotes restriction to $\mathfrak{p}$.

Theorem 3.15. *Let L_X and $L_{\mathfrak{p}}$ denote the Laplace–Beltrami operators on X and $\mathfrak{p}$, respectively. Then the image $(L_X)^{\mathrm{Exp}^{-1}}$ of L_X under* Exp^{-1} (cf. §2) *satisfies the relation*

$$L_X^{\mathrm{Exp}^{-1}} F = (L_{\mathfrak{p}} + \operatorname{grad}(\log J))F \tag{70}$$

for each K-invariant C^∞ function F on $\mathfrak{p}$. The formula can also be written

$$L_X^{\mathrm{Exp}^{-1}} F = (J^{-1/2}L_{\mathfrak{p}} \circ J^{1/2})F - J^{-1/2}L_{\mathfrak{p}}(J^{1/2})F. \tag{71}$$

Proof. Let δ denote the density function for K acting on X and δ_0 the density function for K acting on $\mathfrak{p}$. It suffices to verify the relation above on $\mathfrak{a}$. The functions δ and δ_0 can be taken as

$$\delta(\mathrm{Exp}\,H) = \prod_{\alpha\in\Sigma^+} (\sinh \alpha(H))^{m_\alpha}, \qquad \delta_0(H) = \prod_{\alpha\in\Sigma^+} \alpha(H)^{m_\alpha}$$

for $H \in \mathfrak{a}$ and $J = \delta/\delta_0$. Now if $f = F \circ \mathrm{Exp}^{-1}$,

$$(L_X^{\mathrm{Exp}^{-1}} F)(H) = (L_X f)(\mathrm{Exp}\,H) = (\Delta(L_X)f)(\mathrm{Exp}\,H),$$

which by Theorem 3.7 equals

$$(\delta^{-1/2}L_A \circ \delta^{1/2})f - \delta^{-1/2}L_A(\delta^{1/2})f$$

at Exp H. On the other hand, the right-hand side of (71) has restriction to $\mathfrak{a}$ given by

$$\begin{aligned} &J^{-1/2}[(\delta_0^{-1/2}L_{\mathfrak{a}} \circ \delta_0^{1/2})(J^{1/2}F) - (\delta_0^{-1/2}L_{\mathfrak{a}}(\delta_0^{1/2}))J^{1/2}F] \\ &\quad -J^{-1/2}[(\delta_0^{-1/2}L_{\mathfrak{a}} \circ \delta_0^{1/2})(J^{1/2})F - \delta_0^{-1/2}L_{\mathfrak{a}}(\delta_0^{1/2})(J^{1/2}F)]. \end{aligned}$$

Since the second and fourth terms cancel, this proves (71). Formula (70) now follows by (17), §2.

§4. Invariant Differential Operators on Lie Groups and Homogeneous Spaces

1. Introductory Remarks. Examples. Problems

Let M be a manifold and $\phi: M \to M$ a diffeomorphism of M onto itself. As before, we put

$$f^\phi = f \circ \phi^{-1}, \qquad f \in \mathscr{E}(M),$$

and if D is a differential operator on M, we define D^ϕ by

$$D^\phi: f \to (Df^{\phi^{-1}})^\phi = (D(f \circ \phi)) \circ \phi^{-1}, \qquad f \in \mathscr{E}(M).$$

By the definition in §2, D^ϕ is another differential operator. The operator D is said to be *invariant under* ϕ if $D^\phi = D$, i.e., if $D(f \circ \phi) = (Df) \circ \phi$ for all f. Note that $(Df)^\phi = D^\phi f^\phi$, justifying the notation.

If T is a distribution on M, we write T^ϕ for the distribution $T^\phi(f) = T(f^{\phi^{-1}})$, $f \in \mathscr{D}(M)$.

Since we want to take advantage of the invariance concept we shall now discuss differential operators invariant under a transitive group of diffeomorphisms. Let G be a Lie group, $H \subset G$ a closed subgroup, G/H the manifold of left cosets gH ($g \in G$) and $\mathbf{D}(G/H)$ the algebra of all differential operators on G/H which are invariant under all the transformations $\tau(g): xH \to gxH$ of G/H onto itself. We write $\mathbf{D}(G)$ instead of $\mathbf{D}(G/\{e\})$. The algebra $\mathbf{D}(G/H)$ will play a central role in the remainder of this book.

We recall briefly the concept of a *representation*. Let L be a locally compact group and V a topological vector space, $\mathrm{Aut}(V)$ the group of linear homeomorphisms of V onto itself. A representation π of L on V is a homomorphism of L into $\mathrm{Aut}(V)$ such that the mapping $(l, v) \to \pi(l)v$ of $L \times V$ into V is continuous; π is said to be *irreducible* (or *topologically irreducible*) if $\{0\}$ and V are the only *closed* subspaces of V invariant under $\pi(L)$.

Now we list some problems which arise naturally in connection with the invariant differential operators on G/H; these were stated at the International Congress in Nice 1970. Considerable progress has been made on these problems since then; in particular, for symmetric G/K, Problems A–D are completely solved (see description below).

A. ***The Algebra*** $\mathbf{D}(G/H)$. Describe the algebra $\mathbf{D}(G/H)$ in terms of the Lie algebras of G and H.

See Proposition 4.11, Exercises A3, C1, C2, C3, and C5 for information about this problem.

B. *Solvability*. Given $D \in \boldsymbol{D}(G/H)$, is the differential equation $Du = f$, for $f \in \mathscr{E}(G/H)$ arbitrary, globally solvable (respectively, locally solvable)? In this case we say that D is *globally solvable* (respectively, *locally solvable*).

C. *Joint Eigenfunctions*. Determine the functions on G/H which are eigenfunctions of each $D \in \boldsymbol{D}(G/H)$. Similar problem can be considered for eigendistributions.

D. *Eigenspace Representations*. Let $\mu: \boldsymbol{D}(G/H) \to \boldsymbol{C}$ be a homomorphism and let E_μ denote the corresponding joint eigenspace, i.e.,

$$E_\mu = \{f \in \mathscr{E}(G/H): Df = \mu(D)f \text{ for all } D \in \boldsymbol{D}(G/H)\}.$$

Let T_μ denote the natural representation of G on this eigenspace, i.e.,

$$(T_\mu(g)f)(xH) = f(g^{-1}xH), \qquad g,\ x \in G$$

for $f \in E_\mu$. For which μ is this "eigenspace representation" T_μ irreducible and what representations of G are so obtained?

Here the closed subspace $E_\mu \subset \mathscr{E}(G/H)$ is given the relative topology. The definition of the topology of $\mathscr{E}(G/H)$ (§2, No. 2) shows easily that T_μ is indeed a representation.

E. *Global Properties of Solutions*. What geometric properties (functional equations, mean value properties, behavior at infinity) do solutions of the invariant differential equations have? It is to be expected that the local character of a $D \in \boldsymbol{D}(G/H)$ (as a differential operator) together with its G-invariance will result in global limitations on the solutions.

These problems will be discussed in this book as well as in another volume. But now we shall illustrate the problems in terms of some simple examples. Here we quote some results from elsewhere and anticipate special cases of results to be proved later.

Examples

(i) *The Euclidean Space.* Consider the space $\boldsymbol{R}^n$ as a homogeneous space under all translations. Here $\boldsymbol{D}(\boldsymbol{R}^n)$ consists of all differential operators of constant coefficients in the standard coordinate system on $\boldsymbol{R}^n$. In fact, $\partial/\partial x_i$ is invariant under all translations, so the invariance condition amounts to all the coefficients being invariant under the translations, hence constant.

Problem B has a positive solution in this case; each $D \in \boldsymbol{D}(\boldsymbol{R}^n)$ maps $\mathscr{E}(\boldsymbol{R}^n)$ onto itself (Ehrenpreis [1954]; Malgrange [1955]). For the case $\boldsymbol{R}^n$ we therefore have the following result.

Theorem 4.1.

(a) *The algebra $\boldsymbol{D}(\boldsymbol{R}^n)$ consists of the differential operators on $\boldsymbol{R}^n$ with constant coefficients.*

(b) *Each* $D \in \boldsymbol{D}(\boldsymbol{R}^n)$ *is globally solvable.*

(c) *The joint eigenfunctions are the exponential functions; the eigenspace representations are one dimensional, hence irreducible.*

For $\boldsymbol{R}^2$ viewed as the homogeneous space $\boldsymbol{M}(2)/\boldsymbol{O}(2)$ see the Introduction, Theorems 2.1 and 2.6.

(ii) *The Poincaré Group.* Consider the space $\boldsymbol{R}^4$ as a homogeneous space G/H, where H is the identity component $\boldsymbol{O}_0(1, 3)$ of the Lorentz group and G is generated by H and the translations (the Poincaré group). Here $\boldsymbol{D}(G/H)$ is generated by the d'Alembertian

$$\Box = \partial_1^2 - \partial_2^2 - \partial_3^2 - \partial_4^2$$

(cf. Exercise A3). This implicit physical significance of the Lorentz group so far as electromagnetic phenomena are concerned is made explicit in Einstein's special theory of relativity. Here the Lorentz group is given an interpretation in terms of pure mechanics.

(iii) *The Heisenberg Group N.* Let N be the group of matrices

$$\sigma = \begin{bmatrix} 1 & x_1 & x_3 \\ 0 & 1 & x_2 \\ 0 & 0 & 1 \end{bmatrix}, \qquad x \in \boldsymbol{R}^3,$$

with the obvious Lie group structure. The differential operator

$$E = \frac{\partial}{\partial x_1} + i\frac{\partial}{\partial x_2} + ix_1\frac{\partial}{\partial x_3} \tag{1}$$

on N is invariant under all left translations $L_\alpha\colon \sigma \to \alpha\sigma$ on N. To see this note that under the correspondence $\sigma \to (x_1, x_2, x_3)$, $\alpha \to (a_1, a_2, a_3)$, we have $\alpha\sigma \to (a_1 + x_1, a_2 + x_2, a_3 + x_3 + a_1x_2)$, so

$$[E(f \circ L_\alpha)](\sigma) = (\partial_1 f)(\alpha\sigma) + i(\partial_2 f)(\alpha\sigma) + ia_1(\partial_3 f)(\alpha\sigma) + ix_1(\partial_3 f)(\alpha\sigma)$$

and

$$[Ef](\alpha\sigma) = (\partial_1 f)(\alpha\sigma) + i(\partial_2 f)(\alpha\sigma) + i(a_1 + x_1)(\partial_3 f)(\alpha\sigma).$$

This proves the stated invariance.

Although E by its left invariance seems a natural analog of constant coefficient operators on $\boldsymbol{R}^n$, we shall see that it is not locally solvable on any open subset $\Omega \subset N$.

Theorem 4.2. *For each open subset* $\Omega \subset N$ *we have*

$$E\mathscr{E}(\Omega) \not\supset \mathscr{D}(\Omega).$$

This nonsolvability of E [or rather of the operator $\partial_1 + i\partial_2 - 2i(x_1 + ix_2)\partial_3$] was proved by Lewy [1957]. Later Hörmander found a more general necessary condition for local solvability which for first-order operators can be described as follows. Replacing ∂_i in the expression of a differential operator by the variable ξ_i, we obtain a polynomial in $\xi_1, \dots, \xi_n$ called the *symbol* of the operator. The symbol of E, denoted $E(x, \xi)$, equals $\xi_1 + i\xi_2 + ix_1\xi_3$.

According to Hörmander's theorem, [1963], Chapter VI, specialized to *first-order operators*, we have the following:

The solvability $E\mathscr{E}(\Omega) \supset \mathscr{D}(\Omega)$ implies

$$(2) \qquad [\bar{E}, E](x, \xi) = 0 \quad \text{if} \quad E(x, \xi) = 0, \qquad x \in \Omega, \quad \xi \in \boldsymbol{R}^3.$$

Here $\bar{E}$ is the operator obtained by conjugating the coefficients in E. A simple computation shows $[\bar{E}, E](x, \xi) = 2i\xi_3$, so taking $\xi_1 = 0$, $\xi_2 = -x_1$, $\xi_3 = 1$, we violate (2). Thus the solvability fails.

It turns out that for the solvability question one should look at N as a homogeneous space under the action of the product group $N \times N$ by

$$(g_1, g_2) \cdot g = g_1 g g_2^{-1}, \quad g \in N.$$

With this action all invariant differential operators are globally solvable (cf. Rais [1971] and Wigner [1977] for, respectively, local and global solvability on any simply connected nilpotent Lie group N).

In the Introduction, Problems A–D have been discussed for the cases $\boldsymbol{R}^2$, $\boldsymbol{S}^2$, and $\boldsymbol{H}^2$. Now we shall briefly describe their solutions for the case of a symmetric space $X = G/K$ of the noncompact type; refer to the original papers for detailed proofs.

A. *The algebra* $\mathbf{D}(G/K)$ *is a* (commutative) *polynomial ring in l algebraically independent generators $D_1, \dots, D_l$ whose degrees $d_1, \dots, d_l$ are canonically determined by G* (Chapter II, Theorem 5.18, and Chapter III, §3, No. 1). *Here* $l = \operatorname{rank}(G/K)$.

B. *Each $D \neq 0$ in $\mathbf{D}(G/K)$ is globally solvable.*

This result was proved by Helgason [1973a]. In order to describe the solution to Problem C we consider the Iwasawa decomposition $G = KAN$ and if $g \in G$, let $g = k \exp H(g) n$ $(k \in K, H(g) \in \mathfrak{a}, n \in N)$. Let $\mathfrak{a}_c^*$ denote the space of complex-valued linear functions on $\mathfrak{a}$. The following result of Harish-Chandra will be proved in Chapter IV, §4 [Theorem 4.3 and (7)].

C(i). *The K-invariant joint eigenfunctions of $\mathbf{D}(G/K)$ are the constant multiples of the functions*

$$\phi_\lambda(gK) = \int_K e^{(i\lambda - \rho)(H(gk))}\, dk, \qquad g \in G.$$

Here $\lambda \in \mathfrak{a}_c^*$, ρ as in Chapter I, §5, No. 1, and dk the normalized Haar measure on K.

Let M denote the centralizer of A in K and $B = K/M$. If $x = gK$ in $X = G/K$, $b = kM$ in $B = K/M$ we put

$$A(x, b) = -H(g^{-1}k) \in \mathfrak{a}.$$

This "vector-valued inner product" generalizes the inner product $\langle z, b\rangle$ considered in the Introduction, §4. No. 1. If $\mu \in \mathfrak{a}_c^*$, $b \in B$, the function

$$e_{\mu,b}: x \to e^{\mu(A(x,b))} \qquad (x \in X)$$

is a joint eigenfunction of $\mathbf{D}(G/K)$ (see Chapter IV, Exercise B15).

Each joint eigenfunction f of $\mathbf{D}(G/K)$ can be expanded as $f = \sum_{\delta \in \hat{K}} f_\delta$, where f_δ is a *K-finite* joint eigenfunction of *type* δ (Corollary 3.4 in Chapter V). We have now the following generalization of $\mathbf{C}$(i) (cf. Helgason [1976]).

C(ii). *The K-finite joint eigenfunctions of* $\mathbf{D}(G/K)$ *are the functions*

$$f(x) = \int_B e^{\mu(A(x,b))} F(b)\, db,$$

where F is a K-finite continuous function on B and $\mu \in \mathfrak{a}_c^*$.

If f is any joint eigenfunction of $\mathbf{D}(G/K)$, the expansion $f = \sum_\delta f_\delta$ above together with $\mathbf{C}$(ii) imply a formula

$$f(x) = \int_B e^{\mu(A(x,b))}\, dT(b),$$

where T is a formal Fourier series $\sum_{\delta \in \hat{K}} F_\delta$ on $B = K/M$. How can these T be characterized? Theorem 4.3 of the Introduction suggests an answer.

The space $\mathscr{A}(B)$ of analytic functions on the manifold $B = K/M$ has a natural topology generalizing that explained in the Introduction, §4, No. 1, for the circle. The members of the dual space $\mathscr{A}'(B)$ are called *analytic functionals* (or *hyperfunctions*).

Then we have the following result generalizing Theorem 4.3 of the Introduction.

C(iii). *The joint eigenfunctions of* $\mathbf{D}(G/K)$ *are the functions*

$$f(x) = \int_B e^{\mu(A(x,b))}\, dT(b), \qquad x \in X,$$

where $\mu \in \mathfrak{a}_c^*$ *and* $T \in \mathscr{A}'(B)$ (cf. Kashiwara *et al.* [1978]). A different and a more general proof based on $\mathbf{C}$(ii) above has been given by W. Schmid.

In order to describe the solution to Problem D we consider again the function ϕ_λ above, [see (**C**(i)]. Let $c_\lambda(D)$, $D \in \mathbf{D}(G/K)$, denote the corresponding system of eigenvalues, i.e.

$$D\phi_\lambda = c_\lambda(D)\phi_\lambda, \qquad D \in \mathbf{D}(G/K).$$

For $\lambda \in \mathfrak{a}_c^*$ let $\mathscr{E}_\lambda(X)$ denote the joint eigenspace

$$\mathscr{E}_\lambda(X) = \{f \in \mathscr{E}(X) \colon Df = c_\lambda(D)f \quad \text{for } D \in \mathbf{D}(G/K)\}.$$

Each joint eigenspace has this form for a suitable $\lambda \in \mathfrak{a}_c^*$; in fact, each joint eigenspace contains a K-invariant function which by **C**(i) has the form ϕ_λ $(\lambda \in \mathfrak{a}_c^*)$. For Problem D we have the following answer (Helgason [1970a], Chapter IV, [1976]).

D. *For $\lambda \in \mathfrak{a}_c^*$ let T_λ denote the eigenspace representation of G on $\mathscr{E}_\lambda(X)$. Then*

$$T_\lambda \textit{ is irreducible} \Leftrightarrow 1/\Gamma_X(\lambda) \neq 0.$$

Here the function Γ_X is defined by

$$\Gamma_X(\lambda) = \prod_{\alpha \in \Sigma_0} \Gamma(\tfrac{1}{2}(\tfrac{1}{2}m_\alpha + 1 + \langle i\lambda, \alpha_0\rangle))\Gamma(\tfrac{1}{2}(\tfrac{1}{2}m_\alpha + m_{2\alpha} + \langle i\lambda, \alpha_0\rangle)),$$

where Γ is the usual gamma function, Σ_0 runs over the set of indivisible restricted roots for G/K, m_β denotes the multiplicity of a root β, and $\alpha_0 = \alpha/\langle\alpha, \alpha\rangle$, where $\langle\ ,\ \rangle$ is the bilinear form on $\mathfrak{a}_c^*$ induced by the Killing form of $\mathfrak{g}$. The function Γ_X is the denominator of the function $\mathbf{c}(\lambda)\mathbf{c}(-\lambda)$, where $\mathbf{c}(\lambda)$ is the c-function from Harish-Chandra's expansion of the spherical function ϕ_λ (cf. Chapter IV, §§5–6).

Since the symmetric space G/K is determined by the triple $(\mathfrak{a}, \Sigma, m)$, where Σ is the set of restricted roots and m the multiplicity function (cf., [DS], Chapter X, Exercise F9), it is reasonable that, in principle, Problem D should be answerable in terms of these data. The criterion $\Gamma_X(\lambda)^{-1} \neq 0$ is a confirmation of this.

With this notation we can state the solution to Problem C in a more precise form. Let $\Gamma_X^+(\lambda)$ denote the product above with α only running over the positive elements in Σ_0. Let P_λ denote the *Poisson transform*

$$(P_\lambda T)(x) = \int_B e^{(i\lambda+\rho)(A(x,b))}\, dT(b), \qquad x \in X,\ T \in \mathscr{A}'(B).$$

Then the following refinement of **C**(iii) holds.

C(iv). *Let $\lambda \in \mathfrak{a}_c^*$. Then the following conditions are equivalent:*

(a) $1/\Gamma_X^+(\lambda) \neq 0$.
(b) *The Poisson transform $P_\lambda \colon \mathscr{A}'(B) \to \mathscr{E}_\lambda(X)$ is injective.*
(c) *The Poisson transform $P_\lambda \colon \mathscr{A}'(B) \to \mathscr{E}_\lambda(X)$ is surjective.*

The equivalence (a) $\Leftrightarrow$ (b) is proved in Helgason [1970a] and [1976] and the implication (c) $\Rightarrow$ (a) is a simple corollary of this and the proof of **C**(ii) above. The implication (a) $\Rightarrow$ (c) is proved in Kashiwara *et al.* [1978]. See Schlichtkrull [1984b] for further information.

2. The Algebra $\boldsymbol{D}(G/H)$

Given a coset space G/H, our aim is now to describe the operators in $\boldsymbol{D}(G/H)$. First we consider the case when $H = (e)$ and write $\boldsymbol{D}(G)$ for $\boldsymbol{D}(G/(e))$, the set of left-invariant differential operators on G.

If V is a finite-dimensional vector space over $\boldsymbol{R}$, the *symmetric algebra* $S(V)$ over V is defined as the algebra of complex-valued polynomial functions on the dual space V^*. If $X_1, \ldots, X_n$ is a basis of V, $S(V)$ can be identified with the (commutative) algebra of polynomials

$$\sum_{(k)} a_{k_1 \cdots k_n} X_1^{k_1} \cdots X_n^{k_n}.$$

Let $\mathfrak{g}$ denote the Lie algebra of G (the tangent space to G at e) and $\exp: \mathfrak{g} \to G$ the exponential mapping which maps a line $\boldsymbol{R}X$ through 0 in $\mathfrak{g}$ onto a one-parameter subgroup $t \to \exp tX$ of G. If $X \in \mathfrak{g}$, let $\tilde{X}$ denote the vector field on G given by

$$(3) \qquad (\tilde{X}f)(g) = X(f \circ L_g) = \left\{\frac{d}{dt} f(g \exp tX)\right\}_{t=0}, \qquad f \in \mathscr{E}(G),$$

where L_g denotes the left translation $x \to gx$ of G onto itself. Then $\tilde{X}$ is a differential operator on G, and if $h \in G$, then

$$(\tilde{X}^{L_h} f)(g) = (\tilde{X}(f \circ L_h))(h^{-1}g) = (\tilde{X}f)(g),$$

so $\tilde{X} \in \boldsymbol{D}(G)$. Moreover, the bracket on $\mathfrak{g}$ is by definition given by

$$[X, Y]^{\sim} = \tilde{X}\tilde{Y} - \tilde{Y}\tilde{X}, \qquad X, Y \in \mathfrak{g},$$

the multiplication on the right-hand side being composition of operators.

The following result, which connects $S(\mathfrak{g})$ and $\boldsymbol{D}(G)$, shows in particular that $\boldsymbol{D}(G)$ is generated by the $\tilde{X}$ $(X \in \mathfrak{g})$. Thus it coincides with the algebra introduced in [DS], (Chapter II, §1, No. 4), with the same notation, except that now we are allowing complex scalars.

Theorem 4.3. *Let G be any Lie group with algebra $\mathfrak{g}$. Let $S(\mathfrak{g})$ denote the symmetric algebra over the vector space $\mathfrak{g}$. Then there exists a unique linear bijection*

$$\lambda: S(\mathfrak{g}) \to \boldsymbol{D}(G)$$

such that $\lambda(X^m) = \tilde{X}^m$ $(X \in \mathfrak{g},\ m \in \mathbf{Z}^+)$. *If* $X_1, \ldots, X_n$ *is any basis of* $\mathfrak{g}$ *and* $P \in S(\mathfrak{g})$, *then*

$$(4) \qquad (\lambda(P)f)(g) = \{P(\partial_1, \ldots, \partial_n)f(g \exp(t_1X_1 + \cdots + t_nX_n))\}_{t=0},$$

where $f \in \mathscr{E}(G)$, $\partial_i = \partial/\partial t_i$, *and* $t = (t_1, \ldots, t_n)$.

Proof. Fix a basis $X_1, \ldots, X_n$ of $\mathfrak{g}$. Then the mapping

$$g \exp(t_1X_1 + \cdots + t_nX_n) \to (t_1, \ldots, t_n)$$

is a coordinate system on a neighborhood of g in G, so formula (4) defines a differential operator† $\lambda(P)$ on G. Clearly $\lambda(P)$ is left invariant, and by (3) $\lambda(X_i) = \tilde{X}_i$, so by linearity $\lambda(X) = \tilde{X}$ for $X \in \mathfrak{g}$. Also,

$$\begin{aligned}(\tilde{X}^2f)(g) = \tilde{X}(\tilde{X}f)(g) &= \left\{\frac{d}{dt}(\tilde{X}f)(g \exp tX)\right\}_{t=0} \\ &= \left\{\frac{d}{dt}\left\{\frac{d}{ds}f(g \exp tX \exp sX)\right\}_{s=0}\right\}_{t=0} \\ &= \left\{\frac{d^2}{dt^2}f(g \exp tX)\right\}_{t=0},\end{aligned}$$

which, writing $X = \sum_1^n x_iX_i$, equals

$$\sum x_ix_j(\lambda(X_iX_j)f)(g) = (\lambda(X^2)f)(g).$$

By the same argument†

$$(5) \qquad \lambda(X^m) = \tilde{X}^m, \qquad X \in \mathfrak{g}, \quad m \in \mathbf{Z}^+.$$

For a fixed $m \in \mathbf{Z}^+$, the powers X^m $(X \in \mathfrak{g})$ span the subspace $S^m(\mathfrak{g}) \subset S(\mathfrak{g})$ of homogeneous elements of degree m (cf. Chapter I, Exercise D1). Thus (5) shows that although λ is defined by means of a basis, it is actually independent of this basis.

Next we prove that λ is one-to-one. In fact suppose $\lambda(P) = 0$ where $P \neq 0$. With respect to a "lexicographic ordering," let $aX_1^{m_1} \cdots X_n^{m_n}$ be the leading term in P. Let f be a smooth function on a neighborhood of e in G such that

$$f(\exp(t_1X_1 + \cdots + t_nX_n)) = t_1^{m_1} \cdots t_n^{m_n}$$

for small t. Then $(\lambda(P)f)(e) \neq 0$, contradicting $\lambda(P) = 0$.

Finally, λ maps $S(\mathfrak{g})$ onto $\mathbf{D}(G)$. In fact, if $u \in \mathbf{D}(G)$, there exists a polynomial P such that

$$(uf)(e) = \{P(\partial_1, \ldots, \partial_n)f(\exp(t_1X_1 + \cdots + t_nX_n))\}_{t=0}.$$

Then by the left invariance of u, $u = \lambda(P)$ so λ is surjective.

† See "Some Details," p. 611.

Definition. The mapping λ is usually called *symmetrization.*

The mapping λ has the following property. If $Y_1, \ldots, Y_p \in \mathfrak{g}$, then

$$\lambda(Y_1 \cdots Y_p) = \frac{1}{p!} \sum_{\sigma \in \mathfrak{S}_p} \tilde{Y}_{\sigma(1)} \cdots \tilde{Y}_{\sigma(p)} \tag{5'}$$

where $\mathfrak{S}_p$ is the symmetric group on p letters. This follows from (5) used on $(t_1 Y_1 + \cdots + t_p Y_p)^p$ by equating the coefficients to $t_1 \cdots t_p$.

We now recall some facts concerning the *adjoint representation* Ad of G (or Ad_G) and the *adjoint representation* ad of $\mathfrak{g}$ (or $\mathrm{ad}_{\mathfrak{g}}$). If $g \in G$, the mapping $x \to gxg^{-1}$ is an automorphism of G; the corresponding automorphism of $\mathfrak{g}$ is denoted $\mathrm{Ad}(g)$. Thus

$$\exp \mathrm{Ad}(g)X = g \exp Xg^{-1}, \qquad X \in \mathfrak{g}, \quad g \in G. \tag{6}$$

Then the mapping $g \to \mathrm{Ad}(g)$ is a representation of G on $\mathfrak{g}$. By general theory it induces a representation of $\mathfrak{g}$ on $\mathfrak{g}$, denoted ad (cf., [DS], Chapter II, §5). Thus by definition

$$\mathrm{Ad}(\exp X) = e^{\mathrm{ad}\, X}, \qquad X \in \mathfrak{g}, \tag{7}$$

where for a linear transformation A, e^A denotes $\sum_0^\infty (1/n!)A^n$. From (6) and (7) one can deduce (cf., *loc. cit.*)

$$\mathrm{ad}\, X(Y) = [X, Y], \qquad X, Y \in \mathfrak{g}. \tag{8}$$

These operations can now be extended to differential operators. Let us calculate $(\mathrm{Ad}(g)X)^{\sim}$. Recalling the translations

$$L_g\colon x \to gx, \qquad R_g\colon x \to xg,$$

we have for $f \in \mathcal{E}(G)$

$$\begin{aligned}
[(\mathrm{Ad}(g)\,X)^{\sim} f](x) &= \left\{\frac{d}{dt} f(x \exp t\, \mathrm{Ad}(g)X)\right\}_{t=0} \\
&= \left\{\frac{d}{dt} f(xg \exp tXg^{-1})\right\}_{t=0} = \left\{\frac{d}{dt} f^{R_g}(xg \exp tX)\right\}_{t=0} \\
&= (\tilde{X} f^{R_g})(xg) = (\tilde{X} f^{R_g})^{R_{g^{-1}}}(x),
\end{aligned}$$

so

$$(\mathrm{Ad}(g)\,X)^{\sim} = \tilde{X}^{R_{g^{-1}}}.$$

Thus we *define* for $D \in \boldsymbol{D}(G)$

$$\mathrm{Ad}(g)D = D^{R_{g^{-1}}}. \tag{9}$$

Then $\mathrm{Ad}(g)$ is an automorphism of $\boldsymbol{D}(G)$.

Next we observe that

$$(\mathrm{ad}(X)(Y))^{\sim} = \tilde{X}\tilde{Y} - \tilde{Y}\tilde{X};$$

hence we *define* for $D \in \boldsymbol{D}(G)$

$$(\mathrm{ad}\, X)(D) = \tilde{X}D - D\tilde{X}, \tag{10}$$

and then ad X is a derivation of the algebra $\boldsymbol{D}(G)$. We can also define

$$e^{\mathrm{ad}\, X}(D) = \sum_0^\infty \frac{1}{n!}(\mathrm{ad}\, X)^n(D), \quad D \in \boldsymbol{D}(G), \tag{11}$$

because $(\mathrm{ad}\, X)^n(D)$ by (10) is a differential operator of order $\leq$order of D; thus all the terms in the series (11) lie in a finite-dimensional vector space, so there is no convergence problem. Now using Leibniz's formula for the power of a derivation applied to a product, we have for $D_1, D_2 \in \boldsymbol{D}(G)$

$$\begin{aligned} e^{\mathrm{ad}\, X}(D_1 D_2) &= \sum_0^\infty \frac{1}{n!}(\mathrm{ad}\, X)^n(D_1 D_2) \\ &= \sum_0^\infty \frac{1}{n!} \sum_{0 \leq i, j, i+j=n} \frac{n!}{i!\,j!}(\mathrm{ad}\, X)^i(D_1)(\mathrm{ad}\, X)^j(D_2), \end{aligned}$$

so

$$e^{\mathrm{ad}\, X}(D_1 D_2) = e^{\mathrm{ad}\, X}(D_1)e^{\mathrm{ad}\, X}(D_2).$$

Thus $\mathrm{Ad}(\exp X)$ and $e^{\mathrm{ad}\, X}$ are automorphisms of $\boldsymbol{D}(G)$; they coincide on $\tilde{\mathfrak{g}}$, hence on all of $\boldsymbol{D}(G)$, since by Theorem 4.3, $\tilde{\mathfrak{g}}$ generates $\boldsymbol{D}(G)$. Consequently,

$$\mathrm{Ad}(\exp X)D = e^{\mathrm{ad}\, X}(D), \qquad D \in \boldsymbol{D}(G). \tag{12}$$

Lemma 4.4. *Let $X \in \mathfrak{g}$, $D \in \boldsymbol{D}(G)$. Then $\tilde{X}D = D\tilde{X}$ if and only if*

$$D^{R_{\exp tX}} = D \qquad \text{for} \quad t \in \boldsymbol{R}.$$

In fact we have by (9)–(12)

$$\lim_{t \to 0} \frac{1}{t}\left(D^{R_{\exp(-tX)}} - D\right) = \tilde{X}D - D\tilde{X},$$

so the "if" part is immediate. On the other hand, if $\tilde{X}D = D\tilde{X}$, we have $D^{R_{\exp tX}} = D$ by (9) and (12).

Corollary 4.5. *Assume G is connected. Let $\boldsymbol{Z}(G)$ denote the center of $\boldsymbol{D}(G)$ and $I(\mathfrak{g}) \subset S(\mathfrak{g})$ the set of* Ad(G)*-invariants. Then*

$$\lambda(I(\mathfrak{g})) = \boldsymbol{Z}(G). \tag{13}$$

Moreover $\mathbf{Z}(G)$ *consists of the right-invariant differential operators in* $\mathbf{D}(G)$, *in other words, of the bi-invariant differential operators on* G.

The last statement is immediate from the lemma (since G is connected). Since

$$\lambda(\mathrm{Ad}(g)\,P) = \mathrm{Ad}(g)\,\lambda(P), \qquad P \in S(\mathfrak{g})$$

statement (13) follows immediately.

Now suppose G is a connected Lie group and $H \subset G$ a closed subgroup. Let $\mathfrak{g} \supset \mathfrak{h}$ be their respective Lie algebras and $\mathfrak{m}$ a complementary subspace, $\mathfrak{g} = \mathfrak{m} \oplus \mathfrak{h}$ (direct sum). We now use $\mathfrak{m}$ to introduce coordinates on G/H (cf. [DS], Chapter II, §4). Let $(X_1, \ldots, X_r)$ and $(X_{r+1}, \ldots, X_n)$ be bases of $\mathfrak{m}$ and $\mathfrak{h}$, respectively, and $\pi\colon G \to G/H$ the natural projection. Then, if $g \in G$, the mapping

$$(x_1, \ldots, x_r) \to \pi(g \exp(x_1 X_1 + \cdots + x_r X_r)) \tag{14}$$

is a diffeomorphism of a neighborhood of 0 in $\mathfrak{m}$ onto a neighborhood of $\pi(g)$ in G/H. The inverse of (14) is a local coordinate system near $\pi(g)$, turning G/H into a manifold.

The mapping $\pi\colon G \to G/H$ has a differential $d\pi$ which maps $\mathfrak{g}$ onto the tangent space $(G/H)_o$ to G/H at the origin $o = \{H\}$. The kernel of $d\pi$ is $\mathfrak{h}$. The translation $\tau(g)\colon xH \to gxH$ satisfies

$$\pi \circ L_g = \tau(g) \circ \pi,$$

and since $\pi \circ R_h = \pi$, $\mathrm{Ad}_G(g)\,X = dR_{g^{-1}} \circ dL_g(X)$, we have for the differentials,

$$d\pi \circ \mathrm{Ad}_G(h)\,X = d\tau(h)_o \circ d\pi(X), \qquad X \in \mathfrak{g}.$$

Thus under the isomorphism

$$\mathfrak{g}/\mathfrak{h} \simeq (G/H)_o \tag{15}$$

the linear transformation $\mathrm{Ad}_G(h)$ of $\mathfrak{g}/\mathfrak{h}$ corresponds to the linear transformation $d\tau(h)_o$ of $(G/H)_o$.

The coset space G/H is called *reductive* if the subspace $\mathfrak{m} \subset \mathfrak{g}$ can be chosen such that

$$\mathfrak{g} = \mathfrak{m} \oplus \mathfrak{h}, \qquad \mathrm{Ad}_G(h)\mathfrak{m} \subset \mathfrak{m} \qquad (h \in H).$$

If H is compact [or if just $\mathrm{Ad}_G(H)$ is compact], then G/H is reductive. In fact, $\mathfrak{g}$ will then have a positive definite quadratic form invariant under $\mathrm{Ad}_G(H)$ and we can take for $\mathfrak{m}$ the orthogonal complement of $\mathfrak{h}$ in $\mathfrak{g}$. Let

$$\mathbf{D}_H(G) = \{D \in \mathbf{D}(G) : D^{R_h} = D \qquad \text{for all} \quad h \in H\},$$

and if f is a function on G/H, we put $\tilde{f} = f \circ \pi$. We have now a Lie-algebraic description of $\boldsymbol{D}(G/H)$ for a reductive coset space.

Theorem 4.6. *Assume G/H reductive. Then the mapping $\mu: u \to D_u$, where*

$$(D_u f)^{\sim} = u\tilde{f}, \qquad f \in \mathscr{E}(G/H),$$

is a homomorphism of $\boldsymbol{D}_H(G)$ onto $\boldsymbol{D}(G/H)$. The kernel is

$$\boldsymbol{D}_H(G) \cap \boldsymbol{D}(G)\mathfrak{h},$$

so we have the isomorphism

$$\boldsymbol{D}_H(G)/\boldsymbol{D}_H(G) \cap \boldsymbol{D}(G)\mathfrak{h} \cong \boldsymbol{D}(G/H).$$

Proof. Let $u \in \boldsymbol{D}_H(G)$ and $f \in \mathscr{D}(G/H)$. Then $u\tilde{f}$ is right invariant under H, thus of the form $\tilde{f}_1$, $f_1 \in \mathscr{D}(G/H)$; D_u is the map $f \to f_1$. It decreases supports, so D_u is a differential operator. It is G-invariant because

$$\begin{aligned}(D_u^{\tau(g)}f)^{\sim} &= ((D_u f^{\tau(g)^{-1}})^{\tau(g)})^{\sim} = ((D_u f^{\tau(g^{-1})})^{\sim})^{L_g} \\ &= (u(f^{\tau(g^{-1})})^{\sim})^{L_g} = u\tilde{f} = (D_u f)^{\sim},\end{aligned}$$

so that $D_u^{\tau(g)} = D_u$, whence $D_u \in \boldsymbol{D}(G/H)$. Also, $u \to D_u$ is a homomorphism

Next we prove that the mapping is surjective. Let $E \in \boldsymbol{D}(G/H)$. We express E at o in terms of $x_1, \ldots, x_r$; there exists a polynomial P such that

$$(Ef)(o) = \left[P\left(\frac{\partial}{\partial x_1}, \ldots, \frac{\partial}{\partial x_r}\right) f(\pi(\exp(x_1 X_1 + \cdots + x_r X_r)))\right](0).$$

By the G-invariance,

$$\begin{aligned}(Ef)(g \cdot o) &= Ef^{\tau(g^{-1})}(o) \\ &= \left[P\left(\frac{\partial}{\partial x_1}, \ldots, \frac{\partial}{\partial x_r}\right) f^{\tau(g^{-1})}(\pi(\exp(x_1 X_1 + \cdots + x_r X_r)))\right](0) \\ &= \left\{P\left(\frac{\partial}{\partial x_1}, \ldots, \frac{\partial}{\partial x_r}\right) \tilde{f}(g \exp(x_1 X_1 + \cdots + x_r X_r))\right\}_{x_i = 0}.\end{aligned}$$

In particular, take $g = h \in H$. Then

$$(Ef)(o) = \left\{P\left(\frac{\partial}{\partial x_1}, \ldots, \frac{\partial}{\partial x_r}\right) \tilde{f}(\exp \operatorname{Ad}(h)(x_1 X_1 + \cdots + x_r X_r))\right\}_{x_i = 0};$$

hence we conclude that P is $\operatorname{Ad}(H)$-invariant. Put $u = \lambda(P) \in \boldsymbol{D}(G)$. Then

$$u^{R_{h^{-1}}} = \operatorname{Ad}(h)\, u = \lambda(\operatorname{Ad}(h)\, P) = \lambda(P) = u,$$

so $u \in \boldsymbol{D}_H(G)$. Also,

$$(u\tilde{f})(g) = (\lambda(P)\tilde{f})(g) = \left\{P\left(\frac{\partial}{\partial x_1}, \ldots, \frac{\partial}{\partial x_r}\right)\tilde{f}(g \exp(x_1X_1 + \cdots + x_rX_r)\right\}_{x_i = 0}$$

$$= (Ef)^{\sim}(g),$$

so $D_u = E$. Thus our map is surjective.

It remains to prove $D_u = 0 \Leftrightarrow u \in \boldsymbol{D}_H(G) \cap \boldsymbol{D}(G)\mathfrak{h}$. For this we insert the following lemma. For $d \geq 0$ let $\boldsymbol{D}^d(G) = \lambda(\sum_{e \leq d} S^e(\mathfrak{g}))$.

Lemma 4.7. *$\boldsymbol{D}(G) = \boldsymbol{D}(G)\mathfrak{h} \oplus \lambda(S(\mathfrak{m}))$ (direct sum). Moreover, if $D \in \boldsymbol{D}^d(G)$ decomposes $D = D_1 + D_2$, then $D_1, D_2 \in \boldsymbol{D}^d(G)$.*

Proof. Given $P \in S(\mathfrak{g})$, we claim that there exists a $Q \in S(\mathfrak{m})$ of degree $\leq \deg(P)$ such that $\lambda(P - Q) \in \boldsymbol{D}(G)\mathfrak{h}$. This is clear if P has degree 1. We now assume that the claim is true for $P \in S(\mathfrak{g})$ of degree $<d$. We must prove it holds for P of degree d. We may assume $P = X_1^{e_1} \cdots X_n^{e_n}$ in terms of the bases $X_1, \ldots, X_r$ of $\mathfrak{m}$, $X_{r+1}, \ldots, X_n$ of $\mathfrak{h}$. If $e_{r+1} + \cdots + e_n = 0$, we can take $Q = P$. If $e_{r+1} + \cdots + e_n > 0$, $\lambda(P)$ is a linear combination of terms $\tilde{X}_{\alpha_1} \cdots \tilde{X}_{\alpha_d}$, where $X_{\alpha_i} \in \mathfrak{h}$ for some i. Then

$$\tilde{X}_{\alpha_1} \cdots \tilde{X}_{\alpha_d} - \tilde{X}_{\alpha_1} \cdots \tilde{X}_{\alpha_{i-1}}\tilde{X}_{\alpha_{i+1}} \cdots \tilde{X}_{\alpha_d}\tilde{X}_{\alpha_i} \in \boldsymbol{D}^{d-1}(G),$$

so that

$$\lambda(P) - D \in \boldsymbol{D}(G)\mathfrak{h} \qquad \text{for some} \quad D \in \boldsymbol{D}^{d-1}(G).$$

By the induction hypothesis, there exists a $Q \in S(\mathfrak{m})$ of degree $\leq d - 1$ such that

$$\lambda(Q) - D \in \boldsymbol{D}(G)\mathfrak{h},$$

whence $\lambda(P - Q) \in \boldsymbol{D}(G)\mathfrak{h}$. This gives the decomposition; it remains to prove the directness. Let $P \in S(\mathfrak{m})$, $P \neq 0$. Then there exists a function $f^*(x_1, \ldots, x_r)$ such that

$$\left(P\left(\frac{\partial}{\partial x_1}, \ldots, \frac{\partial}{\partial x_r}\right)f^*\right)(0) \neq 0.$$

Choose $f \in C^\infty(G/H)$ such that

$$f(\pi(\exp(x_1X_1 + \cdots + x_rX_r))) = f^*(x_1, \ldots, x_r)$$

for x_i sufficiently small. Then

$$\lambda(P)(f \circ \pi)(e) \neq 0,$$

so $\lambda(P) \notin \boldsymbol{D}(G)\mathfrak{h}$. Here G/H does not have to be reductive.

Since both summands are stable under $\mathrm{Ad}_G(H)$, we deduce

Corollary 4.8. *Let $I(\mathfrak{m})$ denote the set of $\mathrm{Ad}_G(H)$-invariants in $S(\mathfrak{m})$. Then*

$$\mathbf{D}_H(G) = (\mathbf{D}_H(G) \cap \mathbf{D}(G)\mathfrak{h}) \oplus \lambda(I(\mathfrak{m})).$$

We can now finish the proof of Theorem 4.6. Let $u \in \mathbf{D}_H(G)$ such that $D_u = 0$. Let

$$u = u_1 + u_2,$$

as in Corollary 4.8. Then $D_{u_1} = 0$, so that $D_{u_2} = 0$. But $u_2 = \lambda(P_2)$, $P_2 \in I(\mathfrak{m})$. We claim $u_2 = 0$. If not, then as we saw above, there exists $f \in \mathscr{E}(G/H)$ with $u_2 \tilde{f} \neq 0$, so that $D_{u_2} \neq 0$, which is a contradiction. Thus $u_2 = 0$, so that $u \in \mathbf{D}_H(G) \cap \mathbf{D}(G)\mathfrak{h}$.

Combining Theorem 4.6 and Corollary 4.8, we obtain the following result.

Theorem 4.9. *Let G/H be a reductive homogeneous space. The mapping $Q \to D_{\lambda(Q)}$ is a linear bijection of $I(\mathfrak{m})$ onto $\mathbf{D}(G/H)$. Explicitly, if $Q \in I(\mathfrak{m})$, then*

$$(D_{\lambda(Q)}f)(g \cdot o) = \left[Q\left(\frac{\partial}{\partial x_1}, \ldots, \frac{\partial}{\partial x_r}\right)\tilde{f}(g \exp(x_1 X_1 + \cdots + x_r X_r))\right](0).$$

While the mapping $Q \to D_{\lambda(Q)}$ is not in general multiplicative [even when $\mathbf{D}(G/H)$ is commutative], we have

$$D_{\lambda(P_1 P_2)} = D_{\lambda(P_1)} D_{\lambda(P_2)} + D_{\lambda(Q)},$$

where $Q \in I(\mathfrak{m})$ has degree $< \deg P_1 + \deg P_2$.

By induction we obtain,

Corollary 4.10. *If $I(\mathfrak{m})$ has a finite system of generators $P_1, \ldots, P_l$ and we put $D_i = D_{\lambda(P_i)}$, then each $D \in \mathbf{D}(G/H)$ can be written*

$$D = \sum_{(n)} a_{n_1 \cdots n_l} D_1^{n_1} \cdots D_l^{n_l}.$$

3. The Case of a Two-Point Homogeneous Space. The Generalized Darboux Equation

We consider now the case in which M is a two-point homogeneous space (Chapter I, §4). Here the Laplace–Beltrami operator can be characterized in terms of its invariance.

Proposition 4.11. *Let M be a two-point homogeneous space, $M = G/K$, where $G = I(M)$. Then $\mathbf{D}(G/K)$ consists of the polynomials in the Laplace–Beltrami operator.*

Proof. The two-point homogeneity implies that $\mathrm{Ad}_G(K)$ acts transitively on the unit sphere in the tangent space $(G/K)_o$; thus it is clear that $I(\mathfrak{m})$ is generated by $X_1^2 + \cdots + X_r^2$ if (X_i) is an orthonormal basis of $\mathfrak{m}$. This implies the result.

Remark. It is useful to observe that the result holds also for $\mathbf{D}(G'/K')$ if G' is a subgroup of $I(M)$ doubly transitive on $M = (G'/K')$.

Proposition 4.12. *Let G/K be a symmetric space of rank one, L the Laplacian, and M^r the mean value operator*

$$(M^r f)(x) = \frac{1}{A(r)} \int_{S_r(x)} f(s)\, d\omega(s) \qquad (r \geq 0),$$

where $d\omega$ is the area element of $S_r(x)$ and $A(r)$ the total area. Then

$$M^r L = L M^r, \tag{16}$$

$$L_x(F(x, y)) = L_y(F(x, y)), \tag{17}$$

where

$$F(x, y) = (M^{d(o, y)} f)(x).$$

Remark. This result is the analog to Lemma 2.14 in Chapter I, and (17) is the generalized "Darboux equation."

Proof. Let Ω denote the Laplacian on the semisimple Lie group G, the pseudo-Riemannian structure defined by means of the Killing form. Let $\pi\colon G \to G/K$ be natural projection and $\tilde{f} = f \circ \pi$. Then the operator $\mu(\Omega)$ from Theorem 4.6 is G-invariant, and by Proposition 4.11 (or directly) $\mu(\Omega)$ is proportional to L. If $h \in G$ is such that $r = d(o, h \cdot o)$, then

$$(M^r f)(g \cdot o) = \int_K \tilde{f}(gkh)\, dk.$$

By the bi-invariance of Ω,

$$\Omega_g\left(\int_K \tilde{f}(gkh)\, dk\right) = \int_K (\Omega \tilde{f})(gkh)\, dk = \Omega_h\left(\int_K \tilde{f}(gkh)\, dk\right),$$

and this proves both (16) and (17).

§5. Invariant Differential Operators on Symmetric Spaces

1. The Action on Distributions and Commutativity

Let G be a separable unimodular Lie group, dy a fixed Haar measure on G, and $D \to D^*$ the corresponding adjoint operation on $\boldsymbol{E}(G)$, the set of all differential operators on G. The convolution

(1) $$(f * g)(x) = \int_G f(xy^{-1})\, g(y)\, dy = \int_G f(y) g(y^{-1}x)\, dy$$

is well-defined for f, $g \in C(G)$ if at least one of them has compact support. More generally, we define the convolution $t_1 * t_2$ of two distributions on G, at least one of compact support, as the distribution

(2) $$(t_1 * t_2)(\phi) = \int_G \int_G \phi(xy)\, dt_1(x)\, dt_2(y), \qquad \phi \in \mathscr{D}(G).$$

Then we have the associativity $(t_1 * t_2) * t_3 = t_1 * (t_2 * t_3)$ if at least two of the t_i have compact support (cf. Chapter I, §2, No. 8).

A locally integrable function F on G can be viewed as a distribution $\phi \to \int \phi F$ on G. With this identification we have for $f \in \mathscr{E}(G)$, $t \in \mathscr{D}'(G)$, at least one of compact support,

(3) $$(f * t)(x) = \int f(xy^{-1})\, dt(y), \qquad (t * f)(x) = \int f(y^{-1}x)\, dt(y).$$

If τ is a diffeomorphism of G leaving dg invariant, we have for $D \in \boldsymbol{E}(G)$

(4) $$(D^*)^\tau = (D^\tau)^*, \qquad (DT)^\tau = D^\tau(T^\tau), \qquad T \in \mathscr{D}'(G).$$

The mapping $\check{}: x \to x^{-1}$ is a diffeomorphism and, given $D \in \boldsymbol{D}(G)$, the functional

$$\varepsilon_D \colon \phi \to (D\check{\phi})(e) \qquad \phi \in \mathscr{D}(G)$$

is a distribution with support $\{e\}$. We have

(5) $$(t_1 * t_2)\check{} = \check{t}_2 * \check{t}_1, \qquad Dt = t * \varepsilon_D,$$

whence

(6) $$\varepsilon_{D_1 D_2} = \varepsilon_{D_2} * \varepsilon_{D_1}.$$

Let $X \in \mathfrak{g}$, the Lie algebra of G, and let $\tilde{X}$ (resp. $\bar{X}$) denote the left-invariant (resp. right-invariant) vector field on G such that $\tilde{X}_e = X$ (resp. $\bar{X}_e = X$). Since

$$\int_G (\tilde{X}f)(x \exp tX)\, dx = \int \frac{d}{dt} f(x \exp tX)\, dx = \frac{d}{dt} \int f(x \exp tX)\, dx = 0,$$

we have $\int_G \tilde{X} f = 0$. Using this on a product fg, we obtain

$$(7) \qquad (\tilde{X})^* = -\tilde{X}, \qquad (\tilde{X})^{\check{}} = -\bar{X}.$$

Also, if $D_1, D_2 \in \boldsymbol{E}(G)$, we have

$$(8) \qquad (D_1 D_2)^* = D_2^* D_1^*, \qquad (D_1 D_2)^{\check{}} = D_1^{\check{}} D_2^{\check{}}.$$

Writing $D \in \boldsymbol{D}(G)$ as a polynomial in the $\tilde{X}$ $(X \in \mathfrak{g})$, we deduce from (5)–(8)

$$(9) \qquad \varepsilon_{D^*} = (\varepsilon_D)^{\check{}}.$$

Clearly if D is a bi-invariant operator, i.e., $D \in \boldsymbol{Z}(G)$, then ε_D is invariant under all inner automorphisms $g \to xgx^{-1}$; moreover, $t * \varepsilon_D = \varepsilon_D * t$ for each $t \in \mathscr{D}'(G)$ in this case.

Let $K \subset G$ be a compact subgroup with Haar measure dk normalized by $\int_K dk = 1$. Let $\pi: G \to G/K$ be the natural mapping, put $\tilde{F} = F \circ \pi$, and let dx be the measure on G/K defined by

$$(10) \qquad \int_{G/K} F(x)\, dx = \int_G \tilde{F}(g)\, dg, \qquad F \in C_c(G/K).$$

With $\mathfrak{k}$ denoting the Lie algebra of K, let μ: $\boldsymbol{D}_K(G) \to \boldsymbol{D}(G/K)$ denote the homomorphism from Theorem 4.6 with kernel $\boldsymbol{D}_K(G) \cap \boldsymbol{D}(G)\mathfrak{k}$. Since $(\mu(E)F)^{\sim} = E(\tilde{F})$, formula (10) implies $\mu(E^*) = (\mu(E))^*$.

Extending the operation $F \to \tilde{F}$, we define for $T \in \mathscr{D}'(G/K)$ the distribution $\tilde{T}$ on G by

$$(11) \qquad \tilde{T}(f) = T(\dot{f}), \qquad f \in \mathscr{D}(G),$$

where

$$\dot{f}(gK) = \int_K f(gk)\, dk.$$

Then

$$\tilde{T}(\tilde{F}) = T(F) \qquad \text{for} \quad F \in \mathscr{D}(G/K),$$

so $\tilde{T}$ is determined by its values on the space $\mathscr{D}(G/K)^{\sim}$. If T_1 and T_2 are two distributions on G/K, at least one of compact support, their "convolution" $T_1 \times T_2$ is the distribution on G/K defined by

$$(12) \qquad (T_1 \times T_2)(F) = (\tilde{T}_1 * \tilde{T}_2)(\tilde{F}).$$

Then

$$(T_1 \times T_2)^{\sim} = \tilde{T}_1 * \tilde{T}_2,$$

so $\times$ satisfies the associative law. If δ is the delta distribution $f \to f(o)$ at the origin o in G/K, we have

$$(13) \qquad T \times \delta = T, \qquad \delta \times T = \int_K T^{\tau(k)}\,dk.$$

If $D \in \mathbf{D}(G/K)$, we have

$$(14) \qquad D(T_1 \times T_2) = T_1 \times DT_2.$$

To verify this we select $E \in \mathbf{D}_K(G)$ such that $\mu(E) = D$. Then

$$(15) \qquad (\mu(E)T)^{\sim} = E(\tilde{T}), \qquad T \in \mathscr{D}'(G/K),$$

because both sides have the same value on each $\tilde{F}$ $(F \in \mathscr{D}(G/K))$. By combining this with the fact that E commutes with left convolution $t \to s * t$ on G, (14) follows easily.

We shall now see that in the case when G/K is symmetric there is an analog of (14) with D operating on the *first factor*. Hence, in addition to our assumption that G is a separable unimodular Lie group and K compact, we assume now that (G, K) is a *symmetric pair*, that is, there exists an involutive automorphism θ of G such that $(K_\theta)_0 \subset K \subset K_\theta$, where K_θ is the fixed point set of θ and $(K_\theta)_0$ its identity component. Let s denote the symmetry $gK \to \theta(g)K$ of G/K onto itself.

The automorphism θ induces an automorphism of $\mathfrak{g}$ and of $\mathbf{D}(G)$; these will also be denoted θ. We then have for each $E \in \mathbf{D}(G)$,

$$E^\theta = \theta E.$$

In fact, the maps $E \to E^\theta$ and $E \to \theta E$ are both automorphisms, so it suffices to check this for a left-invariant vector field X in G. But then

$$(X^\theta f)(g) = (Xf^\theta)(\theta g) = \left\{\frac{d}{dt} f^\theta(\theta g \exp tX)\right\}_{t=0}$$
$$= \left\{\frac{d}{dt} f(g \exp t\,\theta X)\right\}_{t=0} = ((\theta X)f)(g).$$

Also, if $E \in \mathbf{D}_K(G)$, we have

$$\mu(\theta E) = (\mu(E))^s$$

as an immediate consequence of

$$(\tilde{F})^\theta = (F^s)^{\sim} \qquad \text{for} \quad F \in \mathscr{E}(G/K).$$

Lemma 5.1. *Let (G, K) be a symmetric pair, K compact, and $D \in \mathbf{D}(G/K)$. Then if $J \in \mathscr{D}'(G/K)$ is K-invariant and $T \in \mathscr{E}'(G/K)$, we have*

$$D(J \times T) = DJ \times T$$

$$D(T \times J) = ((D^*)^s T) \times J.$$

Proof. Select $E \in \mathbf{D}_K(G)$ such that $\mu(E) = D$ and put $j = \tilde{J}$, $t = \tilde{T}$. By the symmetry, $\theta(g) \subset Kg^{-1}K$ $(g \in G)$, so putting

$$f^{\natural}(g) = \iint f(kgk')\, dk\, dk', \qquad f \in \mathscr{D}(G),$$

we have $(f^{\theta})^{\natural} = (\check{f})^{\natural}$. Since $j(f) = j(f^{\natural})$, we deduce

$$j^{\theta} = \check{j}. \tag{16}$$

In particular, since $j * t$ is bi-invariant under K, $(j * t)^{\theta} = (j * t)\check{}$. Now θ is involutive and therefore leaves dy invariant, so

$$\begin{aligned} (E(j * t))^{\theta} &= E^{\theta}(j * t)^{\theta} = E^{\theta}(j * t)\check{} = E^{\theta}(\check{t} * j^{\theta}) \\ &= \check{t} * (Ej)^{\theta} = (Ej * t)\check{} = (Ej * t)^{\theta}. \end{aligned}$$

Combining this with (15), we get the first relation of the lemma.

Since j is bi-invariant under K and ε_E invariant under the conjugation $g \to kgk^{-1}$, it is clear that $j * \varepsilon_E$ is bi-invariant under K. Thus by (9) and (16)

$$\varepsilon_E * j = (j^{\theta} * \varepsilon_{E^*})\check{} = (j^{\theta} * \varepsilon_{E^*})^{\theta} = j * (\varepsilon_{E^*})^{\theta}.$$

But $(\varepsilon_{E^*})^{\theta} = \varepsilon_{(E^*)^{\theta}}$, so this proves

$$\varepsilon_E * j = j * \varepsilon_{(E^*)^{\theta}}.$$

Taking left convolution with t and using (5), we get $(Et) * j = (E^*)^{\theta}(t * j)$. Also, $\mu((E^*)^{\theta}) = (\mu(E^*))^s = (D^*)^s$, and now (15) implies the second relation of the lemma.

Let

$$\mathscr{D}'(G)^{\natural} = \{T \in \mathscr{D}'(G) : T \quad \text{bi-invariant under } K\}$$

and define $\mathscr{E}'(G)^{\natural}$, $C_c^{\natural}(G)$, $\mathscr{D}^{\natural}(G), \ldots$ similarly.

Corollary 5.2. *$\mathscr{E}'(G)^{\natural}$ and $C_c^{\natural}(G)$ are commutative algebras under convolution.*

This follows from (16), since

$$(t_1 * t_2)^{\theta} = t_1^{\theta} * t_2^{\theta}, \qquad (t_1 * t_2)\check{} = \check{t}_2 * \check{t}_1.$$

Corollary 5.3. *If $D \in \mathbf{D}(G/K)$, then*

$$D^* = D^s.$$

This follows by taking $J = \delta$ in Lemma 5.1.

Corollary 5.4. *For (G, K) a symmetric pair, K compact, the algebra $\mathbf{D}(G/K)$ is commutative.*

In fact, if $D_1, D_2 \in \mathbf{D}(G/K)$, we have by Corollary 5.3,

$$D_2^* D_1^* = (D_1 D_2)^* = (D_1 D_2)^s = D_1^s D_2^s = D_1^* D_2^*.$$

Theorem 5.5. *Let (G, K) be a symmetric pair, K compact and $D \in \mathbf{D}(G/K)$. Then*

$$D(S \times T) = DS \times T = S \times DT,$$

for all $S, T \in \mathscr{D}'(G/K)$, at least one of compact support.

For this we need another lemma.

Lemma 5.6. *Let $\mathscr{D}^\natural(G/K)$ be the space of K-invariant functions in $\mathscr{D}(G/K)$. As ϕ runs through $\mathscr{D}^\natural$, and g through G the linear combinations of the functions $\phi^{\tau(g)}$ are dense in $\mathscr{D}(G/K)$.*

Proof. Suppose to the contrary that there were a distribution $T \neq 0$ on G/K such that $T(\phi^{\tau(g)}) = 0$ for all $\phi \in \mathscr{D}^\natural$ and all $g \in G$. This translates quickly into the condition $\tilde{T} * \psi = 0$ for all $\psi \in \mathscr{D}^\natural(G)$. This implies $\tilde{T} - 0$, whence the lemma.

We can now prove Theorem 5.5. If $g \in G$, the mapping $\tau(g)$ of G/K, extended to functions, commutes with D as well as with the right convolution $f \to f \times T$ $(T \in \mathscr{E}'(G/K))$. Hence the first relation of Lemma 5.1 implies

$$D(\phi^{\tau(g)} \times T) = D(\phi^{\tau(g)}) \times T, \qquad T \in \mathscr{E}'(G/K). \tag{17}$$

By Lemma 5.6 and by the continuity of the map $f \to f \times T$ on $\mathscr{D}(G/K)$, (17) implies

$$D(f \times T) = (Df) \times T, \qquad f \in \mathscr{D}(G/K). \tag{18}$$

However, $\mathscr{D}(G/K)$ is dense in $\mathscr{D}'(G/K)$ for the strong topology, and D as well as the convolution $S \to S \times T$ are continuous on $\mathscr{D}'(G/K)$ (Schwartz [1966], pp. 75, 170). This, by (18),

$$D(S \times T) = DS \times T, \qquad S \in \mathscr{D}', \quad T \in \mathscr{E}'.$$

For $S \in \mathscr{E}'$, $T \in \mathscr{D}'$ the proof is similar, so Theorem 5.5, is proved.

Theorem 5.7.

(i) *Let (G, H) be a symmetric pair, G semisimple, H connected. Then $\mathbf{D}(G/H)$ is commutative.*

(ii) *Let L be a connected Lie group, L^* the diagonal in $L \times L$. Under the bijection*

$$(l_1, l_2)L^* \to l_1 l_2^{-1}$$

of $(L \times L)/L^*$ onto L we have the identification $\mathbf{D}((L \times L)/L^*) = \mathbf{Z}(L)$.

Proof. (i) *Let*

$$\mathfrak{g} = \mathfrak{h} + \mathfrak{q} \tag{19}$$

be the decomposition of $\mathfrak{g}$ into the eigenspaces of $d\sigma$ for the eigenvalues $+1$ and -1, respectively. Here $\mathfrak{h}$ is the Lie algebra of H. There exists a Cartan involution θ of $\mathfrak{g}$ commuting with σ ([DS], Chapter III, Exercise B4). Let

$$\mathfrak{g} = \mathfrak{k} + \mathfrak{p} \tag{20}$$

be the corresponding Cartan decomposition. Let $\mathfrak{g}^c$ be the complexification of $\mathfrak{g}$, $\mathfrak{h}^c$ and $\mathfrak{q}^c$ the subspaces of $\mathfrak{g}^c$ generated by $\mathfrak{h}$ and $\mathfrak{q}$. Put $\mathfrak{u} = \mathfrak{k} + i\mathfrak{p}$. Then $\mathfrak{u}$ is a compact real form of $\mathfrak{g}^c$, and since $\sigma\theta = \theta\sigma$, we have

$$\mathfrak{u} = \mathfrak{u} \cap \mathfrak{h}^c + \mathfrak{u} \cap \mathfrak{q}^c,$$

$$\mathfrak{h}^c = \mathfrak{u} \cap \mathfrak{h}^c + i(\mathfrak{u} \cap \mathfrak{h}^c).$$

The subspace

$$\mathfrak{s} = \mathfrak{u} \cap \mathfrak{h}^c + i(\mathfrak{u} \cap \mathfrak{q}^c) \tag{21}$$

is a real form of $\mathfrak{g}^c$ and (21) is a Cartan decomposition of $\mathfrak{s}$. Let S denote the adjoint group $\mathrm{Int}(\mathfrak{s})$, put $\mathfrak{h}^* = \mathfrak{u} \cap \mathfrak{h}^c$, and let H^* be the analytic subgroup of S with Lie algebra $\mathfrak{h}^*$. Then H^* is compact and (S, H^*) is a symmetric pair. By Corollary 5.4, $\mathbf{D}(S/H^*)$ is commutative, so by Theorem 4.6,

$$\mathbf{D}_{H^*}(S)/\mathbf{D}_{H^*}(S) \cap \mathbf{D}(S)\mathfrak{h}^* \quad \text{is commutative.} \tag{22}$$

Passing to the complexifications (22) means that

$$Z_{\mathfrak{h}^c}(\mathfrak{g}^c)/Z_{\mathfrak{h}^c}(\mathfrak{g}^c) \cap U(\mathfrak{g}^c)\mathfrak{h}^c \quad \text{is commutative.} \tag{23}$$

Here $U(\mathfrak{g}^c)$ denotes the universal enveloping algebra of $\mathfrak{g}^c$ and $Z_{\mathfrak{h}^c}(\mathfrak{g}^c)$ the centralizer of $\mathfrak{h}^c$ in $U(\mathfrak{g}^c)$. But just as (22) is equivalent to (23) ($\mathfrak{h}^c$ being the complexification of $\mathfrak{h}^*$), property (23) is equivalent to

$$\mathbf{D}_H(G)/\mathbf{D}_H(G) \cap \mathbf{D}(G)\mathfrak{h} \quad \text{is commutative,} \tag{24}$$

which by Theorem 4.6 proves that $\mathbf{D}(G/H)$ is commutative.

(ii) The action $(l_1, l_2)L^* \to (g_1 l_1, g_2 l_2)L^*$ of $L \times L$ on $(L \times L)/L^*$ corresponds to the translation $l \to g_1 l g_2^{-1}$ on L. This proves (ii), since $\boldsymbol{Z}(L)$ consists of the bi-invariant differential operators on L.

2. The Connection with Weyl Group Invariants

Let $X = G/K$ be a symmetric space of the noncompact type, that is, G is a connected semisimple Lie group with finite center and K is a maximal compact subgroup. Let $\mathfrak{g} = \mathfrak{k} + \mathfrak{p}$ be the corresponding Cartan decomposition of $\mathfrak{g}$, let $\mathfrak{a} \subset \mathfrak{p}$ be a maximal abelian subspace, let $\mathfrak{a}^+ \subset \mathfrak{a}$ be a fixed Weyl chamber, and let the set Σ of roots, the set Σ^+ of positive roots and the Weyl group W be defined as in [DS] (Chapter IX, §1).

If E is a Euclidean space, let $\mathscr{E}(E) \supset \mathscr{S}(E) \supset \mathscr{D}(E)$ denote, respectively, the space of C^∞-functions, the space of rapidly decreasing C^∞-functions, and the space of compactly supported C^∞-functions. The subscripts K and W denote K-invariance and W-invariance, respectively. We shall now consider the subspaces of K-invariant and W-invariant functions,

$$\mathscr{D}_K(\mathfrak{p}) \subset \mathscr{S}_K(\mathfrak{p}) \subset \mathscr{E}_K(\mathfrak{p}) \subset \mathscr{E}(\mathfrak{p}),$$

$$\mathscr{D}_W(\mathfrak{a}) \subset \mathscr{S}_W(\mathfrak{a}) \subset \mathscr{E}_W(\mathfrak{a}) \subset \mathscr{E}(\mathfrak{a}).$$

Theorem 5.8. *The restriction to $\mathfrak{a}$ is an isomorphism of $\mathscr{D}_K(\mathfrak{p})$ onto $\mathscr{D}_W(\mathfrak{a})$.*

Only the surjectivity remains to be proved. For this we first prove two regularity theorems for the Laplacian $L_\mathfrak{p}$.

Lemma 5.9. *Let $T \in \mathscr{D}'(\mathfrak{p})$ satisfy $L_\mathfrak{p} T = 0$. Then $T \in \mathscr{E}(\mathfrak{p})$.*

Proof. If $\phi \in \mathscr{D}(\mathfrak{p})$ then $\phi * T$ is a harmonic function. The mean value theorem for harmonic functions implies that if $\alpha \in \mathscr{D}(\mathfrak{p})$ is a radial function with $\int \alpha = 1$, then

$$(\phi * T) * \alpha = \phi * T.$$

Evaluating at 0, we obtain $T * \alpha = T$, so that $T \in \mathscr{E}(\mathfrak{p})$.

Remark. The same argument shows that if $T \in \mathscr{D}'(\mathfrak{p})$, $LT = 0$ near 0 then T is smooth near 0.

As usual, we let $C^k(\mathfrak{p})$ stand for the space of k times continuously differentiable functions on $\mathfrak{p}$; here $C^0(\mathfrak{p})$ is understood to mean $C(\mathfrak{p})$, the space of continuous functions on $\mathfrak{p}$.

Lemma 5.10. *Let T be a distribution on $\mathfrak{p}$ such that $L_\mathfrak{p} T = g \in C^k(\mathfrak{p})$. Then $T \in C^{k+1}(\mathfrak{p})$.*

Proof. Fix $X \in \mathfrak{p}$ and select $\phi \in \mathcal{D}(\mathfrak{p})$ such that $\phi \equiv 1$ in a neighborhood of X. Assuming first $n \neq 2$ ($n = \dim \mathfrak{p}$) we consider the convolution $F = \phi g * S$, where S is the distribution $((2-n)\Omega_n)^{-1}r^{2-n}$ which satisfies

$$L_{\mathfrak{p}} S = \delta$$

(cf., Chapter I, Lemma 2.35). Then $L_{\mathfrak{p}} F = \phi g$, so by Remark following Lemma 5.9, $F - T$ is C^∞ near X. On the other hand,

$$(2-n)\Omega_n F(X) = \int_{\mathfrak{p}} (\phi g)(X - Y)|Y|^{2-n}\, dY.$$

If $D = D^\alpha$ ($|\alpha| = k$) in the notation of §1, we have

$$(2-n)\Omega_n (DF)(X) = \int_{\mathfrak{p}} (D\phi g)(X - Y)|Y|^{2-n}\, dY$$

$$= \int_{\mathfrak{p}} (D\phi g)(Y)|X - Y|^{2-n}\, dY.$$

We write this in the form

$$H(X) = \int_{\mathfrak{p}} h(Y)|X - Y|^{2-n}\, dY, \qquad h \in C_c(\mathfrak{p}),$$

and have then to prove that $H \in C^1(\mathfrak{p})$. Although $D_i(|X - Y|^{2-n})$ (where $D_i = \partial/\partial x_i$) is readily seen to be locally integrable, so that the function

$$K(X) = \int_{\mathfrak{p}} h(Y)\, D_i(|X - Y|^{2-n})\, dY = \int_{\mathfrak{p}} h(X - Y)\, D_i(|Y|^{2-n})\, dY$$

exists and is continuous, we still have to prove $K = D_i H$. For this we replace the function $Y \to |Y|^{2-n}$ by the function

$$\phi_\varepsilon(Y) = \begin{cases} |Y|^{2-n} & \text{for } |Y| \geq \varepsilon \\ (n/2)\varepsilon^{2-n} + [(2-n)/2]\varepsilon^{-n}|Y|^2 & \text{for } |Y| < \varepsilon, \end{cases}$$

and put

$$H_\varepsilon(X) = \int_{\mathfrak{p}} h(Y)\phi_\varepsilon(X - Y)\, dY.$$

The function ϕ_ε is of class C^1 so

$$(D_i H_\varepsilon)(X) = \int_{\mathfrak{p}} h(Y)(D_i \phi_\varepsilon)(X - Y)\, dY.$$

Also,

$$H(X) - H_\varepsilon(X) = \int_{|Y| \leq \varepsilon} h(X - Y)(|Y|^{2-n} - \phi_\varepsilon(Y))\, dY,$$

so

$$\lim_{\varepsilon \to 0} H_\varepsilon(X) = H(X) \qquad \text{uniformly.}$$

Second,

$$K(X) - (D_i H_\varepsilon)(X) = \int_{\mathfrak{p}} h(X - Y)[D_i(|Y|^{2-n}) - (D_i \phi_\varepsilon)(Y)]\, dY$$

$$= (2 - n) \int_{|Y| \le \varepsilon} h(X - Y)[|Y|^{-n} y_i - \varepsilon^{-n} y_i]\, dY,$$

which by polar coordinate introduction is seen to be bounded by a constant multiple of ε. Thus

$$\lim_{\varepsilon \to 0} (D_i H_\varepsilon)(X) = K(X)$$

uniformly. This proves $K = D_i H$, so we have proved that F and therefore T is of class C^{k+1}.

For $n = 2$ we can take $S = (2\pi)^{-1} \log r$, which satisfies $L_{\mathfrak{p}} S = \delta$ (cf., Lemma 2.35 in Chapter I) and proceed as before, replacing $\phi_\varepsilon(Y)$ by the C^1-function

$$\psi_\varepsilon(Y) = \begin{cases} \log|Y| & \text{for} \quad |Y| \ge \varepsilon \\ \log \varepsilon - 1 + \varepsilon^{-1}|Y| & \text{for} \quad |Y| \le \varepsilon. \end{cases}$$

We can now prove Theorem 5.8. Let $f \in \mathcal{D}_W(\mathfrak{a})$. Since K-conjugate elements in $\mathfrak{a}$ are W-conjugate ([DS], Chapter VII, Proposition 2.2), f can be extended to a K-invariant function $\tilde{f}$ on $\mathfrak{p}$. Since the distance between two K-orbits is minimized on $\mathfrak{a}$ (Chapter I, Proposition 5.18), it follows that $\tilde{f}$ is continuous.

Let $\mathfrak{a}' \subset \mathfrak{a}$ be the set of regular elements and let $\Delta(L_{\mathfrak{p}})$ be the radial part of $L_{\mathfrak{p}}$ for K acting on $\mathfrak{p}$ with transversal manifold $\mathfrak{a}'$. According to Proposition 3.13, we have

$$\Delta(L_{\mathfrak{p}})f = L_{\mathfrak{a}} f + \tfrac{1}{2} \sum_{\alpha \in \Sigma} m_\alpha \alpha^{-1} A_\alpha f \tag{25}$$

on $\mathfrak{a}^+$; but both sides are W-invariant, so (25) actually holds on $\mathfrak{a}'$ Since the function f is invariant under the symmetry s_α, the function $g = A_\alpha f$ satisfies $g^{s_\alpha} = -g$, so the function $\alpha^{-1}(A_\alpha f)$ extends over the hyperplane $\alpha = 0$ to a smooth function on $\mathfrak{a}$. Thus $\Delta(L_{\mathfrak{p}})f$ extends from $\mathfrak{a}$ to a member of $\mathcal{D}_W(\mathfrak{a})$, which we denote by the same symbol. We shall now prove that

$$L_{\mathfrak{p}}(\tilde{f}) = (\Delta(L_{\mathfrak{p}})f)^{\sim} \tag{26}$$

in the sense of distributions. Assuming this formula, we obtain by iteration, $L_{\mathfrak{p}}^m(\tilde{f}) = (\Delta(L_{\mathfrak{p}})^m f)^{\sim}$. But since the right-hand side is continuous, successive use of Lemma 5.10 gives $\tilde{f} \in C^{k+1}(\mathfrak{p})$, $k = 0, 1, \ldots$, proving the theorem.

Since both sides of (26) are K-invariant, it suffices to prove that both sides give the same result when applied to any function $\phi \in \mathscr{D}_K(\mathfrak{p})$. Now using Theorem 5.17 in Chapter I, and (49) in §3, we have, bar denoting restriction to $\mathfrak{a}$,

$$\begin{aligned} L_{\mathfrak{p}}(\tilde{f})(\phi) &= \int_{\mathfrak{p}} \tilde{f}(X)(L_{\mathfrak{p}}\phi)(X)\,dX \\ &= \int_{\mathfrak{a}^+} f(H)(\Delta(L_{\mathfrak{p}})\bar{\phi})(H)\delta(H)\,dH \\ &= \int_{\mathfrak{a}^+} f L_{\mathfrak{a}}(\bar{\phi})\delta\,dH + \int_{\mathfrak{a}^+} f(\operatorname{grad}\delta)(\bar{\phi})\,dH. \end{aligned}$$

Now we have the divergence theorem (cf. (26) in Chapter I, §2)

$$\int_D (\operatorname{div} F)(X)\,dX = \int_{\partial D} \langle F, \boldsymbol{n}\rangle\,d\sigma$$

for a vector field F on $\boldsymbol{R}^n$, $D \subset \boldsymbol{R}^n$ a connected bounded open set whose smooth boundary ∂D has a well-defined outgoing unit normal $\boldsymbol{n}$, dX and $d\sigma$ being the Euclidean measures. Using this on the vector field $F = \psi \operatorname{grad} \zeta$, we obtain Green's first formula:

$$\int_D (\psi L\zeta + \langle \operatorname{grad}\psi, \operatorname{grad}\zeta\rangle)\,dX = \int_{\partial D} \psi\langle \operatorname{grad}\zeta, \boldsymbol{n}\rangle\,d\sigma.$$

By putting $\psi = f\delta$, $\zeta = \bar{\phi}$, the first integral on the right above becomes

$$\begin{aligned} \int_{\mathfrak{a}_\varepsilon^+} f\delta L_{\mathfrak{a}}(\bar{\phi})\,dH = &-\int_{\mathfrak{a}_\varepsilon^+} \langle \operatorname{grad}(f\delta), \operatorname{grad}\bar{\phi}\rangle\,dH \\ &+ \int_{\partial\mathfrak{a}_\varepsilon^+} f\delta\langle \operatorname{grad}\bar{\phi}, \boldsymbol{n}\rangle\,d\sigma. \end{aligned}$$

Here $\mathfrak{a}_\varepsilon^+$ is a region approximating $\mathfrak{a}^+ \cap B_R(0)$, where $B_R(0)$ is a ball in $\mathfrak{a}$ containing $\operatorname{supp}(\bar{\phi})$. Since $\delta = 0$ on $\partial\mathfrak{a}^+$ and $\phi \equiv 0$ in a neighborhood of $S_R(0)$, the integral over $\partial\mathfrak{a}_\varepsilon^+$ tends to 0 as $\varepsilon \to 0$ and we get, using $\operatorname{grad}(f\delta) = \delta \operatorname{grad} f + f \operatorname{grad}\delta$,

$$L_{\mathfrak{p}}(\tilde{f})(\phi) = -\int_{\mathfrak{a}^+} \langle \operatorname{grad} f, \operatorname{grad}\bar{\phi}\rangle\delta\,dH.$$

On the other hand,

$$(\Delta(L_{\mathfrak{p}})f)^{\sim}(\phi) = \int_{\mathfrak{a}^+} (\Delta(L_{\mathfrak{p}})f)(H)\phi(H)\delta(H)\,dH$$

$$= \int_{\mathfrak{a}^+} \bar{\phi} L_{\mathfrak{a}}(f)\delta\,dH + \int_{\mathfrak{a}^+} \bar{\phi}(\operatorname{grad}(\delta))(f)\,dH.$$

We can now repeat the Green's formula application above with f and $\bar{\phi}$ interchanged. Then we end up again with the integral

$$-\int_{\mathfrak{a}^+} \langle \operatorname{grad} \bar{\phi}, \operatorname{grad} f\rangle \delta\,dH.$$

This proves (26) and the theorem.

Corollary 5.11. (i) *The restriction to* $\mathfrak{a}$ *induces an isomorphism of* $\mathscr{E}_K(\mathfrak{p})$ *onto* $\mathscr{E}_W(\mathfrak{a})$.

(ii) *The restriction to* $\mathfrak{a}$ *induces an isomorphism of* $\mathscr{S}_K(\mathfrak{p})$ *onto* $\mathscr{S}_W(\mathfrak{a})$.

The first statement is immediate from Theorem 5.8. For the second statement we first recall (Chapter I, Lemma 2.2) that $F \in \mathscr{S}(\mathbf{R}^n)$ if and only if for all integers $k,\ l \geq 0$

$$\sup_x |(1 + |x|^l)(L^k f)(x)| < \infty, \tag{27}$$

L being the Laplacian. Let $f \in \mathscr{S}_W(\mathfrak{a})$ and let $\tilde{f} \in \mathscr{E}_K(\mathfrak{p})$ be its extension given by part (i). Then we have on $\mathfrak{a}'$

$$L_{\mathfrak{p}}\tilde{f} = L_{\mathfrak{a}} f + \tfrac{1}{2}\sum_{\alpha\in\Sigma} m_\alpha \alpha^{-1} A_\alpha f,$$

but as pointed out above, the right-hand side is smooth on all of $\mathfrak{a}$ and is, in fact, a function $f_1 \in \mathscr{S}_W(\mathfrak{a})$. But then $L_{\mathfrak{p}}\tilde{f}$ is the K-invariant extension of this function to $\mathfrak{p}$. It clearly satisfies (27) (for $k = 1$). Repeating this argument with f replaced by f_1 we obtain (27) for all k, so $\tilde{f} \in \mathscr{S}_K(\mathfrak{p})$, as claimed.

Corollary 5.12. *Every* W*-invariant polynomial* P *on* $\mathfrak{a}$ *can be uniquely extended to a* K*-invariant polynomial on* $\mathfrak{p}$.

In fact, we can assume that P is homogeneous. Then $\tilde{P}$ as given by Corollary 5.11 is a homogeneous C^∞-function, hence a polynomial.

Consider now the bijection $\lambda\colon S(\mathfrak{g}) \to \mathbf{D}(G)$ from Theorem 4.3. It identifies the commutative algebras $S(\mathfrak{a})$ and $\mathbf{D}(A)$ and identifies the set $I(\mathfrak{a})$ of W-invariants in $S(\mathfrak{a})$ with the set $\mathbf{D}_W(A)$ of W-invariant differential operators on $A \cdot o$ with constant coefficients.

We recall now the projection $D \to D'$ (§3, No. 1) mapping differential operators on X into differential operators on $A \cdot o$. From Theorem 3.2 we know that $L'_X = L_A$. We shall now extend this result to other invariant differential operators.

Theorem 5.13. *The projection $D \to D'$ is a bijection of $\mathbf{D}(G/K)$ onto $\mathbf{D}_W(A)$.*

Proof. Suppose $g \in G$ maps $A \cdot o$ into itself. Then for some $a \in A$, $g \cdot o = a \cdot o$. Then $a^{-1}g \in K$ and $\operatorname{Ad}(a^{-1}g)$ maps the tangent space $(A \cdot o)_o$ into itself. Hence $a^{-1}g \in M'$, the normalizer of $\mathfrak{a}$ in K. This shows that $M'A$ is the subgroup of G which leaves $A \cdot o$ invariant. Thus, if $D \in \mathbf{D}(G/K)$, then $D' \in \mathbf{D}_W(A)$.

For the surjectivity, let $\mathfrak{q}$ be the orthogonal complement of $\mathfrak{a}$ in $\mathfrak{p}$, so

$$\mathfrak{g} = \mathfrak{k} + \mathfrak{a} + \mathfrak{q}. \tag{27}$$

Let

$$T_1, \ldots, T_p, H_1, \ldots, H_l, X_1, \ldots, X_q \tag{28}$$

be a basis of $\mathfrak{g}$ compatible with the direct decomposition (27) and orthonormal with respect to $-B(X, \theta Y)$. Let $I(\mathfrak{p})$ denote the set of $\operatorname{Ad}_G(K)$-invariants in $S(\mathfrak{p})$ and let $P \in I(\mathfrak{p})$ be homogeneous of degree m. Writing

$$N = (n_1, \ldots, n_l), \qquad |N| = n_1 + \cdots + n_l,$$

$$M = (m_1, \ldots, m_q), \qquad |M| = m_1 + \cdots + m_q,$$

we have

$$P = \sum_{|N| + |M| = m} a_{N,M} H_1^{n_1} \cdots H_l^{n_l} X_1^{m_1} \cdots X_q^{m_q}$$

$$= P_{\mathfrak{a}} + \sum_{|M| > 0,\, N} a_{N,M} H_1^{n_1} \cdots H_l^{n_l} X_1^{m_1} \cdots X_q^{m_q},$$

where $P_{\mathfrak{a}} \in S(\mathfrak{a})$. Letting M' act on this formula and noting that it leaves $\mathfrak{a}$ and $\mathfrak{q}$ invariant, we deduce that $P_{\mathfrak{a}}$ is W-invariant and of degree m. Writing $\tilde{Z} = \lambda(Z)$ for $Z \in \mathfrak{g}$, we have

$$\lambda(P) = \lambda(P_{\mathfrak{a}}) + \sum_{|M| > 0,\, N} a_{N,M} \tilde{H}_1^{n_1} \cdots \tilde{H}_l^{n_l} \tilde{X}_1^{m_1} \cdots \tilde{X}_q^{m_q} + Q, \tag{29}$$

where Q has order $< m$ and $|N| + |M| = m$.

By Theorem 1.4 in [DS] (Chapter VI), we have the bijective decomposition $G = \exp \mathfrak{a} (\exp \mathfrak{q}) K$. For $Q \in T(\mathfrak{p})$ we put $D_Q = \mu(\lambda(Q))$, where $\mu: \mathbf{D}_K(G) \to \mathbf{D}(G/K)$ is the homomorphism from Theorem 4.6. If $F \in \mathscr{E}(A \cdot o)$ and f on G is determined by

$$F(a \cdot o) = f(a \exp Xk), \qquad X \in \mathfrak{q}, \quad k \in K,$$

we have

$$(30) \qquad (D'_P F)(a \cdot o) = (\lambda(P)f)(a).$$

But if $a(h) = a \exp(h_1 H_1 + \cdots + h_l H_l)$, we have

$$(\tilde{H}_1^{n_1} \cdots \tilde{H}_l^{n_l} \tilde{X}_1^{m_1} \cdots \tilde{X}_q^{m_q} f)(a)$$
$$= \left\{ \frac{\partial^{|N|}}{\partial h_1^{n_1} \cdots \partial h_l^{n_l}} (\tilde{X}_1^{m_1} \cdots \tilde{X}_q^{m_q} f)(a(h)) \right\}_{h_i = 0}$$

and

$$\tilde{X}_1^{m_1} \cdots \tilde{X}_q^{m_q} = \lambda(X_1^{m_1} \cdots X_q^{m_q}) + T,$$

where $T \in \boldsymbol{D}(G)$ has order $< |M|$. But by the definition of λ [(4) in §4] we have

$$(\lambda(X_1^{m_1} \cdots X_q^{m_q})f)(a(h)) = 0 \qquad \text{if } |M| > 0.$$

Thus we conclude from (29) and (30) that for a certain $R \in \boldsymbol{D}(G)$ of order $< m$

$$(D'_P F)(a \cdot o) = (P_{\mathfrak{a}} F)(a \cdot o) + (Rf)(a),$$

for all $a \in A$ and all $F \in \mathscr{E}(A \cdot o)$. If the differential operator R is expressed in terms of the coordinate system

$$\exp(h_1 H_1 + \cdots + h_l H_l) \exp(x_1 X_1 + \cdots + x_q X_q) \exp(t_1 T_1 + \cdots + t_p T_p)$$
$$\to (h_1, \ldots, h_l, x_1, \ldots, x_q, t_1, \ldots, t_p),$$

it becomes obvious, since f in these coordinates is independent of (x_i) and (t_j), that the mapping

$$F \to (Rf)|A$$

is a differential operator on $A \cdot o$ of order less than or equal to that of R. Hence

$$(31) \qquad \operatorname{order}(D'_P - P_{\mathfrak{a}}) < m.$$

Suppose now $Q \in I(\mathfrak{a})$, the set of W-invariants in $S(\mathfrak{a})$. We wish to find $D \in \boldsymbol{D}(G/K)$ such that $D' = Q = \lambda(Q)$. For this we may assume that Q is homogeneous. Let $m = \deg(Q)$. By Corollary 5.12 we can find $P \in I(\mathfrak{p})$ such that $P_{\mathfrak{a}} = Q$. But then by (31) $\operatorname{order}(D'_P - Q) < m$. But $D'_P - Q \in I(\mathfrak{a})$, so the desired statement follows by induction on m.

Finally, suppose a $D \neq 0$ in $\boldsymbol{D}(G/K)$ of order m such that $D' = 0$. Let $P \in I(\mathfrak{p})$ be the homogeneous polynomial of degree m such that

$$\operatorname{order}(D - D_P) < m.$$

Then $D'_P = (D_P - D)'$ has order $<m$, whereas $P_{\mathfrak{a}}$ has degree m. This contradicts (31), so the theorem is proved.

While the projection $D \to D'$ is a linear bijection between the commutative algebras $\mathbf{D}(G/K)$ and $\mathbf{D}_W(A)$, it does not preserve multiplication. This will now be remedied by replacing the decomposition (27) by the Iwasawa decomposition

$$\mathfrak{g} = \mathfrak{k} + \mathfrak{a} + \mathfrak{n},$$

which has the advantage over (27) that $\mathfrak{n}$ is an algebra, whereas $\mathfrak{q}$ is not. Since the decomposition $G/K = \exp \mathfrak{n} \exp \mathfrak{a} \cdot o$, which leads to the N-radial part $\Delta(D)$ for $D \in \mathbf{D}(G/K)$, is rather similar to the above decomposition $G/K = \exp \mathfrak{a} \exp \mathfrak{q} \cdot o$, which led to the projection $D \to D'$, one might expect an analog of Theorem 5.13 for the mapping $D \to \Delta(D)$. With a small but important modification this is indeed true (cf. Corollary 5.19). The proof of Theorem 5.13, however, does not generalize because while the decomposition $G/K = \bigcup_{a \in A} a \exp \mathfrak{q} \cdot o$ is preserved by the group $M'A$, the decomposition $G/K = \bigcup_{a \in A} Na \cdot o$ is only preserved by the smaller group MA.

Lemma 5.14. *For each $D \in \mathbf{D}(G)$ there exists a unique element $D_{\mathfrak{a}} \in \mathbf{D}(A)$ such that*

(32) $$D - D_{\mathfrak{a}} \subset \mathfrak{n}\mathbf{D}(G) + \mathbf{D}(G)\mathfrak{k}.$$

Moreover,

(33) $$(D\phi)^{-} = D_{\mathfrak{a}}\bar{\phi}$$

whenever $\phi \in \mathscr{E}(G)$ satisfies $\phi(ngk) = \phi(g)$ $(n \in N, g \in G, k \in K)$, the bar denoting restriction to A.

Proof. For (32) we choose a basis $X_1, \ldots, X_n$ of $\mathfrak{g}$ compatible with the decomposition $\mathfrak{g} = \mathfrak{n} + \mathfrak{a} + \mathfrak{k}$. Because of Corollary 1.10, Chapter II, in [DS], each $D \in \mathbf{D}(G)$ can be expressed uniquely:

(34) $$D = \sum a_{e_1 \cdots e_n} \tilde{X}_1^{e_1} \cdots \tilde{X}_n^{e_n}.$$

Let $D_{\mathfrak{a}}$ be the sum of terms which only involve basis elements of $\mathfrak{a}$. The uniqueness amounts to

$$\mathbf{D}(A) \cap (\mathfrak{n}\mathbf{D}(G) + \mathbf{D}(G)\mathfrak{k}) = \{0\},$$

and this follows from the uniqueness of the representation (34) combined with the fact that $\mathfrak{n}$ and $\mathfrak{k}$ are subalgebras. For (33) one has to prove

$$(DT\phi)(a) = (XD\phi)(a) = 0$$

for $D \in \boldsymbol{D}(G)$, $T \in \mathfrak{k}$, $X \in \mathfrak{n}$, $a \in A$. But $T\phi \equiv 0$, so that $DT\phi \equiv 0$; second,

$$(XD\phi)(a) = \lim_{t \to 0} \frac{1}{t} [(D\phi)(a \exp tX) - (D\phi)(a)].$$

This expression vanishes, since the element $n_t = a \exp tXa^{-1}$ belongs to N, whence $(D\phi)(a \exp tX) = (D\phi)^{L(n_t^{-1})}(a) = (D\phi)(a)$.

Remark. $D_{\mathfrak{a}}$ is just the radial part of D for the action $g \to ngk$ of the product group $N \times K$ on G with transversal manifold $A \subset G$.

For $g \in G$ let $A(g) \in \mathfrak{a}$ denote the unique element such that

$$g \in N \exp A(g)\, K.$$

Since $\boldsymbol{D}(A)$ is a commutative polynomial ring, each linear mapping $\nu: \mathfrak{a} \to \boldsymbol{C}$ extends uniquely to a homomorphism of $\boldsymbol{D}(A)$ into $\boldsymbol{C}$, denoted $D \to D(\nu)$. Let $\mathfrak{a}^*$ denoted the dual of $\mathfrak{a}$ and $\mathfrak{a}_c^*$ the set of all linear functions of $\mathfrak{a}$ into $\boldsymbol{C}$.

Lemma 5.15. *For each linear function $\nu: \mathfrak{a} \to \boldsymbol{C}$ the function*

$$\phi(g) = \int_K e^{\nu(A(kg))}\, dk$$

is an eigenfunction of each $D \in \boldsymbol{D}_K(G)$. More precisely,

$$D\phi = D_{\mathfrak{a}}(\nu)\phi, \qquad D \in \boldsymbol{D}_K(G).$$

Proof. Let $F(g) = e^{\nu(A(g))}$, $D \in \boldsymbol{D}_K(G)$. Then $F(ngk) \equiv F(g)$, so by (33)

$$(DF)(a) = (D_{\mathfrak{a}}\bar{F})(a) = D_{\mathfrak{a}}(\nu)F(a). \tag{35}$$

The functions DF and $D_{\mathfrak{a}}(\nu)F$ are both left invariant under N, right invariant under K, and by (35) they coincide on A. Hence $DF \equiv D_{\mathfrak{a}}(\nu)F$, and

$$D_g\left(\int_K F(kg)\, dk\right) = \int_K (DF)(kg)\, dk = D_{\mathfrak{a}}(\nu) \int_K F(kg)\, dk,$$

which proves the lemma.

Theorem 5.16. *With $\rho = \frac{1}{2} \sum_{\alpha \in \Sigma^+} m_\alpha \alpha$, let*

$$\phi_\nu(g) = \int_K e^{(i\nu + \rho)(A(kg))}\, dk, \qquad \nu \in \mathfrak{a}_c^*.$$

Then

$$\phi_{s\nu} \equiv \phi_\nu \quad \textit{for} \quad \textit{each } s \in W.$$

Here $s\nu(H) = \nu(s^{-1}H)$, $H \in \mathfrak{a}$, as usual.

Proof. It suffices to prove

$$\int_G \phi_{sv}(g) f(g)\, dg = \int_G \phi_v(g) f(g)\, dg \tag{36}$$

for each $f \in C_c(G)$; since ϕ_{sv} and ϕ_v are both bi-invariant under K, it suffices to prove (36) for f bi-invariant under K. Under the decomposition $g = nak$ we have by Chapter I, Corollaries 5.2 and 5.3,

$$dg = e^{-2\rho(\log a)}\, dn\, da\, dk,$$

so that for $f \in C_c^\natural(G)$

$$\begin{aligned}\int_G \phi_v(g) f(g)\, dg &= \int_K dk \int_G e^{(iv+\rho)(A(kg))} f(g)\, dg \\ &= \int_G e^{(iv+\rho)(A(g))} f(g)\, dg \\ &= \int_A da \int_N e^{(iv-\rho)(\log a)} f(na)\, dn,\end{aligned}$$

so in the notation of Theorem 5.7. Chapter I,

$$\int_G \phi_v(g) f(g)\, dg = \int_A e^{iv(\log a)} F_f(a)\, da, \qquad f \in C_c^\natural(G). \tag{37}$$

The result now follows from the functional equations $F_f(a^s) = F_f(a)$ in the theorem cited.

Theorem 5.17. *The mapping*

$$\gamma : D \to e^{-\rho} D_{\mathfrak{a}} \circ e^{\rho}$$

is a homomorphism of $\boldsymbol{D}_K(G)$ *onto* $\boldsymbol{D}_W(A)$ *with kernel* $\boldsymbol{D}_K(G) \cap \boldsymbol{D}(G)\mathfrak{k}$.

Proof. We first remark that under the canonical identification of $\boldsymbol{D}(A)$ and $S(\mathfrak{a})$ the automorphism $D \to e^{-\rho} D \circ e^{\rho}$ of $\boldsymbol{D}(A)$ corresponds to the automorphism $p \to {}'p$ of $S(\mathfrak{a})$ given by ${}'H = H + \rho(H)$. In fact,

$$e^{-\rho} H(e^{\rho} f) = (H + \rho(H)) f.$$

To see that γ is a homomorphism let $D', D'' \in \boldsymbol{D}_K(G)$. Then

$$D'D'' - D'_{\mathfrak{a}} D''_{\mathfrak{a}} = D'_{\mathfrak{a}}(D'' - D''_{\mathfrak{a}}) + (D' - D'_{\mathfrak{a}}) D'',$$

where $D' - D'_{\mathfrak{a}}$ and $D'' - D''_{\mathfrak{a}}$ belong to $\mathfrak{n}\boldsymbol{D}(G) + D(G)\mathfrak{k}$. Since $TD'' = D''T$ for $T \in \mathfrak{k}$, we have

$$(D' - D'_{\mathfrak{a}}) D'' \in \mathfrak{n}\boldsymbol{D}(G) + \boldsymbol{D}(G)\mathfrak{k},$$

and since $[\mathfrak{n}, \mathfrak{a}] \subset \mathfrak{n}$,

$$D'_{\mathfrak{a}}(D'' - D''_{\mathfrak{a}}) \in \mathfrak{n}\boldsymbol{D}(G) + \boldsymbol{D}(G)\mathfrak{k}.$$

Hence $(D'D'')_{\mathfrak{a}} = D'_{\mathfrak{a}}D''_{\mathfrak{a}}$, and it follows that γ is a homomorphism.

Next we prove that for each nonconstant $p \in I(\mathfrak{p})$

$$\text{degree}(\gamma(\lambda(p)) - \bar{p}) < \text{degree}(p). \tag{38}$$

Here $q \to \bar{q}$ denotes the mapping of $S(\mathfrak{p})$ onto $S(\mathfrak{a})$ given by the decomposition $S(\mathfrak{p}) = S(\mathfrak{a}) + S(\mathfrak{p})\mathfrak{q}$. [If $S(\mathfrak{p})$ and $S(\mathfrak{p}^*)$ are identified by means of the Killing form, $\bar{q}$ means the restriction $q|\mathfrak{a}$.] Let d denote the degree of p. In proving (38) we may assume that p is homogeneous. Writing any element $X \in \mathfrak{q}$ in the form

$$X = Z - \theta Z = 2Z - (Z + \theta Z),$$

where $Z \in \mathfrak{n}$, it is clear that $p - \bar{p} \in \mathfrak{n}S^{d-1}(\mathfrak{g}) + S^{d-1}(\mathfrak{g})\mathfrak{k}$, where $S^e(\mathfrak{g})$ denotes the set of homogeneous elements in $S(\mathfrak{g})$ of degree e. Now $\lambda(\bar{p}) = \bar{p}$ and

$$\lambda(q_1 q_2) - \lambda(q_1)\lambda(q_2) \in \sum_{e < d_1 + d_2} \lambda(S^e(\mathfrak{g}))$$

if $q_1 \in S^{d_1}(\mathfrak{g})$, $q_2 \in S^{d_2}(\mathfrak{g})$. It follows that

$$\lambda(p) - \bar{p} \in \mathfrak{n}\boldsymbol{D}(G) + \boldsymbol{D}(G)\mathfrak{k} + \sum_{e<d} \lambda(S^e(\mathfrak{g})),$$

which implies (38).

To see that γ has kernel equal to $\boldsymbol{D}_K(G) \cap \boldsymbol{D}(G)\mathfrak{k}$, first let $D \in \boldsymbol{D}_K(G) \cap \boldsymbol{D}(G)\mathfrak{k}$. Since the function ϕ in Lemma 5.15 is K-invariant, we have $D\phi = 0$, whence $D_{\mathfrak{a}}(\nu) = 0$ for all ν, so $\gamma(D) = 0$. On the other hand, let $D \in \boldsymbol{D}_K(G)$ be such that $\gamma(D) = 0$. By Corollary 4.8 we have

$$D = \lambda(p) + D',$$

where $D' \in \boldsymbol{D}_K(G) \cap \boldsymbol{D}(G)\mathfrak{k}$ and $p \in I(\mathfrak{p})$. Thus $\gamma(\lambda(p)) = 0$. By decomposing p into homogeneous components if is clear that $\bar{p}$ has the same degree as p. Thus by (38) $p = 0$, so $D \in \boldsymbol{D}_K(G) \cap \boldsymbol{D}(G)\mathfrak{k}$.

Finally, we prove $\gamma(\boldsymbol{D}_K(G)) = \boldsymbol{D}_W(A)$. Let $D \in \boldsymbol{D}_K(G)$ and consider the function ϕ_ν from Theorem 5.16. We have

$$D\phi_\nu = D_{\mathfrak{a}}(i\nu + \rho)\phi_\nu = \gamma(D)(i\nu)\phi_\nu, \tag{38'}$$

and now Theorem 5.16 implies $\gamma(D) \in I(\mathfrak{a})$. For the surjectivity, let q be an arbitrary homogeneous polynomial in $I(\mathfrak{a})$ and let d denote its degree. We prove by induction on d that there exists a $D \in \boldsymbol{D}_K(G)$ such that $\gamma(D) = q$. This is obvious if $d = 0$, so suppose $d > 0$. Let $p \in I(\mathfrak{p})$ satisfy $\bar{p} = q$ (Corollary 5.12). By (38) the polynomial $\gamma(\lambda(p)) - \bar{p}$ has degree

$<d$ and, as we just proved, it belongs to $I(\mathfrak{a})$. By the inductive assumption there exists a $D_1 \in \mathbf{D}_K(G)$ such that $\gamma(D_1) = \gamma(\lambda(p)) - \bar{p}$. But then the operator $D = \lambda(p) - D_1$ belongs to $\mathbf{D}_K(G)$ and satisfies $\gamma(D) = q$.

Combining the maps μ from Theorem 4.6 and the map γ above, we obtain the following result.

Theorem 5.18. *The mapping*

$$\Gamma: \mathbf{D}(G/K) \to \mathbf{D}_W(A)$$

given by $\Gamma(\mu(D)) = \gamma(D)$ *for* $D \in \mathbf{D}_K(G)$ *is a surjective isomorphism.*

Corollary 5.19. *Under the action of* N *on* G/K *with transversal manifold* $A \cdot o$, *let* $\Delta_N(D)$ *denote the radial part of* $D \in \mathbf{D}(G/K)$. *Then* Γ *is given by*

$$\Gamma(D) = e^{-\rho}\Delta_N(D) \circ e^{\rho}$$

and is an isomorphism of $\mathbf{D}(G/K)$ *onto* $\mathbf{D}_W(A)$.

In fact, let $E \in \mathbf{D}_K(G)$ be such that $\mu(E) = D$. Then $E\tilde{f} = (Df)^{\sim}$ for $f \in \mathscr{E}(G/K)$. By taking f N-invariant, this implies $E_{\mathfrak{a}}\bar{f} = \Delta_N(D)\bar{f}$ where $\bar{f} = f \,|\, A \cdot o$. Thus $\Gamma(D) = \Gamma(\mu(E)) = \gamma(E) = e^{-\rho}\, E_{\mathfrak{a}} e^{\rho} = e^{-\rho}\Delta_N(D) \circ e^{\rho}$.

Corollary 5.20. *For the Laplacian* L_X *on* $X = G/K$ *we have*

$$\Gamma(L_X) = L_A - \langle \rho, \rho \rangle.$$

In fact, this is immediate if we compare Proposition 3.8 and Corollary 5.19.

Lemma 5.21. *For each* $D \in \mathbf{D}(G/K)$ *we have*

$$\Gamma(D^s) = \theta\Gamma(D),$$

where s *denotes the symmetry* $gK \to \theta(g)K$.

Proof. For $f \in \mathscr{D}^{\natural}(G)$ consider the function

$$F^*(g) = e^{-\rho(A(g))} \int_N f(ng)\, dn$$

and let $E \in \mathbf{D}_K(G)$ satisfy $\mu(E) = D$. Then

$$E_g(e^{\rho(A(g))}F^*(g)) = \int_N (Ef)(ng)\, dn,$$

the subscript in E_g denoting differentiation with respect to g. The function $g \to e^{\rho(A(g))}F^*(g)$ is left invariant under N, right invariant under K. Note also that by Corollary 5.2, in Chapter I, $F^*(a) = F_f(a)$ for each $a \in$

A. Using now Lemma 5.14 on the function $\phi(g) = e^{\rho(A(g))}F^*(g)$, we obtain the basic formula

$$\gamma(E)F_f = F_{Ef}. \tag{39}$$

From (37) we have for the adjoint E^* of E, and the adjoint $\gamma(E)^*$ of $\gamma(E)$ on A:

$$\begin{aligned}\int_G (E^*\phi_\nu)(g)f(g)\,dg &= \int_G \phi_\nu(g)(Ef)(g)\,dg\\ &= \int_A e^{i\nu(\log a)}F_{Ef}(a)\,da\\ &= \int_A e^{i\nu(\log a)}(\gamma(E)F_f)(a)\,da\\ &= \int_A \gamma(E)^*_a(e^{i\nu(\log a)})F_f(a)\,da\end{aligned}$$

so

$$\gamma(E^*)(i\nu)\int_G \phi_\nu(g)f(g)\,dg = \gamma(E)^*(i\nu)\int \phi_\nu(g)f(g)\,dg.$$

It follows that $\gamma(E^*) = \gamma(E)^*$. But $\gamma(E)^* = \theta\Gamma(D)$ and $\gamma(E^*) = \Gamma(\mu(E^*)) = \Gamma(D^*) = \Gamma(D^s)$ (Cor. 5.3), so the lemma is proved.

Corollary 5.22. *Let $D \in \mathbf{D}(G/K)$. Then the projection D' on $A \cdot o$ and the N-radial part $\Delta_N(D)$ have the same highest order terms, i.e.,*

$$\operatorname{order}(D' - \Delta_N(D)) < \operatorname{order}(D).$$

We may assume $D = D_{\lambda(P)}$ in the sense of Theorem 4.9, where $P \in I(\mathfrak{p})$ is homogeneous. By (31) and (38),

$$\operatorname{order}\,(\Gamma(D) - D') < \operatorname{order} D, \tag{40}$$

so our statement follows from Corollary 5.19. Corollary 5.22 is, of course, very plausible because, if $a \in A$, then the manifolds $Na \cdot o$ and $a \exp \mathfrak{q} \cdot o$ have the same tangent space at $a \cdot o$. If $a \in A^+$, this is also true for the orbit $Ka \cdot o$, so that one would expect an analog to Corollary 5.22 for the K-radial part $\Delta_K(D)$. We shall now prove this in a more precise form.

Let $\alpha_1, \ldots, \alpha_l$ be the system of simple roots in Σ^+ ([DS], Chapter VII, §2) and Λ the set of all linear combinations $n_1\alpha_1 + \cdots + n_l\alpha_l$ ($n_i \in \mathbf{Z}^+$).

Let $\mathscr{R}$ denote the ring of functions on $A^+ \cdot o$ which can be expanded in an absolutely convergent series on $A^+ \cdot o$ with zero constant term:

$$f(\exp H \cdot o) = \sum_{\mu \in \Lambda} a_\mu e^{-\mu(H)}, \qquad a_\mu \in \mathbf{C}, \quad a_0 = 0. \tag{41}$$

Note that f is a power series in the variables $x_i = e^{-\alpha_i(H)}$. The ring $\mathscr{R}$ is invariant under the action of each $D \in \mathbf{D}(A)$.

Proposition 5.23. *Let $D \in \mathbf{D}(G/K)$ be arbitrary and let q be its order. Then there exist functions $g_1, \ldots, g_k \in \mathscr{R}$ and operators $E_1, \ldots, E_k \in \mathbf{D}(A)$ of order $\leq q - 1$ such that the K-radial part $\Delta_K(D)$ satisfies*

$$\Delta_K(D) = e^{-\rho}\Gamma(D) \circ e^\rho + \sum_1^k g_i E_i.$$

Proof. By analogy with Lemma 5.14 we have the direct decomposition

$$\mathbf{D}(G) = \mathbf{D}(A) \oplus (\theta(\mathfrak{n})\mathbf{D}(G) + \mathbf{D}(G)\mathfrak{k}). \tag{42}$$

If $E \in \mathbf{D}^q(G)$ and if (X_i) is a basis of $\mathfrak{n}$ compatible with the decomposition $\mathfrak{n} = \sum_{\alpha \in \Sigma^+} \mathfrak{g}_\alpha$ of $\mathfrak{n}$ into root subspaces $\mathfrak{g}_\alpha$, we can select $D_i \in \mathbf{D}^{q-1}(G)$ such that

$$E - E_A - \sum_i (\theta X_i) D_i \in \mathbf{D}(G)\mathfrak{k}, \tag{43}$$

where E_A is the $\mathbf{D}(A)$-component of E according to (42). Let $X_i \in \mathfrak{g}_\alpha$, $X_i = Y_i + Z_i$ ($Y_i \in \mathfrak{p}$, $Z_i \in \mathfrak{k}$), and $h = \exp H$ in A^+. Then

$$\begin{aligned} \mathrm{Ad}(h^{-1}) Z_i &= \tfrac{1}{2}(\mathrm{Ad}(h^{-1}) X_i + \theta \,\mathrm{Ad}(h) X_i) \\ &= \tfrac{1}{2}(e^{-\alpha(H)}(Y_i + Z_i) + e^{\alpha(H)}(-Y_i + Z_i)), \end{aligned}$$

so

$$Y_i = \coth \alpha(H)\, Z_i - (\sinh \alpha(H))^{-1} \mathrm{Ad}(h^{-1}) Z_i.$$

Hence

$$\theta X_i \equiv g_\alpha(h) Z_i \qquad \mathrm{mod}(\mathrm{Ad}(h^{-1})\, \mathfrak{k}),$$

where $g_\alpha(h) = 1 - \coth \alpha(H)$, so $g_\alpha \in \mathscr{R}$. We substitute this into (43) and replace $Z_i D_i$ by $D_i Z_i + [Z_i, D_i]$. Here $D_i Z_i \in \mathbf{D}(G)\mathfrak{k}$ and $[Z_i, D_i] \in \mathbf{D}^{q-1}(G)$. Repeating this process with E replaced by $[Z_i, D_i]$, etc., we obtain by induction elements $g_i \in \mathscr{R}$, $E_i \in \mathbf{D}^{q-1}(A)$ such that

$$E - E_A - \sum_i g_i(h) E_i \in (\mathrm{Ad}(h^{-1})\, \mathfrak{k})\mathbf{D}(G) + \mathbf{D}(G)\mathfrak{k}. \tag{44}$$

In particular, we can take for E an operator in $\boldsymbol{D}_K(G)$ for which $\mu(\theta E) = D$ (cf. Theorem 4.6). Applying θ to (43) and comparing with (32) and Theorem 5.17, we see that $\Gamma(D) = \gamma(\theta E) = e^{-\rho}\,\theta E_A \circ e^{\rho}$. Taking adjoint on A, we see that $\Gamma(D)^* = e^{\rho}\,E_A \circ e^{-\rho}$, i.e.,

(45) $$E_A = e^{-\rho}\,\theta(\Gamma(D)) \circ e^{\rho}.$$

On the other hand, if

$$C \in (\mathrm{Ad}(h^{-1})\mathfrak{k})\boldsymbol{D}(G) + \boldsymbol{D}(G)\mathfrak{k},$$

and if $f \in \mathscr{D}^{\natural}(G/K)$, then

(46) $$(C\tilde{f})(h) = 0.$$

This is obvious if $C \in \boldsymbol{D}(G)\mathfrak{k}$. Also, if $T \in \mathfrak{k}$, $E_0 \in \boldsymbol{D}(G)$, we have

$$((\mathrm{Ad}(h^{-1})\,T)E_0\tilde{f})(h) = (T^{R(h)}E_0\tilde{f})(h) = (T(E_0\tilde{f})^{R(h^{-1})})(e) = 0.$$

Now $\mu(E) = \mu(\theta\theta E) = D^s$, so, by (44) and (46)

$$\begin{aligned}(\Delta_K(D^s)f)(h\cdot o) &= (D^s f)(h\cdot o) = (E\tilde{f})(h)\\ &= (E_A\tilde{f})(h) + \sum_i g_i(h)(E_i\tilde{f})(h).\end{aligned}$$

This proves that

$$\begin{aligned}\Delta_K(D^s) &= E_A + \sum_i g_i E_i\\ &- e^{-\rho}\,\theta(\Gamma(D)) \circ e^{\rho} + \sum_i g_i E_i,\end{aligned}$$

so by replacing D by D^s and using Lemma 5.21, Proposition 5.23 is proved.

3. The Polar Coordinate Form of the Laplacian

We shall now give an expression of the Laplacian L_X on X which corresponds to the polar decomposition $X' = K/M \times A^+ \cdot o$ (cf. [DS], Chapter IX, Cor. 1.2).

Let $\mathfrak{l}$ denote the orthogonal complement of $\mathfrak{m}$ in $\mathfrak{k}$ (with respect to the Killing form B of $\mathfrak{g}$). Diagonalizing the operators $(\mathrm{ad}\,H)^2$ $(H \in \mathfrak{a})$, we have

$$\mathfrak{l} = \sum_{\alpha \in \Sigma^+} \mathfrak{k}_\alpha,$$

where

$$\mathfrak{k}_\alpha = \{T \in \mathfrak{k} : (\mathrm{ad}\,H)^2 T = \alpha(H)^2 T \quad \text{for} \quad H \in \mathfrak{a}\}.$$

Since $X \to X + \theta X$ maps the root subspace $\mathfrak{g}_\alpha$ onto $\mathfrak{k}_\alpha$, we see that

$$\dim \mathfrak{k}_\alpha = m_\alpha,$$

the multiplicity of α. Let $T_1^\alpha, \ldots, T_{m_\alpha}^\alpha$ be a basis of $\mathfrak{k}_\alpha$, orthonormal with respect to $-B$, and put $\omega_\alpha = \sum_{i=1}^{m_\alpha} T_i^\alpha T_i^\alpha$, where the T_i^α are viewed as differential operators on G. Let $\tilde{f} = f \circ \pi$ for $f \in \mathscr{E}(X)$.

Theorem 5.24. *The Laplace–Beltrami operator L_X has the following polar coordinate decomposition:*

$$(L_X f)(ka \cdot o) = \left(L_A + \sum_{\alpha \in \Sigma^+} m_\alpha (\coth \alpha) A_\alpha\right)_a (f(ka \cdot o))$$
$$+ \sum_{\alpha \in \Sigma^+} \sinh^{-2}(\alpha(\log a))(\mathrm{Ad}(a^{-1})\, \omega_\alpha \tilde{f})(ka)$$

for $k \in K$, $a \in A^+$, $f \in \mathscr{E}(X)$, L_A being the Laplacian on $A \cdot o$.

Proof. We know from Theorem 3.5 and (12) §3 that

$$(L_X f)(a \cdot o) = (L_{Ka \cdot o} f^*)(a \cdot o) + (\Delta(L_X)\bar{f})(a \cdot o), \tag{47}$$

where $L_{Ka \cdot o}$ is the Laplace–Beltrami operator on the orbit $Ka \cdot o$, f^* the restriction of f to this orbit, and $\bar{f}$ the restriction of f to $A \cdot o$. Let $O = a \cdot o$ and let $\phi: k \to k \cdot O$ denote the natural mapping of K onto the orbit $K \cdot O$ (which can be identified with K/M).

Lemma 5.25. *Let $X_i^\alpha = (\sinh \alpha (\log a))^{-1} T_i^\alpha$. Then the vectors*

$$d\phi(X_i^\alpha), \qquad 1 \le i \le m_\alpha, \quad \alpha \in \Sigma^+,$$

form an orthonormal basis of the tangent space $(K \cdot O)_O$.

Proof. The differential $d\phi$ maps $\mathfrak{l}$ onto the tangent space, and since

$$\phi(\exp tT_i^\alpha) = a \exp \mathrm{Ad}(a^{-1}) tT_i^\alpha \cdot o,$$

we see that

$$d\phi(T_i^\alpha) = d\tau(a)(\mathrm{Ad}(a^{-1})\, T_i^\alpha)_{\mathfrak{p}},$$

where $\tau(a)$ is the translation $gK \to agK$ of G/K, $\mathfrak{p}$ is identified with $(G/K)_o$, and the subscript $\mathfrak{p}$ refers to the component for the decomposition $\mathfrak{g} = \mathfrak{k} + \mathfrak{p}$. But

$$(\mathrm{Ad}(a^{-1})\, T_i^\alpha)_{\mathfrak{p}} = \tfrac{1}{2}[\mathrm{Ad}(a^{-1})\, T_i^\alpha - \mathrm{Ad}(a)\, T_i^\alpha],$$

so

$$B(d\phi(T_i^\alpha), d\phi(T_j^\beta)) = \tfrac{1}{4}B(T_i^\alpha - \mathrm{Ad}(a^2)\, T_i^\alpha, T_j^\beta - \mathrm{Ad}(a^2)\, T_j^\beta). \tag{48}$$

Expanding this and using

$$\mathrm{Ad}(a^2)\,T_i^\alpha \equiv \cosh 2\alpha\,(\log a)\,T_i^\alpha \qquad (\mathrm{mod}\ \mathfrak{p}),$$

the result follows from (48).

To conclude the proof of Theorem 5.24, consider the differential operator D on the orbit $K \cdot O$ defined by

$$(49) \qquad (Df)(k \cdot O) = \left\{\sum_{i,\alpha} \partial_{i,\alpha}^2 f(k \exp(\textstyle\sum x_{i,\alpha} X_i^\alpha) \cdot O)\right\}_{x_{i,\alpha}=0},$$

where $\partial_{i,\alpha} = \partial/\partial x_{i,\alpha}$. This operator is well-defined; in fact, if $k \cdot O = k_1 \cdot O$, then $k_1 = km$ $(m \in M)$ and $\mathrm{Ad}(m)$ gives an orthogonal transformation of each $\mathfrak{k}_\alpha$. Moreover, D is invariant under the action of K.

For the local coordinates

$$\exp(\textstyle\sum x_{i,\alpha} X_i^\alpha) \cdot O \to x_{i,\alpha}$$

near O on the orbit $K \cdot O$ the tangent vectors $(\partial_{i,\alpha})_O$ are orthonormal. This follows from Lemma 5.25 since

$$(\partial_{i,\alpha})_O(f) = \left[\frac{d}{dt} f(\exp tX_i^\alpha \cdot O)\right]_{t=0} = d\phi(X_i^\alpha)f.$$

It follows that the Laplacian $L' = L_{K \cdot O}$ satisfies

$$(L'f)(O) = \sum_{i,\alpha} (\partial_{i,\alpha}^2 f)(O) + \text{lower-order terms}$$

so $D - L'$ has order <2 at O hence everywhere, by the K invariance.

Assume now $\mathrm{Ad}_K(M)$ leaves no vector $T \neq 0$ in $\mathfrak{l}$ fixed. Then by Theorem 4.9, K/M has no K-invariant differential operator of order 1 so by the above $D = L_{K \cdot O}$. Substituting the definition of X_i^α into the definition of D and putting $O = a \cdot o$, we obtain

$$(L_X f)(a \cdot o) = (\Delta(L_X)\bar{f})(a \cdot o) + \sum_{\alpha \in \Sigma^+} \sinh^{-2}[\alpha(\log a)][\omega_\alpha(\bar{f}^{R(a^{-1})})](e).$$

But

$$\begin{aligned}[\omega_\alpha(\bar{f}^{R(a^{-1})})](e) &= (\omega_\alpha^{R(a)} \bar{f})^{R(a^{-1})}(e)\\ &= (\mathrm{Ad}(a^{-1})\omega_\alpha \bar{f})(a),\end{aligned}$$

so using Proposition 3.9 for the radial part $\Delta(L_X)$ the theorem follows for $k = e$. It then follows for general k by replacing f by the function $x \to f(k \cdot x)$.

We must still deal with the possibility that $\mathrm{Ad}_K(M)$ has a fixed vector in $\mathfrak{l}$ (and this occurs for some nontrivial examples of X). Here the proof above gives Theorem 5.24 only up to a K-invariant vector field. We therefore apply the formula for the Casimir operator Ω in Exercise A8. Since the vector fields $\tilde{Z}_\alpha$ annihilate the functions on G that are right invariant under K and since $\mu(\Omega) = L_X$ (Exercise A4), we obtain

$$(L_X f)(ka \cdot o) = \Big(L_A + \sum_{\alpha \in \Sigma^+} m_\alpha (\coth \alpha) A_\alpha\Big)_a (f(ka \cdot o))$$
$$- 2 \sum_{\beta \in P_+} \sinh^{-2}(\beta(\log a))[\mathrm{Ad}(a^{-1})Z_\beta Z_{-\beta} \tilde{f}](ka).$$

Now it is easy to see that

$$2 \sum_{\bar{\beta} = \alpha} Z_\beta Z_{-\beta} = -\omega_\alpha,$$

where $\bar{\beta}$ denotes the restriction of β to $\mathfrak{a}$. This then completes the proof of Theorem 5.24.

4. The Laplace–Beltrami Operator for a Symmetric Space of Rank One

Let X be a Euclidean space or a Riemannian globally symmetric space of rank one. We shall derive a useful geometric expression for the Laplace–Beltrami operator L_X on X.

For $p \in X$ let, as usual, $S_r(p)$ denote the sphere in X with center p and radius r. If X is noncompact, $S_r(p)$ is the image of a sphere in the tangent space X_p under the diffeomorphism Exp_p. If X is compact, let L denote its *diameter*, which is, the maximum distance between any pair of points in X. If $r < L$, $S_r(p)$ is the image of a sphere in X_p under a diffeomorphism ([DS], Theorem 10.3, Chapter VII). This, of course, fails to hold for $S_L(p)$ (see, e.g., $X = \boldsymbol{S}^2$). However, $S_L(p)$ is a submanifold of X, being an orbit of a compact subgroup of the Lie group $I(X)$ ([DS], Chapter IV, §3).

In any case, $S_r(p)$ is for each r a submanifold of X and has a Riemannian structure induced by that of X. Let $A(r)$ denote the *area* of $S_r(p)$, that is, the total measure of $S_r(p)$ according to the Riemannian measure on $S_r(p)$. [If $\dim X = 1$, so $X = \boldsymbol{R}^1$ or $X = \boldsymbol{S}^1$, $A(r)$ is understood to be the number of points in $S_r(p)$.] By the homogeneity of X, $A(r)$ is independent of p.

Let $(\theta_1, \ldots, \theta_{n-1})$ be Cartesian coordinates on an open subset of the unit sphere $S_1(0)$ in X_p. The inverse of the mapping

(50) $$(\theta_1, \ldots, \theta_{n-1}, r) \to \mathrm{Exp}_p(r\theta_1, \ldots, r\theta_{n-1})$$

is called a system of *geodesic polar coordinates* at $p \in X$. Here r runs through the intervals $0 < r < \infty$ in the noncompact case, $0 < r < L$ in the compact case. If we fix r, the system $\{\theta_1, \ldots, \theta_{n-1}\}$ can be regarded as a system of coordinates on an open subset of $S_r(p)$.

Proposition 5.26. *In geodesic polar coordinates at p as described above, the Laplace–Beltrami operator L_X on X has the form*

$$L_X = \frac{\partial^2}{\partial r^2} + \frac{1}{A(r)} \frac{dA}{dr} \frac{\partial}{\partial r} + L_S,$$

where L_S is the Laplace–Beltrami operator on $S_r(p)$. Here $0 < r < \infty$ if X is noncompact, $0 < r < L$ if X is compact.

Remark. Compare with Theorems 3.5 and 3.7.

Proof. Put $\theta_n = r$ and let

$$ds^2 = \sum_{i,j=1}^{n} g_{ij}(\theta_1, \ldots, \theta_n)\, d\theta_i\, d\theta_j \tag{51}$$

denote the Riemannian structure of X expressed in geodesic polar coordinates at p. As before, let $\bar{g} = \det(g_{ij})$ and $(g^{ij}) = (g_{ij})^{-1}$ [the inverse of the matrix (g_{ij})]. As shown in [DS], Chapter I, the geodesics emanating from p intersect $S_r(p)$ orthogonally. Hence (51) reduces to

$$ds^2 = dr^2 + \sum_{i,j=1}^{n-1} g_{ij}(\theta_1, \ldots, \theta_{n-1}, r)\, d\theta_i\, d\theta_j. \tag{52}$$

If I denotes the identity mapping of $S_r(p)$ into X, we have $I^*(dr^2) = 0$, so that the Riemannian structure of $S_r(p)$ is given by

$$\sum_{i,j=1}^{n-1} g_{ij}(\theta_1, \ldots, \theta_{n-1}, r)\, d\theta_i\, d\theta_j. \tag{53}$$

Using (52) we obtain

$$L_X = \frac{\partial^2}{\partial r^2} + \frac{1}{\sqrt{\bar{g}}} \frac{\partial \sqrt{\bar{g}}}{\partial r} \frac{\partial}{\partial r} + \frac{1}{\sqrt{\bar{g}}} \sum_{i,j=1}^{n-1} \frac{\partial}{\partial \theta_j} \left(g^{ij} \sqrt{\bar{g}} \frac{\partial}{\partial \theta_i} \right). \tag{54}$$

If A denotes the matrix $(g_{ij})_{1 \le i,j \le n}$ and B denotes the matrix

$$(g_{ij})_{1 \le i,j \le n-1},$$

we have

$$A = \begin{pmatrix} B & 0 \\ 0 & 1 \end{pmatrix}, \qquad A^{-1} = \begin{pmatrix} B^{-1} & 0 \\ 0 & 1 \end{pmatrix}, \qquad \det A = \det B.$$

Consequently, the last term in (54) is the Laplace–Beltrami operator L_S on $S_r(p)$.

The isotropy subgroup of $I(X)$ at p acts transitively on the set of geodesics emanating from p. It follows that the function

$$L_X r = \frac{1}{\sqrt{\bar{g}}} \frac{\partial \sqrt{\bar{g}}}{\partial r}$$

is a function of r alone, so

$$\log \sqrt{\bar{g}} = \alpha(r) + \beta(\theta_1, \ldots, \theta_{n-1})$$

and

$$\sqrt{\bar{g}} = e^{\alpha(r)} \, e^{\beta(\theta_1, \ldots, \theta_{n-1})}$$

for suitable functions α and β. Note that $\alpha'(r)$ is independent of the choice of $\{\theta_1, \ldots, \theta_{n-1}\}$. So is $\alpha(r)$ if we, for example, fix $\alpha(L/2)$ by $\alpha(L/2) = 1$. Now

$$A(r) = \int_{S_r(p)} d\omega,$$

where $d\omega$ is the Riemannian measure on $S_r(p)$ which is locally given by $|\det B|^{1/2} \, d\theta_1 \cdots d\theta_{n-1} = \sqrt{\bar{g}} \, d\theta_1 \cdots d\theta_{n-1}$. Remembering that $\alpha(r)$ is independent of the choice of $\{\theta_1, \ldots, \theta_{n-1}\}$ we have

$$A(r) = Ce^{\alpha(r)} \qquad (C = \text{const}),$$

so

$$\frac{1}{\sqrt{\bar{g}}} \frac{\partial \sqrt{\bar{g}}}{\partial r} = \frac{1}{A(r)} \frac{dA}{dr}.$$

This concludes the proof.

The volume $V(r)$ of $B_r(p)$ is given by

$$V(r) = \int_{B_r(p)} dq,$$

where dq is the Riemannian measure on X, which is locally given by $\sqrt{\bar{g}} \, dr \, d\theta_1 \cdots d\theta_{n-1}$. It follows that

$$V(r) = \int_0^r A(t) \, dt.$$

Let us now determine $A(r)$ in group-theoretic terms. We can restrict ourselves to the case when the isometry group $I(X)$ is semisimple. Let G denote its identity component $I_0(X)$ and $K \subset G$ the isotropy subgroup at a fixed point $o \in X$.

First, we assume that X is noncompact and let $\mathfrak{g} = \mathfrak{k} + \mathfrak{p}$ be a Cartan decomposition of the Lie algebra $\mathfrak{g}$ of G. We assume that the Riemannian structure on X is defined by means of the Killing form of $\mathfrak{g}$. Let

$\mathfrak{a} \subset \mathfrak{p}$ be a maximal abelian subspace (which here is one-dimensional). As shown in Chapter I, §4, we have (as a consequence of the formula for the differential of $\mathrm{Exp}: \mathfrak{p} \to G/K$)

$$A(r) = \int_{S_r(0)} \det\left(\sum_0^\infty \frac{T_Y^k}{(2k+1)!}\right) d\omega(Y), \tag{55}$$

where $S_r(0)$ is the sphere $|Y| = r$ in $\mathfrak{p}$ and T_Y denotes the restriction $(\mathrm{ad}\, Y)^2|\mathfrak{p}$. Let $\pm\alpha$ and possibly $\pm 2\alpha$ denote the restricted roots, and for the multiplicities write $p = m_\alpha$, $q = m_{2\alpha}$. Select $H_0 \in \mathfrak{a}$ such that $\alpha(H_0) = 1$. Then

$$B(H_0, H_0) = 2[p\alpha(H_0)^2 + q(2\alpha(H_0))^2] = 2(p + 4q)$$

and the vector $A_\alpha \in \mathfrak{a}$ defined by $B(H, A_\alpha) = \alpha(H)$ satisfies

$$B(A_\alpha, A_\alpha) = 2(p\alpha(A_\alpha)^2 + q(2\alpha(A_\alpha))^2),$$

so

$$\alpha(A_\alpha) = (2p + 8q)^{-1}, \qquad A_\alpha = (2p + 8q)^{-1} H_0.$$

In order to compute the integrand in (55), we can take $Y = H_r$, where

$$H_r = r|A_\alpha|^{-1} A_\alpha = r(2p + 8q)^{1/2} A_\alpha.$$

The eigenvalues of $(\mathrm{ad}\, H_r)^2|\mathfrak{p}$ being 0, $\alpha(H_r)^2$ (with multiplicity p), and $(2\alpha(H_r))^2$ (with multiplicity q), we have

$$\det\left(\sum_0^\infty \frac{T_{H_r}^k}{(2k+1)!}\right) = \left(\frac{\sinh \alpha(H_r)}{\alpha(H_r)}\right)^p \left(\frac{\sinh 2\alpha(H_r)}{2\alpha(H_r)}\right)^q,$$

so if $c = (2p + 8q)^{-1/2}$, then

$$A(r) = \Omega_{p+q+1}\, 2^{-q} c^{-p-q} \sinh^p(cr) \sinh^q(2cr). \tag{56}$$

For the compact case we obtain in the same way

$$A(r) = \Omega_{p+q+1}\, 2^{-q} c^{-p-q} \sin^p(cr) \sin^q(2cr), \tag{57}$$

as already done in Chapter I [Lemma 4.10 and (40) in §4; note that there the roots were labeled differently].

5. The Poisson Equation Generalized

In the Chapter I, Lemma 2.35, we derived the classical solution formula for the Poisson equation $Lu = f$ in $\boldsymbol{R}^n$. We shall now extend this to symmetric spaces of rank one making use of Proposition 5.26.

Theorem 5.27. *Let X be a Euclidean space or a Riemannian globally symmetric space of rank one. Let $f \in \mathcal{D}(X)$. A solution to Poisson's equation*

$$L_X u = f$$

is given as follows:

(i) *If X is noncompact*

$$u(p) = \int_X H(p, q) f(q)\, dq, \tag{58}$$

where

$$H(p, q) = \int_1^{d(p,q)} \frac{1}{A(r)}\, dr. \tag{59}$$

(ii) *If X is compact a solution exists if and only if $\int f(q)\, dq = 0$. If this condition is satisfied, the unique solution to $L_X u = f$ satisfying $\int u(q)\, dq = 0$ is given by*

$$u(p) = \int_X G(p, q) f(q)\, dq,$$

where

$$G(p, q) = \frac{1}{V(L)} \int_L^{d(p,q)} \frac{V(L) - V(r)}{A(r)}\, dr.$$

Proof of (i). We have for u given by (58)

$$u(p) = \int_0^\infty dr \int_{S_r(p)} H(p, q) f(q)\, d\omega(q) = \int_1^\infty \psi(r) A(r) [M^r f](p)\, dr,$$

where

$$\psi(r) = \int_1^r \frac{1}{A(t)}\, dt,$$

and $[M^r f]\,(p)$ is the mean value of f on $S_r(p)$. Using now Lemma 2.14, Chapter I, Proposition 4.12, and Proposition 5.26, we obtain

$$[L_X u]\,(p) = \int_0^\infty \psi(r) A(r) [L_X M^r f](p)\, dr = \int_0^\infty \psi(r) A(r) L_r [M^r f](p)\, dr$$

where

$$L_r = \frac{\partial^2}{\partial r^2} + \frac{1}{A(r)} \frac{dA}{dr} \frac{\partial}{\partial r}.$$

Keeping p fixed and putting $F(r) = [M^r f](p)$, we have

$$[L_X u](p) = \int_0^\infty \{\psi(r)A(r)F''(r) + \psi(r)A'(r)F'(r)\}\,dr$$

$$= \lim_{\varepsilon \to 0}[\psi(r)A(r)F'(r)]_\varepsilon^\infty - \int_0^\infty F'(r)\,dr.$$

Now $\psi(\varepsilon)A(\varepsilon) \to 0$ as $\varepsilon \to 0$ and $F(r)$ vanishes for all sufficiently large r. Hence

$$[L_X u](p) = F(0) = f(p).$$

The choice of $H(p, q) = \psi(d(p, q))$ was of course motivated by the fact that the equation

$$(L_X)_q(H(p, q)) = \delta_p, \tag{60}$$

where δ_p is the delta distribution $f \to f(p)$, leads to the equation

$$\frac{d^2\psi}{dr^2} + \frac{1}{A(r)}\frac{dA}{dr}\frac{d\psi}{dr} = 0.$$

Proof of (ii). In the compact case we have $\int L_X u(q)\,dq = 0$ for each $u \in \mathscr{E}(X)$ (Prop. 2.3). Taking this fact into account, we replace (60) by the equation

$$(L_X)_q(G(p, q)) = \frac{1}{V(L)}(\delta_p - 1),$$

which by $G(p, q) = \phi(d(p, q))$ leads to the equation

$$\frac{d^2\phi}{dr^2} + \frac{1}{A(r)}\frac{dA}{dr}\frac{d\phi}{dr} = -\frac{1}{V(L)}, \qquad 0 < r < L, \tag{61}$$

whose coefficients are in general unbounded near $r = 0$ and $r = L$. The function

$$\phi(r) = \frac{1}{V(L)}\int_L^r \frac{V(L) - V(t)}{A(t)}\,dt$$

is a solution of (61), unbounded near $r = 0$ (if $n > 1$) but bounded near $r = L$ since $V'(t) = A(t)$. With $u(p) = \int G(p, q)f(q)\,dq$, we have

$$u(p) = \int_0^L dr \int_{S_r(p)} G(p, q)f(q)\,d\omega(q) = \int_0^L \phi(r)A(r)[M^r f](p)\,dr.$$

We put $F(r) = [M^r f](p)$ and observe that $\phi(r)A(r) \to 0$ as $r \to L$ and also as $r \to 0$ if $n > 1$. If $n = 1$, we have $F'(0) = 0$. Then we obtain as above

$$\begin{aligned}[L_X u](p) &= \int_0^L \{\phi(r)A(r)F''(r) + \phi(r)A'(r)F'(r)\}\, dr \\ &= [\phi(r)A(r)F'(r)]_0^L - \int_0^L \phi'(r)A(r)F'(r)\, dr \\ &= -\int_0^L \left(1 - \frac{V(r)}{V(L)}\right) F'(r)\, dr \\ &= F(0) - F(L) + \left[F(r)\frac{V(r)}{V(L)}\right]_0^L - \frac{1}{V(L)}\int_0^L A(r)[M^r f](p)\, dr \\ &= f(p) - F(L) + F(L) - \frac{1}{V(L)}\int_M f(q)\, dq = f(p).\end{aligned}$$

Thus $L_X u = f$. Moreover,

$$\int_X u(p)\, dp = \int_X f(q)\, dq \int_X G(p, q)\, dp = 0$$

because $\int G(p, q)\, dp$ is independent of q and $\int f(q)\, dq = 0$.

For the uniqueness, suppose u_1 and u_2 are two solutions to $L_X u = f$ and $\int u(q)\, dq = 0$. Then the difference v is orthogonal to the constants on X and to the subspace $L_X(\mathscr{E}(X))$, so by the existence just proved, v is orthogonal to $\mathscr{E}(X)$; hence $v = 0$.

6. Ásgeirsson's Mean Value Theorem Generalized

In his paper [1937], Ásgeirsson proved the following result. Let $u \in C^2(\boldsymbol{R}^{2n})$ satisfy the ultrahyperbolic equation

$$\frac{\partial^2 u}{\partial x_1^2} + \cdots + \frac{\partial^2 u}{\partial x_n^2} = \frac{\partial^2 u}{\partial y_1^2} + \cdots + \frac{\partial^2 u}{\partial y_n^2}. \tag{62}$$

Then for each $r \geq 0$ and each point

$$(x_0, y_0) = (x_1^0, \ldots, x_n^0, y_1^0, \ldots, y_n^0) \in \boldsymbol{R}^{2n}$$

the following identity holds for the $(n-1)$-dimensional spherical integrals:

$$\int_{S_r(x_0)} u(x, y_0)\, d\omega(x) = \int_{S_r(y_0)} u(x_0, y)\, d\omega(y). \tag{63}$$

The theorem was proved in the paper quoted for u of class C^2 in a suitable region in $\boldsymbol{R}^{2n}$.

It is obvious (taking u constant in the y-variables) that (63) implies Gauss's mean value theorem

$$u(x_0) = \frac{1}{A(r)} \int_{S_r(x_0)} u(x)\, d\omega(x)$$

for the solutions of Laplace's equation $L_{\boldsymbol{R}^n} u = 0$ on $\boldsymbol{R}^n$. Also, (63) can be used to derive a solution formula for the wave equation

$$L_{\boldsymbol{R}^n}\, u = \frac{\partial^2 u}{\partial t^2}$$

on $\boldsymbol{R}^n$ (cf. Exercise F1).

We shall now state and prove a generalization of Ásgeirsson's theorem to a symmetric space X of rank one. Let $L = L_X$ be the Laplace–Beltrami operator on X and as in §4, No. 3 we consider the spherical mean value operator

$$(M^r f)(x) = \frac{1}{A(r)} \int_{S_r(x)} f(s)\, d\omega(s) \qquad (r \geq 0). \tag{64}$$

Theorem 5.28. *Let X be a symmetric space of rank one* (or $\boldsymbol{R}^n$) *and let the function $u \in C^2(X \times X)$ satisfy the differential equation*

$$L_x(u(x, y)) = L_y(u(x, y)) \tag{65}$$

on $X \times X$ Then for each $r > 0$ and each $(x_0, y_0) \in X \times X$ we have

$$\int_{S_r(x_0)} u(x, y_0)\, d\omega(x) = \int_{S_r(y_0)} u(x_0, y)\, d\omega(y). \tag{66}$$

Proof. If $n = 1$, the solutions to (65) have the form

$$u(x, y) = \phi(x + y) + \psi(x - y)$$

and (66) reduces to the obvious relation

$$u(x_0 - r, y_0) + u(x_0 + r, y_0) = u(x_0, y_0 - r) + u(x_0, y_0 + r).$$

Hence we may assume that $n \geq 2$.

First, let $u \in C^2(X \times X)$ be arbitrary. We form the function

$$U(r, s) = (M_1^r M_2^s u)(x, y). \tag{67}$$

The subscripts 1 and 2 indicate that we consider the first and second variables, respectively; for exampe,

$$(M_1^r u)(x, y) = \frac{1}{A(r)} \int_{S_r(x)} u(s, y)\, d\omega(s).$$

Let $o \in X$ be the origin; write $X = G/K$, where $G = I_0(X)$ and K is the isotropy subgroup of G at o. If $g, h \in G$, $r = d(o, h \cdot o)$ (d = distance) we have from §4, No. 3,

$$(M^r f)(g \cdot o) = \int_K f(gkh \cdot o)\, dk, \qquad f \in C(X), \tag{68}$$

where dk is the normalized Haar measure on K. Thus we have by interchanging the order of integration

$$U(r, s) = (M_1^r M_2^s u)(x, y) = (M_2^s M_1^r u)(x, y). \tag{69}$$

Keeping y and s fixed, we have by Propositions 5.26 and 4.12

$$\frac{\partial^2 U}{\partial r^2} + \frac{1}{A(r)} A'(r) \frac{\partial U}{\partial r} = L_x(M_1^r M_2^s u(x, y)$$

$$= (M_1^r L_1 M_2^s u)(x, y) = (M_1^r M_2^s L_1 u)(x, y),$$

the last identity resulting from the fact that L_1 and M_2^s act on different arguments. Similarly, by using (69),

$$\frac{\partial^2 U}{\partial s^2} + \frac{1}{A(s)} A'(s) \frac{\partial U}{\partial s} = (M_2^s M_1^r L_2 u)(x, y).$$

Assuming now u is a solution to (65), we obtain from the above

$$\frac{\partial^2 U}{\partial r^2} + \frac{1}{A(r)} A'(r) \frac{\partial U}{\partial r} = \frac{\partial^2 U}{\partial s^2} + \frac{1}{A(s)} A'(s) \frac{\partial U}{\partial s}$$

Now putting $F(r, s) = U(r, s) - U(s, r)$, we obtain the relations

$$\frac{\partial^2 F}{\partial r^2} + \frac{1}{A(r)} A'(r) \frac{\partial F}{\partial r} - \frac{\partial^2 F}{\partial s^2} - \frac{1}{A(s)} A'(s) \frac{\partial F}{\partial s} = 0, \tag{70}$$

$$F(r, s) = -F(s, r).$$

After multiplication of (70) by $2A(r)\, \partial F/\partial s$ and some manipulation we obtain

$$-A(r) \frac{\partial}{\partial s}\left[\left(\frac{\partial F}{\partial r}\right)^2 + \left(\frac{\partial F}{\partial s}\right)^2\right] + 2 \frac{\partial}{\partial r}\left(A(r) \frac{\partial F}{\partial r} \frac{\partial F}{\partial s}\right) - \frac{2A(r)}{A(s)} \frac{dA}{ds} \left(\frac{\partial F}{\partial s}\right)^2 = 0.$$

Consider the line MN with equation $r + s = \text{const.}$ in the (r, s)-plane and integrate the last expression over the triangle OMN (see Fig. 11). Using the divergence theorem [Chapter I, §2, (26)] we then obtain, if $\boldsymbol{n}$

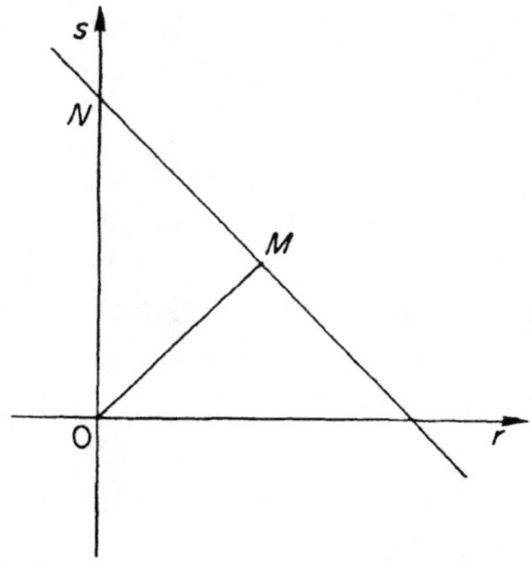

FIG. 11

denotes the outgoing unit normal, dl the element of arc length and dot $\cdot$ the inner product

$$\int_{OMN}\left(2A(r)\frac{\partial F}{\partial r}\frac{\partial F}{\partial s}, -A(r)\left[\left(\frac{\partial F}{\partial r}\right)^2+\left(\frac{\partial F}{\partial s}\right)^2\right]\right)\cdot \boldsymbol{n}dl$$

$$=\iint_{OMN}\frac{2A(r)}{A(s)}\frac{dA}{ds}\left(\frac{\partial F}{\partial s}\right)^2 dr\, ds.$$

On OMN we have $s \geq r$ so that for small r, s

$$\frac{A(r)}{A(s)}A'(s) \leq Cr^{n-1}s^{-1} \leq Cr^{n-2} \qquad (C = \text{const.})$$

so that the last integral does indeed exist since we have assumed $n \geq 2$. We now use the following data:

On OM: $\boldsymbol{n} = (2^{-1/2}, -2^{-1/2})$, $F(r, r) = 0$, so that $\dfrac{\partial F}{\partial r} + \dfrac{\partial F}{\partial s} = 0$.

On MN: $\boldsymbol{n} = (2^{-1/2}, 2^{-1/2})$, on $ON: A(r) = 0$.

The formula above then reduces to

$$(71)\quad 2^{-1/2}\int_{MN} A(r)\left(\frac{\partial F}{\partial r}-\frac{\partial F}{\partial s}\right)^2 dl + \iint_{OMN}\frac{2A(r)}{A(s)}\frac{dA}{ds}\left(\frac{\partial F}{\partial s}\right)^2 dr\, ds = 0.$$

Now we consider separately the compact case and the noncompact case.

I. X *noncompact*. Here we claim $A'(r) \geq 0$ for all $r \geq 0$. If $X = \boldsymbol{R}^n$, this is obvious, so we may assume that G is semisimple. But then the statement is clear from (56). Consequently, both terms in (71) vanish, so we can conclude that $F \equiv 0$. In particular, $U(r, 0) = U(0, r)$, and this is the desired formula (66).

II. *X compact.* In this case (57) shows that $A'(s) \geq 0$ for s in a certain interval $0 \leq s \leq r_0$. As before, we can conclude that $U(r, 0) = U(0, r)$ for r in this interval $0 \leq r \leq r_0$. In order to extend this to all r we approximate the solution u to (65) by analytic solutions. Let ϕ, ψ be analytic functions on the compact Lie group G with Haar measure dg and consider the convolution

$$u_{\phi,\psi}(x, y) = \int_G\int_G u(g_1^{-1} \cdot x, g_2^{-1} \cdot y)\phi(g_1)\psi(g_2)\, dg_1 dg_2. \tag{72}$$

Then

$$\begin{aligned} L_x(u_{\phi,\psi}(x, y)) &= \int_G\int_G (L_1 u)(g_1^{-1} \cdot x, g_2^{-1} \cdot y)\phi(g_1)\psi(g_2)\, dg_1 dg_2 \\ &= \int (L_2 u)(g_1^{-1} \cdot x, g_2^{-1} \cdot y)\phi(g_1)\psi(g_2)\, dg_1\, dg_2 \\ &= L_y(u_{\phi,\psi}(x, y)), \end{aligned}$$

so

$$\int_{S_r(x_0)} u_{\phi,\psi}(x, y_0)\, d\omega(x) = \int_{S_r(y_0)} u_{\phi,\psi}(x_0, y)\, d\omega(y) \tag{73}$$

for $0 \leq r \leq r_0$. For $f \in C(X \times Y)$ define the function $\tilde{f}$ on $G \times G$ by $\tilde{f}(g_1, g_2) = f(g_1 \cdot o, g_2 \cdot o)$; then (73) can be written

$$\int_K \tilde{u}_{\phi,\psi}(g_1 kh, g_2)\, dk = \int_K \tilde{u}_{\phi,\psi}(g_1, g_2 kh)\, dk \tag{74}$$

for all $h \in G$ such that $d(o, h \cdot o) \leq r_0$.

Changing variables in the integral (72), we see that $u_{\phi,\psi}$ is analytic on $X \times X$, so $\tilde{u}_{\phi,\psi}$ is analytic on $G \times G$. Since (74) holds for h varying in an open subset of G, it therefore holds for all $h \in G$. Substituting the definition (72) into (74), we see that for each fixed $h \in G$, the function

$$(z, w) \to \int_K [\tilde{u}(z^{-1}g_1 kh, w^{-1}g_2) - \tilde{u}(z^{-1}g_1, w^{-1}g_2 kh)]\, dk \tag{75}$$

on $G \times G$ is orthogonal to all functions of the form $\phi(z)\psi(w)$ ϕ and ψ being analytic. By the Peter–Weyl theorem (cf. Theorem 4.3 in Chapter V) this remains true for ϕ, $\psi \in \mathscr{E}(G)$. But by the Stone–Weierstrass theorem these functions $\phi(z)\psi(w)$ span a uniformly dense subspace of

$C(G \times G)$. Consequently, the function (75) is identically 0, so (66) is proved for all r.

7. Restriction of the Central Operators in $D(G)$

With the notation as in §5, No. 2 let $\pi: G \to G/K = X$ denote the natural mapping, and as in §4 let $\mu: \boldsymbol{D}_K(G) \to \boldsymbol{D}(G/K)$ be the surjective homomorphism given by $(\mu(D)f) \circ \pi = D(f \circ \pi)$, $f \in \mathscr{E}(G/K)$. We shall now consider the question whether $\boldsymbol{Z}(G)$ is mapped *onto* $\boldsymbol{D}(G/K)$ by the homomorphism μ.

At the same time we consider another mapping ν of $\boldsymbol{D}(G)$ into the algebra $\boldsymbol{E}(X)$ of all differential operators on X. Given $Y \in \mathfrak{g}$, let $\tilde{Y}$ be the corresponding left-invariant vector field on G and define $\nu(\tilde{Y}) \in \boldsymbol{E}(X)$ by

$$(\nu(\tilde{Y})f)(p) = \left\{\frac{d}{dt} f(\exp(-tY)\cdot p)\right\}_{t=0}. \tag{76}$$

According to [DS], Chapter II, Theorem 3.4, ν is a homomorphism of $\mathfrak{g}$ into the Lie algebra $\mathscr{D}^1(X)$ of all vector fields on X; thus by [DS], Chapter II, Proposition 1.1, it extends to a homomorphism of $\boldsymbol{D}(G)$ into $\boldsymbol{E}(X)$, also denoted ν. If $g \in G$, we write also $\nu(g)f = f^{\tau(g)}$ $(f \in \mathscr{E}(X))$.

Lemma 5.29. *Let $D \in \boldsymbol{D}(G)$, $g \in G$. Then*

$$\nu(D)\nu(g) = \nu(g)\nu(D^{R(g)}).$$

Proof. It suffices to verify this for $D = \tilde{Y} \in \mathfrak{g}$ because the maps $D \to D^{R(g)}$ and $D \to \nu(D)$ are homomorphisms. But

$$(\nu(\tilde{Y})\nu(g)f)(p) = \left\{\frac{d}{dt} f(g^{-1}\exp(-tY)\cdot p)\right\}_{t=0}$$

$$= (\nu(\mathrm{Ad}(g^{-1})(\tilde{Y})f)(g^{-1}\cdot p) = (\nu(g)\nu(\tilde{Y}^{R(g)})f)(p),$$

as desired.

Lemma 5.30. *Let $D \in \boldsymbol{D}(G)$, $f \in \mathscr{E}(X)$. Then, superscript $*$ denoting adjoint,*

$$(\nu(D)f)(\pi(g)) = ((D^*)^{R(g)}(f \circ \pi))(g), \qquad g \in G.$$

Proof. If $Y \in \mathfrak{g}$, $g \in G$, we have

$$(\nu(\tilde{Y})f)(\pi(g)) = \left\{\frac{d}{dt} f(\exp(-tY)\cdot \pi(g))\right\}_{t=0}$$

$$= \left\{\frac{d}{dt}((f \circ \pi)(g \exp \mathrm{Ad}(g^{-1})(-tY)))\right\}_{t=0},$$

so

$$(77)\qquad (\nu(\tilde{Y})f)(\pi(g)) = [(\mathrm{Ad}(g^{-1})(-Y))^{\sim}(f \circ \pi)](g).$$

Let $Y_1, \ldots, Y_r \in \mathfrak{g}$. We shall prove by induction on r that

$$(78)\qquad (\nu(\tilde{Y}_1 \cdots \tilde{Y}_r)f)(\pi(g)) = (-1)^r[(\mathrm{Ad}(g^{-1})(\tilde{Y}_r \cdots \tilde{Y}_1))(f \circ \pi)](g).$$

For $r = 1$ this is proved; assuming it for a fixed r we have for $Y_0 \in \mathfrak{g}$

$$(\nu(\tilde{Y}_0 \cdots \tilde{Y}_r)f)(\pi(g))$$

$$= (\nu(\tilde{Y}_0)\nu(\tilde{Y}_1 \ldots \tilde{Y}_r)f)(\pi(g))$$

$$= \left\{\frac{d}{dt}\nu(\tilde{Y}_1 \cdots \tilde{Y}_r)f(\pi(\exp(-tY_0)\,g))\right\}_{t=0}$$

$$= (-1)^r\left\{\frac{d}{dt}[\mathrm{Ad}(g^{-1})\,\mathrm{Ad}(\exp(tY_0))\,(\tilde{Y}_r \cdots \tilde{Y}_1)(f \circ \pi)](\exp(-tY_0)g)\right\}_{t=0}$$

$$= (-1)^r\{\mathrm{Ad}(g^{-1})\,([\tilde{Y}_0, \tilde{Y}_r \cdots \tilde{Y}_1])(f \circ \pi)\}(g)$$

$$+ (-1)^r\{\mathrm{Ad}(g^{-1})\,(-\tilde{Y}_0)\,\mathrm{Ad}(g^{-1})\,(\tilde{Y}_r \cdots \tilde{Y}_1)(f \circ \pi)\}(g),$$

where the unique derivation of $\boldsymbol{D}(G)$ extending the endomorphism $\tilde{Y} \to [\tilde{Y}_0, \tilde{Y}]$ of $\tilde{\mathfrak{g}}$ has been denoted $[\tilde{Y}_0, D]$. However, it is clear that $[\tilde{Y}_0, D] = \tilde{Y}_0 D - D\tilde{Y}_0$, so the two last expressions add up to

$$(-1)^{r+1}\{\mathrm{Ad}(g^{-1})\,(\tilde{Y}_r \cdots \tilde{Y}_1\tilde{Y}_0)(f \circ \pi)\}(g).$$

This proves (78). Since G is unimodular, $\tilde{Y}^* = -\tilde{Y}$ [cf. (7) in §5], so the lemma is proved.

Corollary 5.31. *$\mu(D^*) = \nu(D)$ for each $D \in \boldsymbol{Z}(G)$.*

In fact, $\boldsymbol{Z}(G)$ is invariant under the mapping $D \to D^*$, so $(D^*)^{R(g)} = D^*$.

The image of $\boldsymbol{Z}(G)$ under the mapping μ (or ν) will be denoted $\boldsymbol{Z}(G/K)$. As before, extend $\mathfrak{a}$ to a Cartan subalgebra $\mathfrak{h}$ of $\mathfrak{g}$. The complexification $\mathfrak{h}^C$ is then a Cartan subalgebra of $\mathfrak{g}^C$. Let $\tilde{W}$ denote the corresponding Weyl group.

Denoting the dual of a vector space by a star, we have the following symmetric algebras, with corresponding group action and invariants:

$S(\mathfrak{g}), S(\mathfrak{g}^*)$,	$\mathrm{Ad}(G)$,	$I(\mathfrak{g}), I(\mathfrak{g}^*)$
$S(\mathfrak{p}), S(\mathfrak{p}^*)$,	$\mathrm{Ad}_G(K)$,	$I(\mathfrak{p}), I(\mathfrak{p}^*)$
$S(\mathfrak{a}), S(\mathfrak{a}^*)$,	W,	$I(\mathfrak{a}), I(\mathfrak{a}^*)$
$S(\mathfrak{h}^C), S(\mathfrak{h}^{C*})$,	$\tilde{W}$,	$I(\mathfrak{h}^C), I(\mathfrak{h}^{C*})$.

Let $\widetilde{W}_\theta$ be the subgroup of $\widetilde{W}$ leaving $\mathfrak{a}$ invariant; then the restriction from $\mathfrak{h}^c$ to $\mathfrak{a}$ induces a homomorphism of $\widetilde{W}_\theta$ onto W ([DS]), Proposition 8.10, Chapter VII) and therefore maps $I(\mathfrak{h}^{c*})$ into $I(\mathfrak{a}^*)$.

Proposition 5.32. *The following properties of the symmetric space G/K are equivalent:*

(i) $\mathbf{Z}(G/K) = \mathbf{D}(G/K)$.

(ii) *The restriction from $\mathfrak{g}$ to $\mathfrak{p}$ maps $I(\mathfrak{g}^*)$ onto $I(\mathfrak{p}^*)$.*

(iii) *The restriction from $\mathfrak{h}^c$ to $\mathfrak{a}$ maps $I(\mathfrak{h}^c*)$ onto $I(\mathfrak{a}^*)$.*

Proof. For the equivalence of (ii) and (iii) let $\mathfrak{u}$ be the compact real form $\mathfrak{k} + i\mathfrak{p}$ of $\mathfrak{g}^c$, so $\mathfrak{g}^c = \mathfrak{u} + i\mathfrak{u}$ is a Cartan decomposition. Let $\mathfrak{h}_{\mathbf{R}}$ be the subspace of $\mathfrak{h}^c$, where all roots (of $\mathfrak{g}^c$ with respect to $\mathfrak{h}^c$) take real values. By Corollary 5.12, $I((i\mathfrak{u})^*)$ is isomorphic to $I(\mathfrak{h}_{\mathbf{R}}^*)$ and $I(\mathfrak{p}^*)$ is isomorphic to $I(\mathfrak{a}^*)$ under restriction.

But we have also the identifications $I(\mathfrak{g}^*) = I((\mathfrak{g}^c)^*) = I((i\mathfrak{u})^*)$ (see Lemma 1.4 in the next chapter). Under restrictions from $\mathfrak{g}$ to $\mathfrak{p}$, and from $\mathfrak{h}^c$ to $\mathfrak{a}$, we have the commutative diagram

$$\begin{array}{ccc} I(\mathfrak{g}^*) & \longrightarrow & I((\mathfrak{h}^c)^*) \\ \downarrow & & \downarrow \\ I(\mathfrak{p}^*) & \longrightarrow & I(\mathfrak{a}^*) \end{array}$$

from which the equivalence of (ii) and (iii) follows.

Next consider the mapping $\lambda : S(\mathfrak{g}) \to \mathbf{D}(G)$ from (4) in §4. Let $P \in I(\mathfrak{g})$ and let $\overline{P} \in S(\mathfrak{p})$ be determined by $P - \overline{P} \in S(\mathfrak{g})\mathfrak{k}$. Then $\overline{P} \in I(\mathfrak{p})$. Let $X_1, \ldots, X_n$ be a basis of $\mathfrak{g}$ such that $X_1, \ldots, X_r$ is a basis of $\mathfrak{p}$. Since $\lambda(P) \in \mathbf{D}_K(G)$, there exists by Theorem 4.9 a unique $P' \in I(\mathfrak{p})$ such that

$$(79)\quad (\mu(\lambda(P))f)(\pi(g)) = [P'(\partial_1, \ldots, \partial_r)(f \circ \pi)(g \exp(x_1 X_1 + \cdots + x_r X_r))](0)\,,$$

and in the notation of that theorem, $\mu(\lambda(P)) = D_{\lambda(P')}$, that is, $\mu(\lambda(P)) = \mu(\lambda(P'))$. It follows that

$$(80)\qquad \lambda(\overline{P} - P') + \lambda(P - \overline{P}) = \lambda(P - P') \in \mathbf{D}(G)\mathfrak{k}\,,$$

since μ has kernel $\mathbf{D}_K(G) \cap \mathbf{D}(G)\mathfrak{k}$. Now suppose P is homogeneous of degree $d > 0$. Since $P - \overline{P} \in S(\mathfrak{g})\mathfrak{k}$, there exists an operator $D \in \mathbf{D}(G)$ of degree $< d$ such that $\lambda(P - \overline{P}) - D \in \mathbf{D}(G)\mathfrak{k}$. Thus by (80)

$$(81)\qquad \lambda(\overline{P} - P') + D \in \mathbf{D}(G)\mathfrak{k}\,.$$

Decomposing D according to Lemma 4.7, we may take $D = \lambda(Q)$, where $Q \in S(\mathfrak{p})$ of degree $< d$. But then (81) and the directness in Lemma 4.7 imply that $\overline{P} - P' + Q = 0$; in other words,

$$(82)\qquad \text{degree}\,(\overline{P} - P') < \text{degree}\,(P)\,.$$

The mapping μ maps $\mathbf{Z}(G)$ onto $\mathbf{D}(G/K)$ if and only if the mapping $P \to P'$ maps $I(\mathfrak{g})$ onto $I(\mathfrak{p})$, which by (82) happens if $P \to \overline{P}$ maps $I(\mathfrak{g})$ onto $I(\mathfrak{p})$ since $\deg \overline{P} = d$. Under the bijection between $\mathfrak{g}$ and $\mathfrak{g}^*$ induced by the Killing form, $I(\mathfrak{g})$ and $I(\mathfrak{p})$ correspond to $I(\mathfrak{g}^*)$ and $I(\mathfrak{p}^*)$, respectively, and the mapping $P \to \overline{P}$ corresponds to the restriction of a $p \in I(\mathfrak{g}^*)$ to $\mathfrak{p}$. This shows that (ii) $\Rightarrow$ (i). The converse (i) $\Rightarrow$ (ii) does not follow from (82) since $\deg P'$ might be $< \deg(P)$. The proof that (i) $\Rightarrow$ (ii) is given in *Amer. J. Math.* 114 (1992), 789–811, using classification.

Remark.

(i) As a consequence of the last proof it is shown that $\mathbf{Z}(G/K) = \mathbf{D}(G/K)$ for all irreducible G/K except four, namely $E_6/SO(10)T$, E_6/F_4, E_7/E_6T, $E_8/E_7SU(2)$.

(ii) In the same paper it is proved that given $D \in \mathbf{D}(G/K)$ there exist $Z_1 \neq 0$, Z_2 in $\mathbf{Z}(G/K)$ such that $DZ_1 = Z_2$.

8. Invariant Differential Operators for Complex Semisimple Lie Algebras

Let G be a connected complex semisimple Lie group acting on its Lie algebra $\mathfrak{g}$ by means of the adjoint representation. Let $\mathfrak{h}$ be a Cartan subalgebra and let us adopt the notation from Example (vii) in §3. Let $I(\mathfrak{g}^*)$ denote the set of G-invariant polynomial functions on $\mathfrak{g}$. As p runs through $I(\mathfrak{g}^*)$ the differential operators $\partial(p)$ run through the set of G-invariant differential operators on $\mathfrak{g}$ with constant coefficients. This set is, of course, the algebra $\boldsymbol{D}(G \cdot \mathfrak{g}/G)$ if $G \cdot \mathfrak{g}$ denotes the group of transformations of $\mathfrak{g}$ generated by G and the translations. We can now generalize Proposition 3.14.

Theorem 5.33. *For the adjoint action of G on $\mathfrak{g}$ with transversal manifold $\mathfrak{h}'$ the radial part of the operator $\partial(p)$, $p \in I(\mathfrak{g}^*)$, is given by*

$$\Delta(\partial(p)) = \pi^{-1}\partial(\bar{p}) \circ \pi.$$

Proof. The proof is based on the purely algebraic trick of taking commutators with the Laplacian. As usual, let $E(\mathfrak{g})$, $E(\mathfrak{h}')$ denote the algebras of all differential operators on $\mathfrak{g}$ and $\mathfrak{h}'$, respectively, and let $J(\mathfrak{g})$ [resp. $J(\mathfrak{h}')$] be the subalgebra of the G-invariant ones (resp. the Weyl group-invariant ones). Writing $\{D_1, D_2\} = D_1 \circ D_2 - D_2 \circ D_1$ if D_1 and D_2 are differential operators, we consider the derivation

$$\mu\colon D \to \tfrac{1}{2}\{\partial(\omega), D\}$$

of $E(\mathfrak{g})$ and the derivation

$$\bar{\mu}\colon d \to \tfrac{1}{2}\{\Delta(\partial(\omega)), d\}$$

of $E(\mathfrak{h}')$. Since the operation Δ is a homomorphism of $J(\mathfrak{g})$ into $J(\mathfrak{h}')$, we have

$$(83) \qquad \Delta(\mu(D)) = \bar{\mu}(\Delta(D)), \qquad D \in J(\mathfrak{g}).$$

Now we need a simple lemma.

Lemma 5.34. *Let p be a homogeneous polynomial on $\mathfrak{g}$ of degree m. Considering p as the differential operator $f \to pf$, we have*

$$\mu^m(p) = m!\,\partial(p).$$

Proof. We use induction on m. For $m = 1$, the result is immediate by computation. Let $p = x_1 \cdots x_m$. Since $\mu^2(x_i) = \mu(\partial(x_i)) = 0$, Leibniz's rule for derivations implies

$$(84) \qquad \mu^m(x_1 \cdots x_{m-1}x_m) = \mu^m(x_1 \cdots x_{m-1}) \circ x_m + m\mu^{m-1}(x_1 \cdots x_{m-1}) \circ \mu(x_m).$$

The induction hypothesis gives

$$\mu^{m-1}(x_1 \cdots x_{m-1}) = (m-1)!\,\partial(x_1 \cdots x_{m-1}).$$

Hence the first term on the right in (84) vanishes and the lemma follows.

Combining (83) and the lemma, we have

$$m!\,\Delta(\partial(p)) = \Delta(\mu^m(p)) = \bar{\mu}^m(\Delta(p)) = \bar{\mu}^m(\bar{p}).$$

Now if A is any associative algebra and $a \in A$, put $d_a(b) = \frac{1}{2}(ab - ba)$. Then by induction on k,

$$d_a{}^k(b) = 2^{-k} \sum_0^k \binom{k}{r}(-1)^r a^{k-r} b a^r.$$

If $c \in A$ commutes with b and has an inverse c^{-1} and if $a' = c^{-1}ac$, then

$$(d_{a'})^k(b) = c^{-1} d_a{}^k(b) c.$$

Here we take $A = \mathbf{E}(\mathfrak{h}')$, $a = \partial(\bar{\omega})$, $b = \bar{p}$, $c = \pi$. Then $a' = \pi^{-1}\partial(\bar{\omega}) \circ \pi = \Delta(\partial(\omega))$. By the lemma we have $d_a{}^k(q) = k!\,\partial(q)$ if q is a homogeneous polynomial on $\mathfrak{h}$ of degree k. Hence if $p \in I(\mathfrak{g})$ is homogeneous of degree m

$$m!\,\Delta(\partial(p)) = \bar{\mu}^m(\bar{p}) = (d_{a'})^m(\bar{p}) = \pi^{-1}\, d_a{}^m(\bar{p})\pi = m!\,\pi^{-1}\partial(\bar{p})\pi,$$

proving the theorem.

The theorem can be used to prove an interesting integral formula on a compact semisimple Lie group U. Let $\mathfrak{u}$ denote the Lie algebra of U,

$\mathfrak{t} \subset \mathfrak{u}$ the Lie algebra of a maximal torus $T \subset U$. Let $\mathfrak{g} = \mathfrak{u}^{\mathbf{C}}$, $\mathfrak{h} = \mathfrak{t}^{\mathbf{C}}$ denote the respective complexifications. We retain the notation from Example (vii) in §3 and write uX for $\mathrm{Ad}(u)\, X$ ($u \in U$, $X \in \mathfrak{g}$).

Theorem 5.35. *Let du be a normalized Haar measure on U. Then, if H, $H' \in \mathfrak{h}$,*

$$\pi(H)\pi(H') \int_U e^{\langle uH, H' \rangle}\, du = w^{-1} \partial(\pi)(\pi) \sum_{s \in W} (\det s) e^{\langle sH, H' \rangle},$$

where w is the order of the Weyl group W.

Proof. Let $f \in \mathscr{E}(\mathfrak{u})$ and put $F(X) = \int_U f(u \cdot X)\, du$. Then if $p \in I(\mathfrak{u}^*)$ we have for $H \in \mathfrak{h}'$ $(\partial(p)F)(H) = (\pi^{-1}\partial(\bar{p})(\pi\bar{F}))(H)$, so the function

$$\phi_f(H) = \pi(H) \int_U f(u \cdot H)\, du \qquad (H \in \mathfrak{t})$$

satisfies

$$(\phi_{\partial(p)f})(H) = \pi(H)(\partial(p)F)(H) = (\partial(\bar{p})\phi_f)(H).$$

Thus

$$\phi_{\partial(p)f} = \partial(\bar{p})\phi_f, \qquad p \in I(\mathfrak{u}^*). \tag{85}$$

Now select $H' \in \mathfrak{h}$ such that $\pi(H') \neq 0$ and put $f(x) = e^{\langle X, H' \rangle}$. Then $\partial(q)f = q(H')f$ for each $q \in S(\mathfrak{u}^*)$, so by (85) $\partial(\bar{p})\phi_f = p(H')\phi_f$ for each $p \in I(\mathfrak{u}^*)$. Thus by Corollary 5.12, $\partial(q)\phi_f = q(H')\phi_f$ for all $q \in I(\mathfrak{h}^*)$. Since H' is regular, the simplest instance of Theorem 3.13 in Chapter III shows that

$$\phi_f(H) = \sum_{s \in W} c_s e^{\langle sH, H' \rangle}, \tag{86}$$

where $c_s \in \mathbf{C}$. On the other hand, since $\pi(sH) = (\det s)\pi(H)$, it is clear that $\phi_f(sH) = (\det s)\phi_f(H)$, whence by (86)

$$\phi_f(H) = c \sum_{s \in W} (\det s) e^{\langle sH, H' \rangle}, \tag{87}$$

where $c \in \mathbf{C}$. The constant c can be determined by applying the differential operator $\partial(\pi)$ on both sides and evaluating at $H = 0$. The definition of ϕ_f shows that

$$(\partial(\pi)\phi_f)(0) = \partial(\pi)(\pi)f(0) = \partial(\pi)\pi,$$

whereas the right-hand side of (87) gives the contribution $cw.\pi(H')$. This proves the theorem for $H \in \mathfrak{t}$, $\pi(H') \neq 0$, and hence for all H, $H' \in \mathfrak{h}$ by holomorphic continuation.

We can now use Theorem 5.35 to derive a formula for the order w of the Weyl group quite different from that of [DS], Chapter X, Exercise B9.

Corollary 5.36. *The order w of the Weyl group is given by*

$$w = \partial(\pi)(\pi)/\pi(H_\rho),$$

where 2ρ is the sum of the positive roots in $\Delta(\mathfrak{g}, \mathfrak{h})$.

We put $H' = H_\rho$ in Theorem 5.35. Then by Chapter I, Proposition 5.15, the right-hand side becomes

$$w^{-1}\partial(\pi)(\pi) \prod_{\alpha > 0} (e^{(1/2)\alpha(H)} - e^{-(1/2)\alpha(H)}).$$

Dividing by $\pi(H)$ and letting $H \to 0$, we obtain the result.

Remark. The formula in Theorem 5.35 gives an interesting interpretation of the characters of irreducible representations of the compact group U. This will be discussed in Chapter V, §1, where we shall also prove an analog of Theorem 5.33, generalizing Proposition 3.12.

9. Invariant Differential Operators for $X = G/K$, G Complex

As in Example (iii), §3, No. 4, let δ denote the density function for K acting on the symmetric space $X = G/K$. In the case when G is complex this density function is up to a factor given by

$$\delta(\operatorname{Exp} H) = \prod_{\alpha \in \Sigma^+} (\sinh \alpha(H))^2, \qquad H \in \mathfrak{a}. \tag{88}$$

Let $\Gamma: \boldsymbol{D}(G/K) \to \boldsymbol{D}_W(A)$ be the isomorphism from Theorem 5.18. Then we have the following description of the radial parts generalizing Proposition 3.10.

Theorem 5.37. *Suppose G is complex. The radial parts for K acting on $X = G/K$ are given by*

$$\Delta(D) = \delta^{-1/2}\Gamma(D) \circ \delta^{1/2}, \qquad D \in \boldsymbol{D}(G/K). \tag{89}$$

Since the proof uses some facts from the theory of spherical functions, we postpone it until Chapter IV, §5, No. 2.

There is an analog of Theorem 5.37 concerning the action of K on $\mathfrak{p}$. The corresponding density function is then up to a factor given by

$$\delta_0(H) = \prod_{\alpha \in \Sigma^+} (\alpha(H))^2, \qquad H \in \mathfrak{a}, \tag{90}$$

and we have the following formula generalizing Proposition 3.13 (G complex):

$$\Delta(\partial(p)) = \pi^{-1}\partial(\bar{p}) \circ \pi, \qquad p \in I(\mathfrak{p}^*), \tag{91}$$

where $\bar{p}$ denotes the restriction of p to $\mathfrak{a}$. The proof of this formula is identical to that of Theorem 5.33.

As in Example (viii) in §3, No. 4, let J denote the ratio of the volume elements in X and in $\mathfrak{p}$ (via Exp). Combining the restriction isomorphism $I(\mathfrak{p}^*) \to I(\mathfrak{a}^*)$ with the isomorphism Γ, we obtain a surjective isomorphism σ: $\boldsymbol{D}(G/K) \to \boldsymbol{D}_K(\mathfrak{p})$, where $\boldsymbol{D}_K(\mathfrak{p}) = \boldsymbol{D}(K \cdot \mathfrak{p}/K)$ is the set of K-invariant differential operators on $\mathfrak{p}$ with constant coefficients. By Corollary 5.20 we have $\sigma(L_X) = L_{\mathfrak{p}} - \langle \rho, \rho \rangle$.

Theorem 5.38. *Suppose G is complex. Then if $D \in \boldsymbol{D}(G/K)$, the image $D^{\mathrm{Exp}^{-1}}$ of D under* Exp^{-1} *satisfies the relation*

$$D^{\mathrm{Exp}^{-1}}F = J^{-1/2}\sigma(D)(J^{1/2}F)$$

for all K-invariant C^∞-functions F on $\mathfrak{p}$.

In fact, it suffices to prove this formula on $\mathfrak{a}$, and there it is a direct consequence of (89) and (91) because $J = \delta/\pi^2$.

EXERCISES AND FURTHER RESULTS

A. The Laplace–Beltrami Operator

1. Let M be a Riemannian manifold, $p \in M$ a point, $\{x_1, \ldots, x_m\}$ normal coordinates around p such that $(\partial/\partial x_i)_p$ $(1 \le i \le m)$ is an orthonormal basis of the tangent space M_p. Then the Laplace–Beltrami operator L is given at p by

$$(Lf)(p) = \sum_i \frac{\partial^2 f}{\partial x_i^2}(p).$$

2. Let M be an oriented manifold with a Riemannian structure g and define the $*$ operator on the Grassmann algebra $\mathfrak{A}(M)$ following [DS], Chapter II, §7. Let

$$\delta\omega = (-1)^{np+n+1} * d * \omega, \qquad \omega \in \mathfrak{A}_p(M),$$

and extending δ by linearity put

$$\Delta = -(d\delta + \delta d) \qquad \text{on} \quad \mathfrak{A}(M).$$

Let $X \in \mathscr{D}^1(M)$ and let ω_X be the 1-form $Y \to g(X, Y)$. Show that

(i) $\delta(\omega_X) = -\operatorname{div} X, \ \omega_{\operatorname{grad} f} = df, \ f \in \mathscr{E}(M)$,

and deduce that

(ii) $\Delta f = L_M f$,

where L_M is the Laplace–Beltrami operator.

3. Let G/H be a pseudo-Riemannian manifold of constant curvature represented in accordance with Chapter I, Theorem 6.1, in the form

$$\boldsymbol{O}(p, q+1)/\boldsymbol{O}(p, q), \qquad \boldsymbol{O}(p+1, q)/\boldsymbol{O}(p, q), \qquad \boldsymbol{R}^{p+q} \cdot \boldsymbol{O}(p, q)/\boldsymbol{O}(p, q).$$

Then the algebra $\boldsymbol{D}(G/H)$ of invariant differential operators consists of the polynomials in the Laplace–Beltrami operator L (cf. Helgason [1959], Faraut [1979]; the latter paper contains a detailed study of the H-invariant eigendistributions of L).

4. Let G be a connected semisimple Lie group with Lie algebra $\mathfrak{g}$. Let B denote the Killing form of $\mathfrak{g}$. Let $X_1, \ldots, X_n$ be a basis of $\mathfrak{g}$, put $g_{ij} = B(X_i, X_j)$ $(1 \le i, j \le n)$, and let (g^{ij}) denote the inverse of the matrix (g_{ij}). Show that

(i) The differential operator $\Omega = \sum_{i,j} g^{ij} \tilde{X}_i \tilde{X}_j$ is independent of the choice of the basis (X_i) and belongs to the center $\boldsymbol{Z}(G)$ of $\boldsymbol{D}(G)$. The operator Ω is called the *Casimir operator*. It coincides with the Laplace–Beltrami operator for the bi-invariant pseudo-Riemannian structure on G given by B.

(ii) Assuming G noncompact, let G/K be the symmetric space associated with G. The operator $\mu(\Omega)$ given by Theorem 4.6 coincides with the Laplace–Beltrami operator on G/K.

5. (Conformal Invariance) Let M be a manifold with a pseudo-Riemannian structure g. Let $n = \dim M$. A vector field X on M is said to be *conformal* if the Lie derivative $\theta(X)$ has the property

$$\theta(X)g = \rho(X)g, \qquad \text{where} \quad \rho(X) \in \mathscr{E}(M).$$

(i) Show that if X is conformal, then

$$\rho(X) = \frac{2}{n} \operatorname{div} X.$$

(ii) The action

$$\sigma: \quad z \to \frac{az+b}{cz+d}, \qquad z \in \boldsymbol{C},$$

of the group $\boldsymbol{SL}(2, \boldsymbol{C})$ on $\boldsymbol{R}^2$ is conformal, i.e., $\sigma^* \cdot g = \tau^2 g$ if g is the flat Riemannian structure and $\tau \in \mathscr{E}(\boldsymbol{R}^2)$. Deduce from [DS], Chapter I, Exercise B3, that for each $X \in \mathfrak{sl}(2, \boldsymbol{C})$ the induced vector field X^+ on $\boldsymbol{R}^2$ is conformal. Note also (Introduction, Exercise A4) that $L = L_{\boldsymbol{R}^2}$ satisfies

$$L^\sigma = h(\sigma)L,$$

where $h(\sigma)(z) = |cz - a|^4$. Deduce that

$$[L, X^+] = \rho(X)L, \qquad X \in \mathfrak{sl}(2, \boldsymbol{C}), \tag{1}$$

where [,] denotes commutator and

$$\rho(X) = \left\{\frac{d}{di} h(\exp tX)\right\}_{t=0}$$

(iii)* Let r denote the Ricci curvature tensor of M, i.e., $r_{ji} = -\sum_k R^k{}_{ijk}$ (negative of [DS], Chapter VIII) and K the *scalar curvature*

$$K = \sum_{i,j} g^{ij} r_{ij}.$$

Let M_1 and M_2, respectively, have pseudo-Riemannian structures g_1 and g_2, scalar curvatures K_1 and K_2, and Laplace–Beltrami operators L_1 and L_2. Let

$$T: M_1 \to M_2$$

be a *conformal diffeomorphism*, i.e.,

$$T^* g_2 = \tau^2 g_1 \qquad \text{where} \quad \tau \in \mathscr{E}(M_1).$$

Then if $f \in \mathscr{E}(M_2)$, $c_n = (n-2)/4(n-1)$, $n = \dim M_1 = \dim M_2$,

$$(L_1 - c_n K_1)(\tau^{n/2-1}(f \circ T)) = \tau^{n/2+1}((L_2 - c_n K_2 f) \circ T,$$

so that the null spaces of the operators $L_1 - c_n K_1$ and $L_2 - c_n K_2$ correspond under T.

6*. Let M be a compact Riemannian manifold, g the Riemannian structure, d the distance, L the Laplace–Beltrami operator, $m = \dim M$, ($m \geq 2$), and

$$\langle f_1, f_2 \rangle = \int_M f_1(x) f_2(x)^- \, dx, \qquad f_1, f_2 \in L^2(M),$$

where dx is the volume element.

(i) Let

$$\mathscr{E}_0(M) = \left\{u \in \mathscr{E}(M): \int_M u(x)\, dx = 0\right\}.$$

Then $L: \mathscr{E}_0(M) \to \mathscr{E}_0(M)$ is a bijection and the inverse is given by an integral operator

$$(Gf)(x) = \int_M \gamma(x, y) f(y)\, dy,$$

where $\gamma(x, y) = \gamma(y, x)$ satisfies

(a) γ is C^∞ for $x \neq y$.

(b) $d(x, y)^{m-2}\gamma(x, y)$ is bounded. [If $m = 2$, this property should be replaced by "$(\log d(x, y))^{-1}\gamma(x, y)$ is bounded".]

A proof can be found in de Rham [1955], Chapter V. Theorem 5.27 gives a more explicit version for the rank-one case.

(ii) For $\lambda \in \boldsymbol{C}$ let E_λ be the eigenspace

$$E_\lambda = \{u \in \mathscr{E}(M) : Lu = \lambda u\}$$

and Λ the spectrum

$$\Lambda = \{\lambda \in \boldsymbol{C} : E_\lambda \neq 0\}.$$

Then

(a) Λ is discrete and $\lambda \leq 0$ for each $\lambda \in \Lambda$.

(b) $\dim E_\lambda < \infty$ for each λ.

(c) In accordance with (a) and (b) let $\phi_0, \phi_1, \phi_2, \ldots$ be an orthonormal system in $L^2(M)$ such that each E_λ is spanned by some of the ϕ_i. Then if $f \in L^2(M)$,

$$f = \sum_0^\infty \langle f, \phi_n \rangle \phi_n \qquad \text{[convergence in } L^2(M)\text{].}$$

(d) If $f \in \mathscr{E}(M)$, this expansion converges absolutely and uniformly.

[A proof of the results in (ii) can be found, for example, in F. Warner [1970], Chapter 6.]

7. Let M be a manifold and μ a measure on M equivalent to the Lebesgue measure. Let X be a vector field on M that is symmetric, i.e.,

$$\int_M (Xf)(x) g(x)\, d\mu(x) = \int_M f(x)(Xg)(x)\, d\mu(x), \qquad f, g \in \mathscr{D}(M).$$

Prove that $X = 0$.

8. Let $\mathfrak{g}$ be a semisimple Lie algebra and $\mathfrak{g} = \mathfrak{k} + \mathfrak{p}$ a Cartan decomposition. Let $\mathfrak{a} \subset \mathfrak{p}$ be a maximal abelian subspace, $\mathfrak{m}$ the centralizer of $\mathfrak{a}$ in $\mathfrak{k}$, and $\mathfrak{h}$ a Cartan subalgebra of $\mathfrak{g}$ containing $\mathfrak{a}$. Let superscript c denote complexification and let P_+ denote the set of roots $\beta > 0$ of $\mathfrak{g}^c$ with respect to $\mathfrak{h}^c$ that do not vanish identically on $\mathfrak{a}$. Let $T_1, \ldots, T_m$ be a basis of $\mathfrak{m}$ orthonormal with respect to $-B$, $H_1, \ldots, H_l$ a basis of $\mathfrak{a}$ orthonormal with respect to B. For each $\beta \in P_+$ choose $X_\beta \in (\mathfrak{g}^c)^\beta$, $X_{-\beta} \in (\mathfrak{g}^c)^{-\beta}$ such that $B(X_\beta, X_{-\beta}) = 1$. Because of the decomposition

$$\mathfrak{g}^c = \mathfrak{m}^c + \mathfrak{a}^c + \sum_{\beta \in P_+} (\mathfrak{g}^c)^\beta + \sum_{-\beta \in P_+} (\mathfrak{g}^c)^\beta$$

the Casimir operator Ω (Exercise A4) can be written

$$\Omega = \Omega_{\mathfrak{m}} + \Omega_{\mathfrak{a}} + \sum_{\beta \in P_+} (X_\beta X_{-\beta} + X_{-\beta} X_\beta),$$

where $\Omega_{\mathfrak{m}} = \sum_i T_i^2$, $\Omega_{\mathfrak{a}} = \sum_j H_j^2$. (Here we write X_β for $\tilde{X}_\beta$, etc., for simplicity.)

Let $\mathfrak{a}'$ be the set of regular elements in $\mathfrak{a}$ and fix $a = \exp H$ $(H \in \mathfrak{a}')$. If $\beta \in P_+ \cup (-P_+)$, we write $X_\beta = Y_\beta + Z_\beta (Y_\beta \in \mathfrak{p}^c, Z_\beta \in \mathfrak{k}^c)$ and determine $A_\beta \in \mathfrak{a}$ such that $B(H, A_\beta) = \beta(H)$ for $H \in \mathfrak{a}$. If $g \in G$, $X \in \mathfrak{g}^c$, we put $X^g = \mathrm{Ad}(g)X$. From the expression above show that

$$\begin{aligned} \Omega = \Omega_{\mathfrak{m}} + \Omega_{\mathfrak{a}} &+ \sum_{\beta \in P_+} \coth(\beta(H)) A_\beta \\ &- 2 \sum_{\beta \in P_+} \sinh^{-2}(\beta(H)) [(Z_\beta Z_{-\beta})^{a^{-1}} + Z_\beta Z_{-\beta} \\ &- e^{\beta(H)} Z_{-\beta}^{a^{-1}} Z_\beta - e^{-\beta(H)} Z_\beta^{a^{-1}} Z_{-\beta}] \end{aligned}$$

(Harish-Chandra [1960]; cf. Warner [1972, Vol. II], Proposition 9.1.2.11).

B. The Radial Part

1*. Let H be a Lie transformation group of a manifold V. Then a submanifold $W \subset V$ satisfying the transversality condition [(29) in §3]

$$V_w = (H \cdot w)_w \oplus W_w \qquad \text{(direct sum)}, \quad w \in W,$$

always exists.

2. If f is a radial function on a symmetric space of rank one, then

$$Lf = \frac{\partial^2 f}{\partial r^2} + \frac{1}{A(r)} A'(r) \frac{\partial f}{\partial r}$$

by Proposition 5.26. Derive this formula from Theorem 3.7 as well.

3. Theorem 3.7 remains true if V is a manifold with a pseudo-Riemannian structure g provided the following additional conditions are satisfied:

(i) For each $w \in W$ the orbit $H \cdot w$ is closed and g is nondegenerate on it.

(ii) For each w the isotropy group H^w is compact.

Example. Let (G, H) be a symmetric pair, relative to an involutive automorphism σ of G, where G is assumed connected, semisimple with finite center. Let $K \subset G$ be a maximal compact subgroup invariant under σ ([DS], Chapter III, Exercise B4). Then the K-radial part of $L_{G/H}$ can be determined by the above variation of Theorem 3.7 (cf. Chang [1979a], Flensted-Jensen [1978], §4, Hoogenboom [1983]).

4. The K-radial part of L_X on the symmetric space $X = G/K$ (Proposition 3.9) can also be written

$$\Delta(L_X) = \delta^{-1} \sum_{1 \leq i, j \leq l} g^{ij} H_i \circ \delta H_j.$$

Here δ is the density function $\prod_{\alpha > 0} (\sinh \alpha)^{m_\alpha}$, $H_1, \ldots, H_l$ is any basis of $\mathfrak{a}$, and (g^{ij}) is the inverse of the matrix $g_{ij} = \langle H_i, H_j \rangle$ $(1 \leq i, j \leq l)$ (Harish-Chandra [1958a]).

5*. (Extension of Radial Part to Distributions)

(i) Let M and N be two manifolds and let $\pi: M \to N$ be a surjective C^∞-mapping such that $d\pi_m: M_m \to N_{\pi(m)}$ is surjective for each $m \in M$. Let dm and dn be measures on M and N, respectively, equivalent to Lebesgue measure. Then for each $\alpha \in \mathcal{D}(M)$ there exists a unique function $f_\alpha \in \mathcal{D}(N)$ such that

$$\int_M (F \circ \pi)\alpha \, dm = \int_N F f_\alpha \, dn, \qquad F \in \mathcal{D}(N).$$

By partition of unity the proof can be reduced to the case when M and N are Euclidean rectangles and π is the projection $(x_1, \ldots, x_m) \to (x_1, \ldots, x_n)$ (cf. [DS], Chapter I, Theorem 15.5). Then if $dm \llcorner \pi^*(dn)$ is the "ratio" of the forms dm and $\pi^*(dn)$, one can put

$$f_\alpha(n) = \int_{\pi^{-1}(n)} \alpha(dm \llcorner \pi^* \, dn)$$

(cf. Harish-Chandra [1964a]; also Stoll [1952] and Schwartz [1966]).

(ii) Let H be a Lie transformation group of the manifold V and $W \subset V$ a submanifold satisfying the transversality condition (29):

$$V_w = (H \cdot w)_w \oplus W_w, \qquad w \in W.$$

Fix a left-invariant measure dh on H and measures dv on V, dw on W, equivalent to Lebesgue measure. Let $V^* \subset V$ denote the open set $H \cdot W$. By (i) we can to each $\alpha \in \mathcal{D}(H \times W)$ associate an $f_\alpha \in \mathcal{D}(V)$ such that

$$\int_{H \times W} F(h \cdot w)\alpha(h, w) \, dh \, dw = \int_V F(v) f_\alpha(v) \, dv$$

for all $F \in \mathcal{D}(V^*)$.

(a) Assume the measure dv invariant under H and let $T \in \mathscr{D}'(W^*)$ be H-invariant. Show that there exists a unique distribution $\bar{T}$ on W such that

$$T(f_\alpha) = \bar{T}(A_\alpha) \qquad \text{for} \quad \text{all } \alpha \in \mathscr{D}(H \times W),$$

where

$$A_\alpha(w) = \int_H \alpha(h, w)\, dh.$$

(b) If T is an H-invariant continuous function f on V^*, show that

$$\bar{T}(A_\alpha) = \int_W f(w) A_\alpha(w)\, dw,$$

so $\bar{T} = \bar{f}\, dw$, where $\bar{f} = f | W$.

(c) Suppose H is unimodular and D a differential operator on V invariant under H. Then

$$(DT)^{-} = \Delta(D)\bar{T},$$

where $\Delta(D) \in \boldsymbol{E}(W)$ is the radial part of D. (cf., Helgason [1972a], Chapter I, §2).

6*. The K-radial parts of the operators in $\boldsymbol{D}(\boldsymbol{SU}(p, q)/\boldsymbol{S}(\boldsymbol{U}_p \times \boldsymbol{U}_q)) = \boldsymbol{D}(G/K)$ $(0 < p \leq q)$ can be described as follows. In accordance with [DS] (Chapter VII, Exercise 5), the space $\mathfrak{a} \subset \mathfrak{p}$ can be chosen as

$$\mathfrak{a} = \{H_t = t_1 H_1 + \cdots + t_p H_p : t_i \in \boldsymbol{R}\},$$

where $H_i = E_{i\,p+i} + E_{p+i\,i}$ and the restricted roots are given by

$$\pm f_i,\ \pm 2f_i \qquad (1 \leq i \leq p), \qquad \pm f_i \pm f_j \qquad (1 \leq i < j \leq p),$$

where $f_i(H_t) = t_i$. Put (with ch for coth)

$$L_i = \frac{\partial^2}{\partial t_i^2} + 2[(q - p)\operatorname{ch} t_i + \operatorname{ch} 2t_i]\frac{\partial}{\partial t_i},$$

$$\omega(H_t) = \prod_{i<j} (\operatorname{ch} 2t_i - \operatorname{ch} 2t_j),$$

$$D_i = \omega^{-1} S_i(L_1, \ldots, L_p) \circ \omega,$$

where S_i is the ith elementary symmetric function. Then the algebra $\Delta(\boldsymbol{D}(G/K))$ of K-radial parts is generated by $D_1, \ldots, D_p$.

(For details, as well as formulas for the spherical function ϕ_λ in the same spirit, see Berezin and Karpelevič [1958], Takahashi [1977a], and Hoogenboom [1982].)

C. Invariant Differential Operators

1. Determine explicitly the invariant differential operators on the symmetric space $\boldsymbol{GL}(n, \boldsymbol{R})/\boldsymbol{O}(n)$.

2*. Determine explicitly the algebra

$$\boldsymbol{D}(\boldsymbol{M}(n)/\boldsymbol{M}(p) \times \boldsymbol{O}(n-p))$$

of invariant differential operators on the manifold

$$\boldsymbol{G}(p, n) = \boldsymbol{M}(n)/\boldsymbol{M}(p) \times \boldsymbol{O}(n-p)$$

of p-planes in $\boldsymbol{R}^n$ (Chapter I, §2).

To describe the operators, let $\boldsymbol{G}_{p,n}$ denote the space of p-planes through the origin and let

$$\pi\colon \boldsymbol{G}(p, n) \to \boldsymbol{G}_{p,n}$$

denote the mapping obtained by parallel translating a plane to one through 0. If $\sigma \in \boldsymbol{G}_{p,n}$, the fiber $F = \pi^{-1}(\sigma)$ is naturally identified with the orthogonal complement $\sigma^{\perp}$ and has therefore a canonical Riemannian structure. Let L_F denote the corresponding Laplacian. If $f \in \mathscr{E}(\boldsymbol{G}(p, n))$ let $f\,|\,F$ denote its restriction to F. Consider now the linear mapping $\square$ of $\mathscr{E}(\boldsymbol{G}(p, n))$ into itself given by

$$(\square f)\,|\,F = L_F(f\,|\,F).$$

(i) The operator $\square$ is an $\boldsymbol{M}(n)$-invariant differential operator on $\boldsymbol{G}(p, n)$.

(ii) The algebra $\boldsymbol{D}(\boldsymbol{M}(n)/\boldsymbol{M}(p) \times \boldsymbol{O}(n-p))$ of $\boldsymbol{M}(n)$-invariant differential operators on $\boldsymbol{G}(p, n)$ is for $p = n-1$ generated by $\square$. (See Helgason [1965a], §§7–8.) For $0 < p < n-1$ the algebra has $\min(p+1, n-p)$ algebraically independent generators (cf. Gonzalez–Helgason, Invariant differential operators on Grassmann manifolds, *Advan. Math.* 1986). These have orders 2, 4, 6, . . . etc.

3. Determine explicitly the algebra

$$\boldsymbol{D}(G/M)$$

of invariant differential operators on the manifold G/M corresponding to the hyperbolic space $G/K = \boldsymbol{O}_0(1, n)/\boldsymbol{SO}(n)$. [As usual, M denotes the centralizer of A in K; here it can be identified with $\boldsymbol{SO}(n-1)$.] In view of the fibration $G/M \to G/K$, G/M can be viewed as the *unit sphere bundle* of G/K. With the orthogonal decompositions $\mathfrak{k} = \mathfrak{m} + \mathfrak{l}$, $\mathfrak{p} = \mathfrak{a} + \mathfrak{q}$ we write

$$\mathfrak{g} = \mathfrak{m} + (\mathfrak{a} + \mathfrak{l} + \mathfrak{q}),$$

so by Theorem 4.9, the problem is to determine the algebra $I(\mathfrak{a}+\mathfrak{l}+\mathfrak{q})$ of M-invariant polynomials on $\mathfrak{a}+\mathfrak{l}+\mathfrak{q}$. We have $\dim \mathfrak{a} = 1$, $\dim \mathfrak{l} = \dim \mathfrak{q} = n-1$, $[\mathfrak{l}, \mathfrak{a}] = \mathfrak{q}$, and Σ^+ consists of a single root α. Fix $H \in \mathfrak{a}$ such that $\alpha(H) = 1$; let $T_1, \ldots, T_{n-1}$ be an orthogonal basis of $\mathfrak{l}$ such that $-B(T_i, T_i) = B(H, H)$ $(1 \leq i \leq n-1)$. Put $X_i = [T_i, H]$. Then $B(X_i, X_i) = B(H, H)$ $(1 \leq i \leq n-1)$. With the basis $H, T_1, \ldots, T_{n-1}, X_1, \ldots, X_{n-1}$ of $\mathfrak{a}+\mathfrak{l}+\mathfrak{q}$ put

$$|T|^2 = \sum T_i^2, \qquad T \cdot X = \sum_i T_i \cdot X_i, \qquad |X|^2 = \sum_i X_i^2.$$

Prove that

(i) The algebra $I(\mathfrak{a}+\mathfrak{l}+\mathfrak{q})$ is generated by H, $|T|^2$, $T \cdot X$, $|X|^2$. The corresponding operators D_H, $D_{|T|^2}$, $D_{T\cdot X}$, $D_{|X|^2}$ therefore generate $\boldsymbol{D}(G/M)$.

(ii) The generators above satisfy the following commutation relations:

$$[D_H, D_{|X|^2}] = -2D_{T\cdot X},$$

$$[D_H, D_{|T|^2}] = -2D_{T\cdot X},$$

$$[D_H, D_{T\cdot X}] = -D_{|T|^2} - D_{|X|^2}.$$

(For these results and interesting geometric interpretations of the operators see Reimann [1982]; see also Ahlfors [1975].)

4. The homomorphism $\gamma: \boldsymbol{D}_K(G) \to \boldsymbol{D}_W(A)$ from Theorem 5.17 is defined by means of a choice of positive restricted roots. Show, however, that γ is independent of this choice.

5. (i) Let N denote the three-dimensional Heisenberg group (§4, No. 1). The algebra $\boldsymbol{Z}(N)$ of biinvariant differential operators is generated by $\partial/\partial x_3$.

(ii)* Consider the four-dimensional Lie algebra $\mathfrak{n}_4 = \sum_{i=1}^4 \boldsymbol{R}X_i$ with the brackets

$$[X_1, X_2] = X_3, \qquad [X_1, X_3] = X_4,$$

all other brackets being 0. Then $\mathfrak{n}_4$ is nilpotent and the center $\boldsymbol{Z}(\mathfrak{n}_4)$ of its universal enveloping algebra is generated by X_4 and $X_3^2 - 2X_2X_4$ (cf. Dixmier [1958]).

6*. Let G be a simply connected nilpotent Lie group with Lie algebra $\mathfrak{g}$. Under the adjoint action of G on $\mathfrak{g}$, $\mathfrak{g}$ induces an algebra $\mathfrak{g}^+$ of vector fields on $\mathfrak{g}$ ([DS], Chapter II, Theorem 3.4).

Let $D \in \boldsymbol{Z}(G)$ and let ε_D be the corresponding distribution (§5, No. 1). Viewing ε_D as a distribution on $\mathfrak{g}$ (by exp), we obtain a differential operator D_0 on $\mathfrak{g}$ with constant coefficients by

$$D_0 f = f * \varepsilon_D \qquad \text{(convolution on } \mathfrak{g}).$$

If $D^{\exp^{-1}}$ is the image of D under $\exp^{-1}$, then

$$D_0 - D^{\exp^{-1}} \in \boldsymbol{E}(\mathfrak{g})\mathfrak{g}^+$$

(cf. Kashiwara and Vergne [1978a], which has also more general results). Thus the action of D on distributions invariant under all inner automorphisms is completely determined by the constant coefficient operator D_0.

7*. The symmetrization $\lambda: S(\mathfrak{g}) \to \boldsymbol{D}(G)$ maps $I(\mathfrak{g})$ bijectively onto $\boldsymbol{Z}(G)$. Although $I(\mathfrak{g})$ and $\boldsymbol{Z}(G)$ are both commutative algebras, λ is not in general multiplicative on them.

If $\mathfrak{g}$ is semisimple, Harish-Chandra has defined purely algebraically [1951a] a ring isomorphism between $I(\mathfrak{g})$ and $\boldsymbol{Z}(G)$ (cf. also Theorem 1.9 in Chapter V). For $\mathfrak{g}$ arbitrary Duflo [1977] has established a surjective ring isomorphism $\gamma: I(\mathfrak{g}) \to \boldsymbol{Z}(G)$: Let $j(X)$ be the Jacobian determinant of the exponential map $\exp: \mathfrak{g} \to G$, i.e.,

$$j(X) = \det\left(\frac{1 - e^{-\operatorname{ad} X}}{\operatorname{ad} X}\right)$$

([DS], Chapter II, Theorem 1.7). Then $j^{1/2}$ is analytic in a neighborhood V of 0. View an element P of $S(\mathfrak{g})$ as constant coefficient differential operators $\partial(P)$ on $\mathfrak{g}$. Given $P \in S(\mathfrak{g})$, let $P^\natural \in S(\mathfrak{g})$ be determined by

$$(\partial(P^\natural)\phi)(0) = (\partial(P)(j^{1/2}\phi))(0)$$

for all $\phi \in \mathscr{D}(V)$. Then $\gamma(P) = \lambda(P^\natural)$ for $P \in I(\mathfrak{g})$.

For $\mathfrak{g}$ nilpotent an isomorphism was given by Dixmier [1959].

8*. Consider the space $\mathfrak{gl}(n, \boldsymbol{R})$ of all $n \times n$ real matrices $X = (x_{ij})$. Consider the Capelli differential operator D on $\mathfrak{gl}(n, \boldsymbol{R})$ given by

$$(Df)(X) = (\det X)\left[\det\left(\left(\frac{\partial}{\partial x_{ij}}\right)_{1 \le i, j \le n}\right)f\right](X).$$

Then D is invariant under left and right multiplications by members of $\boldsymbol{GL}(n, \boldsymbol{R})$. Thus it gives a biinvariant differential operator on the group $G = \boldsymbol{GL}^+(n, \boldsymbol{R})$ of $n \times n$ matrices of positive determinant. For $\lambda \in \boldsymbol{C}$ define D_λ by

$$(D_\lambda f)(X) = (\det X)^{-\lambda} D((\det X)^\lambda f)(X).$$

Then

$$D_\lambda = \lambda^n + E_{n-1}\lambda^{n-1} + \cdots + E_1\lambda + D,$$

where $E_i \in \boldsymbol{Z}(G)$. Moreover, $\boldsymbol{Z}(G)$ is the polynomial ring

$$\boldsymbol{Z}(G) = \boldsymbol{C}[E_{n-1}, \ldots, E_1, D].$$

For $n = 2$, $Z(G)$ is generated by

$$D = (x_{11}x_{22} - x_{12}x_{21})\left(\frac{\partial^2}{\partial x_{11}\,\partial x_{22}} - \frac{\partial^2}{\partial x_{12}\,\partial x_{21}}\right),$$

$$E_1 = \sum_{1 \le i, j \le 2} x_{ij}\frac{\partial}{\partial x_{ij}} + 1$$

(cf. Raïs [1977] (unpublished), Carter and Lusztig [1974], and Dixmier [1975]).

D. Restriction Theorems

1. This exercise outlines Harish-Chandra's proof of Chevalley's restriction theorem: The restriction $P \to P|\mathfrak{a}$ is a bijection of $I(\mathfrak{p})$ onto $I(\mathfrak{a})$ (Corollary 5.12).

(a) Let $\mathfrak{p}^C$ and $\mathfrak{a}^C$ denote the complexifications and $I(\mathfrak{p}^C)$ [resp., $I(\mathfrak{a}^C)$] the sets of K-invariants (resp. W-invariants) in $S(\mathfrak{p}^C)$ and $S(\mathfrak{a}^C)$. Prove that

$$I(\mathfrak{p}^C) \qquad \textit{is integrally closed.}$$

For this use the fact that $S(\mathfrak{p}^C)$ is integrally closed (Appendix, Lemma 3.4) and that if $x \in S(\mathfrak{p}^C)$ lies in the quotient field $C(I(\mathfrak{p}^C))$ of $I(\mathfrak{p}^C)$, then $x \in I(\mathfrak{p}^C)$.

(b) Let $J = I(\mathfrak{p}^C)|\mathfrak{a}^C$ (restriction to $\mathfrak{a}^C$). Then

$$S(\mathfrak{a}^C) \qquad \textit{is integral over } J.$$

For this, for $X \in \mathfrak{p}^C$, let $T_X = (\operatorname{ad} X)^2|\mathfrak{p}^C$ and consider the characteristic polynomial

$$\det(\lambda^2 I - T_X) = \lambda^{2r} + p_1(x)\lambda^{2r-2} + \cdots + p_{r-l}(x)\lambda^{2l},$$

where $p_i \in I(\mathfrak{p}^C)$. In particular,

$$\det(\lambda^2 I - T_H) = \lambda^{2r} + p_1(H)\lambda^{2r-2} + \cdots + p_{r-l}(H)\lambda^{2l},$$

so that each root $\alpha \in \Sigma$ is integral over J.

(c) If $H_0 \in \mathfrak{a}$, $H_1 \in \mathfrak{a}^C$ satisfy $p(H_0) = p(H_1)$ for all $p \in J$, then $H_1 = sH_0$ for some $s \in W$.

Since

$$\det(\lambda^2 - T_{H_0}) \equiv \det(\lambda^2 - T_{H_1}),$$

it follows that each $\alpha \in \Sigma$ is real on H_1, so $H_1 \in \mathfrak{a}$. Since each $p \in I(\mathfrak{p})$ has the same value on H_0 and H_1, it follows from Weierstrass's approximation theorem on the *real* space $\mathfrak{p}$ that H_0 and H_1 are $\operatorname{Ad}_G(K)$-conjugate. Now use [DS], Chapter VII, Proposition 2.2.

(d) The field extension $C(S(\mathfrak{a}^C))/C(J)$ is normal.

In fact, the quotient field $C(S(\mathfrak{a}^C))$ is obtained by adjoining all the roots of the polynomial

$$F(\lambda) \equiv \lambda^{2r} + p_1\lambda^{2r-2} + \cdots + p_{r-l}\lambda^l,$$

namely, all $\alpha \in \Sigma$, to $C(J)$.

(e) Let σ be an automorphism of $C(S(\mathfrak{a}^C))$ leaving $C(J)$ pointwise fixed. Fix $H_0 \in \mathfrak{a}$. Then there exists an $H_1 \in \mathfrak{a}^C$ such that

$$p^\sigma(H_0) = p(H_1) \qquad \text{for all } p \in S(\mathfrak{a}^C).$$

In fact, σ permutes the roots of $F(\lambda)$ and hence leaves $S(\mathfrak{a}^C)$ invariant; thus $p \to p^\sigma(H_0)$ is a homomorphism of $S(\mathfrak{a}^C)$ into C.

(f) By (c), $H_1 = sH_0$ for some $s \in W$, so if $q \in I(\mathfrak{a}^C)$, then

$$q^\sigma(H_0) = q(H_1) = q(H_0);$$

thus since $H_0 \in \mathfrak{a}$ was arbitrary, $q^\sigma = q$. Hence by Galois theory, $I(\mathfrak{a}^C) \subset C(J)$. But $I(\mathfrak{a}^C)$ is integral over J and J is [by (a)] integrally closed. Thus $I(\mathfrak{a}^C) = J$, as claimed. (For more details see Helgason [1962a], Chapter X.)

2*. Let U/K be a symmetric space of the compact type, U simply connected, K connected. Let the notation be as in [DS], Chapter VII, §8. Let $\mathscr{E}_K(U/K)$ denote the set of K-invariant functions in $\mathscr{E}(U/K)$ and let $\mathscr{E}_{\Gamma_\Sigma}(\mathfrak{a}_*)$ denote the set of Γ_Σ-invariant functions in $\mathscr{E}(\mathfrak{a}_*)$.

Given $F \in \mathscr{E}_K(U/K)$, define $f \in \mathscr{E}(\mathfrak{a}_*)$ by

$$f(H) = F((\exp H)K), \qquad H \in \mathfrak{a}_*.$$

By [DS], loc. cit., the map $F \to f$ maps $\mathscr{E}_K(U/K)$ into $\mathscr{E}_{\Gamma_\Sigma}(\mathfrak{a}_*)$. Prove that the mapping $F \to f$ is surjective (Dadok [1982]).

3*. Let the notation be as in Proposition 5.32, so that $I(\mathfrak{h}^{C*})$ is the set of polynomials on $\mathfrak{h}^C$ invariant under the Weyl group $\tilde{W} = W(\mathfrak{g}^C, \mathfrak{h}^C)$ and $I(\mathfrak{a}^*)$ is the set of polynomials on $\mathfrak{a}$ invariant under the Weyl group $W = W(\mathfrak{g}, \theta)$.

(i) If G/K is a *classical* symmetric space (i.e., neither G nor K exceptional), then the restriction mapping

$$(*) \qquad p \in I(\mathfrak{h}^{C*}) \to \bar{p} \in I(\mathfrak{a}^*)$$

is *surjective.*

(ii) For the symmetric spaces E_6/F_4, $E_7/E_6 \cdot T$, $E_8/E_7 \times SU(2)$ the restriction mapping $(*)$ is *not surjective.* (For the proof see Helgason [1964a]; for a related positive result see Theorem 3.16 in Chapter III.)

4. Let the notation be as in §5, No. 2. Call a function on a Euclidean space $\boldsymbol{R}^n$ *Lipschitzean* if for each $R > 0$ there exists a constant C_R such that

$$|f(x) - f(Y)| \leq C_R|X - Y| \qquad \text{for} \quad X, Y \in \boldsymbol{R}^n, \quad |X| < R, \quad |Y| < R.$$

Prove that each Weyl group-invariant Lipschitzean function ϕ on $\mathfrak{a}$ extends to a K-invariant Lipschitzean function Φ on $\mathfrak{p}$. [*Hint*: Use Prop. 5.18 in Chapter I.]

E. Distributions

1. In the notation of §5, No. 1 show that if $f \in \mathscr{D}(G/K)$, $T \in \mathscr{E}'(G/K)$, then

$$(f \times T)(g \cdot o) = \int_G f(gh^{-1} \cdot o)\, d\tilde{T}(h),$$

$$(T \times f)(g \cdot o) = \int_G f(h^{-1}g \cdot o)\, d\tilde{T}(h).$$

2*. If $T \in \mathscr{D}'(G)$ such that $T * \mathscr{D} \subset \mathscr{D}$, then $T \in \mathscr{E}'(G)$ (cf., Ehrenpreis [1956]).

3. In the notation of §5, No. 1, let $s, t \in \mathscr{E}'(G)$, $D \in \mathbf{D}(G)$. Then

$$s * (D^*)\check{}\, t = Ds * t.$$

Using this for $s = \tilde{S}$, $t = \tilde{T}$, $D \in \lambda(I(\mathfrak{p}))$, derive another proof of Theorem 5.5 (Koornwinder).

F. The Wave Equation

1. Write down the solution of the wave equation

$$\frac{\partial^2 u}{\partial x_1^2} + \cdots + \frac{\partial^2 u}{\partial x_n^2} = \frac{\partial^2 u}{\partial t^2}, \qquad u(x, 0) = u_0(x), \qquad u_t(x, 0) = u_1(x),$$

by using Ásgeirsson's mean value theorem on the function

$$v(x_1, \ldots, x_n, y_1, \ldots, y_n) = u(x_1, \ldots, x_n, t), \qquad y_1 = t,$$

which satisfies the equation

$$\frac{\partial^2 v}{\partial x_1^2} + \cdots + \frac{\partial^2 v}{\partial x_n^2} = \frac{\partial^2 v}{\partial y_1^2} + \cdots + \frac{\partial^2 v}{\partial y_n^2}$$

(Ásgeirsson [1937]).

2*. As in Chapter I, §4 consider the hyperbolic space $\boldsymbol{H}^n$ with Riemannian structure ds^2 of constant sectional curvature -1. Show that the Lorentzian manifold $\boldsymbol{H}^n \times \boldsymbol{R}$ has constant scalar curvature

$$K = -n(n-1)$$

(cf. Exercise A5). As shown there, the natural wave equation on $\boldsymbol{H}^n$ to consider is

$$\left(L + \left(\frac{n-1}{2}\right)^2\right)u = \frac{\partial^2 u}{\partial s^2}, \qquad u(x, 0) = u_0(x), \quad u_s(x, 0) = u_1(x).$$

Using Theorem 5.28, find an explicit solution formula for this equation. Show that for n odd, Huygens's principle holds.

NOTES

§§1–2. The characterization in Theorem 1.4 is from Peetre [1959, 1960] (with a proof by L. Carleson). We have followed the presentation in Narasimhan [1968] of this as well as of the results 1.2 and 1.3. Distributions on manifolds are treated in detail by Schwartz [1966]. The Laplace–Beltrami operator (cf. Beltrami [1864]) has played a prominent role in Riemannian geometry in each of the forms (12), (13), (26) in §2. Its relation to the operator $d\delta + \delta d$ on forms is explained in de Rham [1955], §26; for another proof see Exercise A2 in this chapter. Proposition 2.4 and Theorem 2.7 are from Helgason [1959], Chapter I, and [1970a], Chapter IV.

§3. Nos. 1–3. This material is based on Helgason [1972a], Chapter I, [1965b], and [1972b]. Of the results in §3, No. 4, Examples (ii) and (iii), the noncompact versions of (v) and (vii) are from Harish-Chandra's papers, [1958a, 1956b], and [1957a], respectively, with entirely different proofs. See also Berezin [1957] and Karpelevič [1962]. Theorem 3.15 is from Helgason [1972a], and so is the proof in the text for Example (vii).

§4. Theorem 4.3 is a modification of results of Harish-Chandra and Schwartz (see Harish-Chandra [1953], p. 192, and [1956b], p. 111). For Corollary 4.5 see, in addition, Gelfand [1950b]. The description of $\boldsymbol{D}(G/H)$ (Theorems 4.9–4.11) is from Helgason [1959]. The nonreductive case has been studied by Hole [1974], Koornwinder [1981] and Jacobsen [1982]. Proposition 4.12 occurs in Günther [1957] for harmonic spaces and in Berezin and Gelfand [1956] and Helgason [1959] for symmetric spaces.

§5, No. 1. The convolution calculus on distributions on G/K (in particular, Theorem 5.5) is from Helgason [1973a]. The commutativity of $\mathscr{D}^\natural(G)$ for G/K symmetric (Corollary 5.2) was first found by Gelfand [1950a]. Corollary 5.4 is stated in a different form in Harish-Chandra [1954c], Lemma 1, and is proved in Selberg [1956], where Corollary 5.3 is also stated. The result is indicated in Gelfand [1950a], but the proof described in Berezin *et al.* [1960] is insufficient because it is based on the statement that $\boldsymbol{Z}(G)$ gets mapped *onto* $\boldsymbol{D}(G/K)$ by applying the operators in $\boldsymbol{Z}(G)$ to functions on G which are right invariant under K. In general, this surjectivity fails (example: E_6/F_4; see Proposition 5.32 and Exercise D3), although it happens to hold for all classical G (real or complex). The present proof of part (i) of Theorem 5.7 was written in 1960, following a discussion with Harish-Chandra. In [1963], Lichnèrowicz proved the same result under the weaker assumption that G/H has invariant volume element. A further (algebraic) generalization was given by Duflo [1979a].

§5, Nos. 2–3. Corollary 5.12 is an unpublished result of Chevalley which is utilized in Harish-Chandra [1958a], whose proof of the result is outlined in Exercise D1 in this chapter. The proof in the text, which also gives the more general Theorem 5.8, is due to Dadok [1982]. Theorem 5.13 is from Helgason [1977a], and Theorems 5.16 and 5.17 and Prop. 5.23 from Harish-Chandra [1958a]. Formula (39) is from Helgason [1964a], where it is used to prove local solvability of each $D \in \boldsymbol{D}(G/K)$. Theorem 5.24 is a special case of Harish-Chandra's expression of the Casimir operator on G in terms of the Cartan decomposition $G = KAK$ (cf. Exercise A8).

§5, Nos. 4–8, Proposition 5.26 and the solution to Poisson's equation (Theorem 5.27) are from Helgason [1959] and [1962a], Chapter X. Part (ii) of Theorem 5.27 is also in Allamigeon [1961]. Theorem 5.28 and its generalization to Riemannian homogeneous spaces are from Helgason [1957b], [1959]. The proof of Theorem 5.28 is a generalization of Ásgeirsson's original proof [1937] for the case $\boldsymbol{R}^n$. Theorems 5.33, 5.35, and 5.37 are from Harish-Chandra [1957a] and [1958a], §14, the material in §5, No. 7 and Theorem 5.38 from the author's paper [1964a].

CHAPTER III

INVARIANTS AND HARMONIC POLYNOMIALS

A linear action of a group on a vector space extends to an action on the symmetric algebra, giving rise to invariant polynomials and corresponding harmonic polynomials. In §1 we prove (under general assumptions) that each polynomial is a polynomial in the invariants with harmonics as coefficients. The harmonics are also described in terms of the variety defined by the invariants. For the linear isotropy action in case of a symmetric space the results hold in a strengthened form, discussed in §5.

In §2 we prove one analogous result for the exterior algebra. In §3 we consider the special case of the standard action of the Weyl group. This case is particularly interesting, because the harmonic polynomials, the invariants, and the corresponding eigenfunctions have profound connections with the root structure of the corresponding Lie algebra. In §4 we study the orbits for the complexified isotropy action for a symmetric space. Here the orbits have important relationships to the invariants and the harmonic polynomials.

§1. Decomposition of the Symmetric Algebra. Harmonic Polynomials

Let E be a finite-dimensional vector space over $\boldsymbol{R}$ and G a group of linear transformations of E. The action of G on E induces an action of G on the ring of polynomials on E. The fixed points, the *G-invariants*, form a subring. The *G-harmonic* polynomials h are the common solutions of the differential equations $Dh = 0$ where D is a G-invariant differential operator of constant coefficients annihilating the constants. Under some general assumptions about G we shall prove results of the following kind.

1. Each polynomial on E is a sum $\sum_k i_k h_k$ where i_k is invariant, h_k harmonic.

2. The G-harmonic polynomials are of two types:

(a) Those which vanish identically on the algebraic variety N_G determined by the G-invariants.

(b) The powers of the linear forms given by the points in N_G.

Each G-harmonic polynomial h is a sum $h = h_a + h_b$ where h_a is of type (a) and h_b a sum of harmonic polynomials of type (b) above.

We proceed now with a more detailed treatment. Let E^* denote the dual of E, $S(E^*)$ and $S(E)$ the corresponding *real* symmetric algebras. Then $S(E^*)$ consists of the polynomial functions on E. If $X \in E$ let $\partial(X)$ denote the differential operator given by

$$(\partial(X)f)(Y) = \left(\frac{d}{dt} f(Y + tX)\right)_{t=0}, \qquad f \in \mathscr{E}(E).$$

The mapping $X \to \partial(X)$ extends to an isomorphism of the symmetric algebra $S(E)$ (resp. the complex symmetric algebra $S^c(E) = \boldsymbol{C} \otimes S(E)$) onto the algebra of all differential operators on E with constant real (resp. complex) coefficients. (In Chapter II we dropped the superscript c for simplicity of notation.)

Each g in the general linear group $\boldsymbol{GL}(E)$ acts on E and on E^* by $(g \cdot e^*)(e) = e^*(g^{-1} \cdot e)$. These actions extend to automorphisms of $S(E)$ and $S(E^*)$.

Let $G \subset \boldsymbol{GL}(E)$ be any subgroup and let $I(E)$ denote the set of G-invariants in $S(E)$ and $I_+(E) \subset I(E)$ the set of G-invariants without constant term. Similarly we define $I_+(E^*) \subset I(E^*) \subset S(E^*)$. An element $h \in S^c(E^*)$ is said to be *G-harmonic* if $\partial(J)h = 0$ for all $J \in I_+(E)$. Let $H^c(E^*)$ denote the set of G-harmonic polynomial functions and put $H(E^*) = S(E^*) \cap H^c(E^*)$. Let $I^c(E)$ and $I^c(E^*)$, respectively, denote the subspaces of $S^c(E)$ and $S^c(E^*)$ generated by $I(E)$ and $I(E^*)$. Each $p \in S^c(E^*)$ extends uniquely to a polynomial function on the complexification E^c, also denoted p. Let N_G denote the variety in E^c defined by

$$N_G = \{X \in E^c : j(X) = 0 \text{ for all } j \in I_+(E^*)\}.$$

Remark. The terminology G-harmonic is of course suggested by the case when G is the orthogonal group. In this case the invariants are the polynomials in the square of the distance and the G-harmonic polynomials are the polynomial solutions of Laplace's equation.

Let B be a nondegenerate symmetric bilinear form on $E \times E$; let B also denote its bilinear extension to $E^c \times E^c$. If $X \in E^c$ let X^* denote the linear form $Y \to B(X, Y)$ on E. The mapping $X \to X^*$ $(X \in E)$ extends uniquely to an isomorphism $\mu: P \to p$ of $S^c(E)$ onto $S^c(E^*)$. Under this isomorphism B gives rise to a bilinear form $\langle\ ,\ \rangle$ on $E^* \times E^*$ which in turn extends to a bilinear form $\langle\ ,\ \rangle$ on $S^c(E^*) \times S^c(E^*)$; the formula for $\langle\ ,\ \rangle$ is

(1) $$\langle p, q\rangle = (\partial(P)q)(0).$$

Since $\partial(X)(B(Y, \cdot)) = B(X, Y)$, $\langle\ ,\ \rangle$ does indeed extend B. On the other hand, let $S^k(E^*)$ denote the space of elements in $S(E^*)$ which are homogeneous of degree k. We define a bilinear form $\langle\ ,\ \rangle'$ on $S^k(E^*) \times S^k(E^*)$ by

$$\langle x_1 \cdots x_k, y_1 \cdots y_k\rangle' = \sum_{\sigma \in \mathfrak{S}_k} \prod_{i=1}^{k} \langle x_i, y_{\sigma(i)}\rangle \tag{2}$$

for $x_i, y_j \in E^*$, $\mathfrak{S}_k$ being the symmetric group on k letters. This bilinear form is indeed well defined because the right-hand side is linear in each x_i (and each y_j) and is independent of their order. We then extend $\langle\ ,\ \rangle'$ to a bilinear form on $S^c(E^*) \times S^c(E^*)$. It is clear [say by (4) below] that $\langle x^k, y^k\rangle' = \langle x^k, y^k\rangle$, so since the powers x^k $(x \in E^*)$ span $S^k(E^*)$ (Chapter I, Exercise D1), $\langle\ ,\ \rangle$ and $\langle\ ,\ \rangle'$ are identical. In particular, $\langle\ ,\ \rangle$ is symmetric and now (1) shows that it is nondegenerate.

Moreover, if $p, q, r \in S^c(E^*)$ then

$$\langle p, qr\rangle = (\partial(QR)p)(0) = (\partial(R)\partial(Q)p)(0) = \langle\partial(Q)p, r\rangle,$$

which shows that multiplication by q is the adjoint to the operator $\partial(Q)$.

Now suppose G leaves B invariant; then $\langle\ ,\ \rangle$ is also G-invariant and

$$\mu(I^c(E)) = I^c(E^*). \tag{3}$$

Let P be a homogeneous element in $S^c(E)$ of degree k. If $n \in \mathbf{Z}$, $n \geq k$, then the relation

$$\partial(P)((X^*)^n) = n(n-1)\cdots(n-k+1)p(X)(X^*)^{n-k} \tag{4}$$

can be verified by a simple computation. In particular, if $X \in N_G$ then $(X^*)^n$ is a harmonic polynomial. Let $H_1(E^*)$ denote the vector space over $\mathbf{C}$ spanned by the functions $(X^*)^n$, $(n = 0, 1, 2, \ldots; X \in N_G)$ and let $H_2(E^*)$ denote the set of harmonic polynomial functions which vanish identically on N_G.

If $A \subset S^c(E^*)$ is a subspace and $k \in \mathbf{Z}^+$, A^k denotes the set of homogeneous elements in A of degree k; A is called *homogeneous* if $A = \sum A^k$. The spaces $I(E^*)$, $H(E^*)$, $H_1(E^*)$ and the ideal $I_+(E^*)S(E^*)$ are clearly homogeneous.

If C and D are subspaces of an associative algebra A then CD will denote the subspace of A spanned by the products cd $(c \in C, d \in D)$.

Theorem 1.1. *Let G be a compact group of linear transformations of a vector space E over $\mathbf{R}$. Then*

$$S(E^*) = I(E^*)H(E^*), \tag{5}$$

that is, each polynomial p on E has the form $p = \sum_k i_k h_k$ where i_k is G-invariant and h_k harmonic.

Example. If G is the orthogonal group $\boldsymbol{O}(n)$ acting on $\boldsymbol{R}^n$ the invariants are the polynomials in $x_1^2 + \cdots + x_n^2$ so (5) gives the classical decomposition of any polynomial p,

$$p = \sum_k (x_1^2 + \cdots + x_n^2)^k h_k, \tag{6}$$

where the polynomials h_k are harmonic, i.e., $L_{\boldsymbol{R}^n}(h_k) = 0$ (cf. Introduction, §3, No. 1).

Proof of Theorem 1.1. Let B be a positive definite quadratic form invariant under G. Using a basis of E orthonormal with respect to B it is easy to see from (2) that $\langle\ ,\ \rangle$ is positive definite on $S(E^*) \times S(E^*)$. For a given $k \in \boldsymbol{Z}^+$ consider the direct decomposition

$$S^k(E^*) = (I_+(E^*)S(E^*))^k \oplus V^k, \tag{7}$$

V^k being the orthogonal complement with respect to $\langle\ ,\ \rangle$. Let $l \leq k$ and let $j \in I_+(E^*)^l$. Since multiplication by j and $\partial(J)$ are adjoint operators, (7) implies

$$\langle S^{k-l}(E^*), \partial(J)V^k\rangle = 0,$$

whence $V^k \subset H^k(E^*)$. The converse inclusion follows in the same way, so we have the orthogonal decomposition

$$S^k(E^*) = (I_+(E^*)S(E^*))^k + H^k(E^*). \tag{8}$$

Now (5) follows by iteration.

Theorem 1.2. *Let E be a finite-dimensional vector space over $\boldsymbol{R}$ and let G be a connected semisimple Lie subgroup of $\boldsymbol{GL}(E)$ leaving invariant a nondegenerate symmetric bilinear form B on E. Then*

$$S(E^*) = I(E^*)H(E^*).$$

Proof. The form B has a unique extension (also denoted B) to a bilinear form on $E^c \times E^c$. Let $\mathfrak{g}$ be the Lie algebra of G, $\mathfrak{g}^C$ its complexification, and let $\mathfrak{u}$ be an arbitrary compact real form of $\mathfrak{g}^C$. Since $\mathfrak{g}$ is a subalgebra of $\mathfrak{gl}(E)$, the Lie algebra of $\boldsymbol{GL}(E)$, $\mathfrak{g}^C$ is a subalgebra of $\mathfrak{gl}(E^c)$, the Lie algebra of all linear transformations of E^c, which again is the Lie algebra of $\boldsymbol{GL}(E^c)$. Let U and G^C be the connected Lie subgroups of $\boldsymbol{GL}(E^c)$ (considered as a real Lie group) corresponding to $\mathfrak{u}$ and $\mathfrak{g}^C$, respectively. The elements of G extend uniquely to linear transformations of E^c whereby G becomes a Lie subgroup of G^C leaving B invariant. This implies that

$$B(T \cdot Z_1, Z_2) + B(Z_1, T \cdot Z_2) = 0, \qquad T \in \mathfrak{g}, \quad Z_1, Z_2 \in E^c. \tag{9}$$

However, since $(T_1 + iT_2) \cdot Z = T_1 \cdot Z + iT_2 \cdot Z$ for $T_1, T_2 \in \mathfrak{g}$, $Z \in E^c$, and since B is $\boldsymbol{C}$-bilinear, it follows that (9) holds for all $T \in \mathfrak{g}^C$; hence, by the connectedness of G^C, B is invariant under G^C.

Lemma 1.3. *There exists a real form F of E^c on which B is strictly positive definite and which is left invariant by U.*

Proof. By the usual reduction of quadratic forms the space E is an orthogonal direct sum $E = E^- + E^+$ of subspaces E^- and E^+ on which $-B$ and B, respectively, are strictly positive definite. Let J be the linear transformation of E^c determined by

$$JZ = iZ \quad (Z \in E^-), \qquad JZ = Z \quad (Z \in E^+).$$

Then the bilinear form

$$B'(Z_1, Z_2) = B(JZ_1, JZ_2), \qquad Z_1, Z_2 \in E^c$$

is strictly positive definite on E. Let $\boldsymbol{O}(B)$, $\boldsymbol{O}(B') \subset \boldsymbol{GL}(E^c)$ denote the orthogonal groups of B and B', respectively, and let $\boldsymbol{O}(B'_E)$ denote the subgroup of $\boldsymbol{O}(B')$ which leaves E invariant, i.e., $\boldsymbol{O}(B'_E) = \boldsymbol{O}(B') \cap \boldsymbol{GL}(E)$. Now we have

$$U \subset G^C \subset \boldsymbol{O}(B) = J\boldsymbol{O}(B')J^{-1}.$$

Since $\boldsymbol{SO}(n)$ is a maximal compact subgroup of $\boldsymbol{SO}(n, \boldsymbol{C})$, the identity component of the group $J\boldsymbol{O}(B'_F)J^{-1}$ is a maximal compact subgroup of the identity component of $J\boldsymbol{O}(B')J^{-1}$. By an elementary special case of Cartan's conjugacy theorem ([DS], Theorem 2.1, Chap. VI) this last group contains an element g such that $g^{-1}Ug \subset J\boldsymbol{O}(B'_E)J^{-1}$. Then the real form $F = gJE$ of E^c has the properties stated in the lemma. In fact,

$$UF = UgJE \subset gJ\boldsymbol{O}(B'_E)J^{-1}(JE) \subset F,$$

and if $X \in F$ then since $J^{-1}g^{-1}J \in \boldsymbol{O}(B')$ we have

$$B(X, X) = B'(J^{-1}X, J^{-1}X) = B'(J^{-1}g^{-1}X, J^{-1}g^{-1}X) \geq 0,$$

B' being positive definite on E. This proves the lemma.

Since a polynomial function is determined by its restriction to a real form it is not difficult to reduce Theorem 1.2 to (5) by means of Lemma 1.3. For the sake of clarity we carry this out by explicit maps.

Since B is nondegenerate on $E \times E$, $F \times F$ and on $E^c \times E^c$ there are induced surjective isomorphisms

$$\mu_1 : S^c(E) \to S^c(E^*), \quad \mu_2 : S^c(F) \to S^c(F^*), \quad \mu : S(E^c) \to S((E^c)^*).$$

By restriction of a complex-valued function on E^c to E and F, respectively, we get the surjective isomorphisms

$$\lambda_1 : S((E^c)^*) \to S^c(E^*), \qquad \lambda_2 : S((E^c)^*) \to S^c(F^*).$$

Since $S(E^c) = S(((E^c)^*)^*)$, we obtain by restricting functions on $(E^c)^*$ to E^* and F^*, respectively, the surjective isomorphisms

$$\Lambda_1 : S(E^c) \to S^c(E), \qquad \Lambda_2 : S(E^c) \to S^c(F).$$

We then have the commutative diagram

$$\begin{array}{ccccc} S^c(E) & \xleftarrow{\Lambda_1} & S(E^c) & \xrightarrow{\Lambda_2} & S^c(F) \\ \downarrow{\scriptstyle \mu_1} & & \downarrow{\scriptstyle \mu} & & \downarrow{\scriptstyle \mu_2} \\ S^c(E^*) & \xleftarrow{\lambda_1} & S((E^c)^*) & \xrightarrow{\lambda_2} & S^c(F^*). \end{array}$$

Corresponding to the actions

$$G \text{ on } E, \qquad U \text{ on } F, \qquad G^{\mathbf{C}} \text{ on } E^c$$

we consider the spaces of invariants

$$I^c(E), \quad I^c(E^*), \quad I^c(F), \quad I^c(F^*), \quad I(E^c), \quad I((E^c)^*).$$

Lemma 1.4. Let $\lambda = \lambda_2 \lambda_1^{-1}$, $\Lambda = \Lambda_2 \Lambda_1^{-1}$. Then

$$\lambda(I^c(E^*)) = I^c(F^*), \qquad \Lambda(I^c(E)) = I^c(F).$$

Proof. Since $G \subset G^{\mathbf{C}}$ it is clear that $\lambda_1(I((E^c)^*)) \subset I^c(E^*)$. On the other hand, let $p \subset I^c(E^*)$. If $Z \in \mathfrak{g}$ let d_Z denote the unique derivation of $S^c(E^*)$ which satisfies $(d_Z \cdot e^*)(X) = e^*(Z \cdot X)$ for $e^* \in E^*$, $X \in E$. Then

$$d_Z \cdot p = 0. \tag{10}$$

Let (X_i) be a basis of E, (x_i) the dual basis of E^*, (z_i) the basis of $(E^c)^*$ dual to (X_i) considered as a basis of E^c. Then (10) is an identity in (x_i) which remains valid after the substitution $x_i \to z_i$ (all i). This means that

$$\delta_Z \cdot (\lambda_1^{-1}(p)) = 0, \tag{11}$$

where δ_Z is the derivation of $S((E^c)^*)$ which satisfies

$$(\delta_Z \cdot e^*)(X) = e^*(Z \cdot X)$$

for $e^* \in (E^c)^*$, $X \in E^c$. However, δ_Z can be defined for all $Z \in \mathfrak{g}^{\mathbf{C}}$ by this last condition and (11) remains valid for all $Z \in \mathfrak{g}^{\mathbf{C}}$. Since $G^{\mathbf{C}}$ is connected, this implies $\lambda_1^{-1}(p) \in I((E^c)^*)$. Thus we have proved $\lambda_1(I((E^c)^*)) = I^c(E^*)$; similarly $\lambda_2(I((E^c)^*)) = I^c(F^*)$, so the first statement of the lemma follows. The second statement follows from the first if we take into account (3) and the diagram above.

Lemma 1.5. *Let $P \in S^c(E)$, $q \in S^c(E^*)$. Then*

$$\partial(\Lambda P)(\lambda q) = \lambda(\partial(P)q).$$

Proof. First, suppose $P = X \in E$, $q = \mu_1(Y)$ $(Y \in E)$. Here both sides of the equation reduce to $B(X, Y)$. Therefore, the derivations $q \to \partial(\Lambda X)\lambda q$ and $q \to \lambda(\partial(X)q)$ of $S^c(E^*)$ coincide on E^*, and hence on all of $S^c(E^*)$. Since the mappings $P \to \partial(\Lambda P)$, $P \to \partial(P)$ are isomorphisms, the lemma follows.

Combining the two last lemmas we obtain the following result.

Corollary 1.6. $\lambda(H^c(E^*)) = H^c(F^*)$.

With U acting on F we deduce from Theorem 1.1

$$S^c(F^*) = I^c(F^*)H^c(F^*).$$

Applying the isomorphism λ^{-1}, Lemma 1.4, and Corollary 1.6, we obtain Theorem 1.2.

Theorem 1.7. *Let E be a finite-dimensional vector space over* $\mathbf{R}$, *and $G \subset \mathbf{GL}(E)$ a subgroup leaving invariant a nondegenerate symmetric bilinear form B on E. Then*

(a) *If G is compact and B positive definite,*

$$H^c(E^*) = H_1(E^*) + H_2(E^*) \qquad (\textit{direct sum}). \tag{12}$$

(b) *If G is a connected semisimple Lie subgroup of* $\mathbf{GL}(E)$, (12) *is also valid.*

We begin with a simple lemma.

Lemma 1.8. *The space $H_2(E^*)$ is homogeneous.*

Proof. We recall that the elements of $S^c(E^*)$ are automatically extended to polynomial functions on E^c without change in notation. Let $J = I_+^c(E^*)S^c(E^*)$. Then N_G is the variety of common zeros of elements of the ideal J. By Hilbert's *Nullstellensatz* (see, e.g., Zariski and Samuel [1960], p. 164) the polynomials in $S^c(E^*)$ which vanish identically on N_G consitute the radical $\sqrt{J}$ of J, that is, the set of elements in $S^c(E^*)$ of which some power lies in J. Now let $p \in \sqrt{J}$ and let $p = \sum_0^r p_k$ be the decomposition of p into homogeneous components. Writing this as $p = q + p_r$ we see that, by the homogeneity of J, the relation $p^n \in J$ implies $p_r^n \in J$; in other words, $p_r \in \sqrt{J}$. By linearity of $\sqrt{J}$, $q \in \sqrt{J}$ and now

$p_k \in \sqrt{J}$ follows inductively. Thus $\sqrt{J}$ is homogeneous, and since $H_2(E^*) = H^c(E^*) \cap \sqrt{J}$ the lemma is proved.

For the theorem consider first Case (a). For $k \in \mathbf{Z}^+$ let M be the span of the orthocomplement of $H(E^*) \cap H_1(E^*)^k$ in $H(E^*)^k$, the superscript k indicating the space of homogeneous components of degree k. Let $q \in H^c(E^*)^k$, $Q = \mu^{-1}(q)$. Then

$$\begin{aligned} q \in M &\Leftrightarrow (\partial(Q)h)(0) = 0 \qquad \text{for all } h \in H_1(E^*)^k \\ &\Leftrightarrow \partial(Q)(X^*)^k = 0 \qquad \text{for all } X \in N_G. \end{aligned}$$

In view of (4) this amounts to q vanishing identically on N_G; consequently, $M = H_2(E^*)^k$. This proves formula (12) since all terms in it are homogeneous.

Next we consider Case (b). We use the notions introduced during the proof of Theorem 1.2. Because of Lemma 1.4 the varieties N_U and N_G coincide. It follows that $\lambda(H_i(E^*)) = H_i(F^*)$ $(i = 1, 2)$ and now Case (b) reduces to the previous case.

Example. Again we consider the example $G = \mathbf{O}(n)$ acting on $\mathbf{R}^n$. In this case

$$H_2(\mathbf{R}^n) = 0, \tag{13}$$

so by (12),

$$H^c(\mathbf{R}^n) = \text{spanned by } \left\{(a_1x_1 + \cdots + a_nx_n)^k : \sum_1^n a_r^2 = 0, k \in \mathbf{Z}^+\right\}. \tag{14}$$

This was already proved in the Introduction, §3, No. 1.

Corollary 1.9. *With the notation of Theorem 1.7 suppose the ideal $I_+^c(E^*)S^c(E^*)$ equals its own radical. Then*

$$H^c(E^*) = H_1(E^*).$$

This is clear since

$$H_2(E^*) = H^c(E^*) \cap J = 0. \tag{15}$$

Theorem 1.10. *Let E be a finite-dimensional vector space over $\mathbf{R}$ and let $G \subset \mathbf{GL}(E)$ be a Lie subgroup which is either (a) compact (possibly finite) or (b) connected and semisimple. Then the algebra $I(E^*)$ is finitely generated.*

Proof. Let us first consider the compact case. Again let

$$J = I_+^c(E^*)S^c(E^*).$$

By Hilbert's basis theorem (cf., e.g., Zariski and Samuel [1958], p. 200) the ideal J has a finite basis. It follows that there exist finitely many homogeneous invariants $j_1, \ldots, j_s \in I^c_+(E^*)$ such that each homogeneous invariant $j \in I^c_+(E^*)$ can be written

$$j = p_1 j_1 + \cdots + p_s j_s, \tag{16}$$

where all $p_k \in S^c(E^*)$ are homogeneous and degree(p_k) = degree(j) − degree(j_k). Applying the linear transformation $g \in G$ and integrating over the compact group G we obtain

$$j = i_1 j_1 + \cdots + i_s j_s,$$

where $i_k = \int_G g \cdot p_k \, dg$. Applying (16) to the homogeneous invariant i_k we obtain by induction, $j \in \mathbf{C}[j_1, \ldots, j_s]$ as desired.

Passing to Case (b), let $\mathfrak{g}$, $\mathfrak{g}^{\mathbf{C}}$, $\mathfrak{u}$, G, $G^{\mathbf{C}}$, U be as in the proof of Theorem 1.2. Consider also the isomorphism λ_1 of $S((E^c)^*)$ onto $S^c(E^*)$, which we proved satisfies

$$\lambda_1(I((E^c)^*)) = I^c(E^*). \tag{17}$$

A polynomial function on E^c is U-invariant if and only if it is $G^{\mathbf{C}}$-invariant. This is readily seen by expressing the invariance condition by means of the Lie algebras $\mathfrak{u}$ and $\mathfrak{g}^{\mathbf{C}}$ (cf. proof of Lemma 1.4). Now Case (a) implies that $I((E^c)^*)$, and therefore $I^c(E^*)$, by (17), is finitely generated.

Writing now $S = S((E^c)^*)$, $I = I((E^c)^*)$, let S^k denote the space of homogeneous elements in S of degree k and put $I^k = I \cap S^k$. We shall now find the *generating function* $\sum_0^\infty \dim(I^k)t^k$ of the sequence $\dim(I^k)$ in the case in which G has finite order $N = |G|$.

Theorem 1.11. *The generating function is given by*

$$\sum_0^\infty \dim(I^k)t^k = \frac{1}{|G|} \sum_{g \in G} (\det(I - tg))^{-1}, \tag{18}$$

where I is the identity operator on E^c.

Proof. Let $F^c \subset E^c$ be the subspace fixed by each $g \in G$. Then the operator $P = N^{-1} \sum_{g \in G} g$ satisfies $g_0 P = P$ for each $g_0 \in G$. Thus $PE^c = F^c$ and $P^2 = P$. In particular,

$$\dim(F^c) = \frac{1}{|G|} \sum_{g \in G} \operatorname{Tr}(g), \tag{19}$$

Tr denoting trace. On the other hand, if $g^{(k)}$ denotes the endomorphism of S^k induced by g we have, by diagonalizing g,

$$\frac{1}{\det(1 - tg)} = \sum_{k \geq 0} \mathrm{Tr}(g^{(k)})t^k.$$

Now the theorem follows by applying (19) to the endomorphism $g^{(k)}$.

§2. Decomposition of the Exterior Algebra. Primitive Forms

Let E be a finite-dimensional vector space over $\boldsymbol{R}$ as in §1 and let $\Lambda(E)$ and $\Lambda(E^*)$, respectively, denote the Grassmann algebras over E and its dual. Each $X \in E$ induces a linear mapping $\delta(X)$ of $\Lambda(E^*)$ given by

$$\delta(X)(x_1 \wedge \cdots \wedge x_m) = \sum_{k=1}^{m} (-1)^{k-1} x_k(X)(x_1 \wedge \cdots \wedge \hat{x}_k \wedge \cdots x_m),$$

where $\hat{x}_k$ denotes omission of x_k. Then $\delta(X)$ is an *antiderivation* of $\Lambda(E^*)$ in the sense that

$$\delta(X)(p \wedge q) = \delta(X)p \wedge q + (-1)^r p \wedge \delta(X)q$$

(cf. [DS], Chapter I, Exercises B5–6) if p has degree r. The mapping $X \to \delta(X)$ extends uniquely to a homomorphism of the tensor algebra $T(E)$ over E into the algebra of all endomorphisms of $\Lambda(E^*)$. Since

$$\delta(X \otimes X) = \delta(X)^2 = 0$$

there is induced a homomorphism $P \to \delta(P)$ of $\Lambda(E)$ into the algebra of endomorphisms of $\Lambda(E^*)$. As will be noted below, this homomorphism is actually an isomorphism.

Now suppose B is any nondegenerate symmetric bilinear form on $E \times E$. The mapping $X \to X^*(X^*(Y) = B(X, Y))$ extends to an isomorphism μ of $\Lambda(E)$ onto $\Lambda(E^*)$. We obtain a bilinear form $\langle\ ,\ \rangle$ on $\Lambda(E^*)$ by the formula

(1) $$\langle p, q \rangle = [\delta(\mu^{-1}(p))q](0).$$

If $x_1, \ldots, x_k, y_1, \ldots, y_l \in E^*$ then by direct computation,

$$\langle x_1 \wedge \cdots \wedge x_k, y_1 \wedge \cdots \wedge y_l \rangle = 0$$

or

(2) $$(-1)^{(1/2)k(k-1)} \det[B(\mu^{-1}x_i, \mu^{-1}y_j)],$$

depending on whether $k \neq l$ or $k = l$. It follows that $\langle\ ,\ \rangle$ is a symmetric nondegenerate bilinear form. Also if $Q \in \Lambda(E)$, $q = \mu(Q)$ then the operator $p \to p \wedge q$ on $\Lambda(E^*)$ is the adjoint of the operator $\delta(Q)$. It is also easy to see from (1) and (2) that the mapping $P \to \delta(P)$ $[P \in \Lambda(E)]$ above is an isomorphism.

Now let G be a group of linear transformations of E. Then G acts on E^* as before and acts as a group of automorphisms of $\Lambda(E)$ and $\Lambda(E^*)$. Let $J(E)$ and $J(E^*)$ denote the set of G-invariants in $\Lambda(E)$ and $\Lambda(E^*)$, respectively; let $J_+(E)$ and $J_+(E^*)$ denote the subspaces consisting of all invariants without constant term. An element $p \in \Lambda(E^*)$ is called *G-primitive* if $\delta(Q)p = 0$ for all $Q \in J_+(E)$. Let $P(E^*)$ denote the set of all G-primitive elements.

Theorem 2.1. *Let B be a nondegenerate, symmetric bilinear form on $E \times E$ and let G be a Lie subgroup of $\mathbf{GL}(E)$ leaving B invariant. Suppose that either* (i) *G is compact and B positive definite or* (ii) *G is connected and semisimple. Then*

$$\Lambda(E^*) = J(E^*)P(E^*). \tag{3}$$

The proof is quite analogous to that of Theorems 1.1 and 1.2. For case (i) one first establishes the orthogonal decomposition

$$\Lambda(E^*) = \Lambda(E^*)J_+(E^*) + P(E^*) \tag{4}$$

in the same manner as (8) in §1. Then (3) follows by iteration of (4). The noncompact case (ii) can be reduced to the case (i) by using Lemma 1.3. We omit the details since they are essentially a duplication of the proof of Theorem 1.2.

Example. Let V be an n-dimensional Hilbert space over $\mathbf{C}$. Considering the set V as a $2n$-dimensional vector space E over $\mathbf{R}$ the unitary group $\mathbf{U}(n)$ becomes a subgroup G of the orthogonal group $\mathbf{O}(2n)$. Let $Z_k = X_k + iY_k$ $(1 \leq k \leq n)$ be an orthonormal basis of V, $z_1, \ldots, z_n$ the dual basis of V^*, and put $x_k = \frac{1}{2}(z_k + \bar{z}_k)$, $y_k = -\frac{1}{2}i(z_k - \bar{z}_k)$ $(1 \leq k \leq n)$.

Let F denote the vector space over $\mathbf{C}$ consisting of all $\mathbf{R}$-linear mappings of V into $\mathbf{C}$. The exterior algebra $\Lambda(F)$ is the direct sum

$$\Lambda(F) = \sum_{0 \leq a, b} F_{a,b},$$

where $F_{a,b}$ is the subspace of $\Lambda(F)$ spanned by all multilinear forms of the type

$$(z_{\alpha_1} \wedge \cdots \wedge z_{\alpha_a}) \wedge (\bar{z}_{\beta_1} \wedge \cdots \wedge \bar{z}_{\beta_b}),$$

where $1 \le \alpha_1 < \alpha_2 < \cdots < \alpha_a \le n$, $1 \le \beta_1 < \beta_2 < \cdots < \beta_b \le n$. The space $J(E^*)$ of G-invariants is given by the space J of invariants of $\boldsymbol{U}(n)$ acting on F. It is clear that $J = \sum_{a,b} J_{a,b}$ where $J_{a,b} = J \cap F_{a,b}$. Now if $\rho_i \in \boldsymbol{R}$ $(1 \le i \le n)$ the mapping

$$(Z_1, \ldots, Z_n) \to (e^{i\rho_1}Z_1, \ldots, e^{i\rho_n}Z_n)$$

is unitary. As a consequence one finds that $J_{a,b} = 0$ if $a \neq b$ and that if $f \in J_{a,a}$ then

$$f = \sum A_{\alpha_1,\ldots,\alpha_a}(z_{\alpha_1} \wedge \cdots \wedge z_{\alpha_a}) \wedge (\bar{z}_{\alpha_1} \wedge \cdots \wedge \bar{z}_{\alpha_a}).$$

Now, there always exists a unitary transformation of V mapping $Z_{\alpha_i} \to Z_i$ $(1 \le i \le a)$. It follows that $A_{1,\ldots,a} = A_{\alpha_1,\ldots,\alpha_a}$ so f is a constant multiple of $(\sum_{\alpha=1}^n z_\alpha \wedge \bar{z}_\alpha)^a$. Since $z_\alpha \wedge \bar{z}_\alpha = -2i(x_\alpha \wedge y_\alpha)$ it is clear that $J(E^*)$ is the algebra generated by $u = \sum_{\alpha=1}^n x_\alpha \wedge y_\alpha$. From Theorem 2.1 we can therefore derive the following consequence.

Corollary 2.2. *Each $q \in \Lambda(E^*)$ can be expressed*

$$q = \sum_k u^k \wedge p_k$$

where $u = \sum_{\alpha=1}^n x_\alpha \wedge y_\alpha$ and each p_k is primitive; i.e., $\delta(u)(p_k) = 0$.

§3. Invariants for the Weyl Group

We shall now amplify and extend the results of §§1, 2 in the case in which W is a finite group of linear transformations of E generated by reflections, in short, a *finite reflection group.* By a *reflection* we mean (as in [DS], Chapter X) a linear transformation of order 2 whose fixed points form a hyperplane. Let w or $|W|$ denote the order of W. The Weyl groups associated with semisimple Lie groups and symmetric spaces are finite reflection groups.

1. Symmetric Invariants

Theorem 1.10 can be put in a more precise form for a finite reflection group.

Theorem 3.1. *Let W be a finite reflection group acting on an n-dimensional real vector space E. Then the algebra $I((E^c)^*)$ of invariants is generated by n homogeneous elements which are algebraically independent.*

Example. Let W be the symmetric group acting linearly on C^n by $\sigma e_i = e_{\sigma(i)}$, e_i being the ith coordinate vector. Here the algebra of invariants is generated by the elementary symmetric functions

$$\sum_i z_i, \sum_{i \neq j} z_i z_j, \ldots, z_1 \cdots z_n,$$

and W is generated by reflections.

As before let $S = S((E^c)^*)$, $I = I((E^c)^*)$, $I_+ \subset I$ be the space of invariants without constant term, J the ideal $I_+ S$, and H the space $H((E^c)^*)$ of harmonic polynomials. Let S^k, I^k, and H^k denote the respective subspaces consisting of homogeneous elements of degree k. Let deg stand for degree.

Lemma 3.2. *Let $j_1, \ldots, j_m \in I$ such that $j_1 \notin \sum_2^m j_s I$. If $q_1, \ldots, q_m \in S$ are homogeneous elements such that $\sum_1^m j_s q_s = 0$, then $q_1 \in J$.*

Proof. The proof goes by induction on $\deg(q_1)$. If $\deg(q_1) = 0$ then $q_1 = 0$ because otherwise the relation $\sum_1^m j_i q_i = 0$ gives, by averaging over W, $j_1 \in \sum_2^m j_s I$, contrary to assumption. Let $\deg(q_1) \geq 1$, let $\sigma \in W$ be a reflection, and let the hyperplane $z = 0$ constitute the fixed points of σ. From the assumptions we deduce $\sum_1^m j_i(q_i - \sigma \cdot q_i) = 0$. But $q_i - \sigma q_i$ vanishes on the hyperplane $z = 0$, so $q_i - \sigma q_i = z r_i$, where $r_i \in S$. Hence $\sum_1^m j_i r_i = 0$, and so by the induction hypothesis $r_1 \in J$; i.e., $\sigma \cdot q_1 \equiv q_1 \mod J$. Since these reflections σ generate W this congruence holds for all $\sigma \in W$. Averaging over σ we deduce $q_1 \in J$.

We now prove Theorem 3.1. Let $j_1, \ldots, j_s \in I_+$ be homogeneous elements forming a minimal basis of the ideal J. During the proof of Theorem 1.10 it was shown that $j_1, \ldots, j_s$ generate I. We shall now show that they are algebraically independent. Suppose this were not so. Let $P(y_1, \ldots, y_s)$ be a polynomial of minimal degree >0, such that

$$P(j_1(Z), \ldots, j_s(Z)) = 0 \quad \text{for all} \quad Z \in E.$$

Let $d_i = \deg(j_i)$. Writing P as a sum of monomials $y_1^{a_1} \cdots y_s^{a_s}$ we may assume P only contains such monomials for which the degree $a_1 d_1 + \cdots + a_s d_s$ of the polynomial $Z \to j_1(Z)^{a_1} \cdots j_s(Z)^{a_s}$ is a certain fixed number h. Let

$$P_i(Z) = \frac{\partial P}{\partial y_i}(j_1(Z), \ldots, j_s(Z)) \qquad (1 \leq i \leq s).$$

The P_i are invariant homogeneous polynomials. Since $\deg P > 0$, $\partial P / \partial y_i \neq 0$ for some i, so by the minimality of $\deg P$ the corresponding P_i is $\neq 0$. We relabel the indices such that $P_1, \ldots, P_t$ are precisely the P_i which are $\neq 0$ and such that among $P_1, \ldots, P_r$ none is in the ideal of I

generated by the others whereas $P_{r+1}, \ldots, P_t$ are in the ideal $\sum_1^r IP_i$. Thus

$$P_{r+j} = \sum_{i=1}^{r} v_{ij} P_i, \qquad 1 \leq j \leq t - r,$$

where each nonzero v_{ij} is a homogeneous polynomial of degree

$$h - d_{r+j} - (h - d_i) = d_i - d_{r+j}.$$

With Cartesian coordinates $z_1, \ldots, z_n$ on E^c we differentiate the relation $P(j_1, \ldots, j_s) = 0$ and obtain

$$0 = \sum_{i=1}^{s} P_i \frac{\partial j_i}{\partial z_k} = \sum_{i=1}^{r} P_i \frac{\partial j_i}{\partial z_k} + \sum_{l=1}^{t-r} P_{r+l} \frac{\partial j_{r+l}}{\partial z_k}$$
$$= \sum_{i=1}^{r} P_i \left(\frac{\partial j_i}{\partial z_k} + \sum_{l=1}^{t-r} v_{il} \frac{\partial j_{r+l}}{\partial z_k} \right).$$

However, the polynomial inside the parentheses is homogeneous of degree $d_i - 1$. By Lemma 3.2 we conclude for each i (with $P_1, \ldots, P_r$ playing the role of $j_1, \ldots, j_m$),

$$\frac{\partial j_i}{\partial z_k} + \sum_{l=1}^{t-r} v_{il} \frac{\partial j_{r+l}}{\partial z_k} = \sum_{m=1}^{s} B_m j_m,$$

where each B_m is homogeneous. Considering the degrees we see that $B_i = 0$. We multiply this relation by z_k, sum over k, and use Euler's formula for homogeneous polynomials. This gives

$$d_i j_i + \sum_{l=1}^{t-r} v_{il}\, d_{r+l} j_{r+l} = \sum_{m=1}^{s} A_m j_m,$$

where $A_i = 0$. This shows that j_i is in the ideal $\sum_{m \neq i} j_m S$ contradicting the minimality of the basis $j_1, \ldots, j_s$.

Thus the $j_1, \ldots, j_s$ are algebraically independent.

It remains to prove that $s = n$. Let $K = \mathbf{C}(j_1, \ldots, j_s)$ be the field generated by the j_i, that is, the quotient field of $I = \mathbf{C}[j_1, \ldots, j_s]$. Also let $Q = \mathbf{C}(z_1, \ldots, z_n)$ be the quotient field of $S = \mathbf{C}[z_1, \ldots, z_n]$. If $p \in S$ then the polynomial

(1) $$R(X) = \prod_{\sigma \in W} (X - \sigma \cdot p) = X^w + q_1 X^{w-1} + \cdots + q_w$$

has coefficients in I. Since $R(p) = 0$ it follows that p, and therefore each element in Q, is algebraic over K. Denoting by tr deg the transcendence degree of a field extension, we have by general theory (Zariski and Samuel [1958], p. 100).

$$\text{tr deg}(Q/C) = \text{tr deg}(Q/K) + \text{tr deg}(K/C). \tag{2}$$

Since the extension Q/K is algebraic this implies $n = \text{tr deg}(K/C)$. Since any two transcendence bases of a field extension have the same cardinality, we conclude that $s = n$, so Theorem 3.1 is proved.

Theorem 3.3. *Let W be a finite reflection group acting on the real vector space E of dimension n. Let $j_1, \ldots, j_n$ be homogeneous generators for the algebra $I((E^c)^*)$ of W-invariants. Let $d_1, \ldots, d_n$ be their respective degrees. Then*

$$\prod_{i=1}^{n} d_i = w, \qquad \sum_{i=1}^{n} (d_i - 1) = r,$$

where w is the order of W and r the number of reflections in W. Moreover, the degrees d_i are unique.

Proof. First we observe from (2) that the j_i are algebraically independent. Let $j \in I^k$, the space of homogeneous invariants of degree k. Then j is a linear combination of monomials $j_1^{a_1} \cdots j_n^{a_n}$ for which

$$a_1 d_1 + \cdots + a_n d_n = k. \tag{3}$$

By the algebraic independence of the j_i it follows that $\dim(I^k)$ equals the number of nonnegative integral solutions $(a_1, \ldots, a_n)$ to (3). Hence

$$\sum_0^{\infty} \dim(I^k) t^k = (1 - t^{d_1})^{-1} \cdots (1 - t^{d_n})^{-1}. \tag{4}$$

Combining this with Theorem 1.11 we conclude

$$w \prod_{1 \leq i \leq n} (1 + t + \cdots + t^{d_i - 1})^{-1} = \sum_{\sigma \in W} \prod_{1 \leq j \leq n} \frac{1 - t}{1 - t c_{\sigma_j}} \tag{5}$$

if the c_{σ_j} are the eigenvalues of σ counted with multiplicity. Letting $t \to 1$, only the term $\sigma = 1$ on the right gives a contribution, so we obtain the first formula of the theorem.

Next we differentiate (5) with respect to t and then let $t \to 1$. A term corresponding to σ on the right gives a nonzero contribution exactly if σ has eigenvalue 1 with multiplicity $n - 1$. But this happens exactly if σ is a reflection; the contribution in question is

$$\left[\frac{d}{dt}\left(\frac{1 - t}{1 + t}\right)\right]_{t=0} = -\tfrac{1}{2}.$$

We conclude that

$$-w\sum_{j=1}^{n}\frac{1}{2}\frac{d_j(d_j-1)}{d_j^2}\prod_{i\neq j}d_i^{-1} = -\tfrac{1}{2}r,$$

and now the first formula of the theorem leads to the second one.

For the uniqueness let $J_1, \ldots, J_n$ be another set of homogeneous polynomials of respective degrees $D_1, \ldots, D_n$ generating I. By reordering we may assume $d_1 \leq d_2 \leq \cdots \leq d_n$ and $D_1 \leq D_2 \leq \cdots \leq D_n$. Because of (4) we have $\prod_i (1 - t^{d_i}) \equiv \prod_i (1 - t^{D_i})$. Considering the lowest power of t we get $d_1 = D_1$ and by cancellation, $d_2 = D_2$, etc. This proves the uniqueness.

2. Harmonic Polynomials

We can now strengthen Theorem 1.1 a bit in the case of a finite reflection group.

Theorem 3.4. *Let W be a finite reflection group and let the notation be as above. Then* $\dim H = w$ *and the mapping* $\phi: j \otimes h \to jh$ *extends to a linear bijection of* $I \otimes H$ *onto* S. *Moreover,*

$$\sum_{k\geq 0}(\dim H^k)t^k = \prod_{1\leq i\leq n}(1 + t + \cdots + t^{d_i - 1}). \tag{6}$$

Proof. We know from Theorem 1.1 that ϕ is surjective. To prove that it is injective we must show that if $\sum_{r,s} a_{r,s} i_r h_s = 0$, where $a_{r,s} \in C$ and $\{i_r\}$ and $\{h_s\}$ are homogeneous bases of the vector spaces I and H, respectively, then $a_{r,s} = 0$. We write the relation in the form

$$\sum_s h_s\left(\sum_r a_{r,s} i_r\right) = 0.$$

Put $I_s = \sum_r a_{r,s} i_r$. We have to prove that each $I_s = 0$ and for this it suffices to consider the case in which each I_s is homogeneous and $\deg h_s + \deg I_s$ the same for all s. Suppose there were an $I_s \neq 0$. We write it in the form

$$I_s = \sum a_{m_1,\ldots,m_n,s} j_1^{m_1}\cdots j_n^{m_n}$$

with nonzero coefficients $a_{m_1,\ldots,m_n,s}$. Then

$$\sum_{(m)}\left(\sum_s a_{m_1,\ldots,m_n,s} h_s\right) j_1^{m_1}\cdots j_n^{m_n} = 0,$$

and at least one of the monomials $j_1^{m_1}\cdots j_n^{m_n}$ is not in the ideal in I generated by the others. The corresponding term

$$\sum_s a_{m_1,\ldots,m_n,s} h_s$$

then belongs to J according to Lemma 3.2. But by Eq. (8) of §1 this term will have to vanish and then the linear independence of the h_s gives the contradiction $a_{m_1,\ldots,m_n,s} = 0$.

The identification $I \otimes H = S$ implies the identity

$$\sum_{k\geq 0} (\dim I^k)t^k \sum_{l\geq 0} (\dim H^l)t^l = \sum_{m\geq 0} (\dim S^m)t^m$$

for the generating functions. Since the right-hand side equals $(1-t)^{-n}$ the formula for $\sum (\dim H^l)t^l$ follows from (4). Putting $t = 1$, we obtain the formula $\dim H = w$ from Theorem 3.3.

Corollary 3.5. *The variety N_W reduces to $\{0\}$ so*

$$H = H_2, \qquad H_1 = 0$$

in the notation of Theorem 1.7.

In fact if $Z \neq 0$ is in N_W then by Eq. (4) of §1 all the powers $(Z^*)^k$ would be W-harmonic and this would contradict the finite-dimensionality of H. We can now determine the harmonic polynomials more explicitly.

Theorem 3.6. *Let W be a finite reflection group acting on the finite-dimensional real vector space E. Let $\sigma_1, \ldots, \sigma_r$ be the reflections in W; $\alpha_1 = 0, \ldots, \alpha_r = 0$ the corresponding reflecting hyperplanes; and let $\pi = \prod_1^r \alpha_i$. Then*

(i) $H = \partial(S)\pi$; *that is, the harmonic polynomials constitute the linear span of the partial derivatives of π.*

(ii) *If $f \in C^\infty(E)$ satisfies $\partial(P)f = 0$ for each $P \in I_+$, then $f \in H$. In other words, each W-harmonic function is a polynomial.*

Proof. First we prove (ii). Considering the polynomial (1) for $p = x_i$ we see that $x_i^w = -q_1 x_i^{w-1} - \cdots - q_w$, so $x_i^w \in I_+(E^*)S(E^*)$ $(1 \leq i \leq n)$. It follows that $\partial(P)f = 0$ whenever P is a homogeneous polynomial of degree $\geq nw$ (because then each term will contain a power X_i^a, $a \geq w$). Hence f is a polynomial of degree $< nw$.

For (i) we first prove some general facts about π. As before let $j_1, \ldots, j_n$ be homogeneous algebraically independent generators of I.

Lemma 3.7. *The Jacobian determinant satisfies the following identity*:

$$\frac{\partial(j_1, \ldots, j_n)}{\partial(x_1, \ldots, x_n)} = c\pi,$$

where c is a constant $\neq 0$.

Proof. Let q denote the left-hand side. Then by the chain rule for the Jacobian, q is *skew*; that is, it satisfies $\sigma \cdot q = (\det \sigma) q$ $(\sigma \in W)$. In particular, q vanishes in each hyperplane $\alpha_i = 0$, and so is divisible by the polynomial α_i. Hence π divides q and now the relation follows from the sum formula in Theorem 3.3.

It remains to prove $c \neq 0$. For each i, $1 \leq i \leq n$, the polynomials x_i, $j_1, \ldots, j_n$ are algebraically dependent. We choose a polynomial Q_i $(x_i, z_1, \ldots, z_n)$ of minimal degree $e_i > 0$ in x_i such that $Q_i(x_i, j_1, \ldots, j_n) = 0$. Applying $\partial/\partial x_k$ we obtain

$$\sum_{r=1}^{n} \frac{\partial Q_i}{\partial z_r}(x_i, j_1, \ldots, j_n) \frac{\partial j_r}{\partial x_k} + \delta_{ik} \frac{\partial Q_i}{\partial x_k}(x_i, j_1, \ldots, j_n) = 0.$$

We write this as a matrix identity

$$AB = C,$$

where

$$A_{ir} = \frac{\partial Q_i}{\partial z_r}(x_i, j_1, \ldots, j_r), \qquad B_{rk} = \frac{\partial j_r}{\partial x_k}$$

and C is the diagonal matrix given by

$$C_{ii} = \frac{\partial Q_i}{\partial x_i}(x_i, j_1, \ldots, j_n).$$

By the minimality of e_i, $\det C \neq 0$ so $\det B \neq 0$ as desired.

Corollary 3.8. *π is skew and each skew polynomial is divisible by π.*

Now it is clear that $\pi \in H$ because if $J \in I_+$, $\partial(J)\pi$ is skew, hence divisible by π. Hence we also have $\partial(S)\pi \subset H$. In order to prove the converse inclusion we prove the following result of independent interest.

Lemma 3.9. *If $\partial(Q)\pi = 0$ then $Q \in I_+ S$.*

Proof. Let B be a positive definite bilinear form on $E \times E$ invariant under W. Let $Q \to q$ be the corresponding bijection of $S(E^c)$ onto $S((E^c)^*)$. For the lemma we may assume Q homogeneous. We have seen that $S^{nw} \subset I_+ S$, so clearly the lemma holds if $\deg q \geq nw$. Assuming it holds for $\deg q = m + 1$ we shall show that it holds for $\deg q = m$ and thus by induction in general. Let σ be any reflection in W and $\alpha = 0$ the corresponding hyperplane. Then $\partial(q)\pi = 0$ implies $\partial(\alpha q)\pi = 0$, so since $\deg(\alpha q) = m + 1$, we have

$$\alpha q = \sum_{k=1}^{n} A_k j_k, \qquad A_k \in S. \tag{7}$$

Applying σ and subtracting, we obtain (since α divides $A_k - \sigma A_k$) $q \equiv -\sigma q \pmod{J}$. Since the reflections σ generate W, this implies $q \equiv (\det s)sq \pmod{J}$ for each $s \in W$. Averaging over W we obtain $q \equiv q^*$ (mod J), where q^* is skew. By Corollary 3.8, $q^* = \pi i$, where i is a homogeneous invariant. If $\deg i > 0$ we have $q \in J$ as desired. If $\deg i = 0$ so $i = c$ (a constant), then $0 = \partial(q)\pi = c\partial(\pi)(\pi)$, so $c = 0$. This proves the lemma.

For Theorem 3.6 it remains to prove $H \subset \partial(S)\pi$ or, equivalently,

$$(\partial(S)\pi)^\perp \subset H^\perp,$$

the orthogonal complement taken with respect to $\langle\ ,\ \rangle$. But if

$$\langle \partial(S)\pi, p\rangle = 0 \text{ then } \langle \partial(P)\pi, S\rangle = 0,$$

so $\partial(P)\pi = 0$. By Lemma 3.9, $p \in J = H^\perp$, so the theorem is proved.

3. The Exterior Invariants

Let W again be a finite reflection group acting on E and let

$$j_1(x_1, \ldots, x_n), \ldots, j_n(x_1, \ldots, x_n)$$

denote homogeneous algebraically independent generators for the algebra of invariants. We shall now describe the W-invariant forms on E with polynomial coefficients. As usual d denotes exterior differentiation.

Proposition 3.10. *Each W-invariant p-form on E with polynomial coefficients may be expressed uniquely as*

$$(8) \qquad \sum_{i_1 < \cdots < i_p} a_{i_1, \ldots, i_p}\, dj_{i_1} \wedge \cdots \wedge dj_{i_p}, \qquad a_{i_1, \ldots, i_p} \in I.$$

Proof. Since $\sigma(dj) = dj$ for $j \in I$, $\sigma \in W$, it is clear that each form (8) is invariant. Next we prove that the $\binom{n}{p}$ forms $dj_{i_1} \wedge \cdots \wedge dj_{i_p}$ are linearly independent over the quotient field $Q = \boldsymbol{R}(x_1, \ldots, x_n)$ so that they form a basis over Q of the space of p-forms with polynomial coefficients. Indeed, suppose

$$(9) \qquad \sum_{i_1 < \cdots < i_p} k_{i_1, \ldots, i_p}\, dj_{i_1} \wedge \cdots \wedge dj_{i_p} = 0,$$

where $k_{i_1, \ldots, i_p} \in Q$. Fix a p-tuple $(e_1 < \cdots < e_p)$ and let $e_{p+1}, \ldots, e_n$ be the complementary indices. Multiplying (9) by $dj_{e_{p+1}} \wedge \cdots \wedge dj_{e_n}$ we obtain from Lemma 3.7.

$$k_{e_1, \ldots, e_p}\, dj_1 \wedge \cdots \wedge dj_n = \pm k_{e_1, \ldots, e_p} c\pi\, dx_1 \wedge \cdots \wedge dx_n = 0,$$

whence $k_{e_1, \ldots, e_p} = 0$, proving the linear independence. It follows that each p-form ω with coefficients in Q can be expressed

$$\omega = \sum_{i_1 < \cdots < i_p} a_{i_1, \ldots, i_p} \, dj_{i_1} \wedge \cdots \wedge dj_{i_p}, \qquad a_{i_1, \ldots, i_p} \in Q. \tag{10}$$

Now assume ω is invariant with polynomial coefficients. Again multiplying by $dj_{e_{p+1}} \wedge \cdots \wedge dj_{e_n}$ we deduce

$$\omega \wedge dj_{e_{p+1}} \wedge \cdots \wedge dj_{e_n} = \pm c a_{e_1, \ldots, e_p} \pi \, dx_1 \wedge \cdots \wedge dx_n,$$

so $a_{e_1, \ldots, e_p} \pi$ is a polynomial. On the other hand, applying $\sigma \in W$ to (10) and averaging over W, we see that each $a_{e_1, \ldots, e_p}$ is W-invariant. Since $a_{e_1, \ldots, e_p} \pi$ is a polynomial and skew we conclude from Corollary 3.8 that $a_{e_1, \ldots, e_p} \in I$ and the proposition is proved.

4. Eigenfunctions of Weyl Group Invariant Operators

Let $\mathfrak{g}$ be a semisimple Lie algebra over $\mathbf{R}$, B its Killing form, θ a Cartan involution of $\mathfrak{g}$, and $\mathfrak{g} = \mathfrak{k} + \mathfrak{p}$ the corresponding Cartan decomposition. Let $\mathfrak{a} \subset \mathfrak{p}$ be a maximal abelian subspace and $W = W(\mathfrak{g}, \theta)$ the corresponding Weyl group acting on $\mathfrak{a}$ (cf. [DS], Chapter IX, §1). Then W is generated by the reflections s_α as α runs through the set Σ of restricted roots. We shall now use the previous results of this section, $\mathfrak{a}$ playing the role of E. We shall determine the functions f on $\mathfrak{a}$ which are eigenfunctions of each $\partial(P)$, $P \in I(\mathfrak{a})$. In the language of Chapter II, §4.1, we are looking for the *joint eigenfunctions* of the operators in $\mathbf{D}(W \cdot \mathfrak{a}/W)$, $W \cdot \mathfrak{a}$ denoting the group of transformations of $\mathfrak{a}$ generated by W and the translations.

Lemma 3.11. *The homomorphisms of $I(\mathfrak{a})$ into $\mathbf{C}$ are precisely*

$$\chi_\mu : P \to P(\mu),$$

where μ is an element in $(\mathfrak{a}^{\mathbf{C}})^*$. Two such homomorphisms χ_μ and χ_ν coincide if and only if $\mu \in W \cdot \nu$.

Proof. Equation (1) shows that the ring $S(\mathfrak{a}^{\mathbf{C}})$ is *integral over* $I(\mathfrak{a}^{\mathbf{C}})$, in the sense that each element in $S(\mathfrak{a}^{\mathbf{C}})$ satisfies an algebraic equation with coefficients in $I(\mathfrak{a}^{\mathbf{C}})$ and leading coefficient one. By Theorem 3.5 of the Appendix each homomorphism of $I(\mathfrak{a})$ into $\mathbf{C}$ extends to a homomorphism of $S(\mathfrak{a})$ into $\mathbf{C}$ and is therefore given by evaluation at some point $\mu \in (\mathfrak{a}^{\mathbf{C}})^*$. Furthermore, suppose for two such μ, $\nu \in (\mathfrak{a}^{\mathbf{C}})^*$ we have $P(\mu) = P(\nu)$ for all $P \in I(\mathfrak{a})$. If the orbits $W \cdot \mu$ and $W \cdot \nu$ were disjoint we can, since these orbits are finite, find a polynomial Q such that $Q > 0$ on $W \cdot \mu$ and $Q < 0$ on $W \cdot \nu$. Then the sum $Q^* = \sum_{\sigma \in W} \sigma \cdot Q$ belongs to $I(\mathfrak{a})$, yet takes different values on μ and ν. This proves the lemma.

Corollary 3.12. *If* $j_1, \ldots, j_l$ $(l = \dim \mathfrak{a})$ *are algebraically independent generators of* I, *then the mapping*

$$e \in \mathfrak{a}^{\mathbf{C}} \to (j_1(e), \ldots, j_l(e)) \in \mathbf{C}^l \tag{11}$$

induces a bijection of $\mathfrak{a}^{\mathbf{C}}/W$ *onto* $\mathbf{C}^l$.

In fact, if $(\xi_1, \ldots, \xi_l) \in \mathbf{C}^l$, then the mapping $j_i \to \xi_i$ $(1 \leq i \leq l)$ gives, by the algebraic independence of the j_i, a homomorphism of $I((\mathfrak{a}^{\mathbf{C}})^*)$ into $\mathbf{C}$ which by the lemma is given by the evaluation at a point $e \in \mathfrak{a}^{\mathbf{C}}$.

Given $\mu \in (\mathfrak{a}^{\mathbf{C}})^*$, let $\mathscr{E}_\mu = \mathscr{E}_\mu(\mathfrak{a})$ denote the joint eigenspace

$$\mathscr{E}_\mu(\mathfrak{a}) = \{f \in \mathscr{E}(\mathfrak{a}) : \partial(J)f = J(\mu)f \quad \text{for all } J \in I(\mathfrak{a})\}. \tag{12}$$

According to Lemma 3.11 this is the most general eigenspace of the operators $\partial(J)$, $J \in I(\mathfrak{a})$. Let W_μ denote the subgroup of W leaving μ fixed. According to [DS, Chapter VII, Theorem 2.15], W_μ is a finite reflection group acting on $\mathfrak{a}$. Let $I^\mu(\mathfrak{a}) \subset S(\mathfrak{a})$ denote the set of W_μ-invariants and H^μ the corresponding set of harmonic polynomial functions.

Theorem 3.13. *For each* $\mu \in (\mathfrak{a}^{\mathbf{C}})^*$,

$$\dim \mathscr{E}_\mu(\mathfrak{a}) = |W|.$$

Moreover, if s *runs through a complete set of representatives of the cosets of* W_μ *in* W *and* h *runs through a basis of* $H^{s\mu}$, *the functions*

$$e \to h(e) \exp[s\mu(e)] \tag{13}$$

form a basis of $\mathscr{E}_\mu(\mathfrak{a})$.

Remark. In particular, if μ is regular (cf. Exercise 2) the functions $e^{s\mu}$ form a basis of $\mathscr{E}_\mu$.

We start by proving the weaker result

$$\dim \mathscr{E}_\mu(\mathfrak{a}) \leq |W|. \tag{14}$$

For this let $h_1, \ldots, h_w$ be a basis of H (where $w = |W|$) and let $H_1, \ldots, H_w$ be the corresponding members of $S(\mathfrak{a}^{\mathbf{C}})$ [under the identification of $\mathfrak{a}^{\mathbf{C}}$ and $(\mathfrak{a}^{\mathbf{C}})^*$ by means of B]. Let $f \in \mathscr{E}_\mu(\mathfrak{a})$ and put for $a \in \mathfrak{a}$

$$c_i(f) = (\partial(H_i)f)(a) \quad (1 \leq i \leq w).$$

Since $S = IH$ it is clear that if $c_1(f) = \cdots = c_w(f) = 0$ then $f = 0$. Thus the mapping

$$f \to (c_1(f), \ldots, c_w(f))$$

is a one-to-one linear mapping of $\mathscr{E}_\mu(\mathfrak{a})$ into $\mathbf{C}^w$ so (14) follows.

Lemma 3.14. *For each $P \in I^\mu(\mathfrak{a})$ and each $h \in H^\mu$ the function*

$$f(e) = h(e)\exp[\mu(e)], \qquad e \in \mathfrak{a},$$

satisfies

(15) $$\partial(P)f = P(\mu)f.$$

In particular, $f \in \mathscr{E}_\mu(\mathfrak{a})$.

Proof. Consider the automorphism $P \to P^\mu$ of $S(\mathfrak{a})$ given by $P^\mu(\lambda) = P(\mu + \lambda)$. Then

$$\begin{aligned}\partial(P)f &= \partial(P)(e^\mu h) = e^\mu e^{-\mu}\partial(P)(e^\mu h)\\ &= e^\mu \partial(P^\mu)(h) = e^\mu(P^\mu(0)h) + e^\mu\partial(P^\mu - P^\mu(0))h,\end{aligned}$$

But if $P \in I^\mu$ then $P^\mu \in I^\mu$, so $\partial(P^\mu - P^\mu(0))h = 0$. Thus $\partial(P)f = P(\mu)f$ as desired.

This lemma shows that the functions (13) belongs to $\mathscr{E}_\mu(\mathfrak{a})$. Since by Theorem 3.4, $\dim H^{s\mu} = |W_{s\mu}| = |W_\mu|$, and since we have inequality (14), Theorem 3.13 will follow if we prove that the functions (13) are linearly independent. As s varies the linear forms $s\mu$ are different; hence they are different on a line in $\mathfrak{a}$ through 0 on which the basis elements h are not identically zero. Thus it just remains to prove the following elementary result.

Lemma 3.15. *Let $c_1, \ldots, c_n$ be distinct complex numbers and $p_1, \ldots, p_n$ polynomials on $\boldsymbol{R}$ with complex coefficients. Assume*

(16) $$\sum_{i=1}^{n} p_i(t)e^{c_i t} = 0 \quad \text{for all } t \in \boldsymbol{R}.$$

Then $p_1 = \cdots = p_n \equiv 0$.

This lemma is a consequence of Exercise D5 of Chapter I.

5. Restriction Properties

With the notation of §3, No. 4, let $\mathfrak{h}$ denote a Cartan subalgebra of $\mathfrak{g}$ containing $\mathfrak{a}$ and $\mathfrak{h}^C$ its complexification. Then in addition to the Weyl group W above which acts on $\mathfrak{a}^C$ we have the Weyl group $\tilde{W}$ of the pair $(\mathfrak{g}^C, \mathfrak{h}^C)$. Corresponding to these groups W and $\tilde{W}$ we have the algebras $I(\mathfrak{a}^*)$ and $I(\mathfrak{h}^{C*})$ of invariant polynomial functions on $\mathfrak{a}^C$ and $\mathfrak{h}^C$, respectively. Let $\tilde{W}_\theta$ denote the subgroup of $\tilde{W}$ leaving $\mathfrak{a}$ invariant. Then we know from [DS] (Proposition 8.10, Chapter VII) that the restriction $s \to s|\mathfrak{a}$ is a homomorphism of $\tilde{W}_\theta$ onto W. If $p \in S((\mathfrak{h}^C)^*)$ its restriction to $\mathfrak{a}^C$ will be denoted by $\bar{p}$. By the above

(17) $$p \in I(\mathfrak{h}^{C*}) \Rightarrow \bar{p} \in I(\mathfrak{a}^*).$$

While this restriction mapping is not always surjective (cf. Chapter II, Exercise D3) the following positive result holds for the rational invariants.

Theorem 3.16. *Let $J \subset I(\mathfrak{a}^*)$ be the image of $I(\mathfrak{h}^{C^*})$ under the restriction mapping $p \to \bar{p}$. Then the quotient fields $\boldsymbol{C}(J)$ and $\boldsymbol{C}(I(\mathfrak{a}^*))$ coincide.*

Proof. From Eq. (1) we know that $S(\mathfrak{h}^{C^*})$ is integral over $I(\mathfrak{h}^{C^*})$. The restriction $p \to \bar{p}$ gives a homomorphism of $S(\mathfrak{h}^{C^*})$ onto $S(\mathfrak{a}^*)$ mapping $I(\mathfrak{h}^{C^*})$ onto J. Hence $S(\mathfrak{a}^*)$ is integral over J. Now suppose $C(J)$ were a proper subset of $C(I(\mathfrak{a}^*))$. Then there exists a $\alpha \in I(\mathfrak{a}^*)$ such that $\alpha \notin C(J)$. But as remarked above α is algebraic over $C(J)$. Since $I(\mathfrak{h}^{C^*})$ is finitely generated, the homomorphic image J also has finitely many generators, say $\xi_1, \ldots, \xi_r$. Let

$$p_\alpha(x) = x^n + f_1(\xi)x^{n-1} + \cdots + f_n(\xi) \qquad (n \geq 2)$$

be the polynomial with coefficients in the field $C(J) = C(\xi_1, \ldots, \xi_r)$ of lowest degree having α as a zero and leading coefficient 1. Let $f(\xi)$ denote the product of the dominators of the $f_i(\xi)$ and let $q(\xi)$ denote the product of $f(\xi)$ with the discriminant of the polynomial $f(\xi)p_\alpha(x)$. Then $q(\xi)$ is a polynomial function on $\mathfrak{a}^C$, not identically 0 (because of the minimality).

Consider now a simply connected Lie group U with Lie algebra $\mathfrak{k} + i\mathfrak{p}$. Put $\mathfrak{a}_* = i\mathfrak{a}$ and let $A_* \subset U$ be the corresponding analytic subgroup. Select $H_* \in \mathfrak{a}_*$ such that

$$q(\xi)(H_*) \neq 0, \qquad \exp \boldsymbol{R}H_* \text{ dense in } A_*. \tag{18}$$

Consider the homomorphism $\lambda: p \to p(H_*)$ of J into $\boldsymbol{C}$. The image of $p_\alpha(x)$ under λ is a polynomial with discriminant $\neq 0$ which therefore has n distinct roots, say $\alpha_1, \ldots, \alpha_n \in \boldsymbol{C}$. For each fixed i $(1 \leq i \leq n)$ we wish to extend λ to a homomorphism $\lambda_i: \boldsymbol{C}[\xi_1, \ldots, \xi_r, \alpha] \to \boldsymbol{C}$ by putting $\lambda_i(\alpha) = \alpha_i$. To see that this is possible, suppose $p(\xi, x)$ is a polynomial with coefficients in $J = \boldsymbol{C}[\xi_1, \ldots, \xi_r]$ which has α as a zero. Then $p_\alpha(x)$ divides $p(\xi, x)$; i.e., $p(\xi, x) = p_\alpha(x)q(\xi, x)$; the coefficients of $q(\xi, x)$ are found by long division, so they appear as rational expressions in $\xi_1, \ldots, \xi_r$ with denominators dividing $f(\xi)$. Since $\lambda(f(\xi)) \neq 0$ [by (18)] the image of $p(\xi, x)$ under λ has $x = \alpha_i$ as a zero. Thus λ_i is well defined. Since $S(\mathfrak{a}^*)$ is integral over J and *a fortiori* over $C[\xi_1, \ldots, \xi_r, \alpha]$, there exists a homomorphism $\Lambda_i: S(\mathfrak{a}^*) \to \boldsymbol{C}$ extending λ_i (Appendix §3). Then there exists $H_i \in \mathfrak{a}^C$ such that $\Lambda_i(p) = p(H_i)$ for all $p \in S(\mathfrak{a}^*)$. In particular,

$$p(H_*) = p(H_1) = p(H_2) \qquad \text{for } p \in I(\mathfrak{h}^{C^*}),$$

which (cf. Lemma 3.11) implies that H_*, H_1, H_2 are all conjugate under $\tilde{W}$. Since $H_* \in \mathfrak{a}_*$ and since the subset $\mathfrak{h} \cap \mathfrak{k} + \mathfrak{a}_* \subset \mathfrak{h}^C$, where all roots $\alpha \in \Delta(\mathfrak{g}^C, \mathfrak{h}^C)$ are purely imaginary, is invariant under $\tilde{W}$ we obtain

$$H_1, H_2 \in (\mathfrak{h} \cap \mathfrak{k} + \mathfrak{a}_*) \cap \mathfrak{a}^C = \mathfrak{a}_*.$$

Now select $u_1, u_2 \in U$ such that $\operatorname{Ad}(u_i)H_* = H_i$ $(i = 1, 2)$. Then

$$u_i \exp tH_* u_i^{-1} = \exp tH_i$$

so by (18) u_1 and u_2 belongs to the normalizer of $\mathfrak{a}_*$ in U. Using [DS] Proposition 8.8, Chapter VII describing this normalizer we can therefore conclude that H_1 and H_2 are conjugate under W. But then, since $\alpha \in I(\mathfrak{a}^*)$,

$$\alpha_1 = \Lambda_1(\alpha) = \alpha(H_1) = \alpha(H_2) = \Lambda_2(\alpha) = \alpha_2,$$

which is a contradiction.

Remark. According to Exercise D3 in Chapter II $J \neq I(\mathfrak{a}^*)$ for some of the exceptional symmetric spaces, so by Proposition 5.30 of Chapter II, $\boldsymbol{Z}(G/K) \neq \boldsymbol{D}(G/K)$ in general. The theorem above suggests that this should be remedied by going over to the quotient fields of $\boldsymbol{Z}(G)$ and of $\boldsymbol{D}(G/K)$.

§4. The Orbit Structure of $\mathfrak{p}$

1. Generalities

In this section we make considerable use of the results proved in Appendix §§1, 2. Let $\mathfrak{g}$ be a semisimple Lie algebra over $\boldsymbol{C}$, $\mathfrak{g}_0$ any real form of $\mathfrak{g}$. Let θ be any Cartan involution of $\mathfrak{g}_0$; we extend it to an involutive automorphism of $\mathfrak{g}$, also denoted θ. We now vary $\mathfrak{g}_0$ too.

Lemma 4.1. *All involutive automorphisms η of $\mathfrak{g}$ arise in this fashion.*

Proof. Let $\mathfrak{g}^{\boldsymbol{R}}$ denote $\mathfrak{g}$ considered as a Lie algebra over $\boldsymbol{R}$ and $\eta^{\boldsymbol{R}}$ the automorphism of $\mathfrak{g}^{\boldsymbol{R}}$ corresponding to η. According to [DS] (Exercise B4 for Chapter II), $\eta^{\boldsymbol{R}}$ commutes with a Cartan involution of $\mathfrak{g}^{\boldsymbol{R}}$; in other words, η leaves a compact real form $\mathfrak{v}$ of $\mathfrak{g}$ invariant. Let $\mathfrak{g} = \mathfrak{k} + \mathfrak{p}$, $\mathfrak{g}_0 = \mathfrak{k}_0 + \mathfrak{p}_0$ be the decompositions into eigenspaces of θ, and put $\mathfrak{u} = \mathfrak{k}_0 + i\mathfrak{p}_0$. Then $\mathfrak{v} = \phi\mathfrak{u}$ for a suitable automorphism ϕ of $\mathfrak{g}$. Since $\phi^{-1}\eta\phi$ leaves $\mathfrak{u}$ invariant, $\mathfrak{u} = \mathfrak{u} \cap \mathfrak{g}_1 + \mathfrak{u} \cap \mathfrak{g}_{-1}$, where $\mathfrak{g}_\varepsilon$ is the eigenspace of $\phi^{-1}\eta\phi$ with eigenvalue ε $(= \pm 1)$. But then $\mathfrak{u} \cap \mathfrak{g}_1 + i(\mathfrak{u} \cap \mathfrak{g}_{-1})$ is a Cartan decomposition of a real form $\mathfrak{g}_0'$ of $\mathfrak{g}$. If θ' is the corresponding Cartan involution, η is the extension of $\phi\theta'\phi^{-1}$ to an involution of $\mathfrak{g}$.

Let G denote the adjoint group of $\mathfrak{g}$, i.e., $G = \text{Int}(\mathfrak{g}^{\boldsymbol{R}})$ with the associated complex structure ([DS], Chapter VIII, §2). Let K_0, $U \subset G$ denote the analytic subgroups corresponding to $\mathfrak{k}_0$ and $\mathfrak{u}$, respectively. Let the automorphism of G induced by θ also be denoted by θ, let K_θ denote the fixed point group of θ and K the identity component $(K_\theta)_0$. We are now interested in studying the orbits for the action of K (and K_θ) on $\mathfrak{p}$.

Let $\mathfrak{a}_0 \subset \mathfrak{p}_0$ be a maximal abelian subspace, $\mathfrak{a} \subset \mathfrak{p}$ its complexification, and A ($=\exp \text{ad}_{\mathfrak{g}}(\mathfrak{a})$) the analytic subgroup of G with Lie algebra $\mathfrak{a}$. Let $F = \{a \in A : a^2 = e\}$. It is clear that $a \in F \Rightarrow \theta a = a^{-1} = a$, so $F \subset K_\theta$.

Lemma 4.2. $K_\theta = FK$.

Proof. Let $g \in K_\theta$. Using the Cartan decomposition $\mathfrak{g} = \mathfrak{u} + i\mathfrak{u}$ ($\mathfrak{u} = \mathfrak{k}_0 + i\mathfrak{p}_0$) and the corresponding decomposition $G = U\exp(i\mathfrak{u})$ of G we have $g = up$ where $u \in U$, $p = \exp X$ ($X \in i\mathfrak{u}$). By uniqueness $\theta u = u$, $\theta p = p$. But exp is one-to-one on $i\mathfrak{u}$ so $\theta X = X$. Thus

$$X \in i\mathfrak{u} \cap \mathfrak{k} = i\mathfrak{k}_0$$

so $p \in K$. Also by [DS] (Chapter V, Theorem 6.7), we have $u = k_1 a k_2$ [$k_1, k_2 \in K_0$, $a \in \exp(i\mathfrak{a}_0)$], so $\theta u = k_1 a^{-1} k_2$. But then $\theta u = u$ implies $a \in F$, so $g \in FK$.

Definition. An element $X \in \mathfrak{p}$ is called *semisimple* (resp *nilpotent*) if $\text{ad}_{\mathfrak{g}}(X)$ is a semisimple (resp. nilpotent) endomorphism; $X \in \mathfrak{p}$ is called *quasi-regular* if its orbit under K has maximal dimension. The set of semisimple (resp. nilpotent, quasi-regular) elements in $\mathfrak{p}$ will be denoted by $\mathscr{S}$ (resp. $\mathscr{N}$, and $\mathscr{R}$). If $X \in \mathfrak{p}$ then $(\text{ad } X)^2$ maps $\mathfrak{p}$ into itself. Thus we generalize the definition in [DS] (Chapter III, §3) and put for $X \in \mathfrak{p}$, $\lambda \in \boldsymbol{C}$

$$(1) \qquad \mathfrak{p}(X, \lambda) = \{Y \in \mathfrak{p} : ((\text{ad } X)^2 - \lambda)^m(Y) = 0 \quad \text{for some } m\}.$$

The element $X \in \mathfrak{p}$ is said to be *regular* if

$$\dim \mathfrak{p}(X, 0) = \min_{Y \in \mathfrak{p}} \dim \mathfrak{p}(Y, 0).$$

Let $\mathfrak{r}$ denote the set of regular elements in $\mathfrak{p}$. It will be proved later that the regular elements are the quasi-regular elements which are also semisimple that is,

$$(2) \qquad \mathfrak{r} = \mathscr{R} \cap \mathscr{S}.$$

The elements in $\mathscr{R} \cap \mathscr{N}$ are called *principal nilpotent elements.*

2. Nilpotent Elements

We begin by relating nilpotent elements to K-invariant polynomials on $\mathfrak{p}$. Then we shall investigate nilpotent elements by imbedding them in subalgebras of $\mathfrak{g}$ isomorphic to $\mathfrak{sl}(2, \boldsymbol{C})$.

As in §1 we consider the symmetric algebras $S(\mathfrak{p})$, $S(\mathfrak{p}^*)$, the sets $I(\mathfrak{p})$, $I(\mathfrak{p}^*)$ of the corresponding K-invariants and the sets $I_+(\mathfrak{p})$, $I_+(\mathfrak{p}^*)$ of K-invariants without constant term.

Theorem 4.3. *Let $X \in \mathfrak{p}$. Then*

$$X \in \mathcal{N} \Leftrightarrow f(X) = 0 \quad \textit{for all } f \in I_+(\mathfrak{p}^*)S(\mathfrak{p}^*).$$

Proof. Suppose $f(X) = 0$ for all $f \in I_+ S$. Then in particular, $f(X) = 0$ for all G-invariant polynomial functions on $\mathfrak{g}$. The characteristic polynomial $\det_{\mathfrak{g}}(\lambda I - \operatorname{ad} X)$ therefore reduces to $\lambda^{\dim \mathfrak{g}}$ so ad X is nilpotent ([DS], Chapter III, §1).

Conversely, suppose $X \in \mathcal{N}$. To prove $f(X) = 0$ for all $f \in I_+ S$ we may assume $X \neq 0$. By [DS], Chapter IX, Theorem 7.4, we can select $H, Y \in \mathfrak{g}$ such that

$$(3) \qquad [H, X] = 2X, \qquad [H, Y] = -2Y, \qquad [X, Y] = H.$$

Writing $H = H_{\mathfrak{k}} + H_{\mathfrak{p}}$ ($H_{\mathfrak{k}} \in \mathfrak{k}$, $H_{\mathfrak{p}} \in \mathfrak{p}$), we have

$$2X = [H, X] = [H_{\mathfrak{k}}, X] + [H_{\mathfrak{p}}, X],$$

so since $X \in \mathfrak{p}$, $[H_{\mathfrak{k}}, X] = 2X$. It follows that $\exp(\boldsymbol{C} \operatorname{ad} H_{\mathfrak{k}}) \cdot X = (\boldsymbol{C} - \{0\})X$, which has 0 in its closure. By continuity $f(X) = f(0) = 0$.

Definition. A *$\mathfrak{sl}_2$-triple* in $\mathfrak{g}$ is a triple (H, X, Y) of nonzero elements in $\mathfrak{g}$ satisfying relations (3) above. The $\mathfrak{sl}_2$-triple is called *normal* if $H \in \mathfrak{k}$, $X, Y \in \mathfrak{p}$.

Lemma 4.4. *If (H, X, Y) and (H, X, Y') are two $\mathfrak{sl}_2$-triples in $\mathfrak{g}$ then $Y = Y'$.*

Proof. We have

$$[X, Y - Y'] = H - H = 0,$$

so $Y - Y' \in \operatorname{kernel}(\operatorname{ad}(X))$. Also,

$$(\operatorname{ad} H + 2I)(Y - Y') = -2Y + 2Y' + 2(Y - Y') = 0.$$

But as remarked in the proof of Lemma 7.6, Chapter IX in [DS], ad $H + 2I$ is nonsingular on the kernel of ad X. Hence $Y = Y'$.

Lemma 4.5. *Let $\mathfrak{n} \subset \mathfrak{g}$ be a nilpotent subalgebra. Suppose $H \in \mathfrak{g}$ satisfies* ad $H(\mathfrak{n}) = \mathfrak{n}$. *Then $e^{\operatorname{ad}(\mathfrak{n})}H = H + \mathfrak{n}$.*

Proof. The inclusion $e^{\mathrm{ad}(\mathfrak{n})}(H) \subset H + \mathfrak{n}$ is obvious. Conversely, we shall prove that if $Z \in \mathfrak{n}$ then $H + Z \in e^{\mathrm{ad}(\mathfrak{n})}(H)$. The central descending series $\mathscr{C}^0\mathfrak{n} = \mathfrak{n}, \ldots, \mathscr{C}^{p+1}\mathfrak{n} = [\mathfrak{n}, \mathscr{C}^p\mathfrak{n}]$ being defined as in [DS] (Chapter III, §2), it suffices to prove

$$H + Z \in e^{\mathrm{ad}(\mathfrak{n})}(H) + \mathscr{C}^p\mathfrak{n} \qquad \text{for all } p \geq 0. \tag{4}$$

In fact, $\mathscr{C}^p\mathfrak{n} = 0$ for p sufficiently large. We prove (4) by induction, the case $p = 0$ being obvious. Suppose we have $Y_p \in \mathfrak{n}$, $Z_p \in \mathscr{C}^{p-1}\mathfrak{n}$ such that

$$H + Z = e^{\mathrm{ad}\, Y_p}(H) + Z_p. \tag{5}$$

Since ad H gives a bijection of $\mathfrak{n}$ onto $\mathfrak{n}$ leaving $\mathscr{C}^{p-1}\mathfrak{n}$ invariant, there exists $Z_0 \in \mathscr{C}^{p-1}\mathfrak{n}$ such that $Z_p = [Z_0, H]$. But then

$$e^{\mathrm{ad}(Y_p + Z_0)}(H) = H + [Y_p + Z_0, H] + \tfrac{1}{2}[Y_p + Z_0, [Y_p + Z_0, H]] + \cdots,$$

$$e^{\mathrm{ad}\, Y_p}(H) = H + [Y_p, H] + \tfrac{1}{2}[Y_p, [Y_p, H]] + \cdots.$$

Both expansions are finite. Their difference belongs to

$$[Z_0, H] + \mathscr{C}^p\mathfrak{n},$$

so using (5) we obtain

$$e^{\mathrm{ad}(Y_p + Z_0)}H \in H + Z + \mathscr{C}^p\mathfrak{n}.$$

This proves (4) and the lemma.

The group K acts on the set of $\mathfrak{sl}_2$-triples (H, X, Y), normal or not, by $k \cdot (H, X, Y) = (k \cdot H, k \cdot X, k \cdot Y)$.

Proposition 4.6. *Any $X \in \mathscr{N} - \{0\}$ is contained in a normal $\mathfrak{sl}_2$-triple. Any two such triples are K-conjugate. This induces a bijection between the set of all K-orbits in $\mathscr{N} - \{0\}$ and the set of all K-conjugacy classes of normal $\mathfrak{sl}_2$-triples.*

Proof. By [DS] (Chapter IX, Theorem 7.4) used on $\mathfrak{g}^{\mathbf{R}}$, there exists an $\mathfrak{sl}_2$-triple (H^*, X, Y^*) containing X. We now modify it to make it normal. Let $H^* = H + H_{\mathfrak{p}}$ $(H \in \mathfrak{k}, H_{\mathfrak{p}} \in \mathfrak{p})$. Then

$$2X = [H^*, X] = [H, X] + [H_{\mathfrak{p}}, X],$$

so $[H, X] = 2X$, $[H_{\mathfrak{p}}, X] = 0$. Writing similarly $Y^* = Y^*_{\mathfrak{k}} + Y^*_{\mathfrak{p}}$, we conclude from $[X, Y^*] = H^*$ that $[X, Y^*_{\mathfrak{p}}] = H$. By Lemma 7.6 in [DS] (Chapter IX), the relations $[H, X] = 2X$, $H \in [X, \mathfrak{g}]$ imply that there exists an element $Y \in \mathfrak{g}$ such that relations (3) hold. Decomposing again $Y = Y_{\mathfrak{k}} + Y_{\mathfrak{p}}$ we see from (3), since $H \in \mathfrak{k}$, $X \in \mathfrak{p}$, that $(H, X, Y_{\mathfrak{p}})$ is also an $\mathfrak{sl}_2$-triple. Thus by Lemma 4.4, $Y = Y_{\mathfrak{p}}$, so (H, X, Y) is a normal $\mathfrak{sl}_2$-triple.

Suppose (H', X, Y') is another normal $\mathfrak{sl}_2$-triple containing X. Let $H_0 = H' - H$ so $[X, H_0] = 0$ and $H_0 \in \mathfrak{k}$. The centralizer $\mathfrak{g}^X$ of X in $\mathfrak{g}$ is invariant under ad H. From §1 of the Appendix used on $\boldsymbol{C}H + \boldsymbol{C}X + \boldsymbol{C}Y$ we know that ad H and therefore the restriction $(\operatorname{ad} H)|\mathfrak{g}^X$ is semisimple and also that the eigenvalues of ad H on $\mathfrak{g}^X$ are in $\boldsymbol{Z}^+$. Let $\mathfrak{g}_+^X \subset \mathfrak{g}^X$ be the subspace spanned by the eigenvectors for ad H with strictly positive eigenvalues. By Corollary 1.5 of the Appendix,

$$\mathfrak{g}_+^X = \mathfrak{g}^X \cap [X, \mathfrak{g}]. \tag{6}$$

The centralizer $\mathfrak{k}^X$ of X in $\mathfrak{k}$ is invariant under ad H. Putting $\mathfrak{k}_+^X = \mathfrak{k}^X \cap \mathfrak{g}_+^X$, we have from (6) $\mathfrak{k}_+^X = \mathfrak{k}^X \cap [X, \mathfrak{g}]$, which clearly contains

$$H_0 = [X, Y' - Y].$$

Now by the Jacobi identity $\mathfrak{g}_+^X$ is a subalgebra which is readily seen to be nilpotent. Hence $\mathfrak{k}_+^X$ is nilpotent and we obtain from Lemma 4.5

$$\exp(\operatorname{ad} \mathfrak{k}_+^X)H = H + \mathfrak{k}_+^X. \tag{6'}$$

In particular, there exists a $k \in K$ such that $kX = X$ and $kH = H + H_0 = H'$. Then Lemma 4.4 implies $Y' = kY$.

We have now proved the two first statements of the proposition. The last one follows immediately, the mapping $K \cdot X \to K \cdot (H, X, Y)$ being the desired bijection.

Lemma 4.7. *Let (H, X, Y) and (H', X', Y') be two normal $\mathfrak{sl}_2$-triples. Then they are K-conjugate if and only if H and H' are K-conjugate.*

Proof. We may assume $H = H'$, and have to show that if (H, X, Y) and (H, X', Y') are normal $\mathfrak{sl}_2$-triples they are K-conjugate. Let $\mathfrak{g}^0$, $\mathfrak{k}^0$, and $\mathfrak{p}^0$ denote the centralizers of H in $\mathfrak{g}$, $\mathfrak{k}$, and $\mathfrak{p}$, respectively. Since $\theta H = H$ we have $\mathfrak{g}^0 = \mathfrak{k}^0 + \mathfrak{p}^0$. Let $K^0 \subset G$ denote the analytic subgroup corresponding to $\operatorname{ad}_{\mathfrak{g}}(\mathfrak{k}^0)$, and let $\mathfrak{g}^2 \subset \mathfrak{g}$, $\mathfrak{k}^2 \subset \mathfrak{k}$, $\mathfrak{p}^2 \subset \mathfrak{p}$ be the eigenspaces of ad H for eigenvalue 2. Again $\theta H = H$ implies $\mathfrak{g}^2 = \mathfrak{k}^2 + \mathfrak{p}^2$. Also $[\mathfrak{g}^0, \mathfrak{g}^2] \subset \mathfrak{g}^2$ so $[\mathfrak{k}^0, \mathfrak{p}^2] \subset \mathfrak{p}^2$.

For $Z \in \mathfrak{p}^2$ consider the map $\Phi^Z: K^0 \to \mathfrak{p}^2$ given by $\Phi^Z(k) = \operatorname{Ad}(k)Z$. Its differential is $d\Phi_e^Z(T) = [T, Z]$. Let $V \subset \mathfrak{p}^2$ be the set of Z for which $d\Phi_e^Z$ is surjective; i.e., $[\mathfrak{k}^0, Z] = \mathfrak{p}^2$.

We claim that $X, X' \in V$. From Corollary 1.5 of the Appendix applied to $\boldsymbol{C}H + \boldsymbol{C}X + \boldsymbol{C}Y$ acting on $\mathfrak{g}$ we deduce $[X, \mathfrak{g}^0] = \mathfrak{g}^2$. From the direct sum decompositions of $\mathfrak{g}^0$ and $\mathfrak{g}^2$ we then deduce $[X, \mathfrak{k}^0] = \mathfrak{p}^2$ so that $X \in V$. Similarly, $X' \in V$. Consider now the linear mapping $-d\Phi_e^Z$, that is, ad $Z: \mathfrak{k}^0 \to \mathfrak{p}^2$ in terms of fixed bases of $\mathfrak{k}^0$ and $\mathfrak{p}^2$. Then surjectivity amounts to at least one of the minors of order $\dim \mathfrak{p}^2$ being $\neq 0$. Thus V is the union of open, intersecting, connected sets ([DS],

Chapter VII, Cor. 12.5); hence V is connected and open in $\mathfrak{p}^2$. However, V is invariant under K^0 and for each $Z \in V, K^0 \cdot Z$ is open in V. Thus V consists of just one K^0-orbit. Hence $X' = k \cdot X$ for some $k \in K^0 \subset K$, so that $k(H, X, Y) = (H, X', kY)$. But then $kY = Y'$ by Lemma 4.4.

Theorem 4.8. *There are only finitely many K-orbits in $\mathcal{N}$.*

Proof. By Proposition 4.6 it suffices to show that there are only a finite number of K-conjugacy classes of normal $\mathfrak{sl}_2$-triples. If $\mathscr{H}$ denotes the set of $H \in \mathfrak{k}$ for which there exist X, $Y \in \mathfrak{p}$ such that (H, X, Y) is a normal $\mathfrak{sl}_2$-triple it suffices by Lemma 4.7 to show that $\mathscr{H}$ (which is clearly stable under K) consists of finitely many K-orbits.

Now by [DS] (Chapter II, Proposition 6.6), we have the direct decompositions

$$\mathfrak{k}_0 = \mathfrak{z}_0 + [\mathfrak{k}_0, \mathfrak{k}_0], \qquad \mathfrak{k} = \mathfrak{z} + [\mathfrak{k}, \mathfrak{k}], \tag{7}$$

where $\mathfrak{z}_0 \subset \mathfrak{k}_0$, $\mathfrak{z} \subset \mathfrak{k}$ are the respective centers. Let $\mathfrak{h}_{\mathfrak{k}_0}$ be a maximal abelian subalgebra and let $\mathfrak{h}_{\mathfrak{k}}$ be the subalgebra of $\mathfrak{k}$ generated by $\mathfrak{h}_{\mathfrak{k}_0}$. Then $\mathfrak{h}_{\mathfrak{k}}$ is a *Cartan subalgebra* of $\mathfrak{k}$ (i.e., the direct sum of $\mathfrak{z}$ with a Cartan subalgebra of $[\mathfrak{k}, \mathfrak{k}]$). Let $\mathfrak{h}_0 \subset \mathfrak{g}_0$ be a maximal abelian subalgebra containing $\mathfrak{h}_{\mathfrak{k}_0}$ and let $\mathfrak{h} = \mathfrak{h}_0 + i\mathfrak{h}_0$. Then $H_0 \in \mathfrak{h}_0$ implies $[\theta H_0 + H_0, \mathfrak{h}_{\mathfrak{k}_0}] = 0$ so $\theta H_0 + H_0 \in \mathfrak{h}_{\mathfrak{k}_0}$, whence $\theta H_0 \in \mathfrak{h}_0$. Thus $\mathfrak{h}_0 = \mathfrak{h}_{\mathfrak{k}_0} \oplus (\mathfrak{h}_0 \cap \mathfrak{p}_0)$, so $\mathfrak{h}_0$ consists of semisimple elements in $\mathfrak{g}_0$. Thus $\mathfrak{h}$ is a Cartan subalgebra of $\mathfrak{g}$. By [DS], Chapter V, Theorem 6.5, all Cartan subalgebras of $[\mathfrak{k}, \mathfrak{k}]$, and hence all Cartan subalgebras of $\mathfrak{k}$, are K-conjugate. By Corollary 1.5 of the Appendix each $H \in \mathscr{H}$ is semisimple and so is each component of H in the decomposition (7) (since each element in $\mathfrak{z}$ is semisimple and commutes with H). By [DS] (Chapter IX, Proposition 4.6) for $[\mathfrak{k}, \mathfrak{k}]^R$ we conclude that H is conjugate to an element in $\mathfrak{h}_{\mathfrak{k}}$. Also, by §1 of the Appendix, the eigenvalues of H are integers with norm less than dim $\mathfrak{g}$. In particular, $0 \leq |\alpha(H)| \leq \dim \mathfrak{g}$ for each root α of $\mathfrak{g}$ with respect to $\mathfrak{h}$. But each $\alpha(H)$ is an integer and an element in $\mathfrak{h}$ is determined by the values the roots take on it. Thus the intersection $K \cdot \mathscr{H} \cap \mathfrak{h}_{\mathfrak{k}}$ is finite and the theorem is proved.

3. Regular Elements

We next turn to the set $\mathfrak{r}$ of regular elements in $\mathfrak{p}$ as defined in §4. No. 1. For $X \in \mathfrak{p}$, $(\text{ad } X)^2$ maps $\mathfrak{p}$ into itself and we consider the characteristic polynomial of the restriction $(\text{ad } X)^2 | \mathfrak{p}$,

$$\det(\lambda I - (\text{ad } X)^2 | \mathfrak{p}) = \lambda^n + a_{n-1}(X)\lambda^{n-1} + \cdots + a_l(X)\lambda^l, \tag{8}$$

where $a_l \not\equiv 0$.

Proposition 4.9. *The element $X \in \mathfrak{p}$ is regular if and only if $a_l(X) \neq 0$. In particular, $\mathfrak{r}$ is connected.*

Proof. Using [DS] (Proposition 1.1, Chapter III) on $(\mathrm{ad}\, X)^2|\mathfrak{p}$, we have

$$\det(\lambda I - (\mathrm{ad}\, X)^2|\mathfrak{p}) = \lambda^{d_1(X)}(\lambda^{n-d_1(X)} + \cdots + b(X)),$$

where $d_1(X) = \dim \mathfrak{p}(X, 0)$, $b(X) \neq 0$. Thus $d_1(X) \geq l$ with equality holding if and only if $a_l(X) \neq 0$. But $d_1(X) = l$ if and only if X is regular, so the proposition is proved (cf. [DS], Chapter VII, Corollary 12.5 for the connectedness).

For $X \in \mathfrak{p}$, $(\mathrm{ad}\, X)^2$ maps $\mathfrak{k}$ into itself, so, in analogy with $\mathfrak{p}(X, \lambda)$, we define for $\lambda \in \boldsymbol{C}$

(9) $$\mathfrak{k}(X, \lambda) = \{T \in \mathfrak{k};\ ((\mathrm{ad}\, X)^2 - \lambda)^m(T) = 0 \quad \text{for some } m\}.$$

Proposition 4.10. *An element $X \in \mathfrak{p}$ is regular if and only if*

$$\dim \mathfrak{k}(X, 0) = \min_{Y \in \mathfrak{p}} \dim \mathfrak{k}(Y, 0).$$

Proof. Consider the eigenspace decompositions of $(\mathrm{ad}\, X)^2$ according to [DS] (Chapter III, §1):

(10) $$\mathfrak{p} = \mathfrak{p}(X, 0) + \mathfrak{p}(X, \lambda_1) + \cdots + \mathfrak{p}(X, \lambda_k) \qquad (\lambda_i \neq 0),$$

(11) $$\mathfrak{k} = \mathfrak{k}(X, 0) + \mathfrak{k}(X, \mu_1) + \cdots + \mathfrak{k}(X, \mu_m) \qquad (\mu_j \neq 0).$$

Since ad X maps $\mathfrak{k}$ into $\mathfrak{p}$ and $\mathfrak{p}$ into $\mathfrak{k}$, it follows immediately from the definitions that ad X maps $\mathfrak{p}(X, \lambda_i)$ into $\mathfrak{k}(X, \lambda_i)$ and maps $\mathfrak{k}(X, \mu_j)$ into $\mathfrak{p}(X, \mu_j)$. Hence $(\lambda_1, \ldots, \lambda_k) = (\mu_1, \ldots, \mu_m)$ $(m = k)$. Also, if $v \neq 0$ in $\mathfrak{p}(X, \lambda_i)$ (resp. $\mathfrak{k}(X, \lambda_i)$), then ad $X(v) \neq 0$, so the mapping

$$\mathrm{ad}\, X\colon \mathfrak{p}(X, \lambda_i) \to \mathfrak{k}(X, \lambda_i)$$

is a bijection. Thus (10) and (11) show that $\dim \mathfrak{p}(X, 0)$ and $\dim \mathfrak{k}(X, 0)$ reach their minimum value for the same X, so the proposition is proved.

In analogy with the definition of a Cartan subalgebra in [DS] (Chapter III, §3), we make the following definition.

Definition. *A Cartan subspace* of $\mathfrak{p}$ is a subspace $\mathfrak{a} \subset \mathfrak{p}$ satisfying the conditions

(i) $\mathfrak{a}$ is a maximal abelian subspace of $\mathfrak{p}$.
(ii) For each $X \in \mathfrak{a}$, $\mathrm{ad}_{\mathfrak{g}} X$ is semisimple.

If $\mathfrak{a}_0$ is a maximal abelian subspace of $\mathfrak{p}_0$ its complexification $\mathfrak{a}_0 + i\mathfrak{a}_0$ is a Cartan subspace. Following the method of [DS] (Chapter III, §3), we shall now give a more general construction.

Theorem 4.11. *Let $H \in \mathfrak{p}$ be regular. Then H is semisimple, $\mathfrak{p}(H, 0)$ equals the centralizer $\mathfrak{p}^H$ of H in $\mathfrak{p}$ and is a Cartan subspace of $\mathfrak{p}$.*

Proof. The proof is similar to that of [DS] (Chapter III, Theorem 3.1). Let $\lambda_1, \ldots, \lambda_k$ be the nonzero eigenvalues of $(\operatorname{ad} H)^2$; put

$$\mathfrak{p}' = \sum_1^k \mathfrak{p}(H, \lambda_i), \qquad \mathfrak{a} = \mathfrak{p}(H, 0), \tag{12}$$

and for any subspace $\mathfrak{s} \subset \mathfrak{g}$ and an element $Y \in \mathfrak{g}$ let $\mathfrak{s}^Y$ denote the centralizer of Y in $\mathfrak{s}$. Let $\operatorname{ad} H = S + N$ be the decomposition of $\operatorname{ad} H$ of $\mathfrak{g}$ into its semisimple part S and its nilpotent part N. Then it is known (Appendix, Prop. 2.7) that $S = \operatorname{ad} H_s$, $N = \operatorname{ad} H_n$, where $H_s, H_n \in \mathfrak{g}$. Since $\theta H = -H$ the uniqueness of the decomposition implies $H_s, H_n \in \mathfrak{p}$. Also, $(\operatorname{ad} H)^2 = (\operatorname{ad} H_s)^2 + N_0$, where

$$N_0 = 2 \operatorname{ad} H_s \operatorname{ad} H_n + (\operatorname{ad} H_n)^2.$$

Since this is a nilpotent operator commuting with $(\operatorname{ad} H_s)^2$, $(\operatorname{ad} H)^2$ has semisimple part $(\operatorname{ad} H_s)^2$. Thus by [DS] (Prop. 1.1, Chapter III) and the semisimplicity of $\operatorname{ad} H_s$,

$$\mathfrak{a} = \mathfrak{p}(H, 0) = \mathfrak{p}(H_s, 0) = \mathfrak{p}^{H_s}. \tag{13}$$

Similarly, since $(\operatorname{ad} H_s)^2 - \lambda$ is the semisimple part of $(\operatorname{ad} H)^2 - \lambda$ we have

$$\mathfrak{p}(H, \lambda) = \mathfrak{p}(H_s, \lambda). \tag{14}$$

But if $Y \in \mathfrak{p}^{H_s}$ then $(\operatorname{ad} Y)^2$ maps $\mathfrak{p}(H_s, \lambda)$ into itself, so as in [DS] (Chapter III, §3) we can for each $Y \in \mathfrak{p}(H, 0)$ consider

$$d(Y) = \det((\operatorname{ad} Y)^2 | \mathfrak{p}').$$

By (12) and (14) $d(H_s) \neq 0$, so, since d is a polynomial function on $\mathfrak{a}$, we have $d(Y) \neq 0$ for all Y in a dense subset of $\mathfrak{a}$. For all such Y, $(\operatorname{ad} Y)^2$ is nonsingular on $\mathfrak{p}'$ and maps each $\mathfrak{p}(H, \lambda_i)$ into itself. Thus if $Z \in \mathfrak{p}(Y, 0)$ and we decompose Z by (10) (with $X = H$) we find $Z \in \mathfrak{p}(H, 0)$. Hence $\mathfrak{p}(Y, 0) \subset \mathfrak{p}(H, 0)$, so by the regularity of H, $\mathfrak{p}(Y, 0) = \mathfrak{p}(H, 0)$. Thus $(\operatorname{ad} Y)^2 | \mathfrak{a}$ is nilpotent; by continuity this is so for all $Y \in \mathfrak{a}$.

The same reasoning with $\mathfrak{k}$ instead of $\mathfrak{p}$ gives (using Proposition 4.10)

$$(\operatorname{ad} Y)^2 | \mathfrak{k}^{H_s} \text{ is nilpotent for each } Y \in \mathfrak{a}.$$

Combining these results and the relation $\mathfrak{g}^{H_s} = \mathfrak{k}^{H_s} + \mathfrak{p}^{H_s}$, we find

$$(\operatorname{ad} Y) | \mathfrak{g}^{H_s} \text{ is nilpotent for } Y \in \mathfrak{a}. \tag{15}$$

The semisimple element H_s lies in a Cartan subalgebra $\mathfrak{h}_s$ of $\mathfrak{g}$, so the centralizer $\mathfrak{g}^{H_s}$ is reductive and has the form

(16) $$\mathfrak{g}^{H_s} = \mathfrak{g}' + \text{center}(\mathfrak{g}^{H_s}), \qquad \mathfrak{g}' = [\mathfrak{g}^{H_s}, \mathfrak{g}^{H_s}],$$

where the ideal $\mathfrak{g}'$ is semisimple (cf., [DS], Chapter IX, §4; Chapter III, Exercise B.1). If $Y \in \mathfrak{g}' \cap \mathfrak{a}$ then by (15) ad Y is nilpotent on $\mathfrak{g}'$, so for the Killing form B' on $\mathfrak{g}'$, $B'(Y, Y) = 0$. The orthogonal decomposition $\mathfrak{g}' = \mathfrak{g}' \cap \mathfrak{k}^{H_s} + \mathfrak{g}' \cap \mathfrak{a}$ therefore implies $B'(\mathfrak{g}', \mathfrak{g}' \cap \mathfrak{a}) = 0$. Hence $\mathfrak{g}' \cap \mathfrak{a} = 0$ and $\mathfrak{g}' \subset \mathfrak{k}$; so, since the decomposition (16) is θ-stable, $\mathfrak{a} \subset \text{center}(\mathfrak{g}^{H_s})$, which in turn is contained in $\mathfrak{h}_s$. This proves the theorem.

We shall now prove, in analogy to the real situation ([DS], Chapter V, Lemma 6.3) that all Cartan subspaces of the complex space $\mathfrak{p}$ are conjugate under K.

Let $\mathfrak{a} \subset \mathfrak{p}$ be an arbitrary Cartan subspace. In analogy with [DS] (Chapter VI, Lemma 3.6, and Chapter VII, Lemma 11.3), we consider the joint eigenspace decompositions of the operators $(\text{ad}_{\mathfrak{g}}(H))^2$ $(H \in \mathfrak{a})$. For $\lambda \in \mathfrak{a}^*$, put

$$\mathfrak{g}_\lambda = \{Z : [H, Z] = \lambda(H)Z \text{ for } H \in \mathfrak{a}\}.$$

Then $\theta\mathfrak{g}_\lambda = \mathfrak{g}_{-\lambda}$, $\mathfrak{g} = \sum_\lambda \mathfrak{g}_\lambda$, and $\mathfrak{g}_\lambda + \mathfrak{g}_{-\lambda} = \mathfrak{k}_\lambda + \mathfrak{p}_\lambda$, where

$$\mathfrak{k}_\lambda = \{T \in \mathfrak{k} : (\text{ad } H)^2 T = \lambda(H)^2 T \text{ for } H \in \mathfrak{a}\},$$

$$\mathfrak{p}_\lambda = \{X \in \mathfrak{p} : (\text{ad } H)^2 X = \lambda(H)^2 X \text{ for } H \in \mathfrak{a}\}.$$

Let $\Delta(\mathfrak{a}) = \{\lambda \in \mathfrak{a}^* : \lambda \neq 0, \mathfrak{g}_\lambda \neq 0\}$ and put

$$\mathfrak{a}' = \{H \in \mathfrak{a} : \lambda(H) \neq 0 \text{ for } \lambda \in \Delta(\mathfrak{a})\}.$$

Since $H \in \mathfrak{a}$ is semisimple, ad H and $(\text{ad } H)^2$ have the same null space. Hence $\mathfrak{k}_0 = \mathfrak{m}$, the centralizer of $\mathfrak{a}$ in $\mathfrak{k}$, and $\mathfrak{p}_0 = \mathfrak{a}$. Also, if $\lambda \in \Delta(\mathfrak{a})$, $H \in \mathfrak{a}'$, then ad H interchanges $\mathfrak{k}_\lambda$ and $\mathfrak{p}_\lambda$ bijectively. Thus

(17) $$\mathfrak{k} = \mathfrak{m} + \sum_{\lambda \in \Delta(\mathfrak{a})} \mathfrak{k}_\lambda, \qquad \mathfrak{p} = \mathfrak{a} + \sum_{\lambda \in \Delta(\mathfrak{a})} \mathfrak{p}_\lambda.$$

Comparing now with (8), we see that the restriction $a_l|\mathfrak{a}$ equals the product of the members of $\Delta(\mathfrak{a})$ with multiplicities. This gives the following description of the set $\mathfrak{r} \cap \mathfrak{a}$.

Lemma 4.12. Let

$$\mathfrak{a}' = \{H \in \mathfrak{a} : \lambda(H) \neq 0 \text{ for all } \lambda \in \Delta(\mathfrak{a})\}.$$

Then if $H \in \mathfrak{a}'$, *we have* $\mathfrak{p}^H = \mathfrak{a}$,

$$\mathfrak{a}' = \mathfrak{r} \cap \mathfrak{a} = \left\{X \in \mathfrak{a} : \dim \mathfrak{p}^X = \min_{Y \in \mathfrak{a}} \dim \mathfrak{p}^Y\right\}$$

$$= \left\{X \in \mathfrak{a} : \dim \mathfrak{k}^X = \min_{Y \in \mathfrak{a}} \dim \mathfrak{k}^Y\right\}.$$

Corollary 4.13. *The set* $K \cdot \mathfrak{a}'$ *is open in* $\mathfrak{p}$.

For this it suffices to verify that the mapping $(k, H) \to k \cdot H$ of $K \times \mathfrak{a}'$ into $\mathfrak{p}$ has a surjective differential at (e, H) and hence at $(k, H) \in K \times \mathfrak{a}'$. For this we must show $[\mathfrak{k}, H] + \mathfrak{a} = \mathfrak{p}$. However, this is obvious from (17) since $\operatorname{ad} H(\mathfrak{k}_\lambda) = \mathfrak{p}_\lambda$.

Theorem 4.14. *Any two Cartan subspaces of* $\mathfrak{p}$ *are conjugate under* K. *Moreover,*

(18) $$\mathfrak{r} = K \cdot \mathfrak{a}'.$$

Proof. Each $X \in \mathfrak{r}$ lies in a Cartan subspace so if $\mathfrak{a}_i$ $(i \in I)$ are the Cartan subspaces of $\mathfrak{p}$, we have $\mathfrak{r} = \bigcup_i K \cdot \mathfrak{a}_i'$. If two summands, say $K \cdot \mathfrak{a}_i'$ and $K \cdot \mathfrak{a}_j'$ intersect we can select $H_i \in \mathfrak{a}_i'$, $H_j \in \mathfrak{a}_j'$, which are conjugate under K. But then, by Lemma 4.12, $\mathfrak{p}^{H_i}$ $(=\mathfrak{a}_i)$ is conjugate to $\mathfrak{p}^{H_j}$ $(=\mathfrak{a}_j)$, so the two summands coincide. Using Corollary 4.13 and the connectedness of $\mathfrak{r}$ we deduce that all summands coincide and that all $\mathfrak{a}_i$ are K-conjugate. The argument also proves that $\mathfrak{r} = K \cdot \mathfrak{a}'$.

Corollary 4.15. *The regular elements in* $\mathfrak{p}$ *are the semisimple quasi-regular elements; i.e.,*

(19) $$\mathfrak{r} = \mathscr{S} \cap \mathscr{R}.$$

For this let $X \in \mathfrak{r}$. We have seen that $X \in \mathscr{S}$ and that $\mathfrak{k}(X, 0) = \mathfrak{k}^X$ has dimension $\dim \mathfrak{m}$ [cf. (17), (18)]. By its definition, $\mathscr{R}$ is the set of points $X \in \mathfrak{p}$ for which the map $k \to k \cdot X$ of K into $\mathfrak{p}$ has maximal rank; expressing this by rank of Jacobians we see that $\mathscr{R}$ is open in $\mathfrak{p}$; thus by Proposition 4.9, $\mathfrak{r} \cap \mathscr{R} \neq \emptyset$. Also, since $\mathfrak{k}^Z$ is the Lie algebra of the isotropy subgroup of K at Z we see that $\mathscr{R}$ is the set of $Z \in \mathfrak{p}$ minimizing $\dim \mathfrak{k}^Z$.

Select $Z \in \mathfrak{r} \cap \mathscr{R}$. Then

$$\dim \mathfrak{k}^X = \dim \mathfrak{m} = \dim k^Z = \min_{Y \in \mathfrak{p}} \dim \mathfrak{k}^Y,$$

so $X \in \mathscr{R}$. This proves $\mathfrak{r} \subset \mathscr{S} \cap \mathscr{R}$. On the other hand, if $X \in \mathscr{S} \cap \mathscr{R}$ then

$$\dim \mathfrak{k}(X, 0) = \dim \mathfrak{k}^X = \min_{Y \in \mathfrak{p}} \dim \mathfrak{k}^Y$$
$$\leq \min_{Y \in \mathfrak{p}} \dim \mathfrak{k}(Y, 0),$$

so $X \in \mathfrak{r}$.

4. Semisimple Elements

Proposition 4.16. *Let $\mathfrak{a} \subset \mathfrak{p}$ be a Cartan subspace and let $\mathfrak{p}_s = \mathscr{S}$ denote the set of semisimple elements in $\mathfrak{p}$. Then*

$$\mathscr{S} = K \cdot \mathfrak{a}.$$

This statement is similar to [DS] (Chapter IX, Proposition 4.6) about Cartan subalgebras, and the proof is also quite analogous. Let $H \in \mathscr{S}$. Then ad H maps $\mathfrak{p}$ into $\mathfrak{k}$ and $\dim \mathfrak{p} = \dim \mathfrak{p}^H + \dim \operatorname{ad} H(\mathfrak{p})$.

Consider now the mapping

$$\tau : \mathfrak{p}^H \times (\operatorname{ad} H(\mathfrak{p})) \to \mathfrak{p}$$

given by $\tau(X, T) = e^{\operatorname{ad} T}(X + H)$. Its differential at $(0, 0)$ is given by

$$d\tau_{(0,0)}(X, T) = X + [T, H].$$

If this vanishes then $(\operatorname{ad} H)^2(T) = 0$. Since $T = \operatorname{ad} H(Y)$, $(Y \subset \mathfrak{p})$ we have $(\operatorname{ad} H)^3(Y) = 0$ so the semisimplicity of ad H implies $T = 0$ whence $X = 0$. Thus the image $\tau(\mathfrak{p}^H \times (\operatorname{ad} H)\mathfrak{p})$ contains an open subset of $\mathfrak{p}$ and therefore a member of $\mathfrak{r}$. Thus $\mathfrak{p}^H$ contains an element X such that $Z = X + H \in \mathfrak{r}$. Then $\mathfrak{p}^Z$ is a Cartan subspace containing H. Because of Theorem 4.14 this proves the proposition.

Lemma 4.17. *Let $X \in \mathfrak{p}$ and let $X = X_s + X_n$ be its additive Jordan decomposition so that $[X_s, X_n] = 0$, $X_s \in \mathscr{S}$, $X_n \in \mathscr{N}$ (since $\theta X = -X$). Then*

$$X_s \in \overline{K \cdot X} \qquad \text{(closure)}.$$

Proof. We may assume $X_n \neq 0$. Then $X_n \in \mathscr{N} \cap \mathfrak{g}^{X_s}$. By [DS] (Chapter III, Exercise B1), $\mathfrak{g}^{X_s}$ is reductive and its center consists of elements which are semisimple in $\mathfrak{g}$. By Propositions 2.12 and 2.16 of the Appendix, $\mathfrak{g}^{X_s}$ is reductive in $\mathfrak{g}$, and X_n is a nilpotent element in the semisimple algebra $\mathfrak{g}' = [\mathfrak{g}^{X_s}, \mathfrak{g}^{X_s}]$. But then by [DS] (Chapter IX, Theorem 7.4), there exists an element $H \in \mathfrak{g}'$ such that $[H, X_n] = X_n$. Decomposing

H according to $\mathfrak{g}^{X_s} = \mathfrak{k}^{X_s} + \mathfrak{p}^{X_s}$, we may take $H \in \mathfrak{g}' \cap \mathfrak{k}$. But then the group $U = \exp(\operatorname{ad} \mathbf{C}H)$ satisfies

$$U \subset K, \qquad U \cdot X = X_s + \mathbf{C}^* \cdot X_n, \qquad \mathbf{C}^* = \mathbf{C} - \{0\}.$$

Hence $X_s \in \overline{U \cdot X} \subset \overline{K \cdot X}$.

From this lemma and Theorem 4.3 we have

Corollary 4.18. *Let j be a K-invariant polynomial on $\mathfrak{p}$ without constant term. Then*

$$j(X) = j(X_s), \qquad j(X_n) = 0.$$

Theorem 4.19. *Let $X \in \mathfrak{p}$. Then the orbit $K \cdot X$ is closed if and only if X is semisimple.*

Proof. From Lemma 4.17 we know that a closed orbit necessarily consists of semisimple elements.

On the other hand, suppose $X \in \mathfrak{p}$ is semisimple. Let $X' = X'_s + X'_n$ be in closure $(K \cdot X)^-$. With j as in Corollary 4.18 we have $j(X'_s) = j(X') = j(X) = j(X_s)$. Because of Proposition 4.16, Corollary 3.12, and Chapter II, Corollary 5.12 (extended in an obvious way to complex $\mathfrak{a}$) we have $X_s = k \cdot X'_s$ for some $k \in K$. Then $k \cdot X'_n$ belongs to the centralizer $\mathfrak{g}^{X_s}$ which is reductive in $\mathfrak{g}$. By Prop. 2.16 and Cor. 2.10 in Appendix, $k \cdot X'_n \in \mathfrak{g}' = [\mathfrak{g}^{X_s}, \mathfrak{g}^{X_s}]$ and an element in $\mathfrak{g}'$ is nilpotent if and only if it is nilpotent as an element in $\mathfrak{g}$. Thus, by Theorem 4.8, we can fix elements $Z_1, \ldots, Z_s \in \mathcal{N} \cap \mathfrak{g}^{X_s}$ such that each nilpotent (in $\mathfrak{g}$) element in $\mathfrak{g}^{X_s} \cap \mathfrak{p}$ is conjugate to some Z_i under the subgroup K^{X_s} of K fixing X_s. It follows that $k \cdot X'_n = u \cdot Z_i$ for some $u \in K^{X_s}$ and some i. This proves that $X' = X'_s + X'_n = k^{-1}X_s + k^{-1}uZ_i = k^{-1}u(X_s + Z_i)$ so

$$(K \cdot X)^- = \bigcup_{i=1}^{r} K \cdot (X_s + Z_i).$$

The space $(K \cdot X)^-$ is locally compact, so, by the Baire category theorem ([DS], Chapter II, §3), one of the orbits $K \cdot (X_s + Z_i)$, say $O_1 = K \cdot (X_s + Z_1)$, has an interior point in $(K \cdot X)^-$. By homogeneity, O_1 is open in $(K \cdot X)^-$. Then the complement $(K \cdot X)^- - O_1$ contains an open orbit, etc. Thus $(K \cdot X)^- = O_1 \cup \cdots \cup O_p$ where for each i, $O_i \cup \cdots \cup O_p$ is closed and contains the orbit O_i as an open subset. In particular O_p is closed. By the first part of the proof it consists of semisimple elements. On the other hand, the relation $X_s = kX'_s$ above showed that the semisimple elements in $(K \cdot X)^-$ form a single orbit which thus must be O_p. In particular, $X \in O_p$, so $K \cdot X = O_p$, which is closed.

5. Algebro-Geometric Results on the Orbits

In this subsection we describe two basic results from Kostant and Rallis [1971] about the K_θ orbits in $\mathfrak{p}$. The proofs will be omitted since they require several tools from algebraic geometry. These results as well as those in §5 which rely on them will not be needed in the present volume.

As in §4, No. 2, consider the symmetric algebras $S(\mathfrak{p})$, $S(\mathfrak{p}^*)$ and the subalgebras $I_+(\mathfrak{p}) \subset I(\mathfrak{p}) \subset S(\mathfrak{p})$ and $I_+(\mathfrak{p}^*) \subset I(\mathfrak{p}^*) \subset S(\mathfrak{p}^*)$. Similarly, let $I(\mathfrak{a})$ and $I(\mathfrak{a}^*)$ denote the spaces of Weyl group invariants acting on $S(\mathfrak{a})$ and $S(\mathfrak{a}^*)$, respectively. By Corollary 5.12 of Chapter II, the restriction $p \to p|\mathfrak{a}$ induces an isomorphism of $I(\mathfrak{p}^*)$ onto $I(\mathfrak{a}^*)$. By Lemma 4.2 the groups K_θ and K induce the same transformations of $\mathfrak{a}$; consequently each $p \in I(\mathfrak{p}^*)$ is K_θ-invariant. In accordance with the stated isomorphism and Theorem 3.1, let $j_1, \ldots, j_l$ be homogeneous algebraically independent generators of $I(\mathfrak{p}^*)$ and let $j: \mathfrak{p} \to \mathbf{C}^l$ denote the mapping $j(x) = (j_1(x), \ldots, j_l(x))$ $(x \in \mathfrak{p})$. According to Corollary 3.12 and Theorem 4.3, j is surjective and $j^{-1}(0) = \mathcal{N}$. As before, let $\mathcal{R}$ denote the set of quasiregular elements in $\mathfrak{p}$.

Theorem 4.20. *The intersection $\mathcal{R} \cap \mathcal{N}$ is a dense open subset of $\mathcal{N}$ and is a single K_θ-orbit.*

More generally, if $\xi \in \mathbf{C}^l$, $\mathcal{R} \cap j^{-1}(\xi)$ is a single K_θ-orbit and is open and dense in $j^{-1}(\xi)$.

For $\xi = (\xi_1, \ldots, \xi_l) \in \mathbf{C}^l$ let $I_\xi = I_\xi(\mathfrak{p}^*)$ denote the ideal in $I(\mathfrak{p}^*)$ generated by $(j_k - \xi_k)$ $(k = 1, \ldots, l)$. Then $I_\xi S(\mathfrak{p}^*)$ is the ideal in $S(\mathfrak{p}^*)$ generated by I_ξ. The corresponding set of common zeros in $\mathfrak{p}$ is just $j^{-1}(\xi)$. According to Hilbert's *Nullstellensatz* a polynomial $f \in S(\mathfrak{p}^*)$ vanishes identically on $j^{-1}(\xi)$ if and only if some power of it belongs to $I_\xi S(\mathfrak{p}^*)$. Here a stronger result holds.

Theorem 4.21. *Let $\xi \in \mathbf{C}^l$ and $f \in S(\mathfrak{p}^*)$. Then*

$$f|j^{-1}(\xi) \equiv 0 \Leftrightarrow f \in I_\xi S(\mathfrak{p}^*).$$

In other words, the ideal $I_\xi S(\mathfrak{p}^*)$ is equal to its own radical.

§5. Harmonic Polynomials on $\mathfrak{p}$

We recall that the elements in $I(\mathfrak{p}^*)$ are also K_θ invariant. Just as in §1 we call a polynomial h on $\mathfrak{p}$ *harmonic* (or *K-harmonic*) if $\partial(j)h = 0$ for each $j \in I_+(\mathfrak{p})$. Let $H(\mathfrak{p}^*)$ denote the space of harmonic polynomials.

From Corollary 1.9 and Theorem 4.21 (for $\xi = 0$) we now have the following result.

Theorem 5.1. *The harmonic polynomials on $\mathfrak{p}$ are spanned by $(X^*)^k$ $(k = 0, 1, \ldots)$ where X runs through $X \in \mathcal{N}$.*

We next prove an analog of a statement in Theorem 3.4, sharpening Theorem 1.1.

Theorem 5.2. *The linear mapping*

$$\phi: I(\mathfrak{p}^*) \otimes H(\mathfrak{p}^*) \to S(\mathfrak{p}^*)$$

given by $\phi(j \otimes h) = jh$ is a bijection.

The surjectivity of ϕ being clear from Theorem 1.1, we have to prove that if $h_1, \ldots, h_p \in H(\mathfrak{p}^*)$ are linearly independent and $i_1, \ldots, i_q \in I(\mathfrak{p}^*)$ are linearly independent and if for some $a_{sr} \in \boldsymbol{C}$,

$$\phi(\textstyle\sum a_{sr} i_s \otimes h_r) = \sum a_{sr} i_s h_r = 0,$$

then all $a_{rs} = 0$. Equivalently, we must show that if $j_1, \ldots, j_p \in I(\mathfrak{p}^*)$ satisfy $\sum_1^p h_r j_r \equiv 0$ then $j_r \equiv 0$ for each r. First we observe that the restrictions $h_i|\mathcal{N}$ $(1 \leq i \leq p)$ are linearly independent. This in fact follows from Theorem 4.21 (for $\xi = 0$) and the directness of the decomposition $S(\mathfrak{p}^*) = H(\mathfrak{p}^*) + I_+(\mathfrak{p})S(\mathfrak{p}^*)$ [cf. Eq. (8) of §1]. Using (by Theorem 4.20 for $\xi = 0$) an element $X \in \mathcal{N}$ for which $K_\theta \cdot X$ is dense in $\mathcal{N}$ it follows that the functions h_i' on K_θ $(1 \leq i \leq p)$ given by $h_i'(k) = h_i(k \cdot X)$ are linearly independent. This means that as k runs through K_θ the vectors $(h_1(k^{-1} \cdot X), \ldots, h_p(k^{-1} \cdot X))$ span $\boldsymbol{C}^p$. Thus we can select $k_1, \ldots, k_p \in K_\theta$ such that the translated polynomials $p_{rs} = k_r \cdot h_s$ $(1 \leq r, s \leq p)$ satisfy $\det(p_{rs}(X)) \neq 0$. By continuity, $\det(p_{rs}(Y)) \neq 0$ for all Y in some neighborhood U of X. But then the functions $h_r|(K_\theta \cdot Y)(1 \leq r \leq p)$ are linearly independent for each $Y \in U$. If the polynomials $j_1, \ldots, j_p$ were not all 0 there would exist a $Y \in U$ such that the vector $(j_1(Y), \ldots, j_p(Y)) \neq 0$. Since each j_r is constant on $K_\theta \cdot Y$ it follows that $\sum_{r=1}^p j_r h_r | K_\theta \cdot Y \not\equiv 0$, contradicting the relation $\sum_1^p j_r h_r \equiv 0$.

Corollary 5.3. *For each $\xi \in \boldsymbol{C}^l$ the direct decomposition*

$$S(\mathfrak{p}^*) = I_\xi S(\mathfrak{p}^*) \oplus H(\mathfrak{p}^*)$$

holds.

Proof. Let $i \in I(\mathfrak{p}^*)$ be a homogeneous element. Writing

$$i = \sum_{(k)} a_{k_1 \cdots k_l} j_1^{k_1} \cdots j_l^{k_l} = \sum_{(k)} a_{k_1 \cdots k_l} (j_1 - \xi_1)^{k_1} \cdots (j_l - \xi_l)^{k_l} + i'. \tag{1}$$

where $\deg(i') < \deg(i)$ we obtain by iteration the direct decomposition $I(\mathfrak{p}^*) = I_\xi \oplus \boldsymbol{C}$. Consequently,

$$I(\mathfrak{p}^*) \otimes H(\mathfrak{p}^*) = (I_\xi \otimes H(\mathfrak{p}^*)) \oplus H(\mathfrak{p}^*). \tag{2}$$

Applying the mapping ϕ in Theorem 5.2, we obtain

$$S(\mathfrak{p}^*) = I_\xi H(\mathfrak{p}^*) \oplus H(\mathfrak{p}^*), \tag{3}$$

the directness coming from the injectivity of ϕ. But since $S(\mathfrak{p}^*) = I(\mathfrak{p}^*)H(\mathfrak{p}^*)$ we have $I_\xi S(\mathfrak{p}^*) \subset I_\xi H(\mathfrak{p}^*) \subset I_\xi S(\mathfrak{p}^*)$ so (3) gives the corollary.

Remark. So far we have only used Theorems 4.20 and 4.21 for the case $\xi = 0$. In the next result we shall, however, need the general ξ.

Theorem 5.4. *Let Γ be a K_θ-orbit in $\mathfrak{p}$ of maximal dimension. If*

$$h \in H(\mathfrak{p}^*), \qquad h | \Gamma \equiv 0, \tag{4}$$

then $h \equiv 0$ on $\mathfrak{p}$.

Proof. Pick $X \in \Gamma$ and put $\xi = j(X)$, the map $j: \mathfrak{p} \to \boldsymbol{C}^l$ being as in §4. Thus $\Gamma \subset \mathscr{R} \cap j^{-1}(\xi)$. But then Theorem 4.20 implies that Γ equals $\mathscr{R} \cap j^{-1}(\xi)$ and is dense in $j^{-1}(\xi)$. Thus (4) implies that h vanishes identically on $j^{-1}(\xi)$, whence by Theorem 4.21 and the directness in Corollary 5.3, $h \equiv 0$.

EXERCISES AND FURTHER RESULTS

1. Let $Q(X) = x_1^2 + \cdots + x_p^2 - x_{p+1}^2 - \cdots - x_{p+q}^2$,

$$\square = \frac{\partial^2}{\partial x_1^2} + \cdots + \frac{\partial^2}{\partial x_p^2} - \frac{\partial^2}{\partial x_{p+1}^2} - \cdots - \frac{\partial^2}{\partial x_{p+q}^2},$$

and C the quadric $Q(X) = 1$. Show that for each polynomial P on $\boldsymbol{R}^{p+q}$ there exists a polynomial H on $\boldsymbol{R}^{p+q}$ such that

$$\square H = 0; \qquad H = P \text{ on } C.$$

2. With the notation of §3, No. 4, let G/K be the symmetric space corresponding to the Cartan involution θ of $\mathfrak{g}$. Let M be the centralizer of $\mathfrak{a}$ in K. Given $\lambda \in \mathfrak{a}_c^*$ let $A_\lambda \in \mathfrak{a}^{\boldsymbol{C}}$ be determined by $B(H, A_\lambda) = \lambda(H)$ $(H \in \mathfrak{a})$ and let K_λ denote the isotropy subgroup of K at A_λ. The element λ is said to be *regular* if $\alpha(A_\lambda) \neq 0$ for each restricted root $\alpha \neq 0$; otherwise λ is said to be *singular*. Prove the following statements:

(i) $\lambda \in \mathfrak{a}_c^*$ is regular $\Leftrightarrow K_\lambda = M \Leftrightarrow \sigma\lambda \neq \lambda$ for each $\sigma \neq e$ in W.
(ii) Let $\mathscr{E}_\lambda(\mathfrak{p})$ denote the joint eigenspace

$$\mathscr{E}_\lambda(\mathfrak{p}) = \{f \in \mathscr{E}(\mathfrak{p}) : \partial(p)f = p(i\lambda)f \text{ for } p \in I(\mathfrak{p})\}.$$

The K-invariant functions in $\mathscr{E}_\lambda(\mathfrak{p})$ are the constant multiples of the function

$$\psi_\lambda(X) = \int_K e^{iB(k \cdot A_\lambda, X)}\, dk \qquad (X \in \mathfrak{p}),$$

dk being the normalized Haar measure (cf. Prop. 4.8, Chapter IV). Show that if h is a K_λ-harmonic polynomial on $\mathfrak{p}$ then the function

$$f(X) = h(X)e^{iB(A_\lambda, X)} \qquad (X \in \mathfrak{p})$$

belongs to $\mathscr{E}_\lambda(\mathfrak{p})$.
(iii) Let $h \in S(\mathfrak{p})$ be an M-harmonic polynomial. Then, if $\lambda \in \mathfrak{a}_c^*$,

$$\int_K h(k \cdot X)e^{iB(A_\lambda, k \cdot X)}\, dk = h(0)\psi_\lambda(X).$$

3. In the notation of §3, No. 4 the natural representation of W on the space H of harmonic polynomials is equivalent to the regular representation of W (Chevalley [1955a]).

4*. In the notation of Theorem 3.3 let for k $(0 \leq k \leq n)$ W_k denote the set of $\sigma \in W$ which fix some k-dimensional subspace of E but fix no subspace of higher dimension. Let $w_k = \text{Card}(W_k)$. Then

$$\sum_0^n w_k t^k = \prod_{k=1}^n (t + d_k - 1),$$

generalizing both identities of Theorem 3.3 (cf., Solomon [1963]).

5*. Let $\sigma \to l(\sigma)$ be the length function on the Weyl group (cf. Chapter IV, §6). Then

$$\sum_{\sigma \in W} t^{l(\sigma)} = \prod_{1 \leq i \leq n} (1 + t + \cdots + t^{d_i - 1})$$

(cf., Bott [1956] and Solomon [1966]). Deduce from Theorem 3.4 that

$$\dim H^k = \text{Card}\{\sigma \in W : l(\sigma) = k\}.$$

6*. Let R be an irreducible reduced root system, $B = \{\alpha_1, \ldots, \alpha_l\}$ any basis. For each $\alpha \in R^+$, $\alpha = \sum_1^l m_i\alpha_i$, put $m(\alpha) = \sum_1^l m_i$. Then

$$\sum_{\alpha \in R^+} t^{m(\alpha)} = \sum_{i=1}^{l} (t + t^2 + \cdots + t^{d_i - 1})$$

(cf. Kostant [1959a]).

7. Show that the group F from Lemma 4.2 has cardinality 2^l if $l = \dim_{\mathbf{C}} \mathfrak{a}$.

8. With the notation of §4 show that $p \in S(\mathfrak{p})$ is K-invariant if and only if it is K_θ-invariant (use Lemma 4.2 and Corollary 5.12 of Chapter II).

9*. Let (H, X, Y) be a normal $\mathfrak{sl}_2$-triple where $X, Y \in \mathcal{N}$. Then the plane $Y + \mathfrak{p}^X$ intersects each K_θ-orbit in $\mathcal{R}$ exactly once. Furthermore, the restriction mapping

$$u \to u|(Y + \mathfrak{p}^X)$$

is a bijection of $I(\mathfrak{p}^*)$ onto $S(Y + \mathfrak{p}^X)$ (cf. Kostant and Rallis [1971], Chapter II, §3; here $S(Y + \mathfrak{p}^X) = S(\mathfrak{p}^*)|(Y + \mathfrak{p}^X)$).

NOTES

§1–§2. Theorems 1.1, 1.2, 1.7, and 2.1 are from Helgason [1962, 1963]. Theorem 1.1 was proved independently by Kostant [1963], and Corollary 1.9 was proved in a different way by Maass [1959]. An example [for $G = \mathbf{SO}(2n + 1)$] for which the ideal $J = I_+^c(E^*)S^c(E^*)$ is not its own radical $\sqrt{J}$ was given by Hesselink [1979]. If ρ is a finite-dimensional irreducible representation of $\mathbf{SL}(2, \mathbf{C})$, combined results of Popov and Dixmier [1981] show that $J = \sqrt{J}$ if and only if $\dim \rho \leq 5$. Theorem 1.11 is from Molien [1898].

Corollary 2.2 (Lepage decomposition) is proved in a different way in Weil [1958] (Chapter I); it is a "flat" analog of the Hodge decomposition of harmonic forms on Kähler manifolds.

§3. Theorems 3.1 and 3.4 are due to Chevalley [1955a], and Theorem 3.3 is used in his paper [1952]. A generalization is given by Solomon [1963] (cf. Exercise 4), where Proposition 3.10 is also proved. Theorems 3.6 and 3.13 are proved in Steinberg [1964]; the latter result was also proved in Harish-Chandra's unpublished paper [1960]. Corollary 3.12 occurs in Kostant [1963] (Proposition 3.10) with a different proof. Theorem 3.16 was proved by the author [1964a] and independently by Harish-Chandra. In §3 we have used some proofs from Flatto [1978].

§4 is mostly due to Kostant and Rallis [1971]. We have sometimes [say, for (6′)] given more elementary proofs in order to make the exposition more self-contained. Theorem 4.19 is from Borel and Harish-Chandra [1962], §10 (in the group case) and Kostant and Rallis [1971], Prop. 16, in the general case. The proof of the "if" part is a generalization of the proof of Proposition 7 of §1 in Varadarajan [1977].

CHAPTER IV

SPHERICAL FUNCTIONS AND SPHERICAL TRANSFORMS

In this chapter we begin a detailed study of spherical functions and their applications to analysis on a semisimple Lie group. Since these functions arise as representation coefficients [see (8) in §4] we begin in §1, No. 1 with a review of some standard notions and results in representation theory. In §1, No. 2 we review some facts about representations of compact groups K and relate these to representations of a bigger group $G \supset K$.

§§2, 3 deal with elementary properties of spherical functions on arbitrary Riemannian homogeneous spaces. In §4 we specialize to the three types of symmetric spaces, prove specific integral formulas for the spherical functions, and exhibit them as representation coefficients. The case of a complex group G has some special features, exemplified by Theorem 4.7 and Proposition 4.10. In §§5, 6 we give an explicit expansion of the spherical function into ordinary exponential functions. Not only does this expansion describe the behavior of the spherical function at infinity, but in addition it can be used to relate the theory of the spherical transform to ordinary Fourier analysis. This serves as the basis of the Paley–Wiener theorem in §7, where, in addition, detailed information is required about the coefficients in the indicated expansion. In §7 we also prove the inversion formula and the Plancherel formula for the spherical transform.

§8 is devoted to the determination of the bounded spherical functions, or, equivalently, the maximal ideal space of the algebra of K-bi-invariant integrable functions. §9 deals with the Paley–Wiener theorem for the spherical transform on the flat space $\mathfrak{p}$ where the spherical function expansion is not available. In §10 we prove geometric properties of the Cartan and Iwasawa decompositions which are then applied to spherical functions.

§1. Representations

1. Generalities

In the preceding chapters representations have played a rather peripheral role; they will now be involved in a more significant way. In this subsection we therefore develop some of the basic notions from the theory of infinite-dimensional representations; occasional results from topological vector space theory will be used without proof.

In Chapter II, §4, we defined the concept of a *representation* π of a locally compact group G on a locally convex topological vector space V

over $\boldsymbol{C}$; we also defined the concept of *irreducibility* of π. It is sometimes convenient to have the following reduction in the assumptions.

Lemma 1.1. *Let G be a locally compact group and V a Fréchet space. Let $\pi: G \to \operatorname{Aut}(V)$ be a mapping satisfying the conditions*

(i) $\pi(xy) = \pi(x)\pi(y), \quad (x, y) \in G, \quad \pi(e) = I.$
(ii) *For each $v \in V$, the mapping $x \to \pi(x)v$ is continuous from G to V.*

Then the mapping $(x, v) \to \pi(x)v$ from $G \times V$ into V is continuous so π is a representation of G on V.

Remark. For the proof, which is a variation of that of the Banach–Steinhaus theorem, see, e.g., Bourbaki [Intégr. Chapter VIII, §2, Proposition 1]. The lemma actually holds for barreled topological vector spaces; in particular, it holds if V is the inductive limit of Fréchet spaces, and also if V is a Montel space. Thus Lemma 1.1 applies to the function spaces $\mathscr{D}(M)$, $\mathscr{E}(M)$ and their duals $\mathscr{D}'(M)$, $\mathscr{E}'(M)$, M being a manifold with a countable base.

Let F be any function space on G invariant under left translations such that the mapping $(g, f) \to f^{L(g)}$ is continuous from $G \times F$ into F. The representation l of G defined by $l(g)f = f^{L(g)}$ is called the *left regular representation* of G on F. Similarly, one defines *right regular representations* r by $r(g)f = f^{R(g^{-1})}$.

Let V' denote the dual space of V. If $v \in V$ and $\lambda \in V'$ the function $g \to \lambda(\pi(g)v)$ is called a *representation coefficient*. In harmonic analysis on G one investigates these functions as well as how they can serve as building blocks for "arbitrary" function spaces on G.

Two representations (π_1, V_1) and (π_2, V_2) of G are said to be *equivalent* if there exists a linear homeomorphism A of V_1 onto V_2 satisfying

$$A\pi_1(x) = \pi_2(x)A, \qquad x \in G. \tag{1}$$

A continuous linear mapping A of V_1 into V_2 satisfying (1) is called an *intertwining operator*.

If V is a Hilbert space and π a representation of G on V with each $\pi(x)$ unitary then π is called a *unitary representation*.

Suppose now G has a left and right invariant Haar measure dx and that the locally convex space V is complete. For the representation π of G we define for each $f \in C_c(G)$ the continuous linear transformation $\pi(f): V \to V$ by

$$\pi(f)v = \int_G f(x)\pi(x)v\,dx, \qquad v \in V. \tag{2}$$

Then $\pi(f * g) = \pi(f)\pi(g)$ if $f * g$ is the *convolution*

$$(f * g)(x) = \int_G f(y)g(y^{-1}x)\,dy$$

Suppose now G is a Lie group with Lie algebra $\mathfrak{g}$. A vector $v \in V$ is said to be *differentiable* if the mapping $\tau_v: x \to \pi(x)v$ of G into V is differentiable. Let V^∞ be the space of differentiable vectors in V. It is clearly invariant under $\pi(G)$. Moreover, if Y is a vector field on G the mapping $x \to (Y\tau_v)(x)$ of G into V is also differentiable. If $X \in \mathfrak{g}$, $v \in V^\infty$, we put

$$\pi(X)v = \lim_{t \to 0} \frac{1}{t}\,[\pi(\exp tX)v - v], \tag{3}$$

knowing that this limit exists (as an element of V).

Lemma 1.2. *For $X \in \mathfrak{g}$ we have $\pi(X)V^\infty \subset V^\infty$ and the mapping $X \to \pi(X)$ is a representation of $\mathfrak{g}$ on V^∞.*

Proof. Let $\tilde{X}$ be the left invariant vector field on G determined by X. Then if $v \in V^\infty$

$$\pi(x)\pi(X)v = (\tilde{X}\tau_v)(x), \tag{4}$$

so $\pi(X)v \in V^\infty$. Let $Y \in \mathfrak{g}$ and replace x in (4) by $x \exp tY$; then we obtain

$$(\pi(X)\pi(Y) - \pi(Y)\pi(X))v = ((\tilde{X}\tilde{Y} - \tilde{Y}\tilde{X})\tau_v)(e). \tag{5}$$

But if λ is any continuous linear functional on V we have [since $\lambda \circ \tau_v$ is a scalar-valued function]

$$\begin{aligned}\lambda \circ ([X, Y]^\sim(\tau_v)) &= [X, Y]^\sim(\lambda \circ \tau_v) = (\tilde{X}\tilde{Y} - \tilde{Y}\tilde{X})(\lambda \circ \tau_v)\\ &= \lambda \circ ((\tilde{X}\tilde{Y} - \tilde{Y}\tilde{X})\tau_v).\end{aligned}$$

By the Hahn–Banach theorem, we conclude

$$[X, Y]^\sim(\tau_v) = (\tilde{X}\tilde{Y} - \tilde{Y}\tilde{X})(\tau_v),$$

so by (4) and (5) $\pi(X)\pi(Y) - \pi(Y)\pi(X) = \pi([X, Y])$, proving the lemma.

Lemma 1.3. *Let $f \in \mathcal{D}(G)$. Then*

$$\pi(f)V \subset V^\infty.$$

Proof. We have for $v \in V$

$$\pi(x)\pi(f)v = \int_G f(y)\pi(x)\pi(y)v\,dy = \int_G f(x^{-1}z)\pi(z)v\,dz,$$

so

$$\pi(x)\pi(f)v = \pi(f^{L(x)})v. \tag{5'}$$

Thus the mapping $\tau_{\pi(f)v}$ is the composite of the differentiable mapping $x \to f^{L(x)}$ of G into $\mathscr{D}(G)$ and the continuous linear mapping $f \to \pi(f)v$ of $\mathscr{D}(G)$ into V; hence $\tau_{\pi(f)v}$ is differentiable as desired.

Corollary 1.4. *V^∞ is dense in V.*

Proof. Let $f \in \mathscr{D}(G)$ be positive, $\int_G f(x)\,dx = 1$, and $v \in V$, $\varepsilon > 0$. Then

$$\pi(f)v - v = \int_G f(x)(\pi(x)v - v)\,dx, \tag{6}$$

and if q is one of the seminorms defining the topology of V there exists a neighborhood U of e in G such that $q(\pi(x)v - v) < \varepsilon$ for $x \in U$. But then if f has support contained in U, (6) implies $q(\pi(f)v - v) \leq \varepsilon$, whence the corollary.

Definition. The representation of $\mathfrak{g}$ on V^∞ defined by Lemma 1.2 is called the *differential* of π. It is denoted by π (and sometimes by $d\pi$).

In view of [DS] (Chapter II, Proposition 1.1), π induces a representation of the universal enveloping algebra $U(\mathfrak{g})$ on V^∞. The representation π of G on V is said to be *quasisimple* if π maps the center of G into scalars and if the differential $d\pi$ maps the center of $U(\mathfrak{g})$ into scalars. (This definition differs slightly from that in Harish-Chandra [1953] (§10).)

Again let V be an arbitrary locally convex space. Let V' denote the dual of V and let V' be given the *strong topology* (that is, the topology of uniform convergence on bounded subsets of V). The mapping which sends a vector $v \in V$ into the continuous linear form $\tilde{v}: \lambda \to \lambda(v)$ on V' is an injective map of V into the second dual space $(V')'$. If this map is surjective, that is, if $\tilde{V} = (V')'$, then V is called *semireflexive*. It is known that any closed subspace W of a semireflexive space V is again semireflexive, (Köthe [1969]).

If A is a continuous linear transformation of V its *transpose*

$${}^tA: V' \to V'$$

is defined by $({}^tA(\lambda))(v) = \lambda(Av)$; it is again continuous. If A is a linear homeomorphism of V onto itself then $({}^tA)^{-1} = {}^t(A^{-1})$.

Let π be a representation of a locally compact group G on a locally convex space V. The mapping $\check{\pi}: g \to {}^t(\pi(g^{-1}))$ is a homomorphism of G into $\mathrm{Aut}(V')$. *If V is semireflexive*, it is known that the mapping $(g, \lambda) \to \check{\pi}(g)\lambda$ of $G \times V'$ into V' is continuous (Bruhat [1956], §2) so $\check{\pi}$ is a representation; it is called the representation *contragredient* to π.

If V is a Hilbert space with inner product $\langle\ ,\ \rangle$ then the mapping which to each v assigns the linear form $v': w \to \langle w, v\rangle$ is a *conjugate linear* bijection of V onto V'. If $A: V \to V$ is a continuous linear mapping with adjoint A^*, we have ${}^tA(v') = (A^*v)'$. We shall now establish an important connection between unitary representations and positive definite functions.

Definition. A complex-valued continuous function ϕ on a locally compact group G is called *positive definite* if

$$\sum_{i,j=1}^{n} \phi(x_i^{-1}x_j)\alpha_i\bar{\alpha}_j \geq 0$$

for all finite sets $x_1, \ldots, x_n$ of elements in G and any complex numbers $\alpha_1, \ldots, \alpha_n$.

A positive definite function ϕ satisfies the conditions

$$\phi(e) \geq 0, \qquad \phi(x^{-1}) = \overline{\phi(x)}, \quad |\phi(x)| \leq \phi(e),$$

which are easily derived from the definition. In particular, a positive definite function is necessarily bounded.

Let π be a unitary representation of a locally compact group G on a Hilbert space $\mathfrak{H}$. Let $\langle\ ,\ \rangle$ denote the scalar product in $\mathfrak{H}$. For each fixed vector $\boldsymbol{e} \in \mathfrak{H}$, the function $x \to \langle \boldsymbol{e}, \pi(x)\boldsymbol{e}\rangle$ is a positive definite function on G. The continuity is obvious, so let $x_1, \ldots, x_n$ be any finite set of elements in G and $\alpha_1, \ldots, \alpha_n$ any complex numbers. Then

$$\sum_{i,j=1}^{n} \langle \boldsymbol{e}, \pi(x_i^{-1}x_j)\boldsymbol{e}\rangle\alpha_i\bar{\alpha}_j = \sum_{i,j=1}^{n} \langle \pi(x_i)\boldsymbol{e}, \pi(x_j)\boldsymbol{e}\rangle\alpha_i\bar{\alpha}_j$$

$$= \left\langle \sum_i \alpha_i\pi(x_i)\boldsymbol{e}, \sum_j \alpha_j\pi(x_j)\boldsymbol{e}\right\rangle \geq 0.$$

On the other hand, a positive definite function $\phi \not\equiv 0$ on G gives in a canonical way rise to a unitary representation of G as follows: Let V_ϕ denote the set of all complex linear combinations of left translates $\phi^{L(x)}$ $(x \in G)$ of ϕ. We define a scalar product in V_ϕ by the formula

$$\langle f, g\rangle = \sum_{i,j} \alpha_i\bar{\beta}_j\phi(x_i^{-1}y_j) \tag{7}$$

if $f = \sum_i \alpha_i\phi^{L(x_i)}$, $g = \sum_j \beta_j\phi^{L(y_j)}$. Now $\langle f, g\rangle = \sum \alpha_i\overline{g(x_i)} = \sum \bar{\beta}_j f(y_j)$ so it is clear that (7) depends only on f and g but not on their special expressions in terms of ϕ. Since $\langle f, f\rangle \geq 0$ we have the Schwarz inequality $|\langle f, g\rangle|^2 \leq \langle f, f\rangle\langle g, g\rangle$. Since $f(x) = \langle f, \phi^{L(x)}\rangle$ it shows that $\langle f, f\rangle = 0$ if and only if $f \equiv 0$. The completion of V_ϕ in the norm $\|f\| = \langle f, f\rangle^{1/2}$ is a Hilbert space $\mathscr{H}_\phi$. Each $x \in G$ gives rise to an endomorphism

$f \to f^{L(x)}$ of V_ϕ; this endomorphism preserves the inner product $\langle\ ,\ \rangle$ and so extends uniquely to a unitary operator $\pi(x)$ of $\mathfrak{H}_\phi$. The mapping $x \to \pi(x)$ satisfies $\pi(xy) = \pi(x)\pi(y)$ for all $x, y \in G$. Moreover, if $f \in V_\phi$ we have

$$\langle f^{L(x)} - f, f^{L(x)} - f \rangle = 2\langle f, f\rangle - \langle f^{L(x)}, f\rangle - \langle f, f^{L(x)}\rangle$$

$$= 2\sum_{i,j} \alpha_i \bar{\alpha}_j \phi(x_i^{-1}x_j) - \sum_{i,j} \alpha_i \bar{\alpha}_j \phi(x_i^{-1}x^{-1}x_j) - \sum_{i,j} \alpha_i \bar{\alpha}_j \phi(x_i^{-1}xx_j),$$

which tends to 0 for $x \to e$. Thus for a dense set of vectors $\boldsymbol{a} \in \mathfrak{H}_\phi$, the mapping $x \to \pi(x)\boldsymbol{a}$ is continuous for $x = e$, hence for all $x \in G$. From the inequality

$$\|\pi(x)\boldsymbol{b} - \boldsymbol{b}\| \leq \|\pi(x)(\boldsymbol{a} - \boldsymbol{b})\| + \|\pi(x)\boldsymbol{a} - \boldsymbol{a}\| + \|\boldsymbol{a} - \boldsymbol{b}\|$$

it follows that for each $\boldsymbol{b} \in \mathfrak{H}_\phi$, the mapping $x \to \pi(x)\boldsymbol{b}$ is continuous. Thus π is a unitary representation of G on $\mathfrak{H}_\phi$. Finally, if $\boldsymbol{e}$ denotes the vector in $\mathfrak{H}_\phi$ which corresponds to the vector $\phi \in V_\phi$ we have

$$\phi(x) = \langle \boldsymbol{e}, \pi(x)\boldsymbol{e}\rangle.$$

We shall call π the unitary representation *associated to* ϕ. Summarizing the results above we have

Theorem 1.5. *Let π be a unitary representation of a locally compact group G on a Hilbert space $\mathfrak{H}$. For each vector $\boldsymbol{e} \in \mathfrak{H}$ the function $\langle \boldsymbol{e}, \pi(x)\boldsymbol{e}\rangle$ is a positive definite function on G. Conversely, to any positive definite function $\phi \not\equiv 0$ on G corresponds a unitary representation π of G such that $\phi(x) = \langle \boldsymbol{e}, \pi(x)\boldsymbol{e}\rangle$ for a suitable vector $\boldsymbol{e}$.*

2. Compact Groups

In this subsection we develop some basic general facts about harmonic analysis on compact groups and their homogeneous spaces. The results are based on Schur's lemma, Schur's orthogonality relations, and the Peter–Weyl theorem, all of which will be stated without proof. This analysis will be developed further in the next chapter.

We recall that a compact group of complex matrices leaves invariant a positive definite Hermitian form; this means that any finite-dimensional representation of a compact group K is unitary for a suitable positive definite inner product. *Schur's lemma* asserts that *such a representation δ of K is irreducible if and only if each operator A commuting with all $\delta(k)$ $(k \in K)$ is a scalar.* In particular, if K is abelian any finite-dimensional irreducible representation of K is one-dimensional.

Let V' and V'' be two finite-dimensional vector spaces over $\boldsymbol{C}$. Their tensor product $V' \otimes V''$ is by definition the dual space of the space of all bilinear forms on $V' \times V''$. If $v' \in V'$, $v'' \in V''$, the element

$$v' \otimes v'' \in V' \otimes V''$$

is defined by $v' \otimes v''(B) = B(v', v'')$ if B is any bilinear form on $V' \times V''$. If $E' \in \operatorname{Hom}(V', V')$, $E'' \in \operatorname{Hom}(V'', V'')$ then $E' \otimes E''$ is the endomorphism of $V' \otimes V''$ defined by

$$E' \otimes E''(v' \otimes v'') = E'v' \otimes E''v''.$$

Let π' and π'' be representations of the group K on V' and V'', respectively. Then the mapping $k \to \pi'(k) \otimes \pi''(k)$ is a representation of K on $V' \otimes V''$, denoted $\pi' \otimes \pi''$ and called the *tensor product* of π' and π''.

On the other hand, we can assign to each $k \in K$ the endomorphism $(\pi' + \pi'')(k)$ of $V' \times V''$ defined by

$$(\pi' + \pi'')(k) = (\pi'(k), \pi''(k));$$

then $\pi' + \pi''$ is a representation of K on $V' \times V''$ called the *direct sum* of π' and π''.

Let K be a compact group and dk the Haar measure on K, normalized by $\int_K dk = 1$. Let δ be a unitary representation of K on a finite-dimensional Hilbert space V_δ with inner product $\langle\ ,\ \rangle$. Let χ_δ and $d(\delta)$, respectively, denote the character and dimension of δ and put

$$\alpha_\delta = d(\delta)\operatorname{conj}(\chi_\delta).$$

For u, $v \in V_\delta$ the representation coefficients $k \to \langle\delta(k)u, v\rangle$ satisfy the *orthogonality relations of Schur*: If δ and δ' are irreducible then (bar denoting complex conjugation)

$$(8) \qquad \int_K \langle\delta(k)u, v\rangle(\langle\delta'(k)u', v'\rangle)^-\, dk = \begin{cases} d(\delta)^{-1}\langle u, u'\rangle(\langle v, v'\rangle)^- \text{ if } \delta = \delta', \\ 0 \text{ if } \delta \nsim \delta', \end{cases}$$

where $\delta \nsim \delta'$ means that δ and δ' are not equivalent. If $v_1, \ldots, v_{d(\delta)}$ is an orthonormal basis of V_δ and the matrix $\delta_{ij}(k)$ is determined by $\delta(k)v_j = \sum_i \delta_{ij}(k)v_i$, then (8) means that *the functions* $d(\delta)^{1/2}\delta_{ij}$ *are orthonormal in* $L^2(K)$. According to the *Peter–Weyl theorem they form a complete orthonormal basis of* $L^2(K)$. This can be expressed by the orthogonal Hilbert space decomposition

$$(9) \qquad L^2(K) = \bigoplus_{\delta \in \hat{K}} H_\delta,$$

where $\hat{K}$ is the set of equivalence classes of finite-dimensional unitary irreducible representations and H_δ is the subspace spanned by the functions (representation-coefficients) on K:

$$k \to \langle \delta(k)u, v \rangle, \quad u, v \in V_\delta.$$

We can also write

(10) $$H_\delta = \{F \in C(K) \colon F(k) = \operatorname{Tr}(\delta(k)C),\ C \in \operatorname{Hom}(V_\delta, V_\delta)\}$$

and, as the notation in (9) indicates, H_δ depends only on the equivalence class of δ. The mapping τ_δ which to the element $(u, v) \in K \times K$ associates the endomorphism $f \to f^{L(u)R(v^{-1})}$ of H_δ is a representation of $K \times K$ on H_δ.

With the orthonormal basis (v_i) of V_δ fixed, consider for each i the subspace

(11) $$H_{i,\delta} = \sum_{j=1}^{d(\delta)} C\delta_{ij}, \quad \delta_{ij}(k) = \langle \delta(k)v_j, v_i \rangle = \overline{\delta_{ji}(k^{-1})},$$

of H_δ; $H_{i,\delta}$ is the subspace spanned by the ith *row* of δ. Similarly, we consider the subspace

(12) $$H_{\delta,j} = \sum_{i=1}^{d(\delta)} C\delta_{ij}$$

of H_δ, i.e., the subspace spanned by the jth *column* of δ.

Theorem 1.6. *Let $\delta \in \hat{K}$.*

(i) *For each i, $H_{i,\delta}$ is right invariant under K and the right regular representation of K on it is equivalent to δ.*

(ii) *For each j, $H_{\delta,j}$ is left invariant under K and the left regular representation of K on it is equivalent to the contragredient representation $\check{\delta}$.*

(iii) *The representation τ_δ of $K \times K$ on H_δ is irreducible. Also,*

$$\delta_{ij} * \delta_{kl} = \varepsilon_{jk} \frac{1}{d(\delta)} \delta_{il} \qquad (\varepsilon_{jk} = \textit{Kronecker delta}).$$

(iv) *The normalized character $\psi = \chi_\delta/d(\delta)$ satisfies the functional equation*

(13) $$\int_K \psi(kuk^{-1}v)\, dk = \psi(u)\psi(v).$$

On the other hand, if $\psi \not\equiv 0$ is a continuous function satisfying (13), *then $\psi = \chi_\delta/d(\delta)$ for some $\delta \in \hat{K}$.*

Proof. (i) For the right regular representation r we have

$$r(k)\delta_{ij} = \delta_{ij}^{R(k^{-1})} = \sum_{p=1}^{d(\delta)} \delta_{pj}(k)\delta_{ip},$$

proving (i).

(ii) The contragredient $\check{\delta}$ operates on the dual space $V'_\delta = V_{\check{\delta}}$. We define the inner product $\langle\ ,\ \rangle$ on $V_{\check{\delta}}$ by $\langle v', w'\rangle = \langle w, v\rangle$. For a linear transformation L of V_δ let L' denote the linear transformation $v' \to (Lv)'$ of $V_{\check{\delta}}$. If L has matrix expression (l_{ij}) in the basis $v_1, \ldots, v_{d(\delta)}$, then L' has matrix expression $(\overline{l_{ij}})$ in the basis $v'_1, \ldots, v'_{d(\delta)}$. Since $\check{\delta}(k) = {}^t\delta(k^{-1})$ we find

$$\begin{aligned}(\check{\delta}(k)v'_j)(w) &= v'_j(\delta(k^{-1})w) = \langle w, \delta(k)v_j\rangle = (\delta(k)v_j)'(w)\\ &= \sum_i (\overline{\delta_{ij}(k)}v'_i)(w),\end{aligned}$$

so

$$\check{\delta}(k) = \delta(k)', \quad \check{\delta} \text{ is unitary}, \quad \check{\delta}_{ij}(k) = \overline{\delta_{ij}(k)}, \tag{14}$$

the last relation holding in terms of the bases (v_i) of V_δ, (v'_i) of $V_{\check{\delta}}$.

For the left regular representation l we have with fixed j

$$(l(k)\delta_{ij})(u) = \delta(k^{-1}u)_{ij} = \delta_{ji}(u^{-1}k)^- = \sum_{p=1}^{d(\delta)} \overline{\delta(k)}_{pi}\delta_{pj}(u),$$

so, comparing with (14), we see that l on $H_{\delta,j}$ is equivalent to $\check{\delta}$.

(iii) Let $f \in H_\delta - \{0\}$ and V_f the K-bi-invariant subspace generated by f. Then V_f contains the function

$$(\delta_{ij} * f)(k) = \sum_p \left(\int_K \delta_{pj}(u^{-1})f(u)\,du\right)\delta_{ip}(k),$$

which for some j is $\neq 0$. Thus $H_{i,\delta} \cap V_f \neq 0$ for each i. Similarly, $H_{\delta,j} \cap V_f \neq 0$ for each j. Now (i) and (ii) imply that V_f contains each $H_{i,\delta}$ and each $H_{\delta,j}$, so (iii) follows [using (8)].

For (iv) consider the integral

$$A(u) = \int_K \delta(kuk^{-1})\,dk \in \mathrm{Hom}(V_\delta, V_\delta).$$

By the invariance of dk, $A(u)$ commutes with each $\delta(l)$ $(l \in K)$. By Schur's lemma, $A(u)$ is a scalar multiple of the identity $I_{d(\delta)}$. Taking traces, we obtain

$$\int_K \delta(kuk^{-1})\,dk = \frac{\chi(u)}{d(\delta)}\, I_{d(\delta)}.$$

Multiplying by $\delta(v)$ on the right and taking traces again we obtain (13).

For the converse we select by the Peter–Weyl theorem a $\sigma \in \hat{K}$ such that $\int \sigma(k)\psi(k)\,dk \neq 0$. Then by (13)

$$\psi(u) \int \psi(v)\sigma(v)\,dv = \iint \psi(kuk^{-1}v)\sigma(v)\,dv\,dk,$$

which by successive change of variables becomes

$$\iint \psi(v)\sigma(ku^{-1}k^{-1}v)\,dk\,dv = \frac{\chi_\sigma(u^{-1})}{d(\sigma)} \int \psi(v)\sigma(v)\,dv.$$

It follows that $\psi = \chi_{\check{\sigma}}/d(\sigma)$ as stated. This finishes the proof.

The characters form a natural description of $\hat{K}$. Two equivalent representations of K have the same character. The converse is a simple consequence of (8) if we recall that each finite-dimensional representation of K can be decomposed into irreducible ones.

Let M be a closed subgroup of K. We shall write out (9) more explicitly for $L^2(K/M)$ viewed as the subspace of $L^2(K)$ consisting of functions right invariant under M. If $f \in C(K)$ is right invariant under M then so is each term in the decomposition $f = \sum_\delta f_\delta$ from (9). Thus we have to describe the subspace H_δ^M of functions in H_δ which are right invariant under M. Let

$$V_\delta^M = \{v \in V_\delta : \delta(m)v = v \text{ for } m \in M\}.$$

and put $l(\delta) = \dim V_\delta^M$. If $F \in H_\delta^M$ then $F(k) = \operatorname{Tr}(\delta(k)C)$ where $C \in \operatorname{Hom}(V_\delta, V_\delta)$ is unique. Since $F(km) \equiv F(k)$ we deduce $\delta(m)C = C$ $(m \in M)$ so $CV_\delta \subset V_\delta^M$. The mapping $F \to C$ is a linear injection of H_δ^M into $\operatorname{Hom}(V_\delta, V_\delta^M)$. The latter space has dimension $d(\delta)l(\delta)$ whereas H_δ^M contains the $l(\delta)d(\delta)$ linearly independent functions

$$\langle \delta(k)v_j, v_i \rangle \qquad (1 \leq j \leq l(\delta),\ 1 \leq i \leq d(\delta))$$

provided the basis (v_i) of V_δ is chosen such that the first $l(\delta)$ elements belong to V_δ^M. The mapping $F \to C$ is therefore a bijection and

$$H_\delta^M = \{F : F(k) = \operatorname{Tr}(\delta(k)C),\ C \in \operatorname{Hom}(V_\delta, V_\delta^M)\}. \tag{15}$$

With a basis (v_i) of V_δ compatible with the direct decomposition $V_\delta^M \oplus (V_\delta^M)^\perp = V_\delta$ the functions

$$\delta_{ij}(k) = \langle \delta(k)v_j, v_i \rangle \qquad (1 \leq i \leq d(\delta),\ 1 \leq j \leq l(\delta))$$

form a basis of H_δ^M. Moreover, we have the orthogonal Hilbert space decomposition

$$L^2(K/M) = \bigoplus_{\delta \in \hat{K}_M} H_\delta^M, \tag{16}$$

where $\hat{K}_M$ is the set of elements in $\hat{K}$ for which $V_\delta^M \neq 0$.

Given a representation π of K on an arbitrary vector space V, a vector $v \in V$ is said to be *K-finite* if the vector space $\{\pi(K)v\}$ spanned by the orbit $\pi(K)v$ is finite-dimensional. The vector v is said to be *K-finite of type δ* if the representation of K on $\{\pi(K)v\}$ given by π decomposes into finitely many copies of δ. Let $V(\delta)$ denote the space of K-finite vectors of type δ. Suppose now V is *locally convex* and *complete* and recall definition (2).

Lemma 1.7. *The mapping $\pi(\alpha_\delta)$ is a continuous projection of V onto $V(\delta)$. In particular, $V(\delta)$ is a closed subspace of V.*

Proof. The orthogonality relations (8) imply $\alpha_\delta * \alpha_\delta = \alpha_\delta$, so $\pi(\alpha_\delta)$ [defined by (2)] is a continuous projection. Also, if $v \in V$ $u \in K$ we have

$$\pi(u)\pi(\alpha_\delta)v = d(\delta)\int_K \overline{\chi_\delta}(u^{-1}k)\pi(k)v\,dk \tag{17}$$

$$= d(\delta)\int_K \chi_{\check{\delta}}(u^{-1}k)\pi(k)v\,dk.$$

But the members of $H_{\check{\delta}}$, in particular $\chi_{\check{\delta}}$, are K-finite of type δ under the left regular representation of K. Hence by (17), $\pi(\alpha_\delta)v \in V(\delta)$. On the other hand, suppose $v \in V(\delta)$. We shall prove $\pi(\alpha_\delta)v = v$. For this we may assume that the action of $\pi(K)$ on $\{\pi(K)v\}$ is irreducible. It is then equivalent to δ and we can take the space V_δ as $\{\pi(K)v\}$ and v as a basis element v_j. Then $\pi(k)v_j = \sum_i \delta_{ij}(k)v_i$, so

$$\pi(\alpha_\delta)v_j = d(\delta)\int_K \overline{\chi_\delta}(k)\sum_i \delta_{ij}(k)v_i\,dk = v_j$$

by (8).

Remark. The same argument shows that if $\delta \not\sim \delta'$ and if $v' \in V(\delta')$ then $\pi(\alpha_\delta)v' = 0$.

Corollary 1.8. *If π is the left regular representation of K on $V = L^2(K)$, then*

$$V(\check{\delta}) = H_\delta \tag{18}$$

and the decomposition (9) *equals*

$$f = \sum_{\delta \in \hat{K}} d(\delta)\chi_\delta * f. \tag{19}$$

More generally, if $M \subset K$ is a closed subgroup and $W = L^2(K/M)$, then for $\delta \in \hat{K}_M$,

$$W(\check{\delta}) = H_\delta^M.$$

In fact, we have remarked that $H_\delta \subset V(\check{\delta})$. But if $f \in V = L^2(K)$,

$$(\pi(\alpha_{\check{\delta}})f)(u) = d(\delta) \int_K \chi_\delta(k) f(k^{-1}u)\, dk$$

$$= d(\delta) \int_K \chi_\delta(uk^{-1}) f(k)\, dk,$$

so $\pi(\alpha_{\check{\delta}})f \in H_\delta$ by (10).

We can now prove an analog of (9) for the general representation π of K on an arbitrary locally convex complete space V.

Lemma 1.9. *The sum $\sum_{\delta \in \hat{K}} V(\delta)$ is direct and it is dense in V.*

Proof. The directness is clear from the remark following Lemma 1.7. Next suppose $\lambda \in V'$ vanishes on each $V(\delta)$. Then for each $k \in K$, $v \in V$, we have $\lambda(\pi(\alpha_\delta)\pi(k)v) = 0$. With $\chi_\delta(k) = \overline{\chi_\delta}(k^{-1})$ this equation reduces to

$$\chi_\delta * \lambda_v = 0, \tag{20}$$

where λ_v is the function $k \to \lambda(\pi(k)v)$. Hence by (19) $\lambda_v \equiv 0$, whence $\lambda = 0$, so the lemma follows via the Hahn–Banach theorem.

Corollary 1.10. *An irreducible representation of a compact group on a complete locally convex space V is finite-dimensional.*

In fact, by Lemma 1.9 V contains a K-finite vector v. The space $\{\pi(K)v\}$ is finite-dimensional, hence closed and so equals V. Thus $\dim V < \infty$.

Suppose now the group K is a compact topological transformation group of a locally compact Hausdorff space X with a countable base. If f and ϕ are continuous functions on X and K, respectively, we write

$$(\phi * f)(x) = \int_K \phi(k) f(k^{-1} \cdot x)\, dk. \tag{21}$$

Putting $\mathscr{K}(X) = C_c(X)$, we consider for each compact subset $S \subset X$ the subspace $\mathscr{K}_S(X) \subset \mathscr{K}(X)$ of functions with support contained in S. This space is a Banach space in the topology of uniform convergence; writing $X = \bigcup_i S_i$, where $S_1, S_2, \ldots$ is an increasing sequence of compact sets with nonempty interior, we have $\mathscr{K}(X) = \bigcup_i \mathscr{K}_{S_i}(X)$ and accordingly give $\mathscr{K}(X)$ the inductive limit topology. This topology does not depend on the choice of the S_i. Furthermore, this topology is the *strict inductive limit* topology in the sense that for each i the topology of

$\mathscr{K}_{S_i}(X)$ is here the relative topology of $\mathscr{K}_{S_{i+1}}(X)$. It is often convenient to take the S_i to be K-invariant.

With $\delta \in \hat{K}$ acting on V_δ we consider also the space

$$\mathscr{K}(X, \operatorname{Hom}(V_\delta, V_\delta)) \tag{22}$$

of compactly supported continuous functions on X with values in $\operatorname{Hom}(V_\delta, V_\delta)$. In the same manner as for $\mathscr{K}(X)$ we consider the subspace $\mathscr{K}_S(X, \operatorname{Hom}(V_\delta, V_\delta))$ of $\mathscr{K}(X, \operatorname{Hom}(V_\delta, V_\delta))$ and give this latter space the strict inductive limit topology given by the sequence of Banach spaces $\mathscr{K}_{S_i}(X, \operatorname{Hom}(V_\delta, V_\delta))$.

We consider now the space

$$\mathscr{K}^\delta(X) = \{F \in \mathscr{K}(X, \operatorname{Hom}(V_\delta, V_\delta)) : F(k \cdot x) \equiv \delta(k)F(x)\}, \tag{23}$$

and for the natural representation of K on $\mathscr{K}(X)$ [assigning to $k \in K$ the linear transformation $f(x) \to f(k^{-1} \cdot x)$] let $\mathscr{K}_\delta(X)$ denote the subspace of K-finite vectors of type δ. The spaces $\mathscr{K}^\delta(X)$ and $\mathscr{K}_\delta(X)$ are given the topologies induced from (22) and $\mathscr{K}(X)$, respectively.

For any continuous function f on X we put

$$f^\delta(x) = d(\delta) \int_K f(k^{-1} \cdot x)\delta(k)\, dk. \tag{24}$$

Proposition 1.11. *The mapping $Q: F(x) \to \operatorname{Tr}(F(x))$ is a homeomorphism of $\mathscr{K}^\delta(X)$ onto $\mathscr{K}_{\check{\delta}}(X)$. The inverse is given by $f \to f^\delta$. Moreover, the maps*

$$p: f \in \mathscr{K}(X) \to d(\delta)\chi_\delta * f \in \mathscr{K}_{\check{\delta}}(X),$$

$$q: f \in \mathscr{K}(X) \to f^\delta \in \mathscr{K}^\delta(X)$$

are continuous open surjections and the subspace $\mathscr{K}_{\check{\delta}}(X)$ of $\mathscr{K}(X)$ is closed.

Proof. Let $F \in \mathscr{K}^\delta(X)$ and let (v_i) be an orthonormal basis of V_δ. Then

$$F(x)v_j = \sum_{i=1}^{d(\delta)} F_{ij}(x)v_i,$$

so $F(k^{-1} \cdot x) \equiv \delta(k^{-1})F(x)$ implies

$$F_{ij}(k^{-1} \cdot x) = \sum_l \overline{\delta_{li}}(k)F_{lj}(x),$$

which in turn shows that each F_{ij}, and therefore $\mathrm{Tr}(F)$, belongs to $\mathscr{K}_{\check{\delta}}(X)$. Also, if $f(x) = \mathrm{Tr}(F(x))$, we see by writing the orthogonality relations (8) in the form

$$d(\delta) \int_K \overline{\delta_{ij}(k)}\delta(k)\, dk = E_{ij} \tag{25}$$

that

$$f^\delta(x) = d(\delta) \int_K \mathrm{Tr}(\delta(k^{-1})F(x))\delta(k)\, dk = F(x).$$

In particular, Q is one-to-one. It is also surjective because if $g \in \mathscr{K}_{\check{\delta}}(X)$ then $g^\delta \in \mathscr{K}^\delta(X)$ and $\mathrm{Tr}(g^\delta) = d(\delta)\chi_\delta * g = g$ [using (25)].

We know from Lemma 1.7 that p is a continuous projection of $\mathscr{K}(X)$ onto the closed subspace $\mathscr{K}_{\check{\delta}}(X)$. This implies that p is open. In fact, let N be a neighborhood of 0 in $\mathscr{K}(X)$ and $M \subset N$ another such, satisfying $p(M) \subset N$. Putting $N_0 = M \cup p(M)$, we have

$$p(N_0) = \{f \in N_0 : p(f) = f\} = N_0 \cap \mathscr{K}_{\check{\delta}}(X),$$

so $p(N) \supset p(N_0)$ is a neighborhood of O in $\mathscr{K}_{\check{\delta}}(X)$. Since $Q \circ q = p$ it follows that q is a continuous open surjection.

We recall that an LF-*space* is by definition the countable strict inductive limit of Fréchet spaces. Thus $\mathscr{K}(X)$ and $\mathscr{K}(X, \mathrm{Hom}(V_\delta, V_\delta))$ are LF-spaces. In case the separable space X is a manifold, $\mathscr{D}(X)$ is also an LF-space, the strict inductive limit of the Fréchet spaces $\mathscr{D}_{S_i}(X)$.

Proposition 1.12. *The spaces $\mathscr{K}^\delta(X)$ and $\mathscr{K}_{\check{\delta}}(X)$ are* LF-*spaces.*

This is a consequence of Proposition 1.11 and the following general lemma (which is necessary since a closed subspace of an LF-space is not necessarily an LF-space).

Lemma 1.13. *Let $E = \bigcup_{i=1}^\infty E_i$ be a strict inductive limit of Fréchet spaces $E_1 \subset E_2 \subset \cdots$, $M \subset E$ a closed topological subspace and $M_i = M \cap E_i$. Assume $p: E \to M$ is a continuous open surjection which for each i maps E_i onto M_i. Then M is the strict inductive limit of $(M_i)_{i=1,2,\ldots}$, each M_i having the topology induced by E.*

Proof. By general theory, E_i carries the relative topology of E, so by completeness it is closed. The restriction $p_i = p|E_i$ gives a continuous surjection of E_i onto M_i. We must prove that a convex subset $V \subset M$ is a neighborhood of 0 if and only if for each i, $V \cap M_i$ is a neighborhood of 0 in M_i $(i = 1, 2, \ldots)$. The "only if" being obvious, we assume $V \cap M_i$ is a neighborhood of 0 in M_i $(i = 1, 2, \ldots)$. Then $p_i^{-1}(V \cap M_i) = W_i$ is a convex neighborhood of 0 in E_i and $W_i \subset W_{i+1}$, so $W = \bigcup_{i=1}^{\infty} W_i$ is convex. Also, $E_i \cap W_{i+j} = W_i$ $(j \geq 0)$ because if $e_i \in E_i$, $p(e_i) \in V \cap M_{i+j}$ then $p(e_i) \in V \cap M_i$ and so $e_i \in W_i$. Consequently, $E_i \cap W = W_i$, so by the definition of an inductive limit, W is a neighborhood of 0 in E. Finally, $p(W) = V$, so, since p is an open map, V is a neighborhood of 0 in M as desired.

§2. Spherical Functions: Preliminaries

1. Definition

Let G be a connected Lie group, K a compact subgroup. Let π denote the natural mapping of G onto $X = G/K$ and as usual we put $o = \pi(e)$ and $\tilde{f} = f \circ \pi$ if f is any function on X. As in Chapter II let $\mathbf{D}(G)$ denote the set of all left invariant differential operators on G, $\mathbf{D}_K(G)$ the subspace of those which are also right invariant under K and $\mathbf{D}(G/K)$ the algebra of differential operators on G/K invariant under all the translations $\tau(g)$: $xK \to gxK$ of G/K.

Definition. Let ϕ be a complex-valued function on G/K of class C^∞ which satisfies $\phi(\pi(e)) = 1$; ϕ is called a *spherical function* if

(i) $\phi^{\tau(k)} = \phi$ for all $k \in K$,
(ii) $D\phi = \lambda_D \phi$ for each $D \in \mathbf{D}(G/K)$,

where λ_D is a complex number.

It is sometimes convenient to consider the function $\tilde{\phi} = \phi \circ \pi$ on G instead of ϕ. We say that $\tilde{\phi}$ is a *spherical function* on G if and only if ϕ is a spherical function on G/K. Then a spherical function $\tilde{\phi}$ on G is characterized by being an eigenfunction of each operator in $\mathbf{D}_K(G)$ and in addition satisfying the relations $\tilde{\phi}(e) = 1$, $\tilde{\phi}(kgk') = \tilde{\phi}(g)$ for all $g \in G$ and all $k, k' \in K$. The last condition will be called *bi-invariance under* K.

We shall now see that spherical functions can also be characterized by means of an integral equation. Let dk denote the invariant measure on K, normalized by $\int_K dk = 1$.

Lemma 2.1. *Let U be an open subset of G such that $Uk \subset U$ for each $k \in K$. Let F be an analytic function on U. Then the function*

$$x \to \int_K F(xk)\, dk$$

is also analytic on U.

Proof. For simplicity, let us assume $U = G$. Let $x_0 \in G$ and let $\{x_1, \ldots, x_n\}$ be a system of coordinates valid in an open neighborhood V of x_0. There exists a finite set of coordinate neighborhoods $U_\alpha \subset K$ whose union equals K and a neighborhood N of x_0 in V such that the function $F(xk)$ is given by a power series

$$F(xk) = P_\alpha(x_1, \ldots, x_n, k_1, \ldots, k_p), \quad x \in N, k \in U_\alpha,$$

where $\{k_1, \ldots, k_p\}$ is a system of local coordinates on U_α. Consider a partition of unity $1 = \sum_\alpha \phi_\alpha$ subordinate to the covering $\{U_\alpha\}$ of K. Then

$$F(xk) = \sum_\alpha \phi_\alpha(k) P_\alpha(x_1, \ldots, x_n, k_1, \ldots, k_p), \quad k \in K, x \in N,$$

which on integration over K gives a power series valid on N.

Proposition 2.2. *Let f be a complex-valued continuous function on G, not identically* 0. *Then f is a spherical function if and only if*

$$\int_K f(xky)\, dk = f(x)f(y) \tag{1}$$

for all $x, y \in G$.

Proof. Let $D \in \mathbf{D}(G)$. Then all the differential operators $\mathrm{Ad}(k)D$, $(k \in K)$, lie in a finite-dimensional subspace of $\mathbf{D}(G)$ and we can form the integral

$$D_0 = \int_K \mathrm{Ad}(k)D\, dk, \tag{2}$$

which is an operator in $\mathbf{D}_K(G)$. The mapping $D \to D_0$ is a linear mapping of $\mathbf{D}(G)$ onto $\mathbf{D}_K(G)$. Let F be a function in $\mathscr{E}(G)$, bi-invariant under K. Then $[DF](k) = [DF](e)$, so

$$[D_0 F](e) = \int_K [D^{R(k^{-1})}F](e)\, dk = \int_K [(DF^{R(k)})^{R(k^{-1})}](e)\, dk,$$

and

$$[D_0 F](e) = [DF](e). \tag{3}$$

Now suppose f is a spherical function on G. Then f has the form $f = \phi \circ \pi$ where ϕ is a spherical function on G/K. The space G/K has a Riemannian structure invariant under the action of G. It is not difficult to see that this Riemannian structure is necessarily analytic. The Laplace–Beltrami operator with respect to this structure has analytic coefficients when expressed in terms of analytic local coordinates on G/K. In addition, this operator is elliptic and consequently, by a theorem of S. Bernstein, its eigenfunctions are analytic (see F. John [1955], p. 57). In particular, the function ϕ, and therefore also the function f, is analytic.

Now, let x be a fixed element in G and consider the function

$$F(y) = \int_K f(xky)\,dk \qquad (y \in G),$$

which is clearly bi-invariant under K. Let D and D_0 be as in (3). Since $D_0 f = \lambda_D f$ we have

$$[D_0 F](y) = \int_K [D_0 f](xky)\,dk = \lambda_D F(y)$$

and consequently

$$[D_0(f(e)F - F(e)f)](e) = 0.$$

Combining this with (3), we obtain

$$[D(f(e)F - F(e)f)](e) = 0$$

for all $D \in \mathbf{D}(G)$. Since $f(e)F - F(e)\,f$ is an analytic function (Lemma 2.1), we conclude from Taylor's formula ([DS], Chapter II, §1, (6)) that

$$f(e)F = F(e)f.$$

Since $f(e) = 1$, this is just (1).

On the other hand, let f be a continuous function on G, not identically 0, satisfying (1). Select $x_0 \in G$ such that $f(x_0) \neq 0$. Now (1) implies that $f(xk)f(x_0) = f(x_0)f(x) = f(x_0)f(kx)$ for all $x \in G$, $k \in K$. Thus f is bi-invariant under K. Putting $y = e$ in (1) gives $f(e) = 1$. In order to see that $f \in \mathscr{E}(G)$, select $\rho \in \mathscr{D}(G)$ such that $\int_G \rho(y)f(y)\,dy \neq 0$, dy denoting a left invariant measure on G. Then

$$\begin{aligned} f(x)\int_G f(y)\rho(y)\,dy &= \int_G \rho(y)\left(\int_K f(xky)\,dk\right)dy \\ &= \int_K\left(\int_G \rho(y)f(xky)\,dy\right)dk \\ &= \int_K\left(\int_G \rho(k^{-1}x^{-1}z)f(z)\,dz\right)dk \\ &= \int_G\left(\int_K \rho(kx^{-1}z)\,dk\right)f(z)\,dz, \end{aligned}$$

which shows that $f \in \mathscr{E}(G)$. For each $D_0 \in \mathbf{D}_K(G)$ we get from (1)

$$f(x)[D_0 f](y) = \int_K [D_0 f](xky)\,dk.$$

Putting $y = e$ we get

$$[D_0 f](x) = [D_0 f](e) f(x), \tag{4}$$

which shows that f is spherical.

From (4) it follows that a spherical function is determined by its system of eigenvalues. More precisely, we have

Corollary 2.3. *Let ϕ and ϕ_1 be two spherical functions on G such that $D_0\phi = \lambda_D \phi$ and $D_0 \phi_1 = \lambda_D \phi_1$ for all $D_0 \in \mathbf{D}_K(G)$. Then $\phi = \phi_1$.*

In fact, Equation (3) implies that $[D\phi](e) = [D\phi_1](e)$ for each $D \in \mathbf{D}(G)$; being analytic, ϕ and ϕ_1 must coincide.

2. Joint Eigenfunctions

We recall (Chapter II, §4) that a *joint eigenfunction* on $X = G/K$ is an eigenfunction of each of the operators $D \in \mathbf{D}(G/K)$. Let $\mu: \mathbf{D}(G/K) \to \mathbf{C}$ be a homomorphism and let E_μ denote the corresponding joint eigenspace; i.e.,

$$E_\mu = \{f \in \mathscr{E}(G/K) : Df = \mu(D) f \text{ for each } D \in \mathbf{D}(G/K)\}.$$

Of course each spherical function belongs to a unique E_μ.

Proposition 2.4. *The joint eigenfunctions on G/K are characterized by the following integral equation: Each joint eigenspace $E_\mu \neq 0$ contains exactly one spherical function ϕ. The members f of E_μ are characterized by the equation*

$$\int_K f(xkyK)\, dk = f(xK)\phi(yK), \qquad x, y \in G. \tag{5}$$

Proof. The first statement is immediate from Corollary 2.3. Next, let $f \in \mathscr{E}(G/K)$ and $x \in G$. Putting

$$F_x(y) = \int_K f(xkyK)\, dk,$$

we find for $D \in \mathbf{D}(G/K)$

$$(DF_x)(y) = \int_K D_y(f(xkyK))\, dk = \int_K (Df)(xkyK)\, dk. \tag{6}$$

If $f \in E_\mu$ this last integral equals $\mu(D)F_x(y)$. Thus F_x is a K-invariant member of E_μ and is therefore proportional to ϕ. Evaluating at $y = e$, we obtain (5).

On the other hand, suppose f is a continuous function satisfying (5). The argument of Proposition 2.2 shows that f is necessarily of class C^∞. By (5) $F_x = f(xK)\phi$, so putting $y = e$ in (6) gives

$$f(xK)(D\phi)(e) = (Df)(xK),$$

which shows that $f \in E_\mu$.

A function u on G/K is said to be *harmonic* if $Du = 0$ for all $D \in \mathbf{D}(G/K)$ which annihilate the constants. In the case in which G/K is two-point homogeneous $\mathbf{D}(G/K)$ is generated by the Laplace–Beltrami operator L (Chapter II, Corollary 4.11), so the harmonic functions are the solutions to the equation $Lu = 0$.

In general the harmonic functions are the members of the eigenspace E_{μ_0} containing the spherical function $\phi \equiv 1$. From Proposition 2.4 we obtain the following result.

Proposition 2.5. (Godement) *The harmonic functions on G/K are characterized by the mean-value property*

$$f(xK) = \int_K f(xkyK)\, dk, \qquad x, y \in G. \tag{7}$$

In the case in which G is the group $\mathbf{M}(n)$ of rigid motions of $\mathbf{R}^n$ and $K = \mathbf{O}(n)$ this result reduces to Gauss's mean-value theorem for solutions of Laplace's equation in $\mathbf{R}^n$.

3. Examples

We shall now determine the spherical function for some simple examples including the three in the Introduction.

I. $X = \mathbf{R}^2, G = \mathbf{R}^2, K = \{e\}$.

Here $\mathbf{D}(G/K)$ consists of all differential operators with constant coefficients. We have obviously

Proposition 2.6. *The spherical functions on G/K are the exponential functions*

$$(x, y) \to e^{\alpha x + \beta y},$$

α, β being any complex numbers.

II. $X = \mathbf{R}^2$, $G = \mathbf{M}_0(\mathbf{R}^2)$ (the e-component of $\mathbf{M}(2)$), $K = \mathbf{SO}(2)$.

Here we have as a special case of Lemma 2.4 of the Introduction

Proposition 2.7. *The spherical functions on $\mathbf{R}^2 = G/K$ are the functions*

$$(8) \qquad \phi(x, y) = \frac{1}{2\pi} \int_0^{2\pi} e^{i\lambda(x\cos\theta + y\sin\theta)}\, d\theta,$$

λ being an arbitrary complex number.

Of course $\phi(x, y)$ can be written $\psi((x^2 + y^2)^{1/2})$ and then the functional equation (1) can be expressed in terms of ψ. If $g, h \in G$ denote translations in the direction of the x-axis of distance r and s, respectively, and k is a rotation around the origin of angle θ then the point $gkh \cdot o$ has distance $(r^2 + s^2 + 2rs\cos\theta)^{1/2}$ from o. The functions (8) are therefore characterized by

$$\frac{1}{2\pi} \int_0^{2\pi} \psi((r^2 + s^2 + 2rs\cos\theta)^{1/2})\, d\theta = \psi(r)\psi(s),$$

where $\phi(x, y) = \psi((x^2 + y^2)^{1/2})$. As shown in the Introduction, §2, we have $\psi(r) = J_0(\lambda r)$.

III. $X = \mathbf{S}^2$, $G = \mathbf{SO}(3)$, $K = \mathbf{SO}(2)$.

Here we consider K as the subgroup of rotation of $\mathbf{S}^2$ leaving the north pole $o = (0, 0, 1)$ fixed. The circle

$$\{(\sin r\cos\phi, \sin r\sin\phi, \cos r) : 0 \le \phi \le 2\pi\}$$

is the K-orbit of a point at geodesic distance r from o. According to Theorem 3.1 of the Introduction the function

$$(r, \phi) \to (a_1 \sin r\cos\phi + a_2 \sin r\sin\phi + a_3\cos r)^n$$

on $\mathbf{S}^2$ is an eigenfunction of the Laplacian L with eigenvalue $-n(n + 1)$ provided the constants a_i satisfy $a_1^2 + a_2^2 + a_3^2 = 0$. Thus we have the following result.

Proposition 2.8. *The spherical functions on $\mathbf{S}^2 = G/K$ are precisely the functions*

$$\phi_n(p) = P_n(\cos(d(o, p)))$$

where P_n is the Legendre polynomial

$$P_n(\cos r) = \frac{1}{2\pi} \int_0^{2\pi} (\cos r + i\sin r\cos u)^n\, du.$$

The functional equation (1) can now be expressed in terms of $P_n(\cos r)$. Consider the rectangular xyz-coordinate system in which $\mathbf{S}^2$

is given by $x^2 + y^2 + z^2 = 1$ and o is the point $(0, 0, 1)$. In the equation

$$\int_K f(gkh)\, dk = f(g)f(h)$$

let g and h denote rotations around the y-axis through the angles r and s, respectively, and let k denote a rotation of angle u around the z-axis. Then the point $kh \cdot o$ is given in coordinates

$$kh \cdot o = (\sin s \cos u, \sin s \sin u, \cos s),$$

and since the rotation g has coordinate expression

$$\begin{Bmatrix} x \\ y \\ z \end{Bmatrix} \to \begin{Bmatrix} \cos r & 0 & \sin r \\ 0 & 1 & 0 \\ -\sin r & 0 & \cos r \end{Bmatrix} \begin{Bmatrix} x \\ y \\ z \end{Bmatrix}$$

the coordinates of $gkh \cdot o$ are

$$(\cos r \sin s \cos u + \sin r \cos s, \sin s \sin u, \ -\sin r \sin s \cos u + \cos r \cos s).$$

Equation (1) therefore takes the form

$$(9) \qquad \frac{1}{2\pi} \int_0^{2\pi} P_n(\cos r \cos s - \sin r \sin s \cos u)\, du = P_n(\cos r) P_n(\cos s).$$

IV. $G = \boldsymbol{SL}(2, \boldsymbol{R})$, $K = \boldsymbol{SO}(2)$, $X = G/K$.

The group G acts transitively on the upper half-plane $\operatorname{Im} z > 0$ by means of the mappings

$$z \to g \cdot z = \frac{az + b}{cz + d} \qquad \text{if } g = \begin{pmatrix} a & b \\ c & d \end{pmatrix} \in \boldsymbol{SL}(2, \boldsymbol{R}).$$

The isotropy subgroup of G at i is K, so X can be identified with the upper half-plane. Let $\mathfrak{g}$ and $\mathfrak{k}$ denote the Lie algebras of G and K, respectively, and as usual we have $\mathfrak{g} = \mathfrak{k} + \mathfrak{p}$, where $\mathfrak{p}$ is the orthogonal complement of $\mathfrak{k}$ in $\mathfrak{g}$ with respect to the Killing form B of $\mathfrak{g}$. Since $\mathfrak{g}$ has complexification $\mathfrak{sl}(2, \boldsymbol{C})$ we have by [DS] (Chapter III, §8),

$$B(Z, Y) = 4 \operatorname{Tr}(ZY), \qquad Z, Y \in \mathfrak{g}.$$

The restriction of $\frac{1}{2}B$ to $\mathfrak{p} \times \mathfrak{p}$ gives rise to a G-invariant Riemannian structure on X with respect to which X has constant curvature (G is transitive). According to [DS], Chapter V, §3, this curvature is

$$\tfrac{1}{2}B([Z, Y], [Z, Y]).$$

if Z, Y is an orthogonal basis of $\mathfrak{p}$. For example, we can take

$$Z = \frac{1}{2}\begin{pmatrix} 0 & 1 \\ 1 & 0 \end{pmatrix}, \qquad Y = \frac{1}{2}\begin{pmatrix} 1 & 0 \\ 0 & -1 \end{pmatrix}.$$

Then

$$[Y, Z] = \frac{1}{2}\begin{pmatrix} 0 & 1 \\ -1 & 0 \end{pmatrix},$$

and so the curvature of X is -1. In geodesic polar coordinates (θ, r), say at the point $i \in X$, the Riemannian structure on X is given by

$$dr^2 + (\sinh r)^2\, d\theta^2,$$

(cf. [DS], Chapter I, Exercise G) and the Laplace–Beltrami operator is

$$L = \frac{\partial^2}{\partial r^2} + \frac{\cosh r}{\sinh r}\frac{\partial}{\partial r} + \frac{1}{(\sinh r)^2}\frac{\partial^2}{\partial \theta^2}.$$

Each spherical function ϕ on G/K has the form $\phi(p) = \psi(d(i, p))$ where the function $\psi(r)$ satisfies the differential equation

$$\frac{d^2\psi}{dr^2} + \frac{\cosh r}{\sinh r}\frac{d\psi}{dr} = \alpha\psi \quad (r > 0), \tag{10}$$

for some complex constant α. As shown in the Introduction, §4, No. 2, the general solution to this equation is given by the *Legendre function*

$$\psi(r) = P_\rho(\cosh r) = \frac{1}{2\pi}\int_0^{2\pi} (\cosh r + \sinh r \cos u)^\rho\, du, \tag{11}$$

$\rho(\rho + 1) = \alpha$, and the following result holds.

Proposition 2.9. *The spherical functions on G/K are given by*

$$\phi(p) = P_\rho(\cosh d(i, p)),$$

where P_ρ is given by (11) *and ρ is an arbitrary complex number.*

Denoting by A the group of diagonal matrices

$$a = \begin{pmatrix} d & 0 \\ 0 & d^{-1} \end{pmatrix}, \qquad d > 0,$$

we have $G = KAK$. In order to find the explicit form of the functional equation (1), let

$$a_i = \begin{pmatrix} e^{(1/2)r_i} & 0 \\ 0 & e^{-(1/2)r_i} \end{pmatrix} \ (i = 1, 2), \qquad k = \begin{pmatrix} \cos u & \sin u \\ -\sin u & \cos u \end{pmatrix}.$$

Then $a_1 k a_2$ can be written $k_1 a k_2$ where $k_1, k_2 \in K$ and

$$a = \begin{pmatrix} e^{r/2} & 0 \\ 0 & e^{-r/2} \end{pmatrix},$$

$$\cosh r = \cosh r_1 \cosh r_2 + \sinh r_1 \sinh r_2 \cos 2u.$$

Hence the functions (11) are characterized by the integral formula

(12)

$$\frac{1}{2\pi}\int_0^{2\pi} P(\cosh r \cosh s + \sinh r \sinh s \cos u)\,du = P(\cosh r)P(\cosh s).$$

V. Compact groups.

In this example let U be an arbitrary compact connected Lie group. Let G denote the product group $U \times U$ and let K denote the diagonal in $U \times U$. Then G/K is identified with U under the mapping $(u_1, u_2)K \to u_1u_2^{-1}$. Under this identification, the mapping $\tau(u_1, u_2)$ of G/K corresponds to the mapping $u \to u_1uu_2^{-1}$ of U. Consequently, a differential operator D on G/K belongs to $\boldsymbol{D}(G/K)$ if and only if, when considered as a differential operator on U, it is invariant under all left and right translations. Thus $\boldsymbol{D}(G/K)$ is identified with the center $\boldsymbol{Z}(U)$ of $\boldsymbol{D}(U)$, (Chapter II, Corollary 4.5). If we interpret functions on G/K as functions on U, the functional equation (1) becomes

$$\int_U \phi(xuyu^{-1})\,du = \phi(x)\phi(y), \tag{13}$$

which characterizes those eigenfunctions of the operators in $\boldsymbol{Z}(U)$ which are invariant under all inner automorphisms of U and satisfy $\phi(e) = 1$.

From Theorem 1.6 we know that the solutions to (13) are precisely $\phi = \chi_\delta/d(\delta)$ where χ_δ is the character of an irreducible representation δ of U and $d(\delta)$ denotes its dimension.

§3. Elementary Properties of Spherical Functions

Let G be a Lie group and dg a left invariant measure on G. The space $C_c(G)$ of complex-valued continuous functions of compact support can be turned into an associative algebra over $\boldsymbol{C}$, the multiplication being the convolution product

$$(f_1 * f_2)(x) = \int_G f_1(g)f_2(g^{-1}x)\,dg$$

and the addition being the pointwise addition of functions. This algebra is called the *group algebra* of G. If θ is an analytic automorphism of G which preserves the measure dg, then the mapping $f \to f^\theta$ is an automorphism of the group algebra. Let $\check{f}$ denote the function $x \to f(x^{-1})$.

Let K be a compact subgroup of G and dk the normalized Haar measure on K. As in Chapter II, §5, No. 1, we put

$$f^{\natural}(x) = \int_K \int_K f(kxk')\, dk\, dk'$$

whenever this integral exists. Then the image $C_c^{\natural}(G) = (C_c(G))^{\natural}$ is a subalgebra of $C_c(G)$ and consists of the functions in $C_c(G)$ that are bi-invariant under K.

Theorem 3.1. (i) *If (G, K) is a Riemannian symmetric pair, then $C_c^{\natural}(G)$ is commutative (under convolution).*

(ii) *If $C_c^{\natural}(G)$ is commutative, then G is unimodular.*

Proof. (i) Let σ be an involutive automorphism of G, identity on K. Then σ leaves dg invariant. Then, since each $x \in G$ can be written $x = kp$ where $\sigma(k) = k$, and $\sigma(p) = p^{-1}$, it follows that $f^{\sigma} = \check{f}$ for $f \in C_c^{\natural}(G)$. For any $f, g \in C_c(G)$ we have

$$\check{f} * \check{g}(x) = \int_G f(y^{-1})g(x^{-1}y)\, dy = \int_G g(z)f(z^{-1}x^{-1})\, dz$$

so that

$$\check{f} * \check{g} = (g * f)\check{}\,.$$

Since $f^{\sigma} * g^{\sigma} = (f * g)^{\sigma}$, we obtain $f * g = g * f$ for $f, g \in C_c^{\natural}(G)$.

(ii) Since $\det(\mathrm{Ad}(x))$ is bi-invariant under K it suffices by Chapter I, Cor. 1.5 to show

$$\int_G f(x^{-1})\, dx = \int_G f(x)\, dx, \qquad f \in C_c^{\natural}(G).$$

If $g \in C_c^{\natural}(G)$ is identically 1 on $\mathrm{Supp}(f) \cup \mathrm{Supp}(\check{f})$ this relation amounts to $(g * f)(e) = (f * g)(e)$, which is guaranteed by the assumption.

Lemma 3.2. *Let ϕ be a continuous complex-valued function on G, bi-invariant under K. Then ϕ is a spherical function if and only if the mapping*

$$L: f \to \int_G f(x)\phi(x)\, dx$$

is a homomorphism of $C_c^{\natural}(G)$ onto $\mathbf{C}$.

Proof. For $f \in C_c(G)$ we put $L(f) = L(f^{\natural})$. We have, by a simple computation,

$$(f^{\natural} * g)^{\natural} = f^{\natural} * g^{\natural}. \tag{1}$$

For a suitably normalized measure dy_K on G/K we have

$$\int_G (f * g)(x)\phi(x)\,dx = \int_G\left(\int_G f(y)g(y^{-1}x)\,dy\right)\phi(x)\,dx$$

$$= \int_G\int_G f(y)g(z)\phi(yz)\,dy\,dz$$

$$= \int_G g(z)\int_{G/K}\left(\int_K f(yk)\phi(ykz)\,dk\right)dy_K\,dz$$

so

$$L(f^\natural * g) = \int_G\int_G f^\natural(y)g(z)\left(\int_K \phi(ykz)\,dk\right)dy\,dz.$$

Since ϕ is bi-invariant and since dy is invariant under the mappings $y \to kyk'$ it follows that

$$L(f^\natural * g) = \int_G\int_G f(y)g(z)\left(\int_K \phi(ykz)\,dk\right)dy\,dz. \tag{2}$$

Moreover,

$$L(f)L(g) = \int_G\int_G f(y)g(z)\phi(y)\phi(z)\,dy\,dz. \tag{3}$$

Considering (1), the lemma follows immediately.

The norm $\|f\| = \int_G |f(x)|dx$ turns the group algebra into a normed vector space. Owing to the additional property $\|f * g\| \leq \|f\|\,\|g\|$ the group algebra is a *normed algebra*. The algebra $C_c^\natural(G)$ is a closed subalgebra.

Theorem 3.3. *The continuous homomorphisms of the algebra $C_c^\natural(G)$ onto $\mathbf{C}$ are the mappings*

$$f \to \int_G f(x)\phi(x)\,dx,$$

where ϕ is a bounded spherical function on G.

Proof. Let L be a continuous homomorphism of $C_c^\natural(G)$ onto $\mathbf{C}$. Then the mapping

$$f \to L(f^\natural)$$

is a continuous linear function on the group algebra. Hence there exists a bounded measurable function ϕ on G such that

$$L(f^\natural) = \int_G f^\natural(x)\phi(x)\,dx$$

for all $f \in C_c(G)$. Here we may assume that ϕ is bi-invariant under K because otherwise it can be replaced by $\phi^\natural$. Since L is a homomorphism, the relations (2) and (3) imply [by approximation in $C_c(G \times G)$] that

$$\int_K \phi(xky)\, dk = \phi(x)\phi(y)$$

except for a set of $(x, y) \in G \times G$ of measure 0. In order to see that ϕ is equal to a continuous function almost everywhere select $\rho \in C_c(G)$ such that $\int_G \rho(y)\phi(y)\, dy \neq 0$. Then for almost all $x \in G$,

$$\begin{aligned}\phi(x) \int_G \phi(y)\rho(y)\, dy &= \int_G \rho(y)\left(\int_K \phi(xky)\, dk\right) dy \\ &= \int_K \left(\int_G \rho(k^{-1}x^{-1}z)\phi(z)\, dz\right) dk \\ &= \int_G \left(\int_K \rho(kx^{-1}z)\, dk\right) \phi(z)\, dz\end{aligned}$$

and this last expression is continuous in x. This proves the theorem.

Remark. For a similar characterization of arbitrary spherical functions see Exercise B16.

A representation π of G on a locally convex topological vector space V will be said to be *spherical* if V contains a nonzero vector fixed under all of $\pi(K)$. We shall now prove that under the correspondance in Theorem 1.5 the positive definite spherical functions correspond to the irreducible unitary spherical representations of G.

Theorem 3.4. *Assume $C_c^\natural(G)$ commutative. Let $\phi \not\equiv 0$ be a positive definite spherical function on G and let π be the unitary representation of G associated to ϕ. Then π is irreducible and spherical.*

On the other hand, if π is an irreducible, unitary, spherical representation of G and $\boldsymbol{e}$ a unit vector left fixed by all $\pi(k)$ $(k \in K)$ then the function $\langle \boldsymbol{e}, \pi(x)\boldsymbol{e}\rangle$ is a positive definite spherical function on G.

Proof. Let ϕ be a positive definite spherical function, let V_ϕ, $\mathfrak{H}_\phi$, and π be as defined in §1, No. 1. Let $\boldsymbol{e}$ be the vector in $\mathfrak{H}_\phi$ which corresponds to $\phi \in V_\phi$. Since ϕ is bi-invariant under K it follows that $\pi(k)\boldsymbol{e} = \boldsymbol{e}$ for all $k \in K$. In order to prove that π is irreducible, consider for each pair $\boldsymbol{a}, \boldsymbol{b} \in \mathfrak{H}_\phi$ the integral

$$B(\boldsymbol{a}, \boldsymbol{b}) = \int_K \langle \pi(k)\boldsymbol{a}, \boldsymbol{b}\rangle\, dk.$$

Since $|B(\boldsymbol{a}, \boldsymbol{b})| \leq \|\boldsymbol{a}\| \|\boldsymbol{b}\|$ there exists a bounded operator P on $\mathfrak{H}_\phi$ such that

$$\langle P\boldsymbol{a}, \boldsymbol{b}\rangle = B(\boldsymbol{a}, \boldsymbol{b})$$

for all $\boldsymbol{a}, \boldsymbol{b} \in \mathfrak{H}_\phi$. Now, since ϕ is spherical it follows from the definition of the scalar product in V_ϕ that

$$\int_K \langle \phi^{L(kx)}, \psi\rangle \, dk = \phi(x^{-1})\langle \phi, \psi\rangle$$

for all $\psi \in V_\phi$. Consequently we have in $\mathfrak{H}_\phi$

$$\int_K \langle \pi(kx)\boldsymbol{e}, \boldsymbol{b}\rangle \, dk = \phi(x^{-1})\langle \boldsymbol{e}, \boldsymbol{b}\rangle$$

for all $\boldsymbol{b} \in \mathfrak{H}_\phi$. This means that

$$P\pi(x)\boldsymbol{e} = \phi(x^{-1})\boldsymbol{e}. \tag{4}$$

Since the space $\mathfrak{H}_\phi$ is generated by the complex linear combinations of the vectors $\pi(x)\boldsymbol{e}$ $(x \in G)$, it follows that $P^2 = P$ and

$$P(\mathfrak{H}_\phi) = C\boldsymbol{e}. \tag{5}$$

Let $\mathfrak{H}'$ denote the closure of the sum of all closed subspaces $U' \subset \mathfrak{H}_\phi$ which are invariant under π and which satisfy $PU' = \{0\}$. Then $\mathfrak{H}'$ and its orthogonal complement $\mathfrak{H}''$ in $\mathfrak{H}_\phi$ are invariant under π. Let U be any closed subspace of $\mathfrak{H}''$ invariant under $\pi (U \neq \{0\})$. Then $PU \neq \{0\}$ so by (5), $\boldsymbol{e} \in U$. But then $\pi(x)\boldsymbol{e} \in U$ for all $x \in G$ so $U = \mathfrak{H}_\phi$. This proves first that $\mathfrak{H}_\phi = \mathfrak{H}''$ and second that π is irreducible.

On the other hand, let π be an irreducible unitary representation of G on a Hilbert space $\mathfrak{H}$ such that π is spherical. Let $\boldsymbol{e}$ be a unit vector in $\mathfrak{H}$ such that $\pi(k)\boldsymbol{e} = \boldsymbol{e}$ for all $k \in K$. Before proving that the function $\langle \boldsymbol{e}, \pi(x)\boldsymbol{e}\rangle$ is spherical we establish a few facts about π.

Let f be a continuous function on G with compact support. Then the Hermitian form

$$C(\boldsymbol{a}, \boldsymbol{b}) = \int_G f(x)\langle \pi(x)\boldsymbol{a}, \boldsymbol{b}\rangle \, dx \qquad (\boldsymbol{a}, \boldsymbol{b} \in \mathfrak{H}),$$

satisfies

$$|C(\boldsymbol{a}, \boldsymbol{b})| \leq \left(\int_G |f(x)| \, dx\right)\|\boldsymbol{a}\| \|\boldsymbol{b}\|,$$

so there is a bounded operator $\pi(f)$ on $\mathfrak{H}$ such that $\langle \pi(f)(\boldsymbol{a}), \boldsymbol{b} \rangle = C(\boldsymbol{a}, \boldsymbol{b})$. We write symbolically

$$\pi(f) = \int_G f(x)\pi(x)\,dx.$$

Then the mapping $f \to \pi(f)$ is a homomorphism of the group algebra of G into the algebra of bounded operators on $\mathfrak{H}$; in other words, we have a representation of the group algebra on $\mathfrak{H}$.

Let $\mathfrak{N}$ be the subspace of $\mathfrak{H}$ consisting of all vectors $\boldsymbol{a} \in \mathfrak{H}$ which are left fixed by each $\pi(k)$, $k \in K$.

Lemma 3.5. *The subspace $\mathfrak{N}$ is invariant under each operator $\pi(f)$, $f \in C_c^\natural(G)$.*

Proof. If $\boldsymbol{a} \in \mathfrak{N}$ and $f \in C_c^\natural(G)$ we have for $k \in K$

$$\pi(k)\pi(f)\boldsymbol{a} = \int_G f(x)\pi(kx)\boldsymbol{a}\,dx = \int_G f(x)\pi(x)\boldsymbol{a}\,dx = \pi(f)\boldsymbol{a}$$

where we have used $f(kx) = f(x)$.

Lemma 3.6. *The space $\mathfrak{N}$ is one-dimensional.*

Proof. For $f \in C_c^\natural(G)$ let A_f denote the restriction of $\pi(f)$ to the Hilbert space $\mathfrak{N}$. Let f^* denote the function

$$f^*(x) = \overline{f(x^{-1})}, \qquad x \in G.$$

Then $f^* \in C_c^\natural(G)$ so by Theorem 3.1(ii)

$$\langle \boldsymbol{a}, \pi(f^*)\boldsymbol{b} \rangle = \int_G f(x^{-1})\langle \pi(x^{-1})\boldsymbol{a}, \boldsymbol{b} \rangle\, d(x^{-1}).$$

Hence $\langle \boldsymbol{a}, A_f * \boldsymbol{b} \rangle = \langle A_f \boldsymbol{a}, \boldsymbol{b} \rangle$ so the operator $A_f *$ is the adjoint of A_f. The operators A_f therefore constitute a commutative family of normal bounded operators. Consequently they have a common spectral resolution

$$A_f = \int p_f(\lambda)\,dE_\lambda,$$

where λ varies over some (unspecified) space, p_f is a complex-valued function, and dE_λ is a measure whose values are projection operators on $\mathfrak{N}$. These operators $E(S)$ commute with each A_f and the range $E(S)$ $\mathfrak{N}$ is therefore invariant under each A_f.

Now suppose that dim $\mathfrak{N}$ were >1. Then either all A_f are scalar multiples of I or $E(S)\,\mathfrak{N}$ is for some S different from 0 and $\mathfrak{N}$. In both of these cases $\mathfrak{N}$ can be decomposed $\mathfrak{N} = \mathfrak{N}_1 + \mathfrak{N}_2$ where $\mathfrak{N}_1$ and $\mathfrak{N}_2$ are nonzero, closed mutually orthogonal subspaces of $\mathfrak{N}$, invariant under each A_f. Select $\boldsymbol{a}_1 \neq 0$ in $\mathfrak{N}_1$ and let $\mathfrak{M}_1$ denote the set of vectors $\pi(f)\boldsymbol{a}_1$ as f runs through the group algebra. Then $\mathfrak{M}_1$ is not $\{0\}$ and is invariant under each $\pi(x)$, $x \in G$. We shall now show that $\mathfrak{M}_1$ and $\mathfrak{N}_2$ are orthogonal. This would imply that the closure of $\mathfrak{M}_1$ is different from $\mathfrak{H}$, which in turn contradicts the assumed irreducibility of π.

Let $f \in C_c(G)$ and $\boldsymbol{a}_2 \in \mathfrak{N}_2$. Then

$$\langle \pi(f)\boldsymbol{a}_1, \boldsymbol{a}_2 \rangle = \int_G f(x)\langle \pi(x)\boldsymbol{a}_1, \boldsymbol{a}_2 \rangle \, dx = \int_G f^{\natural}(x)\langle \pi(x)\boldsymbol{a}_1, \boldsymbol{a}_2 \rangle \, dx$$

so

$$\langle \pi(f)\boldsymbol{a}_1, \boldsymbol{a}_2 \rangle = \langle A_{f^\natural}\boldsymbol{a}_1, \boldsymbol{a}_2 \rangle.$$

This last expression vanishes since $\mathfrak{N}_1$ is invariant under $A_{f^\natural}$. This concludes the proof.

We return now to the proof of Theorem 3.4. It remains to prove that the function $\phi(x) = \langle \boldsymbol{e}, \pi(x)\boldsymbol{e} \rangle$ is spherical. Let $f \in C_c^{\natural}(G)$. In view of Lemma 3.5 and 3.6 the vector $\pi(f)\boldsymbol{e}$ is a scalar multiple of $\boldsymbol{e}$ and since

$$\langle \pi(f)\boldsymbol{e}, \boldsymbol{e} \rangle = \int_G f(x)\overline{\phi(x)} \, dx$$

it is clear that

$$\pi(f)\boldsymbol{e} = \left(\int_G f(x)\overline{\phi(x)} \, dx \right) \boldsymbol{e}.$$

Since the mapping $f \to \pi(f)$ is a representation of $C_c^{\natural}(G)$ on $\mathfrak{H}$, the mapping

$$f \to \int_G f(x)\overline{\phi(x)} \, dx$$

is a homomorphism of $C_c^{\natural}(G)$ onto $\boldsymbol{C}$. Lemma 3.2 now shows that ϕ is a spherical function on G.

The connection between positive definite spherical functions and spherical representations established in Theorem 3.4 can be made more precise by using unitary equivalence of representations.

Definition. Let G be a locally compact group. Two unitary representations π and π' of G on Hilbert spaces $\mathfrak{H}$ and $\mathfrak{H}'$ are called *unitarily equivalent* if there exists a linear mapping A of $\mathfrak{H}$ onto $\mathfrak{H}'$ preserving scalar products such that $\pi'(x) \circ A = A \circ \pi(x)$ for all $x \in G$.

Theorem 3.7. *Assume $C_c^\natural(G)$ commutative. For each irreducible, unitary, spherical representation π of G on a Hilbert space $\mathfrak{H}$ let $\mathfrak{N}_\pi$ denote the (one-dimensional) space of vectors in $\mathfrak{H}$ which are fixed under each $\pi(k)$, $k \in K$. Let Ω denote the set of all unitary equivalence classes ω of such representations of G. Then Ω is in a natural one-to-one correspondence with the set $\mathfrak{P}$ of all positive definite spherical functions ϕ on G. This correspondence $\omega \to \phi$ has the properties*

(i) *If $\pi \in \omega$ and $\boldsymbol{e}$ is any unit vector in $\mathfrak{N}_\pi$, then*

$$\phi(x) = \langle \boldsymbol{e}, \pi(x)\boldsymbol{e} \rangle.$$

(ii) *ω contains the representation associated to ϕ.*

Proof. First we note that $\langle \boldsymbol{e}, \pi(x)\boldsymbol{e} \rangle$ is independent of the choice of the unit vector $\boldsymbol{e} \in \mathfrak{N}_\pi$ and of the choice of π in ω. Thus we have a mapping of Ω into $\mathfrak{P}$. This mapping is onto because

$$\phi(x) = \langle \boldsymbol{e}_\phi, \pi_\phi(x)\boldsymbol{e}_\phi \rangle \qquad (x \in G)$$

if $\phi \in \mathfrak{P}$, π_ϕ is the representation associated to ϕ, and $\boldsymbol{e}_\phi$ any unit vector in $\mathfrak{N}_{\pi_\phi}$. In order to prove that the mapping is one-to-one it suffices to prove that if $\omega \in \Omega$, $\pi \in \omega$ and if we put $\phi(x) = \langle \boldsymbol{e}, \pi(x)\boldsymbol{e} \rangle$ where $\boldsymbol{e}$ is a unit vector in $\mathfrak{N}_\pi$, then $\pi_\phi \in \omega$. But the desired mapping from a Hilbert space $\mathfrak{H}$ (on which π acts) onto $\mathfrak{H}_\phi$ is given by

$$A: \sum_{i=1}^{r} a_i \pi(x_i)\boldsymbol{e} \to \sum_{i=1}^{r} a_i \pi_\phi(x_i)\boldsymbol{e}_\phi$$

where $x_1, \ldots, x_r$ are arbitrary in G and $a_1, \ldots, a_r$ are arbitrary complex numbers. This proves the theorem.

Consider now the Banach space $L^1(G)$ of complex-valued integrable functions on G with the norm

$$\|f\| = \int_G |f(x)|\,dx.$$

Let $L^\natural(G)$ denote the closed subspace of functions in $L^1(G)$ which are bi-invariant under K. With the convolution product $L^\natural(G)$ is a Banach algebra.

Definition. Let f be a function on G, bi-invariant under K. The function $\tilde{f}$ defined by

$$\tilde{f}(\phi) = \int_G f(x)\phi(x)\,dx \tag{6}$$

on a subset of the set Φ of all spherical functions ϕ on G is called the *spherical transform* of f.

According to Theorem 3.3 the continuous homomorphisms of $L^\natural(G)$ onto C are given by $h_\phi: f \to \tilde{f}(\varphi)$, φ being a bounded spherical function.

Proposition 3.8. *Assume $C_c^\natural(G)$ commutative. Then the algebra $L^\natural(G)$ is semisimple; that is, if $f_0 \in L^\natural(G)$ and*

$$\tilde{f}_0(\phi) = 0 \quad \text{for all bounded } \phi \in \Phi, \tag{7}$$

then $f_0 = 0$.

Proof. Assume (7) and put $f^*(x) = \overline{f(x^{-1})}$. Replacing f_0 by a suitable convolution $f_0 * h$ ($h \in C_c^\natural(G)$) we may assume f_0 bounded. But then $f_0 * f_0^*$ is continuous and if $f \in L^\natural(G)$ the convolution $(f_0 * f_0^* * f)(x)$ exists for all $x \in G$ (not just almost all $x \in G$). The linear functional $F: L^\natural(G) \to C$ defined by

$$F(f) = (f_0 * f_0^* * f)(e)$$

satisfies

$$F(f * f^*) \geq 0$$

and by Schwarz inequality

$$|F(f * g)|^2 \leq F(f * f^*)F(g * g^*), \qquad f, g \in L^\natural(G). \tag{8}$$

Letting ϕ approximate the delta function on G and putting $g = \phi^\natural$, we deduce from (8)

$$|F(f)|^2 \leq M \; F(f * f^*),$$

where M is a constant. Using this successively we obtain

$$\begin{aligned} |F(f)|^2 &\leq MF(f * f^*) \leq M^{1+1/2}F(f * f^* * f * f^*)^{1/2} \\ &\leq M^{1+1/2+\cdots+1/2^n}\|F\|^{1/2^n}\|(f * f^*)^{2^n}\|^{1/2^n} \end{aligned}$$

Using the formula

$$\sup_{\phi} |\tilde{g}(\phi)| = \lim_{m \to \infty} \|g^m\|^{1/m}$$

(valid for any commutative Banach algebra) we deduce

$$|F(f)|^2 \leq M^* \sup_{\phi} |(f * f^*)^{\sim}(\phi)| \quad (M^* = \text{const}).$$

In particular, $F(f_0 * f_0^*) = 0$, whence $f_0 = 0$.

Remark. See Exercise C1 for a strengthening of this result.

§4. Integral Formulas for Spherical Functions. Connections with Representations

1. The Compact Type

Let (G, K) be a Riemannian symmetric pair, G compact. We shall express the spherical functions on G/K by means of characters of irreducible spherical representations of G.

Let ϕ be a spherical function on G. Then $\phi(e) = 1$, $\phi(kgk') \equiv \phi(g)$, and

$$D\phi = \lambda_D \phi \tag{1}$$

for each $D \in \mathbf{D}_K(G)$, λ_D being a complex number. Let V_ϕ be the closed subspace of $\mathscr{E}(G/K)$ generated by the translated functions $gK \to \phi(xg)$ $(x \in G)$. Let $\pi = \pi_\phi$ denote the representation of G on V_ϕ given by $\pi(g)f = f^{\tau(g)}$ $(g \in G)$ where, as usual, $\tau(g)xK = gxK$.

Lemma 4.1. *The representation π_ϕ is irreducible and consequently* $\dim V_\phi < \infty$.

Proof. Let $U \subset V_\phi$ be a nonzero closed invariant subspace. Then there exists an $f \in U$ such that $f(o) \neq 0$ (o is the origin in G/K). Since U is complete the integral $\int_K f^{\tau(k)}\,dk$ of the U-valued function $k \to f^{\tau(k)}$ is an element $f^\natural$ of U. Evaluation at a point is a continuous linear functional on U so (by definition) $f^\natural(x) = \int_K f^{\tau(k)}(x)\,dk$ whence $f^\natural \neq 0$. Because of Corollary 2.3, $f^\natural$ is a constant multiple of the spherical function $\phi_0(gK) = \phi(g)$. Hence $U = V_\phi$, proving the irreducibility. The rest follows from Corollary 1.10.

Since π is finite-dimensional the operator $P = \int_K \pi(k)\,dk$ is well defined. We have $P^2 = P$ and the vectors in PV_ϕ are fixed under K, so again by Corollary 2.3, $PV_\phi = \mathbf{C}\phi_0$. The mapping $P\pi(g)P$ maps V_ϕ onto $\mathbf{C}\phi_0$ and the vector ϕ_0 into $\phi(g^{-1})\phi_0$. Hence

$$\phi(g^{-1}) = \mathrm{Tr}(P\pi(g)P) = \mathrm{Tr}(\pi(g)P) = \mathrm{Tr} \int_K \pi(g)\pi(k)\, dk = \int_K \chi(gk)\, dk$$

if χ denotes the character of π.

Theorem 4.2. *The spherical functions on G are precisely the functions of the form*

$$\phi(g) = \int_K \chi(g^{-1}k)\, dk,$$

where χ is the character of a finite-dimensional irreducible, spherical representation π of G. Here ϕ is positive definite and π is the representation associated to ϕ.

Proof. It remains to prove that if π is a unitary, irreducible finite-dimensional representation of G on a Hilbert space $\mathfrak{H}$ such that π is spherical, then the integral above is a spherical function. Let e be a unit vector in $\mathfrak{H}$ which is left fixed by each $\pi(k)$, $k \in K$. Then if we put $\psi(g) = \langle e, \pi(g)e \rangle$ we know from Theorem 3.4 that ψ is spherical. Now put as before

$$P = \int_K \pi(k)\, dk.$$

Then $P^2 = P$, so

$$\mathrm{Tr}(P\pi(g)P) = \mathrm{Tr}(\pi(g)P) = \int_K \mathrm{Tr}\, \pi(gk)\, dk = \int_K \chi(gk)\, dk \tag{2}$$

if χ denotes the character of π. Now for each $x \in G$, the vector $P\pi(x)e$ is left fixed by each $\pi(k)$, $k \in K$. Using Lemma 3.6 we conclude that

$$P\mathfrak{H} = Ce.$$

Let $e_1 = e$, $e_2, \ldots, e_n$ be an orthonormal basis of $\mathfrak{H}$. Since $Pe_i = 0$ for $i \geq 2$ it follows that

$$\mathrm{Tr}(P\pi(g)P) = \langle \pi(g)e, e \rangle = \psi(g^{-1}). \tag{3}$$

From (2) and (3) it now follows that the integral $\int_K \chi(g^{-1}k)\, dk$ is a spherical function.

Remark. For G semisimple a more explicit integral representation is contained in Theorem 4.4 of Chapter V.

2. The Noncompact Type

Let $X = G/K$ be a symmetric space of the noncompact type; that is, G is a connected semisimple Lie group with finite center and K is a maximal compact subgroup. Let $G = NAK$, $\mathfrak{g} = \mathfrak{n} + \mathfrak{a} + \mathfrak{k}$ be corresponding Iwasawa decompositions, and for $g \in G$, let $A(g) \in \mathfrak{a}$, $u(g) \in K$ denote the unique elements such that $g \in N \exp A(g)u(g)$. Let Σ denote the set of roots of $\mathfrak{g}$ with respect to $\mathfrak{a}$, let Σ^+ denote the subset of positive roots (for the ordering corresponding to $\mathfrak{n}$) and let as usual $\rho = \frac{1}{2}\sum_{\alpha\in\Sigma^+} m_\alpha \alpha$, m_α denoting the multiplicity of α. Let $\mathfrak{a}_c^*$ denote the set of complex-valued linear functions on $\mathfrak{a}$.

Theorem 4.3. (Harish-Chandra) *As λ runs through $\mathfrak{a}_c^*$ the functions*

$$\phi_\lambda(g) = \int_K e^{(i\lambda+\rho)(A(kg))}\, dk, \qquad g \in G,$$

exhaust the class of spherical functions on G. Moreover, two such functions ϕ_ν and ϕ_λ are identical if and only if $\nu = s\lambda$ for some s in the Weyl group.

Proof. From Chapter II, §5, No. 2 we know that each ϕ_λ is a spherical function and $\phi_\lambda = \phi_{s\lambda}$. Now let ψ be any spherical function on G/K. Then $D\psi = \lambda_D\psi$ for each $D \in \boldsymbol{D}(G/K)$ where the mapping $D \to \lambda_D$ is a homomorphism of $\boldsymbol{D}(G/K)$ into $\boldsymbol{C}$. Because of Theorem 5.18 of Chapter II and Lemma 3.11 of Chapter III this homomorphism has the form $D \to \Gamma(D)(i\lambda)$ for some $\lambda \in \mathfrak{a}_c^*$. But then ψ and the spherical function ϕ_λ have the same system of eigenvalues; by Corollary 2.3 we conclude $\phi_\lambda(g) = \psi(gK)$. We shall often write ϕ_λ instead of ψ.

Finally, suppose $\phi_\nu \equiv \phi_\lambda$. Using again Lemma 3.11 of Chapter III, we conclude that ν and λ are conjugate under the Weyl group.

We shall now prove a certain symmetry property of the spherical function which will play an important role later.

Lemma 4.4. *The spherical function ϕ_λ satisfies the identity*

$$\phi_\lambda(g^{-1}h) = \int_K e^{(-i\lambda+\rho)(A(kg))}e^{(i\lambda+\rho)(A(kh))}\, dk, \qquad g, h \in G.$$

Proof. Writing $kgh = n \exp A(kg)u(kg)h$ we obtain, since A normalizes N,

$$A(kgh) = A(kg) + A(u(kg)h). \tag{4}$$

Hence

$$\phi_\lambda(g^{-1}h) = \int_K e^{(i\lambda+\rho)(A(kg^{-1}))}e^{(i\lambda+\rho)(A(u(kg^{-1})h))}\, dk. \tag{5}$$

But by (4),

$$A(kg^{-1}) = -A(u(kg^{-1})g). \tag{6}$$

The relations $g = n_1 \exp A(g) u(g) = k(g) \exp H(g) n_2$ imply $k(g) = u(g^{-1})^{-1}$, $A(g) = -H(g^{-1})$. Lemma 5.19 of Chapter I can therefore be written

$$\int_K F(u(kg^{-1}))\, dk = \int_K F(k) e^{2\rho(A(kg))}\, dk.$$

If we now substitute (6) into (5) and use the formula above, the lemma follows.

In particular, we have, putting $h = e$,

$$\phi_\lambda(g^{-1}) \equiv \phi_{-\lambda}(g). \tag{7}$$

We can now prove an analog of Theorem 3.4 for the case of spherical functions which are not necessarily positive definite. Let π be a representation of G on a complete locally convex vector space V. Then the restriction $\pi_K = \pi | K$ is a representation of K on V and for each $\delta \in \hat{K}$ we can consider the space $V(\delta)$ defined in §1, No. 2.

Definition. The representation π of G on V is said to be *K-finite* if $\dim V(\delta) < \infty$ for each $\delta \in \hat{K}$.

The customary terminology is "admissible"; however, we feel that the above term may be more suggestive.

Theorem 4.5. *Let ϕ be a spherical function on $X = G/K$ and $\mathscr{E}_\phi$ the closed subspace of $\mathscr{E}(X)$ generated by the translates of ϕ under G. Let T_ϕ denote the natural representation of G on $\mathscr{E}_\phi$, $\check{T}_\phi$ the contragredient representation of G on the dual space $\mathscr{E}'_\phi$ (strong topology). Then*

(i) *T_ϕ is quasisimple and K-finite.*

(ii) *T_ϕ and $\check{T}_\phi$ are both irreducible; $T_\phi(K)$ and $\check{T}_\phi(K)$ admit fixed vectors ϕ and δ, respectively (unique up to scalar multiples); and, with a suitable normalization,*

$$\phi(g \cdot o) = \langle T_\phi(g^{-1})\phi, \delta \rangle. \tag{8}$$

(iii) *Conversely, suppose T is an irreducible, quasisimple, K-finite, spherical representation of G on a complete semireflexive locally convex space V. Then for a suitable spherical function ϕ, T is weakly equivalent to T_ϕ.*

Remarks. (a) As a closed subspace of the semireflexive space $\mathscr{E}(X)$, $\mathscr{E}_\phi$ is semireflexive, so $\check{T}_\phi$ is well defined (cf. §1, No. 1).

(b) "Weak equivalence" means algebraic equivalence of the representations restricted to appropriate dense invariant subspaces.

(c) There are certain nontrivial implications between the properties irreducible, quasisimple, K-finite (Harish-Chandra [1953]; Hirai [1968], Appendix; Borel [1972]). However, we shall not need them here (although they would reduce a little the assumptions in Theorem 4.5).

Proof. (i) Since each $f \in \mathscr{E}_\phi$ is an eigenfunction of each $D \in \mathbf{D}(G/K)$ the quasisimplicity of T_ϕ follows from Corollary 5.31 in Chapter II. Second, let $\lambda \in \mathfrak{a}_c^*$ be such that $\phi(hK) \equiv \phi_\lambda(h)$. Then the translated function $\phi^{\tau(g)}$ satisfies for each $\delta \in \hat{K}$

$$(T_\phi(\alpha_\delta)\phi^{\tau(g)})(hK) = \int_K \alpha_\delta(u)(T_\phi(u)\phi^{\tau(g)})(hK)\,du$$

$$= d(\delta)\int_K \chi_\delta(u)\phi_\lambda(g^{-1}uh)\,du,$$

which by Lemma 4.4 equals

$$d(\delta)\int_K e^{(i\lambda+\rho)(A(kh))}\left(\int_K e^{(-i\lambda+\rho)(A(ug))}\chi_\delta(u^{-1}k)\,du\right)dk.$$

But for g fixed the inner integral, as a function of k, lies in the finite-dimensional space H_δ. Thus by the continuity of $T_\phi(\alpha_\delta)$ the range $T_\phi(\alpha_\delta)\mathscr{E}_\phi$ has finite dimension, so T_ϕ is K-finite.

(ii) By the Hahn–Banach theorem each $\sigma \in \mathscr{E}'_\phi$ can be extended to a continuous linear form on $\mathscr{E}(X)$, i.e., to a distribution S_σ on X of compact support (Proposition 2.1, Chapter II). If σ is K-invariant we can also take S_σ to be K-invariant. Then, if $k \in K$,

$$S_\sigma(\phi^{\tau(g)}) = \int_X \phi(g^{-1}\cdot x)\,dS_\sigma(x) = \int_X \phi(g^{-1}k\cdot x)\,dS_\sigma(x),$$

which by integrating over K becomes

$$\phi(g^{-1}\cdot o)\int_X \phi(x)\,dS_\sigma(x) = S_\sigma(\phi)\delta(\phi^{\tau(g)}),$$

where δ is the delta distribution at o. Hence

$$\sigma(f) = \sigma(\phi)\delta(f), \qquad f \in \mathscr{E}_\phi,$$

which proves the existence and uniqueness of the $T_\phi^\vee(K)$-fixed vector and proves formula (8). The irreducibility of T_ϕ follows just as Lemma 4.1 (the proof did not use the compactness of G). For the irreducibility of $T_\phi^\vee$ let $W \subset \mathscr{E}'_\phi$ be a closed invariant subspace and W^0 its annihilator in $\mathscr{E}_\phi$. By the irreducibility of T_ϕ, either W^0 equals $\mathscr{E}_\phi$ or $W^0 = 0$. In the first case $W = 0$; in the second case the semireflexivity of $\mathscr{E}_\phi$ implies $W = \mathscr{E}'_\phi$. Hence $T_\phi^\vee$ is irreducible. The uniqueness is clear from Corollary 2.3.

(iii) We now come to the main part. Let T be as in the converse of the theorem. Let $V_K = \bigoplus_{\delta \in \hat{K}} V(\delta)$ denote the space of K-finite vectors in V. Let T_K denote the restriction of T to K. The operator $T_K(\alpha_\delta)$ is a continuous mapping of V onto the finite-dimensional space $V(\delta)$. Since V^∞ is dense in V, (Corollary 1.4) it follows that $T_K(\alpha_\delta)V^\infty$ is dense in $V(\delta)$, whence by the finite-dimensionality, $V(\delta) = T_K(\alpha_\delta)V^\infty$. Let $v \in V^\infty$. Then the map $x \to T(x)v$ of G into V and the map $(x, k) \to T(xk)v$ of $G \times K$ into V are both differentiable, so the mapping

$$x \to \int_K \alpha_\delta(k)T(xk)v\,dk = T(x)\int_K \alpha_\delta(k)T(k)v\,dk$$

is also differentiable. Hence $T_K(\alpha_\delta)V^\infty \subset V^\infty$, so

(9) $$V(\delta) \subset V^\infty.$$

Lemma 4.6. *Let $v \in V$ be fixed under $T(K)$ and let $\lambda \in V'$. Then the function*

$$\lambda_v : g \to \lambda(T(g)v) \qquad (g \in G)$$

is analytic.

Proof. Let $w \in V^\infty$ and λ_w be the function $g \to \langle T(g)w, \lambda\rangle$. Then if $D \in \boldsymbol{D}(G)$

(10) $$(D\lambda_w)(g) = \langle T(g)\,dT(D)w, \lambda\rangle = \lambda_{dT(D)w}(g),$$

as we see by writing D as a linear combination of products of the $\tilde{X}$ ($X \in \mathfrak{g}$). By (9) we have $v \in V^\infty$, so by (10) and the quasisimplicity,

$$(Z\lambda_v)(g) = \lambda(T(g)\,dT(Z)v) = c_Z\lambda_v(g)$$

if $Z \in \boldsymbol{Z}(G)$, c_Z being a scalar. Thus the function $gK \to \lambda_v(g)$ on G/K is an eigenfunction of the operator $\mu(Z) \in \boldsymbol{D}(G/K)$ (μ is defined in Chapter II, §4). But among the operators $\mu(Z)$ occurs the Laplace–Beltrami operator on G/K (cf. Exercise A.4 of Chapter II). Hence λ_v is analytic by the ellipticity of the Laplacian (cf. proof of Proposition 2.2).

The lemma implies that if $v \neq 0$ then $dT(\boldsymbol{D}(G))v$ is dense in V. Otherwise some $\lambda \not\equiv 0$ in V' would annihilate it; then (10) with $w = v$, $g = e$ implies that $\lambda_v \equiv 0$, contradicting the irreducibility of T.

Because of the filtration $\boldsymbol{D}(G) = \sum_{d \geq 0} \boldsymbol{D}^d(G)$ by degrees (Chapter II, §4) we have

$$\boldsymbol{D}(G) = \bigoplus_{\delta \in \hat{K}} \boldsymbol{D}_\delta(G),$$

where $\boldsymbol{D}_\delta(G)$ is the set of elements in $\boldsymbol{D}(G)$ which under $\mathrm{Ad}_G(K)$ transform according to δ. Since $dT(\mathrm{Ad}(k)D) = T(k)\,dT(D)T(k^{-1})$ for $k \in K$,

$D \in \boldsymbol{D}(G)$ and v is fixed under $T(K)$ the mapping $D \to dT(D)v$ of $\boldsymbol{D}(G)$ into V intertwines $\mathrm{Ad}_G|K$ and T. Since every element in $\boldsymbol{D}(G)$ is K-finite [under $\mathrm{Ad}_G(K)$] it follows that $dT(\boldsymbol{D}(G))v \subset V_K$. More specifically,

$$dT(\boldsymbol{D}_\delta(G))v \subset V(\delta).$$

But since $T_K(\alpha_\delta)V = V(\delta)$ and $dT(\boldsymbol{D}(G))v$ is dense in V it follows that $T_K(\alpha_\delta)(dT(\boldsymbol{D}(G))v)$, which equals $dT(\boldsymbol{D}_\delta(G))v$, is dense in $V(\delta)$. By the finite-dimensionality we have $dT(\boldsymbol{D}_\delta(G))v = V(\delta)$. In particular, if δ_0 denotes the one-dimensional identity representation of K we have $dT(\boldsymbol{D}_K(G)) = V(\delta_0)$, the space of fixed points of $T(K)$ in V ($\boldsymbol{D}_K(G)$ was defined in Chapter II, §4). Since $v \in V(\delta_0)$ was arbitrary this shows that $\boldsymbol{D}_K(G)$ acts (under dT) irreducibly on $V(\delta_0)$. Since $(\boldsymbol{D}(G)\mathfrak{k}) \cap \boldsymbol{D}_K(G)$ acts trivially on $V(\delta_0)$, dT induces an irreducible representation τ of

$$\boldsymbol{D}_K(G)/\boldsymbol{D}(G)\mathfrak{k} \cap \boldsymbol{D}_K(G) = \boldsymbol{D}(G/K)$$

on $V(\delta_0)$. Since $\boldsymbol{D}(G/K)$ is commutative and $V(\delta_0)$ finite-dimensional, the operators $\tau(D)$ $(D \in \boldsymbol{D}(G/K))$ have a common eigenvector $v \neq 0$. Hence, by the irreducibility,

$$\dim V(\delta_0) = 1. \tag{11}$$

Fix $v_0 \neq 0$ in $V(\delta_0)$ and select $\lambda \in V'$ such that $\langle v_0, \lambda \rangle = 1$. Let λ_0 denote the linear form $w \to \int_K \langle T(k)w, \lambda \rangle\, dk$ on V. Then $\lambda_0 \in V'$, $\langle v_0, \lambda_0 \rangle = 1$, and we can define the function ϕ on G/K by

$$\phi(gK) = \langle T(g^{-1})v_0, \lambda_0 \rangle. \tag{12}$$

We shall prove that ϕ is a spherical function. Let $f \in C_c^\natural(G)$ and as usual put

$$T(f)v = \int_G f(g)T(g)v\, dg \qquad v \in V.$$

Then the vector $T(f)v_0$ is fixed under $T(K)$, so by (11), $T(f)v_0 = c_f v_0$ $(c_f \in \boldsymbol{C})$. Applying λ_0, we get

$$\int_G f(g)\phi(g^{-1}K)\, dg = c_f. \tag{13}$$

Since $T(f_1 * f_2) = T(f_1)T(f_2)$ the mapping $f \to \int f(g)\phi(g^{-1}K)\, dg$ is a homomorphism of $C_c^\natural(G)$ onto $\boldsymbol{C}$. But then by Lemma 3.2, the function $g \to \phi(g^{-1}K)$, and therefore also ϕ, is spherical. Let T_ϕ be the associated representation of G on $\mathscr{E}_\phi$. Let $V^0 \subset V$, $\mathscr{E}_\phi^0 \subset \mathscr{E}_\phi$ be the subspaces generated (algebraically) by $T(g)v_0$ $(g \in G)$ and $T_\phi(g)\phi$ $(g \in G)$, respectively. These subspaces are dense and the mapping

$$\sum_i a_i T(g_i)v_0 \to \sum_i a_i T_\phi(g_i)\phi, \quad a_i \in \boldsymbol{C}, g_i \in G, \tag{14}$$

sets up the desired equivalence. In fact, the map is well defined because of (12); it is one-to-one because if $\sum a_i \phi(g_i^{-1}xK) \equiv 0$, then

$$(15) \qquad \langle \sum a_i T(g_i)v_0, T^\vee(x)\lambda_0 \rangle \equiv 0,$$

where $T^\vee$ is contragredient to T. But as in part (ii) the semireflexivity of V implies irreducibility of $T^\vee$, so (15) implies $\sum a_i T(g_i)v_0 = 0$. This concludes the proof of Theorem 4.5.

As usual, if $\lambda \in \mathfrak{a}_c^*$ let $A_\lambda \in \mathfrak{a}^c$ be determined by $B(A_\lambda, H) = \lambda(H)$ $(H \in \mathfrak{a})$. Since $\phi_{s\lambda}(x) = \phi_\lambda(x)$ $(\lambda \in \mathfrak{a}_c^*, s \in W, x \in X)$ there exists a unique K-invariant function $Z \to \phi_Z(x)$ on $\mathfrak{p}$ such that $\phi_{A_\lambda}(x) = \phi_\lambda(x)$. Denoting by G_0 the semidirect product $K \times_s \mathfrak{p}$ ($\mathfrak{p}$ as a translation group being the normal subgroup, the action of K on $\mathfrak{p}$ given by $\mathrm{Ad}_G(K)$) we can view $\mathfrak{p}$ as the quotient G_0/K which is a symmetric space of the Euclidean type ([DS], Chapter V). As in Chapter II, §3, No. 4 let $J(X)$ denote the ratio of the volume elements in $\mathfrak{p}$ and G/K.

Theorem 4.7. *Suppose G is complex. Then the function $\phi_Z(x)$ is given by*

$$(16) \qquad \phi_Z(\operatorname{Exp} X) = J^{-1/2}(X) \int_K e^{iB(Z, \operatorname{Ad}(k)X)}\, dk \qquad (X \in \mathfrak{p}).$$

Moreover, $(Z, x) \to \phi_Z(x)$ is a bi-spherical function on $G_0/K \times G/K$, in the sense that in each variable it is proportional to a spherical function.

For the proof let $\Psi_Z \circ \operatorname{Exp}$ be the function on $\mathfrak{p}$ given by the right-hand side of (16). Then by Theorem 5.38 of Chapter II

$$D\Psi_Z = \Gamma(D)(i\lambda)\Psi_Z, \qquad D \in \mathbf{D}(G/K),$$

if $A_\lambda \in \mathfrak{a}^c$ is conjugate to Z under K. By Corollary 2.3, $\Psi_Z = \phi_\lambda = \phi_Z$, proving (16). On the other hand, the G_0-invariant differential operators on G_0/K are the K-invariant differential operators on $\mathfrak{p}$ with constant coefficients. The function $Z \to \phi_Z(x)$ is a K-invariant eigenfunction of these operators, so the theorem follows.

Remarks. **1.** We recall that (16) relies on Theorem 5.37 of Chapter II, which will be proved in the following section [§5, Eq. (29)].

2. Formula (16) can be used to define $\phi_Z(x)$ for any $Z \in \mathfrak{p}^c$. By holomorphic continuation we have then $\phi_{kZ}(x) = \phi_Z(x)$ not only for $k \in K$ but even for $k \in K^c$ (the analytic subgroup of $\mathrm{Int}(\mathfrak{g}^c)$ corresponding to $\mathfrak{k}^c$). The function $x \to \phi_Z(x)$ is still a spherical function, so for each $Z \in \mathfrak{p}^c$ there exists a $\lambda \in \mathfrak{a}_c^*$ such that

$$(17) \qquad \phi_Z = \phi_\lambda$$

even if Z and A_λ are not necessarily conjugate under K.

Consider for example an arbitrary nilpotent $Z \in \mathfrak{p}^{\mathbf{C}}$. By Theorem 4.3 of Chapter III the function $X \to e^{iB(Z,X)}$ is harmonic and (17) takes the form

$$\int_K e^{iB(Z,\operatorname{Ad}(k)X)}\,dk = 1, \qquad X \in \mathfrak{p},\ Z \in \mathcal{N}. \tag{18}$$

This is of course also a special case of Proposition 2.5.

3. The spherical functions as defined in this chapter are often called *elementary spherical functions*, *zonal spherical functions*, or *spherical functions of class one*. More generally, suppose $\delta \in \hat{K}$ such that the group $\delta(M)$ (M being the centralizer of A in K) has a nonzero fixed vector. A *spherical function of class* δ is by definition a joint eigenfunction of $\mathbf{D}(G/K)$ which is K-finite of type δ. It is known that the spherical functions of class δ are precisely the integrals

$$\Phi(gK) = \int_{K/M} e^{(i\lambda+\rho)(A(k^{-1}g))} F(kM)\,dk_M,$$

where F is a K-finite function on K/M of type δ. While this theorem generalizes Theorem 4.3, new methods are required for its proof (cf. Helgason, [1976], §7).

3. The Euclidean Type

The notation of §4, No. 2, being preserved we shall now discuss the spherical functions on the Euclidean-type symmetric space $\mathfrak{p} = G_0/K$. The group G_0 is often called the *Cartan motion group* associated with X.

As in Chapter II let $S(\mathfrak{p})$ denote the symmetric algebra over $\mathfrak{p}$ (with complex coefficients), and if $p \in S(\mathfrak{p})$ let $\partial(p)$ denote the corresponding constant-coefficient differential operator on $\mathfrak{p}$. As p runs through the set $I(\mathfrak{p})$ of K-invariants in $S(\mathfrak{p})$, the operators $\partial(p)$ run through the algebra $\mathbf{D}(G_0/K)$ of G_0 invariant differential operators on $\mathfrak{p} = G_0/K$.

Proposition 4.8. *The spherical functions on G_0/K are the functions*

$$\psi_\lambda(Y) = \int_K e^{iB(A_\lambda, kY)}\,dk \qquad (Y \in \mathfrak{p})$$

as $\lambda \in \mathfrak{a}_c^*$; moreover $\psi_\lambda \equiv \psi_\mu$ if and only if λ and μ are conjugate under W.

The proof is the same as that of Theorem 4.3. Similarly, we have the following result.

Proposition 4.9. *Theorem 4.5 holds with $\mathfrak{p} = G_0/K$ replacing $X = G/K$.*

Combining Proposition 4.8 and Theorem 4.7, we have

Proposition 4.10. *Assume G complex. Then the spherical functions on G/K are given by*

$$\phi_\lambda(\operatorname{Exp} X) = J^{-1/2}(X)\psi_\lambda(X) \qquad (\lambda \in \mathfrak{a}_c^*)$$

where the ψ_λ are the spherical functions on G_0/K.

Remark. Because of Theorem 5.35, Chapter II, these spherical functions can also be expressed as finite sums.

§5. Harish-Chandra's Spherical Function Expansion

1. The General Case

In this section we shall investigate the spherical function ϕ_λ in more detail. We follow the notation of §4, No. 2. Because of Theorem 4.3 we rewrite the *spherical transform* in the form

$$\tilde{f}(\lambda) = \int_G f(x)\phi_{-\lambda}(x)\,dx \qquad (f\ K\text{-bi-invariant}) \tag{1}$$

for all $\lambda \in \mathfrak{a}_c^*$ for which the integral is defined. We know from Proposition 3.8 that the spherical transform $f \to \tilde{f}$ is injective on $L^\natural(G)$. For the deeper theory of this transform we need information about ϕ_λ for beyond Theorem 4.3. Since $G = K\overline{A^+}K$ it suffices to know ϕ_λ on the Weyl chamber $A^+ = \exp(\mathfrak{a}^+)$. There ϕ_λ satisfies the differential equations

$$\Delta(D)\phi_\lambda = \Gamma(D)(i\lambda)\phi_\lambda, \qquad D \in \mathbf{D}(G/K), \tag{2}$$

where $\Delta(D)$ is the radial part of D for the action of K on G/K. According to Proposition 5.23 of Chapter II, the operators $\Delta(D)$ have coefficients belonging to the ring $\mathscr{R}$. It is therefore to be hoped that the joint eigenfunctions of the operators $\Delta(D)$, in particular the function ϕ_λ, should have expansions on $A^+ \cdot o$ resembling those for functions in the ring $\mathscr{R}$. It will turn out that while the entire system (2) is needed to show that such an expansion for ϕ_λ exists, the operator $D = L_X$ alone is sufficient for its actual determination.

We begin by proving the important fact that each $u \in \mathbf{D}(A)$ can be expressed in terms of the radial parts $\Delta(D)$ $(D \in \mathbf{D}(G/K))$ and elements from the ring $\mathscr{R}$. If $\mu \in \mathfrak{a}_c^*$ then e^μ denotes as before the function on A given by $e^\mu(a) = e^{\mu(\log a)}$.

Let $H(\mathfrak{a})$ denote the set of W-harmonic polynomials on $\mathfrak{a}$ and $\boldsymbol{H}(A)$ the set of corresponding constant-coefficient operators on $A \cdot o$ or A. By Chapter III, Theorem 3.4, we have dim $\boldsymbol{H}(A) = w$, the order of W. We shall now apply Proposition 5.23 of Chapter II.

Proposition 5.1. *Let $u_1, \ldots, u_w$ be a homogeneous basis of $\boldsymbol{H}(A)$. Let $u \in \boldsymbol{D}(A)$. Then there exist unique operators $D_1, \ldots, D_w \in \boldsymbol{D}(G/K)$ such that*

$$u = \sum_1^w u_i(e^{-\rho}\Gamma(D_i)e^{\rho}). \tag{3}$$

Furthermore, there exist elements $g_{ij} \in \mathscr{R}$, $D_{ij} \in \boldsymbol{D}(G/K)$ such that

$$u = \sum_i u_i \circ \Delta(D_i) + \sum_{ij} g_{ij} u_i \circ \Delta(D_{ij}). \tag{4}$$

Proof. The mapping $u \to u' = e^{+\rho} u \circ e^{-\rho}$ is an automorphism of $\boldsymbol{D}(A)$ leaving $\boldsymbol{H}(A)$ invariant. Thus u_i' $(1 \le i \le w)$ is a basis of $\boldsymbol{H}(A)$, and so by Chapter III, Theorem 3.4 and Chapter II, Theorem 5.18,

$$u' = \sum_{i=1}^{w} u_i' \Gamma(D_i)$$

for unique elements $D_i \in \boldsymbol{D}(G/K)$. This proves (3).

In order to prove (4) we may assume u homogeneous. Let $q = \deg(u)$, $q_i = \deg(u_i)$. Then D_i has degree $q - q_i$. Using now Proposition 5.23 of Chapter II we have

$$e^{-\rho}\Gamma(D_i) \circ e^{\rho} = \Delta(D_i) + \sum_i h_{ij} E_{ij},$$

where $h_{ij} \in \mathscr{R}$ and $E_{ij} \in \boldsymbol{D}(A)$ has order $\le q - q_i - 1$. Substituting into (3) we get

$$u = \sum_i u_i \circ \Delta(D_i) + \sum_{i,j} u_i \circ (h_{ij} E_{ij}),$$

so

$$u = \sum_i u_i \circ \Delta(D_i) + \sum_k h_k E_k, \tag{5}$$

where $h_k \in \mathscr{R}$ and $E_k \in \boldsymbol{D}(A)$ has order $<q$. Applying (5) to E_k, we obtain formula (4) by iteration.

The proposition has the following important consequence (which is an analog of Theorem 3.13 of Chapter III).

Corollary 5.2. *Let $\chi: \boldsymbol{D}(G/K) \to \boldsymbol{C}$ be a homomorphism and let E_χ denote the set of functions ϕ on $A^+ \cdot o$ satisfying*

$$\Delta(D)\phi = \chi(D)\phi \qquad \textit{for all} \quad D \in \boldsymbol{D}(G/K). \tag{6}$$

Then dim $E_\chi \le w$.

Fix $a_0 \in A^+ \cdot o$. If $\dim E_\chi > w$ it is clear by linear algebra that there exists a $\phi \in E_\chi$, $\phi \neq 0$, such that $(u_i \phi)(a_0) = 0$ $(1 \leq i \leq w)$. But then by Proposition 5.1, $(u\phi)(a_0) = 0$ for all $u \in \boldsymbol{D}(A)$. But this gives the desired contradiction $\phi \equiv 0$ because, by the ellipticity of $\Delta(L_X)$ (Chapter II, Proposition 3.9), ϕ is analytic.

Because of Corollary 5.20 of Chapter II and (2) above the function ϕ_λ satisfies the differential equation

$$\Delta(L_X)\phi_\lambda = -(\langle \lambda, \lambda \rangle + \langle \rho, \rho \rangle)\phi_\lambda. \tag{7}$$

From Proposition 3.9 of Chapter II we have

$$\Delta(L_X) = (L_A + 2A_\rho) + \sum_{\alpha \in \Sigma^+} m_\alpha(\coth \alpha - 1)A_\alpha \tag{8}$$

and

$$\coth \alpha - 1 = 2 \sum_{k \geq 1} e^{-2k\alpha} \in \mathscr{R}. \tag{9}$$

Thus the second term on the right in (8) can be regarded as a perturbation term, the principal term being the constant-coefficient operator $L_A + 2A_\rho$. Since

$$(L_A + 2A_\rho)e^{i\lambda - \rho} = -(\langle \lambda, \lambda \rangle + \langle \rho, \rho \rangle)e^{i\lambda - \rho},$$

we are led to search for a solution Φ_λ of the perturbed equation

$$\Delta(L_X)\Phi_\lambda = -(\langle \lambda, \lambda \rangle + \langle \rho, \rho \rangle)\Phi_\lambda \tag{10}$$

of the perturbed form

$$\Phi_\lambda(\exp H \cdot o) = e^{(i\lambda - \rho)(H)} \sum_{\mu \in \Lambda} \Gamma_\mu(\lambda)e^{-\mu(H)} \qquad (H \in \mathfrak{a}^+) \tag{11}$$

Here Λ has the same meaning as in Chapter II, §5, No. 2 and the $\Gamma_\mu(\lambda)$ are coefficients to be determined.

Ignoring questions of convergence temporarily, we substitute the series (11) into (10) and equate coefficients to $e^{i\lambda - \rho - \mu}$. We have

$$(L_A + 2A_\rho + \langle \lambda, \lambda \rangle + \langle \rho, \rho \rangle)e^{i\lambda - \rho - \mu} = (\langle \mu, \mu \rangle - 2\langle i\mu, \lambda \rangle)e^{i\lambda - \rho - \mu}$$

and

$$\sum_{\alpha \in \Sigma^+} 2m_\alpha \sum_{k \geq 1} e^{-2k\alpha} A_\alpha \left(\sum_{\mu \in \Lambda} \Gamma_\mu(\lambda)e^{i\lambda - \rho - \mu} \right)$$
$$= \sum_{\alpha \in \Sigma^+} 2m_\alpha \sum_{\mu \in \Lambda} e^{i\lambda - \rho - \mu} \sum_{k \geq 1} \Gamma_{\mu - 2k\alpha}(\lambda)\langle i\lambda - \rho - \mu + 2k\alpha, \alpha \rangle,$$

where k runs over all integers ≥ 1 for which $\mu - 2k\alpha \in \Lambda$. This leads to a recursion formula for the function Γ_μ:

$$\{\langle \mu, \mu \rangle - 2i\langle \mu, \lambda \rangle\}\Gamma_\mu(\lambda) = 2 \sum_{\alpha \in \Sigma^+} m_\alpha \sum_{k \geq 1} \Gamma_{\mu - 2k\alpha}(\lambda)\{\langle \mu + \rho - 2k\alpha, \alpha \rangle - i\langle \alpha, \lambda \rangle\}. \tag{12}$$

For $\mu \in \Lambda$ we write $\mu = m_1\alpha_1 + \cdots + m_l\alpha_l$ ($m_i \in \mathbf{Z}^+$) and put $m(\mu) = \sum_i m_i$. Putting $\Gamma_0 \equiv 1$, (12) defines Γ_μ by induction on $m(\mu)$. Then Γ_μ is a rational function on $\mathfrak{a}_c^*$. We now need an estimate for Γ_μ. Let $\tilde{\Lambda} = \mathbf{Z}\alpha_1 + \cdots + \mathbf{Z}\alpha_l$. For $\mu \in \Lambda - \{0\}$, $\nu \in \tilde{\Lambda}$, $s \neq t \in W$, let σ_μ and $\tau_\nu(s, t)$ denote the hyperplanes in $\mathfrak{a}_c^*$ given by

(13)
$$\sigma_\mu = \{\lambda \in \mathfrak{a}_c^*: \langle \mu, \mu \rangle = 2i\langle \mu, \lambda \rangle\},$$
$$\tau_\nu(s, t) = \{\lambda: i(s\lambda - t\lambda) = \nu\}$$

[Recall that $(s\lambda)(H) = \lambda^s(H) = \lambda(s^{-1}H)$ so $st(\lambda) = s(t\lambda)$; we also write a^s for $\exp(s(\log a))$.]

Lemma 5.3. *Suppose $\lambda \in \mathfrak{a}_c^*$ does not lie on any hyperplane σ_μ and let $H \in \mathfrak{a}^+$. Then there exists a constant $K_{\lambda, H}$ such that*

$$|\Gamma_\mu(\lambda)| \leq K_{\lambda, H} e^{\mu(H)} \qquad \textit{for all} \quad \mu \in \Lambda.$$

Proof. Select constants c_1, c_2 such that for all $\mu \in \Lambda$, $\alpha \in \Sigma^+$,

$$|\langle \mu + \rho, \alpha \rangle - \langle i\alpha, \lambda \rangle| \leq c_1[m(\mu) + 1],$$
$$|\langle \mu, \mu \rangle - 2i\langle \mu, \lambda \rangle| \geq c_2 m(\mu)^2.$$

Then (12) implies

(14)
$$|\Gamma_\mu(\lambda)| \leq c m(\mu)^{-1} \sum_{\alpha \in \Sigma^+} m_\alpha \sum_{k \geq 1} |\Gamma_{\mu - 2k\alpha}(\lambda)|,$$

where $c = 4c_1c_2^{-1}$ (dependent on λ). Let N_0 be an integer such that

$$c \sum_{\alpha \in \Sigma^+} m_\alpha \sum_{k=1}^{\infty} e^{-2k\alpha(H)} \leq N_0$$

and select $K = K_{\lambda, H}$ such that

(15)
$$|\Gamma_\nu(\lambda)| \leq K e^{\nu(H)},$$

for $\nu \in \Lambda$, $m(\nu) \leq N_0$. We shall prove (15) for all $\nu \in \Lambda$ by induction on $m(\nu)$. Let $N \in \mathbf{Z}^+$, $N > N_0$, and suppose (15) holds for all $\nu \in \Lambda$ with $m(\nu) < N$. Let $\mu \in \Lambda$ be such that $m(\mu) = N$. Then

$$|\Gamma_\mu(\lambda)| \leq cN^{-1} \sum_{\alpha \in \Sigma^+} m_\alpha \sum_{k=1}^{\infty} K e^{(\mu - 2k\alpha)(H)} \leq K e^{\mu(H)}$$

and the lemma is proved.

The lemma shows (with $\frac{1}{2}H$ for H) that if $\lambda \notin \sigma_\mu$ ($\mu \in \Lambda - \{0\}$) then the series (11) converges absolutely for each $H \in \mathfrak{a}^+$. Being a power series in $x_1 = e^{-\alpha_1(H)}, \ldots, x_l = e^{-\alpha_l(H)}$, the series converges uniformly in a subchamber $\alpha_i(H) > c > 0$ ($1 \leq i \leq l$), $c > 0$ being arbitrary. The series

(11) can also be differentiated term-by-term and so represents a solution to (10). Remarkably enough, Φ_λ is an eigenfunction of all the other operators in $\Delta(\boldsymbol{D}(G/K))$ as well.

Proposition 5.4. *The function Φ_λ is an eigenfuncion of each $\Delta(D)$ $(D \in \boldsymbol{D}(G/K))$; in fact,*

$$\Delta(D)\Phi_\lambda = \Gamma(D)(i\lambda)\Phi_\lambda.$$

Proof. Since $\boldsymbol{D}(G/K)$ is commutative and the map $D \to \Delta(D)$ $(D \in \boldsymbol{D}(G/K))$ is a homomorphism the operators $\Delta(L_X)$ and $\Delta(D)$ commute. Thus the function $\Delta(D)\Phi_\lambda$ satisfies (10). But it is clear from Proosition 5.23 of Chapter II that $\Delta(D)\Phi_\lambda$ has an expansion of the form

$$\Delta(D)\Phi_\lambda = e^{(i\lambda-\rho)(H)} \sum_{\mu \in \Lambda} \Gamma_\mu^*(\lambda) e^{-\mu(H)}.$$

But then the coefficients Γ_μ^* must satisfy the recursion formula (12); consequently they are proportional to the Γ_μ; that is,

$$\Gamma_\mu^*(\lambda) = \Gamma_0^*(\lambda)\Gamma_\mu(\lambda), \qquad \mu \in \Lambda,$$

whence

$$\Delta(D)\Phi_\lambda = \Gamma_0^*(\lambda)\Phi_\lambda. \tag{16}$$

In order to evaluate $\Gamma_0^*(\lambda)$ we take the formula for $\Delta(D)$ from Proposition 5.23 of Chapter II and apply to the expansion for Φ_λ. Because of Lemma 5.3 the term-by-term differentiation is permissible. In the resulting formula

$$e^{(-i\lambda+\rho)(H)}(\Delta(D)\Phi_\lambda)(\exp H \cdot o) = \Gamma_0^*(\lambda)e^{(-i\lambda+\rho)(H)}\Phi_\lambda(\exp H \cdot o)$$

we let $H \to \infty$ along a ray in $\mathfrak{a}^+$. Viewing the expansions of functions in $\mathscr{R}$ as power-series expansions in $x_i = e^{-\alpha_i(H)}$ there is no difficulty in passing to the limit termwise. This gives

$$\Gamma(D)(i\lambda) = \Gamma_0^*(\lambda),$$

proving the proposition.

Now if $s \in W$ the function $\Phi_{s\lambda}$ will also satisfy (6) (with $\chi(D) = \Gamma(D)(i\lambda)$). Suppose now $\lambda \in \mathfrak{a}_c^*$ is such that λ lies on none of the planes $s\sigma_\mu$ $(s \in W, \mu \in \Lambda - \{0\})$ and that in addition $i(s\lambda - \sigma\lambda) \notin \tilde{\Lambda}$ for all $s \neq \sigma$ in W. Then the series $\Phi_{s\lambda}$ are all defined and the exponents $is\lambda - \rho - \mu$ $(\mu \in \Lambda)$ in the series for $\Phi_{s\lambda}$ have no overlap with the exponents $i\sigma\lambda - \rho - \nu$ $(\nu \in \Lambda)$ in the series for $\Phi_{\sigma\lambda}$. Thus there exists a $H \in \mathfrak{a}^+$ such that for $s \neq \sigma \in W$, μ, $\nu \in \Lambda$

$$(is\lambda - \rho - \mu)(H) \neq (i\sigma\lambda - \rho - \nu)(H).$$

The series for $\Phi_{s\lambda}(\exp tH \cdot o)$ converges uniformly in t for $1 \leq t < \infty$. A relation $\sum_{s \in W} a_s \Phi_{s\lambda}(\exp tH \cdot o) = 0$ (a_s constants not all 0) would therefore contradict Exercise D5 of Chapter I. Consequently, the functions $\Phi_{s\lambda}$ ($s \in W$) on A^+ are linearly independent. Thus, by Corollary 5.2, ϕ_λ is a linear combination

$$\phi_\lambda = \sum_{s \in W} c_s(\lambda)\Phi_{s\lambda} \quad \text{on } A^+ \cdot o.$$

Replacing here λ by $\sigma\lambda$ ($\sigma \in W$) we see that $c_s(\lambda) = \mathbf{c}(s\lambda)$ where $\mathbf{c}(\lambda) = c_e(\lambda)$. Recalling now the λ which had to be excluded in order that the functions $\Phi_{s\lambda}$ should be defined and linearly independent we have obtained the following basic result of Harish-Chandra, where the remarkable $\mathbf{c}$-function first enters.

Theorem 5.5. *Suppose $\lambda \in \mathfrak{a}_c^*$ is such that $i(s\lambda - \sigma\lambda) \notin \tilde{\Lambda}$ for $s \neq \sigma$ in W and that $\langle \mu, \mu \rangle \neq 2i\langle \mu, s\lambda \rangle$ for all $\mu \in \Lambda - \{0\}$, $s \in W$. Then*

$$\phi_\lambda = \sum_{s \in W} \mathbf{c}(s\lambda)e^{is\lambda - \rho} \sum_{\mu \in \Lambda} \Gamma_\mu(s\lambda)e^{-\mu} \quad \text{on } A^+ \cdot o, \tag{17}$$

where Γ_μ is given by (12) *and $\Gamma_0 \equiv 1$.*

Moreover, the series (17) *converges absolutely at each point in $A^+ \cdot o$ and uniformly on each subchamber*

$$\{a \cdot o \in A^+ \cdot o : \alpha_i(\log a) > c > 0 \quad (1 \leq i \leq l)\}.$$

This result shows that on $A^+ \cdot o$ ϕ_λ is a perturbation of the elementary function $\sum_s \mathbf{c}(s\lambda)e^{is\lambda - \rho}$. It is therefore an important problem to determine the $\mathbf{c}$-function.

We now prove a variation of Lemma 5.3 which will be needed later. Let $\mathfrak{a}_+^*$ be the Weyl chamber in $\mathfrak{a}^*$ which corresponds to $\mathfrak{a}^+$ under the mapping $\lambda \to A_\lambda$.

Lemma 5.6. *Let $H \in \mathfrak{a}^+$. Then there exists a constant K_H such that*

$$|\Gamma_\mu(\lambda)| \leq K_H e^{\mu(H)} \quad \textit{for} \quad \mu \in \Lambda, \lambda \in \mathfrak{a}^* + i\,\mathrm{Cl}(\mathfrak{a}_+^*).$$

Proof. Consider the radial part $\Delta(L_X)$ in the form (Theorem 3.7 of Chapter II)

$$\Delta(L_X) = \delta^{-1/2} L_A \circ \delta^{1/2} - \delta^{-1/2} L_A(\delta^{1/2}), \tag{18}$$

where

$$\delta(\exp H) = \prod_{\alpha \in \Sigma^+} [\tfrac{1}{2}(e^{\alpha(H)} - e^{-\alpha(H)})]^{m_\alpha}.$$

The functions $\delta^{1/2}$, $\delta^{-1/2}$, and $d = \delta^{-1/2}L_A(\delta^{1/2})$ have expansions on A^+ of the form

$$\delta^{-1/2}(\exp H) = e^{-\rho(H)} \sum_{\mu \in \Lambda} b_\mu e^{-\mu(H)}, \quad \delta^{1/2}(\exp H) = e^{\rho(H)} \sum_{\mu \in \Lambda} c_\mu e^{-\mu(H)},$$

$$d = \sum_{\mu \in \Lambda} d_\mu e^{-\mu(H)}, \qquad d_0 = \langle \rho, \rho \rangle,$$

where the coefficients b_μ, c_μ, and d_μ have at most polynomial growth in μ. Now (10) and (18) imply that the function

$$\Psi_\lambda(\exp H) = \delta^{1/2}(\exp H)\Phi_\lambda(\exp H \cdot o) = \sum_{\mu \in \Lambda} a_\mu(\lambda) e^{(i\lambda - \mu)(H)}$$

where

$$a_\mu(\lambda) = \sum_{\nu, \mu - \nu \in \Lambda} \Gamma_{\mu - \nu}(\lambda) c_\nu$$

satisfies the differential equation

$$L_A \Psi_\lambda = (-\langle \lambda, \lambda \rangle - \langle \rho, \rho \rangle + d)\Psi_\lambda. \tag{19}$$

Using $d_0 = \langle \rho, \rho \rangle$ this gives the recursion formula

$$\{\langle \mu, \mu \rangle - 2i\langle \mu, \lambda \rangle\} a_\mu(\lambda) = \sum_{0 \neq \nu, \mu - \nu \in \Lambda} d_\nu a_{\mu - \nu}(\lambda), \tag{20}$$

which is analogous to (12), but more manageable because the coefficients d_ν on the right do not involve λ. If $\lambda \in \mathfrak{a}^* + i\,\mathrm{Cl}(\mathfrak{a}^*_+)$ then $\mathrm{Im}(\langle \mu, \lambda \rangle) \geq 0$, so $|\langle \mu, \mu \rangle - 2i\langle \mu, \lambda \rangle| \geq \langle \mu, \mu \rangle \geq c^{-1}m(\mu)$, where c is a constant (independent of $\lambda \in \mathfrak{a}^* + i\,\mathrm{Cl}(\mathfrak{a}^*_+)$). It follows that

$$|a_\mu(\lambda)| \leq c\,m(\mu)^{-1} \sum_{0 \neq \nu, \mu - \nu \in \Lambda} |d_\nu||a_{\mu - \nu}(\lambda)|. \tag{21}$$

Let $H \in \mathfrak{a}^+$; repeating (with $N_0 \geq c \sum_{\nu \in \Lambda} |d_\nu| e^{-\nu(H)}$) the proof that (14) $\Rightarrow$ (15) we deduce from (21) that

$$|a_\mu(\lambda)| \leq C_H e^{\mu(H)}, \quad \mu \in \Lambda, \lambda \in \mathfrak{a}^* + i\,\mathrm{Cl}(\mathfrak{a}^*_+), \tag{22}$$

where $C_H \in \mathbf{R}$ only depends on H. But since $\Phi_\lambda = \delta^{-1/2}\Psi_\lambda$ we have

$$\Gamma_\mu(\lambda) = \sum_{\nu, \mu - \nu \in \Lambda} a_{\mu - \nu}(\lambda)\, b_\nu,$$

whence by (22)

$$|\Gamma_\mu(\lambda)| \leq C_H \Big(\sum_{\nu \in \Lambda} b_\nu e^{-\nu(H)} \Big) e^{\mu(H)} \leq K_H e^{\mu(H)},$$

since b_ν has at most polynomial growth in ν. This proves the lemma.

2. The Complex Case

In the case in which G is complex, Theorem 5.5 can be put into much simpler form.

Theorem 5.7. *Suppose G is complex. Then*

$$\phi_\lambda(a) = \frac{\pi(\rho)}{\pi(i\lambda)} \frac{\sum (\det s) e^{is\lambda(\log a)}}{\sum (\det s) e^{s\rho(\log a)}}, \qquad a \in A; \tag{23}$$

the summations extend over the Weyl group and $\pi(\lambda) = \prod_{\alpha \in \Sigma^+} \langle \alpha, \lambda \rangle$. *Moreover,*

$$c(\lambda) = \pi(\rho)/\pi(i\lambda) \tag{24}$$

and

(25) $\Gamma_\mu \equiv$ *the number of partitions of* $\frac{1}{2}\mu$ *into a sum of positive roots.*

Proof. We know from Proposition 5.15 of Chapter I and Proposition 3.10 of Chapter II that with

$$\delta^{1/2}(a) = \sum_s (\det s) e^{s\rho(\log a)} \tag{26}$$

we have

$$\Delta(L_X) = \delta^{-1/2}(L_A - \langle \rho, \rho \rangle) \circ \delta^{1/2}. \tag{27}$$

Consider now the function F on $A' \cdot o$ given by

$$F(a) = \delta^{-1/2}(a) e^{i\lambda(\log a)}.$$

Then F has an expansion

$$F(\exp H \cdot o) = e^{(i\lambda - \rho)(H)} \sum_{\mu \in \Lambda} b_\mu e^{-\mu(H)}, \qquad b_0 = 1,$$

and because of (27) it satisfies

$$\Delta(L_X)F = -(\langle \lambda, \lambda \rangle + \langle \rho, \rho \rangle)F.$$

Thus it must coincide with the function Φ_λ in (11), so by Theorem 5.5

$$\phi_\lambda(a \cdot o) = \delta^{-1/2}(a) \sum_s c(s\lambda) e^{is\lambda(\log a)}, \qquad (a \in A^+).$$

By analyticity, we have for all $a \in A$,

$$\sum_s (\det s) e^{s\rho(\log a)} \phi_\lambda(a \cdot o) = \sum_s c(s\lambda) e^{is\lambda(\log a)}. \tag{28}$$

Replacing here a by a^s we obtain $\mathbf{c}(s\lambda) = (\det s)\mathbf{c}(\lambda)$. Also, applying the differential operator $\partial(\pi)$ on A to (28) and putting $a = e$ we obtain $\pi(\rho) = \mathbf{c}(\lambda)\pi(i\lambda)$. This proves (23) and (24). Finally, $\Gamma_\mu(\lambda) \equiv b_\mu$ as already observed and

$$\sum_{\mu \in \Lambda} b_\mu e^{-\mu} = \prod_{\alpha \in \Sigma^+} (1 + e^{-2\alpha} + e^{-4\alpha} + \cdots),$$

which proves (25).

We can now prove Theorem 5.37 of Chapter II, that for G complex the radial parts $\Delta(D)$ of $D \in \mathbf{D}(G/K)$ are given by

$$\Delta(D) = \delta^{-1/2}\Gamma(D) \circ \delta^{1/2}. \tag{29}$$

We note first that both sides of (29) are differential operators on $A^+ \cdot o$ which agree on all spherical functions. Now let $f \in \mathcal{D}(A^+ \cdot o)$ and extend it to a W-invariant function $\tilde{f} \in \mathcal{D}(A \cdot o)$. Then the function

$$\tilde{f}(a \cdot o) \sum_s (\det s) e^{s\rho(\log a)}$$

has a Fourier integral expansion

$$\tilde{f}(a \cdot o) \sum_s (\det s) e^{s\rho(\log a)} = \int_{\mathfrak{a}^*} F(\lambda) e^{i\lambda(\log a)}\, d\lambda.$$

Since the left-hand side is skew under W so is the function $F(\lambda)$. The right-hand side can therefore be written as the average

$$\frac{1}{w} \int_{\mathfrak{a}^*} F(\lambda) \sum_s (\det s) e^{is\lambda(\log a)}\, d\lambda.$$

This gives a decomposition of $\tilde{f}$ into spherical functions ϕ_λ so (29) follows in general.

Example. Let G denote the group $\mathbf{SL}(2, \mathbf{C})$ considered as a real Lie group. Here

$$A = \left\{ a_t = \begin{pmatrix} e^t & 0 \\ 0 & e^{-t} \end{pmatrix} : t \in \mathbf{R} \right\}$$

and Σ^+ consists of the single root α given by $\alpha(\log a_t) = 2t$ and α has multiplicity 2. Writing $\lambda \in \mathfrak{a}_{\mathbf{C}}^*$ in the form $\lambda = l\alpha$ $(l \in \mathbf{C})$, Theorem 5.7 reduces to

$$\phi_{l\alpha}\begin{pmatrix} e^t & 0 \\ 0 & e^{-t} \end{pmatrix} = \frac{\sin(2lt)}{l \sinh(2t)}. \tag{30}$$

§6. The *c*-Function

1. The Behavior of ϕ_λ at ∞

The function $c(\lambda)$ has been defined on the set

(1) $'\mathfrak{a}_c^* = \{\lambda \in \mathfrak{a}_c^* : \lambda \notin s\sigma_\mu(\mu \in \Lambda - 0), \lambda \notin \tau_\nu(s, t), (\nu \in \tilde{\Lambda}, s \neq t \in W)\}.$

This set is the complement in $\mathfrak{a}_c^*$ of the union of countably many hyperplanes. Each compact subset of $\mathfrak{a}_c^*$ intersects at most finitely many of these hyperplanes. From [DS] (Chapter VII, Theorem 12.2 and Proposition 12.4) we know that $'\mathfrak{a}_c^*$ is a connected open dense subset of $\mathfrak{a}_c^*$. It is clearly W-invariant.

Lemma 6.1. *The function $c(\lambda)$ is holomorphic on $'\mathfrak{a}_c^*$.*

Proof. For $\lambda \in {'\mathfrak{a}_c^*}$ we have the equation

(2) $$\phi_\lambda(h) = \sum_{s \in W} c(s\lambda)\Phi_{s\lambda}(H), \qquad H \in \mathfrak{a}^+, h = \exp H$$

and for each $D \in \mathbf{D}(A)$,

(3) $$(D\phi_\lambda)(h) = \sum_s c(s\lambda)(D\Phi_{s\lambda})(H).$$

Let $s_1, \ldots, s_w$ be the elements of W and let $u_1, \ldots, u_w$ be a homogeneous basis of $H(A)$. If $\lambda_0 \in {'\mathfrak{a}_c^*}$ and $H_0 \in \mathfrak{a}^+$ are fixed we shall prove that

$$\det(\{(u_j\Phi_{s_i\lambda_0})(H_0)\}_{1 \le i, j \le w}) \neq 0.$$

Otherwise there would be complex numbers a_s $(s \in W)$ not all 0 such that $\sum_s a_s(u_j\Phi_{s\lambda_0})(H_0) = 0$ $(1 \le j \le w)$. Thus the function $f = \sum_s a_s\Phi_{s\lambda_0}$ on $\mathfrak{a}^+$ would satisfy $(u_jf)(H_0) = 0$ $(1 \le j \le w)$, and by Proposition 5.4 it is an eigenfunction of each $\Delta(D)$ $D \in \mathbf{D}(G/K)$. But then Proposition 5.1 implies $(Df)(H_0) = 0$ for each $D \in \mathbf{D}(A)$. By the analyticity of f this would imply $f \equiv 0$, contradicting the linear independence of the $\Phi_{s\lambda_0}$.

Putting $D = u_j$ $(1 \le j \le w)$ in (3) we can solve the resulting system with respect to $c(s\lambda)$ and deduce the conclusion of the lemma.

Remark. For a more elementary proof of Lemma 6.1 see Exercise B17.

Next we relate the c-function to the behavior of ϕ_λ at ∞ in A^+. If $\lambda = \xi + i\eta$ $(\xi, \eta \in \mathfrak{a}^*)$ we put $\operatorname{Re}\lambda = \xi$, $\operatorname{Im}\lambda = \eta$. We also denote by $\mathfrak{a}_+^*$ the positive Weyl chamber in $\mathfrak{a}^*$; i.e.,

$$\mathfrak{a}_+^* = \{\lambda \in \mathfrak{a}^* : \langle\lambda, \alpha\rangle > 0 \quad \text{for } \alpha \in \Sigma^+\}.$$

Lemma 6.2. *Let* $\lambda \in {}'\mathfrak{a}_c^*$ *be such that* $\operatorname{Re}(i\lambda) \in \mathfrak{a}_+^*$. *Then if* $H \in \mathfrak{a}^+$

$$\lim_{t \to +\infty} e^{(-i\lambda + \rho)(tH)} \phi_\lambda(\exp tH) = \mathbf{c}(\lambda).$$

Proof. This is a fairly direct consequence of Theorem 5.5. In fact,

$$e^{(-i\lambda+\rho)}\phi_\lambda = \mathbf{c}(\lambda) \sum_\mu \Gamma_\mu(\lambda) e^{-\mu} + \sum_{s \neq e} \mathbf{c}(s\lambda) e^{i(s\lambda - \lambda)} \sum_\mu \Gamma_\mu(s\lambda) e^{-\mu}. \tag{4}$$

Since $\operatorname{Re}(i\lambda) = -\eta$ we know from the proof of [DS] (Theorem 2.22, Chapter VII) that

$$-\eta(H) > s(-\eta)H, \qquad s \neq e.$$

Consequently, if $s \neq e$

$$e^{i(s\lambda - \lambda)(tH)} \to 0 \qquad \text{as} \quad t \to +\infty,$$

so the lemma follows by evaluating (4) for tH and letting $t \to +\infty$.

Remark. Later (in Theorem 6.14) we shall drop the restriction $\lambda \in {}'\mathfrak{a}_c^*$.

For further study of the **c**-function we shall use Theorem 5.20 of Chapter I to transfer the integration over K in Theorem 4.3 to integration over $\bar{N}$. For the Iwasawa decompositions $G = KAN = NAK$ we have

$$g = k(g) \exp H(g) n(g) = n_1(g) \exp A(g) u(g)$$

whence

$$H(g^{-1}) = -A(g), \qquad k(g^{-1}) = u(g)^{-1}. \tag{5}$$

Using Eq. (7) of §4, we therefore have

$$\phi_\lambda(g) = \int_K e^{(i\lambda - \rho)(H(gk))}\, dk$$

and since $H(gkm) = H(gk)$ ($g \in G$, $k \in K$, $m \in M$, the centralizer of A in K) we have by Theorem 5.20 of Chapter I

$$\phi_\lambda(g) = \int_{K/M} e^{(i\lambda - \rho)(H(gk))}\, dk_M$$

$$= \int_{\bar{N}} e^{(i\lambda - \rho)(H(gk(\bar{n})))} e^{-2\rho(H(\bar{n}))}\, d\bar{n}.$$

Put $g = a \in A$ and note that since $\bar{n} \in k(\bar{n}) \exp H(\bar{n}) N$ we have

$$H(ak(\bar{n})) = H(a\bar{n}) - H(\bar{n}) = \log a + H(a\bar{n}a^{-1}) - H(\bar{n}). \tag{6}$$

We have thus proved the following formula for the spherical function.

Proposition 6.3. *Let $\lambda \in \mathfrak{a}_c^*$ and $a \in A$. Then*

$$\phi_\lambda(a) = e^{(i\lambda - \rho)(\log a)} \int_{\bar{N}} e^{(i\lambda - \rho)(H(a\bar{n}a^{-1})) - (i\lambda + \rho)(H(\bar{n}))}\, d\bar{n}.$$

Here the Haar measure on $\bar{N}$ is normalized such that

$$\int_{\bar{N}} e^{-2\rho(H(\bar{n}))}\, d\bar{n} = 1.$$

Each $\bar{n} \in \bar{N}$ can be written uniquely $\bar{n} = \exp X$ where $X \in \theta\mathfrak{n}$; moreover, $X = \sum_{\alpha \in \Sigma^+} x_\alpha X_{-\alpha}$ where $X_{-\alpha} \in \mathfrak{g}_{-\alpha}$. If $a_t = \exp tH$, $H \in \mathfrak{a}^+$, then $a_t \bar{n} a_t^{-1} = \exp(\sum_{\alpha \in \Sigma^+} x_\alpha e^{-t\alpha(H)} X_{-\alpha})$. Hence $\lim_{t \to +\infty} a_t \bar{n} a_t^{-1} = e$, so, ignoring questions of convergence, we would *expect* from Proposition 6.3 and Lemma 6.2,

$$c(\lambda) = \int_{\bar{N}} e^{-(i\lambda + \rho)(H(\bar{n}))}\, d\bar{n} \qquad \text{if} \quad \mathrm{Re}(i\lambda) \in \mathfrak{a}_+^*. \tag{7}$$

After some preparation we shall prove this formula and, on the basis of it, obtain an even more explicit expression for $c(\lambda)$.

2. The Rank-One Case

In this subsection we assume that G has real rank one. Let α denote the single positive indivisible root. Then $\bar{\mathfrak{n}} = \mathfrak{g}_{-\alpha} + \mathfrak{g}_{-2\alpha}$ and the mapping $(X, Y) \to \exp(X + Y) = \bar{n}$ is a diffeomorphism of $\mathfrak{g}_{-\alpha} \times \mathfrak{g}_{-2\alpha}$ onto $\bar{N}$. This is clear from [DS] (Chapter VI, Corollary 4.4 and Chapter IX, Lemma 3.9) ((t, z) is determined by $\bar{n}$) and also from the upcoming Lemma 6.7. From [DS] (Chapter IX, Theorem 3.8) we have

$$e^{\rho(H(\bar{n}))} = [(1 + c|X|^2)^2 + 4c|Y|^2]^{(1/4)(m_\alpha + 2m_{2\alpha})}. \tag{8}$$

Here m_α and $m_{2\alpha}$ are the multiplicities of α and 2α, respectively, $c^{-1} = 4(m_\alpha + 4m_{2\alpha})$ and $|Z|^2 = -\langle Z, \theta Z\rangle$, $\langle\ ,\ \rangle$ being the Killing form.

Theorem 6.4. (*G of real rank one*). *The c-function extends from $\mathfrak{a}_c^*$ to the meromorphic function*

$$c(\lambda) = c_0 \frac{2^{-\langle i\lambda, \alpha_0\rangle}\Gamma(\langle i\lambda, \alpha_0\rangle)}{\Gamma(\frac{1}{2}(\frac{1}{2}m_\alpha + 1 + \langle i\lambda, \alpha_0\rangle))\Gamma(\frac{1}{2}(\frac{1}{2}m_\alpha + m_{2\alpha} + \langle i\lambda, \alpha_0\rangle))}, \tag{9}$$

where $\alpha_0 = \alpha/\langle \alpha, \alpha\rangle$ and

$$c_0 = \Gamma(\tfrac{1}{2}(m_\alpha + m_{2\alpha} + 1))2^{(1/2)m_\alpha + m_{2\alpha}}.$$

Moreover, formula (7) *holds, the integral being absolutely convergent if and only if* $\mathrm{Re}(i\lambda) \in \mathfrak{a}_+^*$.

Proof. It is clear from Theorem 1.14, Chapter I and [DS] (Chapter VI, Corollary 4.4) that $\exp^*(d\bar{n})$ is a constant multiple of the Euclidean measure $dX\,dY$. Let $z = x + iy \in \boldsymbol{C}$ be determined by $\lambda = z\rho$. Then we have for $\operatorname{Re}(iz) > 0$, $P = \frac{1}{4}(m_\alpha + 2m_{2\alpha})$, $c_1 = \text{const.}$,

$$\int_{\bar{N}} e^{-(i\lambda+\rho)(H(\bar{n}))}\,d\bar{n}$$

$$= c_1 \int_{\mathfrak{g}_{-\alpha}} \int_{\mathfrak{g}_{-2\alpha}} [(1 + c|X|^2)^2 + 4c|Y|^2]^{-P(iz+1)}\,dX\,dY$$

$$= c_2 \int_0^\infty \int_0^\infty [(1 + r)^2 + s]^{-P(iz+1)} r^{(1/2)m_\alpha - 1} s^{(1/2)m_{2\alpha} - 1}\,dr\,ds,$$

where c_2 is a constant. Here we first evaluate the integral over s by introducing the variable t by $s = (1 + r)^2 t$. Then we get

$$(10)\qquad \int_{\bar{N}} e^{-(i\lambda+\rho)(H(\bar{n}))}\,d\bar{n}$$

$$= c_2 \int_0^\infty r^{(1/2)m_\alpha - 1}(1 + r)^{-2P(iz+1)+m_{2\alpha}}\,dr \int_0^\infty t^{(1/2)m_{2\alpha} - 1}(1 + t)^{-P(iz+1)}\,dt,$$

and the convergence statement is now obvious. Using the formula

$$(11)\qquad \int_0^\infty v^{x-1}(1 + v)^{-x-y}\,dv = \frac{\Gamma(x)\Gamma(y)}{\Gamma(x + y)}$$

wc obtain (with a new constant c_3)

$$\int_{\bar{N}} e^{-(i\lambda+\rho)(H(\bar{n}))}\,d\bar{n} = c_3 \frac{\Gamma(\frac{1}{2}(\langle i\lambda + \rho, \alpha_0\rangle - m_{2\alpha}))\Gamma(\langle i\lambda, \alpha_0\rangle)}{\Gamma(\frac{1}{2}\langle i\lambda + \rho, \alpha_0\rangle)\Gamma(\langle i\lambda + \rho, \alpha_0\rangle - m_{2\alpha})},$$

which by the duplication formula

$$(12)\qquad \Gamma(2z) = 2^{2z-1}\pi^{-1/2}\Gamma(z)\Gamma(z + \tfrac{1}{2})$$

and the given normalization of $d\bar{n}$ reduces to the right-hand side of (9), the constants being determined by $\boldsymbol{c}(-i\rho) = 1$.

It remains to prove (7) for $\operatorname{Re}(iz) = -y > 0$. Select ε $(0 < \varepsilon < 1)$ so small that $1 + \varepsilon y > 0$ and put $a = \exp tH$ $(t > 0,\ H \in \mathfrak{a}^+)$ in Proposition 6.3. By (8), $0 \le \rho(H(a\bar{n}a^{-1})) \le \rho(H(\bar{n}))$, so the integrand in Proposition 6.3 is majorized by

$$e^{-(y+1)\rho(H(a\bar{n}a^{-1}))} e^{(y-1)\rho(H(\bar{n}))} \le e^{(-y+\varepsilon y)\rho(H(a\bar{n}a^{-1}))} e^{(y-1)\rho(H(\bar{n}))}$$

$$\le e^{(-y+\varepsilon y)\rho(H(\bar{n}))} e^{(y-1)\rho(H(\bar{n}))} = e^{(\varepsilon y-1)\rho(H(\bar{n}))},$$

which as we saw is integrable. Combining Lemma 6.2 and Proposition 6.3 we can let $t \to +\infty$ under the integral sign and derive (7). This proves Theorem 6.4.

3. Properties of $H(\bar{n})$

For the justification of (7) in general we need more information about $H(\bar{n})$. Given the positive Weyl chamber $\mathfrak{a}^+$ let ${}^+\mathfrak{a}$ denote the dual cone defined by

$$^+\mathfrak{a} = \{H \in \mathfrak{a} : B(H, H') > 0 \text{ for all } H' \in \mathfrak{a}^+\},$$

and let $\overline{\mathfrak{a}^+}$ and $\overline{^+\mathfrak{a}}$ denote their respective closures. We know from [DS] (Lemma 2.20, Chapter VII) that $\mathfrak{a}^+ \subset {}^+\mathfrak{a}$. Also,

$$(13) \qquad \overline{^+\mathfrak{a}} = \left\{\sum_{j=1}^{l} a_j A_{\alpha_j} \neq 0 : a_j \geq 0,\ 1 \leq j \leq l\right\},$$

$\alpha_1, \ldots, \alpha_l$ being the simple restricted roots. Thus ${}^+\mathfrak{a} = \overline{^+\mathfrak{a}} - \{0\}$.

Lemma 6.5. *Let $a \in A^+$. Then*

$$\log a - H(ak) \in \overline{^+\mathfrak{a}} \qquad \textit{for each } k \in K.$$

Proof. Extend $\mathfrak{a}$ as usual to a Cartan subalgebra $\mathfrak{h}^C$ of $\mathfrak{g}^C$ (C indicating complexification) let $\mathfrak{h}_R \subset \mathfrak{h}^C$ be the subset where all roots $\alpha \in \Delta = \Delta(\mathfrak{g}^C, \mathfrak{h}^C)$ are real and consider an ordering of the dual of $\mathfrak{h}_R$ compatible with our ordering of the dual $\mathfrak{a}^*$ of $\mathfrak{a}$ ([DS], Chapter VI, §3 and Chapter IX, §1). Let Δ^+ denote the set of positive roots.

For the proof of Lemma 6.5 we may assume that G is the analytic subgroup of the simply connected complex group G^C. In fact, $H(akz) = H(ak)$ for all elements z of the center of G. Let U be the analytic subgroup of G^C with Lie algebra $\mathfrak{u} = \mathfrak{k} + i\mathfrak{p}$. Then U is compact, so we are automatically preserving our assumption that K is compact. Our choice of G permits us to identify finite-dimensional representations π of G on a complex vector space V with such representations of $\mathfrak{g}$, G^C, $\mathfrak{g}^C$, U, and $\mathfrak{u}$. We can always introduce a norm $|\ |$ on V such that $\pi(u)$ is unitary for each $u \in U$.

Now we fix such an irreducible π and let Λ denote its highest weight (cf. Chapter V, §1). Every other weight Λ' then has the form

$$\Lambda' = \Lambda - m_1\beta_1 - \cdots - m_p\beta_p \qquad (m_i \in \mathbf{Z}^+, \beta_i \in \Delta^+).$$

By the compatibility of the orderings we have $\Lambda(H) \geq \Lambda'(H)$ for $H \in \mathfrak{a}^+$. Since det $\pi(g) = 1$ for all g in the commutator group of G^C, hence for all $g \in G^C$ we know that $\pi(H)$ is not only self-adjoint but has trace 0. Thus its eigenvalues are real and add up to 0; hence $\Lambda(H) \geq 0$ for $H \in \mathfrak{a}^+$.

Let $v \in V$ be a unit vector of weight Λ. Let $a \in A^+$. Since $e^{\Lambda(\log a)}$ is the maximal eigenvalue of $\pi(a)$ we have

$$(14) \qquad |\pi(ak)v| = |\pi(a)\pi(k)v| \leq e^{\Lambda(\log a)}|\pi(k)v| = e^{\Lambda(\log a)}.$$

On the other hand, $ak = k_1 \exp H(ak)n_1$ and $\pi(n_1)v = v$. Thus

$$|\pi(ak)v| = e^{\Lambda(H(ak))}, \tag{15}$$

so by (14)

$$\Lambda(\log a - H(ak)) \geq 0, \qquad a \in A^+,\ k \in K. \tag{16}$$

According to Theorem 4.1 of Chapter V we can for each i ($1 \leq i \leq l$) choose $\pi = \pi_i$ such that its highest weight $\Lambda = \Lambda_i$ satisfies

$$\Lambda_i(\mathfrak{h}^{\mathbf{C}} \cap \mathfrak{k}) = 0, \qquad \Lambda_i(A_{\alpha_j}) = c_j \delta_{ij}, \tag{17}$$

where $c_j > 0$ and A_{α_j} as in (13). Writing

$$\log a - H(ak) = \sum_j b_j A_{\alpha_j},$$

we deduce from (17) and (16) that $b_j \geq 0$, and this proves the lemma.

Corollary 6.6. *Let $\bar{n} \in \bar{N}$ and $a \in A^+$. Then*

$$H(\bar{n}) \in \overline{{}^+\mathfrak{a}}, \qquad H(\bar{n}) - H(a\bar{n}a^{-1}) \in \overline{{}^+\mathfrak{a}}.$$

The second relation is an immediate consequence of (6) and Lemma 6.5. The first relation follows by taking $a = a_t = \exp tH$ in the second one and letting $t \to \infty$.

4. Integrals over Nilpotent Groups

As before let M and M' denote the centralizer and normalizer, respectively, of A in K and $W = M'/M$ the Weyl group. Fix $s \in W$ and select a representative $m_s \in M'$. Consider the groups

$$N^s = m_s N m_s^{-1}, \qquad \bar{N}^s = m_s \bar{N} m_s^{-1}, \qquad \bar{N}_s = \bar{N} \cap N^{s^{-1}}$$

whose Lie algebras, respectively, are given by

$$\mathfrak{n}^s = \sum_{s^{-1}\beta > 0} \mathfrak{g}_\beta, \qquad \bar{\mathfrak{n}}^s = \sum_{s^{-1}\beta < 0} \mathfrak{g}_\beta, \qquad \bar{\mathfrak{n}}_s = \sum_{\beta < 0,\, s\beta > 0} \mathfrak{g}_\beta,$$

where $\mathfrak{g} = \mathfrak{g}_0 + \sum_{\beta \in \Sigma} \mathfrak{g}_\beta$ is the usual root-space decomposition of $\mathfrak{g}$ (cf. [DS], Chapter IX, §1). In order to study the integral (7) it is convenient (for the sake of induction proofs) to consider the more general integral

$$c_s(\lambda) = \int_{\bar{N}_s} e^{-(i\lambda + \rho)(H(\bar{n}_s))}\, d\bar{n}_s \tag{18}$$

where $d\bar{n}_s$ is a Haar measure on $\bar{N}_s$.

We need some simple lemmas about nilpotent groups.

Lemma 6.7. *Let N be a connected nilpotent Lie group with Lie algebra $\mathfrak{n}$. Then* $(\mathfrak{n}, \exp)$ *is a covering space of N. If N is simply connected, the map* $\exp: \mathfrak{n} \to N$ *is a diffeomorphism.*

Proof. We know from [DS] (Chapter VI, Corollary 4.4) that

$$\exp: \mathfrak{n} \to N$$

is a regular mapping of $\mathfrak{n}$ onto N. Let $\mathfrak{c}$ denote the center of $\mathfrak{n}$ and put $\mathfrak{n}_{\mathfrak{c}} = \{X \in \mathfrak{c}: \exp X = e\}$. Clearly

$$\exp(X + Z) = \exp X \exp Z = \exp X \qquad \text{if} \quad X \in \mathfrak{n},\ Z \in \mathfrak{n}_{\mathfrak{c}},$$

and since $\exp_{\mathfrak{n}} | \mathfrak{c} = \exp_{\mathfrak{c}}$, $\mathfrak{n}_{\mathfrak{c}}$ is a discrete subgroup of $\mathfrak{n}$.

On the other hand, if $X, Y \in \mathfrak{n}$ such that $\exp X = \exp Y$ then $e^{\operatorname{ad} X} = e^{\operatorname{ad} Y}$, so by [DS] (Chapter VI, Lemma 4.5), $\operatorname{ad} X = \operatorname{ad} Y$. Thus $Y = X + Z$, $Z \in \mathfrak{c}$, so $\exp X = \exp X \exp Z$, and thus $Z \in \mathfrak{n}_{\mathfrak{c}}$. Thus

$$\exp^{-1}(\exp X) = \{X + Z : Z \in \mathfrak{n}_{\mathfrak{c}}\}$$

and the lemma follows.

Lemma 6.8. *Let N be a connected, simply connected nilpotent Lie group with Lie algebra $\mathfrak{n}$, $\mathfrak{n}^{(i)}$ a strictly decreasing sequence of ideals of $\mathfrak{n}$ such that $[\mathfrak{n}, \mathfrak{n}^{(i)}] \subset \mathfrak{n}^{(i+1)}$, $\mathfrak{n}_1$ and $\mathfrak{n}_2$ two mutually complementary subspaces of $\mathfrak{n}$ such that*

$$\mathfrak{n}^{(i)} = \mathfrak{n}_1 \cap \mathfrak{n}^{(i)} + \mathfrak{n}_2 \cap \mathfrak{n}^{(i)} \qquad (i = 0, 1, \dots ; \mathfrak{n}^{(0)} = \mathfrak{n}).$$

Then the mapping

$$\phi: (X, Y) \to \exp X \exp Y$$

is an analytic diffeomorphism of $\mathfrak{n}_1 \times \mathfrak{n}_2$ onto N.

Proof. Let s be the biggest index such that $\mathfrak{n}^{(s)} \neq 0$. Then $\mathfrak{n}^{(s)}$ belongs to the center of $\mathfrak{n}$ and the corresponding analytic subgroup $N^{(s)}$ of N is, by Lemma 6.7, a closed and simply connected central subgroup of N; the factor group $N/N^{(s)}$ is simply connected. We shall first prove, by induction on s, that the mapping ϕ is a bijection. By the induction hypothesis we may assume this holds for the complementary subspaces $\mathfrak{n}_1/(\mathfrak{n}_1 \cap \mathfrak{n}^{(s)})$ and $\mathfrak{n}_2/(\mathfrak{n}_2 \cap \mathfrak{n}^{(s)})$ of the Lie algebra $\mathfrak{n}/\mathfrak{n}^{(s)}$ of $N/N^{(s)}$. This implies that if $\mathfrak{n}_1'$ and $\mathfrak{n}_2'$ are complementary subspaces of $\mathfrak{n}_1 \cap \mathfrak{n}^{(s)}$ and $\mathfrak{n}_2 \cap \mathfrak{n}^{(s)}$ in $\mathfrak{n}_1$ and $\mathfrak{n}_2$, respectively, then the mapping $(X, Y) \to \exp X \exp Y$ is a bijection of $\mathfrak{n}_1' \times \mathfrak{n}_2'$ onto the set $N' = \exp \mathfrak{n}_1' \exp \mathfrak{n}_2'$. A second implication is that the mapping $(n', z) \to n'z$ is a bijection of $N' \times N^{(s)}$ onto N. But $N^{(s)}$ being a vector group, the mapping $(X, Y) \to \exp X \exp Y$ is a bijection of $(\mathfrak{n}_1 \cap \mathfrak{n}^{(s)}) \times (\mathfrak{n}_2 \cap \mathfrak{n}^{(s)})$ onto $N^{(s)}$. Putting these statements together, we find that ϕ is a bijection.

Finally, ϕ is regular. For this let $X_0 \neq 0$ in $\mathfrak{n}_1$, $Y_0 \neq 0$ in $\mathfrak{n}_2$. Then

$$d\phi_{(X,Y)}(X_0, Y_0) = d\phi_{(X,Y)}(X_0, 0) + d\phi_{(X,Y)}(0, Y_0).$$

Since $d\phi_{(X,Y)}(X_0, 0)$ is the tangent vector to the curve

$$t \to \exp(X + tX_0) \exp Y$$

at $t = 0$, we have from [DS] (Chapter II, Theorem 1.7) and the relation $dR_x = dL_x \circ Ad(x^{-1})$

$$d\phi_{(X,Y)}(X_0, 0) = dL_{\exp X \exp Y} e^{-\operatorname{ad} Y}\left(\frac{1 - e^{-\operatorname{ad} X}}{\operatorname{ad} X}\right)(X_0). \tag{19}$$

Similarly,

$$d\phi_{(X,Y)}(0, Y_0) = dL_{\exp X \exp Y}\left(\frac{1 - e^{-\operatorname{ad} Y}}{\operatorname{ad} Y}\right)(Y_0). \tag{20}$$

If $d\phi_{(X,Y)}(X_0, Y_0) = 0$, we thus conclude

$$\frac{1 - e^{-\operatorname{ad} X}}{\operatorname{ad} X}(X_0) = \frac{1 - e^{\operatorname{ad} Y}}{\operatorname{ad} Y}(Y_0). \tag{21}$$

Let i be the largest index such that $X_0, Y_0 \in \mathfrak{n}^{(i)}$. Then (21) implies

$$X_0 - Y_0 \in \mathfrak{n}^{(i+1)} = \mathfrak{n}_1 \cap \mathfrak{n}^{(i+1)} + \mathfrak{n}_2 \cap \mathfrak{n}^{(i+1)},$$

so we get the contradiction $X_0, Y_0 \in \mathfrak{n}^{(i+1)}$.

5. The Weyl Group Acting on the Root System

Next we need some results about the Weyl group W. As before, Σ denotes the set of roots, Σ_0 the set of indivisible roots ([DS], Chapter IX, §2), Σ^+ the set of positive roots, $\Sigma^- = -\Sigma^+$, and

$$\Sigma_0^+ = \Sigma_0 \cap \Sigma^+, \qquad \Sigma_0^- = \Sigma_0 \cap \Sigma^-.$$

If $\alpha \in \Sigma_0^+$ put $(\alpha) = \Sigma^+ \cap \{\alpha, 2\alpha\}$ (i.e., the set of positive roots proportional to α), and if $\beta \in \Sigma_0^-$ put $(\beta) = \Sigma^- \cap \{\beta, 2\beta\}$. By [DS] (Chapter X, Lemma 3.11) we have

$$s_{\alpha_i}(\Sigma_0^+ - \{\alpha_i\}) = \Sigma_0^+ - \{\alpha_i\} \qquad \text{if } \alpha_i \text{ is simple.} \tag{22}$$

We shall now generalize this result. For $s \in W$ let $\Sigma_s^+ = \Sigma_0^+ \cap s^{-1}\Sigma_0^-$ and let $n(s)$ denote the cardinality of the set Σ_s^+. Note that Σ_s^+ consists of the elements in Σ_0^+ which separate $\mathfrak{a}^+$ and $s^{-1}\mathfrak{a}^+$.

Lemma 6.9. *Let $s \in W$, let $\alpha_i \in \Sigma$ be a simple root, and put $s_i = s_{\alpha_i}$. Then*

(i) $s_i(\Sigma_s^+ - \{\alpha_i\}) = \Sigma_{ssi}^+ - \{\alpha_i\}$.

(ii) *α_i is contained in exactly one of the sets Σ_s^+, $\Sigma_{ss_i}^+$.*

(iii) $n(s) - 1 = n(ss_i)$ *if* $\alpha_i \in \Sigma_s^+$,
$n(s) + 1 = n(ss_i)$ *if* $\alpha_i \notin \Sigma_s^+$.

Proof. (i) If $\alpha \in \Sigma_s^+ - \{\alpha_i\}$ then by (22), $s_i\alpha \in \Sigma^+$. Since $ss_is_i\alpha = s\alpha \in \Sigma^-$ we have $s_i\alpha \in \Sigma_{ss_i}^+$. Also, $s_i\alpha \neq \alpha_i$ because $\alpha \neq -\alpha_i = s_i\alpha_i$. Thus we have $s_i(\Sigma_s^+ - \{\alpha_i\}) \subset \Sigma_{ss_i}^+ - \{\alpha_i\}$. Replacing here s by ss_i we get

$$s_i(\Sigma_{ss_i}^+ - \{\alpha_i\}) \subset \Sigma_s^+ - \{\alpha_i\}.$$

This proves (i).

(ii) Suppose $\alpha_i \in \Sigma_s^+$, $\alpha_i \in \Sigma_{ss_i}^+$. Then $s\alpha_i \in \Sigma_0^-$, $-s\alpha_i = ss_i\alpha_i \in \Sigma_0^-$, which is a contradiction. The possibility $\alpha_i \notin \Sigma_s^+$, $\alpha_i \notin \Sigma_{ss_i}^+$ leads to a similar contradiction. Thus if $\alpha_i \in \Sigma_s^+$ we have $\alpha_i \notin \Sigma_{ss_i}^+$, so $\Sigma_{ss_i}^+ = \Sigma_{ss_i}^+ - \{\alpha_i\}$, whence $n(s) - 1 = n(ss_i)$ by (i). Similarly, if $\alpha_i \notin \Sigma_s^+$ then $\alpha_i \in \Sigma_{ss_i}^+$ and (i) implies $n(s) + 1 = n(ss_i)$. This proves (ii), (iii) and the lemma.

Each $s \in W$ can be written $s = s_{\alpha_1} \cdots s_{\alpha_p}$ where each α_i is a simple root. This expression is said to be *reduced* if p is as small as possible. This minimal p is called the *length* $l(s)$ of s (cf. [DS], Chapter VII, §7, where such a notion is considered for the affine Weyl group).

Proposition 6.10. *We have*

$$n(s) = l(s) \qquad \textit{for} \quad s \in W.$$

Proof. Let $k = n(s)$. If $k = 0$ then $s = e$ so $l(s) = 0 = n(s)$. Proceeding by induction on k we assume $n(s) = l(s)$ holds for $s \in W$, $n(s) \leq k - 1$. Let $s \in W$, $n(s) = k > 0$. Then $s \neq e$ so $s\alpha_i \in \Sigma^-$ for some simple root α_i; that is, $\alpha_i \in \Sigma_s^+$. Let $s' = ss_i$. By 6.9(iii) above $n(s') = n(s) - 1 = k - 1$. By the inductive assumption, $n(s') = l(s')$. Since $s = s's_i$ and $l(s') = k - 1$ we get $l(s) \leq k = n(s)$. Thus it remains to prove $n(s) \leq l(s)$. Writing $p = l(s)$, $s = s_1 \cdots s_p$ we have by (iii) $n(\sigma s_i) \leq n(\sigma) + 1$ for each $\sigma \in W$. In particular,

$$n(s_1 \cdots s_p) \leq n(s_1 \cdots s_{p-1}) + 1 \leq \cdots \leq n(s_1) + p - 1 = p$$

by (22); thus $n(s) \leq l(s)$ as desired.

Corollary 6.11. *Let $s \neq e$ in W and let $s = s_{\alpha_1} \cdots s_{\alpha_p}$ be a reduced expression. Then $\beta_1 = -s^{-1}\alpha_1$ is positive. More generally, if*

$$s^{(q)} = s_{\alpha_{q+1}} \cdots s_{\alpha_p}, \qquad \beta_{q+1} = -s^{(q)^{-1}}\alpha_{q+1}\ (= s_{\alpha_p} \cdots s_{\alpha_{q+2}}\alpha_{q+1}) \tag{23}$$

where $0 \leq q \leq p - 1$, then

$$\Sigma_s^+ = \{\beta_1, \ldots, \beta_p\}. \tag{24}$$

Moreover, $s^{-1} = s_{\beta_1} \cdots s_{\beta_p}$ *and*

$$s_{\beta_p} \cdots s_{\beta_j}\beta_i > 0 \qquad \textit{for} \quad 1 \leq i < j \leq p.$$

Proof. We have $\beta_1 = s^{-1}s_{\alpha_1}\alpha_1 = s_{\alpha_p} \cdots s_{\alpha_2}\alpha_1$ and the element $s_{\alpha_p} \cdots s_{\alpha_2}$ has length $p-1$. Thus by Proposition 6.10 $n(s^{-1}s_{\alpha_1}) = p-1$, whence by Lemma 6.9(iii), $\alpha_1 \in \Sigma^+_{s^{-1}}$, so $\beta_1 > 0$. Since the expression (23) for $s^{(q)}$ is reduced we have, more generally, $\beta_{q+1} > 0$. We also have by (22), since $s^{-1} = (s^{(1)})^{-1}s_{\alpha_1}$,

$$s^{-1}(\Sigma_0^- - \{-\alpha_1\}) = (s^{(1)})^{-1}(\Sigma_0^- - \{-\alpha_1\});$$

that is,

$$s^{-1}\Sigma_0^- - \{\beta_1\} = (s^{(1)})^{-1}\Sigma_0^- - \{-\beta_1\}.$$

Intersecting with Σ_0^+ we obtain the disjoint union

$$\Sigma_s^+ = \Sigma_{s^{(1)}}^+ \cup \{\beta_1\}, \tag{25}$$

and now (24) follows by iteration.

Next, the relation $\beta_{q+1} = s_{\alpha_p} \cdots s_{\alpha_{q+2}}\alpha_{q+1}$ implies

$$s_{\beta_{q+1}} = s_{\alpha_p} \cdots s_{\alpha_{q+2}}s_{\alpha_{q+1}}s_{\alpha_{q+2}} \cdots s_{\alpha_p},$$

whence

$$s_{\beta_1}s_{\beta_2} \cdots s_{\beta_p} = s_{\alpha_p} \cdots s_{\alpha_2}s_{\alpha_1} = s^{-1}.$$

Finally,

$$\begin{aligned} s_{\beta_p} \cdots s_{\beta_j}\beta_i &= s_{\alpha_p} \cdots (s_{\alpha_p} \cdots s_{\alpha_{j+1}}s_{\alpha_j}s_{\alpha_{j+1}} \cdots s_{\alpha_p})s_{\alpha_p} \cdots s_{\alpha_{i+1}}\alpha_i \\ &= s_{\alpha_{j-1}}s_{\alpha_{j-2}} \cdots s_{\alpha_{i+1}}\alpha_i \end{aligned}$$

is positive, as we see by using the first part of the proof on the reduced expression $\sigma_{i,j} = s_{\alpha_i}s_{\alpha_{i+1}} \cdots s_{\alpha_{j-1}}$. This proves the corollary.

Remark. By (24) the planes $\beta_p = 0, \ldots, \beta_1 = 0$ are exactly the root planes separating $\mathfrak{a}^+$ and $s^{-1}\mathfrak{a}^+$. Consider a line segment from a point in $\mathfrak{a}^+$ to a point in $s^{-1}\mathfrak{a}^+$ such that its intersections with the planes $\beta_p = 0, \ldots, \beta_1 = 0$ are all different. The relations

$$s^{-1} = s_{\beta_1} \cdots s_{\beta_p},$$

$$\beta_i(s_{\beta_j} \cdots s_{\beta_p}(H)) > 0 \qquad \text{for} \quad H \in \mathfrak{a}^+,\ i < j,$$

express geometrically the passage from $\mathfrak{a}^+$ to $s^{-1}\mathfrak{a}^+$ (cf. [DS], Chapter VII, Lemma 7.3).

6. The Rank-One Reduction. The Product Formula of Gindikin–Karpelevic

With $s \neq e$ in W as in Corollary 6.11 we write now β for β_1 and consider as in [DS] (Chapter IX, §2) the subalgebra $\mathfrak{g}_{(\beta)}$ of $\mathfrak{g}$ generated by the root spaces $\mathfrak{g}_\beta$ and $\mathfrak{g}_{-\beta}$. Let $G_\beta \subset G$ denote the analytic subgroup corresponding to $\mathfrak{g}_{(\beta)}$. Then G_β is semisimple and if $\mathfrak{k}_\beta = \mathfrak{k} \cap \mathfrak{g}_{(\beta)}$, $\mathfrak{p}_\beta = \mathfrak{p} \cap \mathfrak{g}_{(\beta)}$ then $\mathfrak{g}_{(\beta)} = \mathfrak{k}_\beta + \mathfrak{p}_\beta$ is a Cartan decomposition. The space $\mathfrak{a}_\beta = \boldsymbol{R}A_\beta$ is a maximal abelian subspace of $\mathfrak{p}_\beta$. Let K_β and $A_{(\beta)}$ be the analytic subgroups of G_β corresponding to $\mathfrak{k}_\beta$ and $\mathfrak{a}_\beta$, respectively, and let $G_\beta = K_\beta A_{(\beta)} N_\beta$ be the Iwasawa decomposition corresponding to $\boldsymbol{R}^+ A_\beta$ as a positive Weyl chamber in $\mathfrak{a}_\beta$. The group N_β has Lie algebra $\mathfrak{n}_\beta = \mathfrak{g}_\beta + \mathfrak{g}_{2\beta}$. We put $\overline{N}_\beta = \theta N_\beta$, $\bar{\mathfrak{n}}_\beta = \theta \mathfrak{n}_\beta$, and let M_β denote the centralizer of $A_{(\beta)}$ in K_β. Note that the decompositions $G = KAN$ and $G_\beta = K_\beta A_{(\beta)} N_\beta$ are compatible.

Because of (25) we have the direct vector space decomposition of subalgebras

$$\bar{\mathfrak{n}}_s = \bar{\mathfrak{n}}_{s'} + \bar{\mathfrak{n}}_\beta, \tag{26}$$

where for simplicity we write s' for $s^{(1)}$, β for β_1. If $\gamma_1 < \gamma_2 < \cdots$ are the elements of Σ^+ in increasing order and we write $\mathfrak{n}^{(i)} = \bar{\mathfrak{n}}_s \cap (\sum_{j \geq i} \mathfrak{g}_{-\gamma_j})$ then the assumptions of Lemma 6.8 are satisfied with the subspaces $\mathfrak{n}_1 = \bar{\mathfrak{n}}_{s'}$, $\mathfrak{n}_2 = \bar{\mathfrak{n}}_\beta$. Hence we obtain the global decomposition

$$\overline{N}_s = \overline{N}_{s'} \overline{N}_\beta = \overline{N}_\beta \overline{N}_{s'}, \tag{27}$$

the product maps being diffeomorphisms. Moreover, the Haar measures $d\bar{n}_s$, $d\bar{n}_{s'}$, and $d\bar{n}_\beta$, suitably normalized, match under the product maps. This is a simple consequence of Chapter I, Lemma 1.11, because (19) and (20) show that the product map has Jacobian of determinant one.

Similarly, we use Lemma 6.8 on the ideals

$$\mathfrak{n}^{(i)} = \mathfrak{n}^{(s')^{-1}} \cap \operatorname{Ad}(m_{s'}^{-1})\Big(\sum_{j \geq i} \mathfrak{g}_{\gamma_j}\Big)$$

in $\mathfrak{n}^{(s')^{-1}}$. Then we obtain† with $\mathfrak{n}_1 = \mathfrak{n}^{(s')^{-1}} \cap \bar{\mathfrak{n}}$, $\mathfrak{n}_2 = \mathfrak{n}^{(s')^{-1}} \cap \mathfrak{n}$

$$N^{(s')^{-1}} = (N^{(s')^{-1}} \cap \overline{N})(N^{(s')^{-1}} \cap N) = \overline{N}_{s'}(N^{(s')^{-1}} \cap N) \tag{28}$$

in the sense that the product map is a diffeomorphism under which the Haar measures correspond. Thus we have the manifold identification

$$\overline{N}_{s'} = N^{(s')^{-1}}/(N^{(s')^{-1}} \cap N), \tag{29}$$

under which the Haar measure on the left corresponds to the invariant measure on the right. We shall now use the decompositions (27) and

†See "Some Details," p. 611.

(28) on the integral $c_s(\lambda)$ in (18). Let $\mathfrak{a}_s^*$ denote the set of $\lambda \in \mathfrak{a}_c^*$ for which (18) converges absolutely. If $\lambda \in \mathfrak{a}_s^*$ we have by (27)

$$c_s(\lambda) = \int_{\bar{N}_s} e^{-(i\lambda+\rho)(H(\bar{n}_s))}\, d\bar{n}_s$$

$$= \int_{\bar{N}_\beta \bar{N}_{s'}} e^{-(i\lambda+\rho)(H(\bar{n}_\beta \bar{n}_{s'}))}\, d\bar{n}_\beta\, d\bar{n}_{s'} = \int_{\bar{N}_\beta \bar{N}_{s'}} e^{-(i\lambda+\rho)(H(a_\beta n_\beta \bar{n}_{s'}))}\, d\bar{n}_\beta\, d\bar{n}_{s'},$$

where in accordance with the decomposition $G_\beta = K_\beta A_{(\beta)} N_\beta$ we have written $\bar{n}_\beta = k_\beta a_\beta n_\beta$. Since $s'\beta > 0$ we have $N_\beta \subset N^{(s')^{-1}}$, so by (28) we have a unique decomposition

$$n_\beta \bar{n}_{s'} n_\beta^{-1} = (\bar{n}_{s'})^* n. \tag{30}$$

The mapping $\bar{n}_{s'} \to (\bar{n}_{s'})^*$ of $\bar{N}_{s'}$ onto itself corresponds via (29) to the diffeomorphism of the quotient space (29) induced by the inner automorphism $n' \to n_\beta n' n_\beta^{-1}$ of $N^{(s')^{-1}}$ leaving $N^{(s')^{-1}} \cap N$ invariant. By unipotency of $\mathrm{Ad}(n_\beta)$ this diffeomorphism preserves the quotient measure and thus the map $\bar{n}_{s'} \to (\bar{n}_{s'})^*$ is measure preserving. Therefore, since

$$H(a_\beta n_\beta \bar{n}_{s'}) = H(a_\beta (\bar{n}_{s'})^*), \tag{31}$$

we can remove n_β in the last integral. Thus

$$c_s(\lambda) = \int_{\bar{N}_\beta \bar{N}_{s'}} e^{-(i\lambda+\rho)(H(a_\beta \bar{n}_{s'}))}\, d\bar{n}_\beta\, d\bar{n}_{s'},$$

which, since $H(\bar{n}_\beta) = \log a_\beta$, equals

$$\int_{\bar{N}_\beta \bar{N}_{s'}} e^{-(i\lambda+\rho)(H(a_\beta \bar{n}_{s'} a_\beta^{-1}))} e^{-(i\lambda+\rho)(H(\bar{n}_\beta))}\, d\bar{n}_\beta\, d\bar{n}_{s'}.$$

However, by Chapter I, §5 (2), we have

$$\frac{d(a_\beta \bar{n}_{s'} a_\beta^{-1})}{d\bar{n}_{s'}} = e^{2\nu(\log a_\beta)}, \qquad \text{where} \quad 2\nu = \sum_{\alpha<0,\, s'\alpha>0} \alpha, \tag{32}$$

so we obtain

$$c_s(\lambda) = c_{s'}(\lambda) \int_{\bar{N}_\beta} e^{-(i\lambda+2\nu+\rho)(H(\bar{n}_\beta))}\, d\bar{n}_\beta. \tag{33}$$

Lemma 6.12. *The restriction of $2\nu + \rho$ to $\mathbf{R}A_\beta$ equals ρ_β, the ρ-function for the rank-one space G_β/K_β.*

Proof. By the definition of ν we have $2\nu = (s')^{-1}\rho - \rho$ so

$$(2\nu + \rho)(A_\beta) = \rho(A_{\alpha_1}).$$

Since α_1 is simple, s_{α_1} permutes $\Sigma^+ - (\alpha_1)$ so if

$$\rho^{\alpha_1} = \frac{1}{2} \sum_{\gamma \in (\alpha_1)} m_\gamma \gamma$$

$$(\rho - \rho_{\alpha_1})(A_{\alpha_1}) = s_{\alpha_1}(\rho - \rho^{\alpha_1})(A_{\alpha_1}) = (\rho - \rho_{\alpha_1})(-A_{\alpha_1}).$$

Consequently,

(34) $$\rho(A_{\alpha_1}) = \rho_{\alpha_1}(A_{\alpha_1}).$$

Since $\rho_{\alpha_1}(A_{\alpha_1}) = \rho_\beta(A_\beta)$ the lemma is proved. We can now state the basic result for the integral $c_s(\lambda)$.

Theorem 6.13. *Let $s \in W$ and let*

(35) $$\Sigma_0^+ \cap s^{-1}\Sigma_0^- = \{\beta_1, \ldots, \beta_p\}, \qquad \bar{N}_s = \exp\left(\sum_{\beta \in \Sigma^+ \cap s^{-1}\Sigma^-} \mathfrak{g}_{-\beta}\right).$$

Then the integral

(36) $$c_s(\lambda) = \int_{\bar{N}_s} e^{-(i\lambda + \rho)(H(\bar{n}_s))} \, d\bar{n}_s$$

converges absolutely if and only if

(37) $$\langle \text{Re}(i\lambda), \beta \rangle > 0 \qquad \textit{for} \quad \beta \in \Sigma^+ \cap s^{-1}\Sigma^-,$$

and if $d\bar{n}_s$ is suitably normalized,

(38) $$c_s(\lambda) = \prod_{\alpha \in \Sigma_0^+ \cap s^{-1}\Sigma_0^-} c_\alpha(\lambda_\alpha),$$

where c_α is the c-function for the rank-one space G_α/K_α and λ_α denotes the restriction of λ to $\mathfrak{a}_\alpha = \boldsymbol{R}A_\alpha$.

Proof. Using Lemma 6.12, formula (33) reads $c_s(\lambda) = c_{s'}(\lambda) c_\beta(\lambda_\beta) a_\beta$, where a_β is a constant. The product formula (38) now follows by iteration. The convergence statement follows by the Fubini theorem if we take the analogous statement of Theorem 6.4 into account.

If we substitute into (38) the value of $c_\alpha(\lambda_\alpha)$ given by Theorem 6.4 we obtain a product formula for $c_s(\lambda)$. If $\langle \ , \ \rangle_\alpha$ is the Killing form for G_α we shall encounter the expression

$$\left\langle i\lambda_\alpha, \frac{\alpha_\alpha}{\langle \alpha_\alpha, \alpha_\alpha \rangle_\alpha} \right\rangle_\alpha,$$

which, however, equals $\langle i\lambda, \alpha \rangle / \langle \alpha, \alpha \rangle$. Thus we obtain

(39)

$$c_s(\lambda) = a_s \prod_{\alpha \in \Sigma_0^+ \cap s^{-1}\Sigma_0^-} \frac{2^{-\langle i\lambda, \alpha_0 \rangle} \Gamma(\langle i\lambda, \alpha_0 \rangle)}{\Gamma(\frac{1}{2}(\frac{1}{2}m_\alpha + 1 + \langle i\lambda, \alpha_0 \rangle)) \Gamma(\frac{1}{2}(\frac{1}{2}m_\alpha + m_{2\alpha} + \langle i\lambda, \alpha_0 \rangle))}$$

where $\alpha_0 = \alpha/\langle\alpha, \alpha\rangle$ and the constant a_s is given by

$$(40) \qquad a_s = \prod_{\alpha \in \Sigma_0^+ \cap s^{-1}\Sigma_0^-} 2^{(1/2)m_\alpha + m_{2\alpha}}\Gamma(\tfrac{1}{2}(m_\alpha + m_{2\alpha} + 1)).$$

Thus $c_s(\lambda)$ extends to a meromorphic function on $\mathfrak{a}_c^*$.

We still must justify (7). Note that $\bar{N} = \bar{N}_{s^*}$ if s^* is the Weyl group element which maps $\mathfrak{a}^+$ to $-\mathfrak{a}^+$. Thus by (37), the integral

$$\int_{\bar{N}} e^{-(i\lambda+\rho)(H(\bar{n}))}\, d\bar{n}$$

converges absolutely for $\operatorname{Re}(i\lambda) = -\eta \in \mathfrak{a}_+^*$. Proceeding as in the rank-one case, select ε $(0 < \varepsilon < 1)$ such that $\rho + \varepsilon\eta \in \mathfrak{a}_+^*$. If

$$a = \exp tH \quad (H \in \mathfrak{a}^+, t > 0)$$

we see from Corollary 6.6 that the integrand in Proposition 6.3 is majorized by

$$(41) \quad \begin{aligned} e^{-(\eta+\rho)(H(a\bar{n}a^{-1}))}e^{(\eta-\rho)(H(\bar{n}))} &\leq e^{(1-\varepsilon)(-\eta)(H(a\bar{n}a^{-1}))}e^{(\eta-\rho)(H(\bar{n}))} \\ &\leq e^{(1-\varepsilon)(-\eta)(H(\bar{n}))}e^{(\eta-\rho)(H(\bar{n}))} = e^{(\varepsilon\eta-\rho)(H(\bar{n}))}, \end{aligned}$$

which is integrable. This justifies our heuristic derivation of (7) and we have obtained the following result.

Theorem 6.14. *The c-function for the semisimple group G is given by the absolutely convergent integral*

$$(42) \qquad c(\lambda) = \int_{\bar{N}} e^{-(i\lambda+\rho)(H(\bar{n}))}\, d\bar{n}, \qquad \operatorname{Re}(i\lambda) \in \mathfrak{a}_+^*,$$

the Haar measure $d\bar{n}$ being normalized by $\int_{\bar{N}} \exp(-2\rho(H(\bar{n}))\, d\bar{n} = 1$. Also, $c(\lambda)$ is given by the formula

$$(43) \quad c(\lambda) = c_0 \prod_{\alpha \in \Sigma_0^+} \frac{2^{-\langle i\lambda, \alpha_0\rangle}\Gamma(\langle i\lambda, \alpha_0\rangle)}{\Gamma(\frac{1}{2}(\frac{1}{2}m_\alpha + 1 + \langle i\lambda, \alpha_0\rangle))\Gamma(\frac{1}{2}(\frac{1}{2}m_\alpha + m_{2\alpha} + \langle i\lambda, \alpha_0\rangle))},$$

where $\alpha_0 = \alpha/\langle\alpha, \alpha\rangle$ and the constant c_0 is given by $c(-i\rho) = 1$.

Finally, if $H \in \mathfrak{a}^+$, $\operatorname{Re}(i\lambda) \in \mathfrak{a}_+^*$,

$$(44) \qquad \lim_{t \to +\infty} e^{(-i\lambda+\rho)(tH)}\phi_\lambda(\exp tH) = c(\lambda)$$

Remarks. **1.** Formula (43) gives an extension of $c(\lambda)$ from a holomorphic function on the set $'\mathfrak{a}_c^*$ to a meromorphic function on $\mathfrak{a}_c^*$.

2. If G is complex, we have for each α, $m_\alpha = 2$, $m_{2\alpha} = 0$. Using $\Gamma(x + 1) = x\Gamma(x)$, the duplication formula (12), and (43), we recover the formula $c(\lambda) = \pi(\rho)/\pi(i\lambda)$ from Theorem 5.7.

3. Suppose G has all its Cartan subgroups conjugate. By [DS] (Chapter IX, Theorem 6.1) this happens if and only if $\operatorname{rank}(G) = \operatorname{rank}(K) + \operatorname{rank}(G/K)$; equivalently, if and only if all roots $\alpha \in \Sigma$ have even multiplicity. We also know from [DS] (Chapter X, Exercise F4) that in this case $m_{2\alpha} = 0$ for all α. Using again the duplication formula (12) we deduce the following result.

Corollary 6.15. *Suppose G has all its Cartan subgroups conjugate. Then*

$$c(\lambda) = \prod_{\alpha \in \Sigma^+} q_\alpha(\langle \rho, \alpha_0 \rangle)/q_\alpha(\langle i\lambda, \alpha_0 \rangle)$$

where $q_\alpha(x)$ *is the polynomial*

$$q_\alpha(x) = \frac{\Gamma(\frac{1}{2}m_\alpha + x)}{\Gamma(x)}$$

of degree $\frac{1}{2} m_\alpha$.

4. We shall see later that every detail in the formula for the c-function is conceptually significant: its behavior at infinity is significant for the spherical transform, the location of its zeros significant for the proof of the Paley–Wiener theorem in the next section, the numerator enters into the theory of conical distributions and intertwining operators (cf. Helgason [1970a]) and the denominator decides the irreducibility of the eigenspace representations for G/K and the main properties of the Poisson transform (cf. Chapter II, §4, No. 1).

In view of these examples, the c-function gives a graphic illustration of the gamma function's appearance "in nature."

§7. The Paley–Wiener Theorem and the Inversion Formula for the Spherical Transform

According to the classic Paley–Wiener theorem (Introduction, Theorem 2.9) the Fourier–Laplace transform

$$f(x) \to f^*(\zeta) = \int_{\mathbf{R}^n} f(x) e^{-i(x,\zeta)}\, dx, \tag{1}$$

where $(x, \zeta) = x_1\zeta_1 + \cdots + x_n\zeta_n$ $(x \in \mathbf{R}^n, \zeta \in \mathbf{C}^n)$ maps $\mathcal{D}_A(\mathbf{R}^n)$ onto the space $\mathcal{H}^A(\mathbf{C}^n)$ of entire functions F of exponential type A, that is, entire functions satisfying for each $N \in \mathbf{Z}^+$ the inequality

$$\sup |F(\zeta)|(1 + |\zeta|)^N e^{-A|\operatorname{Im}\zeta|} < \infty. \tag{2}$$

We shall now establish an analogous result for the spherical transform

$$(3)\qquad \tilde{f}(\lambda) = \int_G f(g)\phi_{-\lambda}(g)\,dg \qquad (f\ K\text{-bi-invariant})$$

for a connected, noncompact, semisimple Lie group G. In other words, we shall give an intrinsic characterization of the functions $\tilde{f}(\lambda)$ as f runs through the space $\mathscr{D}^\natural(G)$ of functions in $\mathscr{D}(G)$ which are bi-invariant under K. For this we make extensive use of the results in §§5, 6. In general if S is a function space on G (say $S = \mathscr{E}(G)$, $L^p(G)$, etc.) $S^\natural$ denotes the subspace of functions in S which are bi-invariant under K.

1. Normalization of Measures

It is convenient now to establish some conventions about the normalization of certain invariant measures.

The groups G, K, A, N, $\bar{N}$, M, and their Lie algebras $\mathfrak{g}$, $\mathfrak{k}$, $\mathfrak{a}$, $\mathfrak{n}$, $\bar{\mathfrak{n}}$, $\mathfrak{m}$ being as in §§4–6, let as usual $l = \dim \mathfrak{a}$. The Killing form on $\mathfrak{g}$, denoted by B or $\langle\ ,\ \rangle$, induces Euclidean measures on A, $\mathfrak{a}$, and $\mathfrak{a}^*$; multiplying these by the factor $(2\pi)^{-(1/2)l}$, we obtain invariant measures da, dH, and $d\lambda$ and the inversion formula for the Euclidean Fourier transform

$$(4)\qquad f^*(\lambda) = \int_A f(a)e^{-i\lambda(\log a)}\,da, \qquad \lambda \in \mathfrak{a}^*,$$

holds without any multiplicative constant,

$$(5)\qquad f(a) = \int_{\mathfrak{a}^*} f^*(\lambda)e^{i\lambda(\log a)}\,d\lambda, \qquad f \in \mathscr{D}(A).$$

These choices of da, dH, and $d\lambda$ are called *regular normalizations*.

We normalize the Haar measures dk and dm on the compact groups K and M, respectively, such that the total measure is 1. The Haar measures on the nilpotent groups N, $\bar{N}$ are normalized such that

$$\theta(dn) = d\bar{n}, \qquad \int_{\bar{N}} e^{-2\rho(H(\bar{n}))}\,d\bar{n} = 1.$$

The Haar measure dg is then normalized such that

$$\int_G f(g)\,dg = \int_{KAN} f(kan)e^{2\rho(\log a)}\,dk\,da\,dn, \qquad f \in \mathscr{D}(G),$$

(cf. Chapter I, §5). The measures dg and dk define the invariant measure $dx = dg_K$ on $X = G/K$ (Chapter I, Theorem 1.9).

2. The Image of $\mathscr{D}^{\natural}(G)$ under the Spherical Transform. The Paley–Wiener Theorem

As before we put

$$F_f(a) = e^{\rho(\log a)} \int_N f(an)\, dn, \qquad f \in \mathscr{D}^{\natural}(G). \tag{6}$$

Then we know from Chapter II, §5, Eq. (37), that F_f is W-invariant and

$$\tilde{f}(\lambda) = \int_A F_f(a) e^{-i\lambda(\log a)}\, da. \tag{7}$$

Let $R > 0$ be such that the function $gK \to f(g)$ on $X = G/K$ vanishes outside the ball $B_R(o)$. Then by [DS] (Chapter VI, Exercise B2 (iv))

$$F_f(a) = 0 \qquad \text{if} \quad d(o, a \cdot o) > R, \tag{8}$$

d denoting distance. But then (7) and (2) imply that for each $N \in \mathbf{Z}^+$, there exists a constant C_N for which

$$|\tilde{f}(\lambda)| \leq C_N(1 + |\lambda|)^{-N} e^{R|\text{Im}\,\lambda|}, \qquad \lambda \in \mathfrak{a}_{\mathbf{C}}^*. \tag{9}$$

Furthermore, $\tilde{f}$ is W-invariant (since F_f is).

Let $\mathscr{H}^R(\mathfrak{a}_{\mathbf{C}}^*)$ denote the set of entire functions on $\mathfrak{a}_{\mathbf{C}}^*$ satisfying (9) and as before put $\mathscr{H}(\mathfrak{a}_{\mathbf{C}}^*) = \bigcup_{R>0} \mathscr{H}^R(\mathfrak{a}_{\mathbf{C}}^*)$. We write $\mathscr{H}_W(\mathfrak{a}_{\mathbf{C}}^*)$ and $\mathscr{H}_W^R(\mathfrak{a}_{\mathbf{C}}^*)$ for the respective subspaces of W-invariant elements.

Theorem 7.1. *The spherical transform $f \to \tilde{f}$ is a bijection of $\mathscr{D}^{\natural}(G)$ onto $\mathscr{H}_W(\mathfrak{a}_{\mathbf{C}}^*)$. Moreover, the function $gK \to f(g)$ has support in the closed ball $B_R(o)^-$ if and only if $\tilde{f} \in \mathscr{H}_W^R(\mathfrak{a}_{\mathbf{C}}^*)$.*

Proof. It remains to prove the surjectivity. For this we start with a simple estimate of the $\mathbf{c}$-function.

Proposition 7.2. *For suitable constants C_1 and C_2,*

$$|\mathbf{c}(\lambda)|^{-1} \leq C_1 + C_2|\lambda|^p \qquad \text{if} \quad \text{Re}(i\lambda) \in \text{Cl}(\mathfrak{a}_+^*).$$

Here $p = \frac{1}{2}\dim N$.

Proof. By a well-known property of the Gamma function (see, e.g., Titchmarsh [1939], Chapter IV),

$$\lim_{|z| \to \infty} \frac{\Gamma(z + a)}{\Gamma(z)} e^{-a \log z} = 1, \qquad |\arg z| \leq \pi - \delta, \tag{10}$$

where $a \in \mathbf{C}$ is any constant, log is the principal value of the logarithm and $\delta > 0$. Putting $z_\alpha = \langle i\lambda, \alpha_0 \rangle$ and using Eq. (12) of §6, we have

$$\mathbf{c}(\lambda)^{-1} = c_0^{-1} \prod_{\alpha \in \Sigma_0^+} C_\alpha(z_\alpha),$$

where

$$C_\alpha(z) = \frac{2^z\Gamma(\frac{1}{4}m_\alpha + \frac{1}{2} + \frac{1}{2}z)\Gamma(\frac{1}{4}m_\alpha + \frac{1}{2}m_{2\alpha} + \frac{1}{2}z)}{2^{z-1}\pi^{-1/2}\Gamma(\frac{1}{2}z + \frac{1}{2})\Gamma(\frac{1}{2}z)}.$$

It follows from (10) that

$$\lim_{|z|\to\infty} C_\alpha(z)|\tfrac{1}{2}z|^{-(1/2)(m_\alpha+m_{2\alpha})} = 2\pi^{1/2}$$

in the half plane $\operatorname{Re} z \geq 0$. Since $\operatorname{Re}(\langle i\lambda, \alpha_0\rangle) \geq 0$ for each $\alpha \in \Sigma^+$, the proposition follows from the last formula.

Next, let F be a W-invariant holomorphic function on $\mathfrak{a}_c^*$ satisfying (9); i.e., for each $N \in \boldsymbol{Z}^+$,

(11) $$|F(\lambda)| \leq C_N(1 + |\lambda|)^{-N}e^{R|\operatorname{Im}\lambda|}, \qquad \lambda \in \mathfrak{a}_c^*.$$

We write the expansion (17) of §5, in the form

(12) $$\phi_\lambda(\exp H) = \sum_{\mu\in\Lambda} \psi_\mu(\lambda, H), \qquad \lambda \in {}'\mathfrak{a}_c^*,\ H \in \mathfrak{a}^+,$$

where

(13) $$\psi_\mu(\lambda, H) = \sum_{s\in W} c(s\lambda)\Gamma_\mu(s\lambda)e^{(is\lambda-\rho-\mu)(H)}, \qquad \lambda \in {}'\mathfrak{a}_c^*,\ H \in \mathfrak{a}^C.$$

Motivated by the inversion formula [Theorem 7.5(i)] we consider the K-bi-invariant function

(14) $$f(g) = \int_{\mathfrak{a}^*} F(\lambda)\phi_\lambda(g)|c(\lambda)|^{-2}\,dx, \qquad g \in G,$$

the convergence being assured by Proposition 7.2 and (11). Using Theorem 4.3 we see that for each $D \in \boldsymbol{D}(G)$, $(D\phi_\lambda)(g)$ is bounded by a polynomial in λ, uniformly for g varying in a compact set. It follows that $f \in \mathscr{E}(G)$. We wish to prove $f(g) = 0$ for $gK \notin B_R(o)$, or, equivalently,

$$f(\exp H) = 0 \qquad \text{for} \quad H \in \mathfrak{a}^+,\ |H| > R.$$

For this we use the expansion (12) and have for $H \in \mathfrak{a}^+$, $h = \exp H$

(15)

$$f(h) = \int_{\mathfrak{a}^*} F(\lambda)\sum_\mu \psi_\mu(\lambda, H)|c(\lambda)|^{-2}\,d\lambda = \sum_{\mu\in\Lambda}\int_{\mathfrak{a}^*} F(\lambda)\psi_\mu(\lambda, H)|c(\lambda)|^{-2}\,d\lambda,$$

the term-by-term integration being justified by Lemma 5.6. Note also that for $\lambda \in \mathfrak{a}^*$

(16) $$|c(\lambda)|^2 = c(\lambda)c(-\lambda) = c(s\lambda)c(-s\lambda) \qquad (s \in W),$$

and the integrand in the last integral is well-defined.

The following result is crucial both for Theorem 7.1 and for the proof of Theorem 7.5.

Theorem 7.3. *Let $F \in \mathscr{H}_W^R(\mathfrak{a}_C^*)$. Then for each $\mu \in \Lambda$*

$$\int_{\mathfrak{a}^*} F(\lambda)\psi_\mu(\lambda, H)|c(\lambda)|^{-2}\, d\lambda = 0$$

if $H \in \mathfrak{a}^+$, $|H| > R$.

Proof. Because of (13), (16), and the W-invariance of F it suffices to prove

$$\int_{\mathfrak{a}^*} F(-\xi)c(\xi)^{-1}e^{-i\xi(H)}\Gamma_\mu(-\xi)\, d\xi = 0 \tag{17}$$

for $H \in \mathfrak{a}^+$, $|H| > R$. To do this we shift the integration in (17) into the complex space $\mathfrak{a}_C^*$ in such a way that the singularities of the functions $c(\lambda)^{-1}$ and $\Gamma_\mu(-\lambda)$ are simultaneously avoided, Cauchy's theorem applicable, resulting in estimates for (17). By the recursive definition of the rational function $\lambda \to \Gamma_\mu(-\lambda)$ it is clear that it is holomorphic outside the set where $\langle \nu, \nu \rangle = 2i\langle \nu, -\lambda \rangle$ for some $\nu \in \Lambda - \{0\}$, that is, outside the set

$$\{\lambda \in \mathfrak{a}_C^* : \langle \nu, \lambda \rangle = \tfrac{1}{2}i\langle \nu, \nu \rangle \text{ for some } \nu \in \Lambda - \{0\}\}. \tag{18}$$

On the other hand, the function $\lambda \to c(\lambda)^{-1}$ is holomorphic on the complement in $\mathfrak{a}_C^*$ of the set

$$\{\lambda \in \mathfrak{a}_C^* : \langle \lambda, \alpha_0 \rangle \in \tfrac{1}{2}i\,(\mathbf{Z}^+ - \{0\}) \text{ for some } \alpha \in \Sigma_0^+\} \tag{19}$$

Fortunately, both sets (18) and (19) are contained in the set

$$\{\lambda \in \mathfrak{a}_C^* : i\langle \nu, \lambda \rangle < 0 \text{ for some } \nu \in \Lambda - \{0\}\}. \tag{20}$$

The domain of integration in (17) will now be shifted away from this set.

Consider any $\lambda \in \mathfrak{a}_C^*$ for which $\operatorname{Re}(i\lambda) \in \mathfrak{a}_+^*$. This means that if $\lambda = \xi + i\eta$ ($\xi, \eta \in \mathfrak{a}^*$) and $A_\lambda = A_\xi + iA_\eta$ then $-A_\eta \in \mathfrak{a}^+$. Let $H_1, \ldots, H_l$ be the basis of $\mathfrak{a}$ dual to the basis $\alpha_1, \ldots, \alpha_l$ (of simple roots). For $\lambda \in \mathfrak{a}_C^*$ we write $A_\lambda = \sum_j \lambda_j H_j$ ($\lambda_j = \xi_j + i\eta_j \in \mathbf{C}$) and identify a function ϕ on a subset of $\mathfrak{a}_C^*$ with a function on a subset of $\mathbf{C}^l$ by the convention $\phi(\lambda) = \phi(\lambda_1, \ldots, \lambda_l)$. All ξ_j, η_k are real and $\eta_k < 0$ ($1 \leq k \leq l$). The function

$$G(\lambda) = G(\lambda_1, \ldots, \lambda_l) = F(-\lambda)c(\lambda)^{-1}e^{-i\lambda(H)}\Gamma_\mu(-\lambda)$$

is holomorphic at each point

$$\lambda = (\xi_1 + is_1\eta_1, \ldots, \xi_l + is_l\eta_l) \qquad (0 \leq s_1 \leq 1, \ldots, 0 \leq s_l \leq 1)$$

because condition (20) is violated. In fact, if $\nu \in \Lambda - \{0\}$, $\nu = n_1\alpha_1 + \cdots + n_l\alpha_l$ then $\langle \nu, \lambda \rangle = \sum_j n_j(\xi_j + is_j\eta_j)$ has a nonpositive imaginary part.

If the function $G(\lambda)$ behaves suitably at infinity we can therefore conclude from Cauchy's theorem that

$$\int_{\boldsymbol{R}^l} G(\xi_1, \dots, \xi_l)\, d\xi = \int_{\boldsymbol{R}^l} G(\xi_1 + i\eta_1, \dots, \xi_l + i\eta_l)\, d\xi. \tag{21}$$

In order to estimate G we look at the recursion formula (12) of §5 for Γ_μ and observe that, by induction on $m(\mu)$,

$$\Gamma_\mu(-(\xi + i\eta)) = p_\mu(\xi + i\eta)(q_\mu(\xi + i\eta))^{-1},$$

where p_μ and q_μ are polynomials and

$$q_\mu(\xi + i\eta) = \prod_{0 \neq \nu,\, \mu - \nu \in \Lambda} (\langle \nu, \nu \rangle - 2\langle \nu, \eta \rangle + 2i\langle \nu, \xi \rangle).$$

Consequently, we have for a constant c_μ

$$|\Gamma_\mu(-(\xi + i\eta))| \leq c_\mu |p_\mu(\xi + i\eta)|, \qquad \xi \in \mathfrak{a}^*, \ -\eta \in \mathfrak{a}^*_+. \tag{22}$$

The estimates (11) and (22) together with Proposition 7.2 show that the application (21) of Cauchy's theorem is legitimate, so for each $-\eta \in \mathfrak{a}^*_+$,

$$\int_{\mathfrak{a}^*} F(-\xi)\boldsymbol{c}(\xi)^{-1} e^{-i\xi(H)} \Gamma_\mu(-\xi)\, d\xi$$

$$= \int_{\mathfrak{a}^*} F(-\xi - i\eta)\boldsymbol{c}(\xi + i\eta)^{-1} e^{-i\xi(H) + \eta(H)} \Gamma_\mu(-\xi - i\eta)\, d\xi.$$

Choosing the integer N in (11) equal to

$$l + 1 + \deg(p_\mu) + p,$$

we have

$$|Q(H)| \leq C e^{R|\eta|} e^{\eta(H)} \qquad \text{for all } H \in \mathfrak{a}^+, \ -\eta \in \mathfrak{a}^*_+, \tag{23}$$

C being a constant. Now fix $H \in \mathfrak{a}^+$ with $|H| > R$. Let η be such that $A_\eta = -tH$ $(t > 0)$. Then by (23)

$$|Q(H)| \leq C e^{t|H|(R - |H|)}.$$

Letting $t \to +\infty$, we obtain $Q(H) = 0$, and the theorem is proved.

Now (15) shows that

$$f(g) = 0 \qquad \text{if } \ gK \notin B_R(o). \tag{24}$$

To conclude the proof of Theorem 7.1 it remains to prove that $\tilde{f}$ is a constant multiple of F. For this we anticipate Theorem 7.5(i) below. This result [and (14)] imply that the function $\psi(\lambda) = \tilde{f}(\lambda) - cF(\lambda)$ satisfies

$$\int_{\mathfrak{a}^*} \psi(\lambda)\phi_\lambda(g)|\boldsymbol{c}(\lambda)|^{-2}\, d\lambda = 0, \qquad g \in G. \tag{25}$$

Integrating this against any $h \in \mathscr{D}^{\natural}(G)$, we obtain

$$\int_{\mathfrak{a}^*} \psi(\lambda)\tilde{h}(-\lambda)|\mathbf{c}(\lambda)|^{-2}\, d\lambda = 0. \tag{26}$$

By Theorem 4.3, $\phi_\lambda \equiv \phi_\mu$ implies that λ and μ are W-conjugate; hence the functions $\tilde{h}$ $[h \in \mathscr{D}^{\natural}(G)]$ separate points in the space $\mathfrak{a}^*/W$. They vanish at ∞ and form an algebra of functions on $\mathfrak{a}^*/W$ closed under complex conjugation because of the formulas

$$(h_1 * h_2)^{\sim}(\lambda) = \tilde{h}_1(\lambda)\tilde{h}_2(\lambda), \tag{27}$$

$$(h^*)^{\sim}(\lambda) = (\tilde{h}(\lambda))^{-} \qquad \text{if} \quad h^*(g) = \overline{h(g^{-1})}, \tag{28}$$

the latter formula coming from Eq. (7) of §4. Now the Stone–Weierstrass theorem on $\mathfrak{a}^*/W$, formula (26), and the W-invariance of ψ imply that $\psi \equiv 0$ on $\mathfrak{a}^*$, and hence on $\mathfrak{a}_c^*$ by holomorphy. This concludes the proof of Theorem 7.1 [modulo Theorem 7.5(i)].

Corollary 7.4. *Let $\mathscr{D}_W(A)$ denote the subspace of W-invariants in $\mathscr{D}(A)$. Then $f \to F_f$ is a linear homeomorphism of $\mathscr{D}^{\natural}(G)$ onto $\mathscr{D}_W(A)$. Moreover, $*$ denoting convolution on G and on A,*

$$F_{f_1 * f_2} = F_{f_1} * F_{f_2}. \tag{29}$$

In fact, because of (6) and (7) the map $f \to F_f$ maps $\mathscr{D}^{\natural}(G)$ continuously into $\mathscr{D}_W(A)$. Because of (7), Theorem 7.1, and the Paley–Wiener theorem for the Euclidean space A we see that the map is injective and that the image equals $\mathscr{D}_W(A)$. Because of Lemma 1.13, $\mathscr{D}^{\natural}(G)$ and $\mathscr{D}_W(A)$ are LF-spaces (cf. Proposition 1.12 and preceding remarks). Being a one-to-one continuous mapping of one LF-space onto another, the map $f \to F_f$ is a homeomorphism.

3. The Inversion Formula

Preserving the notation above, we shall now prove the inversion formula and the Plancherel formula for the spherical transform.

Theorem 7.5. *The spherical transform satisfies*

(i) $c f(g) = \int_{\mathfrak{a}^*} \tilde{f}(\lambda)\phi_\lambda(g)|\mathbf{c}(\lambda)|^{-2}\, d\lambda, \quad g \in G, \quad f \in \mathscr{D}^{\natural}(G),$

where c is a constant independent of f;

(ii) $c \int_G |f(g)|^2\, dg = \int_{\mathfrak{a}^*} |\tilde{f}(\lambda)|^2 |\mathbf{c}(\lambda)|^{-2}\, d\lambda,$

and the image $\mathscr{D}^{\natural}(G)^{\sim}$ is dense in $L^2(\mathfrak{a}^*/W, |\mathbf{c}(\lambda)|^{-2}\, d\lambda)$.

Here the normalizations of $d\lambda$ and dg can be arbitrary.

Proof. (i) For $\psi \in \mathscr{D}(X)$ define $\psi^\natural \in \mathscr{D}^\natural(G)$ by

$$\psi^\natural(g) = \int_K \psi(kgK)\,dk \tag{30}$$

and consider the linear functional

$$T:\psi \to \int_{\mathfrak{a}^*} (\psi^\natural)^\sim(\lambda)|c(\lambda)|^{-2}\,d\lambda, \qquad \psi \in \mathscr{D}(X). \tag{31}$$

Then T is a distribution because it is composed of the maps

$$\psi \in \mathscr{D}(X) \to \psi^\natural \in \mathscr{D}(G),$$

$$f \in \mathscr{D}^\natural(G) \to F_f \in \mathscr{D}(A),$$

$$g \in \mathscr{D}(A) \to \int_{\mathfrak{a}^*} g^*(\lambda)|c(\lambda)|^{-2}\,d\lambda \in \mathbf{C},$$

the first two being obviously continuous, and the third being continuous because of (4) and Proposition 7.2. We shall now prove that the distribution T has support at the origin $o = \{K\}$ in X. For this fix a function $f_0 \in \mathscr{D}(A)$ with support in the unit ball $B_1(e)^-$ such that in addition the function $F(\lambda) = f_0^*(\lambda)$ satisfies $F(0) = 1$, $F \geq 0$ on $\mathfrak{a}^*$, F is radial (in particular, W-invariant). Let $\psi \in \mathscr{D}(X)$, $f = \psi^\natural$. By dominated convergence, and the Fubini theorem

$$\begin{aligned} T(\psi) &= \int_{\mathfrak{a}^*} \tilde{f}(\lambda)|c(\lambda)|^{-2}\,d\lambda = \lim_{\varepsilon\to 0}\int_{\mathfrak{a}^*} \tilde{f}(\lambda)F(\varepsilon\lambda)|c(\lambda)|^{-2}\,d\lambda \\ &= \lim_{\varepsilon\to 0}\int_{\mathfrak{a}^*}\left(\int_G f(g)\phi_{-\lambda}(g)\,dg\right)F(\varepsilon\lambda)|c(\lambda)|^{-2}\,d\lambda \\ &= \lim_{\varepsilon\to 0}\int_X \psi(x)h_\varepsilon(x)\,dx, \end{aligned}$$

where

$$h_\varepsilon(gk) = \int_{\mathfrak{a}^*} F(\varepsilon\lambda)\phi_{-\lambda}(g)|c(\lambda)|^{-2}\,d\lambda. \tag{32}$$

Thus, viewing h_ε as a distribution on X,

$$T(\psi) = \lim_{\varepsilon\to 0} h_\varepsilon(\psi). \tag{33}$$

Because of (2), (15), and Theorem 7.3 h_ε has support in the closed ball $B_\varepsilon(o)^-$. Hence $\operatorname{supp}(T) \subset \{o\}$.

Next we show that T can be extended to a measure. For this it suffices to prove that the L^1-norm $\|h_\varepsilon\|_1$ remains bounded as $\varepsilon \to 0$ because then by (33),

$$|T(\psi)| \leq \limsup_{\varepsilon \to 0} \int_X |\psi(x)h_\varepsilon(x)|\, dx$$

$$\leq \limsup_{\varepsilon \to 0} \|\psi\,|B_\varepsilon(o)\|_\infty \|h_\varepsilon\|_1 \leq \text{const}\, |\psi(o)|,$$

$\|\ \|_\infty$ denoting the uniform norm. Now

$$\text{supp}(h_\varepsilon) \subset B_\varepsilon(o)^-, \text{ so } \|h_\varepsilon\|_1 \leq \text{vol}(B_\varepsilon(o))\|h_\varepsilon\|_\infty.$$

But using Proposition 7.2, and the boundedness $|\phi_\lambda(g)| \leq C$ for $\lambda \in \mathfrak{a}^*$, $gK \in B_\varepsilon(o)$

$$\|h_\varepsilon\|_\infty \leq C \int_{\mathfrak{a}^*} F(\varepsilon\lambda)|c(\lambda)|^{-2}\, d\lambda = C\varepsilon^{-\dim A} \int_{\mathfrak{a}^*} F(\lambda)|c(\varepsilon^{-1}\lambda)|^{-2}\, d\lambda$$

$$\leq C\varepsilon^{-\dim A} \int_{\mathfrak{a}^*} F(\lambda)(C_1 + C_2|\varepsilon^{-1}\lambda|^p)^2\, d\lambda,$$

so

$$\|h_\varepsilon\|_\infty = O(\varepsilon^{-\dim X}).$$

On the other hand, X being a Riemannian manifold,

$$\text{vol}(B_\varepsilon(o)) = O(\varepsilon^{\dim X}) \qquad \text{for } \varepsilon \text{ small}.$$

These inequalities show that $\|h_\varepsilon\|_1$ is bounded as $\varepsilon \to 0$ so T can be extended to a measure.

We have now proved that $T(\psi) = c\psi(o)$ where $c = \text{const}$; in other words,

$$cf(e) = \int_{\mathfrak{a}^*} \tilde{f}(\lambda)|c(\lambda)|^{-2}\, d\lambda, \qquad f \in \mathscr{D}^\natural(G). \tag{34}$$

It is clear from (27) and (28) that $c \neq 0$.

To prove Theorem 7.5(i) for each $g \in G$ consider the function

$$h(g_1) = \int_K f(gkg_1)\, dk, \qquad g_1 \in G.$$

Then $h \in \mathscr{D}^\natural(G)$ and using Proposition 2.2 we have

$$\begin{aligned}\tilde{h}(\lambda) &= \int_K \left(\int_G f(gkg_1)\phi_{-\lambda}(g_1)\,dg_1 \right) dk \\ &= \int_G f(g_1)\phi_{-\lambda}(g^{-1}g_1)\,dg_1 = \int_G f(g_1)\left(\int_K \phi_{-\lambda}(g^{-1}kg_1)\,dk \right) dg_1 \\ &= \phi_{-\lambda}(g^{-1})\tilde{f}(\lambda),\end{aligned}$$

so by Eq. (7) of §4,

$$\tilde{h}(\lambda) = \phi_\lambda(g)\tilde{f}(\lambda).$$

Now (i) follows by using (34) on the function h.

To derive (ii) we use (34) on the function $g_1 \to \int f(g_1 g)\overline{f(g)}\,dg$, taking (27) and (28) into account. The last statement of the theorem is clear from Theorem 7.1.

Remarks. **1.** The reader will have noticed that Theorems 7.1 and 7.5 are curiously intermingled. Both theorems use Theorem 7.3, which in turn relies on properties of the expansion for ϕ_λ and the c-function.

2. With specific normalizations of $d\lambda$ and dg the constant c in Theorem 7.5 can be determined (see Exercise C4).

3. Theorems 7.1 and 7.5 deal with functions on G, bi-invariant under K, that is, with functions f on $X = G/K$ that are K-invariant: $f(k \cdot x) \equiv f(x)$. For general analysis on X it is of course essential to drop this invariance condition and to consider arbitrary functions F on X. Writing $A(gk, kM) = A(k^{-1}g)$ $(k \in K, g \in G)$, we define the *Fourier transform* $\tilde{F}$ of F by

$$\tilde{F}(\lambda, b) = \int_X F(x)e^{(-i\lambda+\rho)(A(x,b))}\,dx, \qquad \lambda \in \mathfrak{a}_c^*, \quad b \in B = K/M.$$

This was introduced in the author's paper [1965*b*]; a Plancherel- and a Paley–Wiener-type theorem were given there and by the author ([1970*a*] and [1973a]). See the Introduction (Theorem 4.2) for the simplest case.

4. With the density function

$$\delta(H) = \prod_{\alpha \in \Sigma^+} (e^{\alpha(H)} - e^{-\alpha(H)})^{m_\alpha} \tag{35}$$

we have with suitable normalizations

$$\int_G f(g)\,dg = \int_{\mathfrak{a}^+} f(\exp H)\delta(H)\,dH, \qquad f \in \mathscr{D}^\natural(G).$$

If δ_α is the density function δ for the rank-one space G_α/K_α $(\alpha \in \Sigma_0^+)$ formula (35) can be written

$$\delta(H) = \prod_{\alpha \in \Sigma_0^+} \delta_\alpha(H_\alpha) \tag{36}$$

where H_α is the orthogonal projection of H on the space $\mathfrak{a}_\alpha = \boldsymbol{R}A_\alpha$. In view of the Plancherel formula (with $d\lambda$ properly normalized)

$$\int_{\mathfrak{a}^+} |f(\exp H)|^2 \delta(H)\, dH = \int_{\mathfrak{a}_+^*} |\tilde{f}(\lambda)|^2 |\boldsymbol{c}(\lambda)|^{-2}\, d\lambda \qquad (f \in \mathscr{D}^\natural(G))$$

the product formula (36) makes the Gindikin–Karpelevič product formula

$$\boldsymbol{c}(\lambda) = c_0 \prod_{\alpha \in \Sigma_0^+} \boldsymbol{c}_\alpha(\lambda_\alpha) \qquad (c_0 = \text{const.})$$

look quite reasonable; it is an intriguing question whether it can be proved by such simple considerations.

§8. The Bounded Spherical Functions

1. Generalities

According to Theorem 3.3 the continuous homomorphisms of the convolution algebra $L^\natural(G)$ $(=L^1(G)^\natural)$ into $\boldsymbol{C}$ are the maps

$$f \to \int f(g)\phi(g)\, dg, \qquad f \in L^\natural(G), \tag{1}$$

where ϕ is a bounded spherical function on G. When G is connected, noncompact semisimple each spherical function ϕ has the form ϕ_λ $(\lambda \in \mathfrak{a}_c^*)$ and we are thus led to the question

(2) *For which $\lambda \in \mathfrak{a}_c^*$ is ϕ_λ bounded?*

By Banach algebra theory the kernels of the maps (1) are the regular maximal ideals of $L^\natural(G)$, so an answer to (2) amounts to a parametrization of the maximal ideal space of $L^\natural(G)$.

Theorem 8.1. *The spherical function ϕ_λ is bounded if and only if λ belongs to the tube $\mathfrak{a}^* + iC(\rho)$, $C(\rho)$ being the convex hull of the points $s\rho$ $(s \in W)$.*

Let $C^0(\rho)$ denote the interior of $C(\rho)$.

Corollary 8.2. *The spherical transforms*

$$\tilde{f}(\lambda) = \int_G f(g)\phi_{-\lambda}(g)\,dg, \qquad f \in L^{\natural}(G),$$

are holomorphic in the open tube $\mathfrak{a}^* + iC^0(\rho)$.

In fact, the integral of $\tilde{f}(\lambda)$ over any plane closed curve in the tube $\mathfrak{a}^* + iC^0(\rho)$ is 0, $\phi_{-\lambda}$ being bounded and holomorphic in λ. By Morera's theorem, $\tilde{f}$ is holomorphic.

The proof of Theorem 8.1 will be based on a few lemmas. The first is a geometric result which is only needed for the point $H = A_\rho$, but the proof works for any $H \in \overline{\mathfrak{a}^+}$.

2. Convex Hulls

Consider the positive Weyl chamber $\mathfrak{a}^+$ and as in §6. No. 3 the dual cone

$$^+\mathfrak{a} = \{H \in \mathfrak{a} : \langle H, H' \rangle > 0 \text{ for all } H' \in \mathfrak{a}^+\}.$$

Let $\mathfrak{a}^- = -\mathfrak{a}^+$, $^-\mathfrak{a} = -(^+\mathfrak{a})$ and let C^+, C^-, ^+C, and ^-C denote the closures of $\mathfrak{a}^+$, $\mathfrak{a}^-$, $^+\mathfrak{a}$, and $^-\mathfrak{a}$, respectively. For $H \in \mathfrak{a}$ let $C(H)$ denote the convex hull of the orbit $\{sH : s \in W\}$.

Lemma 8.3. *Let* $H \in C^+$. *Then*

(i) $C(H) = \bigcup_{s \in W} s\{C^+ \cap (H + {^-C})\} = \bigcap_{s \in W} s(H + {^-C})$.

(ii) $C^+ \cap C(H) = C^+ \cap (H + {^-C})$.

Proof. We first prove (ii). Given $\lambda \in \mathfrak{a}^*$ such that $A_\lambda \in C^+$, consider the minimal $a \in \mathbf{R}$ such that the half space $\lambda - a \leq 0$ contains $C(H)$. If $H_0 \in C(H)$ satisfies $\lambda(H_0) = a$ and if we write $H_0 = \sum_{s \in W} a_s(sH)$ $(a_s \geq 0, \sum_s a_s = 1)$ then some sH satisfies $\lambda(sH) = a$. Hence $\lambda(sH - H') \geq 0$ for all $H' \in C(H)$. Since $\lambda(H - sH) \geq 0$ ([DS], Chapter VII, Theorem 2.22) we obtain by addition $\lambda(H - H') \geq 0$. Thus $H' - H \in {^-C}$; that is,

$$C(H) \subset H + {^-C}. \tag{3}$$

On the other hand, suppose $H' \in C^+ \cap (H + {^-C})$. To prove $H' \in C(H)$ it suffices to prove that for each $\lambda \in \mathfrak{a}^*$, there exists a $\sigma \in W$ such that $\lambda(\sigma H) \geq \lambda(H')$ [so H' and $C(H)$ cannot be separated by a hyperplane]. For this we select $\sigma \in W$ such that $\sigma^{-1}A_\lambda \in C^+$. Then since the relations

$$H - H' \in {^+C}, \quad H' - \sigma^{-1}H' \in {^+C} \qquad \text{imply} \quad H - \sigma^{-1}H' \in {^+C}$$

we have $\langle \sigma^{-1}A_\lambda, H - \sigma^{-1}H' \rangle \geq 0$; that is, $\lambda(\sigma H) \geq \lambda(H')$. This proves (ii).

Let $C(H)$, C_1, and C_2 denote the three terms in (i). By (ii), $C(H) \subset C_1$; by (ii) and (3) we have for any $s, \sigma \in W$,

$$s(C^+ \cap (H + {}^-C)) = sC^+ \cap C(H) \subset C(H) \subset \sigma(H + {}^-C),$$

whence $C_1 \subset C_2$. Finally, if $H' \in C_2$ then $sH' \in H + {}^-C$ for all $s \in W$. Choosing s such that $sH' \in C^+$, (ii) implies $sH' \in C(H)$ so $H' \in C(H)$. Thus $C_2 \subset C(H)$ and the lemma is proved.

We prove now the first half of Theorem 8.1.

Lemma 8.4. *ϕ_λ is bounded if $\lambda \in \mathfrak{a}^* + iC(\rho)$.*

Proof. As usual we write $\lambda = \xi + i\eta$ ($\xi, \eta \in \mathfrak{a}^*$). Because of the invariance $\phi_{s\lambda} \equiv \phi_\lambda$ and because of Lemma 8.3(i) we may assume

$$-A_\eta \in C^+ \cap (A_\rho + {}^-C).$$

Let Haar measures be normalized as in §7. No. 1. If $f \in L^{\natural}(G)$, we have by the Fubini theorem

$$\int_G f(g)\, dg = \int_{AN} f(an) e^{2\rho(\log a)}\, da\, dn,$$

the function

$$F_f(a) = e^{\rho(\log a)} \int_N f(an)\, dn$$

exists for almost all $a \in A$, and

(4) $$\int_A e^{\rho(\log a)} F_{|f|}(a)\, da < \infty.$$

Since $-A_\eta \in C^+$ we have $-s\eta(H) \leq -\eta(H)$ for $H \in \mathfrak{a}^+$, $s \in W$, so if $A^+ = \exp(\mathfrak{a}^+)$,

$$\int_{A^+} e^{-s\eta(\log a)} F_{|f|}(a)\, da \leq \int_{A^+} e^{-\eta(\log a)} F_{|f|}(a)\, da$$
$$\leq \int_{A^+} e^{\rho(\log a)} F_{|f|}(a)\, da$$

since $\rho + \eta$ is nonnegative on $\mathfrak{a}^+$. Using (4) and the Weyl group invariance of F_f we have, if $|W|$ denotes the order of W,

(5) $$|W|^{-1} \int_A e^{-\eta(\log a)} F_{|f|}(a)\, da \leq \int_{A^+} e^{\rho(\log a)} F_{|f|}(a) < \infty.$$

However, calculating with the positive functions $\phi_{i\eta}$ and $|f|$, we have by the Fubini theorem

$$\begin{aligned}\int_G \phi_{i\eta}(g)|f(g)|\,dg &= \int_G\int_K e^{-(\eta+\rho)(H(gk))}\,dk\,|f(g)|\,dg \\ &= \int_G e^{-(\eta+\rho)(H(g))}|f(g)|\,dg \\ &= \int_{AN} |f(an)|e^{-(\eta+\rho)(H(a))}e^{2\rho(\log a)}\,da\,da \\ &= \int_A e^{-\eta(\log a)}F_{|f|}(a)\,da.\end{aligned}$$

Thus by (5)

$$\int_G \phi_{i\eta}(g)|f(g)|\,dg < \infty,$$

so $\phi_{i\eta}$, and therefore ϕ_λ, is bounded.

Lemma 8.5. *Suppose ϕ_λ is bounded and that $-A_\eta \in \mathfrak{a}^+$. Then $\eta + \rho$ is nonnegative on $\mathfrak{a}^+$.*

Proof. Let $H \in \mathfrak{a}^+$. By the assumption, $\operatorname{Re}(i\lambda) \in \mathfrak{a}^*$, so by Theorem 6.14

$$\lim_{t\to+\infty} e^{(-i\lambda+\rho)(tH)}\phi_\lambda(\exp tH)$$

exists and is $\neq 0$. Since ϕ_λ is bounded we conclude by taking absolute values, $(\eta + \rho)(H) \geq 0$. This proves the lemma.

In order to generalize this lemma to the case $-A_\eta \in \overline{\mathfrak{a}^+}$ we use the concept of a boundary component of the symmetric space. This tool was introduced by Harish-Chandra [1958b] in a different context.

3. Boundary Components

Fix an element $H^0 \in \overline{\mathfrak{a}^+}$ and let

$$\Sigma^0 = \{\alpha \in \Sigma : \alpha(H^0) = 0\}.$$

Let $\mathfrak{g}'$ denote the subalgebra of $\mathfrak{g}$ generated by the root subspaces $\mathfrak{g}_\alpha$ $(\alpha \in \Sigma^0)$.

Lemma 8.6. *The algebra $\mathfrak{g}'$ is semisimple, has a Cartan decomposition*

$$\mathfrak{g}' = \mathfrak{k}' + \mathfrak{p}' \qquad \text{where} \quad \mathfrak{k}' = \mathfrak{g}' \cap \mathfrak{k}, \quad \mathfrak{p}' = \mathfrak{g}' \cap \mathfrak{p},$$

and the space $\mathfrak{a}' = \mathfrak{g}' \cap \mathfrak{a}$ is a maximal abelian subspace of $\mathfrak{p}'$. Furthermore, in the orthogonal decomposition

$$\mathfrak{a} = \mathfrak{a}' + \mathfrak{a}^0, \tag{6}$$

with respect to the Killing form of $\mathfrak{g}$ we have

$$\mathfrak{a}' = \sum_{\alpha \in \Sigma^0} \boldsymbol{R} A_\alpha, \quad \mathfrak{a}^0 = \{H \in \mathfrak{a} : \alpha(H) = 0 \quad \text{for } \alpha \in \Sigma^0\}. \tag{7}$$

Proof. The proof is analogous to that of Proposition 2.1, Chapter IX in [DS]. Let $\mathfrak{g}^c$ denote the complexification of $\mathfrak{g}$ and extend $\mathfrak{a}$ to a maximal abelian subalgebra $\mathfrak{h}$ of $\mathfrak{g}$; then $\mathfrak{h}^c$ is a Cartan subalgebra of $\mathfrak{g}^c$ ([DS], Chapter VI, §3). Then

$$\mathfrak{g}_\alpha = \left(\sum_{\bar{\beta} = \alpha} (\mathfrak{g}^c)^\beta \right) \cap \mathfrak{g}, \tag{8}$$

where β is a root of $(\mathfrak{g}^C, \mathfrak{h}^C)$ and $\bar{\beta}$ denotes the restriction $\beta | \mathfrak{a}$. The subalgebra $\tilde{\mathfrak{g}}$ of $\mathfrak{g}^C$ generated by $(\mathfrak{g}^c)^\beta$ $(\bar{\beta} \in \Sigma^0)$ is semisimple and has the subspace $\tilde{\mathfrak{h}} = \sum_{\bar{\beta} \in \Sigma^0} CH_\beta$ as a Cartan subalgebra ([DS], Chapter III, Exercise B1). The real form $\tilde{\mathfrak{g}} \cap \mathfrak{g}$, which equals $\mathfrak{g}'$, is also semisimple. Invariant under the Cartan involution θ, $\mathfrak{g}'$ has the indicated Cartan decomposition ([DS], Chapter IX, Lemma 2.2). The space $\sum_{\alpha \in \Sigma^0} \boldsymbol{R} A_\alpha$ equals $\tilde{\mathfrak{h}} \cap \mathfrak{p}'$ and is thus a maximal abelian subspace of $\mathfrak{p}'$. This space equals $\mathfrak{g}' \cap \mathfrak{a}$ because

$$\tilde{\mathfrak{h}} \cap \mathfrak{p}' \subset \mathfrak{a} \cap \mathfrak{g}' = \mathfrak{a} \cap \mathfrak{p}' \subset \tilde{\mathfrak{h}} \cap \mathfrak{p}'.$$

This proves (7) and the lemma.

The restrictions $\alpha | \mathfrak{a}'$ $(\alpha \in \Sigma^0)$ form the set Σ' of restricted roots of $\mathfrak{g}'$ with respect to $\mathfrak{a}'$. The subset

$$(\mathfrak{a}')^+ = \{H \in \mathfrak{a}' : \alpha(H) > 0 \text{ for } \alpha \in \Sigma^0 \cap \Sigma^+\} \tag{9}$$

is a maximal connected subset of $(\mathfrak{a}')'$, hence a Weyl chamber in $\mathfrak{a}'$; let $(\Sigma')^+$ denote the corresponding set of positive roots. We have the compatible Iwasawa decompositions

$$\mathfrak{g} = \mathfrak{k} + \mathfrak{a} + \mathfrak{n}, \qquad \mathfrak{g}' = \mathfrak{k}' + \mathfrak{a}' + \mathfrak{n}', \tag{10}$$

where

$$\mathfrak{n}' = \sum_{\alpha \in \Sigma^0 \cap \Sigma^+} \mathfrak{g}_\alpha = \sum_{\beta \in (\Sigma')^+} \mathfrak{g}'_\beta. \tag{11}$$

Moreover, if G', K', A', N' are the corresponding subgroups of G we have

$$(12) \qquad K' = G' \cap K, \qquad A' = G' \cap A, \qquad N' = G' \cap N;$$

in fact, the inclusion $\subset$ is obvious in each case, so (12) follows from $G' = K'A'N'$.

The symmetric space $X' = G'/K'$ is called a *boundary component* of $X = G/K$. Since $\mathfrak{p}'$ is clearly a Lie triple system, X' is a totally geodesic submanifold of X (cf. [DS], Chapter IV, §7).

Lemma 8.7. *Let W' denote the subgroup of W fixing each point in $\mathfrak{a}^0$ and $|W'|$ its order. Then*

(i) *The mapping*

$$\pi^0 : H \to \frac{1}{|W'|} \sum_{s \in W'} sH \qquad (H \in \mathfrak{a})$$

is the orthogonal projection of $\mathfrak{a}$ onto $\mathfrak{a}^0$.

(ii) *If $H \in \mathfrak{a}^+$, then*

$$\alpha(\pi^0(H)) > 0 \qquad \textit{for each } \alpha \in \Sigma^+ - \Sigma^0 \cap \Sigma^+.$$

Proof. First, if $\alpha \in \Sigma^0$ we have $s_\alpha \in W'$ so $s_\alpha(\pi^0 H) = \pi^0 H$; thus $\alpha(\pi^0 H) = 0$ so $\pi^0 H \in \mathfrak{a}^0$. Second, if $A^0 \in \mathfrak{a}^0$ is arbitrary we have $sA^0 = A^0$ for each $s \in W'$ so

$$B(H - \pi^0 H, A^0) = B(H, A^0) - \frac{1}{|W'|} \sum_{s \in W'} B(sH, A^0) = 0,$$

proving (i).

For part (ii) let $\pi' : \mathfrak{a} \to \mathfrak{a}'$ be the orthogonal projection and let $H \in \mathfrak{a}^+$. Since

$$\alpha(H - \pi' H) = 0 \quad \text{for } \alpha \in \Sigma^0,$$

we have

$$(13) \qquad \alpha(\pi' H) > 0 \quad \text{for} \quad \alpha \in \Sigma^0 \cap \Sigma^+.$$

Let B' denote the Killing form of $\mathfrak{g}'$ and, given the restriction $\bar{\alpha} = \alpha | \mathfrak{a}'$ $(\alpha \in \Sigma^0)$, let $A_{\bar{\alpha}} \in \mathfrak{a}'$ be determined by $B'(H, A_{\bar{\alpha}}) = \alpha(H)$ $(H \in \mathfrak{a}')$. Then if $X_\alpha \in \mathfrak{g}_\alpha$ we have

$$[X_\alpha, \theta X_\alpha] = B(X_\alpha, \theta X_\alpha) A_\alpha = B'(X_\alpha, \theta X_\alpha) A_{\bar{\alpha}}$$

(cf. [DS], Chapter IX, §1). It follows that

$$(14) \qquad A_\alpha = c_\alpha A_{\bar{\alpha}}, \qquad c_\alpha > 0.$$

As noted in §6, No. 3,

$$(15) \qquad (\mathfrak{a}')^+ \subset {}^+\mathfrak{a}' = \left\{ \sum_{\substack{\bar{\alpha}_i \in (\Sigma')^+ \\ \text{simple}}} a_i A_{\bar{\alpha}_i} \neq 0 : a_i \geq 0 \right\}.$$

But if $\bar{\alpha}_i \in (\Sigma')^+$ is simple, α_i itself is simple. In fact, if $\alpha_i = \beta + \gamma$, $\beta, \gamma \in \Sigma^+$, then $0 = \alpha_i(H^0) = \beta(H^0) + \gamma(H^0)$ so $\beta, \gamma \in \Sigma^0$ and $\bar{\alpha}_i = \bar{\beta} + \bar{\gamma}$, contradicting the simplicity. By (13), $\pi'(H) \in (\mathfrak{a}')^+$, so by (14) and (15)

$$(16) \qquad \pi'(H) = \sum_{\substack{\alpha_i \in \Sigma^0 \\ \text{simple}}} b_i A_{\alpha_i} \qquad (b_i \geq 0).$$

Now $H = \pi'(H) + \pi^0(H)$. Suppose $\beta \in \Sigma^+ - \Sigma^0 \cap \Sigma^+$ is simple. Then by (16) and [DS] (Chapter III, Lemma 5.6), we have $\beta(\pi'(H)) \leq 0$ so $\beta(\pi^0(H)) = \beta(H) - \beta(\pi'(H)) > 0$. Finally, if $\alpha \in \Sigma^+ - \Sigma^0 \cap \Sigma^+$ is arbitrary we write $\alpha = \sum_i n_i \beta_i$ ($n_i > 0$, β_i simple). Then since $\alpha(H^0) > 0$ and $\beta_i(H^0) \geq 0$ for all i, we have $\beta_i \in \Sigma^+ - \Sigma^0 \cap \Sigma^+$ for *some* i). Writing

$$\alpha = \sum_{\beta_j \in \Sigma^0} n_j \beta_j + \sum_{\beta_k \notin \Sigma^0} n_k \beta_k,$$

we therefore conclude that

$$\alpha(\pi^0(H)) = 0 + \sum_{\beta_k \notin \Sigma^0} n_k \beta_k(\pi^0(H)) > 0$$

and the lemma is proved.

Remark. By general theory ([DS], Chapter VII, Theorem 2.15) W' is generated by the s_α for which $\alpha \in \Sigma$ vanishes identically on $\mathfrak{a}^0$; that is, $\alpha \in \Sigma^0$. Thus the restriction $W'|\mathfrak{a}'$ constitutes the Weyl group of $X' = G'/K'$.

In order to make use of Theorem 6.13 later we insert here the following lemma.

Lemma 8.8. *There exists an* $s \in W$ *such that* $\Sigma^+ - \Sigma^0 \cap \Sigma^+ = \Sigma^+ \cap s^{-1}\Sigma^-$.

Proof. Let $\beta_1, \ldots, \beta_p, \beta_{p+1}, \ldots, \beta_l$ be the simple roots such that $\beta_1(H^0) = \cdots = \beta_p(H^0) = 0$ but $\beta_{p+1}(H^0) > 0, \ldots, \beta_l(H^0) > 0$. We change H^0 slightly to a regular element H^ε satisfying

$$(17) \quad \beta_1(H^\varepsilon) < 0, \ldots, \beta_p(H^\varepsilon) < 0; \qquad \beta_{p+1}(H^\varepsilon) > 0, \ldots, \beta_l(H^\varepsilon) > 0.$$

We claim now that if $|H^\varepsilon - H^0|$ is sufficiently small,

$$(18) \qquad \{\beta \in \Sigma^+ : \beta(H^0) > 0\} = \{\beta \in \Sigma^+ : \beta(H^\varepsilon) > 0\}.$$

The set on the left is contained in the set on the right if H^ε is sufficiently close to H^0. On the other hand, suppose $\beta \in \Sigma^+$, $\beta(H^\varepsilon) > 0$. Writing $\beta = \sum_{i=1}^{l} n_i \beta_i$ $(n_i \in \mathbf{Z}^+)$ we see from (17) that $n_i > 0$ for some $i > p$. But then $\beta(H^0) = \sum_{p+1}^{l} n_i(\beta_i(H^0) > 0$, so β belongs to the left-hand side of (18). Thus

$$\Sigma^+ - \Sigma^0 \cap \Sigma^+ = \{\beta \in \Sigma^+ : \beta(H^\varepsilon) > 0\},$$

which equals $\Sigma^+ \cap s^{-1}\Sigma^-$ if s is the Weyl group element which sends H^ε into $-\mathfrak{a}^+$.

Lemma 8.9. *Suppose ϕ_λ is bounded and that $-A_\eta \in \overline{\mathfrak{a}^+} - \mathfrak{a}^+$ (the boundary of $\mathfrak{a}^+$). Then $\eta + \rho$ is nonnegative on $\mathfrak{a}^+$.*

Proof. Let $H^0 = -A_\eta$ and construct $\mathfrak{g}'$, etc., as above. Put

$$\Delta' = \Sigma^0 \cap \Sigma^+, \qquad \Delta'' = \Sigma^+ - \Sigma^0 \cap \Sigma^+$$

$$\text{(19)} \qquad \bar{\mathfrak{n}}' = \sum_{\alpha \in \Delta'} \mathfrak{g}_{-\alpha}, \qquad \bar{\mathfrak{n}}'' = \sum_{\alpha \in \Delta''} \mathfrak{g}_{-\alpha},$$

and let $\overline{N}'$ and $\overline{N}''$ denote the corresponding analytic (nilpotent) subgroups of G'. Let $2\rho'$, $2\rho''$ denote the sums of the roots in Δ' and Δ'', respectively. Because of Lemma 6.8 and Eqs. (19) and (20) of §6 we can select Haar measures dv, dw on $\overline{N}'$ and $\overline{N}''$, respectively, such that the product measure $dv \times dw$ corresponds to $d\bar{n}$ under the diffeomorphism $(v, w) \to vw$ of $\overline{N}' \times \overline{N}''$ onto $\overline{N}$.

Let us use this change of variables in the formula for $\phi_\lambda(a)$ in Proposition 6.3, and let $a \in A^0 = \exp(\mathfrak{a}^0)$. Putting $\mu = i\lambda - \rho$, $\nu = -(i\lambda + \rho)$, we have, since a centralizes $\overline{N}'$,

$$\phi_\lambda(a) = e^{\mu(\log a)} \int_{\overline{N}} e^{\mu(H(a\bar{n}a^{-1}))} e^{\nu(H(\bar{n}))}\, d\bar{n}$$

$$= e^{\mu(\log a)} \int_{\overline{N}'} \int_{\overline{N}''} e^{\mu(H(vawa^{-1}))\nu(H(vw))}\, dv\, dw.$$

Here we decompose v according to $G' = K'A'N'$, $v = k'a'n'$, and use the fact that G' normalizes $\overline{N}''$ [if $\alpha \in \Sigma$, $\alpha(H^0) > 0$, then $\alpha \in \Delta''$]. Then the last expression becomes

$$e^{\mu(\log a)} \int_{\overline{N}'} \int_{\overline{N}''} e^{\mu(H(aa'wa^{-1}))} e^{\nu(H(a'w))}\, dv\, dw$$

$$= e^{\mu(\log a)} \int_{\overline{N}'} e^{(\mu+\nu+2\rho'')(H(v))}\, dv \int_{\overline{N}''} e^{\mu(H(awa^{-1}))} e^{\nu(H(w))}\, dw,$$

where we have used the Jacobian formula $d(a'w(a')^{-1})/dw = e^{-2\rho''(\log a')}$. Since $\mu + \nu + 2\rho'' = -2\rho'$ this proves the formula

$$(20) \quad \phi_\lambda(a) = ce^{(i\lambda-\rho)(\log a)} \int_{\bar{N}''} e^{(i\lambda-\rho)(H(awa^{-1}))-(i\lambda+\rho)(H(w))}\,dw \qquad (a \in A^0)$$

where the constant c equals $\int_{\bar{N}'} e^{-2\rho'(H(v))}\,dv$.

Now let $H \in \mathfrak{a}$ be such that

$$(21) \qquad \alpha(H) = 0 \quad \text{for } \alpha \in \Delta', \qquad \alpha(H) > 0 \quad \text{for } \alpha \in \Delta'',$$

and put $a = \exp tH$ $(t > 0)$ in (20). Select ε $(0 < \varepsilon < 1)$ such that $\rho + \varepsilon\eta \in \mathfrak{a}_+^*$. Then [cf. Eq. (41) of §6] the integrand in (20) is majorized by $e^{(\varepsilon\eta-\rho)(H(w))}$, which is integrable on $\bar{N}''$ (Theorem 6.13 and Lemma 8.8). Therefore

$$(22) \qquad \lim_{t\to+\infty} e^{(-i\lambda+\rho)(tH)}\phi_\lambda(\exp tH) = c\int_{\bar{N}''} e^{-(i\lambda+\rho)(H(w))}\,dw,$$

which by Theorem 6.13 and Eq. (39) of §6 is $\neq 0$. Taking absolute values and using the boundedness of ϕ_λ we deduce that

$$(23) \qquad (\eta + \rho)(H) \geq 0 \qquad \text{if} \quad H \text{ satisfies (21).}$$

Finally, let $H \in \mathfrak{a}^+$ be arbitrary. We use $\rho(H - sH) \geq 0$ $(s \in W)$, the orthogonality $\eta(H - \pi^0(H)) = -\langle H^0, H - \pi^0(H)\rangle = 0$, and the fact that $\pi^0(H)$ satisfies (21) because of Lemma 8.7(ii). Then, by (23)

$$\begin{aligned}(\eta + \rho)(H) &= (\eta + \rho)(\pi^0(H)) + (\eta + \rho)(H - \pi^0(H)) \\ &= (\eta + \rho)(\pi^0(H)) + \rho(H - \pi^0(H)) \geq 0.\end{aligned}$$

This proves the lemma.

Now Theorem 8.1 follows easily. Suppose ϕ_λ is bounded. Select $s \in W$ such that $A_{-s\eta} \in C^+$. Since $\phi_{s\lambda} = \phi_\lambda$, Lemmas 8.7 and 8.9 imply $A_{s\eta+\rho} \in {}^+C$; i.e., $A_{-s\eta} \in C^+ \cap (A_\rho + {}^-C)$, so by Lemma 8.3, $\eta \in C(\rho)$.

We restate (22) as a corollary generalizing the limit relation of Theorem 6.14.

Corollary 8.10. *Let* $\mathrm{Re}(i\lambda) \in \mathrm{Cl}(\mathfrak{a}_+^*)$, *put* $H^0 = A_{\mathrm{Re}(i\lambda)}$ *and define* $\mathfrak{a}^0$ *and* s *as in Lemmas* 8.6 *and* 8.8. *Then if* $H \in \mathfrak{a}^0 \cap \mathfrak{a}^+$,

$$\lim_{t\to+\infty} e^{(-i\lambda+\rho)(tH)}\phi_\lambda(\exp tH) = c_s(\lambda),$$

provided the measure on N'' *is appropriately normalized.*

§9. The Spherical Transform on $\mathfrak{p}$, the Euclidean Type

We consider now the symmetric space $G_0/K = \mathfrak{p}$ of the Euclidean type and the corresponding spherical functions

$$\psi_\lambda(Y) = \int_K e^{iB(A_\lambda, k\cdot Y)}\,dk, \qquad Y \in \mathfrak{p}, \quad \lambda \in \mathfrak{a}_c^*,$$

as given by Proposition 4.8. The corresponding *spherical transform* is given by

$$\tilde{f}(\lambda) = \int_{\mathfrak{p}} f(Y)\psi_{-\lambda}(Y)\,dY, \tag{1}$$

f being a K-invariant function on $\mathfrak{p}$ and $\lambda \in \mathfrak{a}_c^*$ being such that the integral exists. We shall now establish the analogs of Theorems 7.1 and 7.5 for this case.

According to Chapter I, Theorem 5.17, we have for $F \in \mathscr{D}(\mathfrak{p})$,

$$\int_{\mathfrak{p}} F(Y)\,dY = \int_{\mathfrak{a}^+}\left(\int_K F(\mathrm{Ad}(k)H)\,dk\right)\delta(H)\,dH, \tag{2}$$

where dY and dH are the Euclidean measures, dk the normalized Haar measure, and

$$\delta(H) = c_0\left|\prod_{\alpha\in\Sigma^+}\alpha(H)^{m_\alpha}\right|, \qquad H \in \mathfrak{a}, \tag{3}$$

c_0 being a constant.

We consider now the function spaces $\mathscr{D}(\mathfrak{p})$, $\mathscr{S}(\mathfrak{p})$, $\mathscr{D}(\mathfrak{a}^*)$, $\mathscr{S}(\mathfrak{a}^*)$ and the subspaces $\mathscr{D}_K(\mathfrak{p})$, $\mathscr{S}_K(\mathfrak{p})$, $\mathscr{D}_W(\mathfrak{a}^*)$, $\mathscr{S}_W(\mathfrak{a}^*)$ of K-invariants and, respectively, W-invariants (cf. Chapter II, §5, No. 2).

Theorem 9.1. *The spherical transform $f \to \tilde{f}$ is a bijection of $\mathscr{S}_K(\mathfrak{p})$ onto $\mathscr{S}_W(\mathfrak{a}^*)$ and*

$$f(Y) = c\int_{\mathfrak{a}^*} \tilde{f}(\lambda)\psi_\lambda(Y)\delta(\lambda)\,d\lambda, \tag{4}$$

$$\int_{\mathfrak{p}} |f(Y)|^2\,dY = c\int_{\mathfrak{a}^*} |\tilde{f}(\lambda)|^2\delta(\lambda)\,d\lambda, \tag{5}$$

where c is a constant independent of f.

Proof. If f^* is the Euclidean Fourier transform

$$f^*(Z) = \int_{\mathfrak{p}} f(Y)e^{-iB(Z,Y)}\,dY, \tag{6}$$

we see that $f^*(A_\lambda) = \tilde{f}(\lambda)$, so (4) and (5) follow immediately from the inversion formula and the Plancherel formula for the Euclidean Fourier transform if we take (2) into account. For the surjectivity in Theorem 9.1 use Corollary 5.11 of Chapter II as well as the fact that $F \to F^*$ maps $\mathscr{S}(\mathfrak{p})$ onto itself.

Next we would like to characterize the image of the space $\mathscr{D}_K(\mathfrak{p})$ under the spherical transform. Let $\mathfrak{q}$ be the orthogonal complement of $\mathfrak{a}$ in $\mathfrak{p}$ and put

$$\hat{f}(H) = \int_{\mathfrak{q}} f(H+Q)\,dQ, \qquad f \in \mathscr{D}_K(\mathfrak{p}), \tag{7}$$

dQ being the Euclidean measure. Then (1) can be written

$$\tilde{f}(\lambda) = \int_{\mathfrak{a}} e^{-i\lambda(H)}\hat{f}(H)\,dH; \tag{8}$$

this formula is the Euclidean analog of Eq. (7) of §7. If f has support in the ball $|Y| \leq R$, then by Eq. (2) of §7, $\tilde{f} \in \mathscr{H}^R_W(\mathfrak{a}^*_c)$, $\hat{f}$ clearly being W-invariant.

Theorem 9.2. *The spherical transform $f \to \tilde{f}$ on $\mathfrak{p}$ is a bijection of $\mathscr{D}_K(\mathfrak{p})$ onto $\mathscr{H}_W(\mathfrak{a}^*_c)$. The function f has support in the ball $|Y| \leq R$ if and only if $\tilde{f} \in \mathscr{H}^R_W(\mathfrak{a}^*_c)$.*

Although this theorem is an analog to Theorem 7.1 we cannot imitate the proof of that result, not having a series expansion of ψ_λ analogous to that for ϕ_λ. If on the other hand we want to proceed as for Theorem 9.1 and make use of the relation

$$f^*(A_\lambda) = \tilde{f}(\lambda), \tag{9}$$

we are led to asking for the following result.

Theorem 9.3. *Let ϕ be a W-invariant entire function of exponential type R on $\mathfrak{a}^c$. Then ϕ extends uniquely to a K-invariant entire function Φ of exponential type R on the complexification $\mathfrak{p}^c$ of $\mathfrak{p}$.*

Although this result has a certain formal resemblance to the extension theorems in Chapter II, §5. No. 2 (Theorem 5.8 and Corollary 5.11), it requires entirely different tools, partly because $\mathrm{Int}(\mathfrak{k}^c)(\mathfrak{a}^c) \neq \mathfrak{p}^c$, in contrast to $\mathrm{Int}(\mathfrak{k})(\mathfrak{a}) = \mathfrak{p}$ (Int denotes an adjoint group).

As in Chapter III, §3 let $I(\mathfrak{a}_C^*)$ denote the algebra of W-invariants in the symmetric algebra $S(\mathfrak{a}_C^*)$ and let $j_1, \ldots, j_l$ be homogeneous algebraically independent generators of $I(\mathfrak{a}_C^*)$. We know from Corollary 3.12 (loc. cit.) that the mapping

$$j: H \to (j_1(H), \ldots, j_l(H)) \qquad (H \in \mathfrak{a}^C) \tag{10}$$

induces a bijection of $\mathfrak{a}^C/W$ onto C^l.

Lemma 9.4. *The mapping $j: \mathfrak{a}^C \to C^l$ is proper (that is, the inverse image of any compact set is compact).*

Proof. Let $n = \dim \mathfrak{p}$. For any $k > 0$ there exists a positive number r_k such that for any monic polynomial $p(\lambda) = \lambda^n + \sum_0^{n-1} c_i \lambda^i$ where $c_i \in C$ the inequalities $|c_i| \leq k$ $(0 \leq i \leq n-1)$ imply $|\mu| \leq r_k$ for any root μ of the equation $p(\lambda) = 0$. (In fact if $|\mu| \geq 1$ we would have $|\mu|^n \leq kn|\mu|^{n-1}$, i.e., $|\mu| \leq kn$, so we can in any case take $r_k = kn + 1$.)

Now let $H \in \mathfrak{a}^C$ and consider the characteristic polynomial

$$\det(\lambda I - (\operatorname{ad} H)^2 | \mathfrak{p}^C) \equiv \lambda^n + a_{n-1}(H)\lambda^{n-1} + \cdots + a_l(H)\lambda^l, \tag{11}$$

the roots of which are 0, $\alpha(H)^2$ $(\alpha \in \Sigma)$. There exist unique polynomials $p_{n-1}, \ldots, p_l$ such that $a_i = p_i(j_1, \ldots, j_l)$. This means that

$$a_i(H) = p_i(j(H)), \qquad H \in \mathfrak{a}^C. \tag{12}$$

Let $D \subset C^l$ be any compact set. Let

$$k = \sup_{\substack{\xi \in D, \\ \text{all } i}} |p_i(\xi)|.$$

Then by (12), $|a_i(H)| \leq k$ for all $H \in j^{-1}(D)$. But then by the above

$$|\alpha(H)| \leq r_k^{1/2} \qquad \text{for all } H \in j^{-1}(D).$$

This means that the closed set $j^{-1}(D)$ is bounded, hence compact, and so j is proper.

Lemma 9.5. *Each W-invariant holomorphic function ϕ on $\mathfrak{a}^C$ extends uniquely to a K-invariant holomorphic function Φ on $\mathfrak{p}^C$.*

Proof. Let

$$\phi(H) = \sum_{(\alpha)} a_{\alpha_1, \ldots, \alpha_l} h_1^{\alpha_1} \cdots h_l^{\alpha_l}, \qquad H = (h_1, \ldots, h_l) \in \mathfrak{a}^C, \tag{13}$$

be the Taylor expansion of ϕ, which we write as

$$\phi(H) = \sum_{m=0}^{\infty} A_m(H), \tag{14}$$

where

$$A_m(H) = \sum_{\alpha_1 + \cdots + \alpha_l = m} a_{\alpha_1, \ldots, \alpha_l} h_1^{\alpha_1} \cdots h_l^{\alpha_l}.$$

Each A_m is W-invariant and can therefore be expressed

$$A_m = \sum_{(\beta)} b_{\beta_1, \ldots, \beta_l} j_1^{\beta_1} \cdots j_l^{\beta_l},$$

where $\sum_1^l \beta_i \deg(j_i) = m$. Since the mapping $j: \mathfrak{a}^C \to C^l$ is surjective we conclude from the absolute convergence of (14) that the series

$$\phi^*(\xi) = \sum_0^\infty B_m(\xi), \tag{15}$$

where $\xi = (\xi_1, \ldots, \xi_l) \in C^l$ and

$$B_m(\xi) = \sum_\beta b_{\beta_1, \ldots, \beta_l} \xi_1^{\beta_1} \cdots \xi_l^{\beta_l} \qquad \left(\sum_i \beta_i \deg(j_i) = m\right)$$

converges absolutely. We shall now prove that (15) converges uniformly on each compact subset $D \subset C^l$. [Since (15) is not a power series this is not evident.] By Lemma 9.4 the set $\tilde{D} = j^{-1}(D)$ is compact. The series (13) is a power series so it converges uniformly on the compact set $\tilde{D}$; so does the series (14). But $B_m(j(H)) = A_m(H)$, so we conclude that (15) converges uniformly for $\xi \in j(\tilde{D}) = D$. This proves that ϕ^* is holomorphic on C^l. Now each j_i extends to a K-invariant polynomial J_i on $\mathfrak{p}^C$ (Chapter II, Corollary 5.12). The function $\Phi(Z) = \phi^*(J_1(Z), \ldots, J_l(Z))$ has the property required in the lemma.

In order to prove Theorem 9.3 let $\mathfrak{g}^C = \mathfrak{k}^C + \mathfrak{p}^C$ be the complexification of the Cartan decomposition $\mathfrak{g} = \mathfrak{k} + \mathfrak{p}$. Let G^C be the simply connected Lie group with Lie algebra $\mathfrak{g}^C$, K^C, G, and U the analytic subgroups corresponding to $\mathfrak{k}^C$, $\mathfrak{g}$ and $\mathfrak{u} = \mathfrak{k} + i\mathfrak{p}$. For $X \in \mathfrak{p}^C$ we write $X = X' + iX''$ ($X', X'' \in \mathfrak{p}$). Let $\phi \in \mathscr{H}^R(\mathfrak{a}^C)$ be W-invariant and Φ the extension to $\mathfrak{p}^C$ given by Lemma 9.5. If $Z \in \mathfrak{p}^C$ the mapping $k \to \Phi(k \cdot Z) - \Phi(Z)$ is holomorphic on K^C and vanishes for $k \in K$; hence it vanishes on K^C so Φ is K^C-invariant. In order to deduce from the inequality

$$|\phi(H' + iH'')| \leq Ce^{R|H''|}, \qquad H', H'' \in \mathfrak{a}, \tag{16}$$

an inequality

$$|\Phi(X' + iX'')| \leq Ce^{R|X''|}, \qquad X', X'' \in \mathfrak{p}, \tag{17}$$

by means of the K^C-invariance of Φ it would suffice to show that the infimum of $|Z''|$ as Z runs through the orbit $K^C \cdot (X' + iX'')$ is reached for some $Z = H' + iH''$ in $\mathfrak{a}^C$; in fact then

$$|\Phi(X' + iX'')| = |\phi(H' + iH'')| \leq Ce^{R|H''|} \leq Ce^{R|X''|}.$$

Let $\mathfrak{g}_{\boldsymbol{C}}^{\boldsymbol{R}}$ denote the Lie algebra $\mathfrak{g}^{\boldsymbol{C}}$ considered as a real Lie algebra and $B^{\boldsymbol{R}}$ its Killing form. Then $B^{\boldsymbol{R}} = 2B$ on $\mathfrak{g} \times \mathfrak{g}$. We have the Cartan decompositions

$$\mathfrak{g}_{\boldsymbol{C}}^{\boldsymbol{R}} = \mathfrak{u} + i\mathfrak{u}, \qquad G^{\boldsymbol{C}} = U \exp(i\mathfrak{u})$$

and we let η denote the corresponding Cartan involution of $\mathfrak{g}_{\boldsymbol{C}}^{\boldsymbol{R}}$ and $G^{\boldsymbol{C}}$. If $Z = Z' + iZ''$ $(Z', Z'' \in \mathfrak{p})$ then $\eta(Z) = -Z' + iZ''$ so

$$Z'' = -\tfrac{1}{2}i(Z + \eta(Z)).$$

Using the invariance of $B^{\boldsymbol{R}}$ under $\mathrm{Ad}(K^{\boldsymbol{C}})$ and η, and the relation $B^{\boldsymbol{R}}(iZ, iZ) = -B^{\boldsymbol{R}}(Z, Z)$ we obtain for $k \in K^{\boldsymbol{C}}$

(18) $$|(k \cdot Z)''|^2 = -\tfrac{1}{4}\{B^{\boldsymbol{R}}(Z, Z) + B^{\boldsymbol{R}}(\eta(k \cdot Z), k \cdot Z)\}.$$

Now let $X \in \mathfrak{p}^{\boldsymbol{C}}$ be regular (and hence semisimple). Then the orbit $K^{\boldsymbol{C}} \cdot X$ is closed (Chapter III, Theorem 4.19). Since $B^{\boldsymbol{R}}(\eta(Z), Z) \leq 0$ it is clear that the function $k \to -B^{\boldsymbol{R}}(\eta(k \cdot X), k \cdot X)$ has a minimum on $K^{\boldsymbol{C}}$. By (18) the same is the case for the function $k \to |(k \cdot X)''|^2$ on $K^{\boldsymbol{C}}$. If $k_c \in K^{\boldsymbol{C}}$ is a minimum point we have by (18)

$$\left(\frac{d}{dt} B^{\boldsymbol{R}}(\eta((\exp tTk_c) \cdot X), \exp tTk_c \cdot X)\right)_{t=0} = 0$$

for each $T \in \mathfrak{k}^{\boldsymbol{C}}$. Using $\eta^2 = 1$ it follows that

$$B^{\boldsymbol{R}}(\eta(k_c \cdot X), [\mathfrak{k}^{\boldsymbol{C}}, k_c \cdot X]) = 0.$$

But $B^{\boldsymbol{R}}$ is invariant and nondegenerate on $\mathfrak{k}^{\boldsymbol{C}}$ so this implies

(19) $$[\eta(k_c \cdot X), k_c \cdot X] = 0.$$

This shows that the elements $(k_c \cdot X)'$ and $(k_c \cdot X)''$ commute; since they both belong to $\mathfrak{p}$ there exists a maximal abelian subspace $\mathfrak{b}$ of $\mathfrak{p}$ containing them both. Since $\mathfrak{b}$ is conjugate to $\mathfrak{a}$ under K there exists an element $k_0 \in K$ such that the vector $H = k_0(k_c \cdot X)$ is in $\mathfrak{a}^{\boldsymbol{C}}$. Then

$$H'' = (k_0 k_c \cdot X)'' = k_0 \cdot ((k_c \cdot X)''),$$

so

(20) $$|H''| = \min_{k \in K^{\boldsymbol{C}}} |(k \cdot X)''|.$$

This proves that if $\phi \in \mathscr{H}_W^R(\mathfrak{a}^{\boldsymbol{C}})$ then its K-invariant holomorphic extension Φ satisfies

(21) $$|\Phi(X)| \leq Ce^{R|X''|}$$

for all $X \in \mathfrak{p}^{\boldsymbol{C}}$ which are regular and semisimple, C being a constant. But such X are dense in $\mathfrak{p}^{\boldsymbol{C}}$ so (21) holds on all of $\mathfrak{p}^{\boldsymbol{C}}$ by continuity. We have also by the K-invariance of Φ on $\mathfrak{p}$ that for each $N \in Z^+$ there exists a constant C_N such that

$$(22) \qquad |\Phi(X')| \leq C_N(1 + |X'|)^{-N}, \qquad X' \in \mathfrak{p}.$$

Let $\varepsilon > 0$ and let F be an arbitrary entire function on $\mathfrak{p}^C$ of exponential type ε [cf. (2) of §7]. Then ΦF is entire of exponential type $R + \varepsilon$ and so, as recalled at the beginning of §7, the Fourier transform $(\Phi F)^{\sim} = \tilde{\Phi} * \tilde{F}$ has support in the ball $\mathrm{Cl}(B_{R+\varepsilon}(0))$. Since $\tilde{F}$ can be arbitrary in $\mathscr{D}_{\bar{B}_\varepsilon(0)}(\mathfrak{p})$ it follows that $\tilde{\Phi}$ [which by (22) is smooth] vanishes for $Y \in \mathfrak{p}$, $|Y| > R$. Hence $\Phi = \mathscr{H}^R(\mathfrak{p}^C)$ so Theorem 9.3 is proved.

Now Theorem 9.2 follows immediately: If $\psi \in \mathscr{H}^R_W(\mathfrak{a}^*)$ we define $\phi \in \mathscr{H}^R_W(\mathfrak{a}^C)$ by $\phi(A_\lambda) = \psi(\lambda)$ and define the K-invariant $\Phi \in \mathscr{H}^R(\mathfrak{p}^C)$ as above. Then

$$\Phi(X) = \int_{\mathfrak{p}} f(Y) e^{-iB(X,Y)}\, dY,$$

where $f \in \mathscr{D}_K(\mathfrak{p})$ and

$$\Psi(\lambda) = \Phi(A_\lambda) = \int_K \Phi(kA_\lambda)\, dk = \tilde{f}(\lambda),$$

as claimed.

Taking (8) into account we obtain from Theorem 9.2 the following analog of Corollary 7.4.

Corollary 9.6. *The mapping $f \to \tilde{f}$ is a linear homeomorphism of $\mathscr{D}_K(\mathfrak{p})$ onto $\mathscr{D}_W(\mathfrak{a})$.*

§10. Convexity Theorems

Let $\mathfrak{p}_0$ denote the space of all $n \times n$ complex Hermitian matrices and let $\mathfrak{a}_0 \subset \mathfrak{p}_0$ denote the subspace of all diagonal Hermitian matrices. Consider the projection $Q: \mathfrak{p}_0 \to \mathfrak{a}_0$ defined by $(QY)_{ij} = \delta_{ij} Y_{ij}$; in other words, QY is the diagonal in Y.

If $Y \in \mathfrak{p}_0$ let $\lambda_1 \geq \lambda_2 \geq \cdots \geq \lambda_n$ be the eigenvalues of Y and $H \in \mathfrak{a}_0$ the diagonal matrix with entries $\lambda_1, \ldots, \lambda_n$, i.e., $H_{ij} = \delta_{ij}\lambda_i$. Furthermore, if σ is any permutation of $\{1, 2, \ldots, n\}$ let σH be the diagonal matrix with diagonal $\lambda_{\sigma^{-1}(i)}$, i.e., $(\sigma H)_{ij} = \delta_{ij}\lambda_{\sigma^{-1}(i)}$. Let $C(H)$ denote the convex hull of the set of σH as σ runs over the permutation group $\mathfrak{S}_n$, and let $O(H)$ denote the set of all $n \times n$ Hermitian matrices with the eigenvalues $\lambda_1, \ldots, \lambda_n$. In [1954a], Horn proved the following result.

Theorem 10.1.

$$Q(O(H)) = C(H).$$

We shall now state and prove Kostant's generalization of this result to the tangent space $\mathfrak{p}$ to a symmetric space $X = G/K$ of the noncompact type. As usual let $\mathfrak{a} \subset \mathfrak{p}$ be a maximal abelian subspace and let $Q: \mathfrak{p} \to \mathfrak{a}$ denote the orthogonal projection (with respect to the Killing form). For $Y \in \mathfrak{p}$ put $k \cdot Y = \mathrm{Ad}(k)Y$ ($k \in Y$) and as in §8 let, for $H \in \mathfrak{a}$, $C(H)$ denote the convex hull of the orbit $W \cdot H$.

Theorem 10.2. *For each* $H \in \mathfrak{a}$,

$$Q(K \cdot H) = C(H).$$

For the proof we shall use the general notation from this chapter (especially §§8, 9). We begin by proving

$$(1) \qquad Q(K \cdot H) \subset C(H).$$

If this were false there would exist an element $Z \in K \cdot H$ such that $Q(Z) \notin C(H)$. By the convexity, $Q(Z)$ and $C(H)$ can be separated by a hyperplane, so there exists an $H_0 \in \mathfrak{a}$ such that

$$(2) \qquad \langle H_0, Q(Z) \rangle > \langle H_0, \sigma(H) \rangle, \qquad \sigma \in W.$$

By continuity we can take H_0 regular. By the compactness of the orbit $K \cdot H$ the function

$$Y \to \langle H_0, Q(Y) \rangle = \langle H_0, Y \rangle$$

on $K \cdot H$ has a maximum value at a point Z_0. Then

$$\left(\frac{d}{dt} \langle H_0, \mathrm{Ad}(\exp tT)(Z_0) \rangle \right)_{t=0} = 0 \qquad \text{for} \quad T \in \mathfrak{k},$$

so $\langle H_0, [\mathfrak{k}, Z_0] \rangle = 0$. Hence $[H_0, Z_0] = 0$, and so, by the regularity of H_0, $Z_0 \in \mathfrak{a}$. But $K \cdot H \cap \mathfrak{a} = W \cdot H$ ([DS], Chapter VII, Proposition 2.2), so $Z_0 \in W \cdot H$. But then the maximality of Z_0 contradicts (2).

For the converse of (1) we prove two lemmas. If $H_1, H_2 \in \mathfrak{a}$, $\{H_1, H_2\}$ denotes the segment $H_1 - t(H_1 - H_2)$ $(0 \leq t \leq 1)$ and if $\alpha \in \Sigma$, $\mathfrak{d} \subset \mathfrak{a}$, we put

$$(3) \qquad \{\mathfrak{d}, s_\alpha \mathfrak{d}\} = \bigcup_{H \in \mathfrak{d}} \{H, s_\alpha H\}.$$

Lemma 10.3. *Let* $\mathfrak{d} \subset \mathfrak{a}$ *be a convex subset in the half-space* $\alpha \geq 0$. *Then* $\{\mathfrak{d}, s_\alpha \mathfrak{d}\}$ *is convex.*

Proof. Let $H_1, H_2 \in \{\mathfrak{d}, s_\alpha \mathfrak{d}\}$; if $r_1, r_2 \geq 0$, $r_1 + r_2 = 1$, we must show $H = r_1 H_1 + r_2 H_2 \in \{\mathfrak{d}, s_\alpha \mathfrak{d}\}$. Now $H_i \in \{A_i, s_\alpha A_i\}$, $i = 1, 2$, where $A_1, A_2 \in \mathfrak{d}$; thus, since $s_\alpha(H) = H - \alpha(H)A'_\alpha$, where $A'_\alpha = 2A_\alpha/\langle\alpha, \alpha\rangle$, we have

$$H_i = A_i - t_i A'_\alpha \qquad \text{where} \quad 0 \leq t_i \leq \alpha(A_i). \tag{4}$$

Hence

$$H = r_1(A_1 - t_1 A'_\alpha) + r_2(A_2 - t_2 A'_\alpha),$$

which belongs to the segment $\{r_1 A_1 + r_2 A_2,\ s_\alpha(r_1 A_1 + r_2 A_2)\}$ since $r_1 t_1 + r_2 t_2 \leq \alpha(r_1 A_1 + r_2 A_2)$. This proves the lemma.

A sequence $\gamma_1, \ldots, \gamma_n \in \Sigma_0^+$ (the positive indivisible roots) will be called a *strongly positive sequence* (of length n) if for each $H \in \mathfrak{a}^+$ we have

$$\gamma_i(s_{\gamma_j} s_{\gamma_{j-1}} \cdots s_{\gamma_1}(H)) \geq 0, \qquad \text{for all} \quad 1 \leq j < i \leq n. \tag{5}$$

Corollary 6.11 (and the subsequent remark) shows how each $\sigma \neq e$ in W can be written as a product $\sigma = s_{\gamma_n} \cdots s_{\gamma_1}$ where $\gamma_1, \ldots, \gamma_n$ is a strongly positive sequence.

Lemma 10.4. *Let $\mathfrak{d} \subset \mathfrak{a}$ be any subset such that for any $A \in \mathfrak{d}$ and any root $\alpha \in \Sigma$ the segment $\{A, s_\alpha A\}$ lies in $\mathfrak{d}$. Then for each $H \in \mathfrak{d}$ the convex hull $C(H)$ of $W \cdot H$ also belongs to $\mathfrak{d}$.*

Proof. Since $s_\alpha A \in \mathfrak{d}$ for each $A \in \mathfrak{d}$, we see that $\mathfrak{d}$ is W-invariant. Thus it suffices to show $C(H) \subset \mathfrak{d}$ for each $H \in \mathfrak{d} \cap C^+$ where C^+ is the closure of $\mathfrak{a}^+$. Let $H \in \mathfrak{d} \cap C^+$, let $\sigma_1, \ldots, \sigma_{n+1} \in W$, and $c(\sigma_1, \ldots, \sigma_{n+1})$ be the convex hull of $H, \sigma_1 H, \ldots, \sigma_{n+1} H$. We shall prove

$$c(\sigma_1, \ldots, \sigma_{n+1}) \subset \partial \tag{6}$$

by induction assuming $c(s_1, \ldots, s_n) \subset \mathfrak{d}$ for any $s_1, \ldots, s_n \in W$. For this it suffices to prove for any $\sigma_1, \ldots, \sigma_n \in W$,

$$c(\sigma_1, \ldots, \sigma_n, s_{\gamma_m} s_{\gamma_{m-1}} \cdots s_{\gamma_1}) \subset \mathfrak{d} \tag{7}$$

by induction on m, $\gamma_1, \ldots, \gamma_m$ being any strongly positive sequence. With the definition $s_{\gamma_m} \cdots s_{\gamma_1} = e$ if $m = 0$, (7) holds for $m = 0$; assuming it to hold for m we shall prove it for a strongly positive sequence $\gamma_1, \ldots, \gamma_{m+1}$. Put $H_m = s_{\gamma_m} \cdots s_{\gamma_1}(H)$ and $\gamma = \gamma_{m+1}$. Since $H \in C^+$ we have $\gamma(H_m) \geq 0$ and $\gamma(H) \geq 0$. Given j $(1 \leq j \leq n)$ we have $\gamma(\sigma'_j(H)) \geq 0$, where σ'_j is one of the elements σ_j or $s_\gamma \sigma_j$. By induction, the set

$$\mathfrak{d}_1 = c(\sigma'_1, \ldots, \sigma'_n, s_{\gamma_m} \cdots s_{\gamma_1}) \subset \mathfrak{d}.$$

Also $\mathfrak{d}_1$ lies in the half space $\gamma \geq 0$, so by Lemma 10.3, $\{\mathfrak{d}_1, s_\gamma \mathfrak{d}_1\}$ is convex. By assumption, $A \in \mathfrak{d}$, $\alpha \in \Sigma$ implies $\{A, s_\alpha A\} \subset \mathfrak{d}$. Thus

$\{\mathfrak{d}_1, s_\gamma \mathfrak{d}_1\} \subset \mathfrak{d}$. Since σ_j is one of the elements σ'_j or $s_\gamma \sigma'_j$ we have $\sigma_j H \subset \{\partial_1, s_\gamma \partial_1\}$ for each j. Also

$$H_{m+1} = s_\gamma(H_m) \in \{\mathfrak{d}_1, s_\gamma \mathfrak{d}_1\},$$

so by the convexity of $\{\mathfrak{d}_1, s_\gamma \mathfrak{d}_1\}$

$$c(\sigma_1, \ldots, \sigma_n, s_{\gamma_{m+1}} \cdots s_{\gamma_1}) \subset \{\mathfrak{d}_1, s_\gamma \mathfrak{d}_1\} \subset \mathfrak{d}.$$

This proves (7) for all m. Thus (6) holds for all $\sigma_1, \ldots, \sigma_{n+1} \in W$ and the lemma is proved.

Returning now to the proof of Theorem 10.2 we put $\mathfrak{c} = Q(K \cdot H)$. It remains to prove $C(H) \subset \mathfrak{c}$. Because of Lemma 10.4 it suffices to prove

$$\text{(8)} \qquad \{Y, s_\alpha Y\} \subset \mathfrak{c} \qquad \text{whenever} \quad Y \in \mathfrak{c}, \alpha \in \Sigma_0^+.$$

For this we consider as in [DS] (Chapter IX, §2) the (semisimple) subalgebra $\mathfrak{g}^\alpha$ of $\mathfrak{g}$ generated by the root spaces $\mathfrak{g}_\alpha$ and $\mathfrak{g}_{-\alpha}$. It has the Cartan decomposition $\mathfrak{g}^\alpha = \mathfrak{k}^\alpha + \mathfrak{p}^\alpha$ where $\mathfrak{k}^\alpha = \mathfrak{g}^\alpha \cap \mathfrak{k}$, $\mathfrak{p}^\alpha = \mathfrak{p} \cap \mathfrak{g}^\alpha$; the line $\mathfrak{a}^\alpha = \boldsymbol{R} A_\alpha$ is maximal abelian in $\mathfrak{p}^\alpha$ so the corresponding symmetric space G^α/K^α has rank one. We have the orthogonal decomposition

$$\text{(9)} \qquad \mathfrak{a} = \mathfrak{a}^\alpha + \pi_\alpha,$$

where π_α is the hyperplane $\alpha = 0$ in $\mathfrak{a}$. Let $Y = Y_\alpha + Y_\alpha^\perp$ be the corresponding decomposition of Y. Then $Y_\alpha = \frac{1}{2}\alpha(Y) A'_\alpha$, and

$$\text{(10)} \qquad \{Y, s_\alpha Y\} = \{t Y_\alpha + Y_\alpha^\perp : -1 \leq t \leq 1\}.$$

Now $Y = Q(u \cdot H)$ for some $u \in K$. Consider the mapping $\sigma : K^\alpha \to \mathfrak{a}$ defined by $\sigma(k) = Q(ku \cdot H)$. Then

$$\text{(11)} \qquad \sigma(K^\alpha) \subset \mathfrak{c}.$$

If $Y_\alpha = 0$ then $\{Y, s_\alpha Y\}$ is the single point Y so (8) is trivial. Hence we assume $Y_\alpha \neq 0$. Then we have unique continuous functions $\sigma_\alpha : K^\alpha \to \boldsymbol{R}$, $\sigma_\alpha^\perp : K^\alpha \to \pi_\alpha$ such that

$$\sigma(k) = \sigma_\alpha(k) Y_\alpha + \sigma_\alpha^\perp(k), \quad k \in K^\alpha.$$

We shall prove that the function $\sigma_\alpha^\perp$ is identically constant. For this it suffices to show that for each $H_0 \in \pi_\alpha$, $\langle H_0, \sigma(k) \rangle$ is constant in $k \in K^\alpha$. However,

$$\langle H_0, \sigma(k) \rangle = \langle H_0, Q(ku \cdot H) \rangle = \langle H_0, ku \cdot H \rangle = \langle k^{-1} H_0, u \cdot H \rangle$$

and $k^{-1} H_0 = H_0$ since $[\pi_\alpha, \mathfrak{g}^\alpha] = 0$. Thus $\sigma_\alpha^\perp(k) \equiv \sigma_\alpha^\perp(e) = Y_\alpha^\perp$ so $\sigma(k) = \sigma_\alpha(k) Y_\alpha + Y_\alpha^\perp$. Select $k_\alpha \in K^\alpha$ such that $\mathrm{Ad}(k_\alpha) Y_\alpha = -Y_\alpha$. Since

$$\text{(12)} \qquad [\mathfrak{g}^\alpha, \pi_\alpha] = 0$$

we have $\mathrm{Ad}(k_\alpha) = 1$ on π_α. Hence $\mathrm{Ad}(k_\alpha)\mathfrak{a} = \mathfrak{a}$ and $\sigma(k_\alpha) = Q(k_\alpha u \cdot H) = \mathrm{Ad}(k_\alpha)(Q(u \cdot H)) = -Y_\alpha + Y_\alpha^\perp$. This shows that $\sigma_\alpha(k_\alpha) = -1$ so by the connectedness of K^α, $\sigma_\alpha(K^\alpha) \supset [-1, 1]$. But then (10) and (11) show that

$$\{Y, s_\alpha Y\} \subset \sigma(K^\alpha) \subset \mathfrak{c},$$

and the theorem is proved.

We have seen that the Iwasawa decomposition $G = KAN$, in particular the component $H(g)$ in $g = k(g) \exp H(g)n(g)$ plays an important role in the theory of the spherical transform. The Killing form B of $\mathfrak{g}$ defines a norm $|\ |$ on $\mathfrak{p}$ and a distance function d on G/K. We know from [DS] [Chapter VI, Exercise B2(iv)] that if $o = \{K\}$ then

$$(13) \qquad |H(g)| \leq d(o, g \cdot o), \qquad g \in G,$$

with equality holding only if $g \in KA$. In particular,

$$(14) \qquad |H(ak)| \leq |\log a|, \qquad a \in A, \quad k \in K.$$

We shall now prove a result of Kostant [1973] determining the set $\{H(ak : k \in K\}$ quite explicitly. Again the result is closely related to one of Horn [1954b] and Thompson [1971] for $G = \boldsymbol{SL}(n, \boldsymbol{C})$. The relationship is explained in Kostant's paper.

Theorem 10.5. *Let $a \in A$. Then the set $\{H(ak) : k \in K\}$ equals $C(\log a)$, the convex hull of the orbit $W \cdot \log a$.*

Let us interpret this geometrically in terms of the space $X = G/K$. Each point $x \in X$ can be written $x = n_x \exp A(x) \cdot o$ where $n_x \in N$, $A(x) \in \mathfrak{a}$ are unique. Since the orbit $N \cdot \exp A(x) \cdot o = \exp A(x) N \cdot o$ intersects $A \cdot o$ orthogonally at $\exp A(x) \cdot o$ ([DS], Chapter VI, Exercise B.2) it is reasonable to call $A(x)$ the *N-projection* of the point x. Note that for $g \in G$, $A(g)$ in §4. No. 2 equals the present $A(gK)$. Also, $A(gK) = -H(g^{-1})$ so $A(kaK) = -H(a^{-1}k^{-1})$. Theorem 10.5 can therefore be stated

Theorem 10.6. *Let $a \in A$. The N-projection of the orbit $Ka \cdot o$ in X equals the convex hull $C(\log a)$ of $W \cdot \log a$.*

Because of the orthogonality $(Na \cdot o)_{a \cdot o} \perp (A \cdot o)_{a \cdot o}$, Theorem 10.2 can be viewed as an infinitesimal analog of Theorem 10.6.

Lemma 10.7. *Let X have rank one. Let $b \in A$, $n \in N$. Then for any $a \in \exp C(\log b)$ there exist elements $k, v \in K$, $n' \in N$ such that*

$$(15) \qquad kbnv = an'.$$

Proof. As used above

$$d(o, a \cdot o) \leq d(o, b \cdot o) \leq d(o, bn \cdot o).$$

The distance $d(o, an' \cdot o)$ varies with $n' \in N$ from $d(o, a \cdot o)$ to ∞. Thus there exists an $n' \in N$ such that $d(o, an' \cdot o) = d(o, bn \cdot o)$. Hence $kbn \cdot o = an' \cdot o$ for some $k \in K$ so (14) follows.

Corollary 10.8. *Theorem 10.5 holds for X of rank one.*

In fact if we take $n = e$ in (15) we deduce $\{H(bv) : v \in K\} \supset C(\log b)$. The converse is (14).

We shall now prove Theorem 10.5. For $b \in A$ let

$$\mathfrak{s} = \{H(bk) : k \in K\}.$$

Lemma 10.9. *The set $\mathfrak{s}$ is W-invariant. In fact, for each $H \in \mathfrak{s}$ and each $\alpha \in \Sigma$,*

$$\{H, s_\alpha H\} \subset \mathfrak{s}.$$

Proof. Suppose first α is simple and consider $\mathfrak{g}^\alpha$, $\mathfrak{k}^\alpha$, $\mathfrak{p}^\alpha$, $\mathfrak{a}^\alpha = \boldsymbol{R}A_\alpha$ as above. Corresponding to $\boldsymbol{R}^+ A_\alpha$ as a positive Weyl chamber in $\mathfrak{a}^\alpha$ we have the Iwasawa decompositions $\mathfrak{g}^\alpha = \mathfrak{k}^\alpha + \mathfrak{a}^\alpha + \mathfrak{n}^\alpha$, $G^\alpha = K^\alpha A^\alpha N^\alpha$ compatible with $\mathfrak{g} = \mathfrak{k} + \mathfrak{a} + \mathfrak{n}$ and $G = KAN$. As in §6 we put $(\alpha) = \Sigma^+ \cap \{\alpha, 2\alpha\}$ and consider the subspace $\mathfrak{n}' = \sum_{\beta \in \Sigma^+ - (\alpha)} \mathfrak{g}_\beta$. Then $\mathfrak{n} = \mathfrak{n}^\alpha + \mathfrak{n}'$ and since α is simple, we have

$$[\mathfrak{g}^\alpha, \mathfrak{n}'] \subset \mathfrak{n}'. \tag{16}$$

If $\gamma_1 < \gamma_2 < \cdots$ are the elements of Σ^+ in increasing order and we write $\mathfrak{n}^{(i)} = \sum_{j \geq i} \mathfrak{g}_{\gamma_j}$ then Lemma 6.8 applies to the subspaces $\mathfrak{n}^\alpha$ and $\mathfrak{n}'$ so we get the diffeomorphism decomposition

$$N = N^\alpha N' \qquad \text{where} \quad N' = \exp \mathfrak{n}'. \tag{17}$$

Let $H \in \mathfrak{s}$, $a = \exp H$. Then there exist $k, v \in K$, $n \in N$ such that

$$bv = kan. \tag{18}$$

We decompose $H = H_\alpha + H_\alpha^\perp$ by (9), so by (10)

$$\{H, s_\alpha H\} = \{tH_\alpha + H_\alpha^\perp : -1 \leq t \leq 1\} = C_\alpha(H_\alpha) + H_\alpha^\perp, \tag{19}$$

where $C_\alpha(H_\alpha)$ is the convex hull of H_α and $-H_\alpha$. We put $a_\alpha = \exp H_\alpha$, $a_\alpha^\perp = \exp H_\alpha^\perp$ and write $n = n_\alpha n'$ in accordance with (17). Then by (18) and (12) we have

$$k^{-1}bv = a_\alpha a_\alpha^\perp n_\alpha n' = a_\alpha n_\alpha a_\alpha^\perp n'. \tag{20}$$

Let $c_\alpha \in \exp C_\alpha(H_\alpha)$. By Lemma 10.7 there exist k_α, $v_\alpha \in K^\alpha$, $n_\alpha^0 \in N^\alpha$ such that

$$k_\alpha a_\alpha n_\alpha v_\alpha = c_\alpha n_\alpha^0,$$

whence by (20) and (12),

$$k_\alpha k^{-1} bvv_\alpha = c_\alpha n_\alpha^0 v_\alpha^{-1} a_\alpha^\perp n' v_\alpha = c_\alpha a_\alpha^\perp n_\alpha^0 v_\alpha^{-1} n' v_\alpha. \tag{21}$$

Using (16) we see that this last element belongs to $c_\alpha a_\alpha^\perp N$. Thus (21) implies

$$H(bvv_\alpha) = \log(c_\alpha a_\alpha^\perp) = \log c_\alpha + H_\alpha^\perp. \tag{22}$$

Since $\log c_\alpha$ was arbitrary in $C_\alpha(H_\alpha)$ (22) and (19) show

$$\{H, s_\alpha H\} \subset \mathfrak{s} \qquad \text{if} \quad \alpha \text{ is simple.} \tag{23}$$

This proves that $\mathfrak{s}$ is W-invariant since W is generated by the reflections s_α for α simple.

Now if $\alpha \in \Sigma_0$ is arbitrary there exists a $\sigma \in W$ such that $\sigma\alpha$ is simple. Then we have by (23), for each $H \in \mathfrak{s}$,

$$\mathfrak{s} \supset \{H, s_{\sigma\alpha}(H)\} = \{H, \sigma s_\alpha \sigma^{-1} H\} = \sigma\{\sigma^{-1}H, s_\alpha \sigma^{-1} H\},$$

and this relation implies the lemma now that we know that $\mathfrak{s}$ is W-invariant.

Lemmas 10.4 and 10.9 now imply that for each $b \in A$

$$C(\log b) \subset \{H(bk) : k \in K\}. \tag{24}$$

For the converse inclusion we may, since both sides are invariant under the maps $\log b \to s(\log b)$ $(s \in W)$, assume that $\log b \in C^+$. Then by Lemma 6.5,

$$H(bk) \subset \log b + {}^-C \qquad \text{for all} \quad k \in K;$$

that is, $\mathfrak{s} \subset \log b + {}^-C$. But since $\mathfrak{s}$ is W-invariant this implies

$$\mathfrak{s} \subset s(\log b + {}^-C)$$

for each $s \in W$, whence by Lemma 8.3

$$\mathfrak{s} \subset C(\log b). \tag{25}$$

This proves Theorem 10.5.

Corollary 10.10. $G = KNK$.

In fact if $H \in \mathfrak{a}$ the sum $H_0 = \sum_{\sigma \in W} \sigma H$ is fixed under each reflection s_α $(\alpha \in \Sigma)$ so $H_0 = 0$. Thus $0 \in C(H)$ so $H(\exp Hk) = 0$ for some $k \in K$. Thus $\exp Hk \in KN$, proving the corollary (since $G = KAK$).

Theorem 10.11. (i) *Assume $b \in A$ is such that the convex hull $C(\log b)$ satisfies*

$$\dim C(\log b) = \dim \mathfrak{a}. \tag{26}$$

Then the measure μ_b *given by*

$$\int_K f(H(bk))\,dk = \int_{\mathfrak{a}} f(H)\,d\mu_b(H), \qquad f \in C_c(\mathfrak{a}), \tag{27}$$

satisfies

$$\operatorname{supp}(\mu_b) = C(\log b), \tag{28}$$

$$\mu_b = F_b(H)\,dH, \qquad \text{where} \quad F_b \in L^1(\mathfrak{a}). \tag{29}$$

(ii) *Condition* (26) *holds whenever* $\log b \in \mathfrak{a}$ *is regular and more generally if for the decomposition* $X = \prod_i X_i$, $\mathfrak{a} = \sum_i \mathfrak{a}_i$, *into irreducible factors* X_i *each component of* $\log b$ *is* $\neq 0$.

Proof. (i) It is clear that μ_b as defined by (26) is a measure (Chapter I, §1). Also, if $\operatorname{supp}(f) \cap C(\log b) = 0$ we have $\mu_b(f) = 0$ by (25) so the support of μ_b is contained in the compact set $C(\log b)$.

Consider now the map $g_b: K \to \mathfrak{a}$ given by $g_b(k) = H(bk)$. By Theorem 10.5, $g_b(K) = C(\log b)$ whose interior, say $C_0(\log b)$, is by (26) an open subset of $\mathfrak{a}$. Let $K_0 = g_b^{-1}(C_0(\log b))$. By Sard's theorem (see, e.g., Guillemin and Pollak [1974]) for the map $g_b: K_0 \to C_0(\log b)$, almost every point H_0 in $C_0(\log b)$ is a regular value; that is, for each $k \in K_0 \cap g_b^{-1}(H_0)$, g_b is a submersion at k ([DS], Chapter I, §15). Since g_b being a submersion at k amounts to the nonvanishing of certain determinants, we see that g_b is a submersion on a set $K - C$, dense in K, where C is a certain closed null set for the Haar measure m_K on K. Now, by (27) if $S \subset \mathfrak{a}$,

$$\mu_b(S) = m_K(g_b^{-1}(S)) = m_K(g_b^{-1}(S) \cap (K - C)). \tag{30}$$

Let $k_0 \in K - C$. On suitable coordinate neighborhoods of the points $k_0 \in K - C$, $g_b(k_0) \in \mathfrak{a}$, the map g_b is given by the projection $(x_1, \ldots, x_p) \to (x_1, \ldots, x_l)$, ([DS], Chapter I, §15). Since dm_K is there given by $F\,dx_1 \cdots dx_p$ $(F > 0)$, it follows from the Fubini theorem that if $S \subset \mathfrak{a}$ is a null set then $m_K(g_b^{-1}(S) \cap (K - C)) = 0$. Thus by (30) $\mu_b(S) = 0$ so μ_b is absolutely continuous and (29) is proved.

From the above representation of g_b as a projection near k_0 it is obvious that $g_b(k_0) \in \operatorname{supp}(\mu_b)$. Hence $g_b(K - C) \subset \operatorname{supp}(\mu_b)$, and since $K - C$ is dense in K,

$$C(\log b) = g_b(K) \subset \operatorname{supp}(\mu_b).$$

(ii) Suppose $\log b = \sum_i \log b_i$ according to the decomposition $\mathfrak{a} = \sum_i \mathfrak{a}_i$. The Weyl group W_i for X_i acts irreducibly on $\mathfrak{a}_i$ ([DS], Chapter X §3, No. 3), so since $\log b_i \neq 0$ the set $W_i \cdot \log b_i$ spans $\mathfrak{a}_i$. Thus (26) holds for each $\mathfrak{a}_i$, hence also for $\mathfrak{a}$ itself.

Corollary 10.12. *With $b \in A$ as in Theorem 10.11 the function $\lambda \to \phi_\lambda$ is the Fourier transform of $e^{-\rho}F_b \in L^1(\mathfrak{a})$:*

$$\phi_\lambda(b) = \int_K e^{(i\lambda-\rho)(H(bk))}\,dk = \int_{\mathfrak{a}} e^{i\lambda(H)}(e^{-\rho(H)}F_b(H))\,dH.$$

In particular, by the Riemann–Lebesgue lemma,

$$\lim_{\lambda\in\mathfrak{a}^*,\,|\lambda|\to\infty} \phi_\lambda(b) = 0.$$

We shall finally use Theorem 10.11 to obtain a simple result about restrictions of the functions in $C_c^\natural(G)$ to A. Let $C_c^W(A)$ denote the space of W-invariant functions in $C_c(A)$. Viewing A as a vector space, let $C(a)$ denote the convex hull of the orbit $W \cdot a$, a being arbitrary in A.

Proposition 10.13.

(i) *The restriction from G to A induces a bijection of $C_c^\natural(G)$ onto $C_c^W(A)$.*

(ii) *Suppose X is irreducible and let $b, c \in A$, b, c, bc all $\neq e$. Then there exists a unique W-invariant measure $\mu_{b,c}$ on A such that*

$$\int_K f(bkc)\,dk = \int_A f(a)\,d\mu_{b,c}(a), \qquad f \in C_c^\natural(G). \tag{31}$$

Moreover, $\mu_{b,c}$ has support in $C(b)C(c)$.

Proof. The bijection $(k, X) \to k \exp X$ of $K \times \mathfrak{p}$ onto G identifies the functions in $C_c^\natural(G)$ with $\mathrm{Ad}(K)$-invariant functions in $C_c(\mathfrak{p})$ which under restriction to $\mathfrak{a}$ are identified with $C_c^W(\mathfrak{a})$ (cf. Proposition 5.18 in Chapter I). This proves (i).

This identification means that the functional

$$f \to \int_K f(bkc)\,dk \quad (f \in C_c^\natural(G))$$

can be regarded as a positive W-invariant linear functional on $C_c^W(A)$. Thus the measure $\mu_{b,c}$ exists. Also, $\mu_{b,c}$ has compact support; in fact let $D \subset A$ be a compact set such that $bKc \subset KDK$. Then $\operatorname{supp} \mu_{b,c} \subset D$. Hence formula (31) holds for all $f \in C^\natural(G)$. In particular, we can take $f = \phi_\lambda$. Using the functional equation for ϕ_λ we then obtain

$$\phi_\lambda(b)\phi_\lambda(c) = \int_A \phi_\lambda(a)\,d\mu_{b,c}(a). \tag{32}$$

Our assumption $bc \neq e$ implies that $e \notin bKc$; thus if $f \in C_c^\natural(G)$ has support in a suitably small neighborhood of e in G we have $\int f(bkc)\,dk = 0$. In particular, $\mu_{b,c}(\{e\}) = 0$. Thus we can replace A by $A - \{e\}$ in (32) and use Corollary 10.12 on $\phi_\lambda(a)$ $(a \neq e)$, $\phi_\lambda(b)$, $\phi_\lambda(c)$. Writing $f_a(h) = e^{-\rho(H)}F_a(H)$ if $h = \exp H$, we deduce from (32)

$$f_b * f_c(h) = \int_{A-\{e\}} f_a(h)\, d\mu_{b,c}(a), \tag{33}$$

for almost all $h \in A$, $*$ denoting convolution on A.

We now have to prove that each $a_0 \notin C(b)C(c)$, $a_0 \neq e$, has a spherical neighborhood A_0 such that $\mu_{b,c}(A_0) = 0$. Superscript ^{0} denoting interior, we select

$$u_0 \in C(a_0)^0 - C(b)C(c).$$

By continuity we can then find spherical neighborhoods U_0 of u_0, A_0 of a_0, such that

$$U_0 \subset C(a)^0 - C(b)C(c) \qquad \text{for} \quad a \in A_0. \tag{34}$$

Let $\phi \in C_c(U_0)$, $\phi \geq 0$, $\phi \not\equiv 0$. Since $f_b * f_c$ vanishes almost everywhere outside $C(b)C(c)$, (33) implies

$$\int_{A-\{e\}} \left(\int_A f_a(h)\phi(h)\, dh \right) d\mu_{b,c}(a) = 0.$$

By the positivity we have, a fortiori,

$$\int_{A_0} \left(\int_A f_a(h)\phi(h)\, dh \right) d\mu_{b,c}(a) = 0. \tag{35}$$

But the inner integral equals $\mu_a(e^{-\rho}\phi)$, which by (27) depends continuously on a. Also, by (28)

$$U_0 \subset \operatorname{supp}(\mu_a) \qquad \text{if} \quad a \in A_0,$$

so by general measure theory,

$$\mu_a(e^{-\rho}\phi) > 0 \qquad \text{for} \quad a \in A_0.$$

But then (35) implies $\mu_{b,c}(A_0) = 0$ as claimed. Thus $\operatorname{supp}(\mu_{b,c}) \subset C(b)C(c)$ and the proposition is proved.

EXERCISES AND FURTHER RESULTS

A. Representations

1. (i) Let π be a representation of a Lie group G on a complete locally convex space V. If $v \in V$ show (cf. Corollary 1.4) that $\pi(\mathcal{D}(G))v$ is dense in the closed invariant subspace of V generated by v.

(ii)* Assume V is a Fréchet space; let V^∞ be the space of differentiable vectors in V and V_∞ the set of linear combinations of vectors $\pi(f)v$ ($f \in \mathcal{D}(G)$, $v \in V$) (the *Gårding subspace*). Then

$$V^\infty = V_\infty$$

(Dixmier and Malliavin [1978]).

2. Show that if two unitary representations are equivalent they are unitarily equivalent.

3. (i) Let $I \subset \boldsymbol{R}$ be an open interval and $t \to x_t$ a mapping of I into a complex Banach space V, with dual V'. Assume that for each $\lambda \in V'$ the mapping $t \to \langle x_t, \lambda \rangle$ is of class C^2. Then the mapping $t \to x_t$ is C^1.

(ii) Let π be a representation of a Lie group G on a Banach space V. Then a vector $v \in V$ is differentiable if and only if for each $\lambda \in V'$ the function $x \to \langle \pi(x)v, \lambda \rangle$ on G is differentiable (Poulsen [1970]).

4*. Let G be a connected semisimple Lie group with finite center. Let $K \subset G$ be a maximal compact subgroup. Let π be a quasisimple, K-finite representation of G on a Banach space V. Let V_K denote the space of all K-finite vectors in V. Then if $\mathfrak{g}$ is the Lie algebra of G,

(i) $V_K \subset V^\infty$ and $d\pi(X)V_K \subset V_K$ ($X \in \mathfrak{g}$).

(ii) Let $d\pi_K$ denote the representation of $\mathfrak{g}$ on V_K given by $d\pi_K(X) = d\pi(X)|V_K$. Then $d\pi_K$ is irreducible (algebraically) if and only if π is irreducible (cf. Harish-Chandra [1953], §10).

5. Let G be a connected, noncompact simple Lie group, G/K the associated symmetric space, $\mathfrak{g}$ and $\mathfrak{k}$ the corresponding Lie algebras. We assume G imbedded in the simply connected Lie group $G^{\boldsymbol{C}}$ with Lie algebra $\mathfrak{g}^{\boldsymbol{C}}$. Let $\mathfrak{u} = \mathfrak{k} + i\mathfrak{p}$ and U the corresponding analytic subgroup of $G^{\boldsymbol{C}}$.

Let π be a finite-dimensional representation of U on a real vector space V_0 such that $\pi(K)$ has a fixed vector $v \neq 0$ satisfying

$$K = \{u \in U : \pi(u)v = v\}.$$

Extending π to a representation of $\mathfrak{g}^{\boldsymbol{C}}$ and $G^{\boldsymbol{C}}$ on $V = V_0^{\boldsymbol{C}}$, also denoted π, prove the following:

(i) The mapping $I : gK \to \pi(g)v$ is a bijection of G/K onto a (real) submanifold of V.

(ii) The G-finite functions on G/K are the functions $p \circ I$ where p is a (holomorphic) polynomial function on V.

6. Let K be a compact connected Lie group, δ an irreducible representation of K with character χ_δ. Let $\boldsymbol{D}(K)$ denote the set of left-invariant differential operators on K. Prove that

$$\boldsymbol{D}(K)\chi_\delta = H_\delta$$

the space spanned by the representation coefficients of δ (cf. Helgason [1976], Theorem 4.1).

B. Spherical Functions

1. Show through the indicated steps that the spherical function

$$\phi_0(g) = \int_K e^{-\rho(H(gk))}\, dk$$

satisfies the inequalities

$$(*) \qquad e^{-\rho(\log a)} \leq \phi_0(a) \leq c(1 + |\log a|)^d e^{-\rho(\log a)},$$

where $a \in A^+$, $d = \operatorname{Card}(\Sigma_0^+)$, and c is a constant (cf. Harish-Chandra [1958a], §9).

(a) For the left inequality use Lemma 6.5.

(b)* Multiply the expansion for ϕ_λ by the polynomial $\pi(\lambda)$ and observe that the function $\lambda \to \mathbf{c}(s\lambda)\pi(\lambda)$ is smooth at $\lambda = 0$. Viewing $\partial(\pi)$ as a differential operator on $\mathfrak{a}^*$ we have

$$\phi_0 = c\{\partial(\pi)_\lambda[\pi(\lambda)\phi_\lambda]\}_{\lambda=0}$$

where c is a constant. The right-hand side can be estimated by the expansion for ϕ_λ giving the right-hand inequality in $(*)$.

2. With the notation as in Theorem 6.13 prove that for $f \in C_c(\mathfrak{a})$,

$$\int_{\bar{N}_s} f(H(\bar{n}_s))e^{-\rho(H(\bar{n}_s))}\, d\bar{n}_s$$
$$= \int_{\bar{N}_{\beta_1} \times \cdots \times \bar{N}_{\beta_p}} f(H(\bar{n}_1) + \cdots + H(\bar{n}_p))e^{-\Sigma \rho_i(H(\bar{n}_i))}\, d\bar{n}_1 \cdots d\bar{n}_p.$$

Here ρ_i is the ρ-function for G_{β_i}/K_{β_i}.

3. Let $\beta \in \Sigma_0^+$ and ρ_β be the ρ-function for the rank-one symmetric space G_β/K_β in §6, No. 6. Then

$$\rho(A_\beta) \geq \rho_\beta(A_\beta)$$

with equality if β is simple.

4. Let $d = \operatorname{Card}(\Sigma_0^+)$. Prove that if $\varepsilon > 0$ then

$$\int_{\bar{N}} e^{-\rho(H(\bar{n}))}\{1 + \rho(H(\bar{n}))\}^{-d-\varepsilon}\, d\bar{n} < \infty.$$

For G of rank one this follows from Eq. (8) of §6. For G of arbitrary rank it then follows from B2 if we take B3 into account.

The inequality is due to Harish-Chandra [1958a], who gave an instructive but more difficult proof. The inequality plays an important role in the theory of the spherical transform on the "Schwartz space" $\mathscr{I}^2(G)$ (Problem C6).

5. Show that the spherical function ϕ_λ is real-valued if and only if $\bar{\lambda}$ and $-\lambda$ are conjugate under the Weyl group.

6*. The spherical function ϕ_λ is positive if and only if $i\lambda \in \mathfrak{a}^*$ (cf. Fürstenberg [1965]; Karpelevič [1965]).

7. Using the convexity of the function $x \to e^x$ on $\boldsymbol{R}$ show that if $i\lambda \in \mathfrak{a}^*$ then

$$\phi_\lambda(g) \geq \phi_0(g), \qquad g \in G,$$

(De George and Wallach [1978].)

8. Let G have real rank one. Then Σ_0^+ consists of one element α. The function $t\colon h \to \alpha(\log h)$ can be taken as a coordinate on A whereby the differential equation (7) of §5 for ϕ_λ becomes an ordinary second-order differential equation in t. Introducing the new variable $z = -(\sinh t)^2$, this differential equation becomes the *hypergeometric equation*

$$z(z-1)\frac{d^2\phi_\lambda}{dz^2} + [(a+b+1)z - c]\frac{d\phi_\lambda}{dz} + ab\phi_\lambda = 0,$$

where, with $\alpha_0 = \alpha/\langle\alpha, \alpha\rangle$,

$$a = \tfrac{1}{2}(\tfrac{1}{2}m_\alpha + m_{2\alpha} + \langle i\lambda, \alpha_0\rangle),$$

$$b = \tfrac{1}{2}(\tfrac{1}{2}m_\alpha + m_{2\alpha} - \langle i\lambda, \alpha_0\rangle),$$

$$c = \tfrac{1}{2}(m_\alpha + m_{2\alpha} + 1).$$

Thus ϕ_λ is given by the hypergeometric function

$$\phi_\lambda(h) = F(a, b; c; z), \qquad z = -\mathrm{sh}^2(\alpha(\log h)).$$

The behavior of $\phi_\lambda(h)$ at ∞, in particular the formula of Theorem 6.4 for $\boldsymbol{c}(\lambda)$, can now be determined from a classical functional equation for F valid for $a - b \notin \boldsymbol{Z}$:

$$\begin{aligned} &F(a, b; c; z)/\Gamma(c) \\ &\quad = \Gamma(b-a)[\Gamma(b)\Gamma(c-a)]^{-1}(-z)^{-a}F(a, 1-c+a; 1-b+a; z^{-1}) \\ &\qquad + \Gamma(a-b)[\Gamma(a)\Gamma(c-b)]^{-1}(-z)^{-b}F(b, 1-c+b; 1-a+b; z^{-1}) \end{aligned}$$

(cf. Harish-Chandra [1958a]).

9*. Because of Theorem 3.7 it is of interest to determine the set of $\lambda \in \mathfrak{a}_c^*$ for which ϕ_λ is positive definite. The complete answer does not seem to be known, but one has some partial results:

(i) If $\lambda \in \mathfrak{a}^*$ then ϕ_λ is positive definite.

(ii) Assume G has real rank one and α is the single element in Σ_0^+. Then

(a) Suppose $2\alpha \notin \Sigma^+$. Then ϕ_λ is positive definite if and only if either $\lambda \in \mathfrak{a}^*$ or $i\lambda \in \mathfrak{a}^*$ and $|\langle i\lambda, \alpha\rangle| \leq \langle\rho, \alpha\rangle$.

(b) Suppose $2\alpha \in \Sigma^+$. Then apart from the constant $\phi_{\pm i\rho} \equiv 1$, ϕ_λ is positive definite if and only if either $\lambda \in \mathfrak{a}^*$ or $i\lambda \in \mathfrak{a}^*$ and

$$|\langle i\lambda, \alpha\rangle| \leq (\tfrac{1}{2}m_\alpha + 1)\langle \rho, \alpha\rangle$$

(cf. Kostant [1975]; another proof is given in Flensted-Jensen and Koornwinder [1979a], and partial results were proved by Takahashi [1963] and Faraut and Harzallah [1972]).

(iii) For G complex of rank 2, Duflo [1979b] gives an explicit (but complicated) description of the λ for which ϕ_λ is positive definite.

(iv) For a number of special cases, all the irreducible unitary representations of G have been determined, thus answering the question above. For a survey, see Knapp and Speh [1982].

10. Let $G = \boldsymbol{SU}(2, 1)$. Then the groups M and N can be described by $M = \exp \mathfrak{m}$, $N = \exp \mathfrak{n}$ where

$$\mathfrak{m} = \boldsymbol{R}T, \qquad \mathfrak{n} = \boldsymbol{R}X + \boldsymbol{R}Y + \boldsymbol{R}Z;$$

$$T = \begin{pmatrix} i & 0 & 0 \\ 0 & -2i & 0 \\ 0 & 0 & i \end{pmatrix}, \qquad X = \begin{pmatrix} 0 & 1 & 0 \\ -1 & 0 & 1 \\ 0 & 1 & 0 \end{pmatrix},$$

$$Y = \begin{pmatrix} 0 & i & 0 \\ i & 0 & -i \\ 0 & i & 0 \end{pmatrix}, \qquad Z = \begin{pmatrix} i & 0 & -i \\ 0 & 0 & 0 \\ i & 0 & -i \end{pmatrix}$$

([DS], Chapter IX, §3). Determine $\boldsymbol{D}(MN/M)$ and the spherical functions on $MN/M \approx N$.

Note that N is isomorphic to the Heisenberg group (Chapter II, §4, No. 1).

11*. For $\lambda \in \mathfrak{a}^*$ the spherical function satisfies the following inequality (Harish-Chandra [1958b], p. 583): Let $D \in \boldsymbol{D}(G)$. Then there exists a $k \in \boldsymbol{Z}^+$ and a constant C such that

$$|(1 + |\lambda|)^{-k}\pi(\lambda)(D\phi_\lambda)(a)e^{\rho(\log a)}| \leq C$$

for $a \in A^+$, $\lambda \in \mathfrak{a}^*$.

12*. Deduce the boundedness criterion for ϕ_λ (Theorem 8.1) for complex G directly from the formula for ϕ_λ in Theorem 5.7.

13*. Let G be a unimodular connected Lie group, and $K \subset G$ a compact subgroup. The following properties of G/K are equivalent:

(i) The algebra $\boldsymbol{D}(G/K)$ of G-invariant differential operators on G/K is commutative.

(ii) The convolution algebra $C_c^\natural(G)$ is commutative.

14*. Let U be a compact simply connected simple Lie group, and $T \subset U$ a maximal torus. Deduce from **13** and Theorem 3.5, Chapter V that if $\dim T > 1$, then the algebra $\boldsymbol{D}(U/T)$ of invariant differential operators is not commutative.

15. Using (4) in §4 prove the formula

$$A(g \cdot x, g \cdot b) = A(x, b) + A(g \cdot o, g \cdot b) \qquad (g \in G, x \in X, b \in B),$$

which generalizes (34) in Introduction §4. Using this formula and Proposition 2.4, show that for each $\mu \in \mathfrak{a}_c^*$ the function

$$e_{\mu, b}: x \to e^{\mu(A(x, b))}$$

is a joint eigenfunction of $\boldsymbol{D}(G/K)$.

16. With the notation of §3 consider the space $\mathscr{K}(G) = C_c(G)$ with the inductive limit topology (from §1) and the sub-algebra $C_c^\natural(G)$ with the relative topology. In analogy with Theorem 3.3 show that:

The continuous homomorphisms of the algebra $C_c^\natural(G)$ onto $\boldsymbol{C}$ are the mappings

$$f \to \int_G f(x)\phi(x)\, dx$$

where ϕ is a spherical function on G.

17. Let S be a set and $\Phi_1, \ldots, \Phi_w$ complex-valued functions on S. Show that if $\Phi_1, \ldots, \Phi_w$ are linearly independent over $\boldsymbol{C}$ then there exist $x_1, \ldots, x_w \in S$ such that the matrix

$$(\Phi_i(x_j))_{1 \le i, j \le w}$$

is nonsingular. From this and (2) §6 deduce another proof of Lemma 6.1, generalizing the one for $\boldsymbol{SL}(2, \boldsymbol{R})$ in the Introduction, (51) §4 (Schlichtkrull and Flensted-Jensen).

18. Relate Lemma 5.21, Chapter II, to the formula $\phi_{-\lambda}(g) \equiv \phi_\lambda(g^{-1})$ in (7) §4.

C. Spherical Transforms

1. As in §8 let $L^\natural(G)$ denote the space of functions in $L^1(G)$ which are bi-invariant under K.

(i) Suppose $f \in L^\natural(G)$ is such that

$$\tilde{f}(\lambda) = 0 \qquad \text{for all } \lambda \in \mathfrak{a}^* \text{ (real dual of } \mathfrak{a})$$

Then

$$f(g) = 0 \qquad \text{for almost all } g \in G.$$

(ii) If $f \in L^{\natural}(G)$, show that

$$F_f(a) = e^{\rho(\log a)} \int_N f(an)\, dn$$

exists for almost all $a \in A$ and that the map $f \to F_f$ is injective.

These results are proved in Helgason [1970a], Chapter II, §1; Part (i) sharpens Proposition 3.8.

2. Let $\mathscr{E}_K(\mathfrak{p})$ denote the space of K-invariant functions in $\mathscr{E}(\mathfrak{p})$ with the relative topology. The restriction mapping in Chapter II, Proposition 2.1 identifies the dual $\mathscr{E}'_K(\mathfrak{p})$ with the space of K-invariant distributions on $\mathfrak{p}$ of compact support.

Show that the spherical transform $T \to \tilde{T}$ where

$$\tilde{T}(\lambda) = \int_{\mathfrak{p}} \psi_{-\lambda}(Z)\, dT(Z)$$

is a bijection of $\mathscr{E}'_K(\mathfrak{p})$ onto the space of W-invariant holomorphic functions Φ on $\mathfrak{a}_c^*$ satisfying an inequality of the form

$$|\Phi(\lambda)| \leq C(1 + |\lambda|)^N e^{R|\mathrm{Im}\,\lambda|}, \qquad \lambda \in \mathfrak{a}_c^*,$$

for some positive constants C, N, and R.

3. Suppose G has real rank one. Consider the integral

$$F_f(a) = e^{\rho(\log a)} \int_N f(an)\, dn = e^{\rho(\log a)} \int_{\bar{N}} f(\bar{n}a)\, d\bar{n},$$

where $\theta(dn) = d\bar{n}$. Here an explicit inversion of $f \to F_f$ can be given.

(i) From [DS] (Theorem 3.8, Chapter IX) and its proof show that for $a \in A$,

$$\mathrm{ch}^2(\alpha(A^+(\bar{n}a))) = [\mathrm{ch}(\alpha(\log a)) + \tfrac{1}{2}ce^{\alpha(\log a)}|X|^2]^2 + ce^{2\alpha(\log a)}|Y|^2,$$

in the notation of the quoted theorem.

(ii) Note that in the case $G = \boldsymbol{SO}_0(n, 1)$ we can take either α or 2α as the element in Σ^+ [compare Eq. (26′) of Introduction §4].

(iii) Show that $\alpha(H)^2 = \frac{1}{2}(m_\alpha + 4m_{2\alpha})^{-1}|H|^2$ for $H \in \mathfrak{a}$.

(iv) Renormalize the distance d on G/K and the norm $|Z|^2 = -B(Z, \theta Z)$ by

$$\delta = [2(m_\alpha + 4m_{2\alpha})]^{-1/2} d, \qquad \|Z\|^2 = \tfrac{1}{2}(m_\alpha + 4m_{2\alpha})^{-1}|Z|^2.$$

Then (i) shows if $\alpha(H) = 1$, $a_r = \exp rH$,

$$\mathrm{ch}^2(\delta(o, \bar{n}a_r \cdot o)) = (\mathrm{ch}\, r + \tfrac{1}{4}e^r\|X\|^2)^2 + \tfrac{1}{2}e^{2r}\|Y\|^2.$$

Defining ϕ, ψ on $[1, \infty)$ by

$$\phi(\operatorname{ch}^2(\delta(o, g \cdot o))) = f(g), \qquad \psi((\operatorname{ch} r)^{1/2}) = F_f(a_r),$$

show that

$$(*) \qquad \psi((\operatorname{ch} r)^{1/2}) = \int_{\mathfrak{g}_{-\alpha}} \int_{\mathfrak{g}_{-2\alpha}} \phi((\operatorname{ch} r + \|X\|^2)^2 + \|Y\|^2) \, dX \, dY,$$

with a suitable normalization of the Euclidean measures on $\mathfrak{g}_{-\alpha}$ and $\mathfrak{g}_{-2\alpha}$.

(v) Show that the integral transform $f \to F_f$ or, equivalently, the integral transform $\phi \to \psi$ in $(*)$ can be expressed in terms of *two* Abel integral transforms of the type

$$H(u) = \int_u^\infty h(v)(v^2 - u^2)^p \, dv.$$

(vi) By inverting these two Abelian integral equations deduce the following principal content of Corollary 7.4 in the rank-one case:

If

$$F_f(a) = 0 \qquad \text{for} \quad \|\log a\| > R,$$

then

$$f(g) = 0 \qquad \text{for} \quad \delta(o, g \cdot o) > R.$$

For $G = \mathbf{SO}_0(n, 1)$ this inversion of $f \to F_f$ was given by Takahashi [1963]; generalizations to the rank-one case, related to our method above, have also been found by Faraut [1982], Lohoué-Rychener [1982], and Rouvière [1982]. Earlier, Koornwinder [1975], using a different method, had represented $f \to F_f$ as a similar composite of two fractional integrals.

4*. Suppose the measures dH and $d\lambda$ are regularly normalized and dg normalized by

$$\int_G f(g) \, dg = \int_{\mathfrak{a}^+} f(\exp H) \prod_{\alpha \in \Sigma^+} (e^{\alpha(H)} - e^{-\alpha(H)})^{m_\alpha} \, dH, \qquad f \in \mathscr{D}^\natural(G).$$

Then the Plancherel formula for the spherical transform

$$c \int_G |f(g)|^2 \, dg = \int_{\mathfrak{a}^*} |\tilde{f}(\lambda)|^2 |c(\lambda)|^{-2} \, d\lambda$$

holds with $c = |W|$, the order of W.

5. Suppose now the semisimple group G is *complex*. Put $\pi(\lambda) = \prod_{\alpha \in \Sigma^+} \langle \alpha, \lambda \rangle$ ($\lambda \in \mathfrak{a}_c^*$) and let the K-bi-invariant function $\Delta_0 \in \mathscr{E}(G)$ be defined by

$$\Delta_0(a) = \sum_{s \in W} e^{s\rho(\log a)}, \qquad a \in A,$$

(cf. Chapter II, Corollary 5.11). Using the formula in Theorem 5.7 for the spherical function prove that

$$\pi(\lambda)(\Delta_0 f)^{\sim}(\lambda) = \sum_{s \in W} \tilde{f}(\lambda + is\rho)\pi(\lambda + is\rho), \qquad f \in \mathscr{D}^{\natural}(G)$$

(cf. Helgason [1970a], Chapter II, §2).

6*. For $p \geq 0$ let $\mathscr{I}^p(G)$ denote the set of all $f \in \mathscr{E}(G)$ satisfying the following two conditions:

(a) f is bi-invariant under K.
(b) For each $D \in \mathbf{D}(G)$, $q \in \mathbf{Z}^+$,

$$\sup_{g \in G} |(1 + |g|)^q \phi_0(g)^{-2/p}(Df)(g)| < \infty.$$

Here $|g| = |\log a|$ if $g = k_1 a k_2$ ($k_1, k_2 \in K$, $a \in A$). Because of (*) in B1, $\mathscr{I}^p(G)$ is contained in the space $L^p(G)$ of functions f on G for which $|f|^p \in L^1(G)$.

Theorem. *The spherical transform* $f \to \tilde{f}$ *is a bijection of* $\mathscr{I}^2(G)$ *onto* $\mathscr{S}_W(\mathfrak{a}^*)$.

The substance of this theorem is proved in Harish-Chandra's papers [1958a, b]. See Helgason [1964a] for the additional remarks needed. A central step in Harish-Chandra's proof is the inequality in B11, which is proved by induction on dim G. By a refinement of this induction technique, Trombi and Varadarajan [1971] extended the theorem to a characterization of $\mathscr{I}^p(G)^{\sim}$ as the space of W-invariant rapidly decreasing holomorphic functions in the tube $\mathfrak{a}^* + i(2/p - 1)C^0(\rho)$ in $\mathfrak{a}_c^*$. See also Helgason [1970a], Ch. II, §2, for the rank-one case and $p = 1$.

For G complex, $p = 1$, the identity in Problem C5 can be used to prove this by reducing the case $\mathscr{I}^1(G)$ to Harish-Chandra's case $\mathscr{I}^2(G)$ (see Helgason [1970a], Chapter II, §2 for the details). For G real, p arbitrary, Clerc [1980] has given a related reduction of the proof for $\mathscr{I}^p(G)$ to the case $\mathscr{I}^2(G)$.

D. A Reduction to the Complex Case

Flensted-Jensen [1978] has discovered remarkable relations between the spherical functions on a real semisimple Lie group G_0 and the more elementary spherical functions on the corresponding complex semisimple group G. These results are summarized in 1-4 below.

Let $\mathfrak{g}_0$ be a real noncompact semisimple Lie algebra, $\mathfrak{g}$ its complexification, $\mathfrak{g}_0 = \mathfrak{k}_0 + \mathfrak{p}_0$ a Cartan decomposition, and θ the corresponding Cartan involution. Then $\mathfrak{u} = \mathfrak{k}_0 + i\mathfrak{p}_0$ is a compact real form of $\mathfrak{g}$ and $\mathfrak{k} = \mathfrak{k}_0 + i\mathfrak{k}_0$ is a complex subalgebra of $\mathfrak{g}$.

Let G be a Lie group with Lie algebra $\mathfrak{g}$ (considered as a Lie algebra over $\boldsymbol{R}$). Let G_0, K_0, K, and U be the analytic subgroups of G corresponding to $\mathfrak{g}_0$, $\mathfrak{k}_0$, $\mathfrak{k}$, and $\mathfrak{u}$, respectively. Then $K_0 \subset G_0$, $U \subset G$ are maximal compact subgroups whereas K is noncompact. Let $\boldsymbol{D}_R(K\backslash G)$ denote the set of differential operators on the coset space

$$K\backslash G = \{Kg : g \in G\}$$

invariant under all translations $\tau(x)$: $Kg \to Kgx$ $(g, x \in G)$. Let

$$\mathscr{E}(K\backslash G/U) = \{\phi \in \mathscr{E}(G) : \phi(kgu) \equiv \phi(g)\}.$$

1*. There is a one-to-one correspondence $\phi \to \phi^\eta$ between the set of spherical functions ϕ on G_0/K_0 and the set of functions $\psi = \phi^\eta$ on G satisfying

$$\psi \in \mathscr{E}(K\backslash G/U), \qquad \psi(e) = 1;$$

$$D\psi = \lambda_D \psi \qquad \text{for all} \quad D \in \boldsymbol{D}_R(K\backslash G), \quad \lambda_D \in \boldsymbol{C},$$

such that

$$\phi(x\theta(x)^{-1}) = \phi^\eta(x), \qquad x \in G_0.$$

2*. Let $\mathfrak{g} = \mathfrak{u} + \mathfrak{a} + \mathfrak{n}$ be an Iwasawa decomposition of $\mathfrak{g}$. Then $\mathfrak{g}_0 = \mathfrak{k}_0 + \mathfrak{a}_0 + \mathfrak{n}_0$ with $\mathfrak{a}_0 = \mathfrak{a} \cap \mathfrak{g}_0$, $\mathfrak{n}_0 = \mathfrak{n} \cap \mathfrak{g}_0$ is an Iwasawa decomposition of $\mathfrak{g}_0$. By the Killing form identification of $\mathfrak{a}_0$ and $\mathfrak{a}$ with their duals $\mathfrak{a}_0^*$ is imbedded in $\mathfrak{a}^*$. Let $2\rho_0$ and 2ρ denote the sums (with multiplicity) of the positive restricted roots for $(\mathfrak{g}_0, \mathfrak{k}_0)$ and $(\mathfrak{g}, \mathfrak{u})$, respectively. Similarly, let π_0 and π denote the products of the indivisible positive roots. Let ϕ_λ $(\lambda \in (\mathfrak{a}_0^*)_C)$ denote the spherical functions on G_0/K_0 and Φ_Λ $(\Lambda \in \mathfrak{a}_C^*)$ the spherical functions on G/U.

If $\lambda \in (\mathfrak{a}_0^*)_C$ define $\Lambda \in \mathfrak{a}_C^*$ by

$$\Lambda + i\rho = 2(\lambda + i\rho_0).$$

Then

$$\Phi_\Lambda(x) = \int_U \phi_\lambda^\eta(ux)\, du, \qquad x \in G, \tag{1}$$

where du is the normalized Haar measure on U.

This mapping $\phi_\lambda \to \Phi_\Lambda$ is injective whenever the mapping $\mu: \boldsymbol{Z}(G_0) \to \boldsymbol{D}(G_0/K_0)$ (Chapter II, §5, No. 7) is surjective. If G_0 is simple, this fails only for certain real forms $\mathfrak{g}_0$ of $\mathfrak{e}_6, \mathfrak{e}_7, \mathfrak{e}_8$ (cf. Chapter II, Exercise D3).

3*. Suppose $\mathfrak{g}_0$ is a normal real form of $\mathfrak{g}$. Then if the Haar measure dk on K is suitably normalized,

$$\phi_\lambda(x\theta(x)^{-1}) = |c(\lambda)|^2 |\pi_0(\lambda)|^2 \int_K \Phi_{2\lambda}(kx)\, dk, \qquad \lambda \in \mathfrak{a}_0^*, \tag{2}$$

so, in particular, the Plancherel density for G_0/K_0 is given by

$$|c(\lambda)|^{-2} = |\pi_0(\lambda)|^2 \int_K \Phi_{2\lambda}(k)\, dk, \qquad \lambda \in \mathfrak{a}_0^*. \tag{3}$$

For the case $G_0 = \boldsymbol{SL}(2, \boldsymbol{R})$ this formula becomes

$$\int_{-\infty}^{\infty} \frac{\sin \lambda t}{\operatorname{sh} t}\, dt = \pi \tanh\left(\frac{\pi\lambda}{2}\right),$$

which enters into Godement's proof of the inversion formula for the spherical transform on $\boldsymbol{SL}(2, \boldsymbol{R})$ [see (27) in Introduction §4].

4*. Formulas (1)–(3) have interesting applications to spherical transforms. For example, (1) reduces the determination of the bounded spherical functions (Theorem 8.1) to the case of complex G. For this case one can use the simple formula for $\Phi_{2\Lambda}$ from Theorem 5.7; nevertheless, the proof of the boundedness criterion for this case does not seem all that simple (Exercise B12).

Suppose now $\mathfrak{g}_0$ is a normal real form of $\mathfrak{g}$. Then formulas (1)–(3) reduce the Paley–Wiener theorem, the inversion formula, and the Plancherel formula for the spherical transform on G_0/K_0 (Theorems 7.1 and 7.5) to the corresponding theorems for G/U (these are elementary because of Theorem 5.7). For this and further applications we refer to the paper of Flensted-Jensen quoted above.

NOTES

§1. For further information about representations of Lie groups on locally convex spaces see Bruhat [1956], Harish-Chandra [1966], Borel [1972], and Warner [1972].

In finite-dimensional spaces there is a simple correspondence between representations of a Lie group G and of its Lie algebra $\mathfrak{g}$. For infinite-dimensional representations this correspondence requires modification. Thus Gårding [1947] replaced the representation of G on a Banach space V by the representation of $\mathfrak{g}$ on a certain subspace of the space V^∞ of differentiable vectors. (In this context see Exercise A1.) This was further improved by

Harish-Chandra's space of *well-behaved* or *analytic* vectors [1953]; the proof of Lemma 1.2 is from this paper.

For a general locally compact group where Haar measure is the principal analytic concept, the Hilbert space $L^2(G)$ and the irreducible unitary representations become the central objects in analysis on G. However, when G is a Lie group we have [in addition to $L^p(G)$] spaces like $\mathscr{D}(G)$, $\mathscr{E}(G)$, $\mathscr{D}'(G)$, and $\mathscr{E}'(G)$ forming the core of analysis on G. It has accordingly turned out, at least for semisimple G, that the quasisimple representations (introduced by Harish-Chandra [1953]) form a more natural class than the unitary ones (cf. Langlands [1973], Želobenko [1974], Duflo [1975], Borel and Wallach [1980], Vogan [1981], Knapp and Zuckerman [1982]).

Theorem 1.5 goes back to Gelfand and Raikov [1943]; see also Godement [1948].

For references on the orthogonal decompositions (9) and (16) see notes to Chapter V. The functional equation in Theorem 1.6 is from Weyl [1931, Appendix 2] and Weil [1940, §24]. The decomposition in Lemma 1.9 is from Harish-Chandra [1966]; see also Borel [1972, §3].

§§2–3. Spherical functions on compact symmetric spaces U/K were introduced by Cartan [1929] as coefficients $\langle \pi(u)\boldsymbol{e}, \boldsymbol{e}\rangle$ of representations π of U for which $\pi(K)$ has a fixed vector $\boldsymbol{e}$. Gelfand [1950a] showed the commutativity of the algebra $C_c^\natural(G)$ (Theorem 3.1; the unimodularity of G was remarked by Berg [1973]) and introduced the spherical functions as its homomorphisms into $\boldsymbol{C}$ (Lemma 3.1). He also obtained the functional equation (1) of §2 and the characterization by differential equations which we have used as a definition. Some of these results were also obtained by M. G. Krein. The correspondence in Theorem 3.4 is essentially from Gelfand and Naimark [1952] and Godement [1957a]; the first paper gives an explicit formula for the spherical function for $G = \boldsymbol{SL}(n, \boldsymbol{C})$. The mean value theorem in Proposition 2.5 is from Godement [1952a]. For harmonic spaces (in particular for two-point homogeneous spaces) the result is due to Willmore [1950]. The generalization to arbitrary eigenfunctions (Proposition 2.4) and the semisimplicity (Proposition 3.8) are from Helgason [1962a].

§4. The integral formulas in Theorems 4.2 and 4.3 and Eq. (7) are due to Harish-Chandra [1958a]. Lemma 4.4 and the correspondence theorem (Theorem 4.5) were proved by Helgason ([1970a], p. 116 and [1972a], Chapter II). Kostant ([1975a], Theorem 1.3.5) proved an algebraic analog of Theorem 4.5. Proposition 4.8 was observed by Gindikin [1967] and Korányi [1979], Theorem 4.7 and Proposition 4.10 by Helgason [1972a], Chapter II.

§§5–7. A Plancherel-type formula for the Fourier transform on a unimodular locally compact group G was given by Segal [1950] and Mautner [1950]. For the spherical transform corresponding to a "topological" symmetric space G/K a simpler Plancherel formula was given by Mautner [1951]. A more satisfactory version (with the set of positive definite spherical functions playing the role of the dual space) was given by Godement [1951, 1957a]; see also Harish-Chandra [1954c]. For the case of a noncompact semisimple Lie group with its rich structure and explicit parametrization of the set of spherical functions by $\mathfrak{a}_c^*/W$ (cf. Theorem 4.3) it became a natural problem to determine the Plancherel measure explicitly in terms of this parametrization. Harish-Chandra embarked on this project in his papers [1958a, b]. The first of these papers gives the expansion (Theorem 5.5) for ϕ_λ, introduces the $\boldsymbol{c}$-function, and relates it to the behavior of ϕ_λ at ∞. Some improvements in the proof (namely Lemma 5.3 and the proof of Proposition 5.4) come from Helgason [1966b] and [1972a] (Chapter II). On the basis of the spectral theory of ordinary differential equations, Harish-Chandra expected the Plancherel measure to be given by $|\boldsymbol{c}(\lambda)|^{-2}\, d\lambda$. The proof of this result is in his paper [1958b] reduced to two conjectures; the first one is about the $\mathbf{c}$-function and the second concerns the transform $f \to F_f$.

While Harish-Chandra [1958a] had determined the $\mathbf{c}$-function for the case G complex and the case when G has real rank one, Bhanu-Murthy determined $\mathbf{c}(\lambda)$ for all (except one) classical G which are normal real forms (see [1960a, b]). Guided by his formulas, Gindikin and Karpelevič [1962] proved the general product formula (Theorem 6.13), which when combined with Harish-Chandra's rank-one formula gives a general explicit formula for the $\mathbf{c}$-function. This formula implies easily Harish-Chandra's first conjecture mentioned above (see Helgason [1964a], §§2–4). In [1966, §21], Harish-Chandra proved the second conjecture, thereby completing the proof of the Plancherel formula for the spherical transform.

In [1955] Ehrenpreis and Mautner proved a Paley–Wiener theorem for the spherical transform on $G = \mathbf{SL}(2, \mathbf{R})$; this was extended by Takahashi [1963] to $G = \mathbf{SO}(n, 1)$. For general G the Paley–Wiener theorem for the spherical transform is given by Theorem 7.1 and Corollary 7.4. These were proved in Helgason [1966b] except for full justification of the term-by-term integration (15). This justification (i.e., Lemma 5.6) was given by Gangolli [1971]. (Actually Gangolli proved a stronger estimate than that of Lemma 5.6; the proof of Lemma 5.6 in the text is a simplification from the author's paper [1970a]).

While the proof of the Paley–Wiener theorem indicated above used Harish-Chandra's inversion formula for the spherical transform, Rosenberg in [1977] discovered that this is not really necessary; on the contrary, he showed that the proof of Theorem 7.1 together with some additional arguments (§7, No. 3, in the text) gave the inversion formula as well as the Plancherel formula. However, the surjectivity in Harish-Chandra's result on the Schwartz space $\mathscr{I}^2(G)$ (see Exercise C6) does not seem to be accessible by this method.

We now give some references for the remainder of §§5–7. The formula of Theorem 5.7 is the counterpart to the character formula on a compact group (cf. Theorem 1.7, Chapter V). As indicated earlier, it was proved by Gelfand and Naimark [1952] for the case $G = \mathbf{SL}(n, \mathbf{C})$; the general case was given by Harish-Chandra [1954b, 1958a] and Berezin [1957]. Lemma 6.2 and Proposition 6.3 are also from Harish-Chandra [1958a] and so is the rank one formula (Theorem 6.4) for $\mathbf{c}(\lambda)$. His computation of $\mathbf{c}(\lambda)$ using the hypergeometric equation is outlined in Exercise B8. The method of the text is from Schiffmann [1971] and Helgason [1970a]. Lemma 6.8 is from Harish-Chandra [1957b], and the proof in the text is an elaboration of that of Borel and Harish-Chandra [1962]. Lemma 6.9 and Proposition 6.10 are from Iwahori [1964].

If G has real rank one, the inversion formula and the Plancherel formula can be viewed as parts of the spectral theory of the singular ordinary differential operator $\Delta(L_X)$ in Eq. (8) of §5 (also in Exercise B8). Flensted-Jensen [1972] generalized these rank-one results as well as the Paley–Wiener theorem, allowing the multiplicities m_α, $m_{2\alpha}$ to be arbitrary real positive numbers. Further generalizations have been given, for example, by Chèbli [1974] and Koornwinder [1975].

§8. The determination of the bounded spherical function is from Helgason and Johnson [1969]. In Lemma 8.3 we combine convexity arguments from this paper and from Kostant [1973]. Lemma 8.8 has certain relevance for the $\mathbf{c}$-function. Gindikin and Karpelevič [1962] attached a $\mathbf{c}$-function in a natural way to an arbitrary half-space $R \subset \mathfrak{a}$ with 0 on its boundary. Lemma 8.8 shows that this is not more general than their functions $\mathbf{c}_s(\lambda)$ $(s \in W)$ in Theorem 6.13. Morera's theorem in connection with Cor. 8.2 was suggested by R. Kunze.

§9. The main results of this section are from the author's paper [1980]. Some simplifications by N. Bopp have been taken into account. Lemma 9.4 is essentially from Kostant [1963, §3]. For G complex, (17) is in Flensted-Jensen [1978, §3].

§10. Theorems 10.1 and 10.5 are from Kostant's paper [1973]. Special cases had been proved by Horn [1954a, 1954b] and Thompson [1971]. The proof in the text follows that of Kostant except for a variation in the geometric Lemma 10.7 which makes the reduction to the normal form unnecessary. A different approach, leading to some natural generalizations, is given by Heckman [1980]. Theorem 10.11 and Corollary 10.12 are from Flensted-Jensen and Ragozin [1973]. They had been proved by Koornwinder (cf. [1975]) in the rank-one case. For rank one a stronger version of Proposition 10.13 occurs in Flensted-Jensen and Koornwinder [1973].

CHAPTER V

ANALYSIS ON COMPACT SYMMETRIC SPACES

In this last chapter we turn to analysis on compact homogeneous spaces with emphasis on the symmetric ones. For the study of functions on such spaces the representations of compact groups play a fundamental role, analogous to that of exponentials for Fourier series. Through the theory of maximal tori for compact Lie groups these representations can be determined very explicitly. In §1 we discuss this description of the representations through weights and characters. In §2 we extend to compact nonabelian groups a certain sample of well-known theorems about Fourier series on the circle. While these classical theorems seem at first rather unrelated, the group-theoretic generalizations provide certain links between them and thereby throw a little light on some classical results for Fourier series.

In §3 we discuss Fourier decompositions of vector-valued functions with applications to compact homogeneous spaces. In §4 this is brought into more specific form for compact symmetric spaces and the relevant representations are described more explicitly.

§1. Representations of Compact Lie Groups

Let U be a connected, simply connected, compact, semisimple Lie group. Let $\mathfrak{u}$ denote its Lie algebra and $\mathfrak{g}$ the complexification of $\mathfrak{u}$. Let V be a finite-dimensional vector space over $\boldsymbol{C}$. A *representation* of the real Lie algebra $\mathfrak{u}$ on the complex vector space V is of course an $\boldsymbol{R}$-linear mapping of $\mathfrak{u}$ into the complex Lie algebra $\mathfrak{gl}(V)$ preserving brackets. There is then an obvious one-to-one correspondance between the representation of U, $\mathfrak{u}$, and $\mathfrak{g}$ on V. In this section we develop the standard theory of weights and characters and present some applications. While the weight and character theory is customarily developed purely algebraically for the complex Lie algebra $\mathfrak{g}$ we shall for the sake of brevity of proofs take advantage of the compactness of U and will use some analytic tools already developed. A function f (or a distribution) on U is said to be *central* if it is invariant under conjugation, i.e., $f(uxu^{-1}) \equiv f(x)$ $(x, u \in U)$.

1. The Weights

Let T be a maximal torus of U, $\mathfrak{t}_0$ its Lie algebra. Then the subalgebra $\mathfrak{t}$ of $\mathfrak{g}$ generated by $\mathfrak{t}_0$ is a Cartan subalgebra. Let $\mathfrak{t}^*$ denote the dual space of $\mathfrak{t}$.

Definition. Let π be a representation of $\mathfrak{g}$ on a vector space V over $\boldsymbol{C}$. An element $\mu \in \mathfrak{t}^*$ is called a *weight of* π if there exists a vector $v \neq 0$ in V such that

$$\pi(H)v = \mu(H)v, \qquad H \in \mathfrak{t}.$$

In particular, the weights of the adjoint representation of $\mathfrak{g}$ are the roots of $\mathfrak{g}$ with respect to $\mathfrak{t}$.

Now assume $\dim V < \infty$. As mentioned in Chapter IV, §1. No. 2, there exists a positive definite Hermitian form on $V \times V$ such that $\pi(U)$ consists of unitary operators. Hence $\pi(T)$, and therefore also $\pi(\mathfrak{t})$, form a commutative family of semisimple endomorphisms of V and we have a direct decomposition

$$V = \sum_{\mu} V_{\mu}, \tag{1}$$

where μ runs over the weights of π and for each $\lambda \in \mathfrak{t}^*$ we put

$$V_\lambda = \{v \in V : \pi(H)v = \lambda(H)v \text{ for } H \in \mathfrak{t}\}. \tag{2}$$

The spaces $V_\mu \neq 0$ are called *weight subspaces* and $m_\mu = \dim(V_\mu)$ is called the *multiplicity* of the weight μ.

Since $\pi(t)$ is a unitary diagonal matrix for each $t \in T$, we see that each weight is real-valued on the space $\mathfrak{t}_{\boldsymbol{R}} = i\mathfrak{t}_0$ (cf. [DS], Chapter II, §§5, 6).

As usual, let $\mathfrak{t}_e$ denote the *unit lattice* for U,

$$\mathfrak{t}_e = \{H \in \mathfrak{t}_0 : \exp H = e\}. \tag{3}$$

We know from [DS] (Chapter VII, Corollary 7.8) that

$$\mathfrak{t}_e = \{\text{integral linear combinations of } (4\pi i/\langle\alpha, \alpha\rangle)H_\alpha : \alpha \in \Delta\}. \tag{4}$$

Here Δ is the set $\Delta(\mathfrak{g}, \mathfrak{t})$ of roots $\neq 0$ and if $\mu \in \mathfrak{t}^*$, $H_\mu \in \mathfrak{t}$ is determined by $\langle H, H_\mu\rangle \equiv \mu(H)$, $(H \in \mathfrak{t})$ where $\langle\ ,\ \rangle$ denotes the bilinear form on $\mathfrak{t}$ and $\mathfrak{t}^*$ induced by the Killing form of $\mathfrak{g}$. Put $|\cdot| = \langle\cdot, \cdot\rangle^{1/2}$.

Theorem 1.1. *Let $\mathfrak{t}_e$ denote the unit lattice and Λ the set of all weights of all finite-dimensional representations of U. Then*

(i) $\Lambda = \{\lambda \in \mathfrak{t}^* : \lambda(\mathfrak{t}_e) \subset 2\pi i\boldsymbol{Z}\}$,
(ii) $\mathfrak{t}_e = \{H \in \mathfrak{t} : \Lambda(H) \subset 2\pi i\boldsymbol{Z}\}$.

Proof. Let $\lambda \in \Lambda$ and let π be a representation of $\mathfrak{g}$ on a finite-dimensional vector space V such that $\pi(H)v = \lambda(H)v$ for some $v \neq 0$ in

V and all $H \in \mathfrak{t}$. Then $\pi(\exp H)v = e^{\lambda(H)}v$ $(H \in \mathfrak{t})$ so $\lambda(\mathfrak{t}_e) \subset 2\pi i\mathbf{Z}$. This proves the inclusions $\subset$ in the equalities (i) and (ii).

For the converse inclusion in (i) suppose $\lambda \in \mathfrak{t}^*$ such that $\lambda(\mathfrak{t}_e) \subset 2\pi i\mathbf{Z}$. There exists a homomorphism $\chi: T \to \mathbf{C}$ such that $\chi(\exp H) = e^{\lambda(H)}$ $(H \in \mathfrak{t}_0)$. We claim that there exists a finite-dimensional representation π of U on a space V such that $\pi(t)v = \chi(t)v$ for some $v \neq 0$ in V. For this consider the function space

(5) $$C_\chi(U) = \{f \in C(U): f(ut^{-1}) \equiv f(u)\chi(t)\}.$$

The mapping

(6) $$f(u) \to \int_T f(ut)\chi(t)\,dt$$

(dt is the Haar measure) then maps $C(U)$ onto $C_\chi(U)$; this latter space is nonzero as we see by choosing $f \in C(U)$ by the Tietz extension theorem such that the restriction $f|T$ equals $\mathrm{Re}(\chi)$. Considering the left regular representation of U on $C_\chi(U)$ we know from Lemma 1.9 of Chapter IV that there exists an invariant finite-dimensional subspace $V \subset C_\chi(U)$. Pick $g \in V$ such that $g(e) \neq 0$. Then the function

$$g^\natural(u) = \int_T g(su)\chi(s)\,ds$$

belongs to V, $g^\natural(e) = g(e)\int_T |\chi(t)|^2\,dt \neq 0$, and

$$(g^\natural)^{L(t)} = \chi(t)g^\natural.$$

The representation π on V so constructed satisfies $\pi(\exp H)v = e^{\lambda(H)}v$ $(H \in \mathfrak{t}_0)$ where $v = g^\natural$. It follows that $\lambda \in \Lambda$. This proves (i). Finally, suppose $H \in \mathfrak{t}$ satisfies $\lambda(H) \in 2\pi i\mathbf{Z}$ for each $\lambda \in \Lambda$. Then $H \in \mathfrak{t}_0$ and (1) implies that for each finite-dimensional representation π on, say, V, $\pi(\exp H)v = v$ for each $v \in V$. Then the Peter–Weyl theorem implies $f(\exp H) = f(e)$ for each $f \in C(U)$, so $H \in \mathfrak{t}_e$. This concludes the proof of the theorem.

If $\mu \in \Lambda$ we define the function e^μ on T by

$$e^\mu(t) = e^{\mu(H)} \quad \text{if} \quad t = \exp H, \quad H \in \mathfrak{t}_0.$$

Because of (i) above this is a valid definition. Taking (4) into account the proof above gives the following result.

Corollary 1.2. *The mapping $\mu \to e^\mu$ identifies Λ with the character group of T; moreover,*

$$\Lambda = \left\{\lambda \in \mathfrak{t}^*: 2\frac{\langle \lambda, \alpha\rangle}{\langle \alpha, \alpha\rangle} \in \mathbf{Z} \quad \text{for all } \alpha \in \Delta\right\}.$$

This last integrality condition shows again that $\Delta \subset \Lambda$ (cf. [DS], Chapter III, Theorem 4.3). Note also that Λ is the group $\tilde{T}(\Delta)$ considered in [DS] (Chapter X, §3. No. 6). As there we put $\check{\alpha} = 2\alpha/\langle\alpha, \alpha\rangle$ for $\alpha \in \Delta$.

Now we select a Weyl chamber $\mathfrak{t}^+ \subset \mathfrak{t}_{\boldsymbol{R}}$ and let Δ^+ denote the corresponding set of positive roots. Let $\alpha_1, \ldots, \alpha_l$ denote the corresponding simple roots and let $\mathfrak{t}_{\boldsymbol{R}}^*$ be ordered lexicographically with respect to the basis $\alpha_1, \ldots, \alpha_l$. Let $\omega_1, \ldots, \omega_l$ be the "dual basis" given by $\langle\omega_j, \check{\alpha}_i\rangle = \delta_{ij}$. The weights $\omega_1, \ldots, \omega_l$ are called the *fundamental weights* and we have $\Lambda = \sum_{j=1}^{l} \boldsymbol{Z}\omega_j$ (see loc. cit. Proposition 3.31 and its proof). The vectors H_{ω_j} form the edges of the Weyl chamber $\mathfrak{t}^+$. For each $\alpha \in \Delta^+$ we select nonzero vectors X_α, Y_α in the root spaces $\mathfrak{g}^\alpha$ and $\mathfrak{g}^{-\alpha}$, respectively. Let $\rho = \frac{1}{2}\sum_{\alpha\in\Delta^+} \alpha$. We have of course the root-space decomposition

$$\mathfrak{g} = \mathfrak{t} + \sum_{\alpha\in\Delta} \mathfrak{g}^\alpha.$$

Since the elements $\lambda \in \Lambda$ are characterized by $\langle\lambda, \check{\alpha}\rangle \in \boldsymbol{Z}$ $(\alpha \in \Delta)$ they are often called *integral functions.* If $\lambda \in \Lambda$ is such that $\langle\lambda, \check{\alpha}\rangle \in \boldsymbol{Z}^+$ for all $\alpha \in \Delta^+$ then λ is called a *dominant integral function.* Let $\Lambda(+)$ denote the set of all dominant integral functions. In particular, $\rho \in \Lambda(+)$. Also $\Lambda(+) = \sum_i \boldsymbol{Z}^+\omega_i$.

We shall now study the set $\Lambda(\pi)$ of weights for a single finite-dimensional representation π of $\mathfrak{g}$. Clearly $\Lambda(\pi)$ is invariant under the Weyl group $W = W(\mathfrak{u}) = W(\mathfrak{g}, \mathfrak{t})$. As usual (cf. [DS], Chapter II, Proposition 1.1), π extends to a representation of the universal enveloping algebra $U(\mathfrak{g})$. Because of the intervention of the compact group U, π can be decomposed into irreducible representations (cf. [DS], Chapter III, Exercise B3). Thus we can assume irreducibility.

Theorem 1.3. *Let π be a finite-dimensional irreducible representation of $\mathfrak{g}$ on V. Let λ denote its highest weight, V_λ the corresponding weight space. Then*

(i) $\dim V_\lambda = 1$ *and* $\lambda \in \Lambda(+)$.

(ii) *If* $v \neq 0$ *in* V_λ *and* $\Delta^+ = \{\beta_1, \ldots, \beta_r\}$ *then* V *is spanned by* $\pi(Y_{\beta_1}^{n_1} \cdots Y_{\beta_r}^{n_r})v$, $n_i \in \boldsymbol{Z}^+$.

(iii) *Each weight* $\mu \in \Lambda(\pi)$ *has the form*

$$\mu = \lambda - \sum_1^l m_i\alpha_i, \qquad m_i \in \boldsymbol{Z}^+.$$

Moreover,

$$|\mu| \leq |\lambda| \text{ with equality only if } \mu \in W \cdot \lambda; \tag{7}$$

$$|\mu + \rho| < |\lambda + \rho| \quad \text{if } \mu \neq \lambda. \tag{8}$$

(iv) *If* $\mu \in \Lambda(+)$ *and if*

$$\mu = \lambda - \sum_1^l m_i\alpha_i \qquad (m_i \in \mathbf{Z}^+)$$

then $\mu \in \Lambda(\pi)$.

(v) *Each* $\mu \in \Lambda(\pi)$ *belongs to the convex hull* $C(\lambda)$ *of the orbit* $W \cdot \lambda$.

Remark Note that (iv) is not an exact converse of (iii) since μ in (iii) need not be dominant integral. However, (iii) and (iv) show that $0 \in \Lambda(\pi)$ if and only if $\lambda = \sum_1^l m_i\alpha_i$ with all $m_i \in \mathbf{Z}^+$.

Proof. Since $[H, X_\alpha] = \alpha(H)X_\alpha$ $(H \in \mathfrak{t})$ we have $\pi(X_\alpha)v = V_{\lambda+\alpha}$ so $\pi(X_\alpha)v = 0$ if $v \in V_\lambda$, $\alpha \in \Delta^+$. Hence, using the irreducibility,

$$V = \pi(U(\mathfrak{g}))v = \text{span of } \pi(Y_{\beta_1}^{n_1} \cdots Y_{\beta_r}^{n_r})v, \qquad n_i \in \mathbf{Z}^+.$$

However,

$$\pi(Y_{\beta_1}^{n_1} \cdots Y_{\beta_r}^{n_r})v \in V_{\lambda - n_1\beta_1 - \cdots - n_r\beta_r},$$

so (ii), the equality $\dim V_\lambda = 1$, and the formula $\mu = \lambda - \sum_i m_i\alpha_i$ follow. Next select s in the Weyl group $W(\mathfrak{u})$ such that $H_\nu = H_{s\mu} \in \overline{\mathfrak{t}^+}$. Then $\nu \in \Lambda(\pi)$ so $\nu = \lambda - \beta$, where $\beta = \sum_i k_i\alpha_i$ $(k_i \in \mathbf{Z}^+)$. Hence, if $\beta \neq 0$,

$$\langle \lambda, \lambda \rangle = \langle \nu, \nu \rangle + 2\langle \nu, \beta \rangle + \langle \beta, \beta \rangle > \langle \nu, \nu \rangle = \langle \mu, \mu \rangle.$$

This proves (7) and $H_\lambda \in \overline{\mathfrak{t}^+}$. For (8) we have

$$|\lambda + \rho|^2 - |\mu + \rho|^2 = |\lambda|^2 - |\mu|^2 + 2\langle \lambda, \rho \rangle - 2\langle \mu, \rho \rangle$$
$$= |\lambda|^2 - |\mu|^2 + 2\sum_1^l m_i\langle \alpha_i, \rho \rangle > 0$$

since $H_\rho \in \mathfrak{t}^+$.

For statement (iv) we first prove the following lemma (cf. [DS], Chapter III, Theorem 4.3).

Lemma 1.4. *Let* $\nu \in \Lambda(\pi)$ *and let* $\alpha \in \Delta$. *The set*

$$S = \{n \in \mathbf{Z} : \nu + n\alpha \in \Lambda(\pi)\} \tag{9}$$

is an interval $p \leq n \leq q$ *and*

$$-\langle \nu, \check{\alpha} \rangle = p + q.$$

Proof. Select nonzero vectors $X \in \mathfrak{g}^\alpha$, $Y \in \mathfrak{g}^{-\alpha}$, $H \in \mathfrak{t}$ such that

$$[H, X] = 2X, \qquad [H, Y] = -2Y, \qquad [X, Y] = H.$$

Then π gives by restriction a representation of $\mathbf{C}X + \mathbf{C}Y + \mathbf{C}H$ on V. Suppose $p' = \min S$, $q' = \max S$. If $v_0 \neq 0$ in $V_{v+q'\alpha}$ then by Lemma 1.2 of the Appendix, $\pi(Y)^i(v_0) \neq 0$ for

$$0 \leq i \leq (v + q'\alpha)(H) = \langle v, \check{\alpha} \rangle + 2q'.$$

Thus $v + n\alpha \in \Lambda(\pi)$ for $-\langle v, \check{\alpha} \rangle - q' \leq n \leq q'$, so $p' \leq -\langle v, \check{\alpha} \rangle - q'$; i.e., $p' + q' \leq -\langle v, \check{\alpha} \rangle$. Now replace α by $-\alpha$. Then p' is replaced by $-q'$, q' by $-p'$, so $-q' - p' \leq -\langle v, -\check{\alpha} \rangle$. Thus $p' + q' = -\langle v, \check{\alpha} \rangle$ and the lemma is proved.

For (iv) suppose $\mu = \lambda - \sum_i m_i \alpha_i \in \Lambda(+)$. Suppose $v \in \Lambda(\pi)$ satisfies $v - \mu = \sum_{j \in I} n_j \alpha_j$, with $n_j > 0$. Then since $\langle v - \mu, v - \mu \rangle > 0$ we have $\langle \alpha_j, v - \mu \rangle > 0$ for some $j \in I$. But then

$$\langle \alpha_j, v \rangle = \langle \alpha_j, v - \mu \rangle + \langle \alpha_j, \mu \rangle > 0,$$

so by the lemma $v - \alpha_j \in \Lambda(\pi)$. Using this inductively [which we can since $(v - \alpha_j) - \mu = \sum_k p_k \alpha_k$ $(p_k > 0)$] starting with λ, we conclude that $\mu \in \Lambda(\pi)$.

Finally, for (v) let $\mu \in \Lambda(\pi)$ be arbitrary. We choose $s \in W$ such that $s\mu$ is dominant. Using (iii) we have

$$H_{s\mu} = H_\lambda - \sum_{1}^{l} m_i H_{\alpha_i} \subset H_\lambda + {}^{-}C$$

in the notation of Lemma 8.3 of Chapter IV, which now implies $H_{s\mu} \in C(H_\lambda)$. Thus $\mu \in C(\lambda)$ and Theorem 1.3 is proved.

The importance of the highest weight concept is further affirmed by the following fundamental theorem.

Theorem 1.5. (i) *Two irreducible finite-dimensional representations of* $\mathfrak{g}$ *with the same highest weight are equivalent.*

(ii) *Given a* $\lambda \in \mathfrak{t}^*$ *which is dominant integral, there exists an irreducible finite-dimensional representation of* $\mathfrak{g}$ *with highest weight* λ.

This theorem will follow easily from analytic results proved in the next subsection. The theorem shows that the set $\hat{U}$ of irreducible finite-dimensional representations of U is parametrized by

$$\Lambda(+) = \{\lambda \in \Lambda : H_\lambda \in \overline{\mathfrak{t}^+}\} = \Lambda/W = \hat{T}/W,$$

where $\hat{T}$ is the character group of T (cf. Corollary 1.2). The representations π_i which correspond to the fundamental weights $\omega_1, \ldots, \omega_l$ are called the *fundamental representations*.

2. The Characters

Let π be a representation of U on V, $\dim V < \infty$. Then π is a differentiable function on U with values in $\mathrm{Hom}(V, V)$ (cf. [DS], Chapter II, Theorem 2.6). If $X \in \mathfrak{u}$ let $\tilde{X}$ as usual denote the left invariant vector field on U such that $\tilde{X}_e = X$. Then we have $(\mathrm{Ad}(u)X)^{\sim} = \tilde{X}^{R(u^{-1})}$ [Chapter II, §4, Eq. (9)]. On the other hand,

$$(\tilde{X}\pi)(u) = \left(\frac{d}{dt}\pi(u \exp tX)\right)_{t=0} = \pi(u)\pi(X) \tag{10}$$

and more generally, if $D \in \mathbf{D}(U)$,

$$(D\pi)(u) = \pi(u)\pi(D), \quad u \in U. \tag{11}$$

But if $v \in U$, $\pi(\tilde{X}^{R(v)}) = \pi(\mathrm{Ad}(v^{-1})X) = \pi(v^{-1})\pi(X)\pi(v)$, so by (11)

$$(D^{R(v)}\pi)(u) = \pi(u)\pi(v^{-1})\pi(D)\pi(v). \tag{12}$$

Lemma 1.6. *Let π be an irreducible representation of U and let D be a bi-invariant differential operator on U. Then*

(i) *$\pi(D)$ is a scalar multiple of the identity $\pi(D) = c_D I$;*
(ii) *π is an eigenfunction of D,*

$$D\pi = c_D\pi.$$

In fact, $D^{R(v)} = D$ implies by (12) that $\pi(D)$ commutes with $\pi(v)$, so (i) follows from Schur's lemma; then (11) implies (ii).

Theorem 1.7. *Let π be an irreducible representation of the group U (simply connected, semisimple, and compact). Then the character χ of π is given by*

$$\chi(\exp H) = \frac{\sum_s (\det s)e^{s(\lambda+\rho)(H)}}{\sum_s (\det s)e^{s\rho(H)}},$$

if λ is the highest weight of π, the summations extending over the Weyl group.

Proof. We use the radial part of the Laplacian L_U on U as given by Chapter II, Proposition 3.12. Because of (1),

$$\chi(\exp H) = \sum_{\mu \in \Lambda(\pi)} m_\mu e^{\mu(H)}, \qquad H \in \mathfrak{t}_0. \tag{13}$$

On the other hand, by Lemma 1.6, χ is an eigenfunction of L_U, so, using the quoted formula for the radial part, the function

$$(\delta^{1/2}\chi)(\exp H) = \sum_{s,\mu} (\det s)m_\mu e^{(s\rho+\mu)(H)} \tag{14}$$

is an eigenfunction of L_T. But by (8), $|\mu + s\rho| < |s\lambda + s\rho|$ if $\mu \neq s\lambda$, so the identity

$$L_T e^{s\rho+\mu} = -|s\rho + \mu|^2 e^{s\rho+\mu} \tag{15}$$

together with (14) imply

$$(\delta^{1/2}\chi)(\exp H) = \sum_s (\det s) m_{s\lambda} e^{(s\rho + s\lambda)(H)}.$$

But by Theorem 1.3, $m_{s\lambda} = m_\lambda = 1$, so the theorem is proved. We also see from (15) (with $\mu = s\lambda$) that

$$L_U \chi = (\langle \rho, \rho \rangle - \langle \lambda + \rho, \lambda + \rho \rangle)\chi. \tag{16}$$

Proof of Theorem 1.5. Part (i) is immediate from the character formula since we have already observed (Chapter IV, §1. No. 2) that two representations with the same character are equivalent.

Next suppose $\lambda_0 \in \mathfrak{t}^*$ is dominant integral such that the statement of Part (ii) fails. According to Chapter I, Lemma 5.14 and Proposition 5.15, the function

$$h = \frac{\sum_s (\det s) e^{s(\lambda_0 + \rho)}}{\sum_s (\det s) e^{s\rho}} \tag{17}$$

is a continuous function on T and is invariant under the Weyl group. Since U-conjugate elements in T are W-conjugate ([DS], Chapter VII, Lemma 7.10), h extends uniquely to a central function $\tilde{h}$ on U. Clearly $\tilde{h}$ is bounded and by [DS] (Chapter VII, Lemma 6.3), $\tilde{h}$ is continuous on the regular set U_r whose complement $U - U_r$ has measure 0. If χ is as in Theorem 1.7 we have, using Chapter I, Corollary 5.16,

$$\int_U \tilde{h}(u)\overline{\chi(u)}\, du = \frac{1}{|W|} \int_T \sum_{s,\sigma} \det(s\sigma) e^{s(\lambda_0 + \rho) - \sigma(\lambda + \rho)}(t)\, dt = 0$$

because $\lambda \neq \lambda_0$ by assumption, so the elements $H_{\lambda_0 + \rho}$, $H_{\lambda + \rho}$ which lie in $\mathfrak{t}^+$ cannot be W-conjugate. Using the Peter–Weyl theorem we conclude that $h \equiv 0$, which is a contradiction.

Theorem 1.8. *Let π be an irreducible representation of U with highest weight λ. Then*

$$\dim \pi = \prod_{\alpha \in \Delta^+} \frac{\langle \lambda + \rho, \alpha \rangle}{\langle \rho, \alpha \rangle}.$$

Proof. We must calculate $\chi(e)$ on the basis of Theorem 1.7. As in Chapter I, Lemma 5.14, we consider the algebra F over $\boldsymbol{R}$ generated by e^ω, $\omega \in \Lambda$. Let x be an indeterminate. For each $\nu \in \Lambda$ consider the homomorphism h_ν of F into the ring $\boldsymbol{R}[[x]]$ of formal power series in x

defined by $h_\nu(e^\omega) = e^{\langle \nu, \omega \rangle x}$ ($\omega \in \Lambda$). By (13), $\dim \pi$ equals the constant term in the power series $h_\nu(\chi)$. With the linear mapping $A: F \to F$ given by $A(e^\omega) = \sum_s (\det s) e^{s\omega}$, we have

$$h_\nu(A(e^\omega)) = \sum_s (\det s) e^{\langle \nu, s\omega \rangle x} = h_\omega(A(e^\nu))$$

so, using Chapter I, Proposition 5.15,

$$(18) \qquad h_\rho(A(e^\omega)) = h_\omega(A(e^\rho)) = e^{\langle \omega, \rho \rangle x} \prod_{\alpha \in \Delta^+} (1 - e^{-\langle \alpha, \omega \rangle x})$$

$$= \prod_{\alpha \in \Delta^+} (e^{(1/2)\langle \alpha, \omega \rangle x} - e^{-(1/2)\langle \alpha, \omega \rangle x}).$$

The lowest-degree term in this series (in x) is $\prod_{\alpha \in \Delta^+} \langle \alpha, \omega \rangle x$. But then the equation

$$(19) \qquad h_\rho(A(e^{\lambda + \rho})) = h_\rho(\chi) h_\rho(A(e^\rho))$$

implies

$$\prod_{\alpha \in \Delta^+} \langle \alpha, \lambda + \rho \rangle = \dim(\pi) \prod_{\alpha \in \Delta^+} \langle \alpha, \rho \rangle$$

as desired.

Let $U(\mathfrak{g})$ denote the universal enveloping algebra of $\mathfrak{g}$ and $Z(\mathfrak{g})$ its center. Viewing the members of $U(\mathfrak{g})$ as the left invariant differential operators on U, we identify the elements of $Z(\mathfrak{g})$ with the bi-invariant differential operators on U (Chapter II, Corollary 4.5). By considering the radial parts for the action of U on itself by conjugacy we shall set up an isomorphism between the algebra $Z(\mathfrak{g})$ and the algebra $I(\mathfrak{t})$ of Weyl group invariants in the symmetric algebra $S(\mathfrak{t})$. Viewing this algebra as the universal enveloping algebra of $\mathfrak{t}$ its elements can be regarded as translation-invariant differential operators on T. As in Chapter II, Proposition 3.12, we use the submanifold $\exp P_0 \subset T$ as a transversal manifold for the action of U on itself by the automorphism $I(u_0): u \to u_0 u u_0^{-1}$ ($u_0, u \in U$). The density function δ on T has a well-defined square root $\delta^{1/2}$ given by (cf., Chapter I, Proposition 5.15),

$$(20) \qquad \delta^{1/2}(\exp H) = \sum_s (\det s) e^{s\rho(H)} = e^{\rho(H)} \prod_{\alpha \in \Delta^+} (1 - e^{-\alpha(H)})$$

for $H \in \mathfrak{t}_0$.

Theorem 1.9. *Let $D \in Z(\mathfrak{g})$ with $\Delta(D)$ its radial part (on $\exp P_0$). Then there exists a unique $\gamma(D) \in I(\mathfrak{t})$ such that*

$$\Delta(D) = \delta^{-1/2} \gamma(D) \circ \delta^{1/2} \qquad \textit{on } \exp P_0.$$

Moreover, the mapping $D \to \gamma(D)$ is an isomorphism of $Z(\mathfrak{g})$ onto $I(\mathfrak{t})$.

Proof. (i) We begin with a simple remark on divisibility in $\mathscr{E}(T)$. A function $f \in \mathscr{E}(T)$ is said to be *skew* if $f(\exp(sH)) = (\det s)f(\exp H)$ for $s \in W$, $H \in \mathfrak{t}_0$. The function $\delta^{1/2}$ given by (20) is skew. We claim that each skew $f \in \mathscr{E}(T)$ is divisible by $\delta^{1/2}$ in $\mathscr{E}(T)$. For $n \in \mathbf{Z}$, $\alpha \in \Delta^+$ the reflection $\sigma_{\alpha,n}$ in the hyperplane $\alpha = 2\pi in$ in $\mathfrak{t}_0$ is given by

$$\sigma_{\alpha,n}(H) = s_\alpha(H) + n4\pi i H_\alpha/\langle\alpha, \alpha\rangle, \tag{21}$$

so $(f \circ \exp)(\sigma_{\alpha,n}(H)) = -(f \circ \exp)(H)$ (cf. [DS], Chapter VII, §7). Hence $f \circ \exp$ vanishes on the hyperplanes $\sigma_{\alpha,n}$ $(n \in \mathbf{Z})$ and is therefore divisible by the function $H \to 1 - e^{-\alpha(H)}$. Thus by (20), $f \circ \exp = (\delta^{1/2} \circ \exp)g$ where $g \in \mathscr{E}(\mathfrak{t}_0)$. Also, g depends continuously on $f \circ \exp$, and $g(H + H_e) \equiv g(H)$ for $H \in \mathfrak{t}_0$, $H_e \in \mathfrak{t}_e$. Hence $f\delta^{-1/2}$ belongs to $\mathscr{E}(T)$ as claimed and depends continuously on f.

(ii) Next we prove that if $P \in I(\mathfrak{t})$ and we consider P as an invariant differential operator on T then there exists a $E \in Z(\mathfrak{g})$ such that

$$\Delta(E) = \delta^{-1/2}P \circ \delta^{1/2}.$$

For $F \in \mathscr{E}(U)$ put $\bar{F} = F|T$,

$$F^{\natural}(v) = \int_U F(uvu^{-1})\,du$$

and consider the functional

$$S: F \to [\delta^{-1/2}P(\delta^{1/2}(F^{\natural})^-)](e), \qquad F \in \mathscr{E}(U). \tag{22}$$

Since $(F^{\natural})^-$ is W-invariant the remarks above show that S is a distribution. Moreover, S is central and has support contained in $\{e\}$. Hence the convolution $E: F \to F * \check{S}$ (where $\check{S}$ is the image of S under the map $u \to u^{-1}$ of U) is a bi-invariant differential operator on U. To compute the radial part $\Delta(E)$ we note that if F is central then

$$(\Delta(E)F)(t) = (EF)(t) = \int_U F(tu)\,dS(u) = S(F^{L(t^{-1})})$$

$$= \left[\delta^{-1/2}(s)P_s\left(\delta^{1/2}(s)\int_U F(tusu^{-1})\,du\right)\right]_{s=e}.$$

If F is the character χ of an irreducible representation with highest weight λ then by Chapter IV, Theorem 1.6, the above integral equals $\chi(t)\chi(s)/\chi(e)$. Thus by Theorem 1.7 the entire expression equals

$$P(\lambda + \rho)F(t).$$

This, however, is also equal to $(\delta^{-1/2}P(\delta^{1/2}F))(t)$. Thus $\Delta(E)F = \delta^{-1/2}P(\delta^{1/2}F)$ if $F = \chi$, so by the decomposition of central $F \in \mathscr{E}(U)$ into characters (Corollary 3.6 of this chapter) it follows that

$$\Delta(E) = \delta^{-1/2}P \circ \delta^{1/2}$$

as desired.

(iii) For the last step we prove that if $D \in Z(\mathfrak{g})$ then there exists a $Q \in I(\mathfrak{t})$ such that

$$(DF)(e) = (\delta^{-1/2}Q(\delta^{1/2}\bar{F}))(e)$$

for $F \in \mathscr{E}(U)$ central, the bar denoting restriction to T.

For this we construct a W-invariant distribution S on T with support at e such that $S(\bar{F}) = (DF)(e)$ for each central $F \in \mathscr{E}(U)$. To define S consider an open ball V around 0 in $\mathfrak{u}$ on which $\exp\colon \mathfrak{u} \to U$ is a diffeomorphism and $\exp(V \cap \mathfrak{t}_0) = (\exp V) \cap T$. (cf. [DS], Chapter VII, Lemma 6.4). Suppose $\phi \in \mathscr{E}(T)$ is W-invariant and has compact support in $(\exp V) \cap T$. Passing to $\phi \circ \exp$ and using Theorem 5.8 of Chapter II we can extend ϕ to a function Φ on U which is smooth near e and invariant under conjugation. We put $S(\phi) = (D\Phi)(e)$ and if $f \in \mathscr{E}(T)$ has compact support inside $(\exp V) \cap T$ we put $S(f) = |W|^{-1}S(\sum_s f^s)$. We then extend S to a W-invariant distribution on $\mathscr{E}(T)$ with support $\{e\}$. Then there exists a $Q' \in I(\mathfrak{t})$ such that

$$(DF)(e) = S(\bar{F}) = Q'(\bar{F})(e)$$

for all central $F \in \mathscr{E}(U)$. The differential operators Q' and $\delta^{-1/2}Q' \circ \delta^{1/2}$ have the same leading terms so we see inductively that there exists a $Q \in I(\mathfrak{t})$ such that

$$(DF)(e) = (\delta^{-1/2}Q(\delta^{1/2}\bar{F}))(e)$$

for each central $F \in \mathscr{E}(U)$. Since $(DF)(e) = (DF^{\natural})(e)$, it follows that

$$(DF)(e) = (\delta^{-1/2}Q(\delta^{1/2}(F^{\natural})^{-}))(e), \qquad F \in \mathscr{E}(U).$$

The proof of (ii) then shows that

$$\Delta(D) = \delta^{-1/2}Q \circ \delta^{1/2}.$$

Defining $\gamma\colon Z(\mathfrak{g}) \to I(\mathfrak{t})$ by $\gamma(D) = Q$ parts (ii) and (iii) above imply that γ is a surjective homomorphism. Also γ is one-to-one because if $Q = 0$ we have $DF = 0$ for all central $F \in \mathscr{E}(U)$. But then $(DF)(e) = (DF^{\natural})(e) = 0$ for all $F \in \mathscr{E}(U)$ so $D = 0$.

It is of interest to combine Weyl's character formula with Harish-Chandra's integral formula in Chapter II, Theorem 5.35, also taking into account the formula for the order of the Weyl group (Corollary 5.36). The function

$$(23)\qquad \prod_{\alpha\in\Delta^+} \frac{e^{(1/2)\alpha(H)} - e^{-(1/2)\alpha(H)}}{\alpha(H)} \qquad (H \in \mathfrak{t})$$

is holomorphic and W-invariant so by Chapter IV, Lemma 9.5 it extends to a holomorphic function j on $\mathfrak{u}^C$ invariant under $\mathrm{Ad}(U)$.

Theorem 1.10. *Let χ be the character of an irreducible representation of U with highest weight λ. Then*

$$(24)\qquad j(X)\chi(\exp X) = \chi(e)\int_U e^{\langle H_{\lambda+\rho}, u\cdot X\rangle}\, du, \qquad X \in \mathfrak{u},$$

and

$$|j(X)|^2 = \det_{\mathfrak{u}}\left(\frac{1 - e^{-\operatorname{ad} X}}{\operatorname{ad} X}\right), \qquad X \in \mathfrak{u}.$$

In fact, the roots are purely imaginary on $\mathfrak{t}_0$, so if

$$X = u\cdot H \quad (u \in U, H \in \mathfrak{t}_0)$$

we have

$$|j(X)|^2 = j(H)\overline{j(H)} = e^{\rho(H)}\prod_{\alpha>0}\frac{1 - e^{-\alpha(H)}}{\alpha(H)}\, e^{-\rho(H)}\prod_{\alpha>0}\frac{1 - e^{\alpha(H)}}{-\alpha(H)}$$

$$= \prod_{\alpha\in\Delta}\frac{1 - e^{-\alpha(H)}}{\alpha(H)} = \det_{\mathfrak{u}}\left(\frac{1 - e^{-\operatorname{ad} X}}{\operatorname{ad} X}\right)$$

by [DS], Chapter VII, §4. The formula for χ now follows by putting $H' = H_{\lambda+\rho}$ in the cited Theorem 5.35 (Chapter II), taking Theorems 1.7 and 1.8 into account.

Formula (24) has an interesting interpretation in terms of Fourier transforms. The point $iH_{\lambda+\rho}$ belongs to $\mathfrak{t}_0$ and we denote by μ_λ the U-invariant measure on the orbit $O_\lambda = U\cdot(iH_{\lambda+\rho}) \subset \mathfrak{u}$ normalized by $\mu_\lambda(O_\lambda) = 1$. Then μ_λ can be viewed as a distribution on $\mathfrak{u}$ of compact support; as observed in Chapter I, §2. No. 8, its distributional Fourier transform is the function $\tilde{\mu}_\lambda(X) = \int_{\mathfrak{u}} e^{-i\langle Y, X\rangle}\, d\mu_\lambda(Y)$. Formula (24) can therefore be rewritten as follows.

Corollary 1.11. *The function $j(\chi\circ\exp)$ on $\mathfrak{u}$ is a constant multiple of the Fourier transform $\tilde{\mu}_\lambda$, i.e.,*

$$(25)\qquad j(X)\chi(\exp X) = \chi(e)\tilde{\mu}_\lambda(X), \qquad X \in \mathfrak{u}.$$

Let K be the fixed point group of an involutive automorphism of U; K is actually connected (cf. [DS], Chapter VII. Theorem 8.2). For analysis on symmetric spaces we are interested in the representations π of U which are spherical (with respect to K); that is, $\pi(K)$ has a nonzero fixed vector. In §4, No. 1 of this chapter we shall give an explicit characterization of the weights of these representations.

§2. Fourier Expansions on Compact Groups

1. Introduction. $L^1(K)$ versus $L^2(K)$

Let K be an arbitrary compact group with Haar measure dk normalized by $\int_K dk = 1$. Let $\hat{K}$ denote the set of all equivalence classes of irreducible (finite-dimensional) unitary representations of K. If $\lambda \in \hat{K}$ let U_λ be a member of the class λ acting on the d_λ-dimensional Hilbert space $\mathscr{H}_\lambda$. If for some orthonormal basis of $\mathscr{H}_\lambda$, U_λ has the matrix form $U_\lambda = (u_{ij}^\lambda)$ the Schur orthogonality relations together with the Peter–Weyl theorem express the following result (already used in Chapter IV, §1).

Theorem 2.1. *The functions $d_\lambda^{1/2} \cdot u_{ij}^\lambda$ ($\lambda \in \hat{K}$, $i, j = 1, \ldots, d_\lambda$) form a complete orthonormal system in $L^2(K)$.*

To each $f \in L^1(K)$ we associate the *Fourier series* (Tr denotes the trace)

$$f(k) \sim \sum_{\lambda \in \hat{K}} d_\lambda \operatorname{Tr}(A_\lambda U_\lambda(k)), \tag{1}$$

where the *Fourier coefficient* is the endomorphism of $\mathscr{H}_\lambda$ given by

$$A_\lambda = \int_K f(k) U_\lambda(k^{-1})\, dk. \tag{2}$$

If f is a K-finite function on K (under the left regular representation on K) or, equivalently, if f is a finite linear combination of functions u_{ij}^λ then the orthonormality in Theorem 2.1 implies that the sign $\sim$ in (1) can be replaced by an equality sign. If $g \in L^1(K)$ has Fourier coefficients $\{B_\lambda\}$ then

$$(f * g)(k) \sim \sum_{\lambda \in \hat{K}} d_\lambda \operatorname{Tr}(B_\lambda A_\lambda U_\lambda(k)). \tag{2'}$$

Also, if A^* denotes the Hilbert space adjoint the completeness in Theorem 2.1 implies that

$$\int_K |f(k)|^2\, dk = \sum_{\lambda \in \hat{K}} d_\lambda \operatorname{Tr}(A_\lambda A_\lambda^*) \tag{3}$$

for $f \in L^2(K)$. This relation and (2′) imply that $f \in L^1$ is uniquely determined by its Fourier series. Then (3) even holds in the sense that if for a sequence $\{A_\lambda\}_{\lambda \in \hat{K}}$ the right-hand side of (3) is finite then (1) holds for some $f \in L^2(K)$.

If f is a central function then (1) takes the form (with $\chi_\lambda = \operatorname{Tr}(U_\lambda)$)

$$f(k) \sim \sum_{\lambda \in \hat{K}} a_\lambda \chi_\lambda(k), \qquad a_\lambda = \int_K f(k)\chi_\lambda(k^{-1})\, dk$$

and the characters χ_λ form a complete orthonormal system in the space of central functions in $L^2(K)$.

Since f in (1) is uniquely determined by the series (1) it is of interest to study properties of the family $\{A_\lambda\}_{\lambda \in \hat{K}}$. Let $\| \ \|$ denote the operator norm and put $\|f\|_p = (\int_K |f(k)|^p\, dk)^{1/p}$.

Proposition 2.2. *Let $f \in L^1(K)$. Then the function $\lambda \to \|A_\lambda\|$ vanishes at ∞ on the discrete space $\hat{K}$. In other words, for each $\varepsilon > 0$ the set*

$$\{\lambda \in \hat{K} : \|A_\lambda\| > \varepsilon\}$$

is finite.

Proof. If $f \in L^2(K)$, this is obvious from (3) since $\|A\|^2 \leq \operatorname{Tr}(AA^*)$. In general (2) implies $\|A_\lambda\| \leq \|f\|_1$, so $\|A_\lambda - B_\lambda\| \leq \|f - g\|_1$ and the proposition now follows from the fact that $L^2(K)$ is dense in $L^1(K)$.

In this section we shall discuss some properties of the expansion (1) which emphasize the difference between $L^1(K)$ and $L^2(K)$. For motivation we first discuss briefly the case of the circle group.

2. The Circle Group

Let $\boldsymbol{T}$ denote the circle group $\{e^{ix} : 0 < x \leq 2\pi\}$. We recall that the irreducible representations of $\boldsymbol{T}$ are given by the characters $\chi_n : e^{ix} \to e^{inx}$, n varying over $\boldsymbol{Z}$. Identifying functions f on the circle with periodic functions $F : x \to f(e^{ix})$ on $\boldsymbol{R}$, the Fourier series (1) becomes the classical Fourier series

$$F(x) \sim \sum_{n \in \boldsymbol{Z}} a_n e^{inx}, \qquad F \in L^1(\boldsymbol{T}), \tag{4}$$

where

$$a_n = \frac{1}{2\pi} \int_0^{2\pi} F(x) e^{-inx}\, dx. \tag{5}$$

Here Proposition 2.2 becomes the Riemann–Lebesgue lemma: $a_n \to 0$ as $|n| \to \infty$. Moreover, (3) generalizes the Riesz–Fisher theorem: The convergence

$$\sum_n |a_n|^2 < \infty \tag{6}$$

is a necessary and sufficient condition for the Fourier series (4) to represent a function $F \in L^2(\boldsymbol{T})$; in this case we have the Parseval formula,

$$\frac{1}{2\pi}\int_0^{2\pi} |F(x)|^2\, dx = \sum_n |a_n|^2. \tag{7}$$

Condition (6) has two interesting features: It only involves the absolute value of a_n and is independent of the order. Thus if (6) holds and if $\sigma: \boldsymbol{Z} \to \boldsymbol{Z}$ is a permutation the series $\sum_{n \in Z} a_{\sigma(n)} e^{inx}$ is the Fourier series of an L^2 function. Similarly, if $|\gamma_n| = 1$ for all $n \in \boldsymbol{Z}$ the series

$$\sum_{n \in \boldsymbol{Z}} a_n \gamma_n e^{inx}$$

is the Fourier series of an L^2 function. We are therefore led to the following questions where for a subspace $E \subset L^1$ the notation

$$\sum_n a_n e^{inx} \in E \tag{8}$$

means that $\sum_n a_n e^{inx}$ is the Fourier series for a function in E.

I. *Let* $\sum_n a_n e^{inx} \in L^1(\boldsymbol{T})$ *and assume*

$$\sum_n a_n \gamma_n e^{inx} \in L^1(\boldsymbol{T}) \tag{9}$$

for every sequence (γ_n) *such that* $|\gamma_n| = 1$ *for all* $n \in \boldsymbol{Z}$. *Is then* $\sum_n |a_n|^2 < \infty$?

The answer is yes (*Littlewood* [1924]).

II. *Let* $\sum_n a_n e^{inx} \in L^1(\boldsymbol{T})$ *and assume that for every permutation* σ *of the integers we have*

$$\sum_n a_{\sigma(n)} e^{inx} \in L^1(\boldsymbol{T}). \tag{10}$$

Is then $\sum |a_n|^2 < \infty$?

The answer is yes (Helgason [1958]).

We quote three more results with a similar conclusion.

III. (Sidon [1932]). *Let* (a_n) *be a sequence such that for each*

$$\sum_n b_n e^{inx} \in C(\boldsymbol{T})$$

we have $\sum_n |a_n b_n| < \infty$. *Then* $\sum_n |a_n|^2 < \infty$.

IV. (Zygmund [1930]). *Let*

$$\sum_k a_k e^{in_k x} \in L^1(\boldsymbol{T}), \tag{11}$$

where $n_{k+1}/n_k > 2$ *for all* k. *Then*

$$\sum_1^\infty |a_k|^2 < \infty.$$

Fourier series of the form (11) where $n_{k+1}/n_k > q > 1$ are called *lacunary* Fourier series.

V (Kolmogoroff [1928]). *Consider the compact group* $\boldsymbol{T}^\omega = \prod_{k=1}^\infty T_k$ *where* $T_k = \boldsymbol{T}$ *for each* k. *Assume* $f \in L^1(\boldsymbol{T}^\omega)$ *has Fourier series of the form*

$$f(x_1, x_2, \ldots) \sim \sum_1^\infty a_k e^{ix_k}.$$

Then

$$\sum_1^\infty |a_k|^2 < \infty.$$

We shall now generalize these results to compact groups K (abelian or not). We shall define lacunary Fourier series for such K; then IV and V above will be special cases of the same result and II (generalized to abelian groups) will be a corollary. Finally, the generalization of Littlewood's result I to the group K will at the same time give the extension of Sidon's result III.

3. Spectrally Continuous Operators

Let μ be a complex-valued measure on the compact group K. Let L_μ and R_μ denote the bounded operators on $L^p(K)$ $(p \geq 1)$ defined by

$$(L_\mu F)(k) = (\mu * F)(k) = \int_K F(x^{-1}k)\, d\mu(x),$$

$$(R_\mu F)(k) = (F * \mu)(k) = \int_K F(kx^{-1})\, d\mu(x).$$

If $\mu = f(k)\, dk$ where $f \in L^1(K)$, we write L_f for L_μ, R_f for R_μ.

Let f^* denote the function $k \to (f(k^{-1}))^-$ (the bar denoting complex conjugate) and μ^* the measure $f \to (\mu(f^*))^-$. Let J denote the involution $f \to f^*$ Then a routine computation gives

$$L_{\mu^*} = JR_\mu J. \tag{12}$$

Also, L_{μ^*} acting on $L^q(K)$ $(p^{-1} + q^{-1} = 1)$ is the adjoint of L_μ acting on $L^p(K)$; thus

$$L_{\mu^*} = (L_\mu)^*, \qquad R_{\mu^*} = (R_\mu)^*. \tag{13}$$

In the case $p = q = 2$ we see from (12) and (13) that L_μ and R_μ have the same norm.

Definition. Let $f \in L^1(K)$ and consider L_f and R_f as operators on $L^2(K)$. The common value of the operator norms $\|L_f\|$ and $\|R_f\|$ will be called the *spectral norm* of f. We denote it by $\|f\|_{\mathrm{sp}}$. Then

$$\|f\|_{\mathrm{sp}} = \sup_{g \in L^2} \frac{\|f * g\|_2}{\|g\|_2}.$$

Since $\|f * g\|_2 \le \|f\|_1 \|g\|_2$ for all $g \in L^2(K)$ we have

$$\|f\|_{\mathrm{sp}} \le \|f\|_1, \qquad f \in L^2(K). \tag{14}$$

In the case $K = \boldsymbol{T}$, Parseval's formula (7) shows quickly that

$$\|f\|_{\mathrm{sp}} = \max_n |a_n| \qquad \text{if } f \sim \sum_n a_n e^{inx}. \tag{15}$$

This will be generalized in Lemma 2.5.

Because of (14) the spectral norm topology on $L^1(K)$ is weaker than the usual topology. A linear transformation T on $L^1(K)$ is said to be *spectrally continuous* if it is continuous from the spectral norm topology to the L^1 topology. Equivalently, T is spectrally continuous if

$$\sup_{\|f\|_{sp} \le 1} (\|Tf\|_1) < \infty.$$

Of course (14) shows that a spectrally continuous operator is continuous.

Theorem 2.3. *Let K be a compact group. The spectrally continuous operators S on $L^1(K)$ commuting with all right translations on K are precisely the left convolutions $S = L_f$ with $f \in L^2(K)$. Furthermore,*

$$\frac{1}{\sqrt{2}} \|f\|_2 \le \sup_g \frac{\|f * g\|_1}{\|g\|_{\mathrm{sp}}} \le \|f\|_2. \tag{16}$$

Example. Consider the case when K is the circle group $\boldsymbol{T}$. Then Theorem 2.3 implies Littlewood's result I in the preceding subsection. In fact, let $C_0(\boldsymbol{Z})$ denote the Banach space of sequences $\gamma = (\gamma_n)$ such that $\gamma_n \to 0$ for $|n| \to \infty$, the norm being $\|\gamma\| = \sup_n |\gamma_n|$. Each $\gamma \in C_0(\boldsymbol{Z})$ can be written $\gamma = \alpha + \beta$ where α and β are sequences (α_n) and (β_n) where

$|\alpha_n|$ and $|\beta_n|$ are independent of n. Thus assumption (9) implies that for each $\gamma \in C_0(\boldsymbol{Z})$ there exists a function $f_\gamma \in L^1(\boldsymbol{T})$ such that

$$f_\gamma \sim \sum_n a_n \gamma_n e^{inx}.$$

The mapping $S: \gamma \to f_\gamma$ of $C_0(\boldsymbol{Z})$ into $L^1(\boldsymbol{T})$ is linear and has a closed graph $\{(\gamma, f_\gamma) : \gamma \in C_0(\boldsymbol{Z})\}$ in the product space $C_0(\boldsymbol{Z}) \times L^1(\boldsymbol{T})$. By the closed-graph theorem of Banach, S is continuous. But by (15) this means that S is a spectrally continuous operator on $L^1(\boldsymbol{T})$. Since it clearly commutes with all translations on $\boldsymbol{T}$, Theorem 2.3 implies that $\sum |a_n|^2 < \infty$ as claimed. See Exercise B2 for a nonabelian analog.

For the proof of Theorem 2.3 and related results we start with a simple lemma (cf. Wendel [1952], where, however, the continuity of S is *assumed*).

Lemma 2.4. *Let S be a linear mapping of $L^1(K)$ into itself. The following properties are equivalent:*

(i) *S commutes with all left translation $L(k)$, $k \in K$.*
(ii) *S commutes with all left convolutions L_g, $g \in L^1(K)$.*
(iii) *S has the form*

$$Sf = f * \mu,$$

where μ is a complex-valued measure on K.

In particular, (i) *implies that S is continuous.*

Proof. Let $\lambda \in \hat{K}$ and, as in Chapter IV, §1, let H_λ denote the span of the matrix elements $u_{ij}(k)$ of $U_\lambda(k)$. If χ_λ is the character of λ we know from Chapter IV, Corollary 1.8, that $H_\lambda = \chi_\lambda * L^1(K)$. Thus (ii) implies $SH_\lambda \subset H_\lambda$. Since H_λ consists of the K-finite elements of $L^1(K)$ of a given type it is also clear that (i) implies $SH_\lambda \subset H_\lambda$. Assuming either (i) or (ii) we can therefore define the matrix S_λ by $(S_\lambda)_{ij} = (Su_{ij})(e)$.

We shall now prove that each of the properties (i) or (ii) is equivalent to

(iv) $\quad Sf \sim \sum_\lambda d_\lambda \operatorname{Tr}(S_\lambda A_\lambda U_\lambda(k)) \quad$ for $\quad f \sim \sum_\lambda d_\lambda \operatorname{Tr}(A_\lambda U_\lambda(k))$.

Suppose S satisfies (ii). Then by the above $SH_\lambda \subset H_\lambda$. The orthogonality relations imply

$$u_{il} * u_{mj} = (1/d_\lambda)\delta_{ml} u_{ij},$$

so by (ii)

$$u_{il} * Su_{mj} = (1/d_\lambda)\delta_{ml} Su_{ij}.$$

It follows that Su_{ij} is a linear combination of the u_{iq} $(1 \le q \le d_\lambda)$, so,

$$Su_{ij} = \sum_q (S_\lambda)_{qj} u_{iq}. \tag{17}$$

Then, with f as in (iv) we have

$$\begin{aligned}(\chi_\lambda * Sf)(k) &= S(\chi_\lambda * f)(k) = S(\operatorname{Tr}(A_\lambda U_\lambda))(k)\\ &= \operatorname{Tr}(S_\lambda A_\lambda U_\lambda(k)).\end{aligned}$$

This shows that (ii) $\Rightarrow$ (iv). Also, (iv) $\Rightarrow$ (i). Next assume (i). Since

$$u_{ij}(uk) = \sum_l u_{il}(u) u_{lj}(k), \qquad u, k \in K,$$

we deduce (17), which as we saw implies (iv), which in turn implies (ii) because of (2′). Thus (i) $\Leftrightarrow$ (ii) $\Leftrightarrow$ (iv).

Now assume (iv). Then if $f_n \to f$ and $Sf_n \to g$ in $L^1(K)$ we see from the inequality $\|A_\lambda\| \le \|f\|_1$ that $g = Sf$. Thus S has a closed graph in $L^1(K) \times L^1(K)$, so by the closed-graph theorem, S is continuous. The complex adjoint operator $S^* : L^\infty(K) \to L^\infty(K)$ will also commute with left translations. Since

$$(S^*f)^{L(k)} - S^*f = S^*(f^{L(k)} - f),$$

the boundedness of S^* shows that $S^*(C(K)) \subset C(K)$. The functional $\nu : f \to (S^*f)(e)$ is therefore a well-defined measure on K and by the translation invariance, $S^* = R_{\check\nu}$ where $\check\nu$ is the image of ν under the map $k \to k^{-1}$ of K. Hence by (13), $S = R_\mu$ where $\mu = (\check\nu)^*$.

Since (iii) obviously implies (i) the lemma is proved.

Lemma 2.4′. *Lemma 2.4 also holds with $L^1(K)$ replaced by $C(K)$.*

The proof is the same except that the last step (passing to S^*) is now unnecessary.

Lemma 2.5. *Let $f \in L^1(K)$. Then*

$$\|f\|_{sp} = \max_{\lambda \in \hat K} \|A_\lambda\|$$

in terms of the expansion (1).

Proof. Let $g \in L^2(K)$ and consider the expansion (2′). Then

$$\begin{aligned}\|f * g\|_2^2 &= \sum_\lambda d_\lambda \operatorname{Tr}(B_\lambda A_\lambda (B_\lambda A_\lambda)^*)\\ &= \sum_\lambda d_\lambda \operatorname{Tr}(A_\lambda A_\lambda^* B_\lambda^* B_\lambda) = \sum_\lambda d_\lambda \operatorname{Tr}(B_\lambda^* B_\lambda A_\lambda A_\lambda^*).\end{aligned}$$

The last two expressions are majorized by

$$\sum_{\lambda} d_\lambda \|A_\lambda\|^2 \operatorname{Tr}(B_\lambda^* B_\lambda), \qquad \sum_{\lambda} d_\lambda \|B_\lambda\|^2 \operatorname{Tr}(A_\lambda A_\lambda^*),$$

respectively. We see this, say for the first, by diagonalizing $A_\lambda A_\lambda^*$ and using $\|A_\lambda A_\lambda^*\| = \|A_\lambda\|^2$. This proves

$$(18) \qquad \|f * g\|_2 \leq \max_{\lambda} \|A_\lambda\| \, \|g\|_2, \qquad \|f * g\|_2 \leq \max_{\lambda} \|B_\lambda\| \, \|f\|_2,$$

so $\|f\|_{\text{sp}} \leq \max_\lambda \|A_\lambda\|$. On the other hand, suppose $\|A_\lambda\|$ reaches its maximum for $\lambda = \mu$. Then for each endomorphism B_μ of $\mathscr{H}_\mu$

$$\|f\|_{\text{sp}}^2 \geq \operatorname{Tr}(A_\mu A_\mu^* B_\mu^* B_\mu)/\operatorname{Tr}(B_\mu^* B_\mu).$$

Now choose a basis in $\mathscr{H}_\mu$ diagonalizing $A_\mu A_\mu^*$ and then choose the matrix B_μ having all entries 0 except at the place in the diagonal where the largest eigenvalue of $A_\mu A_\mu^*$ occurs. Then $\|f\|_{\text{sp}} \geq \|A_\mu\|$, whence the lemma.

If $f \in L^2(K)$ and if $g \in L^1(K)$ has Fourier coefficients B_λ we have by (18)

$$(19) \qquad \|f * g\|_1 \leq \|f * g\|_2 \leq \max_{\lambda} \|B_\lambda\| \, \|f\|_2,$$

which shows that L_f is spectrally continuous and the right-hand inequality of Theorem 2.3 follows.

To prove the converse (and the nontrivial) part of Theorem 2.3 we assume that S is a spectrally continuous operator on $L^1(K)$ commuting with the right translations. By Lemma 2.4, we have

$$Sg = \mu * g, \qquad g \in L^1(K),$$

where μ is a complex-valued measure. To μ we assign the *Fourier–Stieltjes series*

$$(20) \qquad \mu \sim \sum_{\lambda \in \hat{K}} d_\lambda \operatorname{Tr}(A_\lambda U_\lambda(k)),$$

where

$$A_\lambda = \int_K U_\lambda(k^{-1}) \, d\mu(k).$$

Again Theorem 2.1 implies that μ is uniquely determined by the family A_λ $(\lambda \in \hat{K})$.

Let $\{\lambda_1, \ldots, \lambda_N\}$ be an arbitrary finite subset of $\hat{K}$, $U_1, \ldots, U_N$ corresponding representations of K, and $d_1, \ldots, d_N$ their degrees. Then by

Lemma 2.5 we have if dV_n is the normalized Haar measure on the unitary group $U(d_n)$ and $dV = dV_1 \cdots dV_N$ on $U(d_1) \times \cdots \times U(d_N)$,

$$\begin{aligned}\|S\| &\geq \sup_{\|g\|_{sp} \leq 1} \|\mu * g\|_1 \geq \sup_{V_n \in U(d_n)} \int_K \left| \sum_1^N d_n \operatorname{Tr}(V_n A_n U_n(k)) \right| dk \\ &\geq \int_{U(d_1) \times \cdots \times U(d_N)} dV \int_K \left| \sum_1^N d_n \operatorname{Tr}(V_n A_n U_n(k)) \right| dk \\ &= \int_{U(d_1) \times \cdots \times U(d_N)} \left| \sum_1^N d_n \operatorname{Tr}(A_n V_n) \right| dV,\end{aligned}$$

the last identity the result of interchanging the integrations and using the right invariance of dV. Putting

$$F(V) = \sum_1^N d_n \operatorname{Tr}(A_n V_n), \qquad V = (V_1, \ldots, V_n),$$

we have proved

$$\|S\| \geq \int |F(V)| \, dV. \tag{21}$$

In order to estimate this last integral further from below we use the inequality

$$\int |F(V)| \, dV \geq \left(\int |F(V)|^4 \, dV \right)^{-1/2} \left(\int |F(V)|^2 \, dV \right)^{3/2}, \tag{22}$$

which is obtained from Hölder's inequality $\int |fg| \leq (\int |f|^p)^{1/p} (\int |g|^q)^{1/q}$ by putting $f = |F|^{2/3}$, $g = |F|^{4/3}$, $p = \frac{3}{2}$, $q = 3$. The inequality (22) is chosen because the integrals $\int |F|^2$ and $\int |F|^4$ are more manageable than $\int |F|$. For the first we have

$$\int_{U(d_1) \times \cdots \times U(d_N)} \left| \sum_1^N d_n \operatorname{Tr}(A_n V_n) \right|^2 dV = \sum_1^N d_n \operatorname{Tr}(A_n A_n^*) \tag{23}$$

because the function F on $U(d_1) \times \cdots \times U(d_N)$ has the Fourier series

$$F(V_1, \ldots, V_N) \sim \sum_1^N d_n \operatorname{Tr}(A_n V_n).$$

For $\int |F|^4$ we prove the following result.

Lemma 2.6. *Let A be an arbitrary $n \times n$ complex matrix $(n \geq 2)$. Then*

$$\int_{U(n)} |\operatorname{Tr}(AV)|^4 \, dV = \frac{2}{n^2 - 1} \left([\operatorname{Tr}(AA^*)]^2 - \frac{1}{n} \operatorname{Tr}(AA^*AA^*) \right).$$

Proof. We begin by writing $A = PO_1$, where P is positive definite, O_1 unitary, and then $B = O_2PO_2^{-1}$, where B is a diagonal matrix and O_2 unitary. Clearly $\operatorname{Tr}(AA^*) = \operatorname{Tr}(BB^*)$, so by the invariance of dV it suffices to prove the lemma when A is a diagonal matrix. Let $e_1, \ldots, e_n$ be the corresponding orthonormal basis for the Hilbert space on which A and V act. Then

$$\int_{U(n)} |\operatorname{Tr}(AV)|^4\, dV = \sum_{i,j,k,l} a_i\bar{a}_j a_k \bar{a}_l \int_{U(n)} v_i\bar{v}_j v_k \bar{v}_l\, dV,$$

where $a_i = (Ae_i, e_i)$, $v_j = (Ve_j, e_j)$ (inner products). Then we have

$$\int_{U(n)} v_i\bar{v}_j v_k\bar{v}_l\, dV = 0 \qquad \text{unless} \quad (i, k) = (j, l) \quad \text{or} \quad (l, j). \tag{24}$$

In fact, let V_0 be a diagonal matrix with diagonal elements α_i of modulus one. Writing the integral in (24) as

$$\int_{U(n)} (Ve_i, e_i)(e_j, Ve_j)(Ve_k, e_k)(e_l, Ve_l)\, dV,$$

we find, by replacing V by VV_0 and using the invariance, that this equals

$$\alpha_i\bar{\alpha}_j\alpha_k\bar{\alpha}_l \int_{U(n)} (Ve_i, e_i)(e_j, Ve_j)(Ve_k, e_k)(e_l, Ve_l)\, dV.$$

Now (24) follows, the α_i being arbitrary.

Note that the equation

$$\int_{U(n)} v_i v_j\, dV = 0, \qquad 1 \leq i, j \leq n, \tag{25}$$

follows in the same way. Using (24) we obtain

$$\int_{U(n)} |\operatorname{Tr}(AV)|^4 = \sum_1^n |a_i|^4 \int_{U(n)} |v_i|^4\, dV + 2\sum_{i \neq j} |a_i|^2|a_j|^2 \int_{U(n)} |v_i|^2|v_j|^2\, dV. \tag{26}$$

Thus it suffices to compute the integrals

$$\int_{U(n)} |v_1|^4\, dV, \qquad \int_{U(n)} |v_1|^2|v_2|^2\, dV.$$

Let

$$U_1(n-1) = \{T \in U(n) : Te_1 = e_1\},$$

$$U_{1,2}(n-2) = \{S \in U_1(n-1) : Se_2 = e_2\},$$

$$\Sigma_n = U(n)/U_1(n-1), \qquad \Sigma_{n-1} = U_1(n-1)/U_{1,2}(n-2),$$

and let $d\tilde{V}$ and $d\tilde{T}$ denote the unique normalized invariant measures on the unit spheres $\Sigma_n \subset C^n$, $\Sigma_{n-1} \subset C^{n-1}$, respectively. Then

$$\begin{aligned}
&\int_{U(n)} |v_1|^2 |v_2|^2 \, dV \\
&\quad = \int_{\Sigma_n} d\tilde{V} \int_{U_1(n-1)} |(VTe_1, e_1)|^2 |(VTe_2, e_2)|^2 \, dT \\
&\quad = \int_{\Sigma_n} d\tilde{V} \int_{\Sigma_{n-1}} d\tilde{T} \int_{U_{1,2}(n-1)} |(VTSe_1, e_1)|^2 |(VTSe_2, e_2)|^2 \, dS \\
&\quad = \int_{\Sigma_n} |(Ve_1, e_1)|^2 \, d\tilde{V} \int_{\Sigma_{n-1}} |(Te_2, V^{-1}e_2)|^2 \, d\tilde{T}.
\end{aligned}$$

Writing

$$Ve_1 = w_1 e_1 + \cdots + w_n e_n, \qquad V^{-1}e_2 = b_1 e_1 + \cdots + b_n e_n,$$

$$Te_2 = 0e_1 + z_2 e_2 + \cdots + z_n e_n,$$

since $(Te_2, e_1) = (e_2, T^{-1}e_1) = (e_2, e_1) = 0$, we obtain

$$\begin{aligned}
&\int_{U(n)} |v_1|^2 |v_2|^2 \, dV \\
&\quad = \int_{|w_1|^2 + \cdots + |w_n|^2 = 1} |w_1|^2 \, d\tilde{V} \int_{|z_2|^2 + \cdots + |z_n|^2 = 1} |\bar{b}_2 z_2 + \cdots + \bar{b}_n z_n|^2 \, d\tilde{T} \\
&\quad = (n-1)^{-1} \int_{|w_1|^2 + \cdots + |w_n|^2 = 1} |w_1|^2 (|b_2|^2 + \cdots + |b_n|^2) \, d\tilde{V} \\
&\quad = (n-1)^{-1} \int_{\Sigma_n} |w_1|^2 (1 - |w_2|^2) \, d\tilde{V}
\end{aligned}$$

since

$$\begin{aligned}
|b_2|^2 + \cdots + |b_n|^2 &= 1 - |b_1|^2 = 1 - |(V^{-1}e_2, e_1)|^2 \\
&= 1 - |(e_2, Ve_1)|^2 = 1 - |w_2|^2.
\end{aligned}$$

On the other hand,

$$n(n-1)\int_{\Sigma_n} |w_1|^2|w_2|^2\,d\tilde{V} = \int_{\Sigma_n}\Big(\sum_i |w_i|^2\Big)^2 d\tilde{V} - n\int_{\Sigma_n}|w_1|^4\,d\tilde{V}$$

$$= 1 - n\int_{\Sigma_n}|w_1|^4\,d\tilde{V},$$

and if $d\Omega$ is the invariant measure on S^{2n-1} normalized by $\int d\Omega = 1$ then

$$\int_{\Sigma_n}|w_1|^4\,d\tilde{V} = \int_{S^{2n-1}}(x_1^2 + x_2^2)^2\,d\Omega = 2\int_{S^{2n-1}}x_1^4\,d\Omega + 2\int_{S^{2n-1}}x_1^2x_2^2\,d\Omega.$$

But by Chapter I, Lemma 2.30,

$$\int_{S^{2n-1}}x_1^4\,d\Omega = \frac{3}{4n(n+1)},$$

and since $\int (x_1^2 + \cdots + x_{2n}^2)^2\,d\Omega = 1$,

$$\int_{S^{2n-1}}x_1^2x_2^2\,d\Omega = \frac{1}{4n(n+1)}.$$

Substituting, we get the desired formula

$$\int_{U(n)}|v_1|^2|v_2|^2\,dV = \frac{1}{n^2-1}. \tag{27}$$

Second,

$$\int_{U(n)}|v_1|^4\,dV = \int_{\Sigma_n}d\tilde{V}\int_{U_1(n-1)}|VTe_1, e_1|^4\,d\tilde{T}$$

$$= \int_{\Sigma_n}|(Ve_1, e_1)|^4\,d\tilde{V} = \int_{\Sigma_n}|w_1|^4\,d\tilde{V}$$

so

$$\int_{U(n)}|v_1|^4\,dV = \frac{2}{n(n+1)}. \tag{28}$$

Substituting into (26) we obtain the lemma.

Corollary 2.7. *The function*

$$F(V_1, \ldots, V_N) = \sum_1^N d_n \operatorname{Tr}(A_n V_n)$$

satisfies

$$\int |F(V)|^4\,dV \leq 2\Big(\int |F(V)|^2\,dV\Big)^2. \tag{29}$$

Using Schwarz's inequality $(\sum_i \alpha_i)^2 \le (\sum_i \alpha_i^2)n$ for the eigenvalue α_i of AA^* we obtain a simple estimate for the expression in Lemma 2.6,

$$\text{(30)} \qquad \frac{2}{n^2-1}\left([\mathrm{Tr}(AA^*)]^2 - \frac{1}{n}\mathrm{Tr}(AA^*AA^*)\right) \le \frac{2}{n^2}[\mathrm{Tr}(AA^*)]^2.$$

Using the orthogonality relation as well as (23), (25), and (30) we obtain

$$\begin{aligned}\int |F(V)|^4\,dV &= \sum_1^N d_n^4 \int_{U(d_n)} |\mathrm{Tr}(A_n V_n)|^4\,dV_n \\ &\quad + 2\sum_{i\neq j}\int_{U(d_i)} |d_i\,\mathrm{Tr}(A_i V_i)|^2\,dV_i \int_{U(d_j)} |d_j\,\mathrm{Tr}(A_j V_j)|^2\,dV_j \\ &\le 2\sum_1^N d_n^2[\mathrm{Tr}(A_n A_n^*)]^2 + 2\sum_{i\neq j} d_i d_j\,\mathrm{Tr}(A_i A_i^*)\mathrm{Tr}(A_j A_j^*),\end{aligned}$$

and now (29) follows.

Combining now (21), (22), and (29) we have

$$\text{(31)} \qquad \frac{1}{2}\sum_1^N d_n\,\mathrm{Tr}(A_n A_n^*) \le \|S\|^2.$$

The coefficients $A_1, \ldots, A_N$ were chosen arbitrarily from the Fourier–Stieltjes series (20). But then (31) shows that $S = L_f$ for a function $f \in L^2(K)$ and (16) holds. This proves Theorem 2.3.

4. Absolute Convergence

A Fourier series (1) on K is said to be *absolutely convergent* if

$$\text{(32)} \qquad \sum_{\lambda\in\hat{K}} d_\lambda\,\mathrm{Tr}(|A_\lambda|) < \infty.$$

Here $|A_\lambda|$ denotes the unique positive self-adjoint operator such that $|A_\lambda|^2 = A_\lambda A_\lambda^*$. [If $A = PU$ is a polar decomposition of a matrix A where P is positive definite and U unitary we have $P = |A|$; also $A^*A = U^{-1}(AA^*)U$, so use of $A_\lambda^* A_\lambda$ instead of $A_\lambda A_\lambda^*$ in defining $|A_\lambda|$ would have led to the same trace.] If K is abelian then each $d_\lambda = 1$ so definition (32) amounts to the usual one.

Definition. Let $A(K)$ denote the set of $f \in L^1(K)$ having absolutely convergent Fourier series.

Lemma 2.8.

(i) *$A(K)$ is invariant under left and right translations of K.*
(ii) *$A(K) \subset C(K)$.*
(iii) *$A(K) = \{g * h : g, h \in L^2(K)\}$.*

Proof. Let $\mathscr{H}$ be an n-dimensional Hilbert space with inner product $(\ ,\ )$ A, B two operators on $\mathscr{H}$, $A = PU$ a polar decomposition. Let $e_1, \ldots, e_n$ be orthonormal vectors diagonalizing P, i.e., $Pe_i = c_i e_i$ where $c_i \geq 0$ $(1 \leq i \leq n)$. Then

$$|\mathrm{Tr}(AB)| = |\mathrm{Tr}(PUB)| = \left|\sum_i c_i(UBe_i, e_i)\right|$$
$$\leq \sum_i c_i \|Be_i\| \, \|e_i\|,$$

so

$$(33) \qquad |\mathrm{Tr}(AB)| \leq \mathrm{Tr}(|A|)\|B\|.$$

Since $|AB| = ABV$ where V is unitary this implies (as $\|BV\| \leq \|B\|$)

$$(34) \qquad \mathrm{Tr}(|AB|) \leq \mathrm{Tr}(|A|)\|B\|, \qquad \mathrm{Tr}(|AB|) \leq \mathrm{Tr}(|B|)\|A\|.$$

Second, the mapping $(A, B) \to \mathrm{Tr}(AB^*)$ is a positive definite inner product on the space of complex $n \times n$ matrices. Thus

$$(35) \qquad \mathrm{Tr}(|AB|) = \mathrm{Tr}(ABV) \leq \mathrm{Tr}(AA^*)^{1/2}\,\mathrm{Tr}(BB^*)^{1/2}.$$

Now part (i) follows directly from (34). Also, (33) shows that the Fourier series

$$(36) \qquad f(k) \sim \sum_\lambda d_\lambda \,\mathrm{Tr}(A_\lambda U_\lambda(k))$$

for $f \in A(K)$ converges uniformly on K, so $f \in C(K)$.

For (iii) consider g, $h \in L^2(K)$ with Fourier coefficients $\{B_\lambda\}$ and $\{C_\lambda\}$, respectively. Since

$$(37) \qquad \sum_\lambda [d_\lambda^{1/2}\,\mathrm{Tr}(B_\lambda B_\lambda^*)^{1/2}\, d_\lambda^{1/2}\,\mathrm{Tr}(C_\lambda C_\lambda^*)^{1/2}] \leq \|g\|_2 \|h\|_2,$$

(35) implies $g * h \in A(K)$. On the other hand, if $f \in A(K)$ and we write $A_\lambda = |A_\lambda| V_\lambda$, where V_λ is unitary, then by (32) the series

$$\sum_\lambda d_\lambda \,\mathrm{Tr}(|A_\lambda|^{1/2} U_\lambda(k)), \qquad \sum_\lambda d_\lambda\,\mathrm{Tr}(|A_\lambda|^{1/2} V_\lambda U_\lambda(k))$$

represent ϕ, $\psi \in L^2(K)$ for which $f = \psi * \phi$.

Remark. Writing $|A| = AU$ (U unitary) we have, using (33),

$$\mathrm{Tr}(|A|) \geq \sup_{\|B\| \leq 1} |\mathrm{Tr}(AB)| \geq |\mathrm{Tr}(AU)| = \mathrm{Tr}(|A|)$$

so

$$\mathrm{Tr}(|A|) = \sup_{\|B\| \leq 1} |\mathrm{Tr}(AB)|.$$

Thus $\mathrm{Tr}(|A_1 + A_2|) \leq \mathrm{Tr}(|A_1|) + \mathrm{Tr}(|A_2|)$ and hence $A(K)$ is a linear subspace of $C(K)$.

Theorem 2.9. *Let $S: C(K) \to A(K)$ be a linear mapping commuting with all left translations $L(k)$, $k \in K$. Then there exists a function $F \in L^2(K)$ such that*

$$Sf = f * F, \qquad f \in C(K).$$

Example. Consider the case when K is the circle group T. Suppose (a_n) is a sequence such that for each

$$f \sim \sum b_n e^{inx}, \qquad f \in C(T), \tag{38}$$

we have $\sum |a_n b_n| < \infty$. Putting $f_a(x) = \sum_n a_n b_n e^{inx}$ the map $f \to f_a$ of $C(T)$ into $A(T)$ then satisfies the assumptions of Theorem 2.9, so we obtain Sidon's conclusion $\sum_n |a_n|^2 < \infty$. (Cf. III in subsection 2 above.)

For the proof of Theorem 2.9 we shall construct a suitable spectrally continuous operator and then invoke Theorem 2.3.

Because of Lemma 2.4′, S has the form

$$Sf = f * \mu \tag{39}$$

where μ is a measure on K. We consider the Fourier–Stieltjes series

$$\mu \sim \sum_{\lambda \in \hat{K}} d_\lambda \, \mathrm{Tr}(S_\lambda U_\lambda(k)), \tag{40}$$

where

$$S_\lambda = \int_K U_\lambda(k^{-1}) \, d\mu(k). \tag{41}$$

Then

$$Sf \sim \sum_\lambda d_\lambda \, \mathrm{Tr}(S_\lambda A_\lambda U_\lambda(k)) \qquad \text{if} \quad f \sim \sum d_\lambda \, \mathrm{Tr}(A_\lambda U_\lambda(k)). \tag{42}$$

Also (41) implies

$$(S_\lambda)_{ij} = (Su_{ij})(e) \qquad \text{if} \quad U_\lambda(k) = (u_{ij}(k))$$

so (17) holds in the present context. By our assumption

$$\sum_\lambda d_\lambda \, \mathrm{Tr}(|S_\lambda A_\lambda|) < \infty \tag{43}$$

for all $f \in C(K)$. Consider now a family

$$B = \{B_\lambda\}_{\lambda \in \hat{K}} \qquad \text{such that} \quad \|B\| = \sup_\lambda \|B_\lambda\| < \infty, \tag{44}$$

B_λ being an endomorphism of $\mathscr{H}_\lambda$. Then by (34) and (43),

(45) $$\sum_\lambda d_\lambda \operatorname{Tr}(|B_\lambda S_\lambda A_\lambda|) < \infty$$

and the mapping which maps the function f to the function in $A(K)$ given by

(46) $$\sum_\lambda d_\lambda \operatorname{Tr}(B_\lambda S_\lambda A_\lambda U_\lambda(k))$$

commutes with all left translations $L(k)$ $(k \in K)$. Hence by Lemma 2.4′

$$\sum_\lambda d_\lambda \operatorname{Tr}(B_\lambda S_\lambda A_\lambda U_\lambda(k)) \equiv f * \mu_B(k),$$

where μ_B is a measure. Consider now the mapping $M: B \to \mu_B$ from the Banach space of families (44) to the Banach space of measures on K. Since μ_B has Fourier–Stieltjes series $\sum_\lambda d_\lambda \operatorname{Tr}(B_\lambda S_\lambda U_\lambda(k))$ it is seen as before that M has closed graph, hence is continuous. If the family $B = \{B_\lambda\}$ forms the set of Fourier coefficients of a function $\phi \in L^1(K)$ then $\mu_B = \mu * \phi \in L^1(K)$. Since M is continuous, the map $\tilde{M}: \phi \to \mu * \phi$ of $L^1(K)$ into itself is *spectrally continuous*. Moreover, $\tilde{M}$ commutes with right translations on K. From Theorem 2.3 we now conclude that μ is given by an L^2 function and this proves Theorem 2.9.

5. Lacunary Fourier Series

Let π' and π'' be two finite-dimensional representations of K on V' and V'', respectively. Let $\pi \otimes \pi''$ denote their tensor product as defined in Chapter IV, §1.

Let (e_i') be a basis of V', (e_j'') a basis of V''. Then the elements $e_i' \otimes e_j''$ form a basis of $V' \otimes V''$. Let $d' = \dim V'$, $d'' = \dim V''$, and let E' and E'' be endomorphisms of V' and V'', respectively. Defining the matrices $A = (a_{ki})$, $B = (b_{lj})$ by

$$E'e_i' = \sum_{k=1}^{d'} a_{ki} e_k', \qquad E''e_j'' = \sum_{l=1}^{d''} b_{lj} e_l'',$$

we have

(47) $$(E' \otimes E'')(e_i' \otimes e_j'') = \sum_{k,l} a_{ki} b_{lj} e_k' \otimes e_l''.$$

Putting $f_{i+d'(j-1)} = e_i' \otimes e_j''$ the matrix expression of $E' \otimes E''$ is given by

(48) $$(E' \otimes E'')f_r = \sum_{s=1}^{d'd''} c_{sr} f_s,$$

where

$$c_{i+d'(j-1),k+d'(l-1)} = a_{ik}b_{jl}. \tag{49}$$

Thus $C = (c_{sr})$ is the $d'd'' \times d'd''$ matrix obtained by taking the matrix B and replacing each entry b_{lj} in it by the entire matrix Ab_{lj}.

We also consider the direct sum representation $\pi' + \pi''$ of K on $V' \times V''$ as defined in Chapter IV, §1.

Lemma 2.10. *Denoting the character of a representation π by χ_π we have*

$$\chi_{\pi'+\pi''} = \chi_{\pi'} + \chi_{\pi''}, \qquad \chi_{\pi'\otimes\pi''} = \chi_{\pi'} \cdot \chi_{\pi''}.$$

In fact, the first relation is obvious and the second follows from (49).

Definition. A (discrete) set S is called a *hypergroup* if there is given a mapping $(\alpha, \beta) \to \mu_{\alpha,\beta}$ of $S \times S$ into the set of measures on S. A subset T of the hypergroup S is called a *subhypergroup* if all the measures $\mu_{\alpha,\beta}$ $(\alpha, \beta \in T)$ have support contained in T.

The group K being compact, the set $\hat{K}$ is a hypergroup in the following sense. For each pair α, $\beta \in \hat{K}$ the tensor product $\alpha \otimes \beta$ has a direct decomposition into irreducible unitary components. This decomposition defines a measure $\mu_{\alpha,\beta}$ on the discrete set $\hat{K}$ with values in $\boldsymbol{Z}^+$. Two finite-dimensional representations of K are said to be *disjoint* if no irreducible component of one is equivalent to an irreducible component of the other.

It will be convenient to denote by E the identity transformation of an arbitrary vector space and to denote by e the trivial representation $k \to 1$ of K. For $M \subset K$ put

$$M^\perp = \{\alpha \in \hat{K} : U_\alpha(k) = E \text{ for all } k \in M\}$$

and for $\mathscr{H} \subset \hat{K}$ let

$$\mathscr{H}^\perp = \{k \in K : U_\alpha(k) = E \text{ for all } \alpha \in \mathscr{H}\}.$$

It will also be convenient to call a subhypergroup $\mathscr{H}$ of $\hat{K}$ a *normal subhypergroup* if $e \in \mathscr{H}$ and if $\alpha \in \mathscr{H}$ implies $\bar{\alpha} \in \mathscr{H}$ (the bar denotes complex conjugation).

Proposition 2.11.

(i) *If $M \subset K$, $M^\perp$ is a normal subhypergroup of $\hat{K}$ and $(M^\perp)^\perp$ is the smallest closed normal subgroup of K containing M.*

(ii) *If $\mathscr{H} \subset \hat{K}$, $\mathscr{H}^\perp$ is a closed normal subgroup of K and $(\mathscr{H}^\perp)^\perp$ is the smallest normal subhypergroup of $\hat{K}$ containing $\mathscr{H}$.*

(iii) *If N is a closed normal subgroup of K, $(K/N)\hat{} = N^\perp$.*

Proof. We start with (ii). Let $\mathscr{H} \subset \hat{K}$. Then it is obvious that $\mathscr{H}^{\perp}$ is a closed normal subgroup and that $(\mathscr{H}^{\perp})^{\perp}$ is a normal subhypergroup. The matrix elements from $\mathscr{H}$ can be regarded as a family of continuous functions on the factor group $K/\mathscr{H}^{\perp}$ separating points. Let $\mathscr{H}^*$ denote the normal subhypergroup generated by $\mathscr{H}$ and let $\mathscr{R}$ denote the set of linear combinations of matrix elements from members of $\mathscr{H}^*$. Then $\mathscr{R}$ is an algebra, it is closed under complex conjugation, it separates points on $K/\mathscr{H}^{\perp}$, and it contains the constants. By the Stone–Weierstrass theorem, $\mathscr{R}$ is uniformly dense in $C(K/\mathscr{H}^{\perp})$. Hence, if $\alpha \in (\mathscr{H}^{\perp})^{\perp} - \mathscr{H}^*$ each matrix element $u(k)$ from $U_{\alpha}(k)$ can be uniformly approximated by members of $\mathscr{R}$, but on the other hand, $u(k)$ is orthogonal to $\mathscr{R}$ by the orthogonality relations. This shows that $(\mathscr{H}^{\perp})^{\perp} = \mathscr{H}^*$.

(i) Let $\alpha, \beta \in M^{\perp}$. Corresponding to the decomposition of $\alpha \otimes \beta$ into irreducible components α_i we have for the respective characters

$$\chi_{\alpha}(k)\chi_{\beta}(k) = \chi_{\alpha_1}(k) + \cdots + \chi_{\alpha_n}(k), \tag{50}$$

and the corresponding dimensions satisfy

$$d_{\alpha}d_{\beta} = d_1 + \cdots + d_n. \tag{51}$$

If $k \in M$ we have $\chi_{\alpha}(k) = d_{\alpha}$, $\chi_{\beta}(k) = d_{\beta}$, and since $\max_k |\chi(k)| = d_{\chi}$ we conclude from (50) and (51) that $\chi_{\alpha_i}(k) = d_i$ for $1 \leq i \leq n$. But, U_{α_i} being unitary, this implies that $U_{\alpha_i}(k) = E$, so $\alpha_i \in M^{\perp}$. Thus $M^{\perp}$ is a subhypergroup which is normal. It is obvious that $(M^{\perp})^{\perp}$ is a closed normal subgroup containing M. If N is an arbitrary closed normal subgroup containing M, then $(M^{\perp})^{\perp} \subset N$ because if $l \in (M^{\perp})^{\perp} - N$ we can (by going over to the factor group K/N) find a representation D_{λ} $(\lambda \in \hat{K})$ such that $D_{\lambda}(k) = E$ for $k \in N$ but $D_{\lambda}(l) \neq E$. Thus $\lambda \in N^{\perp}$ whereas $\lambda \notin [(M^{\perp})^{\perp}]^{\perp}$. This last set equals $M^{\perp}$, as we see by applying (ii) to $\mathscr{H} = M^{\perp}$. Thus we have $\lambda \in N^{\perp} - M^{\perp}$, which contradicts $M \subset N$.

(iii) This is implicit in the proof of (ii) if we let $\mathscr{H} = N^{\perp} = \mathscr{H}^*$, because then $\mathscr{H}^{\perp} = (N^{\perp})^{\perp} = N$ by (i).

Definition. A subset $S \subset \hat{K}$ is said to be *lacunary* if the two following conditions are satisfied.

$\mathbf{L}_{\mathrm{I}}$. Whenever $\alpha, \beta, \gamma, \delta \in S$ such that the sets $\{\alpha, \beta\}$, $\{\gamma, \delta\}$ are different the representations $U_{\alpha} \otimes U_{\beta}$ and $U_{\gamma} \otimes U_{\delta}$ are disjoint.

$\mathbf{L}_{\mathrm{II}}$. The measures $\mu_{\alpha,\beta}$ $(\alpha, \beta \in S)$ are uniformly bounded.

This last condition amounts to that the number $n_{\alpha,\beta}$ of irreducible components of $\alpha \otimes \beta$, counted with multiplicity, has a bound:

$$n_{\alpha,\beta} < C \qquad (\alpha, \beta \in S).$$

A Fourier series on K of the form $\sum_{\lambda \in S} d_\lambda \operatorname{Tr}(A_\lambda U_\lambda(k))$ is called *lacunary* if the subset $S \subset \hat{K}$ is lacunary.

Remarks. If K is abelian, condition L_{II} is automatic and a Fourier series $\sum_{\chi \in S} a_\chi \chi(k)$ on K with nonvanishing coefficients is lacunary if and only if there is no nontrivial relation

$$\chi_1 \chi_2 = \chi_3 \chi_4 \tag{52}$$

between the elements of S. In particular, the series (11) on the circle group is a lacunary Fourier series in the sense of the general definition above. But it is also clear that the Fourier series

$$f(x_1, x_2, \ldots) \sim \sum_{k=1}^{\infty} a_k e^{ix_k}$$

on T^ω (cf. V, No. 2) is a lacunary series on T^ω. We shall now consider a more general situation.

Let I be a set and to each $i \in I$ attached an integer $d_i > 0$. We consider the product group

$$U = \prod_{i \in I} U(d_i). \tag{53}$$

The projection D_i of U onto $U(d_i)$ is a unitary representation which clearly is irreducible; it follows that I can be regarded as a subset of $\hat{U}$. We consider Fourier series on U of the form

$$\sum_{i \in I} d_i \operatorname{Tr}(A_i D_i(u)). \tag{54}$$

Theorem 2.12. *Suppose $f \in L^1(U)$ has Fourier series of the form* (54). *Then $f \in L^2(U)$ and $2^{-1/2}\|f\|_2 \leq \|f\|_1 \leq \|f\|_2$. Moreover, the Fourier series* (54) *is a lacunary Fourier series on U.*

Proof. Let $J \subset I$ be any finite subset and consider the sum

$$s(u) = \sum_{j \in J} d_j \operatorname{Tr}(B_j D_j(u)) \tag{55}$$

for arbitrary B_j. It is then clear that

$$\int_U |s(u)|^m \, du = \int_{V_J} \left| \sum_j d_j \operatorname{Tr}(B_j D_j) \right|^m dD_J \tag{56}$$

where du is the normalized Haar measure on U and dD_J is the normalized Haar measure on the product $V_J = \prod_{j \in J} U(d_j)$. But the integral on the right-hand side of (56) is dealt with in (22) and (29), so we conclude

$$\int_U |s(u)| \, du \geq \frac{1}{\sqrt{2}} \left(\int_U |s(u)|^2 \, du \right)^{1/2}. \tag{57}$$

By the general Peter–Weyl theorem f can be approximated in the L^1 norm by functions $s_1(u), s_2(u), \ldots$, with a finite Fourier series of the form (55). By (57) (s_n) is a Cauchy sequence in $L^2(U)$, so we conclude $f \in L^2(U)$ and (57) also holds for f.

Finally, the lacunary property follows from the following statements.

(a) $D_i \otimes D_j$ is irreducible if $i \neq j$.
(b) $D_i \otimes D_i$ is irreducible if $d_i = 1$.
(c) $D_i \otimes D_i$ is disjoint from $D_j \otimes D_j$ if $i \neq j$.
(d) If $d_i \geq 2$ then $D_i \otimes D_i$ decomposes into two irreducible parts [of dimensions $\frac{1}{2}(d_i^2 + d_i)$ and $\frac{1}{2}(d_i^2 - d_i)$].

Here (a) and (b) are obvious. For (c) we remark that the number of irreducible components common to $D_i \otimes D_i$ and $D_j \otimes D_j$ is equal to

$$\int_U (\chi_{D_i})^2 (\bar{\chi}_{D_j})^2 \, du = \int_{U(d_i) \times U(d_j)} (\operatorname{Tr} D_i)^2 (\operatorname{Tr} \bar{D}_j)^2 \, dD_i \, dD_j$$

[dD_m is the Haar measure on $U(d_m)$], and this last integral vanishes by (25). Finally, Lemma 2.6 shows that

$$\int_U |\chi_{D_i \otimes D_i}(u)|^2 \, du = \int_{U(d_i)} |\operatorname{Tr}(D_i)|^4 \, dD_i = 2,$$

so $D_i \otimes D_i$ decomposes into *two* irreducible parts. Since the symmetric and skew-symmetric second-order tensors do form invariant subspaces they give the decomposition of $D_i \otimes D_i$. This proves Theorem 2.12; note that it is the lacunary property of the series (54) which lies behind the estimate (29) of $\|F\|_4$, which in turn was the main ingredient in the proof of Theorem 2.3.

Next, we consider the case of central functions on K; that is, functions invariant under inner automorphisms. If K is abelian this restriction is of course vacuous.

Theorem 2.13. *Let f be a central function in $L^1(K)$ and suppose f has a lacunary Fourier series. Then $f \in L^2(K)$.*

Proof. The Fourier expansion of f has the form $f(k) \sim \sum_{\lambda \in S} a_\lambda \chi_\lambda(k)$ where $a_\lambda \in \mathbf{C}$ and $S \subset \hat{K}$ is a lacunary subset. We consider a finite partial sum $s(k) = \sum_{n=1}^N a_n \chi_n(k)$. Then

$$s^2 = \sum_p a_p^2 \chi_p^2 + 2 \sum_{p > q} a_p a_q \chi_p \chi_q.$$

If we now expand χ_p^2 and $\chi_p\chi_q$ into a sum of characters χ_i of irreducible representations [cf. (50)] at most one pair (p, q) gives rise to the same χ_i because of condition $\mathrm{L_I}$. Using now $\mathrm{L_{II}}$ as well we have

$$\begin{aligned}\int_K |s(k)|^4\,dk &= \sum_1^N |a_p|^4 n_{p,p} + 2\sum_{p>q} |a_p|^2|a_q|^2 n_{p,q}\\ &\leq C\left(\sum_1^N |a_p|^2\right)^2 = C\left(\int |s(k)|^2\,dk\right)^2,\end{aligned}$$

C being a constant. Again by (22) this leads to an estimate

$$\int |s|^2 \leq C\left(\int |s|\right)^2,$$

so we conclude $f \in L^2(K)$ as before.

Remark. Taking K as the circle group $\boldsymbol{T}$ we obtain results IV and V in No. 2.

We can also use Theorem 2.13 to derive the following simple result (cf. II in No. 2).

Corollary 2.14. *Let K be a compact abelian group. Suppose that for each permutation σ of $\hat{K}$ the series*

$$\sum_{\chi \in \hat{K}} a_{\sigma(\chi)}\chi(k) \qquad (k \in K) \tag{58}$$

is the Fourier series for some L^1 function on K. Then

$$\sum_{\chi \in \hat{K}} |a_\chi|^2 < \infty.$$

Proof. We may assume $\hat{K}$ infinite. Then $\hat{K}$ contains a countably infinite lacunary subset S. In fact, if $\hat{K}$ has an element χ_0 of infinite order we can take $S = \{\chi_0, \chi_0^2, \chi_0^4, \ldots\}$. If on the other hand all members of $\hat{K}$ have finite order we pick $\chi_1 \in \hat{K} - \{e\}$ arbitrary,

$$\chi_2 \notin \langle \chi_1 \rangle,\ \chi_3 \notin \langle \chi_1, \chi_2 \rangle,$$

etc., $\langle L \rangle$ denoting the subgroup of $\hat{K}$ generated by the subset L. Then $S = \{\chi_1, \chi_2, \ldots\}$ has the desired property.

Now we can write

$$\sum_\chi a_\chi \chi(k) = \sum_P a_\chi \chi(k) + \sum_Q a_\chi \chi(k),$$

where P and Q are countably infinite subsets of $\hat{K}$ and $\sum_P |a_\chi| < \infty$. It follows that the series $\sum_{\chi \in Q} a_\chi (\sigma \cdot \chi)(k)$ is an L^1 Fourier series for each permutation σ of $\hat{K}$. Choosing this permutation in such a way that this series is a lacunary series Theorem 2.13 gives the desired conclusion.

Example. As the proof above showed, for each infinite compact abelian group K there are infinite lacunary subsets of $\hat{K}$. We shall now look into the possibility of an analog for the simplest nonabelian case $K = \boldsymbol{SU}(2)$. For the torus

$$T = \begin{pmatrix} e^{i\theta} & 0 \\ 0 & e^{-i\theta} \end{pmatrix}_{\theta \in \boldsymbol{R}}$$

the single positive root α is given by

$$\alpha \begin{pmatrix} 1 & 0 \\ 0 & -1 \end{pmatrix} = 2.$$

The set $\Lambda(+)$ of dominant integral functions consists of the multiples $\lambda = \frac{1}{2}l\alpha$ ($l \in \boldsymbol{Z}^+$). Let U_λ denote a representation corresponding to λ and $\chi_\lambda = \chi_l$ its character. By Theorem 1.7 we have

$$\chi_\lambda(t) = \frac{\sin(l+1)\theta}{\sin \theta} \quad \text{if} \quad t = \begin{pmatrix} e^{i\theta} & 0 \\ 0 & e^{-i\theta} \end{pmatrix}. \tag{59}$$

The Haar integral on K of a central function f is given by

$$\int_K f(k)\, dk = \frac{1}{\pi} \int_0^{2\pi} f(t) \sin^2 \theta \, d\theta. \tag{60}$$

Proposition 2.15. *Let $K = \boldsymbol{SU}(2)$. Then*

(i) *$\hat{K}$ has no infinite lacunary subset.*
(ii) *The series*

$$\sum_{l=0}^{\infty} \chi_l$$

is the Fourier series for a central function belonging to $L^p(K)$ for $p < \frac{3}{2}$.

Remark. The example in (ii) shows that the "abelian" condition cannot be dropped in Corollary 2.14; part (i) explains why.

Proof. For (i) we observe that

$$\int_K |\chi_n \chi_m(k)|^2 \, dk = 1 + \min(n, m), \tag{61}$$

which is unbounded as n and m run through any infinite subset of $\boldsymbol{Z}^+$. Since this number equals the number of irreducible components of the tensor product $U_\nu \otimes U_\mu$ (if $\nu = \frac{1}{2}n\alpha$, $\mu = \frac{1}{2}m\alpha$), it follows that $\hat{K}$ has no infinite lacunary subset.

For (ii) consider the function

$$f(k) = \frac{2}{2 - \operatorname{Tr}(k)}, \qquad k \in K.$$

Then in the notation of (59) $f(t) = (1 - \cos\theta)^{-1}$ so by (60), $f \in L^p(K)$ for $p < \frac{3}{2}$. Since

$$\begin{aligned}\int_K f(k)\chi_l(k)\,dk &= \frac{1}{\pi}\int_{-\pi}^{\pi}(1 - \cos\theta)^{-1}\chi_l(t)\sin^2\theta\,d\theta \\ &= \frac{1}{2\pi}\int_{-\pi}^{\pi}(2 + e^{i\theta} + e^{-i\theta})\chi_l(t)\,d\theta = 2\end{aligned}$$

because of the formula

$$\chi_l(t) = e^{il\theta} + e^{i(l-2)\theta} + \cdots + e^{-il\theta},$$

(ii) is verified.

§3. Fourier Decomposition of a Representation

1. Generalities

The usual Fourier series decomposition $f \sim \sum a_n e^{in\theta}$ can be viewed as a "decomposition" of the regular representation of the circle group T into irreducible representations. We shall now generalize this to an arbitrary representation π of a compact, connected Lie group K on a complete, locally convex topological vector space V.

Let $\mathscr{S}$ denote the set of all continuous seminorms on V. Let $\{v_\alpha\}_{\alpha \in A}$ be an indexed family of members of $\mathscr{S}$. The series

(1) $$\sum_{\alpha \in A} v_\alpha$$

is said to be *absolutely convergent* if

(2) $$\sum_{\alpha \in A} \nu(v_\alpha) < \infty$$

for each $\nu \in \mathscr{S}$. For each finite subset $F \subset A$ put $s_F = \sum_{\alpha \in F} v_\alpha$. The series (1) is said to be *convergent* if for each neighborhood V_0 of 0 in V there exists a finite subset $F_0 \subset A$ such that $s_{F_1} - s_{F_2} \in V_0$ for any two finite subsets F_1, F_2 of A containing F_0. By completeness of V the sums

s_F will then have a limit $s \in V$; s is called the *sum* of the series (1) and we write $s = \sum_{\alpha \in A} v_\alpha$. It is clear that absolute convergence implies convergence.

As in Chapter IV, §1 let $\hat{K}$ denote the set of equivalence classes of finite-dimensional unitary irreducible representations of K. For $\delta \in \hat{K}$ let $\alpha_\delta = d(\delta)\,\mathrm{conj}(\chi_\delta)$ where $d(\delta)$ and χ_δ denote, respectively, the dimension and character of δ. Then we have seen (Chapter IV, Lemma 1.7) that $\pi(\alpha_\delta)$ is a continuous projection of V onto the closed subspace $V(\delta)$ of K-finite vectors of type δ. For the cited case of the circle group T if δ is the representation $e^{i\theta} \to e^{in\theta}$ and π the regular representation then $\pi(\alpha_\delta)$ is the map

$$f(e^{i\theta}) \to a_n e^{-in\theta},$$

where

$$a_n = \frac{1}{2\pi} \int_0^{2\pi} f(e^{i\theta}) e^{in\theta}\, d\theta.$$

We recall (Chapter IV, §1) that a vector $v \in V$ is said to be *differentiable* if the mapping $\tau_v : k \to \pi(k)v$ of K into V is differentiable. We shall now prove a sharpened version of Lemma 1.9, Chapter IV.

Theorem 3.1. *Let $v \in V$ be a differentiable vector. Then the "Fourier series"*

$$\sum_{\delta \in \hat{K}} \pi(\alpha_\delta) v$$

converges absolutely to v.

Proof. The Lie algebra $\mathfrak{k}$ of K is the direct sum $\mathfrak{k} = \mathfrak{z} + [\mathfrak{k}, \mathfrak{k}]$ where $\mathfrak{z}$ is the center of $\mathfrak{k}$ and $[\mathfrak{k}, \mathfrak{k}]$ is semisimple. Let Q be a positive definite quadratic form on $\mathfrak{k}$ which on $[\mathfrak{k}, \mathfrak{k}]$ coincides with the negative of the Killing form of $[\mathfrak{k}, \mathfrak{k}]$ and under which $\mathfrak{z}$ and $[\mathfrak{k}, \mathfrak{k}]$ are orthogonal. Let $X_1, \ldots, X_n$ be a basis of $\mathfrak{k}$, orthonormal with respect to Q, and put

$$\Omega = 1 - \tilde{X}_1^2 - \cdots - \tilde{X}_n^2 \in \mathbf{D}(K).$$

Since Q is $\mathrm{Ad}(K)$-invariant, Ω is a bi-invariant differential operator on K. It is clear from Lemma 1.6 (which does not require semisimplicity of K) and Eq. (16) of §1 that

$$\Omega \alpha_\delta = c_\delta \alpha_\delta, \tag{3}$$

where $c_\delta \geq 1$. On the other hand [Chapter IV, §1, Eq. (5′)], we have for $u \in K$

$$\pi(\alpha_\delta)\pi(u)v = \pi(u)\pi(\alpha_\delta)v = \pi(\alpha_\delta^{L(u)})v = \pi(\alpha_\delta^{R(u)})v$$

so

$$\pi(\alpha_\delta)\pi(\Omega)v = \pi(\Omega\alpha_\delta)v = c_\delta\pi(\alpha_\delta)v. \tag{4}$$

Lemma 3.2. *Let $\nu \in \mathscr{S}$. Then there exists a $\nu_0 \in \mathscr{S}$ such that*

$$\nu(\pi(\alpha_\delta)v) \leq c_\delta^{-m}\, d(\delta)^2\, \nu_0(\pi(\Omega^m)v)$$

for all $\delta \in \hat{K}$ and $m \in \mathbf{Z}^+$.

Proof. By the compactness of K we can choose $\nu_0 \in \mathscr{S}$ such that $\nu(\pi(u)w) \leq \nu_0(w)$ for all $w \in V$, $u \in K$. If $f : K \to V$ is a continuous map we have

$$\nu\left(\int f(k)\,dk\right) \leq \int \nu(f(k))\,dk,$$

so, since $\sup|\alpha_\delta| \leq d(\delta)^2$, we have

$$\nu(\pi(\alpha_\delta)w) \leq d(\delta)^2\, \nu_0(w). \tag{5}$$

On the other hand, iterating (4) we have

$$\pi(\alpha_\delta)v = c_\delta^{-m}\pi(\alpha_\delta)\pi(\Omega^m)v, \qquad m \in \mathbf{Z}^+.$$

Using (5) the lemma follows.

In order to use the lemma we observe that if m is large enough

$$\sum_{\delta \in \hat{K}} d(\delta)^2\, c_\delta^{-m} < \infty. \tag{6}$$

If K is semisimple this is clear from Eq. (16) of §1 and Theorem 1.8. If K is not semisimple let Z and K' denote the analytic subgroups of K corresponding to $\mathfrak{z}$ and $[\mathfrak{k}, \mathfrak{k}]$, respectively. If $\delta \in \hat{K}$ then the restriction $\delta' = \delta | K'$ belongs to $(K')\hat{}$, the elements $\delta(z)$ $(z \in Z)$ being scalars. Thus

$$d(\delta) = d(\delta'), \tag{7}$$

$$\chi_\delta(zu') = \mathrm{Tr}(\delta(z)\delta(u')) = \delta(z)\chi_{\delta'}(u'), \tag{8}$$

for $z \in Z$, $u' \in K'$. At the identity element we have $L_K = L_Z + L_{K'}$ for the respective Laplacians, and this combined with (7) and (8) shows that $c_\delta \geq c_{\delta'}$. This proves (6) in general.

Now Lemma 3.2 shows that the series in Theorem 3.1 converges absolutely. Let v_0 denote its sum and put $w = v - v_0$. If $\delta_0 \in \hat{K}$ is arbitrary,

$$\pi(\alpha_{\delta_0})v_0 = \sum_{\delta \in \hat{U}} \pi(\alpha_{\delta_0})\pi(\alpha_\delta)v = \pi(\alpha_{\delta_0})v$$

by the orthogonality relations. Hence $\pi(\alpha_\delta)w = 0$ for each $\delta \in \hat{K}$. Then if $u \in U$,

$$\pi(\alpha_\delta)\pi(u)w = \pi(u)\pi(\alpha_\delta)w = 0.$$

If λ is arbitrary in the dual space V' of V this equation implies

$$\chi_\delta * \lambda_w = 0, \qquad \delta \in \hat{K}, \tag{9}$$

where $*$ denotes convolution on K and λ_w is the function $u \to \lambda(\pi(u)w)$ (cf. proof of Lemma 1.9, Chapter IV). But (9) implies that $\lambda_w \equiv 0$, so, since λ is arbitrary, $w = 0$. This proves Theorem 3.1.

Let X be a manifold with a countable base and suppose K is a compact, connected Lie transformation group of X. The spaces $\mathscr{D}(X) = C_c^\infty(X)$ and $\mathscr{E}(X) = C^\infty(X)$ with their usual topologies (Chapter II, §2) are locally convex and complete. The mapping π which to $k \in K$ assigns the transformation $f(x) \to f(k^{-1} \cdot x)$ of $\mathscr{E}(X)$ is a representation of K on $\mathscr{E}(X)$.

Lemma 3.3. *Let $f \in \mathscr{E}(X)$. Then f is a differentiable vector for the representation π.*

Let $k \in K$, $T \in \mathfrak{k}$. Then the lemma amounts to the existence of the limit

$$\lim_{t \to 0} \frac{1}{t} (f^{\tau(k \exp tT)} - f^{\tau(k)})$$

in the topology of $\mathscr{E}(X)$. This, however, is clear from the definition.

Remark. The analogous result holds also for the natural representation of K on $\mathscr{D}(X)$.

Defining as in Chapter IV, §1

$$(\phi * f)(x) = \int_K \phi(k) f(k^{-1}x)\, dk \tag{10}$$

for $\phi \in C(K)$, $f \in C(X)$, we have $(\pi(\phi)f)(x) = \phi * f$, so Theorem 3.1 implies the following result.

Corollary 3.4. *Let $f \in \mathscr{E}(X)$. Then*

$$f = \sum_{\delta \in \hat{K}} \alpha_\delta * f,$$

the series being absolutely convergent.

2. Applications to Compact Homogeneous Spaces

Let K be a compact, connected Lie group, $M \subset K$ a closed subgroup, and X the manifold K/M. We can apply the results above to the natural representation π of K on the spaces $C(K/M)$, $\mathscr{D}(K/M)$, $L^2(K/M)$, etc., and combine these with the formalism of Chapter IV, §1.

For each $\delta \in \hat{K}$ let V_δ be a vector space (with inner product $\langle\ ,\ \rangle$) on which a representation of class δ is realized and let this representation also be denoted by δ. Let $\hat{K}_M$ denote the set of elements $\delta \in \hat{K}$ for which the subspace

$$V_\delta^M = \{v \in V_\delta : \delta(m)v = v \text{ for } m \in M\}$$

is nonzero. If $\delta \in \hat{K}_M$ then by Eq. (12) of Chapter IV, §1, the contragredient representation $\check{\delta}$ also belongs to $\hat{K}_M$.

Let

$$d(\delta) = \dim V_\delta, \qquad l(\delta) = \dim V_\delta^M,$$

and let $v_1, \ldots, v_{d(\delta)}$ be a basis of V_δ such that $v_1, \ldots, v_{l(\delta)}$ span V_δ^M. If $C \in \operatorname{Hom}(V_\delta, V_\delta^M)$ then $\delta(m)C = C$ $(m \in M)$ so we can define the function F_C on K/M by

$$F_C(kM) = \operatorname{Tr}(\delta(k)C), \qquad k \in K. \tag{11}$$

For $\delta \in \hat{K}$ let $C_\delta(K/M)$ denote the subspace of $C(K/M)$ consisting of the K-finite vectors (under π) of type δ.

Theorem 3.5. *Let $\delta \in \hat{K}$. Then*

(i) *$C_{\check{\delta}}(K/M) \neq 0$ if and only if $\check{\delta} \in \hat{K}_M$. In this case,*

$$C_{\check{\delta}}(K/M) = \{F_C : C \in \operatorname{Hom}(V_\delta, V_\delta^M)\}$$

and the functions

$$kM \to \langle \delta(k)v_j, v_i\rangle = \delta_{ij}(k), \qquad 1 \le i \le d(\delta), \quad 1 \le j \le l(\delta),$$

constitute a basis of $C_{\check{\delta}}(K/M)$.

(ii) *We have the orthogonal Hilbert space decomposition*

$$L^2(K/M) = \bigoplus_{\delta \in \hat{K}_M} C_\delta(K/M), \tag{12}$$

$$f = \sum_{\delta \in \hat{K}_M} d(\delta)\bar{\chi}_\delta * f, \qquad f \in L^2(K/M), \tag{13}$$

$*$ being as in (10).

(iii) *If $f \in \mathscr{E}(K/M)$, the series (13) converges absolutely and uniformly.*

(iv) *Let $L^2_\natural(K)$ denote the set of $f \in L^2(K)$ that are bi-invariant under M. Let*

$$C_\delta^\natural(K) = \sum_{1 \le i, j \le l(\delta)} C\delta_{ij}, \qquad \delta \in \hat{K}_M.$$

Then we have the orthogonal direct sum

$$L^2_\natural(K) = \bigoplus_{\delta \in \hat{K}_M} C_\delta^\natural(K). \tag{13'}$$

The convolution algebra $L^2_\natural(K)$ is commutative if and only if

$$l(\delta) = 1 \qquad \textit{for each} \quad \delta \in \hat{K}_M.$$

Proof. If the space $C_{\check{\delta}}(K/M)$ is $\neq \{0\}$, it contains a function F such that $F(\{M\}) \neq 0$. By averaging over M we may take F M-invariant so $\check{\delta} \in \hat{K}_M$. The rest of (i) and (ii) is already proved in Chapter IV, §1. Part (iii) follows from Theorem 3.1 when used on the Banach space $V = C(K/M)$ with the uniform norm.

It remains to prove (iv). The space $C_{\check{\delta}}(K/M)$ is by (i) spanned by the first $l(\delta)$ columns of the matrix $\delta(k)$. Thus (13′) follows from (12) and the analogous decomposition of $L^2(K/M)$. The last statement now follows from Chapter IV, Theorem 1.6 (iii), which shows that $C^\natural_{\check{\delta}}(K)$ is a convolution algebra with center $\mathbf{C}(\delta_{11} + \cdots + \delta_{l(\delta)l(\delta)})$.

Corollary 3.6. *Let $f \in \mathscr{E}(K)$ be invariant under conjugation, i.e., $f(klk^{-1}) \equiv f(l)$. Then*

$$f = \sum_{\delta \in \hat{K}} a_\delta \chi_\delta, \qquad a_\delta = \int f(k)\bar{\chi}_\delta(k)\, dk, \tag{14}$$

and the series converges absolutely and uniformly.

In fact, take $M = e$ in (13) above and observe that $\chi_\delta * f$ is a function in $C_{\check{\delta}}(K)$ that is invariant under conjugation. Writing $\chi_\delta * f$ in the form $(\chi_\delta * f)(k) = \mathrm{Tr}(\delta(k)C)$, the condition $\mathrm{Tr}(\delta(kl)C) = \mathrm{Tr}(\delta(lk)C)$ implies $\mathrm{Tr}(\delta(k)\delta(l)C) \equiv \mathrm{Tr}(\delta(k)C\delta(l))$, which again implies $\delta(l)C \equiv C\delta(l)$ so C is a scalar operator. Thus $\chi_\delta * f = c\chi_\delta$ $(c \in \mathbf{C})$ and now (13) implies (14).

§4. The Case of a Compact Symmetric Space

We shall now put Theorem 3.5 into a more explicit form in the case of a symmetric space of the compact type (cf. [DS], Chapter V, §1). First we need some results about representations.

1. Finite-Dimensional Spherical Representations

Let G be a connected noncompact semisimple Lie group, $G = KAN$ an Iwasawa decomposition, and M the centralizer of A in K. Let $\mathfrak{g}$, $\mathfrak{k}$, $\mathfrak{a}$, $\mathfrak{n}$, and $\mathfrak{m}$ denote the corresponding Lie algebras. As in [DS] (Chapter VI), we extend $\mathfrak{a}$ to a maximal abelian subalgebra $\mathfrak{h}$ of $\mathfrak{g}$. Then $\mathfrak{h} = \mathfrak{h} \cap \mathfrak{k} + \mathfrak{a}$, $\mathfrak{h}$ is a Cartan subalgebra of $\mathfrak{g}$ and, superscript $\mathbf{C}$ denoting complexification, $\mathfrak{h}^{\mathbf{C}}$ is a Cartan subalgebra of $\mathfrak{g}^{\mathbf{C}}$.

Each representation δ of G on a finite-dimensional vector space V over $\boldsymbol{C}$ induces a representation, also denoted δ, of $\mathfrak{g}$ (and $\mathfrak{g}^C$) on V. If G is simply connected this process can be reversed. In any case we can use the notion of a weight from §1 for representations of G. The weights (like the roots of $\mathfrak{g}^C$ with respect to $\mathfrak{h}^C$) are real valued on the real subspace $\mathfrak{h}_{\boldsymbol{R}} = i(\mathfrak{h} \cap \mathfrak{k}) + \mathfrak{a}$. We put $\mathfrak{h}_{\mathfrak{k}} = \mathfrak{h} \cap \mathfrak{k}$.

The choice of $\mathfrak{n}$ corresponds to an ordering of the dual space $\mathfrak{a}^*$. Let Σ^+ denote the corresponding set of positive restricted roots. Choosing a compatible ordering of the dual of $\mathfrak{h}_{\boldsymbol{R}}$ we have an ordering of the set of all weights. The Killing form $\langle \ , \ \rangle$ of $\mathfrak{g}^C$ induces an inner product on $\mathfrak{a}$, $\mathfrak{h}_{\boldsymbol{R}}$, and their duals. We can now characterize the spherical representations.

Theorem 4.1. *Let δ be an irreducible representation of G on a finite-dimensional vector space V over $\boldsymbol{C}$.*

(i) *$\delta(K)$ has a nonzero fixed vector if and only if $\delta(M)$ leaves the highest-weight vector of δ fixed.*

(ii) *Let λ be a linear form on $\mathfrak{h}_{\boldsymbol{R}}$. Then λ is the highest weight of an irreducible finite-dimensional spherical representation of G if and only if*

$$(1) \qquad \lambda(i(\mathfrak{h} \cap \mathfrak{k})) = 0, \qquad \frac{\langle \lambda, \alpha \rangle}{\langle \alpha, \alpha \rangle} \in \boldsymbol{Z}^+ \qquad \text{for } \alpha \in \Sigma^+.$$

Proof. Let G^C be the simply connected Lie group with Lie algebra $\mathfrak{g}^C$ and $\tilde{G}$, $\tilde{K}$, $\tilde{A}$, $\tilde{N}$ the analytic subgroups corresponding to $\mathfrak{g}$, $\mathfrak{k}$, $\mathfrak{a}$, and $\mathfrak{n}$ and let U be the (compact simply connected) analytic subgroup corresponding to $\mathfrak{u} = \mathfrak{k} + i\mathfrak{p}$. Here $\mathfrak{p}$ denotes as usual the orthogonal complement of $\mathfrak{k}$ in $\mathfrak{g}$. We can then identify finite-dimensional complex representations of $\mathfrak{g}$, $\mathfrak{g}^C$, $\mathfrak{u}$, $\tilde{G}$, G^C, and U.

For the proof of Theorem 4.1 we first assume $G = \tilde{G}$. Suppose first $\delta(K)$ has a fixed vector $v_K \neq 0$. Let λ denote the highest weight of δ and $v_\lambda \neq 0$ a highest-weight vector. Since $G = KAN$, V is spanned by the vectors $\delta(K)v_\lambda$. Put $P = \int_K \delta(k)\, dk$. Then $P\delta(k)v_\lambda = Pv_\lambda$, so $PV = \boldsymbol{C}Pv_\lambda$. Since $v_K = Pv_K$ it follows that

$$(2) \qquad v_K = cPv_\lambda \qquad \text{for some} \quad c \in \boldsymbol{C}.$$

Suppose now $H \in \mathfrak{h} \cap \mathfrak{k}$. Then

$$v_K = \delta(\exp H)v_K = cPe^{\lambda(H)}v_\lambda = e^{\lambda(H)}v_K,$$

so $e^{\lambda(H)} = 1$ for all $H \in \mathfrak{h} \cap \mathfrak{k}$, whence

$$(3) \qquad \lambda(i(\mathfrak{h} \cap \mathfrak{k})) = 0.$$

Next we recall ([DS], Chapter VII, Theorem 8.5) that since U is now simply connected the set

(4) $$\mathfrak{a}_K = \{H \in i\mathfrak{a} : \exp H \in K\}$$

consists of the integral linear combinations of the vectors

(5) $$e_\alpha = \frac{2\pi i}{\langle \alpha, \alpha \rangle} A_\alpha, \qquad \alpha \in \Sigma^+,$$

where $A_\alpha \in \mathfrak{a}$ is determined by $\langle H, A_\alpha \rangle = \alpha(H)$ for $H \in \mathfrak{a}$. Thus by (2)

$$v_K = \delta(\exp e_\alpha)v_K = cP\delta(\exp e_\alpha)v_\lambda = \exp(2\pi i \langle \lambda, \alpha \rangle / \langle \alpha, \alpha \rangle)v_K$$

so (1) is proved.

Now we recall ([DS], Chapter VI, Lemma 3.6 and Chapter IX, Exercise A3) that M has the expression

(6) $$M = M^0(\exp(i\mathfrak{a}) \cap K),$$

where M^0 is the identity component of M and

$$\mathfrak{m}^C = (\mathfrak{h} \cap \mathfrak{k})^C + \sum_{\beta \in P_-} (\mathfrak{g}^C)^\beta + (\mathfrak{g}^C)^{-\beta}.$$

Here P_- is the set of positive roots β of $\mathfrak{g}^C$ with respect to $\mathfrak{h}^C$ which vanish identically on $\mathfrak{a}$. Since $\lambda + \beta$ is not a weight of δ we have

$$\delta((\mathfrak{g}^C)^\beta)v_\lambda = 0.$$

Also, $\delta((\mathfrak{g}^C)^{-\beta})v_\lambda = 0$; in fact, since $\langle \lambda, \beta \rangle = 0$ [by (3)], the linear form $\lambda - \beta$ is the image of $\lambda + \beta$ under the Weyl reflection s_β and thus is not a weight of δ. Now (3) and the formula for $\mathfrak{m}^C$ show that v_λ is fixed under $\delta(M^0)$, and (1) and (4) show it fixed under $\delta(\exp(\mathfrak{a}_K))$. Hence by (6), $\delta(M)$ leaves v_λ fixed. This proves half of (i) and (ii) for the case $G = \tilde{G}$.

On the other hand, suppose $\delta(M)$ leaves v_λ fixed. Then condition (1) follows readily from (6). The vector

$$v_0 = \int_{K/M} \delta(k)v_\lambda \, dk_M$$

is of course $\delta(K)$-fixed. To see that $v_0 \neq 0$ we use Theorem 5.20 of Chapter I, by which

$$v_0 = \int_{K/M} \delta(k)v_\lambda \, dk_M = \int_{\bar{N}} \delta(k(\bar{n}))v_\lambda e^{-2\rho(H(\bar{n}))} \, d\bar{n}$$

$$= \int_{\bar{N}} \delta(\bar{n})v_\lambda e^{-(\lambda + 2\rho)(H(\bar{n}))} \, d\bar{n}.$$

Consider now an inner product (,) on V for which $\delta(U)$ are unitary operators. Then, $*$ denoting adjointness, we have

$$\delta(T)^* = -\delta(T) \quad (T \in \mathfrak{k}), \qquad \delta(X)^* = \delta(X) \quad (X \in \mathfrak{p}),$$

so $\delta(Z)^* = -\delta(\theta Z)$ for $Z \in \mathfrak{g}$, whence $\delta(g)^* = \delta(\theta g^{-1})$ for $g \in G$. Thus $(\delta(\bar{n})v_\lambda, v_\lambda) = (v_\lambda, \delta(\theta\bar{n}^{-1})v_\lambda) = (v_\lambda, v_\lambda) \neq 0$, so, taking the inner product above, we get

$$(7) \qquad (v_0, v_\lambda) = (v_\lambda, v_\lambda) \int_{\bar{N}} e^{-(\lambda+2\rho)(H(\bar{n}))}\, d\bar{n}.$$

In particular, $v_0 \neq 0$ as claimed.

It remains to prove the "if" part of (ii) and then to remove the assumption $G = \tilde{G}$. Let β be a positive root of $(\mathfrak{g}^C, \mathfrak{h}^C)$ and $\bar{\beta}$ the restriction $\beta|\mathfrak{a}$. Then by [DS] (Chapter VII, Lemma 8.4),

$$\langle \beta, \beta \rangle = m\langle \bar{\beta}, \bar{\beta} \rangle, \qquad m = 1, 2, 4.$$

Moreover, if $m = 4$ then $2\bar{\beta} \in \Sigma^+$.

Assuming (1) and writing

$$2\frac{\langle \lambda, \beta \rangle}{\langle \beta, \beta \rangle} = \begin{cases} 0, & \bar{\beta} = 0 \\ 2m^{-1}\langle \lambda, \bar{\beta} \rangle / \langle \bar{\beta}, \bar{\beta} \rangle, & 2\bar{\beta} \notin \Sigma^+ \\ 4m^{-1}\langle \lambda, 2\bar{\beta} \rangle / \langle 2\bar{\beta}, 2\bar{\beta} \rangle, & 2\bar{\beta} \in \Sigma^+, \end{cases}$$

we see that the left-hand side belongs to $\mathbf{Z}^+$. Thus by Theorem 1.5 there exists an irreducible representation δ of G^C with highest weight λ. As we saw earlier in the proof, condition (1) implies that the highest-weight vector v_λ is fixed under $\delta(M)$. This proves the theorem for the case $G = \tilde{G}$.

Consider now the case of a general G. The representation δ induces, via $\mathfrak{g}$ and $\mathfrak{g}^C$, a representation $\tilde{\delta}$ of $\tilde{G}$ with the same highest weight λ. By Schur's lemma the center $\tilde{Z}$ of $\tilde{G}$ is mapped into scalars by $\tilde{\delta}$. This center is contained in $\tilde{M}$, the centralizer of $\tilde{A}$ in $\tilde{K}$, so if $\tilde{\delta}(\tilde{K})$ has a fixed vector or if $\tilde{\delta}(\tilde{M})$ fixes v_λ, $\tilde{\delta}$ equals the identity on the center. Similar remarks apply for δ relative to the center Z of G. By connectivity of K and $\tilde{K}$ we have $\mathrm{Ad}_G(K) = \mathrm{Ad}_{\tilde{G}}(\tilde{K})$, whence $K/Z = \tilde{K}/\tilde{Z}$. Also, if $k \in K$ such that $\mathrm{Ad}(k)\,\mathrm{Ad}(a) = \mathrm{Ad}(a)\,\mathrm{Ad}(k)$ $(a \in A)$ we have $kak^{-1} = az_a$ $(z_a \in Z)$. Since $z_a \in K$ the uniqueness in the Cartan decomposition $G = K \exp \mathfrak{p}$ implies $z_a = e$, so $k \in M$. This shows that $\mathrm{Ad}_G(M) = \mathrm{Ad}_{\tilde{G}}(\tilde{M})$, so $M/Z = \tilde{M}/\tilde{Z}$. Using the above remarks about the centers Z and $\tilde{Z}$ we conclude for $v \in V$,

$$v \text{ is } \delta(K)\text{-fixed} \Leftrightarrow v \text{ is } \tilde{\delta}(\tilde{K})\text{-fixed};$$

$$v_\lambda \text{ is } \delta(M)\text{-fixed} \Leftrightarrow v_\lambda \text{ is } \tilde{\delta}(\tilde{M})\text{-fixed}.$$

This proves the theorem for G. If the vectors v_K and v_λ are taken as unit vectors we have from (2) since $P^* = P = P^2$

$$1 = (v_K, v_K) = (cPv_\lambda, cPv_\lambda) = c(v_\lambda, v_K),$$

and then (7) takes the form (with $v_0 = Pv_\lambda$)

$$|(v_K, v_\lambda)|^2 = \mathbf{c}(-i(\lambda + \rho)). \tag{8}$$

The proof of Theorem 4.1 has the following consequence.

Corollary 4.2. *Let U be a compact simply connected semisimple Lie group, K the fixed point group of an involutive automorphism of U. Then condition* (1) *characterizes the highest weights of the irreducible spherical representations of U.*

2. The Eigenfunctions and the Eigenspace Representations

With U and K as in Corollary 4.2 we shall now write out Theorem 3.5 in a more explicit form. Let $\hat{U}_K$ denote the set of equivalence classes of spherical representations of U. For each $\delta \in \hat{U}_K$ let V_δ be a representation space for δ. We know from Chapter IV, Lemma 3.6, that now V_δ^K is spanned by a single unit vector $\boldsymbol{e}$.

Theorem 4.3. *For the symmetric space U/K we have the Hilbert space decomposition*

$$L^2(U/K) = \bigoplus_{\delta \in \hat{U}_K} C_\delta(U/K), \tag{9}$$

$$f = \sum_{\delta \in \hat{U}_K} d(\delta)\bar{\chi}_\delta * f, \tag{10}$$

and if $f \in \mathscr{E}(U/K)$ this expansion converges absolutely and uniformly. Moreover,

(i) *$C_\delta(U/K)$ consists of the functions*

$$uK \to \langle \boldsymbol{v}, \delta(u)\boldsymbol{e}\rangle, \qquad \boldsymbol{v} \in V_\delta.$$

The natural representation of U on the function space $C_\delta(U/K)$ is of class δ; in particular, it is irreducible.

(ii) *Each $C_\delta(U/K)$ contains a unique spherical function, namely $\phi_\delta(uK) = \langle \boldsymbol{e}, \delta(u)\boldsymbol{e}\rangle$. If the homomorphism c_δ: $\mathbf{D}(U/K) \to \boldsymbol{C}$ is determined by*

$$D\phi_\delta = c_\delta(D)\phi_\delta, \qquad D \in \mathbf{D}(U/K),$$

then we have for the joint eigenspace

(11) $\{f \in C^\infty(U/K): Df = c_\delta(D)f \quad \text{for all} \quad D \in \boldsymbol{D}(U/K)\} = C_\delta(U/K).$

(iii) *The class* $\hat{U}_K$ *coincides with the class of eigenspace representations for* U/K. *In particular, each eigenspace representation is irreducible and finite-dimensional.*

Proof. Part (i) is clear from Theorem 3.5 because if $f(uK) = \langle \boldsymbol{v}, \delta(u)\boldsymbol{e}\rangle$ then $f(u_0^{-1}uK) = \langle \delta(u_0)\boldsymbol{v}, \delta(u)\boldsymbol{e}\rangle$. For Part (ii) we first note from Chapter IV, Theorem 3.4 that the function $\phi_\delta(uK) = \langle \boldsymbol{e}, \delta(u)\boldsymbol{e}\rangle$ is a spherical function in $C_\delta(U/K)$. Because of Lemma 3.6 in Chapter IV, the irreducibility in (i) implies the uniqueness of ϕ_δ as well as the fact that its translates span $C_\delta(U/K)$. This again implies the inclusion $\supset$ in (11). For the converse inclusion as well as part (iii) let f be a joint eigenfunction of $\boldsymbol{D}(U/K)$, i.e., $Df = v(D)f$ for all $D \in \boldsymbol{D}(U/K)$, v being a homomorphism of $\boldsymbol{D}(U/K)$ into $\boldsymbol{C}$. Applying D to (10) we conclude that

$$f = \sum_{c_\delta = v} d(\delta)\bar{\chi}_\delta * f. \tag{12}$$

However, by Corollary 2.3 in Chapter IV, all the c_δ are different, so (12) contains just one term, whence $f \in C_\delta(U/K)$. This concludes the proof.

While Theorem 4.3 gives for U/K an answer to Problems A, B, and C in Introduction, §1, we can still look for a more specific description of the joint eigenfunctions than that provided by (i) above. We shall now obtain such a description by going over to the complexification $U^{\boldsymbol{C}} = G^{\boldsymbol{C}}$. We adopt the notation of §4, No. 1 and choose for G the group $\tilde{G}$. As in Chapter IV, §4 we let, for $g \in G$, the vector $A(g) \in \mathfrak{a}$ be determined by $g \in N \exp A(g)K$. We consider $\mathfrak{u}^{\boldsymbol{C}} = \mathfrak{g}^{\boldsymbol{C}}$ with the usual Hilbert space inner product $(X, Y) = -\langle X, \tau Y\rangle$, τ being the conjugation of $\mathfrak{u}^{\boldsymbol{C}}$ with respect to $\mathfrak{u}$. Because of the vector space direct sum of the complexifications

$$\mathfrak{u}^{\boldsymbol{C}} = \mathfrak{n}^{\boldsymbol{C}} + \mathfrak{a}^{\boldsymbol{C}} + \mathfrak{k}^{\boldsymbol{C}} \tag{13}$$

the map $(X, H, T) \to \exp X \exp H \exp T$ is a holomorphic diffeomorphism of a neighborhood of $(0, 0, 0)$ onto a neighborhood $U_0^{\boldsymbol{C}}$ of e in $U^{\boldsymbol{C}}$. The map

$$\exp X \exp H \exp T \to H \tag{14}$$

is therefore a well-defined holomorphic mapping of $U_0^{\boldsymbol{C}}$ into $\mathfrak{a}^{\boldsymbol{C}}$ extending the map A. We denote this extension also by A. We can take $U_0^{\boldsymbol{C}}$ as the diffeomorphic image (under exp) of an open $B(0) \subset \mathfrak{u}^{\boldsymbol{C}}$ with center 0.

Then $U_0^{\mathbf{C}}$ is invariant under the conjugations $u \to kuk^{-1}$ by elements $k \in K$ and so is the set $U_0 = U_0^{\mathbf{C}} \cap U$.

Theorem 4.4. *Each joint eigenfunction of all $D \in \mathbf{D}(U/K)$ has the form*

$$f(uK) = \int_{K/M} e^{-\mu(A(k^{-1}uk))} F(kM)\, dk_M, \qquad u \in U_0, \tag{15}$$

where $\mu \in \mathfrak{a}^*$ and F satisfy

$$F \in \mathscr{E}(K/M), \qquad \frac{\langle \mu, \alpha \rangle}{\langle \alpha, \alpha \rangle} \in \mathbf{Z}^+ \qquad \text{for} \quad \alpha \in \Sigma^+. \tag{16}$$

Conversely, if μ and F satisfy (16) *then the function f defined by* (15) *extends uniquely to an analytic function on U/K and this function is a joint eigenfunction of all $D \in \mathbf{D}(U/K)$.*

Proof. Let f be a joint eigenfunction. By Theorem 4.3 there exists a $\delta \in \hat{U}_K$, $\mathbf{v} \in V_\delta$, such that $f(uK) \equiv \langle \mathbf{v}, \delta(u)\mathbf{e} \rangle$. The corresponding spherical function $\phi(uK) = \langle \mathbf{e}, \delta(u)\mathbf{e} \rangle$ can be extended to the function

$$\tilde{\phi}(u) = \langle \mathbf{e}, \delta(u)\mathbf{e} \rangle, \qquad u \in U^{\mathbf{C}}, \tag{17}$$

because δ automatically extends to a representation of $U^{\mathbf{C}}$. Let μ denote the restriction to $\mathfrak{a}$ of the highest weight of δ and let v_μ be a corresponding highest-weight vector. Then by (2) since P is self-adjoint,

$$\langle \mathbf{e}, \delta(u)\mathbf{e} \rangle = |c|^2 \langle v_\mu, P\delta(u)Pv_\mu \rangle. \tag{18}$$

Let $k \in K$, $g \in G$. Then $gk = k' \exp(-A(k^{-1}g^{-1}))n$ where $k' \in K$, $n \in N$. Hence

$$\begin{aligned} P\delta(gk)v_\mu &= \exp(-\mu(A(k^{-1}g^{-1})))P\delta(k')v_\mu \\ &= \exp(-\mu(A(k^{-1}g^{-1})))Pv_\mu, \end{aligned}$$

so

$$P\delta(g)Pv_\mu = \int_K e^{-\mu(A(kg^{-1}))}\, dk\, Pv_\mu. \tag{19}$$

But $P^2 = P$ so $|c|^2\langle v_\mu, Pv_\mu \rangle = |c|^2 \langle Pv_\mu, Pv_\mu \rangle = 1$, whence by (17)–(19)

$$\tilde{\phi}(g) = \int_K e^{-\mu(A(kg^{-1}))}\, dk, \qquad g \in G. \tag{20}$$

We write $\mu = i\lambda - \rho$ ($\lambda \in \mathfrak{a}_c^*$, ρ half the sum of the positive restricted roots) and recall from Chapter IV, §4, that $\phi_\lambda(g) = \phi_{-\lambda}(g^{-1})$ and $\phi_{s\lambda} = \phi_\lambda$ for all s in the Weyl group W. Thus by (20)

$$\tilde{\phi}(g) = \phi_\lambda(g) = \int_K e^{(s^*\mu)(A(kg))}\, dk, \qquad g \in G, \tag{21}$$

where s^* is the Weyl group element mapping Σ^+ to $-\Sigma^+$.

Using now the extension $A : U_0^{\mathbf{C}} \to \mathfrak{a}$ we can extend ϕ_λ by the formula

$$\phi_\lambda(u) = \int_K e^{(i\lambda + \rho)(A(k^{-1}uk))}\, dk, \qquad u \in U_0^{\mathbf{C}}.$$

Then the integral formula (Chapter IV, Lemma 4.4)

$$\phi_\lambda(g^{-1}h) = \int_K e^{(-i\lambda + \rho)(A(k^{-1}gk))} e^{(i\lambda + \rho)(A(k^{-1}hk))}\, dk \tag{22}$$

holds by holomorphic continuation for g, h, $g^{-1}h \in U_0^{\mathbf{C}}$.

The vector $\boldsymbol{v}$ above is a linear combination

$$\boldsymbol{v} = \sum_i a_i \delta(u_i)\boldsymbol{e}, \qquad a_i \in \mathbf{C},\ u_i \in U, \tag{23}$$

and here we may assume the elements u_i contained in an arbitrary neighborhood U_e of e in U. In fact, if a linear form β on V vanishes on $\delta(U_e)\boldsymbol{e}$ then the analytic function $u \to \beta(\delta(u)\boldsymbol{e})$ vanishes on U_e, hence on U. Since $\delta(U)\boldsymbol{e}$ spans V we conclude $\beta = 0$. Now by (23)

$$f(uK) = \sum_i a_i \phi(u_i^{-1}uK).$$

If u is in a sufficiently small neighborhood of e we can write $\phi(u_i^{-1}uK) = \phi_{s^*\lambda}(u_i^{-1}u)$ and can substitute from (22). This gives formula (15) with μ replaced by $-(is^*\lambda + \rho) = -s^*\mu$. Since $-s^*\mu$ satisfies the integrality condition (16) formula (15) is proved.

On the other hand, let $\mu \in \mathfrak{a}^*$ satisfy (16) and let by Theorem 4.1 δ be the irreducible finite-dimensional spherical representation of G on a vector space V with highest weight having restriction to $\mathfrak{a}$ given by μ. Let V^* be the dual of V, $v_0 \neq 0$ a vector fixed under $\delta(K)$, and v^* a vector in V^* such that $v^*(v_0) = 1$. Let $v_0^* \in V^*$ be defined by

$$v_0^*(v) = \int_K v^*(\delta(k)v)\, dk, \qquad v \in V,$$

and the function ϕ on G/K by

$$\phi(gK) = v_0^*(\delta(g^{-1})v_0).$$

Comparing with Eq. (12) of Chapter IV, §4 and the ensuing discussion we see that ϕ is a spherical function on G/K and that δ is equivalent to the natural representation δ_ϕ of G on the vector space V_ϕ spanned by the translates of ϕ. Let ψ be a highest-weight vector for δ_ϕ. Then, writing $o = \{K\}$, we have

$$\psi(na \cdot o) = (\delta_\phi(a^{-1}n^{-1})\psi)(o) = e^{-\mu(\log a)}\psi(o),$$

so, taking $\psi(o) = 1$, we have

$$\psi(gK) = e^{-\mu(A(g))}.$$

But δ_ϕ extends to $U^{\mathbf{C}}$, so ψ, being a representation coefficient, extends to a holomorphic function on $U^{\mathbf{C}}$. By analytic continuation this extension $\tilde{\psi}$ satisfies

$$\tilde{\psi}(uK) = e^{-\mu(A(u))}, \qquad u \in U_0. \tag{24}$$

Being a joint eigenfunction of $\mathbf{D}(G/K)$, ψ satisfies

$$\int_K \psi(gkx \cdot o)\, dk = \psi(g \cdot o) \int_K \psi(kx \cdot o)\, dk, \quad g, x \in G$$

(cf. Chapter IV, Proposition 2.4). By holomorphic continuation, $\tilde{\psi}$ satisfies this functional equation on U/K; hence it is a joint eigenfunction of $\mathbf{D}(U/K)$. But then (24) gives the desired conclusion about the integral (15). This concludes the proof.

Note that the formula

$$f(uK) = \int_{K/M} \tilde{\psi}(k^{-1}uK)F(kM)\, dk_M$$

is valid on all of U/K.

3. The Rank-One Case

Suppose now the symmetric space U/K above has rank one. We shall then give more explicit formulas for the spherical function $\phi = \phi_\delta$ corresponding to the spherical representation δ.

As before, let μ denote the restriction to $\mathfrak{a}$ of the highest weight of δ and put $\mu = i\lambda - \rho$. By (16) we have

(i) $\mu = n\alpha, \quad n \in \mathbf{Z}^+ \quad \text{if } \Sigma^+ = \{\alpha\};$
(ii) $\mu = 2n\alpha, \quad n \in \mathbf{Z}^+ \quad \text{if } \Sigma^+ = \{\alpha, 2\alpha\}.$

According to Exercise B8, Chapter IV, the spherical function ϕ_λ in (21) is given by the hypergeometric function

$$\phi_\lambda(h) = F(\tfrac{1}{2}(m_\alpha + n), -\tfrac{1}{2}n, \tfrac{1}{2}(m_\alpha + 1), -sh^2 t)$$

$$\phi_\lambda(h) = F(\tfrac{1}{2}m_\alpha + m_{2\alpha} + n, -n, \tfrac{1}{2}(m_\alpha + m_{2\alpha} + 1), -sh^2 t)$$

in the two respective cases; here $t = \alpha(\log h)$. Using the Gauss–Kummer transformation formula (Erdélyi *et al.*, Vol. I, p. 65)

$$F(a, b; a + b + \tfrac{1}{2}; 4z(1 - z)) = F(2a, 2b; a + b + \tfrac{1}{2}; z),$$

the first formula above can be written

$$\phi_\lambda(h) = F\left(m_\alpha + n, -n, \tfrac{1}{2}(m_\alpha + 1), -sh^2\left(\frac{t}{2}\right)\right).$$

Letting now β denote the larger element in Σ^+ both case (i) (ii) can be combined in the formula

$$(25)\quad \phi_\lambda(\exp H) = F\left(\tfrac{1}{2}m_{\beta/2} + m_\beta + n, -n, \tfrac{1}{2}(m_{\beta/2} + m_\beta + 1), -sh^2\left(\frac{\beta(H)}{2}\right)\right),$$

which by the expansion for F reduces to a polynomial. Both sides are holomorphic on all of $\mathfrak{a}^c$ so in particular we can take $H \in i\mathfrak{a}$ and have then proved the following result.

Theorem 4.5. *Let U/K be a simply connected compact symmetric space of rank one and β the larger element in Σ^+. Let δ be a spherical representation of U and let μ denote the restriction to $\mathfrak{a}$ of its highest weight. Then $\mu = n\beta$ where $n \in \mathbf{Z}^+$ and the spherical function ϕ_δ is given by the* (Jacobi) *polynomial*

$$(26)\quad \phi_\delta(\exp H) = F\left(\tfrac{1}{2}m_{\beta/2} + m_\beta + n, -n; \tfrac{1}{2}(m_{\beta/2} + m_\beta + 1); \sin^2\left(\frac{\beta(H)}{2}\right)\right).$$

Each $n \in \mathbf{Z}^+$ appears for a suitable δ.

EXERCISES AND FURTHER RESULTS

A. Representations

1. If $\mathfrak{g}$ is a simple Lie algebra over $\mathbf{C}$, show, using Eq. (16) of §1 that

$$\langle \delta + \rho, \delta + \rho \rangle - \langle \rho, \rho \rangle = 1,$$

where δ is the highest root, and ρ half the sum of the positive roots.

2. (i) For $s \in W = W(\mathfrak{g}, \mathfrak{t})$ let $n(s)$ denote the number of positive roots which s maps into negative roots. Then

$$\det s = (-1)^{n(s)}.$$

(ii) Show that the orbit

$$\{s\rho : s \in W\}$$

equals the sphere $\{\lambda \in \Lambda : |\lambda| = |\rho|\}$.

3. Let π be as in Theorem 1.7, $\Lambda(\pi)$ the set of weights μ of π, m_μ the multiplicity of μ. Let λ denote the highest weight of π. Writing Eq. (19) of §1, in the form

$$\prod_{\alpha \in \Delta^+} \sinh(\tfrac{1}{2}\langle \alpha, \lambda + \rho \rangle x) = \sum_{\mu \in \Lambda(\pi)} m_\mu e^{\langle \rho, \mu \rangle x} \prod_{\alpha \in \Delta^+} \sinh(\tfrac{1}{2}\langle \alpha, \rho \rangle x),$$

deduce

$$\sum_{\mu \in \Lambda(\pi)} m_\mu \langle \mu, \rho \rangle^2 = \tfrac{1}{24}(\dim \pi)(\langle \lambda + \rho, \lambda + \rho \rangle - \langle \rho, \rho \rangle).$$

(cf. Freudenthal and de Vries [1969], §47).

4. Specializing A3 to the adjoint representation and using A1 deduce

$$\dim_{\boldsymbol{C}} \mathfrak{g} = 24\langle \rho, \rho \rangle$$

for each simple Lie algebra $\mathfrak{g}$ over $\boldsymbol{C}$.

5. Let π and m_μ ($\mu \in \Lambda(\pi)$) be as in A3 and for $\nu \in \Lambda$ (cf. Theorem 1.1) let $\mathscr{P}(\nu)$ be the number of unordered partitions of ν into a sum of positive roots. Because of the analogy between χ_π and ϕ_λ (for G complex) as expressed in Theorem 5.7, Chapter IV and Theorem 1.7, Chapter V, the multiplicity m_μ corresponds to the coefficient Γ_μ. Thus prove the following analogs of (25) and (12) in Chapter IV, §5,

(i) $$m_\mu = \sum_{s \in W} (\det s)\mathscr{P}(s(\lambda + \rho) - (\mu + \rho)),$$

(ii) $$(\langle \lambda + \rho, \lambda + \rho \rangle - \langle \mu + \rho, \mu + \rho \rangle)m_\mu = 2 \sum_{k \geq 1, \alpha \in \Delta^+} m_{\mu + k\alpha}\langle \alpha, \mu + k\alpha \rangle$$

(cf. Kostant [1959b] and Freudenthal [1954], respectively).

Deduce from (i) the recursion formula

$$\mathscr{P}(\mu) = -\sum_{s \in W, s \neq e} (\det s)\mathscr{P}(\mu - (\rho - s\rho)), \qquad \mu \in \Lambda.$$

6. Let U be as in Theorem 1.7 and let $\Lambda(+)$ denote the set of all $\lambda \in \Lambda$ which are dominant integral. For each $\lambda \in \Lambda(+)$ let π_λ denote the corresponding irreducible representation of U and χ_λ its character. Let λ, $\mu \in \Lambda(+)$.

(i) Show that

$$\chi_\lambda \chi_\mu = \sum_{\nu \in \Lambda(+)} m_\nu(\lambda, \mu)\chi_\nu,$$

where

$$m_\nu(\lambda, \mu) = \sum_{s, s' \in W} \det(ss')\mathscr{P}(s(\lambda + \rho) + s'(\mu + \rho) - \nu - 2\rho)$$

with $\mathscr{P}$ having the same meaning as in A5 (cf. Steinberg [1961]; for an earlier formula see Brauer [1937]).

In view of Lemma 2.10 this result gives the direct decomposition of $\pi_\lambda \otimes \pi_\mu$ into irreducibles.

(ii) Show that

$$m_{\lambda+\mu}(\lambda, \mu) = 1;$$

in other words, $\pi_{\lambda+\mu}$ appears exactly once in the decomposition of $\pi_\lambda \otimes \pi_\mu$ into irreducibles. This representation $\pi_{\lambda+\mu}$ is sometimes called the *Cartan product* of π_λ and π_μ.

(iii) Let $V = \sum_{\mu \in \Lambda(\pi_\lambda)} V_\mu$ be the weight space decomposition of V. Let $\mu \in \Lambda(\pi_\lambda)$ and $\alpha \in \Delta^+$ such that $\mu + \alpha \in \Lambda(\pi_\lambda)$. Using Lemma 1.2 in the Appendix show that

$$\pi_\lambda(X_\alpha)V_\mu \neq 0.$$

7. Let $U = \boldsymbol{SU}(2)$. For the torus

$$T = \left\{\exp\begin{pmatrix} i\theta & 0 \\ 0 & -i\theta \end{pmatrix} : \theta \in \boldsymbol{R}\right\},$$

the single positive root α is given by

$$\alpha\begin{pmatrix} 1 & 0 \\ 0 & -1 \end{pmatrix} = 2$$

and $\Lambda(+)$ consists of the multiples $\lambda = \frac{1}{2}l\alpha$ $(l \in \boldsymbol{Z}^+)$. Show that

(i) $$\chi_\lambda\begin{pmatrix} e^{i\theta} & 0 \\ 0 & e^{-i\theta} \end{pmatrix} = \frac{\sin(l+1)\theta}{\sin\theta}, \qquad \dim \pi_\lambda = l + 1.$$

Let $\mu \in \Lambda(+)$, $\mu = \frac{1}{2}m\alpha$. Deduce from Exercise A6 that if $\nu = \frac{1}{2}n\alpha$ $(n \in \boldsymbol{Z}^+)$ then

(ii) $$m_\nu(\lambda, \mu) = \begin{cases} 1, & n = l + m, l + m - 2, \ldots, |l - m| \\ 0, & \text{else,} \end{cases}$$

(Clebsch–Gordan formula). Another proof comes by writing

$$\chi_\lambda\begin{pmatrix} e^{i\theta} & 0 \\ 0 & e^{-i\theta}\end{pmatrix} = e^{il\theta} + e^{i(l-2)\theta} + \cdots + e^{-il\theta}.$$

(iii) Let V_λ denote the set of linear combinations $\sum_{p+q=l} a_{p,q} z^p w^q$. Show that the natural representation of U on V_λ has character χ_λ.

8. Deduce from Lemma 2.6 that if J denotes the canonical representation $u \to u$ of $\boldsymbol{U}(n)$ (for $n > 1$) then $J \otimes J$ decomposes into two irreducible components (the spaces of second-order skew-symmetric and symmetric tensors, respectively).

9. For the simply connected group $\boldsymbol{SU}(n+1)$ let J denote the canonical representation $u \to u$ of $\boldsymbol{SU}(n+1)$ on the vector space $V = \boldsymbol{C}^{n+1}$. The action of $\boldsymbol{SU}(n+1)$ on V induces for each r a representation $\bigwedge^r J$ of $\boldsymbol{SU}(n+1)$ on $\bigwedge^r V$, given by

$$(\textstyle\bigwedge^r J)(u)(v_1 \wedge \cdots \wedge v_r) = u \cdot v_1 \wedge \cdots \wedge u \cdot v_r,$$

(cf. [DS], Chapter I, §2). Show that the representations $\bigwedge^r J$ $(1 \le r \le n)$ are irreducible and constitute the fundamental representations of $\boldsymbol{SU}(n+1)$.

Geometrically, $\bigwedge^r J$ amounts to the action of $\boldsymbol{SU}(n+1)$ on the family of the r-dimensional subspaces of V.

10. Let π be a finite-dimensional representation of $\mathfrak{u}$ and $\check{\pi}$ its contragredient. Show that

(i) $\Lambda(\check{\pi}) = -\Lambda(\pi)$.

(ii) If π has highest weight λ then $\check{\pi}$ has highest weight $-s\lambda$, s being the Weyl group element which interchanges $\mathfrak{t}^+$ and $-\mathfrak{t}^+$. In particular, if $\mathfrak{u}$ is simple and $\neq \mathfrak{a}_l$ $(l > 1)$, $\mathfrak{d}_{2k+1}$, $\mathfrak{e}_6$ then every π is self-contragredient ([DS], Chapter X, Exercise B6).

(iii) Suppose $\mathfrak{u}$ is simple and π irreducible. The diagram of π is obtained from the Dynkin diagram of $\mathfrak{g} = \mathfrak{u}^{\boldsymbol{C}}$ by attaching to each simple root α in the diagram the integer $2\langle\lambda, \alpha\rangle/\langle\alpha, \alpha\rangle$. In the exceptional cases under (ii) π is self-contragredient if and only if its diagram is invariant under the automorphism $-s$ of the Dynkin diagram.

(iv) If π is fundamental so is $\check{\pi}$.

11. Let $\mathfrak{g} = \mathfrak{sl}(2, \boldsymbol{R})$ and consider the Cartan decomposition $\mathfrak{g} = \mathfrak{k} + (\mathfrak{a} + \mathfrak{q})$ where $\mathfrak{k}$, $\mathfrak{a}$, and $\mathfrak{q}$, respectively, are spanned by the vectors

$$X_1 = \begin{pmatrix} 0 & 1 \\ -1 & 0 \end{pmatrix}, \quad X_2 = \begin{pmatrix} 1 & 0 \\ 0 & -1 \end{pmatrix}, \quad X_3 = \begin{pmatrix} 0 & 1 \\ 1 & 0 \end{pmatrix}.$$

The Killing form of $\mathfrak{g}$ is given by

$$B(X, X) = 8(-x_1^2 + x_2^2 + x_3^2) \qquad \text{if} \quad X = \sum_1^3 x_i X_i$$

and we put $\mathfrak{a}^+ = (\boldsymbol{R}^+ - 0)X_2$, $\mathfrak{n} = \boldsymbol{R}(X_1 + X_3)$.

The space $\mathfrak{a}^{\boldsymbol{C}} = \boldsymbol{C}X_2$ is a Cartan subalgebra of $\mathfrak{g}^{\boldsymbol{C}}$ and the highest weights of the finite-dimensional representations of $\mathfrak{g}^{\boldsymbol{C}}$ are $\lambda = \frac{1}{2}n\alpha$ where $n \in \boldsymbol{Z}^+$ and $\Delta^+(\mathfrak{g}^{\boldsymbol{C}}, \mathfrak{a}^{\boldsymbol{C}}) = \{\alpha\}$. The spherical representations of G are (by Theorem 4.1) given by $\lambda = n\alpha$ ($n \in \boldsymbol{Z}^+$).

In particular, the adjoint representation Ad_G of $G = \boldsymbol{SL}(2, \boldsymbol{R})$ is spherical. With K, A, N, and M as in §4 show that

(i) The mapping $gK \to \mathrm{Ad}_G(g)X_1$ is a bijection of G/K onto the quadric

$$B(X, X) = B(X_1, X_1) \quad (x_1 > 0).$$

(ii) The mapping $gMN \to \mathrm{Ad}_G(g)(\frac{1}{2}(X_1 + X_3))$ is a bijection of G/MN onto the null cone

$$(*) \qquad B(Z, Z) = 0 \qquad (x_1 > 0).$$

(iii) The horocycles in G/K are given by the plane sections $B(X, Z) = -1$, Z being a fixed element of the null cone $(*)$.

B. Fourier Series

1. Let K be a compact group. To each $f \in L^1(K)$ we associate the Fourier series

$$(1) \qquad f(k) \sim \sum_{\lambda \in \hat{K}} d_\lambda \operatorname{Tr}(A_\lambda U_\lambda(k))$$

as in §2. No. 1. A *hyperfunction* on $\hat{K}$ is a mapping Γ which assigns to each $\lambda \in \hat{K}$ a linear transformation Γ_λ of $\mathscr{H}_\lambda$ (the representation space of U_λ). A hyperfunction Γ is said to be *unitary* if each Γ_λ is unitary. Let $\mathscr{E}$ and $\mathscr{F}$ be function spaces on K. A hyperfunction Γ on $\hat{K}$ is called an *$(\mathscr{E}, \mathscr{F})$-multiplier* if for each $f \in \mathscr{E}$ with Fourier series (1) the series

$$(2) \qquad \sum_{\lambda \in \hat{K}} d_\lambda \operatorname{Tr}(\Gamma_\lambda A_\lambda U_\lambda(x))$$

represents a function $f_\Gamma \in \mathscr{F}$. Show that

(i) If Γ is an $(L^1(K), L^1(K))$-multiplier [or a $(C(K), C(K))$-multiplier] then

$$f_\Gamma = f * \mu_\Gamma \qquad \text{for all } f,$$

where μ_Γ is a measure on K.

(ii) If Γ is a $(C(K), A(K))$-multiplier then for a certain function $F \in L^2(K)$

$$f_\Gamma = f * F \qquad \text{for} \quad f \in C(K).$$

2. Suppose $f \in L^1(K)$ such that for each unitary hyperfunction Γ the series (2) represents a function $f_\Gamma \in L^1(K)$. Then $f \in L^2(K)$. (Use Theorem 2.3.) Compare with Littlewood's theorem (§2, No. 2).

3*. (Failure of Theorem 2.3 for noncompact groups.) Let G be a non-pact, connected, separable, unimodular locally compact group. Then every spectrally continuous operator T on $L^1(G)$ commuting with the right translations on G is 0, (cf. Helgason [1957]; for a further reduction of the assumptions on G see Sakai [1964]).

4. Let U be a compact, simply connected (semisimple) Lie group. Identifying $\Lambda(+)$ with $\hat{U}$ (Theorem 1.5), we call a hyperfunction Γ on $\Lambda(+)$ *rapidly decreasing* if for each $k \in \boldsymbol{Z}^+$,

$$\lim_{|\lambda| \to \infty} |\lambda|^k \operatorname{Tr}(\Gamma_\lambda \Gamma_\lambda^*)^{1/2} = 0.$$

Then the mapping $f \to \{A_\lambda\}_{\lambda \in \Lambda(+)}$ defined by (1) is a bijection of $\mathscr{E}(U)$ onto the space of rapidly decreasing hyperfunctions on $\hat{U}$ (cf. Sugiura [1971] even for U just assumed to be compact and connected Lie group).

5. Let K be an arbitrary compact group. A subset $S \subset \hat{U}$ is said to be *distinguished* if for each continuous function

$$f(k) \sim \sum_{\lambda \in \hat{K}} d_\lambda \operatorname{Tr}(A_\lambda U_\lambda(k))$$

the subseries

$$\sum_{\lambda \in S} d_\lambda \operatorname{Tr}(A_\lambda U_\lambda(k))$$

also represents a continuous function f_S. Show that

(i) If S is distinguished then

$$f_S = f * \mu_S$$

where μ_S is a central measure on K which is idempotent in the sense that $\mu_S * \mu_S = \mu_S$. Conversely, each central idempotent measure arises in this way from a distinguished set S.

(ii) The distinguished sets which preserve positivity in the sense that $f_S \geq 0$ whenever $f \geq 0$ are precisely the normal subhypergroups of $\hat{K}$.

NOTES

§1. The basic weight theory for finite-dimensional representations of simple Lie algebras over C was founded by Cartan [1913] (later he considered real Lie algebras as well [1914]; see also Iwahori [1959]). Here appears Theorem 1.5 (on the highest weight) and a description of the fundamental representations. While Theorem 1.3 is a well-known standard result I believe inequality (8) and the remark following the theorem are due to Freudenthal [1954, 1956], and I learned part (iv) from Kostant in 1965.

Using integration methods, Schur [1924] developed the representation theory for the orthogonal group $O(n)$ including the character formula (Theorem 1.7); an algebraic treatment of $SO(n)$ is due to Brauer [1926]. Weyl in [1925, 1926] adapted Schur's method to arbitrary compact semisimple Lie groups and proved the character formula (Theorem 1.7) and the dimension formula (Theorem 1.8).

In Weyl's derivation of Theorems 1.5 and 1.8 from Theorem 1.7 we have followed Cartier's exposition in Séminaire S. Lie [1955]. In [1954], Freudenthal gave an algebraic proof of the character formula; this proof can be found in Jacobson [1962]. Cartier's paper quoted gives it also along with Weyl's original proof. The proof in the text is an analytic modification of Freudenthal's proof; it goes back to Sugiura [1960], although I was unaware of this fact in [1972a].

The isomorphism $Z(\mathfrak{g}) \approx I(\mathfrak{t})$ from Theorem 1.9 was first obtained by Harish-Chandra [1951a] (Part III), who gave an algebraic definition and proof. The analytic proof in the text is from Helgason [1972a] (Chapter I, §2).

Theorem 1.10, which is a corollary of Harish-Chandra's integral formula (Theorem 5.35 in Chapter II), represents the character in terms of the Fourier transform of the characteristic function of an orbit. Kirillov [1962] proved a formula of this type for nilpotent groups and in [1968, 1969] he posed the problem of generalization to arbitrary Lie groups. For semisimple groups positive results on this question have been proved by Gutkin [1970], Duflo [1970], Rossmann [1978b, 1980], and Vergne [1979]; for solvable groups by Duflo in Bernat *et al.* [1972]; and for general groups by Khalgui [1982].

§2. Schur's orthogonality relations (Schur [1I]) and the Peter–Weyl theorem (Peter and Weyl [1927]) are proved in Weil [1940] and in many more recent books, including Sugiura [1975], Lang [1975], Schempp and Dreseler [1980], and Wallach [1973]. The results in this section (in particular Lemma 2.6) which deal with spectrally continuous operators, absolute convergence, lacunary Fourier series on groups, and the hypergroup structure on $\hat{K}$ are from the author's papers [1957a, 1958] in which the group-theoretic lacunary condition was introduced. Further work on lacunary series was done by Hewitt and Zuckerman, Figa-Talamanca, Rider, Edwards, Ross, and others. See Hewitt and Ross [1970] for an extensive treatment. However, as Prop. 2.15(i) indicates (Helgason [1958]) and as is shown more generally by Ceccini [1972] and others, the notion, and certain natural variations of it, is primarily of interest for abelian compact groups and for infinite-dimensional compact groups. The example in Proposition 2.15(ii) occurs in Coifman and Weiss [1971, Chapter IV]; it was pointed out to me by Travaglini.

According to Theorem 2.3, the constant

$$\inf_f \sup_g \frac{\|f * g\|_1}{\|f\|_2 \|g\|_{sp}},$$

which only depends on the group K, lies between $2^{-1/2}$ and 1. The exact value is not known even for K abelian; see Edwards and Ross [1973] for work in this direction.

An extensive exposition of Fourier series on compact groups is given in Hewitt and Ross [1970]; see also Edwards [1972], Dunkl and Ramirez [1971], and Rudin [1962]. Further integrals of the type in Lemma 2.6 (Helgason [1957]) can be found in Hewitt and Ross [1970] (§29) and in James [1961].

§§3–4. Theorem 3.1 and its corollaries 3.4, 3.6, and Theorem 3.5(iii) are due to Harish-Chandra [1954b, 1966]. Theorem 4.3 on Fourier expansions on compact symmetric spaces goes back to Cartan [1929] with some amplifications by Sugiura [1960]. The expansion in Theorem 3.5 goes back to Weyl [1934]; see also Weil [1940] (§23).

Theorem 4.1 on the spherical representations is stated and proved in Helgason [1965b] and [1970a] (Chapter III, §3). An informative extension was given by Schlichtkrull [1984a]. The question settled by Corollary 4.2 is discussed in Cartan [1929] (§VI); a weaker version is stated without proof in Sugiura [1962]; the first proof seems to be in Helgason [1970a], Chapter III. A complexified version is in Kostant [1975a] (Chapter II, §6). Formulas (7) and (8) occur in Helgason [1970a] (Chapter III, §3) and Lasalle [1978] (§5), respectively.

The eigenfunction integral representation in Theorem 4.4 was proved by the author in [1977]. The mapping (14) was used by Clerc [1976] and Stanton [1976] in a different context, and it was observed by Sherman [1977] that Lemma 4.4, Chapter IV extends to the present case [as formula (22) of §4]. Also, formulas (18)–(20) are due to Harish-Chandra [1958a] (Lemma 5).

Theorem 4.5 is given in Cartan [1929], §§VII–VIII for $m_{\beta/2} = 0$ or 1; it is stated in Gangolli [1967], p. 179, that a similar proof works in general. For an analytic generalization see Koornwinder [1973, 1974] and for further results on the spherical functions for compact spaces of higher rank see Vretare [1976], Hoogenboom [1983].

SOLUTIONS TO EXERCISES

INTRODUCTION

A. The Spaces R^n and S^n.

A.1. (i) The space spanned by the translates of f can be decomposed into subspaces on which $O(n)$ acts irreducibly. Each of these will contain a function ϕ_k for some k, so by the irreducibility of the spaces E_k these subspaces are among the E_k.

(ii) The group $O(1, 1)$ is generated by the transformations

$$a_t\colon \begin{pmatrix} x_1 \\ x_2 \end{pmatrix} \to \begin{pmatrix} \operatorname{ch}t & \operatorname{sh}t \\ \operatorname{sh}t & \operatorname{ch}t \end{pmatrix} \begin{pmatrix} x_1 \\ x_2 \end{pmatrix}, \qquad \begin{pmatrix} x_1 \\ x_2 \end{pmatrix} \to \begin{pmatrix} \pm x_1 \\ -x_2 \end{pmatrix}.$$

Consider the function f on $x_1^2 - x_2^2 = -1$ defined by

$$f(x_1, x_2) = \sinh^{-1}(x_1).$$

Then $f^{a_t}(x_1, x_2) = f(x_1, x_2) - t$, so f is $O(1, 1)$ finite, yet is not the restriction of a polynomial.

A.2. Consider for $\lambda \neq 0$ the transform $F \to f$ given by

$$f(x) = \int_{S^{n-1}} e^{i\lambda(x,\omega)} F(\omega)\, d\omega, \qquad F \in L^2(S^{n-1}).$$

Generalizing Lemma 2.7, we see that the map $F \to f$ is one-to-one. Using Theorem 2.7 of Chapter II, we let S_λ be a sphere in R^n with center 0 such that each $f \in \mathscr{E}_\lambda(R^n)$ is determined by its restriction $f|S_\lambda$ to S_λ. Because of Theorem 3.1, the space $E_k = E_k(S^{n-1})$ is exactly the space of functions $f \in C(S^{n-1})$ which are $O(n)$-finite and for which the representation of $O(n)$ on the space of translates is equivalent to δ (in Prop. 3.2). The space $E_k(S_\lambda)$ is similarly characterized. The maps

$$F \to f|S_\lambda, \qquad F \to f, \qquad F \in L^2(S^{n-1}),$$

are, by the above, one-to-one and commute with the action of $O(n)$. By the characterization of E_k indicated the first map sends $E_k(S^{n-1})$ injectively into $E_k(S_\lambda)$, hence surjectively. Hence each $O(n)$-finite $f \in \mathscr{E}_\lambda(R^n)$

has the form stated. The proof of the irreducibility criterion for T_λ now proceeds as for Theorem 2.6.

A.3. There is a natural bijection of $\mathscr{H}_k/\mathscr{H}_{k-1}$ onto E_k commuting with the $\boldsymbol{O}(n)$ action. Thus we see that already $\boldsymbol{O}(n)$ acts irreducibly on $\mathscr{H}_k/\mathscr{H}_{k-1}$.

A.4. The Lie algebra $\mathfrak{sl}(2, \boldsymbol{C})$ of $\boldsymbol{SL}(2, \boldsymbol{C})$ consists of the complex 2×2 matrices of trace 0 viewed as a real Lie algebra. Given $X \in \mathfrak{sl}(2, C)$, $u \in \mathscr{E}(\boldsymbol{R}^2)$ we put

$$(X^*u)(z) = \left(\frac{d}{dt}u(\exp(-tX)\cdot z)\right)_{t=0}, \qquad z \in \boldsymbol{R}^2.$$

By the quasi-invariance of L, X^*u is harmonic if u is harmonic. The statement to be proved is that if $A: \mathscr{E}_0(\boldsymbol{R}^2) \to \mathscr{E}_0(\boldsymbol{R}^2)$ is a continuous linear mapping such that $AX^* = X^*A$ for all $X \in \mathfrak{su}(2, \boldsymbol{C})$ then A is a scalar. Taking X as

$$\begin{pmatrix} 0 & 1 \\ 0 & 0 \end{pmatrix} \quad \text{or} \quad \begin{pmatrix} 0 & i \\ 0 & 0 \end{pmatrix}$$

we find that A commutes with $\partial/\partial x$ and $\partial/\partial y$, and therefore maps the subspaces $\mathscr{H}$ and $\overline{\mathscr{H}}$ (of holomorphic and antiholomorphic functions, respectively) into themselves. Taking X as

$$\begin{pmatrix} 1 & 0 \\ 0 & -1 \end{pmatrix}$$

we find that A commutes with the operator $x\partial/\partial x + y\partial/\partial y$, which on $\mathscr{H}$ coincides with $z\partial/\partial z$. This implies easily that A is a scalar on $\mathscr{H}$ and similarly on $\overline{\mathscr{H}}$, hence on $\mathscr{E}_0(\boldsymbol{R}^2)$.

A.5. (i) Let $D_k(n) = \dim P_k$ and $d_k(n) = \dim H_k$. Writing a polynomial in $(x_1, \ldots, x_n)$ as a polynomial in x_n with coefficients in $(x_1, \ldots, x_{n-1})$ we see that

$$D_k(n) = \sum_{p=0}^{k} D_{k-p}(n-1) = D_k(n-1) + D_{k-1}(n),$$

so by induction on $k + n$ we have

$$D_k(n) = \frac{n(n+1)\cdots(n+k-1)}{k!} = \binom{n+k-1}{k}.$$

Since by Eq. (4) of §3 $d_k(n) = D_k(n) - D_{k-2}(n) = D_k(n-1) + D_{k-1}(n-1)$, part (i) follows.

Part (ii) follows from Eq. (3) of §3. For (iii) we have by the mean-value theorem for harmonic functions

$$\int_{\mathbf{R}^n} h(x+y)f(x)\,dx = \left(\int_{\mathbf{R}^n} f(x)\,dx\right)h(y)$$

if f is any rapidly decreasing radial function. Here we can take $f(x) = e^{-(1/2)|x|^2}$ and since h is a polynomial we can replace y by any complex $z \in \mathbf{C}^n$. This gives

$$\int_{\mathbf{R}^n} h(x-iy)e^{-(1/2)(x,x)}\,dx = (-i)^k(2\pi)^{(1/2)n}h(y).$$

Using Cauchy's theorem on the holomorphic function

$$z \to h(z-iy)e^{-(1/2)(z,z)}$$

we can shift the integral on the left to $\mathbf{R}^n + iy$. Then (iii) follows. (The result is given in Bochner [1955] with a proof involving Bessel functions.)

(iv) Let $\phi \in E_k$ and let h_ϕ denote its unique extension to an element of H_k. Note that $\partial(h_\phi)(|x|^{2-n})$ is a harmonic function of homogeneity $2-n-k$. Writing it as $|x|^{2-n-k}\psi(x/|x|)$ we see using Eq. (3) of §3 that ψ is an eigenfunction of L with eigenvalue

$$-(2-n-k)(2-n-k+n-2),$$

so $\psi \in E_k$. The mapping

$$\phi \to \partial(h_\phi)(|x|^{2-n})|_{S^{n-1}} = \psi$$

thus maps E_k into itself. Since it clearly commutes with the rotations, Schur's lemma and Theorem 3.1(ii) show that the mapping is a scalar multiple of the identity. This proves (iv) except for the value of the constant $c_{k,n}$, which can easily be calculated by using h_{k+1} from part (ii).

A computational proof (using induction on k) can easily be given for h of the form $(a, x)^k$ $a \in \mathbf{C}^n$ isotropic.

B. The Hyperbolic Plane

B.1. In fact, if $g \in G$ the point $g \cdot o$ can be written $na \cdot o$ ($n \in N, a \in A$) so $g^{-1}na \in K$.

B.2. Part (i) follows from [DS] (p. 548), which also shows that $\mathbf{SU}(1, 1)$ and the conjugation $z \to \bar{z}$ generate the group of isometries of D. Since the conjugation $z \to \bar{z}$ of D corresponds to the map $w \to 1/\bar{w}$ of the upper half plane part (ii) follows.

(iii) Since c maps Euclidean circles into circles it is clear that as a set $S_r(i)$ is a Euclidean circle. It passes through the points $(0, e^{\pm 2r})$ so the statement is clear.

B.3. This follows from (24) and (32) combined with the classical Paley–Wiener theorem for $\boldsymbol{R}$. Another proof can be given by means of Abel's integral equation

$$\phi(u) = \int_{\boldsymbol{R}} F(u + y^2)\, dy,$$

appearing in the proof of Theorem 4.6.

B.4. (i) Clearly $\phi_{-\lambda}(z)$ is real for all z if and only if $\tilde{f}(\lambda)$ is real for all real $f \in \mathscr{D}^\natural(D)$, which by B3 is equivalent to

$$\int_{-\infty}^{\infty} e^{-i\lambda t}\phi(t)\, dt$$

being real for every real $\phi \in \mathscr{D}^\natural(\boldsymbol{R})$. This in turn is equivalent to $\cos(\lambda t) = \cos(\bar{\lambda}t)$ and (i) follows.

(ii) Let $\lambda = \xi + i\eta$ where $\xi,\ \eta \in \boldsymbol{R}$. Assume first $|\eta| \leq 1$. Then if $f \in L^1(D)$ and is radial

$$\begin{aligned}\int |f(z)\phi_{-\lambda}(z)|\, dz &\leq \int |f(z)|\phi_{i\eta}(z)\, dz \\ &= \int_{\boldsymbol{R}} e^{\eta t} F_{|f|}(t)\, dt = \int_0^{\infty} (e^{\eta t} + e^{-\eta t}) F_{|f|}(t)\, dt \\ &\leq 2\int_0^{\infty} e^t F_{|f|}(t)\, dt < \infty\end{aligned}$$

because we have

$$\int_D |f(z)|\, dz = \int_{\boldsymbol{R}} e^{-t} F_{|f|}(t)\, dt < \infty$$

and $F_{|f|}$ is even. Thus ϕ_λ has a finite integral against any L^1-function and so is bounded.

On the other hand, suppose ϕ_λ bounded. If $\eta = 0$ there is nothing to prove and since $\phi_\lambda = \phi_{-\lambda}$ we can assume $\eta < 0$. Then Theorem 4.5 applies and $e^{(-i\lambda+1)r}\phi_\lambda(a_r \cdot 0)$ has a nonzero limit as $r \to +\infty$. But the boundedness of ϕ_λ then implies $\operatorname{Re}(-i\lambda + 1) \geq 0$ and so $\eta \geq -1$ as desired.

B.5. The calculation is the same as that of Lemma 4.9 (cf. Helgason [1976], p. 203).

B.6. This follows by approximating f by functions $f_n \in \mathscr{D}^\natural(D)$ and using (32).

B.7. As in text expand the function $f \in \mathcal{D}(D)$,

$$f(z) = \sum_{m \in \mathbf{Z}} f_m(z),$$

where $f_m(e^{i\theta}z) = e^{im\theta}f_m(z)$. Then $\tilde{f}_m$ and $\hat{f}_m$ have the form

$$\tilde{f}_m(\lambda, e^{i\theta}) = e^{im\theta}\tilde{f}_m(\lambda, 1),$$
$$\hat{f}_m(e^{i\theta}, t) = e^{im\theta}\hat{f}_m(1, t),$$

and

$$\tilde{f}_m(\lambda, 1) = \int_{\mathbf{R}} e^{-i\lambda t}e^t\hat{f}_m(1, t)\, dt.$$

Also, by (58),

$$\tilde{f}_m(\lambda, 1) = \frac{p_m(-i\lambda)}{p_m(i\lambda)}\tilde{f}_m(-\lambda, 1)$$

and since $p_m(i\lambda)$ and $p_m(-i\lambda)$ are relatively prime this implies

$$\tilde{f}_m(\lambda, 1) = p_m(-i\lambda)h(\lambda),$$

where $h(\lambda)$ even and of exponential type. Now the result follows from the Fourier inversion

$$\hat{f}_m(1, t)e^t = (2\pi)^{-1}\int_{\mathbf{R}} \tilde{f}_m(\lambda, 1)e^{i\lambda t}\, d\lambda.$$

B.8. These are the circular arcs connecting -1 and 1 inside D. They are the curves $x_t: r \to x(r, t)$ in Exercise B9 and consist of points of fixed non-Euclidean distance from the geodesic $A \cdot o$.

B.9. We have $x(r, t) = a_r \cdot x(0, t) = a_r \cdot (\tanh t\, i)$ where $a_r \in \mathbf{SU}(1, 1)$ is as in §4. This gives the formula for $x(r, t)$. Next substitute $\zeta = x(r, t)$ into the formula

$$dx = \tfrac{1}{2}i\frac{d\zeta \wedge d\bar{\zeta}}{(1 - |\zeta|^2)^2}.$$

B.10. We have by (24)–(27)

$$f(\tanh s) = F((\mathrm{ch} s)^2)$$

if $F(x) = x^{-(1/2)(i\lambda + 1)}$. It follows that

$$F_f(t) = \int_{\mathbf{R}} F((\mathrm{ch} t)^2 + y^2)\, dy$$
$$= (\mathrm{ch} t)^{-i\lambda}\int_{\mathbf{R}} (1 + z^2)^{-(1/2)(i\lambda + 1)}\, dz$$
$$= (\mathrm{ch} t)^{-i\lambda}\pi c(\lambda).$$

C. Fourier Analysis on the Sphere

(i) Immediate from §3. No. 1.

(ii) The first relation and the second for $(s, a) > 0$ are clear from the uniqueness of ϕ_k in Prop. 3.2. For $(s, a) < 0$ the second relation holds at least up to a constant factor which however is 1 because by the first relation $C_k^n(-1) = (-1)^k$

(iii) Denote the integral by $F(s, s')$. Since $F(-s, -s') = F(s, s')$ we may assume $(s', a) > 0$. Next observe $F(us, us')$ if u is a rotation of B. Let u_ϕ be a rotation of angle ϕ in the $x_n x_{n+1}$ plane. If we can prove

$$F(\mu_\phi s, u_\phi s') = F(s, s')$$

for all ϕ with $(u_\phi s', a) > 0$ then (iii) is reduced to the case $s' = a$ and then (ii) applies. Thus it suffices to prove $\partial F(u_\phi s, u_\phi s')/\partial\phi = 0$.

Each $b \in B$ can be written

$$b = (c_1 \cos\theta, c_2 \cos\theta, \ldots, c_{n-1} \cos\theta, \sin\theta, 0), \qquad c \in \boldsymbol{S}^{n-2}.$$

Putting

$$g(c, \theta, \phi, s)$$

$$= i \cos\theta \sum_{j=1}^{n-1} c_j s_j + i \sin\theta (s_n \cos\phi - s_{n+1} \sin\phi) + s_n \sin\phi + s_{n+1} \cos\phi$$

we have

$$e_{b,k}(u_\phi s) = [g(c, \theta, \phi, s)]^k,$$

$$f_{b,k}(u_\phi s') = [g(c, \theta, \phi, s')]^{-k-n+1}.$$

For each $m \in \boldsymbol{Z}$ we have the differential equation

$$i \frac{\partial g^m}{\partial\phi} - \cos\theta \frac{\partial g^m}{\partial\theta} = m g^m \sin\theta$$

and the volume element decomposition

$$db = (\text{const}) \cos^{n-2}\theta \, d\theta \, dc \qquad (-\pi/2 \leq \theta \leq \pi/2).$$

Using these facts we get after simple computation

$$\frac{\partial F(u_\phi s, u_\phi s')}{\partial\phi}$$

$$= -i \int_{c \in \boldsymbol{S}^{n-2}} \int_{-\pi/2}^{\pi/2} \frac{\partial}{\partial\theta} [g(c, \theta, \phi, s)^k g(c, \theta, \phi, s')^{-k-n+1} \cos^{n-1}\theta] \, d\theta = 0.$$

(The result is from Sherman [1975], the proof from J.-G. Yang [1983].)

(iv) Let $\phi_k(s) = C_k^n((s, a))$ $(k \in \boldsymbol{Z}^+)$ so by Lemma 3.5

$$f = \sum_{k=0}^{\infty} d(k) f * \phi_k.$$

Then

$$\begin{aligned}\int_B e_{b,k}(s')\tilde{f}(b, k)\, db &= \int_{S^n} f(s)\left(\int_B e_{b,n}(s') f_{b,n}(s)\, db\right) ds \\ &= \int_{S^n} f(s) C_k^n((s, s'))\, ds \\ &= \int_{\mathbf{O}(n+1)} f(u \cdot a) C_k^n((u \cdot a, s'))\, du\end{aligned}$$

so we have the compact analog of Lemma 4.7.

$$f * \phi_k(s) = \int_B e_{b,k}(s)\tilde{f}(b, k)\, db$$

and the result follows using the expansion above. (With a suitable definition of $\tilde{f}$ the result holds for all $f \in \mathscr{E}(\boldsymbol{S}^n)$; cf. Sherman [1975].)

CHAPTER I

A. Invariant Measures

A.1. (i) If H is compact, $|\det(\mathrm{Ad}_G(H))|$ and $|\det(\mathrm{Ad}_H(H))|$ are compact subgroups of the multiplicative groups of the positive reals, hence identically 1.

(ii) G/H has an invariant measure so $|\det \mathrm{Ad}_H(h)| = |\det \mathrm{Ad}_G(h)|$, which by unimodularity of G equals 1.

(iii) Let $G_0 = \{g \in G : |\det \mathrm{Ad}_G(g)| = 1\}$. Then G_0 is a normal subgroup of G containing H. Since $\mu(G/H) < \infty$, Prop. 1.13 shows that the group G/G_0 has finite Haar measure, and hence is compact. Thus the image $|\det \mathrm{Ad}_G(G)|$ is a compact subgroup of the group of positive reals, and hence consists of 1 alone.

A.2. The element $H = \begin{pmatrix} 0 & 1 \\ -1 & 0 \end{pmatrix}$ spans the Lie algebra $\mathfrak{o}(2)$ and $\exp \mathrm{Ad}(g)tH = g \exp tHg^{-1} = \exp(-tH)$.

A.3. We have $\det \mathrm{Ad}(\exp X) = \det(e^{\mathrm{ad}\, X}) = e^{\mathrm{Tr}(\mathrm{ad}\, X)}$, so (i) follows. For (ii) we know that G/H has an invariant measure if and only if

$$\exp(\mathrm{Tr}(\mathrm{ad}_{\mathfrak{g}}\, T)) = \exp(\mathrm{Tr}(\mathrm{ad}_{\mathfrak{h}}\, T)), \quad T \in \mathfrak{h}.$$

Put $T = tX_i$ $(r < i \leq n)$, $t \in \boldsymbol{R}$, and differentiate with respect to t. Then the desired relations follow.

A.4. To each $g \in \boldsymbol{M}(n)$ we associate the translation T_x by the vector $x = g \cdot o$ and the rotation k given by $g = T_x k$. Then $kT_x k^{-1} = T_{k \cdot x}$, so

$$g_1 g_2 = T_{x_1} k_1 T_{x_2} k_2 = T_{x_1 + k_1 \cdot x_2} k_1 k_2 .$$

Since $g_{k_1, x_1} \cdot g_{k_2, x_2} = g_{k_1 k_2, x_1 + k_1 \cdot x_2}$ this shows that the mapping $g \to g_{k,x}$ is an isomorphism. Also

$$\int f(g_{k_0, x_0} g_{k, x})\, dk\, dx = \int f(g_{k_0 k,\, x_0 + k_0 \cdot x})\, dk\, dx$$

$$= \int f(g_{k, x})\, dk\, dx$$

since dx is invariant under $x \to x_0 + k_0 \cdot x$.

A.5. By [DS], Chapter II, §7, the entries ω_{ij} in the matrix $\Omega = X^{-1}\, dX$ constitute a basis of the Maurer–Cartan forms (the left invariant 1-forms) on $\boldsymbol{GL}(n, \boldsymbol{R})$. Writing $dX = X\Omega$ we obtain from [DS] (Chapter I, §2, No. 3) for the exterior products

$$\prod_{i,j} dx_{ij} = (\det X)^n \prod_{i,j} \omega_{ij},$$

so $|\det X|^{-n} \prod_{i,j} dx_{ij}$ is indeed a left invariant measure. The same result would be obtained from the right invariant matrix $(dX)X^{-1}$ so the unimodularity follows.

A.6. Let the subset $G' \subset G$ be determined by the condition $\det X_{11} \neq 0$ and define a measure $d\mu$ on G' by

$$d\mu = |\det X_{11}|^{-1} \prod_{(i,j) \neq (1,1)} dx_{ij}.$$

If dg is a bi-invariant Haar measure on G we have (since $G - G'$ is a null set)

$$\int_G f(g)\, dg = \int_{G'} f(g)\, dg = \int_{G'} f(g) J(g)\, d\mu,$$

where J is a function on G'. Let T be a diagonal matrix with $\det T = 1$ and $t_1, \dots, t_n$ its diagonal entries. Under the map $X \to TX$ the product $\prod_{(i,j) \neq (1,1)} dx_{ij}$ is multiplied by $t_1^{n-1} t_2^n \cdots t_n^n$ and $|\det X_{11}|$ is multiplied by $t_2 t_3 \cdots t_n$. Since $\det T = 1$, these factors are equal, so the set G' and the measure μ are preserved by the map $X \to TX$. If A is a supertriangular matrix with diagonal 1, the mapping $X \to AX$ is supertriangular with diagonal 1 if the elements x_{ij} are ordered lexicographically. Thus $\prod_{(i,j) \neq (1,1)} dx_{ij}$ is unchanged and a simple inspection shows[†] $\det((AX)_{11}) = \det(X_{11})$. It follows that G' and $d\mu$ are invariant under

[†]See "Some Details," p. 611.

each map $X \to UX$ where U is a supertriangular matrix in G. By transposition, G' and $d\mu$ are invariant under the map $X \to XV$ where V is a lower triangular matrix in G. The integral formulas above therefore show that $J(UXV) \equiv J(X)$. Since the products UV form a dense subset of G ([DS], Chapter IX, Exercise A2) μ is a constant multiple of dg. For another, more down-to-earth proof see Gelfand and Naimark [1957] (Chapter I, §4).

A.7. A simple computation shows that the measures are invariant under multiplication by diagonal matrices as well as by unipotent matrices; hence they are invariant under $\boldsymbol{T}(n, \boldsymbol{R})$; cf. Gelfand and Naimark [1957] (Chapter I).

A.8. G leaves invariant the metric $y^{-2}(dx^2 + dy^2)$ on H ([DS], Chapter X, Exercise G1) and the Riemannian measure is $y^{-2}\, dx\, dy$, so D has finite area. Since $G = \Gamma\pi^{-1}(D)$ where $\pi\colon G \to H$ is the natural projection $g \to g \cdot i$, $\mu(G/\Gamma) < \infty$.

Remarks. (i) D is actually a fundamental domain for Γ in the sense that each Γ-orbit in H meets D and if two distinct points of D are in the same Γ-orbit they lie on the boundary of D. Also, T and S generate Γ (see, e.g., Serre [1970]).

(ii) The finiteness $\mu(G/\Gamma) < \infty$ for $G = \boldsymbol{SL}(n, \boldsymbol{R})$, $\Gamma = \boldsymbol{SL}(n, \boldsymbol{Z})$ is classical (Hermite, Siegel; for an elementary proof, see Bourbaki [1963], *Intégration*, Chapter 7, Exercise 7, §3). For an extension to semisimple groups see Borel and Harish-Chandra [1962].

A.9. Part (i) is straightforward. For Part (ii) we just indicate the proof in the case of $\boldsymbol{G}'_{2,4}$. The invariant measure is of the form

$$D(a, b, c, d)\, da\, db\, dc\, dd$$

where the density D is to be determined. Requiring invariance under rotations in the (f_1, f_2)-plane and in the (e_1, e_2)-plane gives, respectively,

$$(1)\quad D(a\cos\phi - b\sin\phi, a\sin\phi + b\cos\phi, c\cos\phi - d\sin\phi, c\sin\phi + d\cos\phi) = D(a, b, c, d),$$

$$(2)\quad D(a\cos\phi - c\sin\phi, b\cos\phi - d\sin\phi, a\sin\phi + c\cos\phi, b\sin\phi + d\cos\phi) = D(a, b, c, d).$$

Invariance under rotations in the (e_1, f_1)-plane gives

$$(3)\qquad D(a', b', c', d') = (a\sin\phi - \cos\phi)^4 D(a, b, c, d)$$

where

$$\begin{pmatrix} a' & b' \\ c' & d' \end{pmatrix} = (\cos\phi - a\sin\phi)^{-1}$$

$$\times \begin{pmatrix} a\cos\phi + \sin\phi & b \\ c & d\cos\phi + (bc - ad)\sin\phi \end{pmatrix}.$$

From (1) we obtain with $r = (a^2 + b^2)^{1/2}$,

$$(4)\qquad D(r, 0, r^{-1}(ac + bd), r^{-1}(ad - bc)) = D(a, b, c, d),$$

and from (3) we derive

$$(5)\qquad D(0, 0, -(a^2 + 1)^{-1}c, d) = (a^2 + 1)^2 D(a, 0, c, d).$$

Combining (4) and (5) we get

$$(6)\quad D(a, b, c, d) = (r^2 + 1)^{-2} D(0, 0, -(r^2 + 1)^{-1/2} r^{-1}(ac + bd), r^{-1}(ad - bc)).$$

Also, by (1) and (5),

$$D(0, 0, c, d) = D(0, 0, 0, (c^2 + d^2)^{1/2},$$

$$D(a, 0, 0, d) = (a^2 + 1)^{-2} D(0, 0, 0, d),$$

whence

$$D(a, 0, 0, 0) = (a^2 + 1)^{-2} D(0, 0, 0, 0).$$

But (1) and (2) imply

$$D(a, 0, 0, 0) = D(0, a, 0, 0) = D(0, 0, a, 0) = D(0, 0, 0, a).$$

The formula for D now follows easily.

A.10. (i) This comes from the usual formula for surface area in $\boldsymbol{R}^n$.

(ii) The invariant measure can be written

$$f(x_1, \ldots, x_{p+q+1})\, dx_1 \cdots dx_{p+q}$$

letting $x_1, \ldots, x_{p+q}$ serve as coordinates. Since $\boldsymbol{O}(p, q)$ acting on $x_1, \ldots, x_{p+q}$ leaves the measure invariant f is a function of

$$x_1^2 + \cdots + x_p^2 - x_{p+1}^2 - \cdots - x_{p+q}^2$$

and x_{p+q+1} only, hence of x_{p+q+1} only. Now use the invariance under the group

$$(x_1, x_2, \ldots, x_{p+q+1}) \to (x_1 \operatorname{ch} t + x_{p+q+1} \operatorname{sh} t, x_2, \ldots, x_1 \operatorname{sh} t + x_{p+q+1} \operatorname{ch} t).$$

This shows that

$$f(x_{p+q+1})|x_{p+q+1}| = \text{const.}$$

A.11. (i)–(ii) follow by routine computation. For (iv) it is sufficient just to verify that given $\alpha, \beta \in \boldsymbol{C}$, $\alpha\beta \neq 0$, $|\alpha|^2 + |\beta|^2 = 1$ there exist unique t, θ, ϕ such that

$$0 \leq t < \pi, \qquad 0 \leq \theta < 2\pi, \qquad -2\pi \leq \phi < 2\pi,$$

and such that (1) holds. For further information concerning this exercise see, e.g., Vilenkin [1968].

B. Radon Transforms

B.1. Let $\phi \in \mathcal{D}_H(\boldsymbol{P}^n)$ and put

$$\Phi(\lambda, \omega) = \int_{\boldsymbol{R}} \phi(\omega, p) e^{-ip\lambda}\, dp.$$

Then Φ is holomorphic in λ. Let $a \in \boldsymbol{C}^n$ be isotropic and consider the expression

$$\left(\frac{d^l}{d\lambda^l} \int_{S^{n-1}} \Phi(\lambda, \omega)(a, \omega)^k\, d\omega\right)_{\lambda=0},$$

where $l, k \in \boldsymbol{Z}^+$, $l < k$. Using $\phi \in \mathcal{D}_H(\boldsymbol{P}^n)$ we see this expression vanish. Now apply Theorem 2.10 of the Introduction to Φ.

B.2. (i, ii) As in Introduction, §3, let S_{km} $[1 \leq m \leq d(k)]$ be an orthonormal basis of the eigenspace E_k of $L = L_{S^{n-1}}$. Then we have the following result.

Theorem. *The range $\mathcal{D}(\boldsymbol{R}^n)^\wedge$ consists of the functions $\psi \in \mathcal{D}(\boldsymbol{P}^n)$ which when expanded,*

$$\psi(\omega, p) = \sum_{k \geq 0,\, 1 \leq m \leq d(k)} \psi_{km}(p) S_{km}(\omega),$$

have the property that, for each $k \in \boldsymbol{Z}^+$, $1 \leq m \leq d(k)$,

$$\psi_{km}(p) = \frac{d^k}{dp^k} \phi_{km}(p)$$

where $\phi_{km} \in \mathcal{D}(\boldsymbol{R})$ is even.

By Theorem 2.10, $\mathcal{D}(\boldsymbol{R}^n)^\wedge$ is characterized by

$$\int_{\boldsymbol{R}} \psi_{km}(p) p^l\, dp = 0 \qquad \text{for} \quad l < k,$$

which (by Fourier transform) is equivalent to the description above. [The evenness of ϕ_{km} comes from the property $\psi(-\omega, -p) = \psi(\omega, p)$.]

B.3. Let $T \in \mathscr{E}'(\boldsymbol{P}^n)$ such that $\check{T} = 0$. Then $T(\hat{f}) = 0$ for each $f \in \mathscr{D}(\boldsymbol{R}^n)$. With notation from B2 define $T_{km} \in \mathscr{E}'(\boldsymbol{R})$ by

$$T_{km}(\phi) = \int \phi(p) S_{km}(\omega)\, dT(p, \omega).$$

Since the expansion for ψ (in B2) converges in $\mathscr{E}(\boldsymbol{P}^n)$ (Introduction, Theorem 3.4 or Chapter V, Corollary 3.4),

$$T(\psi) = \sum_{k,m} T_{km}(\psi_{km}).$$

Thus by B2, $d^k/dp^k(T_{km})$ annihilates all even functions in $\mathscr{D}(\boldsymbol{R})$. On the other hand, it annihilates all odd functions because

$$\int \phi(-p)\, dT_{km}(p) = (-1)^k \int \phi(p)\, dT_{km}(p).$$

Thus $d^k/dp^k(T_{km}) = 0$, so, since T_{km} has compact support, $T_{km} = 0$.

B.4. Because of Eq. (46) of §2 we have

$$\mathscr{N} = \left\{ \phi \in \mathscr{E}(\boldsymbol{P}^n) : \int \hat{f}(\xi)\phi(\xi)\, d\xi = 0 \quad \text{for} \quad f \in \mathscr{D}(\boldsymbol{R}^n) \right\};$$

in other words, $\mathscr{N}$ equals the annihilator $(\mathscr{D}^\wedge)^\perp$ of $\mathscr{D}(\boldsymbol{R}^n)^\wedge$ in $\mathscr{E}(\boldsymbol{P}^n)$. But then (since $d\xi = d\omega\, dp$) Theorem 2.10 implies that $\mathscr{N}$ contains each space $E_k \otimes p^l$ if $k - l > 0$ and even. Denoting now

$$\langle \phi, \psi \rangle = \int_{\boldsymbol{P}^n} \phi(\xi)\psi(\xi)\, d\xi,$$

we shall prove for $\phi \in \mathscr{D}(\boldsymbol{P}^n)$

(∗) $\phi \in \mathscr{D}(\boldsymbol{R}^n)^\wedge \Leftrightarrow \langle \phi, E_k \otimes p^l \rangle = 0 \quad$ for $\quad k - l > 0$, even.

It suffices to verify the implication $\Leftarrow$. For this we expand

$$\phi(\omega, p) = \sum_{r \geq 0} \sum_{1 \leq m \leq d(r)} \phi_{rm}(p) S_{rm}(\omega).$$

The condition $\langle \phi, E_k \otimes p^l \rangle = 0$ then implies

$$\int_{\boldsymbol{R}} \phi_{km}(p) p^l\, dp = 0, \qquad k - l > 0, \text{ even}.$$

But since $\phi_{km}(-p) = (-1)^k \phi_{km}(p)$ this integral vanishes for $k - l$ odd. Hence $\phi \in \mathscr{D}(\boldsymbol{R}^n)^\wedge$ as stated (cf. B2).

Now suppose $\psi \in \mathscr{N}$ and decompose

$$(**) \qquad \psi = \sum_{k \geq 0} \psi_k$$

where (cf. B2)

$$\psi_k(\omega, p) = \sum_{1 \le m \le d(k)} \psi_{km}(p) S_{km}(\omega).$$

The decomposition $\psi = \sum \psi_k$ is the one given by Corollary 3.4 of Chapter V, so the various ψ_k transform under inequivalent representations of $\boldsymbol{O}(n)$. Thus the condition $\check{\psi} = 0$ implies $\check{\psi}_k = 0$ for each k.

In analogy with (**) we decompose each $f \in \mathscr{D}(\boldsymbol{R}^n)$

$$f = \sum_{k \ge 0} f_k$$

[Corollary 3.4, Chapter V or Introduction, Eq. (19), §3.] Then by (*) we have for each k and each $\phi \in \mathscr{D}(\boldsymbol{P}^n)_k$

$$\phi \in (\mathscr{D}(\boldsymbol{R}^n)_k)^\wedge \Leftrightarrow \langle \phi, E_k \otimes p^l \rangle = 0, \qquad k - l > 0, \text{ even}.$$

For a fixed k the spaces $E_k \otimes p^l$ ($l < k$, $k - l$ even) span a finite-dimensional space $\tilde{E}_k$ whose annihilator $(\tilde{E}_k)^\perp$ in $\mathscr{D}(\boldsymbol{P}^n)_k$ is $(\mathscr{D}(\boldsymbol{R}^n)_k)^\wedge$. But $\check{\psi}_k = 0$ implies that ψ_k lies in the double annihilator $((\tilde{E}_k)^\perp)^\perp$, which by the finite dimensionality equals $\tilde{E}_k$. Since (**) converges in the topology of $\mathscr{E}(\boldsymbol{P}^n)$ (Corollary 3.4, Chapter V), ψ lies in the closed subspace generated by the spaces $E_k \otimes p^l$.

B.5. (Sketch) It is easy to prove by approximation that each Σ in the image $(\mathscr{E}')^\wedge$ has the property stated. Conversely, suppose $\Sigma \in \mathscr{E}'(\boldsymbol{P}^n)$ has the stated property. Because of the Remark following Theorem 2.17 we define $S \in \mathscr{D}'(\boldsymbol{R}^n)$ by

$$S(f) = \Sigma(\Lambda \hat{f}).$$

Let dg be a Haar measure on the isometry group $G = \boldsymbol{M}(n)$. If $g \in G$ let S^g denote the image of S under g [i.e., $S^g(f) = \int f(g \cdot x)\, dS(x)$] and define Σ^g similarly. If $F \in \mathscr{D}(G)$ the distribution

$$f \to \int_G F(g) S^g(f) dg, \qquad f \in \mathscr{D}(\boldsymbol{R}^n),$$

is given by a function s on $\boldsymbol{R}^n$. Also, the distribution

$$\phi \to \int_G F(g) \Sigma^g(\phi)\, dg, \qquad \phi \in \mathscr{D}(\boldsymbol{P}^n),$$

is given by a compactly supported function σ on $\boldsymbol{P}^n$. Since Σ vanishes on the kernel $\mathscr{N}$ from B4 we have $\sigma(\mathscr{N}) = 0$ so by Theorem 2.10 $\sigma = \hat{h}$ for some $h \in \mathscr{D}(\boldsymbol{R}^n)$. The Remark following Theorem 2.17 now shows $s(x) = c\, h(x)$ so $s \in \mathscr{D}(\boldsymbol{R}^n)$. Hence, by Exercise E2, Chapter II, $S \in \mathscr{E}'(\boldsymbol{R}^n)$. Since $\hat{s} = c\, \sigma$ a suitable limit argument shows $\hat{S} = c\, \Sigma$ as desired.

A related result has been proved by Ambrose (unpublished) and a generalization by Hertle [1983].

C. Spaces of Constant Curvature

C.1. Let $I: Q_{-1} \to \mathbf{R}^{n+1}$ denote the identity mapping. We shall prove

$$I^*\Phi^*(ds^2) = dx_1^2 + \cdots + dx_n^2 - dx_{n+1}^2,$$
$$I^*\Psi^*(d\sigma^2) = dx_1^2 + \cdots + dx_n^2 - dx_{n+1}^2,$$

which is the substance of C1. But by [DS] (Chapter I, §3. No. 3)

$$\Phi^*(dy_i) = (x_{n+1} + 1)^{-1} dx_i - x_i(x_{n+1} + 1)^{-2}\, dx_{n+1},$$

so

$$\Phi^*(ds^2) = \sum_{i=1}^{n} dx_i^2 + (x_{n+1} + 1)^{-2} \sum_{1}^{n} x_i^2\, dx_{n+1}^2$$
$$- 2(x_{n+1} + 1)^{-1}\left(\sum_{1}^{n} x_i\, dx_i\right) dx_{n+1}.$$

But on Q_{-1} we have

$$\sum_{1}^{n} x_i^2 = x_{n+1}^2 - 1, \qquad \sum_{1}^{n} x_i\, dx_i = x_{n+1}\, dx_{n+1},$$

so

$$I^*(\Phi^*(ds^2)) = \sum_{1}^{n} dx_i^2 + \frac{x_{n+1} - 1}{x_{n+1} + 1}\, dx_{n+1}^2 - \frac{2x_{n+1}}{x_{n+1} + 1}\, dx_{n+1}^2$$
$$= \sum_{1}^{n} dx_i^2 - dx_{n+1}^2.$$

Also

$$\Psi^*(dz_i) = (x_{n+1} - x_n)^{-1}\, dx_i - x_i(x_{n+1} - x_n)^{-2}(dx_{n+1} - dx_n) \qquad (1 \le i < n),$$
$$\Psi^*(dz_n) = -(x_{n+1} - x_n)^{-2}(dx_{n+1} - dx_n).$$

But since

$$I^*\left(\sum_{1}^{n-1} x_i\, dx_i + x_n\, dx_n - x_{n+1}\, dx_{n+1}\right) = 0$$

a computation similar to that made before gives

$$I^*(\Psi^*(d\sigma^2)) = dx_1^2 + \cdots + dx_n^2 - dx_{n+1}^2.$$

C.2. We may use as the hyperbolic plane the unit disk $|z| < 1$ with the metric

$$ds^2 = 4\,\frac{dx^2 + dy^2}{(1 - x^2 - y^2)^2},$$

which gives curvature -1 and the distance

$$d(z_1, z_2) = \log \frac{1 + \delta(z_1, z_2)}{1 - \delta(z_1, z_2)}, \qquad \delta(z_1, z_2) = \frac{|z_2 - z_1|}{|1 - \bar{z}_1 z_2|},$$

[Introduction, §1, Eq. (6)]. This means that

$$\delta = \tanh(d/2)$$

It suffices to verify (2) for a triangle $ABC = oz_1z_2$ with z_1 on the x axis. Then $A = o$, $B = \tanh(c/2)$, $C = \tanh(b/2)e^{iA}$, so

$$\tanh \frac{a}{2} = \delta(B, C) = \frac{|\tanh(b/2)e^{iA} - \tanh(c/2)|}{|1 - \tanh(b/2)e^{iA} \tanh(c/2)|}.$$

Now (2) follows by a simple computation using the formulas

$$\cosh a = \frac{1 + \tanh(a/2)}{1 - \tanh^2(a/2)}, \qquad \sinh a = \frac{2 \tanh(a/2)}{1 - \tanh^2(a/2)}.$$

Formula (1) can be derived directly from (2).

For curvature $-\varepsilon^2$ we use the model D_ε:

$$ds^2 = 4 \frac{dx^2 + dy^2}{(1 - \varepsilon^2(x^2 + y^2))^2} \qquad |z| < \varepsilon^{-1}.$$

Then we find

$$\varepsilon|z| = \tanh[\tfrac{1}{2}\varepsilon d(o, z)].$$

We consider again a triangle $ABC = oz_1z_2$ with z_1 on the x axis. Then

$$A = o, \qquad B = \frac{1}{\varepsilon} \tanh\left(\frac{\varepsilon c}{2}\right), \qquad C = \frac{1}{\varepsilon} \tanh\left(\frac{\varepsilon b}{2}\right) e^{iA}.$$

The mapping

$$z \to (z - z_1)/(1 - \varepsilon^2 \bar{z}_1 z)$$

maps the disk $|z| < \varepsilon^{-1}$ isometrically onto itself so

$$a = d(z_1, z_2) = d\left(o, \frac{z_2 - z_1}{1 - \varepsilon^2 \bar{z}_1 z_2}\right)$$

and

$$\varepsilon(|z_2 - z_1|)/|1 - \varepsilon^2 \bar{z}_1 z_2| = \tanh(\varepsilon a/2).$$

This equation can be written

$$\tanh\left(\frac{\varepsilon a}{2}\right) = \frac{|\tanh(\varepsilon b/2)e^{iA} - \tanh(\varepsilon c/2)|}{|1 - \tanh(\varepsilon b/2)e^{iA} \tanh(\varepsilon c/2)|}.$$

Thus (1) and (2) hold for D_ε if we replace a, b, and c by εa, εb, and εc. Letting $\varepsilon \to 0$, D_ε formally tends to the flat plane and formulas (1) and (2) converge to the standard trigonometric formulas

$$\frac{a}{\sin A} = \frac{b}{\sin B} = \frac{c}{\sin C},$$

$$a^2 = b^2 + c^2 - 2bc \cos A.$$

D. Some Results in Analysis

D.1. As in Introduction §3 consider the bilinear form

$$\langle p, q \rangle = \left[p\left(\frac{\partial}{\partial x_1}, \ldots, \frac{\partial}{\partial x_n}\right) q \right](0)$$

on the space P_k of homogeneous polynomials of degree k. It is nondegenerate and the annihilator of the set of polynomials $x \to (x, \omega)^k$ as ω runs through S^{n-1} is 0.

D.3. It is obvious that

$$\lim_{r=0} r \int_0^\infty F(re^t, re^{-t})e^{-t}\, dt = 0$$

and if $r > 0$

$$r \int_0^\infty F(re^t, re^{-t})e^t\, dt = \int_0^\infty F(\tau, r^2\tau^{-1})\, d\tau \to \int_0^\infty F(\tau, 0)\, d\tau.$$

D.4. Let $\alpha \to 0$ in the formula $L(\alpha^{-1}(r^\alpha - 1)) = \alpha r^{\alpha - 2}$.
Note, however, that while

$$L((\log r)r^\alpha) = 2\alpha r^{\alpha-2} + \alpha^2(\log r)r^{\alpha-2},$$

multiplication by $\log r$ is not continuous in $\mathscr{S}'$ so we cannot state $\lim_{\alpha \to 0} \alpha^2(\log r)r^{\alpha-2} = 0$ (remark by R. Melrose).

D.5. Let $a = \limsup_{t \to +\infty} |f(t)|$; if $\varepsilon > 0$ then $|f(t)| \le a + \varepsilon$ for all sufficiently large t so

$$\lim_{T \to +\infty} \frac{1}{T} \int_0^T |f(t)|^2\, dt \le (a + \varepsilon)^2.$$

Now (*) follows. Next let $l_j = \operatorname{Re} k_j$. If $l_j < 0$ then $\lim_{t \to +\infty} p_j(t)e^{k_j t} = 0$. Suppose $l_j \ge 0$ for $1 \le j \le s$, $l_j < 0$ for $j > s$. Then

$$\limsup_{t \to +\infty} \left| p_0(t) + \sum_{j=1}^{s} p_j(t)e^{k_j t} \right| \le a.$$

Thus we may assume $s = r$. We first claim $l_1 = l_2 = \cdots = l_r = 0$. Otherwise let $l = \max(l_1, \ldots, l_r) > 0$ and suppose $l_1 = \cdots = l_i = l$, $l_j < l$ for $j > i$. Let n be the maximal degree among $p_1, \ldots, p_i$ and put

$$c'_j = \lim_{t \to +\infty} p_j(t)/t^n.$$

Then

$$\begin{aligned} 0 &= \limsup_{t \to +\infty} t^{-n} e^{-lt} |p_0(t) + p_1(t)e^{k_1 t} + \cdots + p_r(t)e^{k_r t}| \\ &= \limsup_{t \to +\infty} |c'_1 e^{(k_1 - l)t} + \cdots + c'_i e^{(k_i - l)t}|, \end{aligned}$$

contradicting (*). Thus $l = 0$.

Next let m be the maximal degree among $p_0, \ldots, p_r$ and put $c_j = \lim_{t \to +\infty} p_j(t)/t^m$. If $m > 0$ our assumption implies

$$\lim_{t \to +\infty} t^{-m} \left| p_0(t) + \sum_{1 \le j \le r} p_j(t) e^{k_j t} \right| = 0,$$

whence

$$\lim_{t \to +\infty} \left| c_0 + \sum_{1 \le j \le r} c_j e^{k_j t} \right| = 0,$$

contradicting (*). Hence $m = 0$, so each p_j is a constant and (*) implies $|p_0| \le a$.

CHAPTER II

A. The Laplace-Beltrami Operator

A.1. The differential equations for geodesics {[DS], Chapter I, §5, Eq. (3)} show that $\Gamma^k_{ij}(p) = 0$. Now use Prop. 2.6.

A.2. The results are local so we may assume M has an orthonormal basis $X_1, \ldots, X_n$ of the vector fields. Let $\omega^1, \ldots, \omega^n$ be the dual basis of 1-forms. Then $\omega^k = \omega_{X_k}$ and $\omega^1 \wedge \cdots \wedge \omega^n$ is the volume element. The Lie derivative $\theta(X_k)$ satisfies

$$\theta(X_k)\omega^i = \sum_j (\Gamma^i_{jk} - \Gamma^i_{kj})\omega^j$$

(use [DS], pp. 540, 44–45), whence by Lemma 2.2,

$$\operatorname{div}(X_k) = \sum_i (\Gamma^i_{ik} - \Gamma^i_{ki}).$$

On the other hand, $*\omega^1 = \omega^2 \wedge \cdots \wedge \omega^n$, etc., and $\delta\omega^k$ can be calculated by means of the first structural equation ([DS], Chapter I, Theorem 8.1) in terms of the Γ^i_{jk}. One finds $\delta\omega^k = -\operatorname{div} X_k$ and now $\delta(\omega_X) = -\operatorname{div} X$ follows by using Eq. (14) of §2.

Finally, if $f \in \mathscr{E}(M)$ we have

$$\operatorname{grad} f = \sum_i (X_i f) X_i, \qquad df = \sum_i (X_i f)\omega_i,$$

$$\operatorname{div}(\operatorname{grad} f) = \sum_i X_i^2 f + \sum_i (X_i f) \operatorname{div} X_i,$$

$$\begin{aligned} \Delta f &= -\delta\, df = -\delta\Big(\sum_i (X_i f)\omega^i\Big) \\ &= *\, d\Big[\Big(\sum_i X_i f\Big) * \omega^i\Big] \\ &= \sum_i (X_i f) * d * \omega^i + * d\Big(\sum_i X_i f\Big) \wedge * \omega^i \\ &= \sum_i (X_i f) \operatorname{div} X_i + * \sum_{i,j} (X_i X_j f)\omega^j \wedge * \omega^i \\ &= \operatorname{div} \operatorname{grad} f. \end{aligned}$$

A.3. The space G/H is in each case symmetric, hence reductive, so $\boldsymbol{D}(G/H)$ is determined by the algebra $I(\mathfrak{m})$ of $\operatorname{Ad}_G(H)$-invariants on $\mathfrak{m}$ (cf. Theorem 6.1 in Chapter I). The space G/H being isotropic, each $q \in I(\mathfrak{m})$ is constant on each "sphere" $\{X \neq 0 \text{ in } \mathfrak{m} : g_e(X, X) = c\}$ in $\mathfrak{m}$; hence $q(X) = F(g_e(X, X))$ where F is a function. But then F is a polynomial; now use Corollary 4.10.

A.4. (i) We have $g^{ij} = g^{ji}$ so by Theorem 4.3

$$\begin{aligned} &((g^{ij}\tilde{X}_i\tilde{X}_j + g^{ji}\tilde{X}_j\tilde{X}_i)f)(e) \\ &\quad = 2(g^{ij}\lambda(X_iX_j)f)(e) = 2g^{ij}\left(\frac{\partial^2}{\partial t_i\, \partial t_j} f(\exp(t_iX_i + t_jX_j))\right)_{t=0} \end{aligned}$$

For the pseudo-Riemannian structure given by B the one-parameter subgroups are the geodesics through e; the Christoffel symbols therefore vanish at e for the canonical coordinates. Thus by Proposition 2.6, Ω coincides with the Laplace-Beltrami operator L_G at e, hence everywhere by their invariance. Now $\Omega \in \boldsymbol{Z}(G)$ by Corollary 4.5.

(ii) Choosing the basis (X_i) compatible with the Cartan decomposition $\mathfrak{g} = \mathfrak{k} + \mathfrak{p}$ the argument above shows that $\mu(\Omega)$ and $L_{G/K}$ coincide at the origin $\{K\}$, hence everywhere by the invariance.

A.5. Using notation from A2 we have $g = \sum_i \omega^i \otimes \omega^i$ so (i) follows from the proof of A2. Formula (1) follows from $L^{\exp tX} = h(\exp tX)L$ by differentiation.

(iii) Let M have pseudo-Riemannian structure g and put $h = \tau^2 g$. Then $h_{ij} = \tau^2 g_{ij}$, $h^{ij} = \tau^{-2} g^{ij}$, $\bar{h} = \tau^{2n}\bar{g}$, and by (22) §2 the new Christoffel symbols are

$$\Gamma^{*k}_{ij} = \Gamma^k_{ij} + \delta^k_i \partial_j(\log \tau) + \delta^k_j \partial_i(\log \tau) - g_{ij} \sum_l g^{lk} \partial_l(\log \tau) \tag{1}$$

Thus by Proposition 2.6 the new Laplacian L^* is given by

$$L^* f = \tau^{-2}[Lf + (n-2)(\operatorname{grad} f)(\log \tau)],$$

which by (17) §2 can also be written

$$L^* f = \tau^{-n/2-1}[L(\tau^{n/2-1} f) - L(\tau^{n/2-1}) f]. \tag{2}$$

Now we need the identity

$$-L(\tau^{n/2-1}) = c_n(K^* \tau^{n/2+1} - K\tau^{n/2-1}), \tag{3}$$

where K^* is the scalar curvature for the Riemannian structure h (see Yamabe [1960], where the scalar curvature has the same sign). The desired result now follows from (2) and (3) because, L and K being invariant under isometries, we can take T as the identity mapping.

This quasi-invariance of L occurs several places in the literatures; see, e.g., Friedlander [1975], Ørsted [1981b], and also Kosman [1975]. To a large extent, it goes back to Cotton [1900].

A.7. Let U be an arbitrary coordinate neighborhood on M and let μ be given on U by

$$d\mu = \delta(x)\, dx_1 \cdots dx_n = \delta(x)\, dx.$$

Then the equations

$$\int (Xf)(x) g(x) \delta(x)\, dx = \int f(x)(Xg)(x)\delta(x)\, dx, \qquad f \in \mathcal{D}(U), \quad g \in \mathcal{E}(U),$$

$$\int [(Xf)g + f(Xg)]\delta(x)\, dx = \int X(fg)1\delta(x)\, dx = \int fg \cdot 0\delta(x)\, dx = 0$$

imply

$$\int (Xf)(x) g(x) \delta(x)\, dx = 0$$

for all f, g, whence $X = 0$.

A.8. (Harish-Chandra [1960]) During the proof of Proposition 5.23 we saw

$$Y_\beta = (\coth c)Z_\beta - (\sinh c)^{-1}Z_\beta^{a^{-1}}, \qquad c = \beta(H),$$

so

$$X_\beta = (\sinh c)^{-1}\{e^c Z_\beta - Z_\beta^{a^{-1}}\}.$$

Hence

$$-X_\beta X_{-\beta} = (\sinh c)^{-2}\{Z_\beta Z_{-\beta} + (Z_\beta Z_{-\beta})^{a^{-1}} - e^c Z_\beta Z_{-\beta}^{a^{-1}} - e^{-c} Z_\beta^{a^{-1}} Z_{-\beta}\}.$$

Using

$$\begin{aligned} Z_\beta Z_{-\beta}^{a^{-1}} &= Z_{-\beta}^{a^{-1}} Z_\beta + [Z_\beta, Z_{-\beta}^{a^{-1}}] \\ &= Z_{-\beta}^{a^{-1}} Z_\beta + (\cosh c)[Z_\beta, Z_{-\beta}] + (\sinh c)[Z_\beta, Y_{-\beta}] \end{aligned}$$

we obtain

$$\begin{aligned} -X_\beta X_{-\beta} = (\sinh c)^{-2}\{&Z_\beta Z_{-\beta} + (Z_\beta Z_{-\beta})^{a^{-1}} - e^c Z_{-\beta}^{a^{-1}} Z_\beta - e^{-c} Z_\beta^{a^{-1}} Z_{-\beta} \\ &- \tfrac{1}{2}(e^{2c} + 1)[Z_\beta, Z_{-\beta}] - \tfrac{1}{2}(e^{2c} - 1)[Z_\beta, Y_{-\beta}]\} \end{aligned}$$

Here we apply θ and note that $\theta(Z_\beta^{a^{-1}}) = Z_\beta^a$, etc. This gives a formula for $\theta(X_\beta X_{-\beta})$ in which we replace β by $-\beta$ and a^{-1} by a. Adding, we obtain

$$\begin{aligned} &-(X_\beta X_{-\beta} + \theta(X_{-\beta} X_\beta)) \\ &\quad = (\sinh c)^{-2}\{Z_\beta Z_{-\beta} + Z_{-\beta} Z_\beta + (Z_\beta Z_{-\beta} + Z_{-\beta} Z_\beta)^{a^{-1}} \\ &\quad\quad - 2e^c Z_{-\beta}^{a^{-1}} Z_\beta - 2e^{-c} Z_\beta^{a^{-1}} Z_{-\beta}\} \\ &\quad\quad + e^c(\sinh c)^{-1}\{[Z_{-\beta}, Y_\beta] - [Z_\beta, Y_{-\beta}]\}. \end{aligned}$$

Put $\Omega_+ = \Sigma_{\beta \in P_+}(X_\beta X_{-\beta} + X_{-\beta} X_\beta)$. Then $\theta\Omega_+ = \Omega_+$ so that $\Omega_+ = \frac{1}{2}(\Omega_+ + \theta\Omega_+)$. Also, $[Z_\beta, Y_{-\beta}] + [Y_\beta, Z_{-\beta}] = A_\beta$ so that the formula above gives

$$\begin{aligned} \Omega_+ = &\sum_{\beta \in P_+} \coth c A_\beta \\ &- \sum_{\beta \in P_+} (\sinh c)^{-2}\{Z_\beta Z_{-\beta} + Z_{-\beta} Z_\beta + (Z_\beta Z_{-\beta} + Z_{-\beta} Z_\beta)^{a^{-1}} \\ &- 2e^c Z_{-\beta}^{a^{-1}} Z_\beta - 2e^{-c} Z_\beta^{a^{-1}} Z_{-\beta}\} \end{aligned}$$

Since the mapping $\beta \to -\beta^\theta = \beta^*$ is a permutation of P_+ we have $X_{\beta^*} = c_\beta \theta X_{-\beta}$, $X_{-\beta^*} = c_{-\beta}\theta X_\beta$ where $c_\beta \in \boldsymbol{C}$. Since θ preserves B and since $B(X_\beta, X_{-\beta}) = 1$, we have $c_\beta c_{-\beta} = 1$. Hence

$$Z_{\beta^*} Z_{-\beta^*} = c_\beta c_{-\beta} Z_{-\beta} Z_\beta = Z_{-\beta} Z_\beta.$$

This gives the desired formula. See also Anderson [1979].

B. The Radial Part

B.1. Let $v_0 \in V$ be such that the orbit $H \cdot v_0$ has maximal dimension. Then $\dim(H \cdot v) = \dim(H \cdot v_0)$ for v in a neighborhood N_0 of v_0 in V. According to Chevalley [1946] (Chapter III, §7, Theorem 1) there exists a coordinate system $\{x_1, \ldots, x_n\}$ near v_0 in V such that

$$x_i(v_0) = 0 \quad (1 \le i \le n)$$

and such that each slice $x_{r+1} = a_{r+1}, \ldots, x_n = a_n$ is a part of an orbit $H \cdot v$. Then the submanifold $x_1 = 0, \ldots, x_r = 0$ can serve as W (remark by R. Palais).

B.2. This is clear from (33′) because here $\delta(r) = A(r)$.

B.3. By (i) the orbit $H \cdot w$ is locally compact and thus homeomorphic to H/H^w. By (ii), H^v is compact for each $v \in V^*$ so no change is required in the proof of Theorem 3.7.

B.4. The right-hand side of the formula is independent of the choice of basis (H_i). But with (H_i) orthonormal the computation is the same as in the proof of Proposition 3.9.

C. Invariant Differential Operators

C.1. The group $G = \mathbf{GL}(n, \mathbf{R})$ acts on the space $\mathbf{P}_n$ of $n \times n$ symmetric positive definite matrices by

$$p \to gp{}^tg, \qquad \text{if} \quad p \in \mathbf{P}_n, \quad g \in G, \quad {}^tg = \text{transpose of } g,$$

and the isotropy group of I is $K = \mathbf{O}(n)$. For the Lie algebras we have $\mathfrak{g} = \mathfrak{k} + \mathfrak{p}$ and $\mathfrak{p}$ is the space of all $n \times n$ real symmetric matrices X. If $k \in K$, then $\mathrm{Ad}_G(k)|\mathfrak{p}$ is given by $X \to kXk^{-1}$. The algebra $I(\mathfrak{p})$ of invariant polynomial functions is generated by

$$\mathrm{Tr}(X),\ \mathrm{Tr}(XX),\ \ldots,\ \mathrm{Tr}(X^n).$$

This boils down to the fact that the symmetric polynomial in n variables $\lambda_1, \ldots, \lambda_n$ (the eigenvalues of X) is a polynomial in

$$\lambda_1 + \cdots + \lambda_n, \ldots, \lambda_1^n + \cdots + \lambda_n^n.$$

By Corollary 4.10 the algebra $\mathbf{D}(G/K)$ is generated by the corresponding operators D_i $(1 \le i \le n)$. With the inner product $\mathrm{Tr}(XY)$ on $\mathfrak{p}$ the matrices $e_i = E_{ii}$, $1 \le i \le n$, $e_{ij} = 2^{-1/2}(E_{ij} + E_{ji})$ $(i > j)$ form an orthonormal basis. Writing the matrix $X = (x_{ij})$ in the form $X = \sum \xi_{ii} e_i + \sum \xi_{ij} e_{ij}$ we have $\xi_{ii} = x_{ii}$, $\xi_{ij} = 2^{1/2} x_{ij}$. If

$$\mathrm{Tr}(X^k) = P_k(\xi_{pq})$$

we have to determine $P_k(\partial/\partial\xi_{pq})$. But for $a = 2^{-1/2}$

$$X = \begin{pmatrix} \xi_{11}, \ldots, a\xi_{n1} \\ \vdots \quad\quad \vdots \\ a\xi_{n1}, \ldots, \xi_{nn} \end{pmatrix}$$

and we have

$$\begin{pmatrix} \dfrac{\partial}{\partial\xi_{11}}, \ldots, a\dfrac{\partial}{\partial\xi_{n1}} \\ \vdots \quad\quad \vdots \\ a\dfrac{\partial}{\partial\xi_{n1}}, \ldots, \dfrac{\partial}{\partial\xi_{nn}} \end{pmatrix} = \frac{\partial}{\partial X}$$

the symmetric matrix with entries $\frac{1}{2}(1 + \delta_{ij})\,\partial/\partial x_{ij}$. Since $\operatorname{Tr}(X^k) = P_k(\xi_{pq})$ we have

$$\operatorname{Tr}\left(\left(\frac{\partial}{\partial X}\right)^k\right) = P_k\left(\frac{\partial}{\partial\xi_{pq}}\right).$$

Thus the operators

$$(D_k f)(g \cdot 0) = \left[\operatorname{Tr}\left(\left(\frac{\partial}{\partial X}\right)^k\right) f(g \exp X \cdot o)\right]_{X=0}$$

generate $\boldsymbol{D}(G/K)$.

A different description of $\boldsymbol{D}(G/K)$ is given Maass [1955] and Selberg [1956].

C.3. (i) Since $\operatorname{Ad}_G(m)$ fixes H each $P \in I(\mathfrak{a} + \mathfrak{l} + \mathfrak{q})$ can be written

$$P(H, T, X) = \sum_k H^k P_k(T, X).$$

But each $P_k(T, X)$ is invariant under the action

$$(T, X) \to (mT, mX), \qquad m \in M = \boldsymbol{SO}(n-1).$$

By a standard result about invariants (cf. Weyl [1939], p. 31) P_k is a polynomial in $|T|^2$, $T \cdot X$, and $|X|^2$. This proves (i).

(ii) We have the commutation relations $X_i = [T_i, H]$, $[X_i, T_j] = \delta_{ij}H$, $[X_i, H] = T_i$ {[DS], Eq. (17) p. 407}. The symmetrization λ satisfies

$$\lambda(H\,|\,T\,|^2) = \frac{1}{6}\sum_{i=1}^{n-1} 2(T_i T_i H + T_i H T_i + H T_i T_i).$$

Using the commutation relations this can be written

$$\sum_{i=1}^{n-1} T_i T_i H - \frac{1}{2}\sum_{i=1}^{n-1}(X_i T_i + T_i X_i) + \frac{n-1}{6} H$$

and also as

$$\sum_{i=1}^{n-1} H T_i T_i + \frac{1}{2}\sum_{i=1}^{n-1}(X_i T_i + T_i X_i) + \frac{n-1}{6} H.$$

By subtraction, $[D_H, D_{|T|^2}] = -2D_{T \cdot X}$; the other formulas from (ii) follow similarly.

C.4. In the notation of Theorem 5.16 we have

$$D\phi_\nu = \gamma(D)(i\nu)\phi_\nu \qquad D \in \mathbf{D}_K(G).$$

Let $s \in W$, let $m_s \in M'$ be a representative and put $N^s = m_s N m_s^{-1}$. Then

$$g = n \exp A^s(g)k, \qquad k \in K,\ A^s(g) \in \mathfrak{a},\ n \in N^s.$$

Let $\gamma^s : \mathbf{D}_K(G) \to \mathbf{D}_W(A)$ be the homomorphism defined by using the Iwasawa decomposition $G = N^s AK$ in place of $G = NAK$ and put

$$\phi_\nu^s(g) = \int_K e^{(i\nu + s\rho)(A^s(kg))}\, dk.$$

Then by Eq. (38′) of §5,

$$D\phi_\nu^s = \gamma^s(D)(i\nu)\phi_\nu^s.$$

Comparing the two Iwasawa decompositions we see that

$$A^s(g) = sA(m_s^{-1}g).$$

As a result $\phi_\nu^s = \phi_\nu$, so $\gamma^s = \gamma$.

The result is given in Schiffmann [1979] with a different proof.

D. Restriction Theorems

D.4. For $R > 0$ select C_R such that

$$|\phi(H) - \phi(H')| \leq C_R |H - H'|$$

for all H, H' in the ball $|H| < R$ in $\mathfrak{a}$. The K-invariant extension Φ satisfies such an inequality on each maximal abelian subspace $\mathfrak{b}$ of $\mathfrak{p}$. Given $X, Y \in \mathfrak{p}$ the distance from X to the orbit $\mathrm{Ad}(K)Y$ is minimized at a point $\mathrm{Ad}(k_0)Y$ for which X and $\mathrm{Ad}(k_0)Y$ commute (Prop. 5.18, Chapter I) and so lie in the same $\mathfrak{b}$. Thus $|\Phi(X) - \Phi(Y)| \leq C_R|X - Y|$ if both $|X|$, $|Y| < R$. (cf. Helgason [1980a], Proposition 2.4).

E. Distributions

E.1. By definition

$$\int_G (f \times T)(g \cdot o)F(g \cdot 0)\, dg = (\check{f} * \tilde{T})(\tilde{F})$$

$$= \int_G \int_G \tilde{F}(gh)\tilde{f}(g)\, dg\, d\tilde{T}(h)$$

$$= \int_G \tilde{F}(g)\left(\int_G f(gh^{-1} \cdot o)\, d\tilde{T}(h)\right) dg,$$

proving the first relation. The second follows in the same way.

E.3. If $F \in \mathcal{D}(G)$ then

$$(s * (D^*)\check{}\, t)(F) = \int\left(\int F(xy)\, ds(x)\right) d((D^*)\check{}\, t)(y)$$

$$= \iint F(xy^{-1})\, ds(x)\, d(D^*\check{t})(y) = \int (\check{F} * s)(y)\, d(D^*\check{t})(y)$$

$$= \check{t}(\check{F} * Ds) = \iint \check{F}(yx^{-1})\, d(Ds)(x)\, d\check{t}(y)$$

$$= \iint F(xy)\, d(Ds)(x)\, dt(y) = (Ds * t)(F).$$

Use this result for $s = \tilde{S}$, $t = \tilde{T}$, $D \in \lambda(I(\mathfrak{p}))$. Then, using (13) §5 and the bi-invariance of $t^\natural$,

$$D(s * t) = D(s * \delta * t) = D(s * t^\natural) = s * Dt^\natural$$

$$= s * (D^*)^\rho t^\natural = s * (D^*)\check{}\, t^\natural = Ds * t^\natural = Ds * t.$$

F. The Wave Equation

F.1. By Eq. (63) of §5 we have for $x_0 \in \boldsymbol{R}^n$, $r \geq 0$,

$$\int_{S^r(x_0)} v(x, 0)\, d\omega(x) = \int_{S^r(0)} v(x_0, y)\, d\omega(y).$$

The function $v(x_0, y)$ is constant in y on hyperplanes perpendicular to the y_1 axis. Proceeding as with Lemma 2.30 in Chapter I we therefore obtain

$$(1) \qquad \Omega_n r^{n-1}(M^r u_0)(x) = \Omega_{n-1} r \int_{-r}^{r} (r^2 - t^2)^{(1/2)(n-3)} u(x_0, t)\, dt.$$

Case I. n odd. Divide Eq. (1) by r, differentiate $\frac{1}{2}(n-3)$ times with respect to r^2, and then apply $\partial/\partial r$. This gives

$$(2) \qquad u(x_0, r) + u(x_0, -r)$$
$$= \frac{\Omega_n}{[\frac{1}{2}(n-3)!\Omega_{n-1}} \frac{\partial}{\partial r}\left(\frac{\partial}{\partial(r^2)}\right)^{(1/2)(n-3)} [r^{n-2}(M^r u_0)(x_0)].$$

Now the function $w(x, t) = u_t(x, t)$ satisfies the wave equation with initial data $w_0 = u_1$, $w_1 = u_{tt}(x, 0) = L_{\mathbf{R}^n} u_0$. We can replace u by w, u_0 by $w_0 = u_1$ in the last formula and get by integration

$$(3) \qquad u(x_0, r) - u(x_0, -r)$$
$$= \frac{\Omega_n}{[\frac{1}{2}(n-3)]!\Omega_{n-1}} \left(\frac{\partial}{\partial(r^2)}\right)^{(1/2)(n-3)} [r^{n-2}(M^r u_1)(x_0)].$$

Adding (2) and (3) an explicit solution formula for n odd is obtained.

Case II. n even. Again divide (1) by r and differentiate $\frac{1}{2}(n-2)$ times with respect to r^2. This gives

$$\frac{n-3}{2}\frac{n-5}{2}\cdots\frac{1}{2}\int_0^r (r^2-t^2)^{-1/2}[u(x_0, t) + u(x_0, -t)]\, dt$$
$$= \frac{\Omega_n}{\Omega_{n-1}}\left(\frac{\partial}{\partial(r^2)}\right)^{(1/2)(n-2)} \{r^{n-2}(M^r u_0)(x_0)\}.$$

The integral on the left is inverted as in Eq. (19) of §2 in Chapter I. Again we can use the identity replacing u by u_t and u_0 by u_1. Here one gets the solution formula

$$u(x_0, \pm r)$$
$$= \frac{1}{[\frac{1}{2}(n-2)]!}\left[\frac{\partial}{\partial r}\int_0^r \frac{t}{(r^2-t^2)^{1/2}}\left(\frac{\partial}{\partial(t^2)}\right)^{(1/2)(n-2)} [t^{n-2}(M^t u_0)(x_0)]\, dt\right.$$
$$\left.\pm \int_0^r \frac{t}{(r^2-t^2)^{1/2}}\left(\frac{\partial}{\partial(t^2)}\right)^{(1/2)(n-2)} [t^{n-2}M^t u_1(x_0)]\, dt\right].$$

F.2. In accordance with Chapter I, Exercise C1, we use the half-space model for $\boldsymbol{H}^n$ and have

$$ds^2 = y_n^{-2}(dy_1^2 + \cdots + dy_n^2), \qquad y_n > 0.$$

Then the Laplace–Beltrami operator is given by

$$L = y_n^2\left(\frac{\partial^2}{\partial y_1^2} + \cdots + \frac{\partial^2}{\partial y_{n-1}^2}\right) + y_n^2\frac{\partial^2}{\partial y_n^2} - (n-2)y_n\frac{\partial}{\partial y_n},$$

which by the substitution $y_n = e^s$ can be written

$$L = e^{2s}\left(\frac{\partial^2}{\partial y_1^2} + \cdots + \frac{\partial^2}{\partial y_{n-1}^2}\right) + e^{(1/2)(n-1)s}\left[\frac{\partial^2}{\partial s^2} - \left(\frac{n-1}{2}\right)^2\right] \circ e^{-(1/2)(n-1)s}.$$

With $y_n = e^s$ put

$$v(x, y) = v(x_1, \ldots, x_n, y_1, \ldots, y_n) = e^{(1/2)(n-1)s}u(x_1, \ldots, x_n, s).$$

Then, using the differential equation for u,

$$L_x(v(x, y)) = e^{(1/2)(n-1)s}\left(\frac{\partial^2 u}{\partial s^2} - \left(\frac{n-1}{2}\right)^2 u\right)$$

$$= L_y(v(x, y)).$$

Now use Theorem 5.28 for the point $(x_0, y_0) = (x, o)$, where o is the origin $(0, \ldots, 0, 1)$ in $\boldsymbol{H}^n$. Since $v(x, o) = u(x, 0)$ we obtain

$$\int_{S_r(x)} u(z, 0)\, d\omega(z) = \int_{S_r(o)} v(x, y)\, d\omega(y).$$

The non-Euclidean sphere $S_r(o)$ is a Euclidean sphere with center $(0, \ldots, 0, \operatorname{ch} r)$ and radius $\operatorname{sh} r$. On the plane $y_n = e^s$ the metric is

$$e^{-2s}(dy_1^2 + \cdots + dy_{n-1}^2)$$

so the intersection with the sphere $S_r(o)$ is a non-Euclidean sphere of Euclidean radius $[2e^s(\operatorname{ch} r - \operatorname{ch} s)]^{1/2}$; its non-Euclidean $(n-2)$-dimensional area is

$$\Omega_{n-1}[2e^{-s}(\operatorname{ch} r - \operatorname{ch} s)]^{(1/2)(n-2)}$$

and the function $y \to v(x, y)$ is constant on it. The arc-element on the circle

$$y_1 = (2e^s(\operatorname{ch} r - \operatorname{ch} s))^{1/2}, \qquad y_2 = \cdots = y_{n-1} = 0, \qquad y_n = e^s$$

is

$$y_n^{-1}(dy_1^2 + dy_n^2)^{1/2} = \operatorname{sh} r[2e^s(\operatorname{ch} r - \operatorname{ch} s)]^{-\frac{1}{2}}\, ds.$$

The integral relation above thus becomes

(4)

$$\Omega_n \operatorname{sh}^{n-1} r(M^r u_0)(x) = \Omega_{n-1} \operatorname{sh} r \int_{-r}^{r} u(x, s)(2 \operatorname{ch} r - 2 \operatorname{ch} s)^{(1/2)(n-3)}\, ds.$$

The function $w(x, s) = u_s(x, s)$ satisfies the same differential equation as $u(x, s)$. This we can in the last formula replace $u(x, s)$ by $u_s(x, s)$ and $u_0(x)$ by $u_1(x)$. Using now the method of the last problem we obtain:

Case I. n odd. The solution $u(x, s)$ is given by

$$(5)\quad u(x, s) = \frac{1}{2(\frac{1}{2}(n-3))!}\frac{\Omega_n}{\Omega_{n-1}}\left[\frac{\partial}{\partial s}\left(\frac{\partial}{\partial(2\operatorname{ch} s)}\right)^{(n-3)/2}[\operatorname{sh}^{n-2}s(M^s u_0)(x)] + \left(\frac{\partial}{\partial(2\operatorname{ch} s)}\right)^{(n-3)/2}[\operatorname{sh}^{n-2}s(M^s u_1)(x)]\right].$$

In particular, Huygens's principle holds.

Case II. n even. From (4) we obtain

$$\frac{n-3}{2}\cdot\frac{n-5}{2}\cdots\frac{1}{2}\int_0^s(2\operatorname{ch} s - 2\operatorname{ch} r)^{-1/2}[u(x, r) + u(x, -r)]\,dr = \frac{\Omega_n}{\Omega_{n-1}}\left(\frac{\partial}{\partial(2\operatorname{ch} s)}\right)^{(1/2)(n-2)}[\operatorname{sh}^{n-2}s(M^s u_0)(x)].$$

This Abel-type integral equation can be solved for $u(x, s) + u(x, -s)$. Using (4) on u_s we get an analogous equation for $u(x, s) - u(x, -s)$, so a solution formula for $u(x, s)$ follows.

For relevant references see Hölder [1938] (for the pure equation $Lu - u_{ss} = 0$ with Huygens's principle absent), Günther [1957, p. 23] (giving (5) for $n = 3$, using a different method), Helgason [1964a, 1977c, 1984], Lax and Phillips [1978], Ørsted [1981b], and Kiprijanov and Ivanov [1981].

CHAPTER III

1. Since the invariants of $\boldsymbol{O}(p, q)$ are the polynomials in Q (compare Exercise A3, Chapter II) this follows from Theorem 1.2 [for $(p, q) \neq (1, 1)$]. For $(p, q) = (1, 1)$ use substitution $y_1 = x_1$, $y_2 = ix_2$ and observe that a polynomial identity $P(\cosh\zeta, \sinh\zeta) = 0$ implies $P(\cos\theta, i\sin\theta) = 0$.

2. (i) If λ is singular then K_λ has a Lie algebra larger than that of M so $K_\lambda \neq M$ and $s_\alpha\lambda = \lambda$ for a certain Weyl reflection $s_\alpha \neq e$ in W. If λ is regular then K_λ and M have the same Lie algebra. If $k \in K_\lambda$ and $\lambda = \xi + i\eta$ $(\xi, \eta \in \mathfrak{a}^*)$ then $k\xi = \xi$, $k\eta = \eta$. For each restricted root α either $\alpha(A_\xi)$ or $\alpha(A_\eta)$ is nonzero. Thus the centralizers $\mathfrak{z}_\xi$ and $\mathfrak{z}_\eta$ of ξ and η, respectively, in $\mathfrak{p}$ have intersection $\mathfrak{a}$. Since k leaves each of these centralizers invariant we have $k \in M'$, the normalizer of $\mathfrak{a}$ in K. Then the Weyl group element $s = \operatorname{Ad}(k)|\mathfrak{a}$ fixes the set $\boldsymbol{R}A_\xi + \boldsymbol{R}A_\eta$ pointwise. But

$$|\alpha(A_\xi)| + |\alpha(A_\eta)| > 0$$

for each restricted root α so this set contains a regular element. Hence $s = e$ so $k \in M$.

(ii) The statement $f \in \mathscr{E}_\lambda(\mathfrak{p})$ is proved just as was Lemma 3.14.

(iii) This follows from (ii) because each M-harmonic polynomial on $\mathfrak{p}$ is K_λ-harmonic.

3. Let $h_1, \dots, h_w$ be a homogeneous basis of H. As in Exercise D1 of Chapter II consider the normal field extension

$$\boldsymbol{C}(S)/\boldsymbol{C}(I) \qquad (S = S(\mathfrak{a}^{\boldsymbol{C}}),\ I = I(\mathfrak{a}^{\boldsymbol{C}})),$$

which has Galois group W. Because of Theorem 3.4 the h_i $(1 \leq i \leq r)$ are linearly independent over $\boldsymbol{C}(I)$. Each element of $\boldsymbol{C}(S)$ can be written as a quotient of a polynomial and an invariant polynomial. Thus the h_i form a basis of $\boldsymbol{C}(S)$ over $\boldsymbol{C}(I)$ so the representation $\sigma \to (a_{ij}(\sigma))$ of W given by

$$\sigma \cdot h_j = \sum_i a_{ij}(\sigma) h_i$$

is equivalent to the regular representation.

7. Let $\alpha_1, \dots, \alpha_l$ be the simple restricted roots; $H_1, \dots, H_l$ the dual basis of $\mathfrak{a}_0$; $a = \exp(\sum_i t_i H_i)$ any member of A. It is now easy to see that $a^2 = e \Leftrightarrow t_j \in \pi i \boldsymbol{Z}$ for each j. Thus $\{0, 1\}^l$ is in a one-to-one correspondence with F.

8. Because of the results quoted it suffices to prove $a \cdot j = j$ for $a \in F$ and $j \in I(\mathfrak{a})$. Since a acts trivially on $\mathfrak{a}$ this is obvious.

CHAPTER IV

A. Representations

A.1. (i) The proof of Corollary 1.4 gives the result.

A.2. Let π, π' be two unitary representations of G on Hilbert spaces V and V', respectively. Assume the linear homeomorphism $A: V \to V'$ satisfies $A\pi(x) \equiv \pi'(x)A$. Passing to adjoints one finds

$$A^*A\pi(x) = \pi(x)A^*A.$$

Let P be a positive bounded self-adjoint operator such that $P^{-2} = A^*A$. Then AP sets up the desired unitary equivalence (cf. Borel [1972], §5).

A.3. (i) Let $v \to \hat{v}$ be the canonical mapping of V into $(V')'$. Consider the functional $X_t: \lambda \to (d/dt)\langle x_t, \lambda\rangle$ on V', which does belong to $(V')'$ (Dunford and Schwartz [1958], Theorem II 1.18). The function $t \to X_t(\lambda)$ is C^1 for each $\lambda \in V'$. Fix $t \in I$ and an $\varepsilon > 0$ such that $t + s \in I$ for $|s| \leq \varepsilon$.

Let

$$V'_n = \{\lambda \in V' : |(1/s)\langle X_{t+s} - X_t, \lambda\rangle| \leq n \text{ for } |s| \leq \varepsilon\}.$$

Then V'_n is closed for $n = 1, 2,$ and $V' = \bigcup_n V'_n$ so by the Baire Category Theorem there exists an $m \in \mathbf{Z}^+$ such that V'_m contains a ball $B_\delta(\lambda_0)$. Thus

$$\|\lambda\| < \delta \Rightarrow |(1/s)\langle X_{t+s} - X_t, \lambda + \lambda_0\rangle| \leq m \qquad \text{for} \quad |s| \leq \varepsilon,$$

which implies that the mapping $t \to X_t$ is continuous from I to $(V')'$. Since

$$\begin{aligned} \frac{1}{h}\langle \hat{x}_{t+h} - \hat{x}_t, \lambda\rangle &= \frac{1}{h}\int_t^{t+h} \frac{d}{ds}\langle \hat{x}_s, \lambda\rangle\, ds \\ &= \frac{1}{h}\int_t^{t+h} \langle X_s, \lambda\rangle\, ds = \left\langle \frac{1}{h}\int_t^{t+h} X_s\, ds, \lambda\right\rangle \end{aligned}$$

we see that

$$\frac{1}{h}(\hat{x}_{t+h} - \hat{x}_t) - X_t = \frac{1}{h}\int_t^{t+h} (X_s - X_t)\, ds \to 0$$

in the norm on $(V')'$ as $h \to 0$. But $\hat{V}$ is closed in $(V')'$, so $X_t \in \hat{V}$. Since the injection $v \to \hat{v}$ of V in $(V')'$ is isometric there is a unique continuous mapping $t \to \xi_t$ of I into V such that $\hat{\xi}_t = X_t$. This shows that $t \to x_t$ is of class C^1.

(ii) Let $X_1, \ldots, X_n$ be a basis for the Lie algebra $\mathfrak{g}$ of G. If N_e is a suitable neighborhood of e in G the inverse of the mapping $(t_1, \ldots, t_n) \to \exp(t_1X_1) \cdots \exp(t_nX_n)$ is a local coordinate system on N_e. Assuming $x \to \langle \pi(x)v, \lambda\rangle$ C^∞ on G for each $\lambda \in V'$ we see from (i) that the map $(t_1, \ldots, t_n) \to \pi(\exp(t_1X_1) \cdots \exp(t_nX_n))v$ of $\mathbf{R}^n$ into V has continuous partial derivatives of any order.

A.5. (i) As an orbit $\pi(G) \cdot v$ the image $I(G/K)$ is a submanifold of V and the injectivity is clear from the maximality of K ([DS], Chapter VI, Exercise A3).

(ii) Let ϕ be a G-finite function on G/K, V_ϕ the space of complex linear combinations of translates $\phi^{\tau(g)}$, T the natural representation of G on V_ϕ, and $T^{\mathbf{C}}$ the extension of T to a representation of $\mathfrak{g}^{\mathbf{C}}$ and $G^{\mathbf{C}}$ on V_ϕ. We have $T(g)\psi = \psi^{\tau(g)}$, $\psi \in V_\phi$. Define $\phi^{\mathbf{C}}$ on $G^{\mathbf{C}}$ by $\phi^{\mathbf{C}}(x) = (T^{\mathbf{C}}(x^{-1})\phi)(o)$ where o is the origin in G/K. Then $\phi^{\mathbf{C}}(g) = \phi(gK)$ for $g \in G$. Also $\phi^{\mathbf{C}}$ is a $G^{\mathbf{C}}$-finite function on $G^{\mathbf{C}}$. [If $\phi = e_1, \ldots, e_n$ is a basis of V_ϕ and $T^{\mathbf{C}}(x)e_j = \sum_i T_{ij}(x)e_i$ then $\phi^{\mathbf{C}}(x) = \sum_i T_{i1}(x^{-1})e_i(o)$.].

Let $k \in K$. Then if $x \in G^{\mathbf{C}}$,

$$\phi^{\mathbf{C}}(xk) = (T^{\mathbf{C}}(k^{-1}x^{-1})\phi)(o) = (T(k^{-1})T^{\mathbf{C}}(x^{-1})\phi)(o),$$

so $\phi^{\mathbf{C}}(xk) = \phi^{\mathbf{C}}(x)$. In particular, we can define ϕ_0 on U/K by $\phi_0(uk) = \phi^{\mathbf{C}}(u)$. Then ϕ_0 is K-finite. Denoting by I_0 the injection $uK \to \pi(u)v$ of U/K into V_0 we have by Introduction, Exercise A1 $\phi_0 = p_0 \circ I_0$ where p_0 is a polynomial function on V_0. If p is the canonical extension of p_0 to a (holomorphic) polynomial on V put $\psi = p \circ I$. Since ψ is a G-finite function on G/K we can repeat the above process, replacing ϕ by ψ. Then $\phi_0 = \psi_0$ so $\phi^{\mathbf{C}} = \psi^{\mathbf{C}}$ by holomorphic continuation.

A.6. If $D \in \boldsymbol{D}(K)$ let as in Chapter II, §5 ε_D denote the distribution with support $\{e\}$ defined by $\varepsilon_D(f) = (D\check{f})(e)$. The mapping $D \to \varepsilon_D$ then identifies $\boldsymbol{D}(K)$ with a subspace of $\mathscr{D}'(K)$. Each element ϕ in the space $\mathscr{A}(K)$ of analytic functions on K gives a linear functional $\tau_\phi: T \to T(\phi)$ on $\mathscr{D}'(K)$; we assign to $\mathscr{D}'(K)$ the topology $\sigma(\mathscr{D}'(K), \mathscr{A}(K))$, i.e., the weakest topology making all these functionals continuous. From general theory it is then known that the dual of $\mathscr{D}'(K)$ for this topology consists just of the functionals τ_ϕ $(\phi \in \mathscr{A}(K))$. Since no such functional annihilates the subspace $\boldsymbol{D}(K) \subset \mathscr{D}'(K)$ (by Taylor's formula [DS], Chapter II, (6) §1) we conclude that $\boldsymbol{D}(K)$ is a dense subspace of $\mathscr{D}'(K)$. Second, by the definition of the topology $\sigma(\mathscr{D}'(K), \mathscr{A}(K))$, the convolution $T \to T * \chi_\delta$ is a continuous mapping of $\mathscr{D}'(K)$ into itself. Thus $\boldsymbol{D}(K)\chi_\delta$ is dense in $\mathscr{D}'(K) * \chi_{\check{\delta}}$ which, however, by Corollary 1.8, equals the finite-dimensional space H_δ.

B. Spherical Functions

B.1. We have by Lemma 6.5 $\rho(H(ak)) \leq \rho(\log a)$ for $a \in A^+$ so $e^{-\rho(\log a)} \leq \phi_0(a)$. Next put $\psi(a) = e^{\rho(\log a)}\phi_0(a)$ $(a \in A)$. Let $b \in A$. Then using Lemma 5.19 of Chapter I we find [since $H(abk) = H(ak(bk)) + H(bk)$]

$$\phi_0(a) = \int_K e^{-\rho(H(ak))}\, dk = \int_K e^{-\rho(H(abk))}e^{-\rho(H(bk))}\, dk$$

so

$$\psi(a) \leq \psi(ab) \qquad \text{if} \quad a \in A, \quad b \in \overline{A^+}. \tag{1}$$

Select $H_0 \in \mathfrak{a}^+$ such that $\alpha_i(H_0) = 1$ $(1 \leq i \leq l)$ and put for $t > 0$

$$F(t) = \psi(\exp tH_0) = ce^{\rho(tH_0)}\{\partial(\pi)_\lambda(\pi(\lambda)\phi_\lambda(\exp tH_0))\}_{\lambda=0}.$$

Then

$$F(t) = \sum_{\mu \in \Lambda} p_\mu(t)e^{-t\mu(H_0)},$$

where p_μ is a polynomial of degree $\leq d$. Using the uniform convergence for $1 \leq t < \infty$; this implies $|F(t)| \leq a_0(1 + t)^d$ for $t \geq 0$, a_0 being a constant.

Let $H \in \mathfrak{a}^+$ and put $\beta(H) = \max_i \alpha_i(H)$. Then $\beta(H)H_0 - H \in \overline{\mathfrak{a}^+}$, so inequality (1) above implies

$$\psi(\exp H) \leq F(\beta(H)) \leq a_0(1 + \beta(H))^d \leq a(1 + |H|)^d$$

as desired.

B.2. If $f \in \mathscr{D}(\mathfrak{a})$ we can write it as a Fourier transform

$$f(H) = \int_{\mathfrak{a}^*} g(\lambda)e^{-i\lambda(H)}\, d\lambda$$

and then our formula comes from Eq. (38) of §6 if we ignore problems of convergence. The problem is that the integral for $c_s(\lambda)$ is not absolutely convergent for $\lambda \in \mathfrak{a}^*$. One way to overcome the difficulty is to retrace the steps in the proof of the product formula (33). A simpler way is to shift the integration in the formula for f and write

$$f(H) = \int_{\mathfrak{a}^* - i\rho} g(\lambda)e^{-i\lambda(H)}\, d\lambda,$$

because then the absolute convergence of (36) is guaranteed by (37).

B.3. We know from Eq. (34) of §6 that $\rho(A_\alpha) = \rho_\alpha(A_\alpha)$ if α is one of the simple roots $\alpha_1, \ldots, \alpha_l$. For $\beta \in \Sigma_0^+$ put as in §5, $\beta = \Sigma_1^l m_i\alpha_i$ and $m(\beta) = \Sigma_i m_i$. Assume now β not simple and select an i such that $\langle\beta, \alpha_i\rangle > 0$. Then the root $\gamma = s_{\alpha_i}\beta = \beta - c_i\alpha_i$ ($c_i \in \mathbf{Z}^+$) satisfies $m(\gamma) < m(\beta)$ and $\gamma \in \Sigma_0^+$. We have $\rho(A_\beta) \geq \rho(A_\gamma)$. After finitely many steps we get $\rho(A_\beta) \geq \rho(A_{s\beta})$ where $s \in W$ and $s\beta$ is simple. By the first part of the proof, $\rho(A_{s\beta}) = \rho_{s\beta}(A_{s\beta}) = \rho_\beta(A_\beta)$ (cf. Kostant [1975a]).

B.4. Using B2 for $\overline{N}_s = \overline{N}$, $p = d$, $f(H) = [1 + \rho(H)]^{d+\varepsilon}$ we have by B3,

$$1 + \rho(H(\bar{n}_1) + \cdots + H(\bar{n}_d)) \geq 1 + \rho_1(H(\bar{n}_1)) + \cdots + \rho_d(H(\bar{n}_d))$$

so

$$[1 + \rho(H(\bar{n}_1) + \cdots + H(\bar{n}_d))]^{d+\varepsilon} \geq \{[1 + \rho_1(H(\bar{n}_1))] \cdots [1 + \rho_d(H(\bar{n}_d))]\}^{1+\varepsilon/d},$$

which reduces to result to the rank-one case.

B.5. A corollary of Theorem 4.3.

B.7. Writing $\eta = -i\lambda$ we have

$$\phi_\lambda(g) = \int_K e^{\rho(A(kg))}e^{-\eta(A(kg))}\, dk.$$

Here we can replace $e^{-\eta(A(kg))}$ by the average $w^{-1}\sum_{s\in W} e^{-s\eta(A(kg))}$. However, if $H \in \mathfrak{a}$, $\sum_{s\in W} s\cdot H$ is fixed under each s_α and hence is 0. By convexity of e^x, $\sum_{i=1}^n e^{a_i} \geq n$ whenever $\sum_i a_i = 0$. Thus

$$\sum_{s\in W} e^{-s\eta(A(kg))} \geq w,$$

giving the desired result.

B.8. We have $2\rho = (m_\alpha + 2m_{2\alpha})\alpha$ and $\langle\alpha, \alpha\rangle = \frac{1}{2}(m_\alpha + 4m_{2\alpha})^{-1}$. From Eq. (8) of §5 we have

$$2(m_\alpha + 4m_{2\alpha})\Delta(L_X) = \frac{d^2}{dt^2} + \{m_\alpha \coth t + 2m_{2\alpha}\coth 2t\}\frac{d}{dt}.$$

Also

$$\langle\lambda, \lambda\rangle + \langle\rho, \rho\rangle = \tfrac{1}{2}[1/(m_\alpha + 4m_{2\alpha})][\langle\lambda, \alpha_0\rangle^2 + (\tfrac{1}{2}m_\alpha + m_{2\alpha})^2]$$

and the rest follows by direct computation.

B.9. (i) Let $f^*(g) = \overline{f(g^{-1})}$. Then since $\phi_{-\lambda}(g) \equiv \phi_\lambda(g^{-1})$ we have $(f^*)^\sim(\lambda) = \overline{\tilde{f}(\lambda)}$ for $\lambda \in \mathfrak{a}^*$. Thus if $\lambda \in \mathfrak{a}^*$

$$\int_G\int_G \phi_\lambda(h^{-1}g)f(h)\overline{f(g)}\,dh\,dg = \int_G \phi_{-\lambda}(g)(f^* * f)(g)\,dg = |\tilde{f}(\lambda)|^2$$

so ϕ_λ is positive definite.

B.10. Considering the group $\mathrm{Ad}_{MN}(M)$ acting on $\mathfrak{n}$ we see from Chapter II, Theorem 4.9, that with the identification $MN/M = N$ the algebra $D(MN/M)$ is generated by the two commuting operators $\tilde{X}^2 + \tilde{Y}^2$ and $\tilde{Z}$ ($\tilde{V}$ being the left invariant vector field on N generated by $V \in \mathfrak{n}$). Putting for $F \in \mathscr{E}(N)$,

$$f(x, y, z) = F(\exp(xX + yY + zZ))$$

we have, by [DS], Chapter II, Lemma 1.8, since

$$[X, Y] = 2Z, \qquad [X, Z] = [Y, Z] = 0,$$

$$(\tilde{X}F)(n) = \frac{\partial f}{\partial x} - y\frac{\partial f}{\partial z}, \qquad (\tilde{Y}F)(n) = \frac{\partial f}{\partial y} + x\frac{\partial f}{\partial z}, \qquad (\tilde{Z}F)(n) = \frac{\partial f}{\partial z},$$

whence

$$(\tilde{X}^2 + \tilde{Y}^2)F(n) = \frac{\partial^2 f}{\partial x^2} + \frac{\partial^2 f}{\partial y^2} + (x^2 + y^2)\frac{\partial^2 f}{\partial z^2} + 2x\frac{\partial^2 f}{\partial y\,\partial z} - 2y\frac{\partial^2 f}{\partial x\,\partial z}.$$

Thus the spherical functions on $N = MN/M$ are given by

$$f(x, y, z) = \phi(x^2 + y^2)e^{\alpha z},$$

where $\alpha \in \boldsymbol{C}$ and ϕ satisfies the differential equation

$$\frac{d^2\phi}{dr^2} + \frac{1}{r}\frac{d\phi}{dr} + r^2\alpha^2\phi = -\lambda^2\phi$$

for some $\lambda \in \boldsymbol{C}$.

This example and its generalization to G of real rank one are given in Korányi [1980]; see also Kaplan and Putz [1977].

B.12. Let $\lambda_0 = \xi_0 + i\eta_0$ $(\xi_0, \eta_0 \in \mathfrak{a}^*)$ be such that $-A_{\eta_0} \in \overline{\mathfrak{a}^+}$ and ϕ_{λ_0} is bounded. We have to prove $\eta_0 + \rho$ nonnegative on $\mathfrak{a}^+$. Let $U \subset V \subset W$ be the subgroups of the Weyl group W leaving fixed λ_0 and η_0, respectively. We can assume that for the lexicographic ordering of $\mathfrak{a}^*$ defined by means of the simple roots $\alpha_1, \ldots, \alpha_l$ we have $\xi_0 \geq s\xi_0$ for $s \in V$. In particular, $\alpha(A_{\xi_0}) \geq 0$ for $\alpha \in \Sigma^+$ satisfying $\alpha(A_{\eta_0}) = 0$.

Lemma. *The subgroup U of W leaving λ_0 fixed is generated by the refections s_{α_i} where α_i is a simple root vanishing at A_{λ_0}.*

For this let U' denote the subgroup of U generated by these s_{α_i}. The group U is generated by s_α as α runs through the roots, vanishing at A_{λ_0} ([DS], Chapter VII, Theorem 2.15). For each such $\alpha > 0$ we shall prove $\alpha = s\alpha_p$ where $s \in U'$ and α_p is a simple root vanishing at A_{λ_0}. We prove this by induction on $\sum m_i$ if $\alpha = \sum m_i\alpha_i$ $(m_i \in \boldsymbol{Z}^+ - \{0\})$. The statement is clear if $\sum m_i = 1$ so assume $\sum m_i > 1$. Since $\langle \alpha, \alpha \rangle > 0$ we have $\langle \alpha, \alpha_k \rangle > 0$ for some k among the indices i above. Then $s_{\alpha_k}\alpha \in \Sigma^+$ so $s_{\alpha_k}\alpha = \sum m_j'\alpha_j$ $(m_j' \in \boldsymbol{Z}^+)$ and $\sum m_j' < \sum m_i$. Now $\alpha(A_{\lambda_0}) = 0$ so $\alpha_i(A_{\eta_0}) = 0$ for each i above, whence $\alpha_i(A_{\xi_0}) \geq 0$ by our convention. But then $\alpha(A_{\lambda_0}) = 0$ implies $\alpha_i(A_{\xi_0}) = 0$. In particular, $s_{\alpha_k} \in U$. Thus the induction assumption applies to $s_{\alpha_k}\alpha$, giving a $s' \in U'$ such that $s_{\alpha_k}\alpha = s'\alpha_p$. Hence $\alpha = s\alpha_p$ with $s \in U'$. But then $s_\alpha = ss_{\alpha_p}s^{-1}$, proving the lemma.

Let π' denote the product of the positive roots vanishing on λ_0, π'' the product of the remaining positive roots. We multiply the formula (Theorem 5.7) for $\phi_\lambda(a)$ by $\pi'(\lambda)$, then apply the differential operator $\partial(\pi')$ on $\mathfrak{a}_c^*$ and evaluate at $\lambda = \lambda_0$. This gives the formula

$$\sum_{s \in w} (\det s)e^{s\rho(\log a)}\phi_{\lambda_0}(a) = c \sum_{s \in W} p_s(\log a)e^{is\lambda_0(\log a)},$$

with $c \in \boldsymbol{C}$, the p_s being certain polynomials. Now write the sum on the right in the form

$$e^{-\eta_0(\log a)}\left(\sum_{V/U} e^{is\xi_0(\log a)} \sum_{t \in U} p_{st}(\log a)\right) + \sum_{s \in W - V} p_s(\log a)e^{is\lambda_0(\log a)}.$$

Using the lemma above we see that $\pi'(sH) = (\det s)\pi'(H)$ for $s \in U$. Examining the definition of p_s, i.e.,

$$p_s(\log a) = \{\partial(\pi')_\lambda[(\det s)\pi''(\lambda)^{-1}e^{is\lambda(\log a)}]\}_{\lambda=\lambda_0}e^{-is\lambda_0(\log a)},$$

we see that the highest-degree term in $\sum_{s \in U} p_s(\log a)$ is a nonzero constant multiple of $\pi'(\log a)$. Hence

$$\sum_{t \in U} p_t(\log a) \not\equiv 0.$$

Thus the expression $q(a)$ inside the large parentheses above is nonzero (cf. Chapter I, Exercise D5). We have

$$\left|\sum_{s \in W} p_s(\log a)e^{is\lambda_0(\log a)}\right|$$

$$\geq e^{-\eta_0(\log a)}\left||q(a)| - \sum_{s \in W-V} |p_s(\log a)|e^{(\eta_0 - s\eta_0)(\log a)}\right|$$

and

$$\left|\sum_{s \in W} (\det s)e^{s\rho(\log a)}\right| \leq e^{\rho(\log a)}\left(1 + \sum_{s \neq e} e^{(s\rho-\rho)(\log a)}\right).$$

Now $(\eta_0 - s\eta_0)(H) < 0$ for $H \in \mathfrak{a}^+$, and $s \in W - V$, and of course $s\rho - \rho < 0$ on $\mathfrak{a}^+$ if $s \neq e$. Fix a unit vector $H \in \mathfrak{a}^+$ such that $(s\xi_0)(H) \neq \xi_0(H)$ for $s \in V - U$ and write

$$q(\exp tH) = \sum_j q_j(t)e^{ir_jt},$$

where the r_j are different real numbers and the q_j are polynomials. The q_j which corresponds to $r_j = \xi_0(H)$ equals $\sum_{s \in U} p_s(\exp tH)$ whose leading term is $\pi'(tH)$ (up to a constant factor). By Chapter I, Exercise D5, the function $t \to q(\exp tH)$ is unbounded. The inequalities above, together with the boundedness of ϕ_{λ_0}, thus imply that $e^{-(\eta_0+\rho)(tH)}$ is bounded for $t \geq 0$; thus $\eta_0 + \rho \geq 0$ on a dense subset of $\mathfrak{a}^+$, hence on all of $\mathfrak{a}^+$.

B.13. For $x \in G$, consider the mean value operator M^x given by

$$(M^x f)(g) = \int_K f(gkx)\,dk$$

for f in the space

$$C_K(G) = \{f \in C(G) : f(gk) = f(g) \quad \text{for} \quad g \in G, k \in K\}.$$

Let $\mathfrak{g}$ be the Lie algebra of G. For $X \in \mathfrak{g}$, $m \in \boldsymbol{Z}^+$, let D_m^X denote the operator

$$D_m^X = \int_K \mathrm{Ad}(k)\tilde{X}^m\,dk \in \boldsymbol{D}_K(G).$$

Let $f \in C_K(G)$ be analytic. Then if $X \in \mathfrak{g}$ is sufficiently small, the formula

$$M^{\exp X} f = \sum_0^\infty \frac{1}{m!} D_m^X f$$

holds on a neighborhood of e in G (cf. Helgason [1957a, 1962a], Chapter X). Assuming (i) we thus derive the relation

$$(M^x M^y f)(e) = (M^y M^x f)(e)$$

for x, y near e, hence for all $x, y \in G$ by analyticity. Using approximation by analytic functions (loc. cit. Chapter X, Lemma 7.9) the relation is seen to hold for all $f \in C_K(G)$. It follows that if $f \in C_c^\natural(G)$, then

$$\int_K f(xky^{-1})\,dk = \int_K f(y^{-1}kx)\,dk.$$

Integrating this relation against $g(y)$ ($g \in C_c^\natural(G)$) and using the unimodularity we get $f * g = g * f$, proving (ii).

Conversely, assume (ii) holds and let $D, E \in \mathbf{D}_K(G)$, $f, g \in \mathcal{D}^\natural(G)$. Then since D is left invariant, $D(f * g) = f * Dg = Dg * f$, so

$$DE(f * g) = D(Eg * f) = Eg * Df = Df * Eg,$$

$$ED(f * g) = ED(g * f) = Eg * Df.$$

Thus E and D commute on the subspace $\mathcal{D}^\natural * \mathcal{D}^\natural$ of $\mathcal{D}^\natural$, hence, by density, on all of $\mathcal{D}^\natural$.

Now let $F \in \mathcal{D}(G) \cap C_K(G)$ and consider

$$F^*(x, y) = (M^x F)(y).$$

Let subscript 1 denote differentiation with respect to the first variable. Since F^* is bi-invariant under K in the first variable, we have

$$(D_1 F^*)(x, y) = \int_K (DF)(ykx)\,dk = (DF)^*(x, y),$$

so

$$(DF)(y) = (D_1 F^*)(e, y).$$

Hence

$$\begin{aligned}(DEF)(y) &= (D_1(EF)^*)(e, y) = (D_1 E_1 F^*)(e, y)\\ &= (E_1 D_1 F^*)(e, y) = (EDF)(y),\end{aligned}$$

so (i) holds. Part (ii) $\Rightarrow$ (i) is from Helgason [1979]; part (i) $\Rightarrow$ (ii) was noticed independently by Thomas [1984].

B.14. The adjoint representation of U is irreducible and the multiplicity of the 0 weight (Chapter V, §1) equals dim T. Now the result follows from B.13 and Theorem 3.5, Chapter V.

B.16. Assuming $L: C_c^\natural(G) \to C$ to be a continuous surjective homomorphism we have, by the definition of a measure, a measure μ on G, bi-invariant under K such that

$$L(f) = \int_G f(x)\,d\mu(x), \qquad f \in C_c^\natural(G).$$

Take $g \in C_c^\natural(G)$ such that $L(g) = 1$. Then

$$L(f) = L(f)L(g) = L(f * g) = \int_G f * g(x)\,d\mu(x)$$

$$= \int_G f(x)\mu * \check{g}(x)\,dx$$

so $\mu = \mu * \check{g}$ is a continuous function which by Lemma 3.2 is spherical.

B.17. Let r be the maximal rank of the matrices $(\Phi_i(x_j))_{i,j}$ for all sets of $x_j \in S$, and choose $x_1, \dots, x_r$ such that $(\Phi_i(x_j))_{i,j}$ has rank r. If $r < w$ there exists a nonzero w-tuple $(\lambda_1, \dots, \lambda_w)$ depending on $x_1, \dots, x_r$ such that if $v_j = (\Phi_1(x_j), \dots, \Phi_w(x_j))$ then

$$\sum_{i=1}^{w} \lambda_i \Phi_i(x_j) = 0, \qquad 1 \le j \le r.$$

Let $x \in S$ be arbitrary and put $v = (\Phi_1(x), \dots, \Phi_w(x))$. Since the matrix formed by the column vectors $v_1, \dots, v_r, v$ has rank r (by the maximality) there exist numbers $a_j(x)$ such that $v = \sum_1^r a_j(x)v_j$. Then, by the above,

$$\sum_{i=1}^{w} \lambda_i \Phi_i(x) = \sum_{i=1}^{w} \lambda_i \sum_{j=1}^{r} a_j(x)\Phi_i(x_j) = 0.$$

Since $x \in S$ is arbitrary, this is a contradiction.

C. Spherical Transforms

C.1. (i) Let $\lambda = \xi + i\eta$ ($\xi, \eta \in \mathfrak{a}^*$) be such that ϕ_λ is bounded. For $F \in L^\natural(G)$ we have

$$\infty > \int_G |F(g)|\,dg = \int_{KAN} |F(kan)| e^{2\rho(\log a)}\,dk\,da\,dn$$

$$= \int_A e^{2\rho(\log a)}\left(\int_N |F(an)|\,dn\right) da,$$

so $\int_N F(an)\,dn$ exists for almost all $a \in A$. Also $\phi_{i\eta}$ is bounded by Theorem 8.1, so

$$\int_K \int_G |F(g)e^{(i\lambda - \rho)(H(gk))}|\,dg\,dk = \int_G |F(g)|\phi_{i\eta}(g)\,dg < \infty,$$

whence by Fubini, whenever ϕ_λ is bounded,

$$(*) \qquad \int_G F(g)\phi_\lambda(g)\,dg = \int_G F(g)e^{(i\lambda-\rho)(H(g))}\,dg = \int_A \phi_F(a)e^{i\lambda(\log a)}\,da,$$

where

$$\phi_F(a) = e^{\rho(\log a)} \int_N F(an)\,dn.$$

Replacing F by $|F|$ and taking $\lambda = 0$ in $(*)$ we conclude that $\phi_{|F|} \in L^1(A)$; hence $\phi_F \in L^1(A)$. The assumption $\tilde{F}(\lambda) = 0$ $(\lambda \in \mathfrak{a}^*)$ and the injectivity of the Fourier transform on $L^1(A)$ then imply $\phi_F \equiv 0$ almost everywhere on A. But then $(*)$ together with Prop. 3.8 implies that $F \equiv 0$ almost everywhere on G. The argument also gives part (ii).

C.2. Clearly $\tilde{T}$ is W-invariant. Also since T is K-invariant

$$\tilde{T}(\lambda) = T^*(A_\lambda)$$

where

$$T^*(Y) = \int_{\mathfrak{p}} e^{-iB(Y,Z)}\,dT(Z).$$

By the Paley–Wiener theorem for $\mathscr{E}'(\mathfrak{p})$ (cf. Hörmander [1963]) T^* and therefore $\tilde{T}$ satisfies the desired inequality.

Conversely, suppose Φ satisfies the stated conditions. Consider the K-invariant distribution T on $\mathfrak{p}$ defined by (cf. (2) §9)

$$T(f) = \int_{\mathfrak{a}^*} \tilde{f}^{\natural}(\lambda)\Phi(\lambda)\delta(\lambda)\,d\lambda, \qquad f \in \mathscr{D}(\mathfrak{p}),$$

where $f^{\natural}(Z) = \int f(k \cdot Z)\,dk$. Let $\eta_\varepsilon \in \mathscr{D}_K(\mathfrak{p})$ have support in $B_\varepsilon(0)$. Then $\Phi(\lambda)\tilde{\eta}_\varepsilon(\lambda)$ is entire of exponential type $\leq R + \varepsilon$ so by Theorem 9.2, $T * \eta_\varepsilon$ has support in $B_{R+\varepsilon}$. Letting $\varepsilon \to 0$, we see that $T \in \mathscr{E}'_K(\mathfrak{p})$.

C.3. (i) Writing $a_r = \exp rH$, $\alpha(H) = 1$, we have as in [DS], p. 417, $\bar{n}a_r = u_1 \exp A^+(\bar{n}a_r)u_2$ $(u_1, u_2 \in K)$, whence

$$a_r\theta(\bar{n})^{-1}\bar{n}a_r = u_2^{-1} \exp 2A^+(\bar{n}a_r)u_2,$$

giving

$$\mathrm{Tr}(a_r\theta(\bar{n})^{-1}\bar{n}a_r) = 2\,\mathrm{ch}(2\alpha[A^+(\bar{n}a_r)]) + 1,$$

and (i) follows by direct computation.

(ii) If α is the root in Σ^+ the formula reads, replacing $\mathrm{ch}\,x$ by $2\,\mathrm{ch}^2(\tfrac{1}{2}x) - 1$,

$$\mathrm{ch}^2(\tfrac{1}{2}\alpha[A^+(\bar{n}a)]) = \mathrm{ch}^2(\tfrac{1}{2}\alpha(\log \alpha)) + (1/16m_\alpha)e^{\alpha(\log a)}|X|^2.$$

If 2α is the root in Σ^+ the formula reads, writing $\beta = 2\alpha$,

$$\mathrm{ch}^2(\tfrac{1}{2}\beta[A^+(\bar{n}a)]) = \mathrm{ch}^2(\tfrac{1}{2}\beta(\log a)) + (1/16m_{2\alpha})e^{\beta(\log a)}|Y|^2.$$

which is the same formula.

(iii) See solution to B8.

(iv) The change of variables $X \to e^{(1/2)r}X$, $Y \to e^rY$ cancels out the factor $e^{\rho(\log a)}$, so $(*)$ follows.

(v) Putting

$$\Phi(u) = \int_{\mathfrak{g}_{-2\alpha}} \phi(u^2 + \|Y\|^2)\,dY,$$

we have

$$\psi(v) = \int_{\mathfrak{g}_{-\alpha}} \Phi(v^2 + \|X\|^2)\,dX.$$

Introducing polar coordinates in $\mathfrak{g}_{-2\alpha}$ and then putting $t^2 = u^2 + \|Y\|^2$, we find

$$\Phi(u) = c_1 \int_u^\infty t\phi(t^2)(t^2 - u^2)^{(1/2)(m_{2\alpha}-2)}\,dt,$$

and similarly

$$\psi(v) = c_2 \int_v^\infty s\Phi(s^2)(s^2 - v^2)^{(1/2)(m_\alpha - 2)}\,ds,$$

c_1 and c_2 being constants.

(vi) These two integral transforms can be inverted by the method used in the proof of Theorem 2.6 of Chapter I. This gives the conclusions

$$\begin{aligned} \psi(v) &= 0 \qquad \text{for } \; v > A,\\ \Rightarrow \Phi(s^2) &= 0 \qquad \text{for } s > A,\\ \Rightarrow \phi(t^2) &= 0 \qquad \text{for } \; t > A^2. \end{aligned}$$

Now assume $F_f(a_r) = 0$ for $\|\log a_r\| > R$. Since $\|\log a_r\| = r$ we have $\psi(\sqrt{\mathrm{ch}\, r}) = 0$ for $r > R$, i.e., $\sqrt{\mathrm{ch}\, r} > \sqrt{\mathrm{ch}\, R}$. By the above, $\phi(t^2) = 0$ for $t > \mathrm{ch}\, R$, that is, $f(g) = 0$ if $\delta(o, g\cdot o) > R$.

C.4. Select $f \in \mathscr{D}^\natural(G)$ such that the function $H \to f(\exp H)$ on $\mathfrak{a}^+$ has support inside the set

$$\mathfrak{a}^+(M) = \{H \in \mathfrak{a}^+ : \alpha(H) > M \text{ for } \alpha \in \Sigma^+\},$$

$M > 0$ being fixed.

With $\delta(H) = \prod_{\alpha>0} 2\sinh\alpha(H)$, consider the function $f_\varepsilon \in \mathscr{D}^\natural(G)$ given by

$$f_\varepsilon(H)\delta^{1/2}(H) = \varepsilon^{(1/2)l} f(\varepsilon H)\delta^{1/2}(\varepsilon H), \qquad H \in \mathfrak{a}^+.$$

Then

$$\int_G |f_\varepsilon(g)|^2\, dg = \int_G |f(g)|^2\, dg.$$

Put $\chi(s, \lambda) = c(-s\lambda)/c(-\lambda)$ and

$$F_\varepsilon(\lambda) = \varepsilon^{-(1/2)l} \sum_{s \in W} \chi(s, \lambda) \int_{\mathfrak{a}^+} f(H)\delta^{1/2}(H)e^{-is\lambda(H)/\varepsilon}\, dH.$$

By the Euclidean Plancherel formula,

(1) $$\int_{\mathfrak{a}^*} |F_\varepsilon(\lambda)|^2\, d\lambda \to |W| \int_G |f(g)|^2\, dg \qquad \text{as} \quad \varepsilon \to 0.$$

On the other hand,

$$\tilde{f}_\varepsilon(\lambda) = \int_{\mathfrak{a}^+} f_\varepsilon(H)\phi_{-\lambda}(H)\, \delta(H)\, dH,$$

and, using the expansion for $\phi_{-\lambda}$ and the Euclidean Paley–Wiener theorem on $\mathfrak{a}$, it can be proved that

(2) $$\int_{\mathfrak{a}^*} |\tilde{f}_\varepsilon(\lambda)c(-\lambda)^{-1} - F_\varepsilon(\lambda)|^2\, d\lambda \to 0 \qquad \text{as} \quad \varepsilon \to 0.$$

Since

$$c \int_G |f(g)|^2\, dg = \int_{\mathfrak{a}^*} |\tilde{f}_\varepsilon(\lambda)c(\lambda)^{-1}|^2\, d\lambda$$

(1) and (2) would imply $c = |W|$.

This proof, with further details, was given by Rosenberg [1977]; in Harish-Chandra [1958b] (§15) the measures $d\lambda$ and da are regularly normalized, the measure dg is normalized by

$$\int f(g)\, dg = \int_{KAN} f(kan)e^{2\rho(\log a)}\, dk\, da\, dn,$$

and the Plancherel formula (Theorem 7.5) is proved with the same constant $c = |W|$. In particular, the two normalizations of dg considered above must agree. See also Mneimné [1983].

C.5. From the inversion formula we have

$$f(\exp H) = \frac{1}{|W|} \int_{\mathfrak{a}^*} \tilde{f}(\lambda)\phi_\lambda(\exp H)|c(\lambda)|^{-2}\, d\lambda.$$

Using Theorem 5.7 we get for $H \in \mathfrak{a}$, $\varepsilon(s) = \det(s)$,

$$i^{-d}\pi(\rho) \sum_{s \in W} \varepsilon(s)e^{s\rho(H)}f(\exp H) = \int_{\mathfrak{a}^*} \tilde{f}(\lambda)\pi(\lambda)e^{i\lambda(H)}\, d\lambda,$$

and in this last integral we can shift the integration to the extremes of the tube $\mathfrak{a}^* + iC(\rho)$ where $C(\rho)$ is the convex hull of the orbit $W \cdot \rho$. This gives

$$i^{-d}\pi(\rho) \sum_{s \in W} \varepsilon(s)e^{s\rho(H)}(\Delta_0 f)(\exp H) = \int_{\mathfrak{a}^*} \psi^*(\lambda)\, \pi(\lambda)e^{i\lambda(H)}\, d\lambda$$

where

$$\psi^*(\lambda) = \pi(\lambda)^{-1} \sum_{s \in W} \tilde{f}(\lambda + is\rho)\pi(\lambda + is\rho).$$

Reintroducing the formula for ϕ_λ this gives the desired formula.

CHAPTER V

A. Representations

A.1. Apply Eq. (16) of §1 to the representation $\pi = \mathrm{Ad}_U$; then $\pi(L_U) = cI$, $\lambda = \delta$ and $\mathrm{Tr}(\pi(L_u)) = -c \dim \mathfrak{u}$ so $c = -1$.

A.2. (i) Use Proposition 6.10, Chapter IV. For (ii) suppose $\lambda \in \Lambda$ and $|\lambda| = |\rho|$. By Chapter I, Proposition 5.15(ii)

$$\sum_{s \in W} (\det s)e^{s\lambda} = \left(\sum_{s \in W} (\det s)e^{s\rho}\right) \sum_{\mu} a_\mu e^{\mu}.$$

By the W-invariance we have $a_\mu \neq 0$ for some μ with $H_\mu \in \overline{\mathfrak{t}^+}$. But then if λ_0 is the highest among $s\lambda$ ($s \in W$) we have $\lambda_0 = \rho + \mu$. Then $\langle \mu, \rho \rangle \geq 0$, so $|\lambda_0|^2 \geq |\rho|^2 + |\mu|^2$, which implies $\mu = 0$.

A.3. While the formula for $\dim(\pi)$ (Theorem 1.8) is obtained by comparing the lowest-degree terms in x, the present formula is obtained by considering the next power of x.

A.4. Clear from A1 and A3 if we also recall

$$\langle \mu, \nu \rangle = \sum_{\alpha \in \Delta} \langle \mu, \alpha \rangle \langle \nu, \alpha \rangle$$

([DS], Chapter III, Exercise C4).

A.5. As in the proof of Eq. (25), Chapter IV, §5, we have

$$\begin{aligned} \sum_{\mu \in \Lambda} \mathscr{P}(\mu)e^{-\mu} &= \prod_{\alpha \in \Delta^+} (1 + e^{-\alpha} + e^{-2\alpha} + \cdots) \\ &= \prod_{\alpha \in \Lambda^+} (1 - e^{-\alpha})^{-1} = e^{\rho}\left(\sum_{s} (\det s)e^{s\rho}\right)^{-1}. \end{aligned}$$

Hence

$$\chi_\pi = \sum_{\mu \in \Lambda(\pi)} m_\mu e^\mu = \sum_s (\det s) e^{s(\lambda+\rho)} \left(\sum_s (\det s) e^{s\rho} \right)^{-1}$$
$$= \sum_s (\det s) e^{s(\lambda+\rho)} \sum_{\nu \in \Lambda} \mathscr{P}(\nu) e^{-\nu-\rho},$$

so (i) follows by putting $\nu = s(\lambda + \rho) - (\mu + \rho)$ (cf. Cartier [1961]).

For (ii) note that by Eq. (16) of §1 and Prop. 3.12 of Chapter II we have

$$(1) \qquad \Delta(L_U)\left(\sum_{\mu \in \Lambda(\pi)} m_\mu e^\mu \right) = (\langle \rho, \rho \rangle - \langle \rho + \lambda, \rho + \lambda \rangle) \sum_{\mu \in \Lambda(\pi)} m_\mu e^\mu.$$

By Theorem 3.7 of Chapter II

$$\Delta(L_U) = L_T + \operatorname{grad}(\log \delta)$$

and

$$\delta = e^{2\rho} \prod_{\alpha \in \Lambda^+} (1 - e^{-\alpha})^2.$$

For $\nu \in \Lambda$ let H_ν be determined by $\langle H, H_\nu \rangle = \nu(H)$. Then as a differential operator on T, $H_\nu e^\mu = \langle \mu, \nu \rangle e^\mu$. Let $H_1, \ldots, H_l$ be a basis of $\mathfrak{t}$ such that $-\langle H_i, H_j \rangle = \delta_{ij}$. Then

$$\operatorname{grad}(\log \delta) = \sum_i H_i(\log \delta) H_i$$
$$= 2\rho(H_i)H_i + 2 \sum_i \sum_{\alpha \in \Delta^+} (1 - e^{-\alpha})^{-1} e^{-\alpha} \alpha(H_i) H_i$$
$$= -H_{2\rho} - 2 \sum_{\alpha \in \Delta^+} e^{-\alpha}(1 - e^{-\alpha})^{-1} H_\alpha,$$

so

$$\Delta(L_U) = L_T - H_{2\rho} - 2\left(\sum_{\alpha \in \Delta^+, k \geq 1} e^{-k\alpha} \right) H_\alpha.$$

Now we have

$$L_T\left(\sum_\mu m_\mu e^\mu \right) = -\sum_\mu m_\mu \langle \mu, \mu \rangle e^\mu,$$
$$-H_{2\rho}\left(\sum_\mu m_\mu e^\mu \right) = -\sum m_\mu \langle \mu, 2\rho \rangle e^\mu,$$
$$-2 \sum_{\alpha, k} e^{-k\alpha} H_\alpha \left(\sum_\mu m_\mu e^\mu \right) = -2 \sum_\mu \left(\sum_{\nu - k\alpha = \mu} m_\nu \langle \alpha, \nu \rangle \right) e^\mu.$$

Substituting this into (1) we obtain the desired formula (ii) in analogy with Eq. (12), Chapter IV, §5.

For the last statement put $\lambda = 0$ in (i).

A.6. (i) Consider the mapping A from the proof of Theorem 1.8. Let $l_\nu \in \boldsymbol{Z}^+$ be determined by $\chi_\lambda \chi_\mu = \sum_{\nu \in \Lambda(+)} l_\nu \chi_\nu$. Then multiplying by $A(e^\rho)$ we get

$$\sum_{\nu \in \Lambda(+)} l_\nu A(e^{\nu+\rho}) = \chi_\lambda A(e^{\mu+\rho})$$

$$= \left(\sum_{\xi \in \Delta} m_\xi e^\xi\right) \sum_{s \in W} (\det s) e^{s(\mu+\rho)}$$

$$= \sum_{\xi \in \Lambda} \left(\sum_{s \in W} (\det s) m_{\xi+\rho-s(\mu+\rho)}\right) e^{\xi+\rho}.$$

Now $H_{\nu+\rho} \in \mathfrak{t}^+$ whereas $H_{s(\nu+\rho)} \notin \mathfrak{t}^+$ for $s \neq e$ in W. Thus l_ν is the coefficient to $e^{\nu+\rho}$ on the left. Hence

$$l_\nu = \sum_{s \in W} (\det s) m_{\nu+\rho-s(\mu+\rho)}$$

and now the result follows from A.5(i).

(ii) We have $\chi_\lambda = \sum_{\nu \in \Lambda(\pi_\lambda)} m_\nu e^\nu$ and a similar formula for χ_μ. Since $m_\lambda = m_\mu = 1$ the result follows by multiplication.

(iii) Let π denote the restriction of π_λ to the algebra $\mathfrak{g}^\alpha + \mathfrak{g}^{-\alpha} + \boldsymbol{C}H_\alpha$, which is isomorphic to $\mathfrak{sl}(2, \boldsymbol{C})$. Now apply Lemma 1.2 of the Appendix.

A.8. In fact $\chi_{J \otimes J} = (\chi_J)^2$ so by Lemma 2.6 for $A = I$,

$$\int_{U(n)} |\chi_{J \otimes J}(u)|^2 = 2$$

and the statement follows.

A.9. The simple roots α_i $(1 \leq i \leq n)$ are given by (cf. [DS], Chapter III, §8)

$$\alpha_i(H) = e_i(H) - e_{i+1}(H),$$

where H is a diagonal matrix of trace 0 with diagonal elements $e_i(H)$. The fundamental weights ω_j satisfy $\langle \omega_j, \check{\alpha}_i \rangle = \delta_{ij}$ so since $\langle e_i, e_i \rangle = \frac{1}{2}(n+1)^{-1}$ we obtain $\omega_j = e_1 + \cdots + e_j$. If $v_1, \ldots, v_{n+1}$ is the standard basis of $V = \boldsymbol{C}^{n+1}$ the elements $v_{i_1} \wedge \cdots \wedge v_{i_r}$ $(i_1 < \cdots < i_r)$ form a basis of $\bigwedge^r V$. For $t \in T$ [maximal torus formed by the diagonal elements in $\boldsymbol{SU}(n+1)$] we have for $t = \exp H$

$$(\textstyle\bigwedge^r J)(t)(v_{i_1} \wedge \cdots \wedge v_{i_r}) = (e_{i_1} + \cdots + e_{i_r})(H) v_{i_1} \wedge \cdots \wedge v_{i_r}.$$

Thus $e_{i_1} + \cdots + e_{i_r}$ is a weight of $\bigwedge^r J$ and $\omega_r = e_1 + \cdots + e_r$ is the highest one. The reflection s_{α_i} interchanges e_i and e_{i+1} so the Weyl group W consists of all permutations of the e_i. Thus W permutes transitively the weights of $\bigwedge^r J$, so it must be irreducible.

The construction of the fundamental representations of the simple Lie algebras goes back to Cartan [1913]. For $\boldsymbol{Sp}(n)$ the results are similar; one has to replace $\bigwedge^r V$ by the subspace of primitive elements (in the sense of Corollary 2.2 of Chapter III); for $\boldsymbol{Spin}(2n + 1)$ [the double covering of $\boldsymbol{SO}(2n + 1)$] one needs in addition to the exterior powers $\bigwedge^r J$, the *spin-representation*; for $\boldsymbol{Spin}(2n)$ one needs in addition to the exterior powers $\bigwedge^r J$, the two *semi-spin-representations*. For detailed description see Weyl [1939]; Boerner [1955]; Bourbaki, *Groupes et algèbres de Lie*, Chapter 8.

A.10. With notation from §1 let $V = \sum V_\mu$, where μ runs through $\Lambda(\pi)$. If $v' \in V'$ is such that $v'(V_\mu) = 0$ for all $\mu \in \Lambda(\pi)$ except μ_0 then $\check{\pi}(H)v' = -\mu_0(H)v'$. Thus $-\mu_0 \in \Lambda(\check{\pi})$ and $\dim V'_{-\mu_0} = \dim V_{\mu_0}$. This proves (i). Part (ii) follows from Theorem 1.3. Part (iv) follows since $-s$ permutes the "co-roots" $\check{\alpha}_j$.

A.11. For $G = \boldsymbol{SL}(2, \boldsymbol{R})$ we have $\mathrm{Ad}(g)X = gXg^{-1}$. Since $M = \pm I$, $N = \begin{pmatrix} 1 & \boldsymbol{R} \\ 0 & 1 \end{pmatrix}$ we find quickly

$$K = \{g \in G : \mathrm{Ad}(g)X_1 = X_1\},$$

$$MN = \{g \in G : \mathrm{Ad}(g)(X_1 + X_3) = X_1 + X_3\},$$

so the two maps are injections. If

$$g = \begin{pmatrix} a & b \\ c & d \end{pmatrix},$$

then

$$g^{-1} = \begin{pmatrix} d & -b \\ -c & a \end{pmatrix},$$

so

$$\mathrm{Ad}(g)X_1 = \tfrac{1}{2}(a^2 + b^2 + c^2 + d^2)X_1 - (bd + ac)X_2 + \tfrac{1}{2}(a^2 + b^2 - c^2 - d^2)X_3,$$

$$\mathrm{Ad}(g)\tfrac{1}{2}(X_1 + X_3) = \tfrac{1}{2}(a^2 + c^2)X_1 - acX_2 + \tfrac{1}{2}(a^2 - c^2)X_3,$$

so (i) and (ii) follow easily. For (iii) consider the horocycle $N \cdot X_1 = \mathrm{Ad}(N)(X_1)$, which by the above is the section of G/K with the plane $x_1 = x_3 + 1$. Writing $Z_0 = \frac{1}{8}(X_1 + X_3)$ this has the form $B(X, Z_0) = -1$. Since G permutes the horocycles transitively, (iii) follows.

B. Fourier Series

B.1. Part (i) is clear from Lemmas 2.4 and 2.4′ and then (ii) follows from Theorem 2.9.

B.2. Let A be an operator on $\mathscr{H}_\lambda$. Then $A = PU$ where U is unitary and P is the unique nonnegative square root of AA^*. Then $P = \sum_1^d c_i P_i$ where $c_i \geq 0$; the c_i^2 are the eigenvalues of AA^* and the P_i are mutually orthogonal projections, $\sum_i P_i = I$. Let $2c = \max c_i$. Then we can write $c_i = \alpha_i + \beta_i$ $(1 \leq i \leq d)$ where $|\alpha_i| = |\beta_i| = c$ for all i (cf. Example to Theorem 2.3). Put

$$V = c^{-1}\Big(\sum_i \alpha_i P_i\Big)U, \qquad W = c^{-1}\Big(\sum_i \beta_i P_i\Big)U.$$

Then V and W are unitary and since $\|A\| = \|AA^*\|^{1/2} = 2c$ we have

$$A = \tfrac{1}{2}\|A\|(V + W).$$

This shows that the Fourier series of f can be multiplied on the left by any hyperfunction B, bounded in the sense of Eq. (44) of §2, giving a function $f_B \in L^1$. By the closed-graph theorem the map $B \to f_B$ is continuous, i.e., $\|f_B\| \leq C \sup_\lambda \|B_\lambda\|$ where C is a constant. Restricting B to Fourier coefficients of functions $g \in L^1$ we see that the mapping $g \to f * g$ is spectrally continuous so by Theorem 2.3, $f \in L^2(K)$.

Remark. This corollary of Theorem 2.3 (Helgason [1957]) has been extended by Figa-Talamanca and Rider [1967] so that the assumption is only required for a set of unitary hyperfunctions Γ of positive measure in the product group $\prod_{\lambda \in K} U(d_\lambda)$. They also proved [1966] a certain "converse" to Exercise B2, namely that if $f \in L^2(K)$ then for any $p < \infty$ there exists a unitary hyperfunction Γ such that $f_\Gamma \in L^p(K)$. For the circle group such results are classical (Littlewood [1924]; Zygmund [1959], Vol. II, §8.14).

B.4. (Sketch) Since $L = L_U$ satisfies

$$L(U_\lambda) = (\langle \rho, \rho \rangle - \langle \lambda + \rho, \lambda + \rho \rangle)U_\lambda$$

[because of Lemma 1.6 and (16)] we see from (3) §2 that the Fourier coefficients of $f \in \mathscr{E}(U)$ form a rapidly decreasing hyperfunction. Conversely, let $\{A_\lambda\}_{\lambda \in \Lambda(\pi)}$ be a rapidly decreasing hyperfunction and define $f \in C(U)$ by

$$f(u) = \sum_{\lambda \in \hat{u}} d_\lambda \operatorname{Tr}(A_\lambda U_\lambda(u)),$$

the series being clearly uniformly convergent. Let dU_λ be the differential of U_λ. If $H \in \mathfrak{t}$ then $dU_\lambda(H)$ is a diagonal matrix with diagonal formed by $\mu(H)$ as μ runs over the weights of U_λ. Thus by Theorem 1.3(iv)

$$\mathrm{Tr}(dU_\lambda(H)(dU_\lambda(H))^*) \leq d_\lambda |\lambda|^2 |H|^2,$$

and by the conjugacy theorem for $\mathfrak{u}$ this holds for H replaced by any $X \in \mathfrak{u}$. Since d_λ is majorized by a polynomial in λ (Theorem 1.8) this inequality implies quickly that the series for f remains uniformly convergent after repeated differentiation. Hence $f \in \mathscr{E}(U)$.

B.5. (i) By Exercise B1 we know that $f_S = f * \mu_S$, where μ_S is a measure with Fourier–Stieltjes series

$$\mu_S \sim \sum_{\lambda \in S} d_\lambda \,\mathrm{Tr}(U_\lambda(k))$$

to μ_S is a central idempotent. The converse is obvious. Under the assumption in (ii) μ_S is a positive measure so by a theorem of Wendel [1954], $\mu_S * \mu_S = \mu_S$ implies that μ_S is the Haar measure on a compact subgroup $H \subset K$. The Fourier–Stieltjes series above then shows that S equals the annihilator $H^\perp$ so by Proposition 2.11 we are finished, (cf. Helgason [1958]).

The central idempotent measures for abelian groups were determined (in increasing generality) by Helson [1953], Rudin [1959], and Cohen [1960]; for classes of nonabelian compact groups by Rider [1970, 1971] and Ragozin [1972] (this latter for all simple compact Lie groups).

APPENDIX

In this Appendix we collect some scattered results which are used in the text but which are nevertheless somewhat extraneous to the main themes. In §§1–2 we have primarily followed the treatment in Dixmier [1974], but we have also used some proofs from Chevalley [1955b], Jacobson [1953], and Iwahori [1959]. Proposition 2.12 is due to Jacobson [1951]. In §3 we follow to some extent the treatment in Zariski and Samuel [1958, 1960]; the proof of Theorem 3.5(ii) was kindly communicated to the author by Arthur Mattuck.

§1. The Finite-Dimensional Representations of $\mathfrak{sl}(2, C)$

In this section we write down for reference the finite-dimensional representations of $\mathfrak{sl}(2, C)$; these are familiar from most books on Lie algebras.

The elements

$$H = \begin{pmatrix} 1 & 0 \\ 0 & -1 \end{pmatrix},$$

$$X = \begin{pmatrix} 0 & 1 \\ 0 & 0 \end{pmatrix},$$

$$Y = \begin{pmatrix} 0 & 0 \\ 1 & 0 \end{pmatrix}$$

form a basis of $\mathfrak{sl}(2, C)$ satisfying

$$(1) \qquad [H, X] = 2X, \qquad [H, Y] = -2Y, \qquad [X, Y] = H.$$

Let $k \in Z^+$. Consider the linear mapping

$$\pi_k: \mathfrak{sl}(2, C) \to \mathfrak{gl}(k + 1, C)$$

given by

$$\pi_k(H) = \begin{pmatrix} k & & & \\ & k-2 & & \\ & & \ddots & \\ & & & -k \end{pmatrix} \qquad \text{(diagonal matrix)}$$

$$\pi_k(X) = \begin{pmatrix} 0 & \lambda_1 & 0 & \cdots & 0 \\ 0 & 0 & \lambda_2 & \cdots & 0 \\ & & & & \\ 0 & 0 & \cdots & 0 & \lambda_k \\ 0 & 0 & \cdots & 0 & 0 \end{pmatrix}, \qquad \pi_k(Y) = \begin{pmatrix} 0 & 0 & \cdots & & 0 \\ 1 & 0 & \cdots & & 0 \\ & & & & \\ 0 & 0 & \cdots & 1 & 0 \end{pmatrix}$$

where $\lambda_j = j(k - j + 1)$, $1 \le j \le k$.

Proposition 1.1. *The mapping π_k is an irreducible representation of $\mathfrak{sl}(2, \boldsymbol{C})$ on $\boldsymbol{C}^{k+1}$.*

Proof. To see that π_k is a representation one just verifies directly that $\pi_k(H)$, $\pi_k(X)$, and $\pi_k(Y)$ satisfy the bracket relations in (1). For the irreducibility we can take $k > 0$. Suppose $0 \neq V \subset \boldsymbol{C}^{k+1}$ is an invariant subspace. Let $v \in V - \{0\}$. If $e_1, \ldots, e_{k+1}$ is the standard basis of $\boldsymbol{C}^{k+1}$ we have

(2) $$\pi_k(H)e_i = (k - 2i + 2)e_i \qquad (1 \le i \le k + 1),$$

(3) $$\pi_k(Y)e_i = e_{i+1} \quad (1 \le i \le k), \qquad \pi_k(Y)e_{k+1} = 0,$$

(4) $$\pi_k(X)e_1 = 0, \qquad \pi_k(X)e_{i+1} = \lambda_i e_i \quad (1 \le i \le k).$$

Thus, if $n \in \boldsymbol{Z}^+$ is suitably chosen, $\pi_k(Y)^n v$ is a nonzero multiple of e_{k+1}, whence $e_{k+1} \in V$. Using (4) successively, we see that each $e_i \in V$, so $V = \boldsymbol{C}^{k+1}$.

Lemma 1.2. *Let π be a representation of $\mathfrak{sl}(2, \boldsymbol{C})$ on a finite-dimensional vector space V. Let $v_0 \neq 0$ be an eigenvector of $\pi(H)$, $\pi(H)v_0 = \lambda v_0$, $\lambda \in \boldsymbol{C}$. Then if $k \in \boldsymbol{Z}^+$,*

(i) $\pi(H)\pi(X)^k v_0 = (\lambda + 2k)\pi(x)^k v_0$,
$\pi(H)\pi(Y)^k v_0 = (\lambda - 2k)\pi(Y)^k v_0$.

(ii) *Suppose $\pi(X)v_0 = 0$. Putting*

$$v_i = \pi(Y)^i v_0, \quad i = 0, 1, 2, \ldots,$$

we have

$$\pi(X)v_i = i(\lambda - i + 1)v_{i-1} \quad (i \ge 1),$$

so if $v_{i_0} = 0$, $v_{i_0 - 1} \neq 0$, we have $\lambda = i_0 - 1$.

Proof. (i)

$$\pi(H)\pi(X)v_0 = \pi(X)\pi(H)v_0 + \pi([H, X])v_0 = (\lambda + 2)\pi(X)v_0$$

and the first formula follows by iteration; the other is proved in the same way. For (ii) we use induction on i. For $i = 1$, $\pi(X)\pi(Y)v_0 = \pi(Y)\pi(X)v_0 + \pi(H)v_0 = \lambda v_0$; assuming formula for i we have by the first part

$$\begin{aligned}\pi(X)v_{i+1} &= \pi(X)\pi(Y)v_i = \pi(Y)\pi(X)v_i + \pi(H)v_i \\ &= i(\lambda - i + 1)v_i + (\lambda - 2i)v_i = (i + 1)(\lambda - i)v_i.\end{aligned}$$

Theorem 1.3. *Let π be an irreducible representation of* $\mathfrak{sl}(2, \mathbf{C})$ *on a complex vector space V of dimension $k + 1$. Then π is equivalent to π_k; that is, $\pi_k(X)Av = A\pi(X)v$ ($v \in V$, $X \in \mathfrak{sl}(2, \mathbf{C})$) for a suitable linear bijection $A: V \to \mathbf{C}^{k+1}$.*

Proof. The endomorphism $\pi(H)$ of the complex vector space V has an eigenvector $v \neq 0$, say $\pi(H)v = \mu v$ ($\mu \in \mathbf{C}$). Then by Lemma 1.2(i)

$$\pi(H)\pi(X)^k v = (\mu + 2k)\pi(X)^k v$$

and since $\pi(H)$ has at most finitely many eigenvalues there exists a $k_0 \in \mathbf{Z}^+$ such that $\pi(X)^{k_0}v \neq 0$, $\pi(X)^{k_0+1}v = 0$. Put $v_0 = \pi(X)^{k_0}v$, $\lambda = \mu + 2k_0$, so

$$\pi(H)v_0 = \lambda v_0, \qquad \pi(X)v_0 = 0.$$

Put $v_i = \pi(Y)^i v_0$ ($i \geq 0$) and let v_l be the last nonzero v_i. The vectors $v_0, \ldots, v_l$ have different eigenvalues for $\pi(H)$ and the space $\sum_0^l \mathbf{C}v_i$ is invariant, hence equals V. Thus $l = k$.

Since $(k + 1)(\lambda - k)v_k = \pi(X)v_{k+1} = 0$ we conclude $\lambda = k$. Finally, using (2)–(4) we see that the linear mapping given by $v_0 \to e_1$, $v_1 \to e_2$, $\ldots$, $v_k \to e_{k+1}$ sets up an equivalence between π and π_k.

Corollary 1.4. *Let π be any finite-dimensional representation of* $\mathfrak{sl}(2, \mathbf{C})$. *Then*

(5) $\quad \text{Range } \pi(X) \cap \text{Kernel } \pi(Y) = 0 = \text{Range } \pi(Y) \cap \text{Kernel } \pi(X).$

For the representation $\pi = \pi_k$ this is obvious from (3) and (4). Since the representation π is semisimple ([DS], Chapter III, Exercise B3) it decomposes into irreducibles which by Theorem 1.3 have the form π_k. This proves (5) in general.

Corollary 1.5. *Let π be any finite-dimensional representation of* $\mathfrak{sl}(2, \mathbf{C})$ *on a vector space V. Then*

(i) *The endomorphism $\pi(H)$ is semisimple and its eigenvalues are all integers. Let V_r be the eigenspace of $\pi(H)$ for the eigenvalue r.*

(ii) Kernel $\pi(X) \cap$ Range $\pi(X) \subset \sum_{r>0} V_r \subset$ Range $\pi(X)$.
(iii) Kernel $\pi(Y) \cap$ Range $\pi(Y) \subset \sum_{r<0} V_r \subset$ Range $\pi(Y)$.

Again it suffices to verify this in the case $\pi = \pi_k$ and there it is immediate from (2)–(4).

§2. Representations and Reductive Lie Algebras

1. Semisimple Representations

Let k be a field, $\mathfrak{g}$ a Lie algebra over k and V a vector space over k. *We assume* $k = \boldsymbol{R}$ *or* $\boldsymbol{C}$ *and that* dim $\mathfrak{g}$ *and* dim V *are both finite.* We shall now recall some notions and results about *representations* of $\mathfrak{g}$ on V, i.e., homomorphisms of $\mathfrak{g}$ into $\mathfrak{gl}(V)$. An injective representation is usually called a *faithful representation.*

Let π_1 and π_2 be representations of $\mathfrak{g}$ on V_1 and V_2, respectively; π_1 and π_2 are said to be *equivalent* if there exists a linear bijection $A: V_1 \to V_2$ such that $\pi_2(X)Av = A\pi_1(X)v$ for $v \in V_1$, $X \in \mathfrak{g}$. If we form the direct sum $V = V_1 \oplus V_2$ we can define a representation π of $\mathfrak{g}$ on V by $\pi(X)(v_1 + v_2) = \pi_1(X)v_1 + \pi_2(X)v_2$; π is then called the *direct sum* of π_1 and π_2. This can obviously be extended to several representations π_i of $\mathfrak{g}$.

Given the representations $\pi_1: \mathfrak{g} \to \mathfrak{gl}(V_1)$, $\pi_2: \mathfrak{g} \to \mathfrak{gl}(V_2)$, the *tensor product* $\pi_1 \otimes \pi_2$ is the representation of $\mathfrak{g}$ on $V_1 \otimes V_2$ given by

$$\pi_1 \otimes \pi_2(X)(v_1 \otimes v_2) = (\pi_1(X)v_1) \otimes v_2 + v_1 \otimes (\pi_2(X)v_2),$$

$X \in \mathfrak{g}$, $v_1 \in V_1$, $v_2 \in V_2$. If K is a Lie group with Lie algebra $\mathfrak{k}$ and ρ_i a representation of K on V_i $(i = 1, 2)$ then the representation $\rho_1 \otimes \rho_2$ as defined in Chapter IV, §1, No. 2 has differential

$$d(\rho_1 \otimes \rho_2) = d\rho_1 \otimes d\rho_2.$$

Let π be a representation of $\mathfrak{g}$ on V; π is said to be *irreducible* if the only $\pi(\mathfrak{g})$-invariant subspaces of V are 0 and V; π is said to be *semisimple* (or *completely reducible*) if for each $\pi(\mathfrak{g})$-invariant subspace $W \subset V$ there exists a complementary invariant subspace W', i.e.,

$$V = W \oplus W', \qquad \pi(\mathfrak{g})W' \subset W'. \tag{1}$$

Proposition 2.1. *The representation π is semisimple if and only if it is the direct sum of irreducible representations.*

Proof. Suppose π is a semisimple representation of $\mathfrak{g}$ on V. If $W \subset V$ is an invariant subspace the *subrepresentation* of $\mathfrak{g}$ on W given by $X \to$

$\pi(X)|W$ is also semisimple. Using this remark on W' in (1) and iterating we obtain a direct decomposition $V = \bigoplus_{i=1}^{r} V_i$ where for each i the representation $X \to \pi(X)|V_i$ is irreducible.

Conversely, suppose we have such a direct decomposition of V. Suppose $W \subset V$ is any invariant subspace. Then either $W = V$ or some V_i, say V_{i_1}, is not contained in W. Put $W_1 = W + V_{i_1}$. By the irreducibility, $W \cap V_{i_1} = 0$ so we have the direct decomposition $W_1 = W \oplus V_{i_1}$. Repeating this argument with W_1 in place of W we obtain if $V_{i_2} \not\subset W_1$, and if we put $W_2 = W_1 + V_{i_2}$,

$$W_2 = W \oplus V_{i_1} \oplus V_{i_2}.$$

The process stops when each V_i belongs to a certain W_j, in which case

$$V = W_j = W \oplus V_{i_1} \oplus V_{i_2} \oplus \cdots \oplus V_{i_j}.$$

Then the sum $W' = \sum_{p=1}^{j} V_{i_p}$ satisfies (1).

Proposition 2.2. *Let π be a semisimple representation of $\mathfrak{g}$ on V. Let $V^{\mathfrak{g}}$ denote the joint null space, i.e.,*

$$V^{\mathfrak{g}} = \{v : \pi(X)v = 0 \quad \text{for all } X \in \mathfrak{g}\},$$

and let $[\pi(\mathfrak{g})V]$ denote the subspace of V generated by the vectors $\pi(X)v$, $X \in \mathfrak{g}$, $v \in V$. Then

$$V = V^{\mathfrak{g}} \oplus [\pi(\mathfrak{g})V]$$

and the subspace $[\pi(\mathfrak{g})V]$ is the only $\pi(\mathfrak{g})$-invariant subspace W complementary to $V^{\mathfrak{g}}$.

Proof. If $V = V^{\mathfrak{g}} \oplus W$, $\pi(\mathfrak{g})W \subset W$, we have for each $v \in V$, $X \in \mathfrak{g}$, $v = v_0 + w$ ($v_0 \in V^{\mathfrak{g}}$, $w \in W$), $\pi(X)v = \pi(X)w \in W$. Thus $[\pi(\mathfrak{g})V] \subset W$. Let $W_1 \subset W$ be a $\pi(\mathfrak{g})$-invariant subspace complementary to $[\pi(\mathfrak{g})V]$. If $w_1 \in W_1$, then $\pi(X)w_1 \in [\pi(\mathfrak{g})V] \cap W_1 = 0$ so $w_1 \in V^{\mathfrak{g}}$ whence $w_1 = 0$. Thus $W_1 = 0$ and $W = [\pi(\mathfrak{g})V]$.

Proposition 2.3. *Let π be a representation of the real Lie algebra $\mathfrak{g}$ on the real vector space V. Let $\pi^{\mathbf{C}}$ denote the induced representation of the complexification $\mathfrak{g}^{\mathbf{C}}$ on the complexified vector space $V^{\mathbf{C}}$, i.e.,*

$$\pi^{\mathbf{C}}(X_1 + iX_2)(v_1 + iv_2) = \pi(X_1)v_1 - \pi(X_2)v_2 + i[\pi(X_2)v_1 + \pi(X_1)v_2],$$

$X_j \in \mathfrak{g}$, $v_j \in V$. Then

(i) *π is semisimple $\Leftrightarrow \pi^{\mathbf{C}}$ is semisimple.*

(ii) *If π is irreducible then either $\pi^{\mathbf{C}}$ is irreducible or is the direct sum of two irreducible components.*

Proof. (i) $\Leftarrow$. Let $\sigma: V^{\mathbf{C}} \to V^{\mathbf{C}}$ denote the conjugation of $V^{\mathbf{C}}$ with respect to V. Let $W \subset V$ be a $\pi(\mathfrak{g})$-invariant subspace. Then $W^{\mathbf{C}} = W + iW$ is a $\pi^{\mathbf{C}}(\mathfrak{g}^{\mathbf{C}})$-invariant subspace of $V^{\mathbf{C}}$ so $W^{\mathbf{C}}$ has an invariant complement $Z' \subset V^{\mathbf{C}}$. Then the space

$$Z = (1 + \sigma)[Z' \cap (1 - \sigma)^{-1}(iW)]$$

satisfies

$$V = W \oplus Z, \qquad \pi(\mathfrak{g})Z \subset Z$$

(cf. [DS], Chapter III, solution to Exercise B3), so π is semisimple.

(i) $\Rightarrow$. For this we may assume, by Proposition 2.1, that π is irreducible and $\pi^{\mathbf{C}}$ not irreducible. Let $Z \subset V^{\mathbf{C}}$ be any $\pi^{\mathbf{C}}(\mathfrak{g}^{\mathbf{C}})$-invariant subspace, $Z \neq \{0\}$, $Z \neq V^{\mathbf{C}}$. Then

$$V^{\mathbf{C}} = Z + \sigma Z, \qquad Z \cap (\sigma Z) = \{0\}. \tag{2}$$

In fact, since $\sigma(Z + \sigma Z) = Z + \sigma Z$ we have $Z + \sigma Z = Y + iY$ where $Y = (Z + \sigma Z) \cap V$. Then $Y \neq 0$ is a $\pi(\mathfrak{g})$-invariant subspace of V so $Y = V$ and $V^{\mathbf{C}} = Z + \sigma Z$. Similarly, $Z \cap (\sigma Z) = \{0\}$ so (2) holds. Also σZ is a complex subspace of $V^{\mathbf{C}}$ since $\alpha\sigma(v) = \sigma(\bar{\alpha}v)$ for $\alpha \in \mathbf{C}$, $v \in V^{\mathbf{C}}$. Finally,

$$\pi^{\mathbf{C}}(X_1 + iX_2)\sigma v = \sigma\pi^{\mathbf{C}}(X_1 - iX_2)v \tag{3}$$

so σZ is $\pi^{\mathbf{C}}(\mathfrak{g}^{\mathbf{C}})$-invariant. Thus (2) implies that $\pi^{\mathbf{C}}$ is semisimple.

(ii) It suffices to show that the space Z in (2) above is irreducible under $\pi^{\mathbf{C}}(\mathfrak{g}^{\mathbf{C}})$. But if $U \subset Z$ is an invariant subspace, $U \neq 0$, $U \neq Z$, then (2) holds for it, which is a contradiction.

2. Nilpotent and Semisimple Elements

Generalizing the notion of nilpotent and semisimple linear transformations we shall now define such notions in a semisimple Lie algebra.

Lemma 2.4. *Let V be a vector space over k and $\mathfrak{g} = \mathfrak{gl}(V)$. Let $X \in \mathfrak{g}$ be nilpotent. Then $\operatorname{ad}_{\mathfrak{g}} X$ is nilpotent.*

Proof. Since $(\operatorname{ad}_{\mathfrak{g}} X)Y = XY - YX$ we have

$$(\operatorname{ad}_{\mathfrak{g}} X)^p(Y) = \sum_{i=0}^{p} (-1)^i \binom{p}{i} X^{p-i} Y X^i,$$

so the lemma follows.

Lemma 2.5. *Let* $\mathfrak{g} = \mathfrak{gl}(n, k)$ *and let* $A \in \mathfrak{g}$ *be a diagonal matrix*

$$A = d_1 E_{11} + \cdots + d_n E_{nn}.$$

Then

$$\mathrm{ad}_{\mathfrak{g}} A(E_{ij}) = (d_i - d_j) E_{ij}$$

so in the basis (E_{ij}) *of* $\mathfrak{g}$, $ad_{\mathfrak{g}}\, A$ *is given by a diagonal matrix.*

Lemma 2.6. *Let V be a vector space over k, A an endomorphism of V, and* $A = S + N$ *its additive Jordan decomposition, S semisimple, N nilpotent, and* $SN = NS$. *Let* $\mathfrak{g} = \mathfrak{gl}(V)$. *Then*

$$\mathrm{ad}_{\mathfrak{g}}\, A = \mathrm{ad}_{\mathfrak{g}}\, S + \mathrm{ad}_{\mathfrak{g}}\, N \tag{4}$$

is the additive Jordan decomposition of $ad_{\mathfrak{g}}\, A$.

Proof. Passing to the complexifications in the case $k = \boldsymbol{R}$ we may assume $k = \boldsymbol{C}$. Then S is diagonalizable and so is $\mathrm{ad}_{\mathfrak{g}}\, S$ by Lemma 2.5. Also Lemma 2.4 implies that $\mathrm{ad}_{\mathfrak{g}}\, N$ is nilpotent. Since in addition

$$[\mathrm{ad}_{\mathfrak{g}}, S, \mathrm{ad}_{\mathfrak{g}}\, N] = \mathrm{ad}_{\mathfrak{g}}([S, N]) = 0$$

(4) is indeed the Jordan decomposition.

Proposition 2.7.* *Let* $\mathfrak{g}$ *be a semisimple subalgebra of* $\mathfrak{gl}(V)$. *Let* $A \in \mathfrak{g}$ *be arbitrary and* $A = S + N$ *is additive Jordan decomposition. Then* $S, N \in \mathfrak{g}$.

Proof. In the case $k \in \boldsymbol{R}$ we pass to the complexification so we may assume $k = \boldsymbol{C}$. Let $\mathscr{V}$ be the set of subspaces of V invariant under $\mathfrak{g}$. For each $W \in \mathscr{V}$ let

$$\mathfrak{g}_W = \{X \in \mathfrak{gl}(V) : X(W) \subset W, \mathrm{Tr}(X \mid W) = 0\}.$$

Then $\mathfrak{g}_W$ is a subalgebra of $\mathfrak{gl}(V)$; it contains $\mathfrak{g}$ because, by semisimplicity, $\mathfrak{g} = [\mathfrak{g}, \mathfrak{g}]$. Next consider the normalizer

$$\mathfrak{n} = \{A \in \mathfrak{gl}(V) : [A, \mathfrak{g}] \subset \mathfrak{g}\}$$

and put

$$\mathfrak{g}_* = \mathfrak{n} \cap \left(\bigcap_{W \in \mathscr{V}} \mathfrak{g}_W \right).$$

Let $A \in \mathfrak{g}_*$, put $\tilde{A} = \mathrm{ad}_{\mathfrak{gl}(V)}(A)$, and let

$$A = S + N, \qquad \tilde{A} = \tilde{S} + \tilde{N}$$

* Cf. [DS], Chapter IX, Exercise A6.

be their additive Jordan decompositions. By (4), $\tilde{S} = \mathrm{ad}_{\mathfrak{gl}(V)}(S)$, $\tilde{N} = \mathrm{ad}_{\mathfrak{gl}(V)}(N)$. Since $\tilde{S}$ and $\tilde{N}$ are polynomials in $\tilde{A}$ we have $S, N \in \mathfrak{n}$. Since S and N are polynomials in A they leave each $W \in \mathscr{V}$ invariant; also $\mathrm{Tr}(N|W) = 0$ and $\mathrm{Tr}(S|W) = \mathrm{Tr}(A|W) = 0$ so $S, N \in \mathfrak{g}_W$. Thus S and N belong to $\mathfrak{g}_*$.

It suffices now to prove $\mathfrak{g}_* = \mathfrak{g}$.

Since $\mathfrak{g}$ is an ideal in $\mathfrak{g}_*$ the Killing form B of $\mathfrak{g}$ is the restriction of the Killing form B_* of $\mathfrak{g}_*$. Hence B_* is nondegenerate on $\mathfrak{g}$ so we have the direct decomposition $\mathfrak{g}_* = \mathfrak{g} \oplus \mathfrak{g}^\perp$ where the orthogonal complement $\mathfrak{g}^\perp$ is an ideal in $\mathfrak{g}_*$. Now since the representation of $\mathfrak{g}$ on V is semisimple (Weyl's theorem) we have $V = \bigoplus_i V_i$ where each V_i is $\mathfrak{g}$-irreducible. Each $X \in \mathfrak{g}^\perp$ commutes elementwise with $\mathfrak{g}$ so by Schur's lemma it acts on V_i by a scalar multiplication. Since $\mathrm{Tr}(X|V_i) = 0$ we conclude $X|V_i = 0$ so $X = 0$ and $\mathfrak{g} = \mathfrak{g}_*$.

Corollary 2.8. *With the notation of Proposition* 2.7, $A \in \mathfrak{g}$ *is semisimple (nilpotent) if and only if* $\mathrm{ad}_{\mathfrak{g}}(A)$ *is semisimple (nilpotent).*

Let $A = S + N$ be the additive Jordan decomposition. Since $\mathrm{ad}_{\mathfrak{g}}(A) = \mathrm{ad}_{\mathfrak{gl}(V)}(A)|\mathfrak{g}$ we deduce from Lemma 2.6 that $\mathrm{ad}_{\mathfrak{g}}(A) = \mathrm{ad}_{\mathfrak{g}}(S) + \mathrm{ad}_{\mathfrak{g}}(N)$ is the additive Jordan decomposition of $\mathrm{ad}_{\mathfrak{g}}(A)$. Now $A = S \Leftrightarrow \mathrm{ad}_{\mathfrak{g}}(A) = \mathrm{ad}_{\mathfrak{g}}(S)$, so the corollary is verified.

Corollary 2.9. *Let* $\mathfrak{g}$ *be a semisimple Lie algebra,* $A \in \mathfrak{g}$. *The following conditions are equivalent:*

(i) $\mathrm{ad}_{\mathfrak{g}}(A)$ *is semisimple (resp. nilpotent).*

(ii) *There exists a faithful representation* π *of* $\mathfrak{g}$ *such that* $\pi(A)$ *is semisimple (resp. nilpotent).*

(iii) *For each representation* π *of* $\mathfrak{g}$ $\pi(A)$ *is semisimple (resp. nilpotent).*

Proof. (iii) $\Rightarrow$ (ii) is obvious; (ii) $\Rightarrow$ (i) by Corollary 2.8. Finally, (i) $\Rightarrow$ (iii) as follows: assuming $\mathrm{ad}_{\mathfrak{g}}(A)$ semisimple, consider a representation π of $\mathfrak{g}$ on V. Let $\mathfrak{g}' = \mathfrak{g}/\mathrm{Kernel}(\pi)$, view $\mathfrak{g}'$ as the subalgebra $\pi(\mathfrak{g}) \subset \mathfrak{gl}(V)$, and put $A' = \rho(A)$ where $\rho: \mathfrak{g} \to \mathfrak{g}'$ is the natural mapping. If $\mathfrak{v} \subset \mathfrak{g}'$ satisfies $[A', \mathfrak{v}] \subset \mathfrak{v}$ then $[A, \rho^{-1}\mathfrak{v}] \subset \rho^{-1}\mathfrak{v}$ so there exists a $\mathfrak{u} \subset \mathfrak{g}$ such that $\rho^{-1}\mathfrak{v} \oplus \mathfrak{u} = \mathfrak{g}$, $[A, \mathfrak{u}] \subset \mathfrak{u}$. But then $\rho(\mathfrak{u})$ satisfies $\mathfrak{v} \oplus \rho(\mathfrak{u}) = \mathfrak{g}'$ and $[A', \rho(\mathfrak{u})] \subset \rho(\mathfrak{u})$. Thus $\mathrm{ad}_{\mathfrak{g}'}(A')$ is semisimple. By Corollary 2.8 used on $\pi(\mathfrak{g}) = \mathfrak{g}' \subset \mathfrak{gl}(V)$ we conclude that $A' = \pi(A)$ is semisimple as desired. The nilpotent case of (i) $\Rightarrow$ (iii) is proved in a similar way.

Definition. Let $\mathfrak{g}$ be a semisimple Lie algebra. An element $A \in \mathfrak{g}$ verifying the conditions of Corollary 2.9 is said to be *semisimple* (resp. *nilpotent*).

Corollary 2.10.

(i) *Let* $\mathfrak{g}$ *be a semisimple Lie algebra. Then each* $A \in \mathfrak{g}$ *has a unique decomposition*

$$A = S + N \tag{5}$$

where $S \in \mathfrak{g}$ *is semisimple,* $N \in \mathfrak{g}$ *is nilpotent.*

(ii) *Let* $\mathfrak{g}_1 \subset \mathfrak{g}$ *be a semisimple subalgebra. If* $A \in \mathfrak{g}_1$, *then* $S, N \in \mathfrak{g}_1$. *An element* $A \in \mathfrak{g}_1$ *is semisimple (resp. nilpotent) in* $\mathfrak{g}_1$ *if and only if it is semisimple (resp. nilpotent) in* $\mathfrak{g}$.

This follows directly from Proposition 2.7.

Definition. Decomposition (5) is called the *additive Jordan decomposition of A.*

3. Reductive Lie Algebras

Reductive Lie algebras are natural generalizations of semisimple Lie algebras; in this subsection we prove for them a few results which are useful primarily in Chapter III, §4. A Lie algebra will mean a Lie algebra over $\boldsymbol{R}$ or $\boldsymbol{C}$ and a representation will mean a finite-dimensional representation.

Let $\mathfrak{g}$ be a Lie algebra, $\mathfrak{h} \subset \mathfrak{g}$ a subalgebra.

Definition. The subalgebra $\mathfrak{h} \subset \mathfrak{g}$ is said to be *reductive in* $\mathfrak{g}$ if the representation $H \to \operatorname{ad}_{\mathfrak{g}}(H)$ of $\mathfrak{h}$ on $\mathfrak{g}$ is completely reducible (i.e., semisimple). A Lie algebra $\mathfrak{g}$ is said to be *reductive* if it is reductive in itself. Obviously, if $\mathfrak{h}$ is reductive in $\mathfrak{g}$ then $\mathfrak{h}$ is reductive.

Proposition 2.11. *A Lie algebra* $\mathfrak{g}$ *is reductive if and only if it is the direct sum*

$$\mathfrak{g} = \mathfrak{a} + \mathfrak{s} \tag{6}$$

of an abelian ideal $\mathfrak{a}$ *and a semisimple ideal* $\mathfrak{s}$. *In that case,*

$$\mathfrak{a} = \text{center}(\mathfrak{g}), \qquad \mathfrak{s} = [\mathfrak{g}, \mathfrak{g}]. \tag{7}$$

Proof. Since the representation $\operatorname{ad}_{\mathfrak{g}}$ is semisimple $\mathfrak{g}$ has a decomposition $\mathfrak{g} = \sum \mathfrak{g}_i$ into subspaces on which $\operatorname{ad}_{\mathfrak{g}}$ acts irreducibly. In particular, each $\mathfrak{g}_i$ is an ideal of $\mathfrak{g}$. Each ideal of $\mathfrak{g}_i$ is an ideal in $\mathfrak{g}$ so ([DS], Chapter II, §6; Chapter III, Exercise B8) $\mathfrak{g}_i$ is either a simple Lie algebra or one-dimensional. Decomposition (6) now follows by taking $\mathfrak{s}$ as the sum of the simple $\mathfrak{g}_i$ and $\mathfrak{a}$ the sum of the 1-dimensional $\mathfrak{g}_i$.

The converse is clear since by the semisimplicity of $\mathfrak{s}$, $\operatorname{ad}_{\mathfrak{g}}(\mathfrak{g}) = \operatorname{ad}_{\mathfrak{g}}(\mathfrak{s})$ is semisimple.

Proposition 2.12. *Let $\mathfrak{g}$ be a reductive Lie algebra (over $\boldsymbol{R}$ or $\boldsymbol{C}$) and π a representation of $\mathfrak{g}$ on a finite-dimensional vector space V. Then π is semisimple if and only if the restriction $\pi|$center $(\mathfrak{g})$ is semisimple.*

Proof. We shall use the familiar fact from linear algebra that a commutative family of endomorphisms is semisimple if and only if each member of the family is semisimple.

Assume first π is semisimple and consider the decomposition (6). Fix $H \in \mathfrak{a}$ and write $\pi(H) = N + S$ where N and S are the nilpotent and semisimple parts of $\pi(H)$, respectively. Let W be the null space of N. Since N is a polynomial in $\pi(H)$ ([DS], Chapter III, §1) N commutes with each element in $\pi(\mathfrak{g})$. Hence $\pi(\mathfrak{g})$ leaves the subspace $W \subset V$ invariant. Let $W' \subset V$ be a complementary invariant subspace. Since $\pi(H)W' \subset W'$ we have $NW' \subset W'$ and the restriction $N' = N|W'$ is nilpotent. But since $W \cap W' = 0$, N' has no null vector $x \neq 0$ in W'. But, N' being nilpotent, this implies $N' = 0$ ([DS], Chapter III, §1). Thus $N = 0$, so $\pi(H)$ is semisimple, whence by the initial remark, $\pi|\mathfrak{a}$ is semisimple.

For the converse we may, by Proposition 2.3, assume that $\mathfrak{g}$ is a Lie algebra over $\boldsymbol{C}$. The commutative family $\pi(\mathfrak{a})$ now assumed semisimple, let $V = \bigoplus_i V_i$ be the direct decomposition into joint eigenspaces of $\pi(\mathfrak{a})$. Then each V_i is $\pi(\mathfrak{g})$-invariant and $\pi(\mathfrak{g})|V_i = \pi(\mathfrak{s})|V_i$. Since $\mathfrak{s}$ is semisimple this last family of operators is semisimple. Thus the representation $\pi_i\colon X \to \pi(X)|V_i$ of $\mathfrak{g}$ on V_i is semisimple. Decomposing this into irreducibles we conclude from Proposition 2.1 that π is semisimple.

Proposition 2.13. *Let $\mathfrak{g}$ be a reductive Lie algebra, π_1 and π_2 two semisimple representations of $\mathfrak{g}$. Then $\pi_1 \otimes \pi_2$ is semisimple.*

Proof. Because of Propositions 2.1 and 2.3 we may assume $\mathfrak{g}$ complex, π_1 and π_2 irreducible. Using (6) we see that $\pi_1(\mathfrak{a})$ commutes elementwise with $\pi_1(\mathfrak{g})$ so by Schur's lemma consists of scalar multiples of the identity. The same holds for $\pi_2(\mathfrak{a})$ and therefore also for $(\pi_1 \otimes \pi_2)(\mathfrak{a})$, whence, by Proposition 2.12, $\pi_1 \otimes \pi_2$ is semisimple.

Remark. The result holds also for $\mathfrak{g}$ nonreductive (see, e.g., Dixmier [1974], Chapter I).

Proposition 2.14. *Suppose the subalgebra $\mathfrak{h}$ of $\mathfrak{g}$ is reductive in $\mathfrak{g}$ and let π be a representation of $\mathfrak{g}$ on V.*

(i) *Let W be the sum of the subspaces W_i of V which are invariant and irreducible under $\pi(\mathfrak{h})$. Then $\pi(\mathfrak{g})W \subset W$.*

(ii) *If π is semisimple, so is $\pi|\mathfrak{h}$.*

Proof. Let $W_i \subset V$ be a subspace which is $\pi(\mathfrak{h})$-invariant and irreducible and consider the representation $\rho: X \to \mathrm{ad}_{\mathfrak{g}}(X)$ of $\mathfrak{h}$ on $\mathfrak{g}$. By Prop. 2.13 the representation $\rho \otimes (\pi|\mathfrak{h})$ of $\mathfrak{h}$ on $\mathfrak{g} \otimes W_i$ is semisimple. Consider now the linear map $A: \mathfrak{g} \otimes W_i \to V$ given by $A(X \otimes w) = \pi(X)w$. Then A commutes with the action of $\mathfrak{h}$; in fact if $Y \in \mathfrak{h}$

$$\begin{aligned} A((\rho \otimes (\pi|\mathfrak{h}))(Y)(X \otimes w)) &= A([Y, X] \otimes w + X \otimes \pi(Y)w) \\ &= \pi([Y, X])\omega + \pi(X)\pi(Y)w = \pi(Y)A(X \otimes w). \end{aligned}$$

Hence we have a semisimple representation of $\mathfrak{h}$ on $A(\mathfrak{g} \otimes W_i)$. Thus $\pi(\mathfrak{g})W_i = A(\mathfrak{g} \otimes W_i) \subset W$. This proves (i). For part (ii) we may assume π irreducible. Then by (i) $W = V$. Thus $V = \sum_i W_i$ and by irreducibility $W_i \cap W_j = 0$ unless $W_i = W_j$. Thus V is the direct sum of certain W_i so $\pi|\mathfrak{h}$ is semisimple.

Corollary 2.15. *If $\mathfrak{h}$ is reductive in $\mathfrak{g}$ and $\mathfrak{g}$ reductive in $\mathfrak{k}$, then $\mathfrak{h}$ is reductive in $\mathfrak{k}$.*

Proposition 2.16. *Let $\mathfrak{k}$ be semisimple and $\mathfrak{g}$ reductive in $\mathfrak{k}$. If $X \in \mathfrak{g}$ and $\mathrm{ad}_{\mathfrak{k}}(X)$ nilpotent, then $X \in [\mathfrak{g}, \mathfrak{g}]$.*

Proof. We may assume $\mathfrak{g}$ and $\mathfrak{k}$ complex. Let $X = C + A$ where $C \in$ center($\mathfrak{g}$) and $A \in [\mathfrak{g}, \mathfrak{g}]$ and let $A = S + N$ be the additive Jordan decomposition of A in $[\mathfrak{g}, \mathfrak{g}]$. Let $\pi = \mathrm{ad}_{\mathfrak{k}}|\mathfrak{g}$. Then π is a faithful semisimple representation of $\mathfrak{g}$ which has a decomposition $\pi = \bigoplus_i \pi_i$ into irreducibles π_i. By 2.9 for $[\mathfrak{g}, \mathfrak{g}]$, $\pi(N)$ is nilpotent, $\pi(S)$ semisimple. By Prop. 2.12, $\pi(C)$ is semisimple, so since $\pi(S)$ and $\pi(C)$ commute, $\pi(S + C)$ is semisimple. Since $\pi(X) = \pi(N) + \pi(S + C)$ is nilpotent and $\pi(N)$ and $\pi(S + C)$ commute it follows that $\pi(S + C) = 0$. For each i $\pi_i(C)$ commutes elementwise with the irreducible $\pi_i(\mathfrak{g})$, so is a scalar, and $\mathrm{Tr}(\pi_i(S)) = \mathrm{Tr}(\pi_i(A)) = 0$. Hence $\pi_i(C) = 0$, so $\pi(C) = 0$ and $C = 0$, as desired.

§3. Some Algebraic Tools

We recall that "field" always means commutative field of characteristic 0. In this section we only consider rings and algebras which are commutative, have an identity element, and no divisors of 0. A module M over a ring A is called *finite* if there exist finitely many elements $x_1, \ldots, x_m \in M$ such that $M = Ax_1 + \cdots + Ax_m$.

Definition. Let B be a ring and A a subring of B (with the same identity). An element $x \in B$ is called *integral* over A if it satisfies an equation

$$x^n + a_1 x^{n-1} + \cdots + a_n = 0 \tag{1}$$

with leading coefficient 1, where $a_i \in A$. If each $x \in B$ is integral over A, the ring B is said to be integral over A.

Lemma 3.1. *Let A be a subring of B and let $x \in B$. The following conditions are equivalent:*

(i) *x is integral over A.*
(ii) *The ring $A[x]$ (the subring of B generated by A and x) is a finite A-module.*
(iii) *There exists an intermediate ring R, $A[x] \subset R \subset B$, such that R is a finite A-module.*

Proof. (i)$\Rightarrow$(ii) There exists an integer $n > 0$ such that $x^n \in \sum_{i=0}^{n-1} Ax^i$. It follows that $x^{n+r} \in \sum_{i=0}^{n-1} Ax^{i+r}$ so by induction on r, $x^{n+r} \in \sum_{i=0}^{n-1} Ax^i$ for each $r > 0$.

(ii)$\Rightarrow$(iii) Take $R = A[x]$.

(iii)$\Rightarrow$(i) Select $y_1, \ldots, y_n \in R$ such that $R = \sum_{i=1}^n Ay_i$. Then we have for suitable elements $a_{ij} \in A$, $xy_i = \sum_{j=1}^n a_{ij}y_j$ $(1 \le i \le n)$. Writing this system of linear equations as

$$\sum_{j=1}^{n} (\delta_{ij}x - a_{ij})y_j = 0 \qquad (1 \le i \le n),$$

we conclude that the determinant $d = \det(\delta_{ij}x - a_{ij})$ satisfies $dy_i = 0$ for each i. Then $dR = 0$ so $d = 0$. But this is an equation of the form (1) so x is integral over A.

By induction we conclude from Lemma 3.1,

Lemma 3.2. *Let $x_1, \ldots, x_n$ be elements of a ring B which are integral over a subring A. Then the ring $A[x_1, \ldots, x_n]$ (the subring of B generated by A and $x_1, \ldots, x_n$) is a finite A-module.*

From this lemma and Lemma 3.1(iii) we obtain the following:

Corollary 3.3. *Let A be a subring of a ring B. The set of elements in B which are integral over A form a subring $\bar{A}$ of B containing A.*

Definition. The subring $\bar{A}$ is called the *integral closure* of A in B.

Definition. A ring A is called *integrally closed* if it coincides with its integral closure in the quotient field $C(A)$ of A.

Lemma 3.4. *A unique factorization domain A is integrally closed.*

Proof. Let $x \in C(A)$ be integral over A. Then $x = \alpha/\beta$ where we may assume that α and β are relatively prime. Now x satisfies an equation

$$x^n + a_1x^{n-1} + \cdots + a_n = 0$$

so

$$\alpha^n + a_1\alpha^{n-1}\beta + \cdots + \alpha_n\beta^n = 0.$$

If β has a prime factor γ, then γ must divide α^n and therefore γ divides α. This contradicts the fact that α and β are relatively prime. Hence $x \in A$.

In particularly, the symmetric algebra $S(V)$ over a finite-dimensional vector space V is integrally closed.

Theorem 3.5. *Let K be an algebraically closed field and A and B finitely generated algebras (with identity) over K. Suppose that $A \subset B$ and that B is integral over A. Then*

(i) *Each homomorphism $\varphi: A \to K$ extends to a homomorphism ψ of B into K.*

(ii) *If each homomorphism $\varphi: A \to K$ extends uniquely to a homomorphism of B into K then the quotient fields $C(A)$ and $C(B)$ coincide.*

Proof of (i). If $\varphi(A) = \{0\}$ we define ψ by $\psi(B) = \{0\}$. If $\varphi(A) \neq \{0\}$ then the kernel of φ is a maximal ideal $\mathfrak{m}$ of A. We first prove the existence of a maximal ideal $\mathfrak{n}$ of B such that $\mathfrak{n} \cap A = \mathfrak{m}$. For this purpose consider the set of all ideals $\mathfrak{p}$ of B satisfying $\mathfrak{p} \cap A \subset \mathfrak{m}$. This set is partially ordered under inclusion and every (totally) ordered subset has an upper bound. By Zorn's lemma there exists a maximal element $\mathfrak{n}$ of the set. If $\mathfrak{n} \cap A$ is a proper subset of $\mathfrak{m}$, let x be an element in $\mathfrak{m}$ which does not belong to $\mathfrak{n}$. Then $\mathfrak{n}$ is a proper subset of the ideal $\mathfrak{n} + Bx$; by the choice of $\mathfrak{n}$, $(\mathfrak{n} + Bx) \cap A$ is not a part of $\mathfrak{m}$. In other words, there exists an element $z \in B$ and an element $y \in A$, not in $\mathfrak{m}$, such that $zx - y \in \mathfrak{n}$. Now z satisfies an equation $z^n + a_1 z^{n-1} + \cdots + a_n = 0$, $a_i \in A$, which after multiplication by x^n and use of the congruence $zx \equiv y \pmod{\mathfrak{n}}$ gives

$$y^n + a_1 x y^{n-1} + \cdots + a_n x^n \equiv 0 \pmod{\mathfrak{m}}.$$

This contradicts $x \in \mathfrak{m}$, $y \notin \mathfrak{m}$. Hence $\mathfrak{n} \cap A = \mathfrak{m}$.

If $\mathfrak{n}$ were not a maximal ideal in B, suppose $\mathfrak{n}'$ is a maximal ideal in B satisfying the proper inclusions $\mathfrak{n} \subset \mathfrak{n}' \subset B$. Then $\mathfrak{n}' \cap A$ is an ideal in A properly containing $\mathfrak{m}$ so $A \subset \mathfrak{n}'$. Let $x \in B$. Since x is integral over $\mathfrak{n}'$ we have $x^l \in \mathfrak{n}'$ for some integer l. Since $B/\mathfrak{n}'$ is a field we conclude that $x \in \mathfrak{n}'$. Thus $\mathfrak{n}' = B$, which is a contradiction. This proves that $\mathfrak{n}$ is a maximal ideal in B so $B/\mathfrak{n}$ is a field. Let 1 denote the identity of A. Then the mapping $\alpha \to \alpha \cdot 1$ $(\alpha \in K)$ is an isomorphism of K into A and the mapping $\alpha \to \alpha \cdot 1 + \mathfrak{n}$ is an isomorphism of K into $B/\mathfrak{n}$. Now select $b_1, \ldots, b_n \in B$ such that $B = K[b_1, \ldots, b_n]$. Then $B = A[b_1, \ldots, b_n]$ so by Lemma 3.2, B is a finite A-module. Writing $B = Ax_1 + \cdots + Ax_m$ $(x_i \in B)$ and using $A = K + \mathfrak{m}$ we see that $B/\mathfrak{n}$ is a finite extension of K. Since K is algebraically closed we have $B/\mathfrak{n} = K$. The natural mapping $\psi: B \to K$ gives the desired extension of φ.

Proof of (ii). In order to prove (ii) we make use of the following theorem (the Noether normalization theorem, see, e.g., Zariski and Samuel [1958], p. 266).

Let $R = k[y_1, \ldots, y_n]$ be a finitely generated algebra over a field k and let d be the transcendence degree of the quotient field $k(y_1, \ldots, y_n)$ over k. There exist d linear combinations $z_1, \ldots, z_d$ of the y_i with coefficients in k, algebraically independent over k, such that R is integral over $k[z_1, \ldots, z_d]$.

Combining this theorem with (i) we see that if k is algebraically closed there exists a homomorphism of $k[y_1, \ldots, y_n]$ onto k.

Suppose now that the quotient fields $C(A)$ and $C(B)$ were different. Let $a_1, \ldots, a_m$ be a set of generators of A so $A = K[a_1, \ldots, a_m]$. Pick any element $\alpha \in B$ which does not belong to $C(A)$. We shall find a homomorphism of $K[a_1, \ldots, a_m]$ into K which has more than one extension to a homomorphism of $K[a_1, \ldots, a_m, \alpha]$ into K. Since B is integral over $K[a_1, \ldots, a_m, \alpha]$, (ii) will then follow from (i). Let

$$p_\alpha(x) = x^n + f_1(a)x^{n-1} + \cdots + f_n(a) = 0$$

be the polynomial with coefficients in the field $C(A) = K(a_1, \ldots, a_m)$ of lowest degree having α as a zero and leading coefficient 1. [Here a stands for the m-tuple $(a_1, \ldots, a_m)$]. Let $q(a)$ denote the product of all the denominators of all the $f_i(a)$ with the discriminant of the polynomial $p_\alpha(x)$. From the remark above we see that there exists a homomorphism φ of $K[a_1, \ldots, a_m, 1/q(a)]$ onto K. The image of the polynomial $p_\alpha(x)$ under φ will then be a polynomial with coefficients in K having n distinct roots, say $\alpha_1, \ldots, \alpha_n$. Fix one α_i. We wish to extend φ (or more precisely, the restriction of φ to $K[a_1, \ldots, a_m]$) to a homomorphism $\psi_i : K[a_1, \ldots, a_m, \alpha] \to K$ by putting $\psi_i(\alpha) = \alpha_i$. The condition for this being possible is that whenever α satisfies a polynomial equation $p(a, x) = 0$ with coefficients in $K[a_1, \ldots, a_m]$ then α_i satisfies the corresponding equation $p(\varphi(a), x) = 0$ with coefficients in K. Since the polynomial $p_\alpha(x)$ has minimum degree it divides $p(a, x)$:

$$p(a, x) = p_\alpha(x)q(a, x).$$

Here the polynomial $q(a, x)$ can be found by long division; its coefficients $g_i(a)$ are rational expressions in $a_1, \ldots, a_m$ whose denominators divide the product of the denominators of the $f_i(a)$. Since φ does not map this product into 0 it is clear that $p(\varphi(a), x)$ vanishes for $x = \alpha_i$. Now $n > 1$ and the homomorphisms ψ_i $(1 \leq i \leq n)$ are all different. This concludes the proof.

SOME DETAILS

As explained in the preface the present short section furnishes detailed explanations of a few statements in the text, the proof of which is only indicated there.

Page 9, line 11^-: As remarked by Stetkær one does not have to use the analyticity here. For a differential equation $y'' + f_1(r)y' + f_0(r)y = 0$ it is a classical result that if $y_1(x)$ is a solution then

$$y_2 = y_1 \int \frac{1}{y_1^2} e^{-\int f_1(r)\,dr}\,dr$$

is another solution. In our case formula (13a) shows that the solution $y_1(r) = \varphi_{\lambda,n}(r)$ is divisible by $r^{|n|}$. As a result, $y_2(r)$ is singular near $r = 0$, also for $n = 0$. This shows that all solutions which are smooth near $r = 0$ are proportional.

Page 19, line 12^+: The remark for Page 9 applies here to show that all smooth solutions are proportional.

Page 27, line 1^+: Combining this formula with Lemma 3.6 we obtain

$$\int_{\mathbf{S}^{n-1}} e^{i(x,w)} p(w)\,dw = \left(\frac{i}{2\pi}\right)^k p(x) \int_{\mathbf{S}^{n+2k-1}} e^{i(x,w)}\,dw$$

Page 28, line 16^-: Since $\Phi_{\lambda,\delta}(re_n) = \Phi_{r\lambda,\delta}(e_n)$ this constant is independent of λ (remark by Stetkær).

Page 30, line 1^+: In fact, putting $t = x(\tau)/|x|$ we get

$$L(\gamma) \geq \int_\alpha^\beta \frac{|x'(\tau)|}{1 - x(\tau)^2}\,d\tau \geq \int_\alpha^\beta \frac{x'(\tau)}{1 - x(\tau)^2}\,d\tau = \int_0^1 \frac{|x|}{1 - t^2x^2}\,dt = L(\gamma_0)$$

Page 50, line 7^+: The remark for Page 9 shows that indeed all the smooth solutions of (45) are proportional.

Page 64, line 8^+, 6^-: The function $e^{(i\lambda+1)\langle z,\zeta\rangle}$ can be written

$$(1-z\bar{z})^{\nu}\bar{z}^{-\nu}\left(\frac{\zeta-z}{\zeta}\right)^{-\nu}\left(\frac{1}{\bar{z}}-\zeta\right)^{-\nu}.$$

If $|\zeta|=1$ then $\frac{\zeta-z}{\zeta}\in S_{|z|}(1)$ so $\log\left(\frac{\zeta-z}{\zeta}\right)$ is holomorphic for ζ in an annulus around $S_1(0)$. Also, $\log\left(\frac{1}{\bar{z}}-\zeta\right)$ is holomorphic for ζ in a slit plane from $\frac{1}{\bar{z}}$ to ∞.

Page 65, line 7^+: Here we give a more detailed proof of the formula

$$(1)\qquad \int_K\left(\int_B f(k,b)\,dT(b)\right)dk=\int_B\left(\int_K f(k,b)\,dk\right)dT(b),$$

where $f(k,b)=e^{(i\lambda+1)\langle gk\cdot z,b\rangle}$, $g\in G$, $z\in D$ fixed, since the text only gives and indication of a proof.

The function $b\to f(k,b)$ extends to a holomorphic function in an annulus U containing B (same for all k). This extension has a Laurent expansion

$$(2)\qquad f(k,\zeta)=\sum_{-\infty}^{\infty}A_n(k)\zeta^n\,,\ A_n(k)=\tfrac{1}{2\pi i}\int_{|\zeta|=R}\frac{f(k,\zeta)}{\zeta^{n+1}}\,d\zeta\,,$$

converging uniformly on each compact subset $C\subset U$. We have

$$(3)\qquad |A_n(k)|\leq\frac{M}{R^n}$$

whenever the circle $|\zeta|=R$ is contained in U, M being a constant. As $N\to\infty$ we have

$$\sum_{-N}^{N}A_n(k)\zeta^n\to f(k,\zeta)$$

uniformly on $K\times C$. After taking $\int_K dk$ we still have uniform convergence for $\zeta\in C$. Since the embedding $\mathcal{H}(U)\to\mathcal{A}(B)$ is continuous this implies

$$(4)\quad \int_B\left(\int_K\sum_{-N}^{N}A_n(k)\,dk\right)\chi_n(b)\,dT(b)\to\int_B\left(\int_K f(k,b)\,dk\right)dT(b),$$

χ_n being as in Proposition 4.11.

On the other hand, the left hand side of (4) equals

$$(5)\qquad \sum_{-N}^{N}\int_K\left(\int_B A_n(k)\chi_n(b)\,dT(b)\right)dk = 2\pi\sum_{-N}^{N}\int_K A_n(k)a_{-n}\,dk\,.$$

We shall prove that

$$(6)\qquad \sum_{-\infty}^{\infty}\int_K |A_n(k)a_{-n}|\,dk < \infty\,.$$

Assuming this, the right hand side of (5) has (by the dominated convergence theorem) the limit

$$(7)\qquad 2\pi\sum_n\int_K A_n(k)a_{-n}\,dk = \int_K\sum_n A_n(k)\int_B \chi_n(b)\,dT(b)\,dk\,.$$

This last expression equals

$$(8)\qquad \int_K\left(\int_B\sum_n A_n(k)\chi_n(b)\,dT(b)\right)dk\,,$$

because the uniform convergence of (2) implies that $\sum_n A_n(k)\chi_n(b)$ converges in $\mathcal{A}(B)$. On the other hand expression (8) equals

$$\int_K\left(\int_B f(k,b)\,dT(b)\right)dk$$

so (1) would follow from (4).

To verify (6) we consider the cases $n \geq 0$ and $n < 0$ separately. We know by Lemma 4.21 that

$$(9)\qquad \sum_n |a_n|r^{|n|} < \infty \qquad 0 \leq r < 1\,.$$

(i) $n \geq 0$. Choose $R > 1$ such that (3) holds and then $r < 1$ such that $Rr > 1$. Then by (9), $|A_n(k)a_{-n}| \leq \text{const.}(Rr)^{-n}$.

(ii) $n < 0$. Choose $R < 1$ such that (3) holds and then $r < 1$ such that $R < r < 1$. Then by (9), $|A_n(k)a_{-n}| \leq \text{const.}\left(\frac{R}{r}\right)^{-n}$. Combining these cases (i) and (ii) relation (6) follows so (1) is proved.

Page 81, line 11^+: Here it is worth while remarking that each measure μ has the form $w - \nu$ where w and ν are positive measures. To see this define for $f \geq 0$, $w(f) = \sup_{0 \leq g \leq f} \mu(g)$. Then if f, f' are ≥ 0 it is easy to see that $w(f) + w(f') \leq w(f + f')$. On the other hand, if $h \geq 0$, $0 \leq h \leq f + f'$ then putting $g = \min(f, h)$, $g' = h - g$ we have $0 \leq g \leq f$, $0 \leq g' \leq f'$ and $h = g + g'$. Hence $\mu(h) = \mu(g) + \mu(g') \leq w(f) + w(f')$ whence $w(f + f') \leq w(f) + w(f')$. Thus $w(f + f') = w(f) + w(f')$ for $f, f' \geq 0$. If $k \in C_c(S)$ is written $k = f - g = f' - g'$ (f, g, f', g' all ≥ 0) then $f + g' = f' + g$ so $w(k)$ is well-defined as $w(f) - w(g)$. Thus w is a positive measure and so is $\nu = w - \mu$.

Page 92, line 6^-: The uniqueness can be proved as follows. If μ and μ' are two left invariant Haar measures, define ν by $\nu(f) = \mu'(\check{f})$ where $\check{f}(g) = f(g^{-1})$, $f \in C_c(G)$. For $g \in C_c(G)$ consider the function $s \to f(s) \int g(ts)\, d\nu(t)$ and integrate it with respect to μ. Since the ν-integral is constant in s the result is

$$\begin{aligned} \mu(f)\nu(g) &= \int f(s) \int g(ts)\, d\nu(t)\, d\mu(s) = \int d\nu(t) \int f(s) g(ts)\, d\mu(s) \\ &= \int d\nu(t) \int f(t^{-1}\sigma) g(\sigma)\, d\mu(\sigma) = \int g(\sigma)\, d\mu(\sigma) \int f(t^{-1}\sigma)\, d\nu(t)\,. \end{aligned}$$

Put $h(\sigma) = \int f(t^{-1}\sigma)\, d\nu(t)/\mu(f)$. Then the formula shows $\nu = h\mu$ (h independent of f). Taking $\sigma = e$ we deduce

$$h(e)\mu(f) = \nu(\check{f}) = \mu'(f)$$

so μ and μ' are proportional.

Page 115, line 1^+: More explicitly, we have for each $F \in \mathcal{S}(R)$,

$$\int_{\mathbf{R}} \operatorname{sgn} s\, e^{ist} \left(\int_{\mathbf{R}} F(p) e^{-isp}\, dp \right) ds = 2\pi (\mathcal{H}F)(t)\,.$$

In fact if we apply both sides to $\widetilde{\psi}(\psi \in \mathcal{S}(\mathbf{R}))$ the left-hand side is (by (40))

$$\begin{aligned} &\int_{\mathbf{R}} \left(\int_{\mathbf{R}} \operatorname{sgn} s \widetilde{F}(s) e^{ist}\, ds \right) \widetilde{\psi}(t)\, dt \\ &\qquad = \int_{\mathbf{R}} (\mathcal{H}F)^{\sim}(s) 2\pi \psi(s)\, ds = 2\pi \int_{\mathbf{R}} (\mathcal{H}F)(t)(\widetilde{\psi})(t)\, dt \end{aligned}$$

Page 176, line 3^+: To verify the surjectivity statement in full we note by Introduction, Theorem 3.4 that

$$f = \sum_{k,m} a_{km} S_{km} \qquad (k \text{ even})$$

and by (32)–(34)

$$\widehat{f} = \Omega_n \sum_{k,m} a_{km}\varphi_k\left(\frac{\pi}{2}\right) S_{km} \qquad (k \text{ even}).$$

But by p. 23

$$\varphi_{2k}\left(\frac{\pi}{2}\right) = \frac{\Omega_{n-1}}{\Omega_n}(-1)^k \int_0^\pi (\cos\varphi)^{2k}(\sin\varphi)^{n-2}\,d\varphi \sim k^{-\frac{n-1}{2}}$$

By the characterization in Introduction, Theorem 3.4(ii), the surjectivity of $f \to \widehat{f}$ follows.

Page 194, line 14^-: The factors $1 - e^{-\alpha}$ and $1 - e^{-\beta}$ are relatively prime.

Proof. Let $\omega_1, \ldots \omega_\ell$ be as above, put $\Lambda = \widetilde{T}(R)$, so $\Lambda = \oplus_1^\ell \mathbf{Z}\omega_i \approx \mathbf{Z}^\ell$.

If $\alpha = \sum_1^\ell n_k\omega_k$ and d the highest common factor of the n_k then $-\alpha = du_1$ where $d > 0$, $u_1 \in \Lambda$, and coordinates $m_j = n_j/d$ of u_1 have no common factor. This being the case, we can extend the row $m_1, \ldots, m_\ell$ to a matrix $(m_{ij}) \in \mathbf{SL}(\ell, \mathbf{Z})$ with $m_{1j} = m_j$. We extend u_1 to a basis $u_1, u_2', \ldots, u_\ell'$ of Λ over $\mathbf{Z}$ by $u_k' = \sum_1^\ell m_{kj}\omega_j (k \geq 2)$. Let $\Lambda' = \oplus_2^\ell \mathbf{Z}u_k'$. We have

$$-\beta = nu_1 + \mu\,, \qquad \mu \neq 0 \text{ in } \Lambda'\,.$$

Then $\mu = mu_2$ where $u_2 \in \Lambda'$ has coordinates relative to (u_k') which have no common factor. We can then by the earlier argument for Λ extend u_2 to a basis (u_k) of Λ'. Then $\Lambda = \oplus_1^\ell \mathbf{Z}u_i$ and

$$-\alpha = du_1\,, \qquad -\beta = nu_1 + mu_2\,, \qquad m \neq 0\,.$$

The algebra F is isomorphic to the ring

$$A = \mathbf{Z}\left[X_1, \ldots, X_\ell, X_1^{-1}, \ldots, X_\ell^{-1}\right] \qquad X_i = e^{u_i}$$

and

$$\begin{aligned} 1 - e^{-\alpha} &= 1 - X_1^d\,, \\ 1 - e^{-\beta} &= 1 - X_1^n X_2^m\,. \end{aligned}$$

Since $1 - X_1^d = \prod_{\nu=0}^{d-1}(1 - \epsilon^\nu X_1)$ where $\epsilon = e^{\frac{2\pi}{d}}$, and since no factor $1 - \epsilon^\nu X_1$ divides $1 - X_1^n X_2^m$ (it does not vanish for $X_1 = e^{-\nu}$) the elements $1 - e^{-\alpha}$, $1 - e^{-\beta}$ are relatively prime.

Page 206, line 1^+: Since $H = \begin{pmatrix} 0(1,q) & 0 \\ 0 & 1 \end{pmatrix}$ is the isotropy group of $(0,\dots,0,1)$ it leaves each plane $y_{q+2} = \text{const.}$ invariant. Thus $\mathbf{S}_r(0)$ is a section of a plane with our quadric (or rather the part of this section which lies in $\mathbf{D}_0$).

Page 250, line 3^+: Proof of (31). Given $x' = (x_1, \dots, x_{n-1})$ put

$$I(x') = \{x_n : (x', x_n) \in B_R\} .$$

By Schwarz's inequality,

$$\begin{aligned} |\varphi(x_1,\dots,x_n)|^2 &\leq \int_{-R}^{x_n} dt \int_{I(x')} |(\partial_n\varphi)(x',t)|^2\,dt \\ &\leq 2R \int_{I(x')} |(\partial_n\varphi)(x',t)|^2\,dt . \end{aligned}$$

The right hand side is independent of x_n so

$$\begin{aligned} \int_{B_R} |\varphi(x)|^2\,dx &= \int_{|x'|\leq R} dx' \int_{I(x')} |\varphi(x',x_n)|^2\,dx_n \\ &\leq \int_{|x'|\leq R} dx' \int_{I(x')} \left(2R\int_{I(x')} |(\partial_n\varphi)(x',t)|^2\,dt\right) dx_n \\ &\leq 4R^2 \int_{|x'|<R} dx' \int_{I(x')} |(\partial_n\varphi)(x',t)|^2\,dt \\ &= 4R^2 \int_{B_R} |(\partial_n\varphi)(x)|^2\,dx , \end{aligned}$$

which proves (31) even in a stronger form.

Page 281, line 16^-: To prove (5) recall that

$$(\widetilde{X}^m f)(g) = \left\{\frac{d^m}{dt^m} f(g \exp tX)\right\}_{t=0} .$$

Put $F(x_1,\dots,x_n) = f(g\exp X)$. By induction

$$\frac{d^m}{dt^m} F(tx_1,\dots,tx_n) = \sum x_{i_1}\dots x_{i_m}(\partial_{i_1}\dots\partial_{i_m}F)(tx_1,\dots,tx_n) ,$$

which for $t = 0$ gives

$$(\widetilde{X}^m f)(g) = \sum x_{i_1}\dots x_{i_m}\lambda(X_{i_1}\dots X_{i_m})f(g) = (\lambda(X^m)f)(g) .$$

Page 281, line 8^+: That $\lambda(P)$ is a differential operator is even easier to see from (5) and the fact that the powers X^m ($X \in \mathfrak{g}, m \in \mathbf{Z}^+$) span $S(\mathfrak{g})$.

Page 444, line 7^-: Actually, Lemma 6.8 gives (with $\sigma = (s')^{-1}$)

$$N^\sigma = \exp(\mathfrak{n}^\sigma \cap \overline{\mathfrak{n}}) \exp(\mathfrak{n}^\sigma \cap \mathfrak{n}) .$$

If there were an element $x \in N^\sigma \cap \overline{N}$ not in $\exp(\mathfrak{n}^\sigma \cap \overline{\mathfrak{n}})$ then the formula gives $x = x_1 x_2$ and $x_2 = \overline{N} \cap N = e$. Hence $\exp(\mathfrak{n}^\sigma \cap \overline{\mathfrak{n}}) = N^\sigma \cap \overline{N}$, proving (28).

Page 558, line 2^-: Note in fact that $(AX)_{11} = A_{11}X_{11}$ if the x_{ij} are ordered by $x_{11}, \dots, x_{1n}, x_{2n}, \dots x_{2n}, \dots x_{n1}, \dots x_{nn}$.

BIBLIOGRAPHY

ADIMURTI, KUMARESAN S.
1979 On the singular support of distributions and Fourier transforms on symmetric spaces. *Ann. Scuola Norm. Sup. Pisa Cl. Sci.* **6** (1979), 143–150.

AHLFORS, L. V.
1975 Invariant operators and integral representations in hyperbolic space. *Math. Scand.* **36** (1975), 27–43.
1981 "Möbius Transformations in Several Variables." Lecture Notes. Univ. of Minnesota, 1981.

ALLAMIGEON, A. C.
1961 Propriétés globales des espaces harmoniques. *C. R. Acad. Sci. Paris* **252** (1961), 1093–1095.

ANDERSON, M. F.
1979 A simple expression for the Casimir operator on a Lie Group. *Proc. Amer. Math. Soc.* **77** (1979), 415–420.

ARAKI, S. I.
1962 On root systems and an infinitesimal classification of irreducible symmetric spaces. *J. Math. Osaka City Univ.* **13** (1962), 1–34.

ÁSGEIRSSON, L.
1937 Über eine Mittelwertseigenschaft von Lösungen homogener linearer partieller Differentialgleichungen 2. Ordnung mit konstanten Koeffizienten. *Math. Ann.* **113** (1937), 321–346.

ASKEY, R.
1975 "Orthogonal Polynomials and Special Functions." Regional Conf. Ser. in Appl. Math. SIAM, Philadelphia, 1975.

ATIYAH, M. *et al.*
1979 "Representation Theory of Lie Groups." London Math. Soc. Lecture Notes 34. Cambridge Univ. Press, London and New York, 1979.

BAGCHI, SOMESH CHANDRA, and SITARAM, A.
1979 Spherical mean periodic functions on semisimple Lie groups. *Pacific J. Math.* **84** (1979), 241–250.

BARBASH, D.
1979 Fourier inversion for unipotent invariant integrals. *Trans. Amer. Math. Soc.* **249** (1979), 51–83.

BARBASH, D., and VOGAN, D.
1980 The local structure of characters. *J. Funct. Anal.* **37** (1980), 27–55.
1982 Primitive ideals and orbital integrals in complex classical groups. *Math. Ann.* **259** (1982), 153–199.
1983 Primitive ideals and orbital integrals in complex exceptional groups. *J. Algebra* **80** (1983), 350–382.

BARGMANN, V.
1947 Irreducible unitary representations of the Lorentz groups. *Ann. of Math.* **48** (1947), 568–640.

BARKER, W.
1975 The spherical Bochner theorem on semisimple Lie groups. *J. Funct. Anal.* **20** (1975), 179–207.

BARUT, A. D., and RACZKA, R.
1977 "Theory of Group Representations and Applications." Polish Scientific Publishers, Warsaw, 1977.

BELTRAMI, E.
1864 Ricerche di analisi applicata alla geometria. *Giornale di Mat.* **2, 3** (1864) (1865), Opera I, p. 107–198.
1869 Zur Theorie des Krümmungsmaasses. *Math. Ann.* **1** (1869), 575–582.

BEREZIN, F. A.
1957 Laplace operators on semi-simple Lie groups. *Trudy Moskov. Mat. Obšč.* **6** (1957), 371–463 [*English transl.*: *Amer. Math. Soc. Transl.* **21** (1962), 239–339].

BEREZIN, F. A., and GELFAND, I. M.
1956 Some remarks on the theory of spherical functions on symmetric Riemannian manifolds. *Trudy Moscov. Mat. Obšč.* **5** (1956), 311–351 [*English transl.*: *Amer. Math. Soc. Transl.* **21** (1962), 193–239].

BEREZIN, F. A., GELFAND I. M., GRAEV, M. A., and NAIMARK, M. A.
1960 Representations of groups. *Usp. Math. Nauk* **11** (1956), 13–40 [*English transl.*: *Amer. Math. Soc. Transl.* **16** (1960), 325–353].

BEREZIN, F. A., and KARPELEVIČ, F. I.
1958 Zonal spherical functions and Laplace operators on some symmetric spaces. *Dokl. Akad. Nauk. USSR* **118** (1958), 9–12.

BERG, C.
1973 Dirichlet forms on symmetric spaces. *Ann. Inst. Fourier* **23** (1973), 135–156.

BERLINE, N., and VERGNE, M.
1981 Équations de Hua et intégrales de Poisson. *C. R. Acad. Sci. Paris Ser. A* **290** (1980), 123–125. *In* "Non-Commutative Harmonic Analysis and Lie Groups," Lecture Notes in Math. No. 880, pp. 1–51. Springer-Verlag, New York, 1981.

BERNAT, P., CONZE, N. *et coll.*
1972 "Représentations des groupes de Lie résolubles." Dunod, Paris, 1972.

BESSE, A.
1978 Manifolds all of whose geodesics are closed. "Ergebnisse der Math.," Vol. 93. Springer-Verlag, New York, 1978.

BHANU MURTHY, T. S.

1960a Plancherel's measure for the factor space **SL**(n, **R**)/**SO**(n, **R**). *Dokl. Akad. Nauk. SSSR* **133** (1960), 503–506.

1960b The asymptotic behavior of zonal spherical functions on the Siegel upper half-plane. *Dokl. Akad. Nauk SSSR* **135** (1960), 1027–1029.

BOAS, R. P. Jr.

1954 "Entire Functions." Academic Press, New York, 1954.

BOCHNER, S.

1955 "Harmonic Analysis and the Theory of Probability." Univ. of California Press, Berkeley, California, 1955.

BOCKWINKEL, H. B. A.

1906 On the propagation of light in a biaxial crystal about a midpoint of oscillation. *Verh. Konink. Acad. V. Wet. Wis-en. Natur.* **14**, (1906), 636.

BOERNER, H.

1955 "Darstellungen von Gruppen mit Berücksichtigung der Bedürfnisse der Modernen Physik." Springer-Verlag, Berlin and New York, 1955; Zweite Auflage, 1967.

BOPP, N.

1981 Distributions de type K-positif sur l'espace tangent. *J. Funct. Anal.* **44** (1981), 348–358.

BOREL, A.

1972 "Représentations de Groupes Localement Compacts," Lecture Notes in Math. No. 276. Springer-Verlag, Berlin and New York, 1972.

BOREL, A., and HARISH-CHANDRA

1962 Arithmetic subgroups of algebraic groups. *Ann. of Math.* **75** (1962), 485–535.

BOREL, A., and WALLACH, N.

1980 "Continuous Cohomology, Discrete Subgroups, and Representations of Reductive Groups," Ann. Math. Studies. Princeton Univ. Press, Princeton, New Jersey, 1980.

BOROVIKOV, W. A.

1959 Fundamental solutions of linar partial differential equations with constant coefficients. *Trudy Moscov Mat. Obšč.* **8** (1959), 199–257.

BOTT, R.

1956 An application of the Morse theory to the topology of Lie groups. *Bull. Soc. Math. France* **84** (1956), 251–286.

BOURBAKI, N.

"Éléments de Mathématique," Vol. VI, Intégration, Chapters 1–8. Hermann, Paris, 1952–1963.

BOURBAKI, N.

Éléments de mathématique. "Groupes et algébres de Lie," Chapters I–VIII. Hermann, Paris, 1960–1975.

BOURBAKI, N.

"Éléments de Mathématique," Vol. V, Espaces vectoriels topologiques, Chapters I–V. Hermann, Paris, 1953–1955.

BRACEWELL, R. N., and RIDDLE, A. C.

1967 Inversion of fan beam scans in radio astronomy, *Astrophys. J.* **150** (1967), 427–434.

BRAUER, R.

1926 Über die Darstellung der Drehungsgruppe durch Gruppen linearer Substitutionen. Dissertation, Berlin, 1926.

1937 Sur la multiplication des caracteristiques des groupes continus et semi-simples. *C. R. Acad. Sci. Paris* **204** (1937), 1784–1786.

BRUHAT, F.

1956 Sur les représentations induites des groupes de Lie. *Bull. Soc. Math. France* **84** (1956), 97–205.

CALDERON A. P., and ZYGMUND, A.

1957 Singular differential operators and differential equations. *Amer. J. Math.* **79** (1957), 901–927.

CARTAN, É.

1896 Le principe de dualité et certaines intégrales multiples de l'espace tangential et de espace reglé. *Bull. Soc. Math. France* **24** (1896), 140–177.

1913 Les groupes projectifs qui ne laissent invariant aucune multiplicité plane. *Bull. Soc. Math. France* **41** (1913), 53–96.

1914 Les groupes projectifs continus réels qui ne laissent invariant aucune multiplicité plane. *J. Math. Pure Appl.* **10** (1914), 149–186.

1922 "Lecons sur les Invariant Integraux." Hermann, Paris, 1922. Reprinted 1958.

1927 Sur certaines formes riemanniennes remarquables des géometries a groupe fondamental simple. *Ann. Sci. École Norm. Sup.* **44** (1927), 345–467.

1929 Sur la detérmination d'un système orthogonal complet dans un espace de Riemann symétrique clos. *Rend. Circ. Mat. Palermo* **53** (1929), 217–252.

1950 "Leçons sur la Géometrie Projective Complexe," 2nd ed. Gauthier-Villars, Paris, 1950.

CARTAN, H.

1957 Fonctions Automorphes. Seminaire 1957–1958, Paris.

CARTER, R. W.

1972 "Simple Groups of Lie Type." Wiley, New York, 1972.

CARTER, R. W., and LUSZTIG, G.

1974 On the representations of the general linear and symmetric group *Math. Z.* **136** (1974), 193–242.

CARTIER, P.

1955 Articles in Séminaire Sophus Lie [1] (1955).

1961 On H. Weyl's character formula. *Bull. Amer. Math. Soc.* **67** (1961), 228–230.

CARTIER, P., and DIXMIER, J.

1958 Vecteurs analytiques dans les répresentations des groupes de Lie. *Amer. J. Math.* **80** (1958), 131–145.

CECCHINI, C.

1972 Lacunary Fourier series on compact Lie groups. *J. Funct. Anal.* **11** (1972), 191–203.

CERÈZO, A., and ROUVIÈRE, F.

1972 Sur certains opérateurs differentiels invariants du groupe hyperbolique. *Ann. Sci. Ecole Norm. Sup.* **5** (1972), 581–597.

1973 Opérateurs differentiels invariants sur un groupe de Lie. Séminaire Goulaouic-Schwartz 1972–1973. École Polytech., Paris, 1973.

CHAMPETIER, C. and DELORME, P.

1981 Sur les représentations des groupes de déplacements de Cartan. *J. Funct. Anal.* **43** (1981), 258–279.

CHANG, W.

1979a Global solvability of the Laplacians on pseudo-Riemannian symmetric spaces. *J. Funct. Anal.* **34** (1979), 481–492.

1979b Global solvability of bi-invariant differential operators on solvable Lie groups. "Non Commutative Harmonic Analysis," Lecture Notes in Math. No. 728, pp. 8–16. Springer-Verlag, Berlin and New York, 1979.

1982 Invariant differential operators and P-convexity of solvable Lie groups. *Adv. in Math.* **46** (1982), 284–304.

CHERN, S. S.

1942 On integral geometry in Klein spaces. *Ann. of Math.* **43** (1942), 178–189.

CHEVALLEY, C.

1946 "Theory of Lie Groups," Vol. I. Princeton Univ. Press, Princeton, New Jersey, 1946.

1952 The Betti numbers of the exceptional simple Lie groups. *Proc. Internat. Congr. Math. 1950* **2** (1952), 21–24.

1955a Invariants of finite groups generated by reflections, *Amer. J. Math.* **77** (1955), 778–782.

1955b "Théorie des groupes de Lie," Vol. III. Hermann, Paris, 1955.

CHÈBLI, H.

1974 Sur un théoreme de Paley-Wiener associé a la décomposition spectrale d'un opérateur de Sturm-Liouville sur $(0, \infty)$. *J. Funct. Anal.* **17** (1974), 447–461.

CLERC, J. L.

1972 Les sommes partielles de la decomposition en harmoniques sphérique ne convergent pas dans $L^p (p \neq 2)$. *C. R. Acad. Sci. Paris Ser. A–B* **274** (1972), A59–A61.

1976 Une formule de type Mehler–Heine pour les zonal d'un espace riemannien symétrique. *Studia Math.* **57** (1976), 27–32.

1979 Multipliers on symmetric spaces. *Proc. Symp. Pure Math.* **35**, Part 2, (1979), 345–353.

1980 Transformation de Fourier sphérique des espaces de Schwartz. *J. Funct. Anal.* **37** (1980), 182–202.

1981 Estimations à l'infini des fonctions des Bessel généraliseés. *C. R. Acad. Sci. Paris Math.* **292** (1981), 429–430.

CODDINGTON, E. A., and LEVINSON, N.

1955 "Theory of Ordinary Differential Equations." McGraw-Hill, New York, 1955.

COHEN, P.

1960 On a conjecture of Littlewood and idempotent measures. *Amer. J. Math.* **82** (1960), 191–212.

COIFMAN, R. R., and WEISS, G.

1971 "Analyse harmonique noncommutative sur certains espaces homogènes," Springer Lecture Notes No. 242. Springer Verlag, Berlin and New York, 1971.

CORMACK, A. M.

1963–1964 Representation of a function by its line integrals, with some radiological application I, II. *J. Appl. Phys.* **34** (1963), 2722–2727; **35** (1964), 2908–2912.

CORMACK, A. M., and QUINTO, T.

1980 A Radon transform on spheres through the origin in $\mathbb{R}^n$ and applications to the Darboux equation. *Trans. Amer. Math. Soc.* **260** (1980), 575–581.

COTTON, E.

1900 Sur les invariant differentiels.... *Ann. Éc. Norm. Sup.* **17** (1900), 211–244.

COURANT, R., and LAX, A.

1955 Remarks on Cauchy's problem for hyperbolic partial differential equations with constant coefficients in several independent variables. *Comm. Pure Appl. Math.* **8** (1955), 497–502.

COEXTER, H. S. M.

1934 Discrete groups generated by reflections. *Ann. of Math.* **35** (1934), 588–621.

1957 "Non-Euclidean Geometry." Univ. of Toronto Press, Toronto, 1957.

CROFTON, M. W.

1868 On the theory of local probability. *Phil. Trans. Roy. Soc. London* **158** (1868), 181–199.

1885 Probability. *Encyclopaedia Britannica*, 9th ed. (1885), 768–788.

DADOK, J.

1980 Solvability of invariant differential operators of principal type on certain Lie groups and symmetric spaces. *J. D'Analyse* **37** (1980), 118–127.

1982 On the C^∞ Chevalley theorem. *Adv. in Math.* **44** (1982), 121–131.

1983 Polar coordinates induced by actions of compact Lie groups (preprint).

DE GEORGE D. L., and WALLACH, N.

1978 Limit formulas for multiplicities in $L^2(G/\Gamma)$. *Ann. of Math.* **107** (1978), 133–150.

DELORME, P.

1981 Homomorphismes de Harish-Chandra et K-types minimaux de series principales des groupes de Lie semisimples reels. *In* "Non-Commutative Harmonic Analysis and Lie Groups." Lecture Notes in Math. No. 880. Springer-Verlag, Berlin and New York, 1981.

1982 Theorème de type Paley–Wiener pour les groupes de Lie semisimples reels avec une seule classe de conjugaison de sons groupes des Cartan. *J. Funct. Anal.* **47** (1982), 26–63.

DE MICHELE, L., and RICCI, F. (eds.)

1982 *Proc. Seminar Topics Mod. Harmonic Anal.* Turin and Milano, 1982.

DE RHAM, G.

1955 "Variétés différentiables." Hermann, Paris, 1955.

DIEUDONNÉ, J.
1980 "Special Functions and Linear Representations of Lie Groups," Conf. Board Math. Sci. Series. No. 42. American Mathematical Society, Providence, Rhode Island, 1980.
1975 "Éléments d'Analyse," Vols. V and VI. Gauthier-Villars, Paris, 1975.

DIXMIER, J.
1958 Sur les représentations unitaires des groupes de Lie nilpotents III. *Canad. J. Math.* **10** (1958), 321–348.
1959 Sur l'algèbre enveloppante d'une algèbre de Lie nilpotente. *Arch. Math.* **10** (1959), 321–326.
1967 Sur le centre de l'algèbre enveloppante d'une algèbre de Lie, *C. R. Acad. Sci. Paris Ser. A* **265** (1967), 408–410.
1974 "Algèbres Enveloppants." Gauthier-Villars, Paris, 1974.
1975 Sur les algèbres enveloppantes de $\mathfrak{sl}(n, C)$ et $\mathfrak{af}(n, C)$. *Bull. Sci. Math* **100** (1975), 57–95.
1981 Sur les invariants des formes binaires. *C. R. Acad. Sci. Paris* **292** (1981), 987–990.

DIXMIER, J., and MALLIAVIN, P.
1978 Factorisation de fonctions et de vecteurs indéfiniment différentiables. *Bull. Sci. Math.* **102** (1978), 305–330.

DOOLEY, A. H.
1979 Norms of characters and lacunarity for compact Lie groups. *J. Funct. Anal.* **32** (1979), 254–267.

DRESELER, B.
1977 "On Summation Processes of Fourier Expansions for Spherical Functions," Lecture Notes in Math. Vol. 571, pp. 65–84. Springer-Verlag, Berlin and New York, 1977.
1981 Norms of zonal spherical functions and Fourier series on compact symmetric spaces. *J. Funct. Anal.* **44** (1981), 74–86.

DRESELER, B., and SCHEMPP, W.
1975 On the convergence and divergence behavior of approximation processes in homogeneous Banach spaces. *Math. Z.* **143** (1975), 81–89.

DUFLO, M.
1970 Fundamental series representations of a semisimple Lie group. *Funct. Anal. Appl.* **4** (1970), 122–126.
1972 Charactère des représentations des groupes résolubles associées a une orbite entière. *In* Bernat *et al.* [1972].
1975 Représentations irréductibles des groupes semi-simples complexes. *In* "Analyse Harmonique sur les Groupes de Lie," Lecture Notes in Math. No. 497, pp. 26–88. Springer-Verlag, Berlin and New York, 1975.
1976 Sur l'analyse harmonique sur les groupes de Lie résolubles. *Ann. Sci. Ecole Norm. Sup.* **9** (1976), 107–144.
1977 Opérateurs differentiels bi-invariants sur un groupe de Lie. *Ann. Sci. Ecole Norm. Sup.* **10** (1977), 265–288.
1979a Opérateurs invariants sur un espace symétrique. *C. R. Acad Sci. Paris Ser. A* **289** (1979), 135–137.

1979b Représentations unitaires irréductibles des groupes simples complexes de rang deux. *Bull. Soc. Math. France* **107** (1979), 55–96.

DUFLO, M., and WIGNER, D.

1978 Sur la résolubilité des équations differentielles invariantes sur un groupe de Lie. *Sém. Goulaouic-Schwartz, 1977–1978*. École-Polytech., Palaisesu, 1978.

DUISTERMATT, J. J., KOLK, J. A. C., and VARADARAJAN, V. S.

1979 Spectra of compact locally symmetric manifolds of negative curvature. *Invent. Math.* **52** (1979), 27–93.

1983 Functions, flows, and oscillatory integrals on flag manifolds and conjugacy classes in real semisimple Lie groups. Composite Math. 49 (1983), 309–398.

DUNFORD, N., and SCHWARTZ, J. J.

1958 "Linear Operators," Part I. Wiley (Interscience), New York, 1958.

DUNKL, C. F.

1977 Spherical functions on compact groups and applications to special functions. *Symposia Mathematica* **22**, (1977), 145–161.

DUNKL, C. F., and RAMIREZ, D.

1971 "Topics in Harmonic Analysis." Appleton, New York, 1971.

DYM, H., and MCKEAN, H. P.

1972 "Fourier Series and Integrals." Academic Press, New York, 1972.

EDWARDS, R. E.

1972 "Integration and Harmonic Analysis on Compact Groups," London Math. Soc. Lecture Notes Series No. 8. Cambridge Univ. Press, London and New York, 1972.

EDWARDS, R. E., and ROSS, K. A.

1973 Helgason's number and lacunarity constants. *Bull. Austral. Math. Soc.* **9** (1973), 187–218.

EGUCHI, M., and OKAMOTO, K.

1977 The Fourier transform of the Schwartz space on a symmetric space. *Proc. Japan Acad.* **53** (1977), 237–241.

EHRENPREIS, L.

1954 Solutions of some problems of division I. *Amer. J. Math.* **76** (1954), 883–903.

1956 Some properties of distributions on Lie groups. *Pacific J. Math.* **6** (1956), 591–605.

1969 "Fourier Analysis in Several Complex Variables." Wiley, New York, 1969.

EHRENPREIS, L., and MAUTNER, F.

1955 Some properties of the Fourier transform on semi-simple Lie groups I. *Ann. of Math.* **61** (1955), 406–443; II, III. *Trans. Amer. Math. Soc.* **84** (1957), 1–55; **901** (1959), 431–484.

ERDÉLYI, A., MAGNUS, W., OBERHETTINGER, F., and TRICOMI, F. G.

"Higher Transcendental Functions" (Bateman Manuscript Project), Vols. I, II, and III. McGraw-Hill, New York, 1953, 1953, 1955.

EYMARD, P.

1977 Le noyau de Poisson et la théorie des groupes. *Symposia Mathematica* **22** (1977), 107–132.

FARAUT, J.

1979 Distributions sphériques sur les espaces hyperboliques. *J. Math. Pures Appl.* **58** (1979), 369–444.

1982 Un théoreme de Paley–Wiener pour la transformation de Fourier sur un espace Riemannien symétrique de rang un. *J. Funct. Anal.* **49** (1982), 230–268.

FARAUT, J., and HARZALLAH, K.

1972 Fonctions sphériques de type positif sur les espaces hyperboliques. *C. R. Acad. Sci. Paris* **274** (1972), A1396–A1398.

FATOU, P.

1906 Séries trigonométriques et séries de Taylor. *Acta Math.* **30** (1906), 335–400.

FIGA-TALAMANCA, A., and RIDER, D.

1966 A theorem of Littlewood and lacunary series for compact groups. *Pacific J. Math.* **16** (1966), 505–514.

1967 A theorem on random Fourier series on noncommutative groups. *Pacific J. Math.* **21** (1967), 487–492.

FLATTO, L.

1978 Invariants of finite reflection groups. *Enseign. Math.* **24** (1978), 237–292.

FLENSTED-JENSEN, M.

1972 Paley–Wiener theorems for a differential operator connected with symmetric spaces. *Ark. Mat.* **10** (1972), 143–162.

1975 "Spherical Functions and Discrete Series," Springer Lecture Notes No. 466, pp. 65–78. Springer-Verlag, Berlin and New York, 1975.

1977a Spherical functions on a simply connected semisimple Lie group. *Amer. J. Math.* **99** (1977), 341–361.

1977b Spherical functions on a simply connected semisimple Lie Group, II. *Math. Ann.* **228** (1977), 65–92.

1978 Spherical functions on a real semi-simple Lie group. A method of reduction to the complex case. *J. Funct. Anal.* **30** (1978), 106–146.

1980 Discrete series for semisimple symmetric spaces. *Ann. of Math.* **111** (1980), 253–311.

1981 K-finite joint eigenfunctions of $U(\mathfrak{g})^K$ on a non-Riemannian semisimple symmetric space G/H. *In* "Non-Commutative Harmonic Analysis and Lie Groups," Lecture Notes in Math. No. 880. Springer-Verlag, Berlin and New York, 1981.

FLENSTED-JENSEN, M., and KOORNWINDER, T.

1973 The convolution structure for Jacobi function expansions. *Ark. Mat.* **11** (1973), 245–262.

1979a Positive-definite spherical functions on a noncompact rank one symmetric space. *In* "Analyse Harmonique sur les Groupes de Lie II," Springer Lecture Notes No. 739. Springer-Verlag, Berlin and New York, 1979.

1979b Jacobi functions: the addition formula and the positivity of the dual convolution structure. *Ark. Math.* **17** (1979), 139–151.

FLENSTED-JENSEN, M., and RAGOZIN, D. L.

1973 Spherical functions are Fourier transforms of L_1-functions. *Ann. Sci. École. Norm. Sup.* **6** (1973), 457–458.

FOLLAND, G. B., and STEIN, E.

1974 Estimates for the ∂_b-complex and analysis on the Heisenberg group. *Comm. Pure Appl. Math.* **27** (1974), 429–522.

FREUDENTHAL, H.

1954 Zur Berechnung der Charaktere der halbeinfachen Lieschen Gruppen. I. *Indag. Math.* **16** (1954), 363–368.

1956 The existence of a vector of weight 0 in irreducible Lie groups without centre. *Proc. Amer. Math. Soc.* **7** (1956), 175–176.

FREUDENTHAL, H., and DE VRIES, H.

1969 "Linear Lie Groups." Academic Press, New York, 1969.

FRIEDLANDER, F. C.

1975 "The Wave Equation in Curved Space." Cambridge Univ. Press, London and New York, 1975.

FROTA-MATTOS, L. A.

1978 Analytic continuation of the Fourier series on connected compact Lie groups. *J. Funct. Anal.* **29** (1978), 1–15.

FUGLEDE, B.

1958 An integral formula. *Math. Scand.* **6** (1958), 207–212.

FUNK, P.

1916 Über eine geometrische Anwendung der Abelschen Integralgleichung. *Math. Ann.* **77** (1916), 129–135.

FURSTENBERG, H.

1963 A Poisson formula for semisimple Lie groups. *Ann. of Math.* **77** (1963), 335–386.

1965 Translation-invariant cones of functions on semi-simple Lie groups. *Bull. Amer. Math. Soc.* **71** (1965), 271–326.

FURSTENBERG, H., and TZKONI, I.

1971 Spherical functions and integral geometry. *Israel Math. J.* **10** (1971), 327–338.

GÅRDING, L.

1947 Note on continuous representations of Lie groups. *Proc. Nat. Acad. Sci. U.S.A.* **33** (1947), 331–332.

1960 Vecteurs analytiques dans les représentations des groups de Lie. *Bull. Soc. Math. France* **88** (1960), 73–93.

1961 Transformation de Fourier des distributions homogènes. *Bull. Soc. Math. France* **89** (1961), 381–428.

GANGOLLI, R.

1967 Positive definite kernels on homogeneous spaces and certain stochastic processes related to Levy's Brownian motion of several parameters. *Ann. Inst. H. Poincaré Sect. B.* **3** (1967), 121–226.

1968 Asymptotic behaviour of spectra of compact quotients of certain symmetric spaces. *Acta Math.* **12** (1968), 151–192.

1971 On the Plancherel formula and the Paley–Wiener theorem for spherical functions on semisimple Lie groups. *Ann. of Math.* **93** (1971), 150–165.

1972 "Spherical functions on semisimple Lie group. Short Courses, Washington Univ." Dekker, New York, 1972.

GELFAND, I. M.

1950a Spherical functions on symmetric spaces. *Dokl. Akad. Nauk USSR* **70** (1950), 5–8. *Amer. Math. Soc. Transl.* **37** (1964), 39–44.

1950b The center of an infinitesimal group algebra. *Mat. Sb.* **26** (1950), 103–112.

1960 Integral geometry and its relation to group representations. *Russ. Math. Surveys* **15** (1960), 143–151.

GELFAND, I. M., and GRAEV, M. I.

1955 Analogue of the Plancherel formula for the classical groups. *Trudy Moscow. Mat. Obšč.* **4** (1955), 375–404, (also in Gelfand and Naimark [1957]).

1968 Complexes of straight lines in the space $\mathbb{C}^n$. *Funct. Anal. Appl.* **2** (1968), 39–52.

GELFAND, I. M., GRAEV, M. I., and SHAPIRO, S. J.

1969 Differential forms and integral geometry. *Funct. Anal. Appl.* **3** (1969), 24–40.

GELFAND, I. M., GRAEV, M. I., and VILENKIN, N.

1966 "Generalized Functions," Vol. 5, Integral Geometry and Representation Theory. Academic Press, New York, 1966.

GELFAND, I. M., and NAIMARK, M. A.

1948 An analog of Plancherel's formula for the complex unimodular group. *Dokl. Akad. Nauk USSR* **63** (1948), 609–612.

GELFAND, I. M., and NAIMARK, M. R.

1952 Unitary representation of the unimodular group containing the identity representation of the unitary subgroup. *Trudy Moscov. Mat. Obšč.* **1** (1952), 423–475.

1957 "Unitäre Darstellungen der Klassischen Gruppen." Akademie Verlag, Berlin, 1957.

GELFAND, I. M., and RAIKOV, D. A.

1943 Irreducible unitary representations of locally compact groups. *Mat. Sb.* **13** (1943), 301–316.

GELFAND, I. M., and SCHILOV, G. E.

1960 "Verallgemeinerte Funktionen," Vol. I. German Transl. VEB, Berlin, 1960.

GELFAND, I. M., and SHAPIRO, S. J.

1955 Homogeneous functions and their applications, *Uspehi Mat. Nauk.* **10** (1955), 3–70.

GINDIKIN, S. G.

1967 Unitary representations of groups of automorphisms of Riemannian symmetric spaces of null curvature. *Funct. Anal. Appl.* **1** (1967), 28–32.

1975 Invariant generalized functions in homogeneous domains. *Funct. Anal. Appl.* **9** (1975), 50–52.

GINDIKIN, S. G., and KARPELEVIČ, F. I.

1962 Plancherel measure of Riemannian symmetric spaces of non-positive curvature, *Dokl. Akad. Nauk SSSR* **145** (1962), 252–255.

1969 On an integral connected with symmetric Riemannian spaces of negative curvature. *Amer. Math. Soc. Transl.* (*2*) **85** (1969), 249–258.

GIULINI, S., SOARDI, P. M., and TRAVAGLINI, G.

1982 Norms of characters and Fourier series on compact Lie groups. *J. Funct. Anal.* **46** (1982), 88–101.

GODEMENT, R.

1948 Les fonctions de type positif et la théorie des groupes. *Trans. Amer. Math. Soc.* **63** (1948), 1–84.

1951 Sur la théorie des représentations unitaires. *Ann. of Math.* **53** (1951), 68–124.

1952a Une généralisation du théorème de la moyenne pour les fonctions harmoniques, *C. R. Acad. Sci. Paris* **234** (1952), 2137–2139.

1952b A theory of spherical functions I. *Trans. Amer. Math. Soc.* **73** (1952), 496–556.

1957a Introduction aux travaux de A. Selberg. *Séminaire Bourbaki* No. 144 *Paris* (1957).

1957b Articles in H. Cartan, Fonctions automorphes, Séminaire, 1957–1958, Paris.

1966 The decomposition of $L^2(G/\Gamma)$ for $\Gamma = SL(2, Z)$, *Proc. Symp. Pure Math.* **9** (1966), 211–244.

GODIN, P.

1982 Hypoelliptic and Gevrey hypoelliptic invariant differential operators on certain symmetric spaces. *Ann. Scuola Norm. Pisa* **IX** (1982), 175–209.

GOODMAN, R.

1969a Analytic domination by fractional powers of a positive operator. *J. Funct. Anal.* **3** (1969), 55–76.

1969b Analytic and entire vectors for representations of Lie groups. *Trans. Amer. Math. Soc.* **143** (1969), 55–76.

GRAY, A., and WILLMORE, T.

1982 Mean-value theorems for Riemmanian manifolds. *Proc. Roy. Soc. Edinburgh Sect. A* **92A** (1982), 343–364.

GRINBERG, E. L.

1983 Spherical harmonics and integral geometry on projective spaces. *Trans. Amer. Math. Soc.* **279** (1983), 187–213.

GROSS, K. I.

1978 On the evolution of noncommutative harmonic analysis. *Amer. Math. Mon.* **85** (1978), 525–548.

GROSS, K. I., HOLMAN, W. J., and KUNZE, R.

1979 A new class of Bessel functions and applications in harmonic analysis. *Proc. Symp. Pure Math.* **35** (1979), Part 2, Harmonic Analysis in Euclidean Spaces. American Mathematical Society, Providence, Rhode Island, 1979.

GÜNTHER, P.

1957 Über einige specielle Probleme aus der Theorie der linearen partiellen Differentialgleichungen 2. Ordnung. *Ber. Verh. Sächs. Akad. Wiss. Leipzig* **102** (1957), 1–50.

GUILLEMIN, V.

1976 Radon Transform on Zoll surfaces. *Advances in Math.* **22** (1976), 85–199.

1979a A Szegö-type theorem for symmetric spaces. *Ann. Math. Stud.* **93** (1979), 63–78.

1979b Some micro-local aspects of analysis on compact symmetric spaces. *Ann. Math. Stud.* **93** (1979), 79–111.

GUILLEMIN, V., and POLLACK, A.

1974 "Differential Topology." Prentice Hall, Englewood Cliffs, New Jersey, 1974.

GUILLEMIN, V., and STERNBERG, S.

1977 "Geometric Asymtotics," Math. Surveys No. 14. American Mathematical Society, Providence, Rhode Island, 1977.

1979 Some problems in integral geometry and some related problems in micro-local analysis. *Amer. J. Math.* **101** (1979), 915–955.

GUTKIN, E. A.

1970 Representations of the principal series of a complex semisimple Lie group. *Funct. Anal.* **4** (1970), 117–121.

HAAR, A.

1933 Der Maassbegriff in der Theorie der Kontinuerlichen Gruppen. *Ann. of Math.* **34** (1933), 147–169.

HARISH-CHANDRA

1951a On some applications of the universal enveloping algebra of a semi-simple Lie algebra. *Trans. Amer. Math. Soc.* **70** (1951), 28–96.

1951b Plancherel formula for complex semisimple Lie groups. *Proc. Nat. Acad. Sci. U.S.A.* **37** (1951), 813–818.

1952 Plancherel formula for the 2 × 2 real unimodular group. *Proc. Nat. Acad. Sci. U.S.A.* **38** (1952), 337–342.

1953 Representations of semisimple Lie groups, I. *Trans. Amer. Math. Soc.* **75** (1953), 185–243.

1954a Representations of semisimple Lie groups, II. *Trans. Amer. Math. Soc.* **76** (1954), 26–65.

1954b Representations of semisimple Lie groups, III. *Trans. Amer. Math. Soc.* **76** (1954), 234–253.

1954c Plancherel formula for the right invariant functions on a semisimple Lie group. *Proc. Nat. Acad. Sci. U.S.A.* **4** (1954), 200–204.

1954d The Plancherel formula for complex semisimple Lie groups. *Trans. Amer. Math. Soc.* **76** (1954), 485–528.

1956a Representations of semisimple Lie groups, VI. *Amer. J. Math.* **78** (1956), 564–628.

1956b The characters of semisimple Lie groups. *Trans. Amer. Math. Soc.* **83** (1956), 98–163.

1957a Differential operators on a semi-simple Lie algebra, *Amer. J. Math.* **79** (1957), 87–120.

1957b A formula for semisimple Lie groups. *Amer. J. Math.* **79** (1957), 733–760.

1958a Spherical functions on a semi-simple Lie group I. *Amer. J. Math.* **80** (1958), pp. 241–310.

1958b Spherical functions on a semisimple Lie group, II. *Amer. J. Math.* **80** (1958), 553–613.

1959 Some results on differential equations and their applications, *Proc. Nat. Acad. Sci. U.S.A.* **45** (1959), 1763–1764.

1960 Differential equations and semisimple Lie groups (1960) (unpublished).

1964a Invariant distributions on Lie algebras. *Amer. J. Math.* **86** (1964), 271–309.

1964b Invariant differential operators and distributions on a semi-simple Lie algebra. *Amer. J. Math.* **86** (1964), 534–564.

1964c Some results on an invariant integral on a semi-simple Lie algebra. *Ann. of Math.* **80** (1964), 551–593.

1966 Discrete series for semisimple Lie groups II. *Acta Math.* **116** (1966), 1–111.

1970 Harmonic analysis on semisimple Lie groups. *Bull. Amer. Math. Soc.* **78** (1970), 529–551

HASHIZUME, M., KOWATA, A., MINEMURA, K., and OKAMOTO, K.

1972 An integral representation of an eigenfunction of the Laplacian on the Euclidean space. *Hiroshima Math. J.* **2** (1972), 535–545.

HECKMAN, G. J.
1980 Projections of orbits and asymtotic behaviour of multiplicities for compact Lie groups. Thesis, Leiden, 1980.

HEINE, H. E.
1878–1880 "Kugelfunktionen I, II." Berlin, 1878–1880.

HEJHAL, D. A.
1976 "The Selberg Trace Formula for PSL(2, **R**)," Lecture Notes in Mathematics No. 548. Springer-Verlag, Berlin and New York, 1976.

HELGASON, S.
1957a Topologies of group algebras and a theorem of Littlewood. *Trans. Amer. Math. Soc.* **86** (1957), 269–283.
1957b Partial differential equations on Lie groups. *Scand. Math. Congr., 13th, Helsinki, 1957*, 110–115.
1958 Lacunary Fourier series on non-commutative groups. *Proc. Amer. Math. Soc.* **9** (1958), 782–790.
1959 Differential operators on homogeneous spaces. *Acta Math.* **102** (1959), 239–299.
1961 Some remarks on the exponential mapping for an affine connection. *Math. Scand.* **9** (1961), 129–146.
1962a "Differential Geometry and Symmetric Spaces." Academic Press, New York, 1962.
1962b Some results in invariant theory. *Bull. Amer. Math. Soc.* **68** (1962), 367–371.
1963 Invariants and fundamental functions. *Acta Math.* **109** (1963), 241–258.
1964a Fundamental solutions of invariant differential operators on symmetric spaces. *Amer. J. Math.* **86** (1964), 565–601.
1964b A duality in integral geometry: some generalizations of the Radon transform. *Bull. Amer. Math. Soc.* **70** (1964), 435–446.
1965a The Radon transform on Euclidean spaces, compact two-point homogeneous spaces and Grassmann manifolds. *Acta Math.* **113** (1965), 153–180.
1965b Radon–Fourier transforms on symmetric spaces and related group representations. *Bull. Amer. Math. Soc.* **71** (1965), 757–763.
1966a A duality in integral geometry on symmetric spaces, *Proc. U.S.-Japan Seminar in Differential Geometry, Kyoto, 1965.* Nippon Hyronsha, Tokyo, 1966.
1966b An analogue of the Paley–Wiener theorem for the Fourier transform on certain symmetric spaces. *Math. Ann.* **165** (1966), 297–308.
1968 Lie groups and symmetric spaces. "Battelle Rencontres," pp. 1–71. Benjamin, New York, 1968.
1970a A duality for symmetric spaces with applications to group representations. *Advan. Math.* **5** (1970), 1–154.
1970b Group representations and symmetric spaces. *Actes Congr. Internat. Math.* **2** (1970), 313–319.
1972a "Analysis on Lie Groups and homogeneous Spaces," Conf. Board Math. Sci. Series, No. 14. American Mathematical Society, Providence, Rhode Island, 1972.
1972b A formula for the radial part of the Laplace–Beltrami operator. *J. Differential Geometry* **6** (1972), 411–419.
1973a The surjectivity of invariant differential operators on symmetric spaces. *Ann. of Math.* **98** (1973), 451–480.
1973b Paley–Wiener theorems and surjectivity of invariant differential operators on symmetric spaces and Lie groups. *Bull. Amer. Math. Soc.* **79** (1973), 129–132.

1974 Eigenspaces of the Laplacian; integral representations and irreducibility. *J. Functional Anal.* **17** (1974), 328–353.

1976 A duality for symmetric spaces with applications to group representations, II. Differential equations and eigenspace representations. *Advan. Math.* **22** (1976), 187–219.

1977a Some results on eigenfunctions on symmetric spaces and eigenspace representations. *Math. Scand.* **41** (1977), 79–89.

1977b Invariant differential equations on homogeneous manifolds. *Bull. Amer. Math. Soc.* **83** (1977), 751–774.

1977c Solvability questions for invariant differential operators. In "Colloquim on Group Theoretical Methods in Physics." Academic Press, New York, 1977.

1978 "Differential Geometry, Lie Groups and Symmetric Spaces." Academic Press, New York, 1978.

1979 Invariant differential operators and eigenspace representations, pp. 236–286 in Atiyah *et al.* [1].

1980a A duality for symmetric spaces with applications to group representations, III. Tangent space analysis. *Advan. Math.* **30** (1980), 297–323.

1980b Support of Radon transforms. *Advan. Math.* **38** (1980), 91–100.

1980c "The Radon Transform." Birkhäuser, Basel and Boston, Massachusetts, 1980.

1980d The X-ray transform on a symmetric space. *Proc. Conf. Diff. Geom. and Global Anal., Berlin, 1979*, Lecture Notes in Math. No. 838. Springer-Verlag, New York, 1980.

1981 "Topics in Harmonic Analysis on Homogeneous Spaces." Birkhäuser, Basel and Boston, Massachusetts, 1981.

1983a Ranges of Radon Transforms. AMS Short Course on Computerized Tomography, January, 1982. *Proc. Symp. Appl. Math.* Amer. Math. Soc. Providence, Rhode Island, 1983.

1983b The Range of the Radon transform on Symmetric Spaces. In *Proc. Conf. Representation Theory of Reductive Lie Groups, Utah, 1982* (P. Trombi, ed.), pp. 145–151. Birkhäuser, Basel and Boston, Massachusetts, 1983.

1984 Wave equations on homogeneous spaces. *In* "Lie Group Representations III." Lecture Notes in Math. No. 1077, Springer-Verlag, New York, 1984, 254–287.

Helgason, S., and Johnson, K.

1969 The bounded spherical functions on symmetric spaces. *Adv. in Math.* **3** (1969), 586–593.

Helgason, S., and Korànyi, A.

1968 A Fatou-type theorem for harmonic functions on symmetric spaces. *Bull. Amer. Math. Soc.* **74** (1968), 258–263.

Helson, H.

1953 A note on harmonic functions. *Proc. Amer. Math. Soc.* **4** (1953), 686–691.

Hertle, A.

1982 A characterization of Fourier and Radon transforms on Euclidean spaces. *Trans. Amer. Math. Soc.* **273** (1982), 595–608.

1983 Continuity of the Radon transform and its inverse on Euclidean space. *Math. Zeitschr.* **184** (1983), 165–192.

1984 On the range of the Radon transform and its dual. *Math. Ann.* **267** (1984), 91–99.

Herz, C.

1970 Functions which are divergences. *Amer. J. Math.* **92** (1970), 641–656.

HESSELINK, W. H.
1979 Desingularization of varieties of null forms. *Invent. Math.* **55** (1979), 141–163.

HEWITT, E., and ZUCKERMAN, H.
1959 Some theorems on lacunary series with extensions to compact groups. *Trans. Amer. Math. Soc.* **93** (1959), 1–19.

HEWITT, W., and ROSS, K. A.
1970 "Abstract Harmonic Analysis, II." Springer-Verlag, Berlin and New York, 1970.

HIRAI, T.
1968 The characters of some induced representations of semisimple Lie groups. *J. Math. Kyoto Univ.* **8** (1968), 313–363.

HIRAOKA, K., and MATSUMOTO, S., and OKAMOTO, K.
1977 Eigenfunctions of the Laplacian on a real hyperboloid of one sheet. *Hiroshima Math. J.* **7** (1977), 855–864.

HOBSON, E. W.
1931 "The Theory of Spherical and Elliptical Harmonics." Cambridge Univ. Press, London and New York, 1931.

HOLDER, E.
1938 Poissonsche Wellenformel in nichtenklidischen Raumen. Ber. Verh. Sachs. *Akad. Wiss. Leipzig* **90** (1938), 55–66.

HOLE, A.
1974 Invariant differential operators and polynomials of Lie transformation groups. *Math. Scand.* **34** (1974), 109–123.
1975 Representations of the Heisenberg group of dimension $2n + 1$ on eigenspaces. *Math. Scand.* **37** (1975), 129–141.

HOOGENBOOM, B.
1982 Spherical functions and differential operators on complex Grassmann manifolds. *Ark. Mat.* **20** (1982), 69–85.
1983 "Intertwining Functions and Compact Lie Groups." Proefschrift, Amsterdam, 1983.

HORN, A.
1954a Doubly stochastic matrices and the diagonal of a rotation matrix. *Amer. J. Math.* **76** (1954), 620–630.
1954b On the eigenvalues of a matrix with prescribed singular values. *Proc. Amer. Math. Soc.* **5** (1954), 4–7.

HOWE, R.
1980 On the role of the Heisenberg group in harmonic analysis. *Bull. Amer. Math. Soc.* **3** (1980), 821–843.

HUA, LO-KENG
1981 "Starting with the Unit Circle; Background to Higher Analysis." Springer-Verlag, Berlin and New York, 1981.

HURWITZ, A.
1897 Über die Erzeugung der Invarianten durch Integration. *Gött. Nachr.* (1897), 71–90.

HÖRMANDER, L.
1963 "Linear Partial Differential Operators." Springer-Verlag, Berlin and New York, 1963.

IWAHORI, N.
1959 On real irreducible representations on Lie algebras. *Nagoya Math. J.* **14** (1959), 59–83.
1964 On the structure of a Hecke ring of a Chevalley group over a finite field. *J. Fac. Sci. Univ. Tokyo* **10** (1964), 215–236.

IWAHORI, N., and MATSUMOTO, H.
1965 On some Bruhat decompositions and the structure of the Hecke rings of p-adic Chevalley groups. *Inst. Hautes Etudes Sci. Publ. Math.* **25** (1965), 5–48.

JACOBSEN, J.
1982 Invariant differential operators on some homogeneous spaces for solvable Lie groups. Preprint No. 34, Aarhus Univ., 1982.
1983 Eigenspace representations of nilpotent Lie groups, II. *Math. Scand.* **52** (1983), 321–333.

JACOBSEN, J., and STETKAER, H.
1981 Eigenspace representations of nilpotent Lie groups. *Math. Scand.* **48** (1981), 41–55.

JACOBSON, N.
1951 Completely reducible Lie algebras of linear transformations. *Proc. Amer. Math. Soc.* **2** (1951), 105–133.
1953 "Lectures in Abstract Algebra," Vol. II. Van Nostrand Reinhold, New York, 1953.
1962 "Lie Algebras." Wiley (Interscience), New York, 1962.

JACQUET, H.
1967 Fonctions de Whittaker associées aux groupes de Chevalley. *Bull. Soc. Math. France* **95** (1967), 243–309.

JAKOBSEN, H. P.
Intertwining differential operators for $Mp(n, R)$ and $SU(n, n)$. *Trans. Amer. Math. Soc.* **246** (1978), 311–337.

JAKOBSEN, H. P., and VERGNE, M.
1977 Wave and Dirac operators and representations of the conformal group. *J. Funct. Anal.* **24** (1977), 52–66.

JAMES, A. T.
1961 Zonal polynomials of the real positive definite symmetric matrices. *Ann. of Math.* **74** (1961), 456–469.

JOHN, F.
1934 Bestimmung einer Funktion aus ihren Integralen über gevisse Mannigfaltigkeiten. *Math. Ann.* **109** (1934), 488–520.
1935 Anhängigheit zwischen den Flächenintegralen einer stetigen Funktion. *Math. Ann.* **111** (1935), 541–559.
1938 The ultrahyperbolic differential equation with 4 independent variables. *Duke Math. J.* **4** (1938), 300–322.
1955 "Plane Waves and Spherical Means." Wiley (Interscience), New York, 1955.

JOHNSON, K. D.
1980 On a ring of invariant polynomials on a Hermitian symmetric space. *J. of Algebra* **67** (1980), 72–81.

JOHNSON, K., and KORÁNYI, A.
1980 The Hua operators on bounded symmetric domains of tube type. *Ann. of Math.* **111** (1980), 589–608.

KAPLAN, A., and PUTZ, R.
1977 Boundary behaviour of harmonic forms on a rank one symmetric space. *Trans. Amer. Math. Soc.* **231** (1977), 369–384.

KARPELEVIČ, F. I.
1962 Orispherical radial parts of Laplace operators on symmetric spaces. *Sov. Math.* **3** (1962), 528–531.
1965 The geometry of geodesics and the eigenfunctions of the Beltrami–Laplace operator on symmetric spaces. *Trans. Moscow Math. Obšč.* **14** (1965), 48–185; *Trans. Moscow Math. Soc.* (1965), 51–199.

KASHIWARA, M., KOWATA, A., MINEMURA, K., OKAMOTO, K., OSHIMA, T., and TANAKA, M.
1978 Eigenfunctions of invariant differential operators on a symmetric space. *Ann. of Math.* **107** (1978), 1–39.

KASHIWARA, M., and VERGNE, M.
1976 Remarque sur la covariance de certains opérateurs differentiels. *In* "Non-Commutative Harmonic Analysis." Lecture Notes in Math. No. 587, 119–137. Springer-Verlag, New York, 1976.
1978a The Campbell–Hausdorff formula and invariant distributions. *Invent. Math.* **47** (1978), 249–272.
1978b On the Segal–Shale–Weil representation and harmonic polynomials. *Invent. Math.* **44** (1978), 1–47.
1979 *K*-types and singular spectrum. *In* "Non-Commutative Harmonic Analysis." Lecture Notes in Math. No. 728. Springer-Verlag, Berlin and New York, 1979.

KHALGUI, M. S.
1982 Sur les characteres des groupes de Lie. *J. Funct. Anal.* **47** (1982), 64–77.

KIPRIJANOV, I. A., and IVANOV, L. A.
1981 The Euler–Poisson–Darboux equation in a Riemannian space. *Soviet Math. Dokl.* **24** (1981), 331–335.

KIRILLOV, A.
1962 Unitary representations of nilpotent Lie groups. *Russian Math. Surveys* **17** (1962), 53–104.
1968 The characters of unitary representations of Lie groups. *Funct. Anal. Appl.* **2** (1968), 40–55; **3** (1969), 36–47.

KNAPP, A. W., and SPEH, B.
1981 Status of classification of irreducible unitary representations. *Proc. Conf. on Harmonic Anal., 1981* Lecture Notes in Math. Vol. 908. Springer, New York, 1982.

KNAPP, A. W., and WILLIAMSON, R. E.
1971 Poisson integrals and semisimple groups. *J. Anal. Math.* **24** (1971), 53–76.

KNAPP, A. W., and ZUCKERMAN, G. J.

1982 Classification of irreducible tempered representations of semisimple groups. *Ann. of Math.* **116** (1982), 389–455.

KOBAYASHI, S., and NOMIZU, K.

1963 "Foundations of Differential Geometry" Vols. I and II. Wiley (Interscience), New York, 1963, 1969.

KÖTHE, G.

1969 "Topological Vector Spaces." Springer-Verlag, New York, 1969.

KOLK, J. A. C.

1977 "The Selberg Trace Formula and Asymtotic Behaviour of Spectra." Proefschrift, Utrecht, 1977.

KOLMOGOROFF, A.

1928 Über die Summen durch den Zufall bestimmter unabhängiger Grössen. *Math. Ann.* **99** (1928), 309–319; **102** (1930), 484–488.

KOORNWINDER, T. H.

1973 The addition formula for Jacobi polynomials and spherical harmonics. *SIAM J. Appl. Math.* **25** (1973), 236–246.

1974 Jacobi polynomials II. An analytic proof of the product formula. *SIAM J. Math. Anal.* **5** (1974), 125–137.

1975 A new proof of a Paley–Wiener theorem for the Jacobi transform. *Ark. Mat.* **13** (1975), 145–159.

1981 "Invariant Differential Operators on Non-reductive Homogeneous Spaces." Publ. Math. Centrum, Amsterdam, Report ZW 153, 1981.

1982 The representation theory of $SL(2\boldsymbol{R})$. A non-infinitesimal approach. *L'Enseign. Math.* **28** (1982), 53–90.

KORÁNYI, A.

1963 On the boundary values of holomorphic functions in wedge domains. *Bull. Amer. Math. Soc.* **69** (1963), 475–480.

1970 Generalizations of Fatou's theorem to symmetric spaces. *Rice Univ. Stud.* **56** (1970), 127–136.

1979 A survey of harmonic functions on symmetric spaces. *Proc. Symp. Pure Math.* Vol. 35, Part 1, Harmonic Analysis in Euclidean Spaces. American Mathematical Society, Providence, Rhode Island, 1979.

1980 Some applications of Gelfand pairs in classical analysis. *In* "Harmonic Analysis and Group Representations." CIME, Cortona, 1980.

1982a On the injectivity of the Poisson transform. *J. Funct. Anal.* **45** (1982), 293–296.

1982b Kelvin transforms and harmonic polynomials on the Heisenberg group. *J. Funct. Anal.* **49** (1982), 177–185.

1982c Geometric properties of Heisenberg type groups. *Advan. Math.* (to appear).

KORÁNYI, A., and MALLIAVIN, P.

1975 Poisson formula and compound diffusion associated to an overdetermined elliptic system on the Siegel halfplane of rank two. *Acta Math.* **134** (1975), 185–209.

KOSMAN, Y.

1975 Sur les degrés conformes des opérateurs differentiels. *C. R. Acad. Sci. Paris Ser. A* 280 (1975), 229–232.

KOSTANT, B.
1959a The principal three-dimensional subgroup and the Betti numbers of a complex simple Lie group. *Amer. J. Math.* **81** (1959), 973–1032.
1959b A formula for the multiplicity of a weight. *Trans. Amer. Math. Soc.* **93** (1959), 53–73.
1963 Lie group representations on polynomial rings. *Amer. J. Math.* **85** (1963), 327–404.
1973 On convexity, the Weyl group and the Iwasawa decomposition. *Ann. Sci. École Norm. Sup.* **6** (1973), 413–455.
1975a On the existence and irreducibility of certain series of representations. *In* "Lie Groups and Their Representations" (I. M. Gelfand, ed.), pp. 231–329. Halsted, New York, 1975.
1975b Verma modules and the existence of quasiinvariant partial differential operators. *In* "Non-Commutative Harmonic Analysis." Lecture Notes in Math. No. 466. Springer-Verlag, New York, 1975.

KOSTANT, B., and RALLIS, S.
1971 Orbits and Lie group representations associated to symmetric spaces. *Amer. J. Math.* **93** (1971), 753–809.

KOWALSKI, O. and VANHECKE, L.
1983 Opérateurs différentiels invariants et symmétries géodesignes préservant le volume. *C. R. Acad. Sci. Paris.* **296** (1983), 1001–1004.

KOWATA, A., and OKAMOTO, K.
1974 Harmonic functions and the Borel–Weil theorem. *Hiroshima Math. J.* **4** (1974), 89–97.

KOWATA, A., and TANAKA, M.
1980 Global solvability of the Laplace operator on a non-compact affine symmetric space. *Hiroshima Math. J.* **10** (1980), 409–417.

KRÄMER, M
1979 Some remarks suggesting an interesting theory of harmonic functions on $SU(2n+1)/Sp(n)$ and $SO(2n+1)/U(n)$. *Arch. Math.* **33** (1979/80), 76–79.

KUBOTA, T.
1973 "Elementary Theory of Eisenstein Series." Halsted, New York, 1973.

KUČMENT, P. A.
1981 Representations of solutions of invariant differential equations on certain symmetric spaces. *Sov. Math. Dokl.* **24** (1981), 104–106.

LANG, S.
1975 "$SL_2(R)$." Addison-Wesley, Reading, Massachusetts, 1975.

LANGLANDS, R. P.
1973 On the classification of irreducible representations of real algebraic groups. Preprint, Princeton, 1973.

LASALLE, M.
1978 Séries de Laurent des fonctions holomorphes dans la complexification d'un espace symétrique compact. *Ann. Sci. Ecole Norm. Sup.* **11** (1978), 167–210.
1982 Transformeés de Poisson, algèbres de Jordan et équations de Hua. *C. R. Acad. Sci. Paris* **294** (1982), 325–328.

LAX, P. D., and PHILLIPS, R. S.
1967 "Scattering Theory." Academic Press, New York, 1967.
1978 An example of Huygens' principle. *Comm. Pure Appl. Math.* **31** (1978), 415–423.
1979 Translation representations for the solution of the non-Euclidean wave equation. *Comm. Pure. Appl. Math.* **32** (1979), 617–667.
1982 A local Paley–Wiener theorem for the Radon transform of L^2 functions in a non-Euclidean setting. *Comm. Pure Appl. Math.* **35** (1982), 531–554.

LEPOWSKY, J.
1975 On the Harish-Chandra homomorphism. *Trans. Amer. Math. Soc.* **208** (1975), 193–218.
1976 Cartan subspaces of symmetric Lie algebras. *Trans. Amer. Math. Soc.* **216** (1976), 217–228.
1977 Generalized Verma modules, the Cartan–Helgason theorem and the Harish-Chandra homomorphism. *J. Algebra* **49** (1977), 470–495.

LEPOWSKY, J., and MCCOLLUM, G. W.
1976 Cartan subspaces of symmetric Lie algebras. *Trans. Amer. Math. Soc.* **216** (1976), 217–228.

LEWIS, J. B.
1970 Eisenstein series on the boundary of the disk. Thesis, MIT, Cambridge, Massachusetts, 1970.
1978 Eigenfunctions on symmetric spaces with distribution-valued boundary forms. *J. Funct. Anal.* **29** (1978), 287–307.

LEWY, H.
1957 An example of a smooth linear partial differential equation without solution. *Ann. of Math.* **66** (1957), 155–158.

LICHNÉROWICZ, A.
1963 Opérateurs differentiels invariants sur un espace symétrique. *C. R. Acad. Sci. Paris* **257** (1963), 3548–3550.
1964 Opérateurs différentiels invariants sur un espace homogène. *Ann. Sci. École Norm. Sup.* **81** (1964), 341–385.

LICHNÉROWICZ, A., and WALKER, A. G.
1945 Sur les espaces Riemanniens harmoniques de type hyperbolique normal. *C. R. Acad. Sci. Paris* **221** (1945), 397–396.

LICHTENSTEIN, W.
1979 Qualitative behavior of special functions on compact symmetric spaces. *J. Funct. Anal.* **34** (1979), 433–455.

LIE, S.
1897 Die Theorie der Integralinvarianten ist ein Corollar der Theorie der Differentialinvarianten. *Ber. Sächs. Gesellsch. Leipzig* (1897), 342–357.

LIMIC, N., NIDERLE, J., and RACZKA, R.
1967 Eigenfunction expansions associated with the second-order invariant operator of hyperboloids and cones, III. *J. Math. Phys.* **8** (1967), 1079–1093.

LITTLEWOOD, J. E.
1924 On the mean value of power series. *Proc. London Math. Soc.* **25** (1924), 328–337.

LOHOUÉ, N. and RYCHENER, T.
1982 Die Resolvente von Δ auf symmetrischen Räumen vom nichtkompakten. *Typ. Comment. Math. Helv.* **57** (1982), 445–468.

LOOMIS, L. H.
1953 "Abstract Harmonic Analysis." Van Nostrand Reinhold, New York, 1953.

LOWDENSLAGER, D. B.
1958 Potential theory in bounded symmetric homogeneous complex domains. *Ann. of Math.* **67** (1958), 467–484.

LUCQUIAND, J.-C.
1978 Expression sous forme covariante des fonctions sphériques zonales attachées aux groupes SO(N) et SH(N). *C. R. Acad. Sci. Paris. Ser. A-B* **287** (1978), A67–A69.

LUDWIG, D.
1966 The Radon transform on Euclidean space. *Comm. Pure Appl. Math.* **23** (1966), 49–81.

MAASS, H.
1955 Die Bestimmung der Dirichletreihen mit Grössencharakteren zu den Modulformen n-ten Grades. *J. Indian Math. Soc.* **19** (1955), 1–23.
1956 Spherical functions and Quadratic Forms. *J. Indian Math. Soc.* **20** (1956), 117–162.
1958 Zur Theorie der Kugelfunktionen einer Matrixvariablen. *Math. Ann.* **135** (1958), 391–416.
1959 Zur Theorie der harmonischen Formen. *Math. Ann.* **137** (1959), 142–149.
1971 "Siegel's Modular Forms and Dirichlet Series," Lecture Notes in Mathematics No. 216. Springer-Verlag, Berlin and New York, 1971.

MACDONALD, I.
1968 Spherical functions on a p-adic Chevalley group. *Bull. Amer. Math. Soc.* **74** (1968), 520–525.
1972 Spherical Functions on Groups of p-adic Type. *Publ. Ramanujan Inst. Adv. Study No. 2. Madras, India* (1972).

MACKEY, G. W.
1952 Induced representations of locally compact groups, I. *Ann. of Math.* **55** (1952), 101–139.
1953 Induced representations of locally compact groups, II. *Ann. of Math,* **58** (1953), 193–221.
1976 "The Theory of Unitary Group Representations." Univ. of Chicago Press, Chicago, Illinois, 1976.
1978 "Unitary Group Representations in Physics, Probability and Number Theory." Benjamin/Cummings Publ., Reading, Massachusetts, 1978.
1980 Harmonic analysis as the exploitation of symmetry—a historical survey. *Bull. Amer. Math. Soc.* **3** (1980), 543–698.

MAGNUS, W., and OBERHETTINGER, R.
1948 "Formeln und Sätze fur die speciellen Funktionen der mathematischen Physik." Springer-Verlag, Berlin and New York, 1948.

Malgrange, B.

1955 Existence et approximation des solutions des équations aux dérivées partielles et des équations de convolution. *Ann. Inst. Fourier Grenoble* **6** (1955–56), 271–355.

Matsumoto, H.

1971 Quelques remarques sur les espaces riemanniens isotropes. *C. R. Acad. Sci. Paris* **272** (1971), 316–319.

Maurin, K.

1968 "General Eigenfunction Expansions and Unitary Representations of Topological Groups." Polish Sci. Publ., Warszawa, 1968.

Mautner, F. I.

1950 Unitary representations of locally compact groups II. *Ann. of Math.* **52** (1950), 528–556.

1951 Fourier analysis and symmetric spaces. *Proc. Nat. Acad. Sci. U.S.A.* **37** (1951), 529–533.

1958, 1964 Spherical functions over p-adic fields, I, II. *Amer. J. Math.* **80** (1958), 441–457; **86** (1964), 171–200.

1969, 1970 Fonctions propres des opérateurs de Hecke, *C. R. Acad. Sci. Paris* **269** (1969), 940–943; **270** (1970), 89–92.

Maxwell, J. C.

1892 "A Treatise on the Electricity and Magnetism," Vol. 1, 3rd ed. Oxford Univ. Press, London and New York, 1892.

Meany, C.

1977 A Cantor–Lebesgue theorem for spherical convergence on a compact Lie group. *Proc. Ann. Sem. Canad. Math. Congr. Queens Univ. Kingston, Ontario* (1977).

Michel, R.

1972 Sur certains tenseurs symétriques de projectifs réels. *J. Math. Pures Appl.* **51** (1972), 275–293.

1973 Problemes d'analyse géometrique liés a la conjecture de Blaschke. *Bull. Soc. Math. France* **101** (1973), 17–69.

Michelson, H. L.

1975 A decomposition for certain real semisimple Lie groups. *Trans. Amer. Math. Soc.* **213** (1975), 177–193.

Miller, J., and Simms, D. J.

1969 Radial limits of the rational functions 'Γ_μ of Harish-Chandra, *Proc. Roy. Irish Acad. Sect. A* **68** (1969), 41–47.

1970 A difference equation satisfied by the functions 'Γ_μ of Harish-Chandra, *Amer. J. Math.* **92** (1970), 362–368.

1973 On the coefficients of an asymptotic expansion of spherical functions on symmetric spaces. *Proc. Amer. Math. Soc.* **37** (1973), 448–452.

Miller, W. Jr.

1977 "Symmetry and Separation of Variables." Addison-Wesley, Reading, Massachusetts, 1977.

MIZONY, M.
1976 Algèbres et noyaux de convolution sur le dual sphérique d'un groupe de Lie semi-simple, noncompact et de rang 1. *Publ. Dep. Math. Lyon.* **13** (1976), 1–14.

MNEIMNE, R.
1983 Équation de chaleur sur un espace Riemannien symétrique et formule de Plancherel. *Bull. Sci. Math.* **107** (1983), 261–287.

MOLČANOV, V. F.
1968 Analogue of the Plancherel formula for hyperboloids. *Sov. Math. Dokl.* **9** (1968), 1385–1387.
1976 Spherical functions on hyperboloids. *Math. Sb.* **99** (141) (1976), 139–161, 295.

MOLIEN, TH.
1898 Über die Invarianten der linearen Substitutionsgruppen. *Berliner Sitznngsberichte* (1898), 1152–1158.

MOORE, C. C.
1964 Compactifications of symmetric spaces II; The Cartan domains. *Amer. J. Math.* **86** (1964), 358–378.

MOORE, C. C., and WOLF, J. A.
1971 Totally real representations and real function spaces. *Pacific J. Math.* **38** (1971), 537–542.

MORIMOTO, M.
1981 "Analytic Functionals on the Sphere and Their Fourier-Borel Transformations." Banach Center Publ., 1981, Warsaw.

MOSTOW, G. D.
1952 On the L^2-space of a Lie group. *Amer. J. Math.* **74** (1952), 920–928.
1962 Homogeneous spaces with finite invariant measure. *Ann. of Math.* **75** (1962), 17–37.

MÜLLER, C.
1966 "Spherical Harmonics," Lecture Notes in Math. No. 17. Springer-Verlag, Berlin and New York, 1966.

MYERS, S. B., and STEENROD, N.
1939 The group of isometries of a Riemannian manifold. *Ann. of Math.* **40** (1939), 400–416.

NACHBIN, L.
1965 "The Haar Integral." Van Nostrand Reinhold, New York, 1965.

NAGANO, T.
1959 Homogeneous sphere bundles and the isotropic Riemannian manifolds. *Nagoya Math. J.* **15** (1959), 29–55.

NARASIMHAN, R.
1968 "Analysis on Real and Complex Manifolds." North-Holland Publ., Amsterdam, 1968.

NELSON, E.
1959 Analytic vectors. *Ann. of Math.* **70** (1959), 572–615.

NEVANLINNA, R.
1936 "Eindeutische Analytische Funktionen." Springer-Verlag, Berlin and New York, 1936.

NUSSBAUM, E. A.
1975 Extension of positive definite functions and representations of functions in terms of spherical functions in symmetric spaces of noncompact type of rank one. *Math. Ann.* **215** (1975), 97–116.

ØRSTED, B.
1976 A note on the conformed quasi-invariance of the Laplacian on a pseudo-Riemannian manifold. *Lett. Math. Phys.* **1** (1976), 183–186.
1981a Conformally invariant differential equations and projective ,geometry. *J. Funct. Anal.* **44** (1981), 1–23.
1981b The conformal invariance of Huygens' principle. *J. Differential Geom.* **16** (1981), 1–9.

OLSANECKII, M. A.
1972 The asymtotic behavior of spherical functions. *Usp. Mat. Nauk* **27** (1972), 211–212.

ORIHARA, A.
1961 Bessel functions and the Euclidean motion group. *Tohoku Math. J.* **13** (1961), 66–74.

ORTNER, N.
1980 Faltung hypersingularer Integraloperatoren. *Math. Ann.* **248** (1980), 19–46.

OSHIMA, T.
1981 Fourier analysis on semisimple symmetric spaces. *In* "Non-Commutative Harmonic Analysis and Lie Groups," Lecture Notes in Math. No. 880, pp. 357–369, Springer-Verlag, Berlin and New York, 1981.

OSHIMA, T., and MATSUKI, T.
1980 Orbits on affine symmetric spaces under its action of the isotropy subgroups. *J. Math. Soc. Japan* **32** (1980), 399–414.

OSHIMA, T., and SEKIGUCHI, J.
1980 Eigenspaces of invariant differential operators on an affine symmetric space. *Invent. Math.* **57** (1980), 1–81.

OVSIANNIKOV, L. V.
1982 "Group Analysis of Differential Equations." Academic Press, New York, 1982.

PALAMODOV, V. P.
1970 "Linear Differential Operators with Constant Coefficients." Springer-Verlag, Berlin and New York, 1970.

PALEY, R., and WIENER, N.
1934 "Fourier Transforms in the Complex Domain." American Mathematical Society, Providence, Rhode Island, 1934.

PEETRE, J.
1959–1960 Une caractérization abstraite des opérateurs différentiels. *Math. Scand.* **7** (1959), 211–218; Rectification, *ibid.* **8** (1960), 116–120.

PENROSE, R.
1967 Twistor algebra. *J. Math. Phys.* **8** (1967), 345–366.

PETER, F., and WEYL, H.
1927 Die Vollständigkeit der primitiven Darstellungen einer geschlossenen kontinuerlichen Gruppe. *Math. Ann.* **97** (1927), 737–755.

PETROV, E. E.
1977 A Paley-Wiener theorem for a Radon complex. *Izv. Vyss. Ucebn. Zaved. Math.* **3**, (1977), 66–77.

POINCARÉ, H.
1887 Sur les residus des integrales doubles. *Acta Math.* **9** (1887), 321–380.

POULSEN, N. S.
1970 Regularity aspects of the theory of infinite dimensional representations of Lie groups. Ph.D. thesis, Massachusetts Institute of Technology, 1970.
1972 On C^∞ vectors and intertwining bi-linear forms for representations of Lie groups. *J. Funct. Anal.* **9** (1972), 87–120.

PUKANSZKY, L.
1964 The Plancherel formula for the universal covering group of **SL**(2, **R**). *Math. Ann.* **156** (1964), 96–143.

QUINTO, E. T.
1978 On the locality and invertibility of the Radon transform. Thesis, Massachusetts Institute of Technology (1978).
1980 The dependence of the generalized Radon transform on defining measures. *Trans. Amer. Math. Soc.* **257** (1980), 331–346.
1981 Topological restrictions on double fibrations and Radon transforms. *Proc. Amer. Math. Soc.* **81** (1981), 570–574.
1982 Null spaces and ranges for the classical and spherical Radon transforms. *J. Math. Anal. Appl.* **90** (1982), 408–420.

RADER, C.
1976 Spherical functions on a semisimple Lie groups. Lecture Notes, Univ. of Chicago, Chicago, Illinois, 1976.

RADON, J.
1917 Über die Bestimmung von Funktionen durch ihre Integralwerte längs gewisser Mannigfaltigkeiten. *Ber. Verh. Sächs. Akad. Wiss. Leipzig. Math.-Nat. kl.* **69** (1917), 262–277.

RAGOZIN, D. L.
1972 Central measures on compact simple Lie groups. *J. Funct. Anal.* **10** (1972), 212–229.
1976 Approximation theory, absolute convergence, and smoothness of random Fourier series on compact Lie groups. *Math. Ann.* **219** (1976), 1–11.

RAIKOV, D. A.
1946 On the theory of normed rings with involution. *Dokl. Akad. Nauk USSR* **54** (1946), 387–390.

RAÏS, M.
1971 Solutions élémentaires des opérateurs différentiels bi-invariants sur un groupe de Lie nilpotent. *C. R. Acad. Sci. Paris* **273** (1971), 495–498.

1977 Invariant Differential Operators on Lie Groups. Lecture Notes, Bielefeld, 1977.
Sur le centre de l'algèbre envelopante de **GL**(n, **R**) (unpublished).
1983 Groupes linéaires compacts et fonctions C^∞ covariantes. *Bull. Sci. Math.* **107** (1983), 93–111.

RALLIS, S.
1968 Lie Group Representations Associated to Symmetric Spaces, Ph.D. Thesis, Massachusetts Institute of Technology, 1968.

RAO, R.
1972 Orbital integrals in reductive Lie groups. *Ann. of Math.* **96** (1972), 505–510.

RAUCH, J., and WIGNER, D.
1976 Global solvability of the Casimir operator. *Ann. of Math.* **103** (1976), 229–236.

REIMANN, H. M.
1982 Invariant differential operators in hyperbolic space. *Comment. Math. Helv.* **57** (1982), 412–444.

REITER, H.
1968 "Classical Harmonic Analysis on Locally Compact Groups." Oxford Univ. Press (Clarendon), London and New York, 1968.

RHEE, H.
1970 A representation of the solutions of the Darboux equation in odd-dimensional spaces. *Trans. Amer. Math. Soc.* **150** (1970), 491–498.

RICCI, F.
1977 Local properties of the central Wiener algebra on the regular set of a compact Lie group. *Bull. Sci. Math.* **101** (1977), 87–95.

RICHARDSON, R. W. JR.
1967 Conjugacy classes in Lie Algebras and algebraic groups, *Ann. of Math.* **86** (1967), 1–15.

RIDER, D.
1970 Central idempotent measures on unitary groups. *Canad. J. Math.* **22** (1970), 719–725.
1971 Central idempotent measures on SIN groups. *Duke Math. J.* **38** (1971), 189–191.

RIESZ, M.
1949 L'intégrale de Riemann-Liouville et le problème de Cauchy. *Acta Math.* **81** (1949), 1–223.

ROBERT, A.
1983 "Introduction to Representation Theory of Compact and Locally Compact Groups," London Math. Soc. Lecture Notes No. 80. Cambridge Univ. Press, London and New York, 1983.

ROSENBERG, J.
1977 A quick proof of Harish-Chandra's Plancherel theorem for spherical functions on a semisimple Lie group. *Proc. Amer. Math. Soc.* **63** (1977), 143–149.

ROSSMANN, W.
1978a Analysis on real hyperbolic spaces. *J. Funct. Anal.* **30** (1978), 448–477.

1978b Kirillov's character formula for reductive Lie groups. *Invent. Math.* **48** (1978), 207–220.

1980 Limit characters of reductive Lie groups. *Invent. Math.* **61** (1980), 53–66.

ROUVIERE, F.

1976 Sur la résolubilité locale des opérateurs bi-invariants. *Ann. Scuola Norm. Superore Pisa* **3** (1976), 231–244.

1976 Solutions distributions de l'opérateur de Casimir. *C. R. Acad. Sci. Paris Ser. A-B* **282** (1976), A853–A856.

1983 Sur la transformation d'Abel groupes de Lie semisimples de rang un. *Ann. Scuola Norm. Pisa* **10** (1983), 263–290.

RUDIN, W.

1959 Idempotent measures on abelian groups. *Pacific J. Math.* **9** (1959), 195–209.

1962 "Fourier Analysis on Groups." (Interscience), New York, 1962.

SAKAI, S.

1964 Weakly compact operators on operator algebras. *Pacific J. Math.* **14** (1964), 659–664.

SAMII, H.

1982 "Les Transformation de Poisson dans la Boule Hyperbolique." Thèse Univ. de Nancy, Nancy, 1982.

SANTALO, L.

1976 "Integral Geometry and Geometric Probability." Addison-Wesley, Reading, Massachusetts, 1976.

SATAKE, I.

1963 Theory of spherical functions on reductive algebraic groups over p-adic fields. *Inst. Hautes Etudes Sci. Publ. Math.* **18** (1963), 5–69.

SCHEMPP, W., and DRESELER, B.

1980 "Einführung in die Harmonische Analyse." Tuebner, Stuttgart, 1980.

SCHIFFMAN, G.

1971 Integrales d'entrelacement et fonctions de Whittaker. *Bull. Soc. Math. France* **99** (1971), 3–72.

1979 Travaux de Kostant sur la série principale. *In* "Analyse Harmonique sur les Groupes de Lie II," Springer Lecture Notes No. 739, pp. 460–510. Springer-Verlag, Berlin and New York, 1979.

SCHLICHTKRULL, H.

1984a One-dimensional K-types in finite-dimensional representations of semisimple Lie groups. A generalization of Helgason's theorem. *Math. Scand.* **54** (1984), 279–294.

1984b "Hyperfunctions and Harmonic Analysis on Symmetric Spaces." Birkhäuser. Boston, 1984 (to appear).

SCHMID, W.

1969 Die Randwerte holomorpher Funktionen auf Hermitesch symmetrischen Räumen, *Invent. Math.* **9** (1969), 61–80.

SCHUR, I.

1924 Neue Anwendungen der Integralrechnung auf Probleme der Invariantentheorie. I, II, III. *Sitz. Ber. Preuss. Akad. Wiss. Phys. Math. Kl.* (1924), 189–208, 297–321, 346–355.

SCHWARTZ, L.
1966 "Théorie des Distributions," 2nd ed. Hermann, Paris, 1966.

SEGAL, G.
1968 The representation ring of a compact Lie group. *Inst. Hautes Etudes Sci. Publ. Math.* **34** (1968), 113–128.

SEGAL, I. E.
1950 An extension of Plancherel's formula to separable unimodular groups. *Ann. of Math.* **52** (1950), 272–292.

SEKIGUCHI, J.
1980 Eigenspaces of the Laplace–Beltrami operator on a hyperboloid. *Nagoya Math. J.* **79** (1980), 151–185.

SELBERG, A.
1956 Harmonic analysis and discontinuous groups in weakly symmetric Riemannian spaces with applications to Dirichlet series. *J. Indian Math. Soc.* **20** (1956), 47–87.
1962 Discontinuous groups and harmonic analysis. *Proc. Internat. Congr. Math., Stockholm* (1962), 177–189.

SEMYANISTY, V. I.
1960 On some integral transforms in Euclidean space. *Sov. Math. Dokl.* **1** (1960), 1114–1117.
1961 Homogeneous functions and some problems of integral geometry in spaces of constant curvature. *Sov. Math. Dokl.* **2** (1961), 59–62.

SERRE, J. P.
1966 "Algèbres de Lie Semi-simple Complexes." Benjamin, New York, 1966.
1970 "Cours d'arithmetique." Presses Univ. de France, Paris, 1970.

SÉMINAIRE "SOPHUS LIE"
1955 1. "Théorie des algèbres de Lie; Topologie des groupes de Lie" École Norm. Sup., Paris, 1955.

SHELSTAD, D.
1979 Characters and inner forms of a quasi-split group over **R**. *Composito Math.* **39** (1979), 11–45.

SHEPHARD, G., and TODD, J.
1954 Finite unitary reflection groups. *Canad. J. Math.* **6** (1954), 274–304.

SHEPP, L. A., *et al.*
1983 AMS short courses on computerized tomography, Cincinnati, January 1982; *Proc. Symp. Appl. Math. Amer. Math. Soc., Providence, Rhode Island* **27** (1983).

SHEPP, L. A., and KRUSKAL, J. B.
1978 Computerized Tomography; the new medical X-ray technology. *Amer. Math. Mon.* (1978), 420–438.

SHERMAN, T.
1975 Fourier analysis on the sphere. *Trans. Amer. Math. Soc.* **209** (1975), 1–31.
1977 Fourier analysis on compact symmetric space. *Bull. Amer. Math. Soc.* **83** (1977), 378–380.

SHINYA, H.
1977 Spherical matrix functions on locally compact groups of a certain type. *J. Math. Kyoto Univ.* **17** (1977), 501–509.

SIDON, S.
1932 Ein Satz über die Fourierschen Reihen stetiger Funktionen. *Math. Z.* **34** (1932), 485–486.

SITARAM, A.
1980 An analogue of the Wiener–Tauberian theorem for spherical transforms on semisimple Lie group. *Pacific J. Math.* **89** (1980), 439–445.

SMITH, K. T., and SOLMON, D. C.
1975 Lower-dimensional integrability of L^2 functions. *J. Math. Anal. Appl.* **51** (1975), 539–549.

SMITH, K. T., SOLMON, D. C., and WAGNER, S. L.
1977 Practical and mathematical aspects of the problem of reconstructing objects from radiographs. *Bull. Amer. Math. Soc.* **83** (1977), 1227–1270; Addendum. *Ibid.* **84** (1978), 691.

SOLMON, D. C.
1976 The X-ray transform, *J. Math. Anal. Appl.* **56** (1976), 61–83.

SOLOMON, L.
1963 Invariants of finite reflection groups, *Nagoya Math. J.* **22** (1963), 57–64.
1966 The orders of the finite Chevalley groups. *J. Algebra* **3** (1966), 376–393.

SPRINGER, T. A.
1968 Weyl's character formula for algebraic groups, *Invent. Math.* **5** (1968), 85–105.

STANTON, R. J.
1976 On mean convergence of Fourier series on compact Lie groups. *Trans. Amer. Math. Soc.* **218** (1976), 61–87.

STANTON, R. J., and THOMAS, P. A.
1978a Polyhedral summability of Fourier series on compact Lie groups. *Amer. J. Math.* **100** (1978), 477–493.
1978b Expansions for spherical functions on noncompact symmetric spaces. *Acta Math.* **140** (1978), 251–276.
1979 Pointwise inversion of the spherical transform on $L^2(G/K)$ $(1 \leq p < 2)$. *Proc. Amer. Math. Soc.* **73** (1979), 398–404.

STEIN, E. M.
1970 Some problems in harmonic analysis suggested by symmetric spaces and semisimple Lie groups. *Actes Congr. Internat. Math.* **1** (1970), 173–189.

STEIN, E. M., and WEISS, G.
1968 Generalization of the Cauchy–Riemann equations and representations of the rotation group. *Amer. J. Math.* **90** (1968), 163–196.

STEINBERG, R.
1961 A general Clebsch–Gordon theorem. *Bull. Amer. Math. Soc.* **67** (1961), 406–407.
1964 Differential equations invariant under finite reflection groups. *Trans. Amer. Math. Soc.* **112** (1964), 392–400.

1965 Regular elements of semi-simple algebraic groups, *Inst. Hautes Etudes Sci. Publ. Math.* **25** (1965), 49–80.

STERN, A. I.
1969 Completely irreducible class I representations of real semi-simple Lie group. *Sov. Math.* **10** (1969), 1254–1257.

STOLL, W.
1952 Mehrfache Integrale auf komplexen Mannigfaltigkeiten. *Math. Z.* **57** (1952), 116–154.

STRICHARTZ, R. S.
1973 Harmonic analysis on hyperboloids. *J. Funct. Anal.* **12** (1973), 341–383.
1974 Multiplier transformations on compact Lie groups and algebras. *Trans. Amer. Math. Soc.* **193** (1974), 99–110.
1975 The explicit Fourier decomposition of $L^2(\mathbf{SO}(n)/\mathbf{SO}(n-m))$. *Canad. J. Math.* **27** (1975), 294–310.
1981 L^p-Estimates for Radon transforms in Euclidean and non-Euclidean spaces. *Duke Math. J.* **48** (1981), 699–727.

SUGIURA, M.
1960 Spherical functions and representations theory of compact Lie groups. *Sci. Papers Coll. Gen. Ed. Univ. Tokyo* **10** (1960), 187–193.
1962 Representations of compact groups realized by spherical functions on symmetric spaces. *Proc. Japan Acad.* **38** (1962), 111–113.
1971 Fourier series of smooth functions on compact Lie groups. *Osaka Math. J.* **8** (1971), 33–47.
1975 "Unitary Representations and Harmonic Analysis." Wiley, New York, 1975.

SULANKE, R.
1966 Croftonsche Formeln in Kleinschen Räumen. *Math. Nachr.* **32** (1966), 217–241.

TAKAHASHI, R.
1961 Sur les fonctions sphériques et la formule de Plancherel dans le groupe hyperbolique, *Japanese J. Math.* **31** (1961), 55–90.
1963 Sur les représentations unitaires des groupe de Lorentz généralisés. *Bull. Soc. Math. France* **91** (1963), 289–433.
1977a Fonctions Sphériques zonales sur $\mathbf{U}(n, n+k; \mathbf{F})$. *In* Séminaire d'Analyse Harmonique (1976–77)." Fac. de Sciences, Tunis, 1977.
1977b Spherical functions in $\mathbf{Spin}(1, d)/\mathbf{Spin}(d-1)$. In "Non-commutative Harmonic Analysis," Lecture Notes in Math. Vol. 587. Springer-Verlag, Berlin and New York, 1977.
1979 Quelque resultats sur l'analyse harmonique dans l'espace symétrique noncompact de rang 1 du type exceptionel. *In* "Analyse Harmonique sur les Groupe de Lie II," Springer Lecture Notes No. 739. Springer-Verlag, Berlin and New York, 1979.

TAMAGAWA, T.
1960 On Selberg's trace formula. *J. Fac. Sci. Univ. Tokyo* **8** (1960), 363–386.

TAYLOR, M.
1968 Fourier series on compact Lie groups. *Proc. Amer. Math. Soc.* **19** (1968), 1103–1105.

TEDONE, O.
1898 Sull' integrazionne dell'equazione $\partial^2 I/\partial t^2 - \sum \partial^2 I/\partial_i^2 = 0$. *Ann. Mat.* **1** (1898), 1–24.

TERRAS, A.
1982 Non-Euclidean harmonic analysis. *SIAM Rev.* **24** (1982), 159–193.

THOMAS, E. G. F.
1984 An infinitesimal characterization of Gelfand pairs. *Proc. Conf. Modern Anal. and Probability*. Yale Univ., 1984.

THOMPSON, C.
1971 Inequalities and partial orders on matrix spaces. *Indiana Univ. Math. J.* **21** (1971), 469–480.

TITCHMARSH, E. C.
1939 "The Theory of Functions," 2nd ed. Oxford Univ. Press, London and New York, 1939.

TITS, J.
1955 Sur certains classes d'espaces homogènes de groupes de Lie. *Acad. Roy. Belg. Cl. Sci. Mem. Coll.* **29** (1955), No. 3.
1967 "Tabellen zu den einfachen Lie Gruppen und ihren Darstellungen," Lecture Notes in Mathematics 40. Springer-Verlag, Berlin and New York, 1967.

TORASSO, P.
1977 Le théorème de Paley–Wiener pour l'espace des fonctions indéfinement differentiables et a support compact sur un espace symétrique de type noncompact. *J. Funct. Anal.* **26** (1977), 201–213.

TRÈVES, F.
1963 Équations aux dérivees partielles inhomogènes a coefficients constants dépendent de paramètres. *Ann. Inst. Fourier Grenoble* **13** (1963), 123–138.
1967 "Topological Vector Spaces, Distributions and Kernels." Academic Press, New York, 1967.

TRICERRI, F., and VANHECKE, L.
1983 "Homogeneous Structures on Riemannian Manifolds." London Math. Soc. Lect. Notes, 83. Cambridge Univ. Press, 1983.

TROMBI, P., and VARADARAJAN, V. S.
1971 Spherical transforms on semi-simple Lie groups. *Ann. of Math.* **94** (1971), 246–303.

VAN DEN BAN, E. P.
1982 "Asymtotic Expansions and Integral Formulas for Eigenfunctions on a Semisimple Lie Group." Proefscrift, Utrecht, 1982.

VAN DIJK, G.
1969a On symmetry of group algebras of motion groups. *Math. Ann.* **179** (1969), 219–226.
1969b Spherical functions on the p-adic group PGL(2). *Indag. Math.* **31** (1969), 213–241.

VARADARAJAN, V. S.
1968 On the ring of the invariant polynomials on a semi-simple Lie algebra. *Amer. J. Math.* **90** (1968), 308–317.

1974 "Lie Groups, Lie Algebras and their Representations." Prentice-Hall, Englewood Cliffs, New Jersey, 1974.

1977 "Harmonic Analysis on Real Reductive Groups," Springer Lecture Notes No. 576. Springer-Verlag, Berlin and New York, 1977.

VERGNE, M.

1979 On Rossmann's character formula for discrete series. *Invent. Math.* **54** (1979), 11–14.

VILENKIN, N.

1957 On the theory of associated spherical functions on Lie groups. *Mat. Sb.* **42** (1957), 485–496.

1968 "Special Functions and the Theory of Group Representations," Translations of Math. Mono., Vol. 22. American Mathematical Society, Providence, Rhode Island, 1968.

VOGAN, D.

1981 "Representations of Real Reductive Groups." Birkhäuser, Basel and Boston, Massachusetts, 1981.

VRETARE, L.

1975 On L_p Fourier multipliers on certain symmetric spaces. *Math. Scand.* **37** (1975), 111–121.

1976 Elementary spherical functions on symmetric spaces. *Math. Scand.* **39** (1976), 343–358.

1977 On a recurrence formula for elementary spherical functions on symmetric spaces and its applications. *Math. Scand.* **41** (1977), 49–112.

WALLACH, N.

1973 "Harmonic Analysis in Homogeneous Spaces." Dekker, New York, 1973.

WANG, H. C.

1952 Two-point homogeneous spaces. *Ann. of Math.* **55** (1952), 177–191

WARNER, F.

1970 "Foundations of Differentiable Manifolds and Lie Groups." Scott Foresman, Glenview, Illinois, 1970.

WARNER, G.

1972 "Harmonic Analysis on Semisimple Lie Groups," Vols. I, II. Springer-Verlag, Berlin and New York, 1972.

WEIL, A.

1940 "L'intégration dans les groupes topologiques et ses applications." Hermann, Paris, 1940.

1958 "Variétés Kähleriennes." Hermann, Paris, 1958.

WEISS, B.

1967 Measures that vanish on half spaces. *Proc. Amer. Math. Soc.* **18** (1967), 123–126.

WEISS, G.

1976 Harmonic analysis on compact groups. *MAA Stud. Math.* **13** (1976), 198–223.

WELLS, R., JR.

1979 Complex manifolds and mathematical physics. *Bull. Amer. Soc.* **1** (1979), 296–336.

WENDEL, J. G.
1952 Left centralizers and isomorphisms of group algebras. *Pacific J. Math.* **2** (1952), 251–261.
1954 Haar measure and the semigroup of measures on a compact group. *Proc. Amer. Math. Soc.* **5** (1954), 923–929.

WEYL, H.
1910 Über gewöhnliche Differentialgleichungen mit Singularitäten und die zurgehörigen Entwicklungen willkürlicher Funktionen. *Math. Ann.* **68** (1910), 220–269.
1925–1926 Theorie der Darstellung kontinuerlicher halbeinfacher Gruppen durch lineare Transformationen I, II, III und Nachtrag. *Math. Z.* **23** (1925), 271–309; **24** (1926), 328–376, 377–395, 789–791.
1931 "The Theory of Groups and Quantum Mechanics." Dover, New York, 1931.
1934 Harmonics on homogeneous manifolds. *Ann. of Math.* **35** (1934), 486–499.
1939 "The Classical Groups." Princeton Univ. Press, Princeton, New Jersey, 1939.

WHITTAKER, E. T., and WATSON, G. N.
1927 "A Course of Modern Analysis." Cambridge Univ. Press, London and New York, 1927.

WIGNER, D.
1977 Bi-invariant operators on nilpotent Lie groups. *Invent. Math.* **41** (1977), 259–264.

WILLMORE, T. J.
1950 Mean value theorems in harmonic Riemann spaces. *J. London Math. Soc.* **25** (1950), 54–57.

WOLF, J. A.
1967 "Spaces of Constant Curvature." McGraw-Hill, New York, 1967.

YAMABE, H.
1960 On the deformation of Riemannian structures on compact manifolds. *Osaka J. Math.* **12** (1960), 21–37.

YANG, J.-G.
1983 A proof of a formula in Fourier analysis on the sphere. *Proc. Amer. Math. Soc.* **88** (1983), 602–604.

YANO, K., and OBATA, M.
1970 On conformal changes of Riemannian metrics. *J. Differential Geometry* **4** (1970), 53–72.

ZALCMAN, L.
1980 Offbeat integral geometry. *Amer. Math. Monthly* **87** (1980), 161–175.
1982 Uniqueness and nonuniqueness for the Radon transforms. *Bull. London Math. Soc.* **14** (1982), 241–245.

ZARISKI, O., and SAMUEL, P.
1958 "Commutative Algebra," Vol. I. Van Nostrand Reinhold, New York, 1958.
1960 "Commutative Algebra," Vol. II. Van Nostrand Reinhold, New York, 1960.

ZELOBENKO, D. P.
1963 On the theory of representations of complex and real Lie groups. *Trudy Moscov. Mat. Obšč.* **12** (1963), 53–98.

1973 "Compact Lie Groups and their Representations." Transl. of Math. Monogr., American Mathematical Society, Providence, Rhode Island, 1973.

1974 "Harmonic Analysis on Complex Semisimple Lie Groups." WAUKA, Moscow, 1974 (in Russian).

ZYGMUND, A.

1930 On the convergence of lacunary trigonometric series. *Fundamental Matem.* **16** (1930), 90–107.

1959 "Trigonometric Series," 2nd ed., I, II. Cambridge Univ. Press, London and New York, 1959.

SYMBOLS FREQUENTLY USED

$\mathscr{E}(X)$: $C^\infty(X)$, 3, 239
$\mathscr{E}^\natural(X)$, $\mathscr{E}'^\natural(X)$: space of K-invariant elements in $\mathscr{E}(X)$, $\mathscr{E}'(X)$, 292
$\mathscr{E}_\lambda$: eigenspace, 12, 35, 279
$E_{a,b}$: space, 4
E': space of entire functionals, 5
$F(a, b; c; z)$: hypergeometric function, 50, 484
ϕ_λ: spherical function, 38, 418
$\Phi_{\lambda,m}$: generalized spherical function, 50
$\boldsymbol{G}(d, n)$, $\boldsymbol{G}_{d,n}$: manifolds of d-planes, 124
$\boldsymbol{GL}(n, \boldsymbol{R})$, $\boldsymbol{GL}(n, C)$: classical groups, 222
$\mathscr{H}(Z)$: space of holomorphic functions on Z, 15
$\mathscr{H}^A(\boldsymbol{C}^n)$: exponential type, 15
Hom(V, W): space of linear transformations of V into W, 392
$\boldsymbol{H}^n$: hyperbolic space, 29, 152, 177
$H(E)$: space of harmonic polynomials on E, 346
$\mathscr{H}$: Hilbert transform, 113
$H(g)$: component in $g = k \exp H(g)n$, 277
Im: imaginary part, 2
$I(E)$: space of invariant polynomials on E, 346
$I_{p,q}$: matrix, 201
I^γ: Riesz potential, 135
$J_n(z)$: Bessel function, 11
$\mathscr{K}(X)$: $C_c(X)$, 396
χ_δ: character of δ, 391
$\mathfrak{k}$: algebra in Cartan decomposition, 186
$L^1(X)$: space of integrable functions on X, 414
$L^p(X)$: space of f with $|f|^p \in L^1(X)$, 510
L: diameter of X, 164
$L = L_X$: Laplace–Beltrami operator on X, 31, 244
$L(g) = L_g$: left translation by g, 85
l: left regular representation, 386
$l(\delta)$: dimension, 394
$l(s)$: length function, 442
λ: symmetrization, 282
Λ: set of integral functions (weights), 498, operator, 113, root lattice, 307
$\Lambda(+)$: set of dominant integral functions, 498
$\Lambda(\pi)$: set of weights of π, 498
$\Lambda(E)$: Grassmann algebra over E, 354
M_p: the tangent space to a manifold M at p, 84
M^r: mean-value operator, 103, 288
$\boldsymbol{M}(n)$: group of isometries of $\boldsymbol{R}^n$, 4
$n(s)$: cardinality of Σ_s^+, 441
$\mathscr{N}$: set of nilpotent elements, 369
$\boldsymbol{O}(p, q)$: orthogonal groups, 200
Ω_n: area of $\boldsymbol{S}^{n-1}$, 18
$\boldsymbol{P}^n$: set of hyperplanes in $\boldsymbol{R}^n$, 97
$\mathfrak{p}$: part of a Cartan decomposition, 181
$\boldsymbol{R}^n$: real n-space, 3
$\boldsymbol{R}^+$: set of reals ≥ 0, 2
Re: real part, 2

R_g or $R(g)$: right translation by g, 85
r: right regular representation, 386
Res: residue, 133
ρ : half sum of roots, 181
$\mathfrak{r}$: set of regular elements, 369
$\mathfrak{R}$: set of quasi-regular elements, 369
S^n: n-sphere, 3
$S_r(p)$: sphere or radius r and center p, 103, 312
$\mathscr{S}(\boldsymbol{R}^n)$: space of rapidly decreasing functions on $\boldsymbol{R}^n$
$\mathscr{S}^*(\boldsymbol{R}^n)$, $\mathscr{S}_0(\boldsymbol{R}^n)$: subspaces of $\mathscr{S}(\boldsymbol{R}^n)$, 104
$\mathscr{S}'(\boldsymbol{R}^n)$: space of tempered distributions, 131, 132
$S(V)$: symmetric algebra over V, 346
sgn(x): signum function, 114
$\mathscr{S}_H(\boldsymbol{P}^n)$: subspace of $\mathscr{S}(\boldsymbol{P}^n)$, 99, 100
sh x: sinh x, 36
Σ, Σ^+, Σ_0^+, Σ_s^+ sets of restricted roots, 441
$\mathscr{S}$: set of semisimple elements, 369
th x: tanh x, 76
${}^tA$: transpose of A, 388
Tr(A): trace of A, 221
$\tau(x)$: translation on G/H, 88
$U(\mathfrak{g})$: universal enveloping algebra of $\mathfrak{g}$, 503
$\boldsymbol{U}(n)$: unitary group, 515
V_δ: representation space of δ, 391
V_δ^M: space of fixed vectors under $\delta(M)$, 394
$V(\delta)$: space of K-finite vectors of type δ, 395
V^∞: space of differentiable vectors, 387
W: Weyl group, 194
$\boldsymbol{Z}$, $\boldsymbol{Z}^+$: the integers, the nonnegative integers, 2
$\boldsymbol{Z}(G)$: center of $D(G)$
$Z(\mathfrak{g})$: center of the universal enveloping algebra $U(\mathfrak{g})$, 503
$\tilde{}$: Fourier transform, 4, 33, 457; spherical transform, 39, 415, 449
$\hat{}$: Radon transform, 76, 97; incidence 140
$\check{}$: Dual Radon transform, 97; incidence, 140
$*$, $\times$: convolutions, 22, 43, 44, 99, 119, 132, 289, 290; adjoint operation, 241, 513; pullback, 83; star operator, 330
$\oplus$: direct sum, 601
$\otimes$: tensor product, 391
$\wedge$: exterior product, 354
$\langle , \rangle$: inner product, 18, 29, 32, 186, 347
$^\natural$, $E^\natural$: space of K-invariants in E, 43, 292
$\| \|_m^S$: norm on $C^\infty(V)$, 234
$\square$: operator 98, 337; wave operator, 208

Set theory. Let A and B be sets. We use the usual symbols $\in$, $\subset$, $\cap$, $\cup$ for being element of, inclusion, intersection, and union. The set $A - B$ is the set of elements in A not in B. If $\mathscr{P}$ is a property and M a set, then $\{x \in M\colon x$ has property $\mathscr{P}\}$ denotes the set of elements $x \in M$ with property $\mathscr{P}$. The composite of two maps $f\colon A \to B$ and $g\colon B \to M$ is denoted $g \circ f\colon A \to M$. The sign $\Rightarrow$ means "implies."

Algebra. If K is a subgroup of a group G, the symbol G/K denotes the set of left cosets gK, $g \in G$. When K is considered as an element in G/K it will sometimes be denoted by $\{K\}$. If $x \in G$, the mapping $gK \to xgK$ of G/K onto itself will be denoted by $\tau(x)$.

Let $A: V \to W$ be a linear mapping of a vector space V into a vector space W. Let V^* and W^* denote the respective duals. The mapping ${}^tA: W^* \to V^*$ defined by ${}^tA(w^*)(v) = w^*(Av)$ is called the *transpose* of A. A linear map $A: V \to V$ will often be called an *endomorphism* of V. If V has finite dimension the *determinant* and *trace* of A will be denoted by $\det(A)$ and $\mathrm{Tr}(A)$, respectively.

Topology. Let X be a topological space, $A \subset X$ a subset. The closure of A is denoted $\bar{A}$ or $Cl(A)$, the interior of A by $\mathring{A}$. If f is a function on X its restriction to A is denoted $f|A$. The Hausdorff separation axiom for a topological space will always be assumed. A coset space G/H where H is a closed subgroup of the topological group G will always be taken with its *natural topology* which is characterized by the fact that the mapping $g \to gH$ of G onto G/H is continuous and open. A topological vector space V (over $\boldsymbol{R}$ or $\boldsymbol{C}$) can be topologized by a family of *seminorms* if and only if it is locally convex. This family can be chosen countable if and only if V is metrizable. If in addition V is complete it is called a *Fréchet space.*

Manifolds. If M is a manifold and $p \in M$, the tangent space to M at p is denoted by M_p. If $\phi: q \to (x_1(q), \ldots, x_m(q))$ is a system of coordinates near p the mappings

$$(\partial/\partial x_i)_p: f \to \left(\frac{\partial(f \circ \phi^{-1})}{\partial x_i}\right)(\phi(p)) \qquad f \in C^\infty(M)$$

form a basis of M_p. Let $\Phi: M \to N$ be a differentiable mapping of M into a manifold N. Its *differential* at p, mapping M_p into $N_{\Phi(p)}$ is denoted $d\Phi_p$. If η is a form on N its *pullback* by Φ is denoted $\Phi^*\eta$; it is a form on M.

Lie groups. Lie groups are usually denoted by capital letters and their Lie algebras by the corresponding lower case German letters. The exponential mapping of a Lie algebra into a corresponding Lie group is denoted exp. The identity component of a Lie group G is denoted G_0. We use the standard notation for the classical groups: $\boldsymbol{GL}(n, \boldsymbol{C})$, $\boldsymbol{GL}(n, \boldsymbol{R})$, $\boldsymbol{SL}(n, \boldsymbol{C})$, $\boldsymbol{SL}(n, \boldsymbol{R})$, $\boldsymbol{U}(p, q)$, $\boldsymbol{SU}(p, q)$, $\boldsymbol{SU}^*(2n)$, $\boldsymbol{SO}(n, \boldsymbol{C})$, $\boldsymbol{O}(p, q)$, $\boldsymbol{SO}^*(2n)$, $\boldsymbol{Sp}(n, \boldsymbol{C})$, $\boldsymbol{Sp}(n, \boldsymbol{R})$, and $\boldsymbol{Sp}(p, q)$. See, for example, [*DS*], Chapter X, §2 for detailed descriptions.

INDEX

E

F

G

H

I

J

K

L

M

N

O

P

Q

R

S

T

U

V

W

X

ERRATA

Page and line in $\begin{cases}\text{above}\\ \text{below}\end{cases}$	Instead of:	Read:
16_2	Cor. 4.11	Proposition 4.11
88_{11}	$\operatorname{Ad} X$	$\operatorname{ad} X$
88_{13}	$r^{\operatorname{Tr} A}$	$e^{\operatorname{Tr} A}$
104_5	2.41	2.34
104_9	$C1$	$D1$
110^9	B_e	B_ϵ
127_2	$\|x\|p$	$(\|x\|p)^k$
145_{17}	$\int_{G/H} dg_\equiv$	$\int_{G/H_\equiv} dg_{H_\equiv}$
151_2	Q_ϵ	B_ϵ
$160^{4,6}$	K	H_X
161^3	p	x
167^3	$u(\breve{0})$	$\mu(\breve{o})$
170_5	(r)	(x)
176_3	Theorem 1.11	Theorem 4.11
178^8	follow	follows
179_{18}	M	X
190^6	uk_1	uK_1
191_1	corrollary	corollary
199_9	$g_y(X, X)$	$g_y(Z, Z)$
258_{8-9}	Proposition 3.1	Theorem 3.2
258_{13}	Proposition 3.1	Theorem 3.2
259_{17}	V_W	V_w
262^{10}	y_1	$y_1(b)$

Page and line in {above / below}	Instead of:	Read:
269^1	By	In
273^1	by analogy	in analogy
278_6	Theorem 4.3	Theorem 4.4
302_{15}	$\subset$	$\in$
303^{11}	denoted	denote
319_2	exampe	example
332_{10}	K_2	$K_2)$
336^1	W^*	V^*
336_{13}	p	$\mathfrak{p}$
362^{14}	$j_r)$	$j_n)$
367^{17}	dominators	denominators
377_1	k^Z	$\mathfrak{k}^Z$
389_1	$\mathcal{H}_\varphi$	$\mathfrak{H}_\varphi$
405_2	orthogonal	orthonormal
425_{16}	for	far
439_9	N^{s-1}	$N^{s^{-1}}$
442^3	ssi	ss_i
448^5	(21)	(12)
455_5	gk	gK
457_{10}	gk	gK
471^{12}	4.21	4.19
473^7	$k \in Y$	$k \in K$
476^{15}	Hak	$H(ak)$
479^9	(26)	(27)
483_{14}	N_{β_p}	$\overline{N}_{\beta_p}$
504_{12}	e	(e)
523^{10}	(49)	(47)
529_8	$\mathcal{S}$	V
551^8	(ii)	(iii)
556^{10}	μ	u

Page and line in {above / below	Instead of:	Read:
566^{13}	$r = 0$	$r \to 0$
569^{7}	$\sum_i$	$\sum_\ell$
572_{3}	(17)	(7)
575^{4}	$[\frac{1}{2}(n-3)!$	$[\frac{1}{2}(n-3)]!$
576_{2}	This	Thus
595_{10}	to	so
599_{16}	$\sum_0^i$	$\sum_0^\ell$
603_{11}	$k \in \mathbf{R}$	$k = \mathbf{R}$
609^{2}	$\alpha_n \beta^n$	$a_n \beta^n$
615_{13}	zonal	zonales
618^{5}	Palaisesu	Palaiseau
618^{6}	Duistermatt	Duistermaat
618^{10}	Composite	Compositio
622_{2}	Asymtotics	Asymptotics
626^{19}	nichtenklischen	nichteuklidischen
630^{20}	geodesignes	geodesiques
635^{1}	Eindeutische	Eindeutige
643_{2}	Amer. Soc.	Amer. Math. Soc.
645^{3}	WAUKA	NAUKA
653_{8}	399	415

Selected Titles in This Series

(Continued from the front of this publication)

For a complete list of titles in this series, visit the
AMS Bookstore at **www.ams.org/bookstore/**.